Prose Writers Before 1867, edited by William B. Thesing (1987)

Fiction Writers, 1914-1945, by James Hardin (1987)

Prose Writers After 1867, edited by William B. Thesing (1987)

...n and Caroline Dramatists, edited by Fredson Bowers (1987)

Literary Critics and Scholars, ...850, edited by John W. ...n and Monica M. Grecu

...an Writers Since 1960, Second ..., edited by W. H. New (1987)

...n Writers for Children Since ...: Poets, Illustrators, and Nonfiction ..., edited by Glenn E. Estes

...than Dramatists, edited by ...n Bowers (1987)

American Critics, 1920-1955, by Gregory S. Jay (1988)

...an Literary Critics and Scholars, ...1880, edited by John W. ...un and Monica M. Grecu

Novelists, 1900-1930, edited ...atharine Savage Brosman

...n Fiction Writers, 1885-1913, 2 edited by James Hardin

...n American Critics Since 1955, by Gregory S. Jay (1988)

...ian Writers, 1920-1959, First Series, edited by W. H. New (1988)

...mporary German Fiction Writers, ...Series, edited by Wolfgang D. ...nd James Hardin (1988)

...h Mystery Writers, 1860-1919, ...d by Bernard Benstock and ...as F. Staley (1988)

...can Literary Critics and Scholars, ...1900, edited by John W. ...bun and Monica M. Grecu

...h Novelists, 1930-1960, edited ...Catharine Savage Brosman

...ican Magazine Journalists, 1741-..., edited by Sam G. Riley (1988)

...ican Short-Story Writers Before ..., edited by Bobby Ellen Kim...with the assistance of William ...rant (1988)

...mporary German Fiction Writers, ...d Series, edited by Wolfgang D. ...and James Hardin (1988)

76 *Afro-American Writers, 1940-1955*, edited by Trudier Harris (1988)

77 *British Mystery Writers, 1920-1939*, edited by Bernard Benstock and Thomas F. Staley (1988)

78 *American Short-Story Writers, 1880-1910*, edited by Bobby Ellen Kimbel, with the assistance of William E. Grant (1988)

79 *American Magazine Journalists, 1850-1900*, edited by Sam G. Riley (1988)

80 *Restoration and Eighteenth-Century Dramatists, First Series*, edited by Paula R. Backscheider (1989)

81 *Austrian Fiction Writers, 1875-1913*, edited by James Hardin and Donald G. Daviau (1989)

82 *Chicano Writers, First Series*, edited by Francisco A. Lomelí and Carl R. Shirley (1989)

83 *French Novelists Since 1960*, edited by Catharine Savage Brosman (1989)

84 *Restoration and Eighteenth-Century Dramatists, Second Series*, edited by Paula R. Backscheider (1989)

85 *Austrian Fiction Writers After 1914*, edited by James Hardin and Donald G. Daviau (1989)

86 *American Short-Story Writers, 1910-1945, First Series*, edited by Bobby Ellen Kimbel (1989)

87 *British Mystery and Thriller Writers Since 1940, First Series*, edited by Bernard Benstock and Thomas F. Staley (1989)

88 *Canadian Writers, 1920-1959, Second Series*, edited by W. H. New (1989)

89 *Restoration and Eighteenth-Century Dramatists, Third Series*, edited by Paula R. Backscheider (1989)

90 *German Writers in the Age of Goethe, 1789-1832*, edited by James Hardin and Christoph E. Schweitzer (1989)

91 *American Magazine Journalists, 1900-1960, First Series*, edited by Sam G. Riley (1990)

92 *Canadian Writers, 1890-1920*, edited by W. H. New (1990)

93 *British Romantic Poets, 1789-1832, First Series*, edited by John R. Greenfield (1990)

94 *German Writers in the Age of Goethe: Sturm und Drang to Classicism*, edited by James Hardin and Christoph E. Schweitzer (1990)

95 *Eighteenth-Century British Poets, First Series*, edited by John Sitter (1990)

96 *British Romantic Poets, 1789-1832, Second Series*, edited by John R. Greenfield (1990)

97 *German Writers from the Enlightenment to Sturm und Drang, 1720-1764*, edited by James Hardin and Christoph E. Schweitzer (1990)

98 *Modern British Essayists, First Series*, edited by Robert Beum (1990)

99 *Canadian Writers Before 1890*, edited by W. H. New (1990)

100 *Modern British Essayists, Second Series*, edited by Robert Beum (1990)

101 *British Prose Writers, 1660-1800, First Series*, edited by Donald T. Siebert (1991)

102 *American Short-Story Writers, 1910-1945, Second Series*, edited by Bobby Ellen Kimbel (1991)

103 *American Literary Biographers, First Series*, edited by Steven Serafin (1991)

104 *British Prose Writers, 1660-1800, Second Series*, edited by Donald T. Siebert (1991)

105 *American Poets Since World War II, Second Series*, edited by R. S. Gwynn (1991)

106 *British Literary Publishing Houses, 1820-1880*, edited by Patricia J. Anderson and Jonathan Rose (1991)

107 *British Romantic Prose Writers, 1789-1832, First Series*, edited by John R. Greenfield (1991)

108 *Twentieth-Century Spanish Poets, First Series*, edited by Michael L. Perna (1991)

109 *Eighteenth-Century British Poets, Second Series*, edited by John Sitter (1991)

110 *British Romantic Prose Writers, 1789-1832, Second Series*, edited by John R. Greenfield (1991)

111 *American Literary Biographers, Second Series*, edited by Steven Serafin (1991)

112 *British Literary Publishing Houses, 1881-1965*, edited by Jonathan Rose and Patricia J. Anderson (1991)

113 *Modern Latin-American Fiction Writers, First Series*, edited by William Luis (1992)

114 *Twentieth-Century Italian Poets, First Series*, edited by Giovanna Wedel De Stasio, Glauco Cambon, and Antonio Illiano (1992)

115 *Medieval Philosophers*, edited by Jeremiah Hackett (1992)

116 *British Romantic Novelists, 1789-1832*, edited by Bradford K. Mudge (1992)

Dictionary of Literary Biogr

1. *The American Renaissance in New England*, edited by Joel Myerson (1978)
2. *American Novelists Since World War II*, edited by Jeffrey Helterman and Richard Layman (1978)
3. *Antebellum Writers in New York and the South*, edited by Joel Myerson (1979)
4. *American Writers in Paris, 1920-1939*, edited by Karen Lane Rood (1980)
5. *American Poets Since World War II*, 2 parts, edited by Donald J. Greiner (1980)
6. *American Novelists Since World War II, Second Series*, edited by James E. Kibler Jr. (1980)
7. *Twentieth-Century American Dramatists*, 2 parts, edited by John MacNicholas (1981)
8. *Twentieth-Century American Science-Fiction Writers*, 2 parts, edited by David Cowart and Thomas L. Wymer (1981)
9. *American Novelists, 1910-1945*, 3 parts, edited by James J. Martine (1981)
10. *Modern British Dramatists, 1900-1945*, 2 parts, edited by Stanley Weintraub (1982)
11. *American Humorists, 1800-1950*, 2 parts, edited by Stanley Trachtenberg (1982)
12. *American Realists and Naturalists*, edited by Donald Pizer and Earl N. Harbert (1982)
13. *British Dramatists Since World War II*, 2 parts, edited by Stanley Weintraub (1982)
14. *British Novelists Since 1960*, 2 parts, edited by Jay L. Halio (1983)
15. *British Novelists, 1930-1959*, 2 parts, edited by Bernard Oldsey (1983)
16. *The Beats: Literary Bohemians in Postwar America*, 2 parts, edited by Ann Charters (1983)
17. *Twentieth-Century American Historians*, edited by Clyde N. Wilson (1983)
18. *Victorian Novelists After 1885*, edited by Ira B. Nadel and William E. Fredeman (1983)
19. *British Poets, 1880-1914*, edited by Donald E. Stanford (1983)
20. *British Poets, 1914-1945*, edited by Donald E. Stanford (1983)
21. *Victorian Novelists Before 1885*, edited by Ira B. Nadel and William E. Fredeman (1983)
22. *American Writers for Children, 1900-1960*, edited by John Cech (1983)
23. *American Newspaper Journalists, 1873-1900*, edited by Perry J. Ashley (1983)
24. *American Colonial Writers, 1606-1734*, edited by Emory Elliott (1984)
25. *American Newspaper Journalists, 1901-1925*, edited by Perry J. Ashley (1984)
26. *American Screenwriters*, edited by Robert E. Morsberger, Stephen O. Lesser, and Randall Clark (1984)
27. *Poets of Great Britain and Ireland, 1945-1960*, edited by Vincent B. Sherry Jr. (1984)
28. *Twentieth-Century American-Jewish Fiction Writers*, edited by Daniel Walden (1984)
29. *American Newspaper Journalists, 1926-1950*, edited by Perry J. Ashley (1984)
30. *American Historians, 1607-1865*, edited by Clyde N. Wilson (1984)
31. *American Colonial Writers, 1735-1781*, edited by Emory Elliott (1984)
32. *Victorian Poets Before 1850*, edited by William E. Fredeman and Ira B. Nadel (1984)
33. *Afro-American Fiction Writers After 1955*, edited by Thadious M. Davis and Trudier Harris (1984)
34. *British Novelists, 1890-1929: Traditionalists*, edited by Thomas F. Staley (1985)
35. *Victorian Poets After 1850*, edited by William E. Fredeman and Ira B. Nadel (1985)
36. *British Novelists, 1890-1929: Modernists*, edited by Thomas F. Staley (1985)
37. *American Writers of the Early Republic*, edited by Emory Elliott (1985)

117 *Twentieth-Century Caribbean and Black African Writers, First Series,* edited by Bernth Lindfors and Reinhard Sander (1992)

118 *Twentieth-Century German Dramatists, 1889–1918,* edited by Wolfgang D. Elfe and James Hardin (1992)

119 *Nineteenth-Century French Fiction Writers: Romanticism and Realism, 1800–1860,* edited by Catharine Savage Brosman (1992)

120 *American Poets Since World War II, Third Series,* edited by R. S. Gwynn (1992)

121 *Seventeenth-Century British Nondramatic Poets, First Series,* edited by M. Thomas Hester (1992)

122 *Chicano Writers, Second Series,* edited by Francisco A. Lomelí and Carl R. Shirley (1992)

123 *Nineteenth-Century French Fiction Writers: Naturalism and Beyond, 1860–1900,* edited by Catharine Savage Brosman (1992)

124 *Twentieth-Century German Dramatists, 1919-1992,* edited by Wolfgang D. Elfe and James Hardin (1992)

125 *Twentieth-Century Caribbean and Black African Writers, Second Series,* edited by Bernth Lindfors and Reinhard Sander (1993)

126 *Seventeenth-Century British Nondramatic Poets, Second Series,* edited by M. Thomas Hester (1993)

127 *American Newspaper Publishers, 1950-1990,* edited by Perry J. Ashley (1993)

128 *Twentieth-Century Italian Poets, Second Series,* edited by Giovanna Wedel De Stasio, Glauco Cambon, and Antonio Illiano (1993)

129 *Nineteenth-Century German Writers, 1841-1900,* edited by James Hardin and Siegfried Mews (1993)

130 *American Short-Story Writers Since World War II,* edited by Patrick Meanor (1993)

131 *Seventeenth-Century British Nondramatic Poets, Third Series,* edited by M. Thomas Hester (1993)

132 *Sixteenth-Century British Nondramatic Writers, First Series,* edited by David A. Richardson (1993)

133 *Nineteenth-Century German Writers to 1840,* edited by James Hardin and Siegfried Mews (1993)

134 *Twentieth-Century Spanish Poets, Second Series,* edited by Jerry Phillips Winfield (1994)

135 *British Short-Fiction Writers, 1880–1914: The Realist Tradition,* edited by William B. Thesing (1994)

136 *Sixteenth-Century British Nondramatic Writers, Second Series,* edited by David A. Richardson (1994)

137 *American Magazine Journalists, 1900-1960, Second Series,* edited by Sam G. Riley (1994)

138 *German Writers and Works of the High Middle Ages: 1170–1280,* edited by James Hardin and Will Hasty (1994)

139 *British Short-Fiction Writers, 1945-1980,* edited by Dean Baldwin (1994)

140 *American Book-Collectors and Bibliographers, First Series,* edited by Joseph Rosenblum (1994)

141 *British Children's Writers, 1880–1914,* edited by Laura M. Zaidman (1994)

142 *Eighteenth-Century British Literary Biographers,* edited by Steven Serafin (1994)

143 *American Novelists Since World War II, Third Series,* edited by James R. Giles and Wanda H. Giles (1994)

144 *Nineteenth-Century British Literary Biographers,* edited by Steven Serafin (1994)

145 *Modern Latin-American Fiction Writers, Second Series,* edited by William Luis and Ann González (1994)

146 *Old and Middle English Literature,* edited by Jeffrey Helterman and Jerome Mitchell (1994)

147 *South Slavic Writers Before World War II,* edited by Vasa D. Mihailovich (1994)

148 *German Writers and Works of the Early Middle Ages: 800–1170,* edited by Will Hasty and James Hardin (1994)

149 *Late Nineteenth- and Early Twentieth-Century British Literary Biographers,* edited by Steven Serafin (1995)

150 *Early Modern Russian Writers, Late Seventeenth and Eighteenth Centuries,* edited by Marcus C. Levitt (1995)

151 *British Prose Writers of the Early Seventeenth Century,* edited by Clayton D. Lein (1995)

152 *American Novelists Since World War II, Fourth Series,* edited by James and Wanda Giles (1995)

153 *Late-Victorian and Edwardian British Novelists, First Series,* edited by George M. Johnson (1995)

154 *The British Literary Book Trade, 1700-1820,* edited by James K. Bracken and Joel Silver (1995)

155 *Twentieth-Century British Literary Biographers,* edited by Steven Serafin (1995)

156 *British Short-Fiction Writers, 1880-1914: The Romantic Tradition,* edited by William F. Naufftus (1995)

157 *Twentieth-Century Caribbean and Black African Writers, Third Series,* edited by Bernth Lindfors and Reinhard Sander (1995)

158 *British Reform Writers, 1789–1832,* edited by Gary Kelly and Edd Applegate (1995)

159 *British Short Fiction Writers, 1800–1880,* edited by John R. Greenfield (1996)

160 *British Children's Writers, 1914–1960,* edited by Donald R. Hettinga and Gary D. Schmidt (1996)

161 *British Children's Writers Since 1960, First Series,* edited by Caroline Hunt (1996)

162 *British Short-Fiction Writers, 1915-1945,* edited by John H. Rogers (1996)

163 *British Children's Writers, 1800–1880,* edited by Meena Khorana (1996)

164 *German Baroque Writers, 1580–1660,* edited by James Hardin (1996)

165 *American Poets Since World War II, Fourth Series,* edited by Joseph Conte (1996)

166 *British Travel Writers, 1837–1875,* edited by Barbara Brothers and Julia Gergits (1996)

167 *Sixteenth-Century British Nondramatic Writers, Third Series,* edited by David A. Richardson (1996)

168 *German Baroque Writers, 1661–1730,* edited by James Hardin (1996)

169 *American Poets Since World War II, Fifth Series,* edited by Joseph Conte (1996)

170 *The British Literary Book Trade, 1475-1700,* edited by James K. Bracken and Joel Silver (1996)

171 *Twentieth-Century American Sportswriters,* edited by Richard Orodenker (1996)

172 *Sixteenth-Century British Nondramatic Writers, Fourth Series,* edited by David A. Richardson (1996)

173 *American Novelists Since World War II, Fifth Series,* edited by James R. Giles and Wanda H. Giles (1996)

174 *British Travel Writers, 1876–1909,* edited by Barbara Brothers and Julia Gergits (1997)

Documentary Series

1 *Sherwood Anderson, Willa Cather, John Dos Passos, Theodore Dreiser, F. Scott Fitzgerald, Ernest Hemingway, Sinclair Lewis,* edited by Margaret A. Van Antwerp (1982)

2 *James Gould Cozzens, James T. Farrell, William Faulkner, John O'Hara, John Steinbeck, Thomas Wolfe, Richard Wright,* edited by Margaret A. Van Antwerp (1982)

3 *Saul Bellow, Jack Kerouac, Norman Mailer, Vladimir Nabokov, John Updike, Kurt Vonnegut,* edited by Mary Bruccoli (1983)

4 *Tennessee Williams,* edited by Margaret A. Van Antwerp and Sally Johns (1984)

5 *American Transcendentalists,* edited by Joel Myerson (1988)

6 *Hardboiled Mystery Writers: Raymond Chandler, Dashiell Hammett, Ross Macdonald,* edited by Matthew J. Bruccoli and Richard Layman (1989)

7 *Modern American Poets: James Dickey, Robert Frost, Marianne Moore,* edited by Karen L. Rood (1989)

8 *The Black Aesthetic Movement,* edited by Jeffrey Louis Decker (1991)

9 *American Writers of the Vietnam War: W. D. Ehrhart, Larry Heinemann, Tim O'Brien, Walter McDonald, John M. Del Vecchio,* edited by Ronald Baughman (1991)

10 *The Bloomsbury Group,* edited by Edward L. Bishop (1992)

11 *American Proletarian Culture: The Twenties and The Thirties,* edited by Jon Christian Suggs (1993)

12 *Southern Women Writers: Flannery O'Connor, Katherine Anne Porter, Eudora Welty,* edited by Mary Ann Wimsatt and Karen L. Rood (1994)

13 *The House of Scribner, 1846–1904,* edited by John Delaney (1996)

14 *Four Women Writers for Children, 1868–1918,* edited by Caroline C. Hunt (1996)

Yearbooks

1980 edited by Karen L. Rood, Jean W. Ross, and Richard Ziegfeld (1981)

1981 edited by Karen L. Rood, Jean W. Ross, and Richard Ziegfeld (1982)

1982 edited by Richard Ziegfeld; associate editors: Jean W. Ross and Lynne C. Zeigler (1983)

1983 edited by Mary Bruccoli and Jean W. Ross; associate editor: Richard Ziegfeld (1984)

1984 edited by Jean W. Ross (1985)

1985 edited by Jean W. Ross (1986)

1986 edited by J. M. Brook (1987)

1987 edited by J. M. Brook (1988)

1988 edited by J. M. Brook (1989)

1989 edited by J. M. Brook (1990)

1990 edited by James W. Hipp (1991)

1991 edited by James W. Hipp (1992)

1992 edited by James W. Hipp (1993)

1993 edited by James W. Hipp, contributing editor George Garrett (1994)

1994 edited by James W. Hipp, contributing editor George Garrett (1995)

1995 edited by James W. Hipp, contributing editor George Garrett (1996)

Concise Series

Concise Dictionary of American Literary Biography, 6 volumes (1988-1989): *The New Consciousness, 1941-1968; Colonization to the American Renaissance, 1640-1865; Realism, Naturalism, and Local Color, 1865-1917; The Twenties, 1917-1929; The Age of Maturity, 1929-1941; Broadening Views, 1968-1988.*

Concise Dictionary of British Literary Biography, 8 volumes (1991-1992): *Writers of the Middle Ages and Renaissance Before 1660; Writers of the Restoration and Eighteenth Century, 1660-1789; Writers of the Romantic Period, 1789-1832; Victorian Writers, 1832-1890; Late Victorian and Edwardian Writers, 1890-1914; Modern Writers, 1914-1945; Writers After World War II, 1945-1960; Contemporary Writers, 1960 to Present.*

Dictionary of Literary Biography® • Volume One Hundred Seventy-Four

British Travel Writers, 1876–1909

Edited by
Barbara Brothers
and
Julia Gergits
Youngstown State University

A Bruccoli Clark Layman Book
Gale Research
Detroit, Washington, D.C., London

Advisory Board for
DICTIONARY OF LITERARY BIOGRAPHY

John Baker
William Cagle
Patrick O'Connor
George Garrett
Trudier Harris

Matthew J. Bruccoli and Richard Layman, Editorial Directors
C. E. Frazer Clark Jr., Managing Editor
Karen Rood, Senior Editor

Printed in the United States of America

Published simultaneously in the United Kingdom
by Gale Research International Limited
(An affiliated company of Gale Research)

The paper used in this publication meets the minimum requirements
of American National Standard for Information Sciences–Permanence
Paper for Printed Library Materials, ANSI Z39.48-1984. ∞™

This publication is a creative work fully protected by all applicable copyright laws, as well as by misappropriation, trade secret, unfair competition, and other applicable laws. The authors and editors of this work have added value to the underlying factual material herein through one or more of the following: unique and original selection, coordination, expression, arrangement, and classification of the information.

All rights to this publication will be vigorously defended.

Copyright © 1997 by Gale Research
835 Penobscot Building
Detroit, MI 48226

All rights reserved including the right of reproduction in
whole or in part in any form.

Library of Congress Cataloging-in-Publication Data

British travel writers, 1876-1909 / edited by Barbara Brothers and Julia Gergits.
    p. cm. – (Dictionary of literary biography; v. 174)
"A Bruccoli Clark Layman book."
Includes bibliographical references and index.
ISBN 0-8103-9937-7 (alk. paper)
1. Travelers' writings, English – Bio-bibliography – Dictionaries. 2. British – Travel – Foreign countries – History – Dictionaries. 3. English prose literature – 19th century – Bio-bibliography – Dictionaries. 4. English prose literature – 20th century – Bio-bibliography – Dictionaries. 5. Great Britain – History – Victoria, 1837–1901 – Biography – Dictionaries. 6. Great Britain – History – Edward VII, 1901-1910 – Biography – Dictionaries. 7. Authors, English – 19th century – Biography – Dictionaries. 8. Authors, English – 20th century – Biography – Dictionaries. 9. Travelers – Great Britain – Biography – Dictionaries. 10. Travel writing – Bio-bibliography – Dictionaries. I. Brothers, Barbara, 1937- . II. Gergits, Julia Marie. III. Series.
PR788.T72B743 1997
820.9'355 – dc21   96-49467
[B]   CIP

---

10 9 8 7 6 5 4 3 2 1

To our fellow travelers, Lawrence Haims and James Schramer

Contents

Plan of the Series ix
Introduction xi

Gertrude Margaret Lowthian Bell
   (1868–1926) 3
   Nancy V. Workman

Hilaire Belloc (1870–1953) 19
   Brian D. Reed

James Theodore Bent (1852–1897) and
   Mabel Virginia Anna Bent (?–?) 31
   Kelly Belanger

Wilfrid Scawen Blunt (1840–1922) and
   Lady Anne Isabella Noel Blunt
   (1837–1917) 41
   Syrine C. Hout

Margaret Brooke, Ranee of Sarawak
   (1849–1936) 54
   Susan Morgan

Samuel Butler (1835–1902) 60
   Keith C. Odom

Constance Gordon Cumming
   (1837–1924) 67
   Susan Schoenbauer Thurin

W. H. Davies (1871–1940) 89
   Ira Grushow

Florence Douglas Dixie
   (1857–1905) 97
   Catherine Barnes Stevenson

Charles M. Doughty (1843–1926) 103
   Stephen E. Tabachnick

Amelia Anne Blandford Edwards
   (1831–1892) 116
   Patricia O'Neill

Matilda Barbara Betham-Edwards
   (1836–1919) 130
   Cynthia Ellen Patton

Douglas W. Freshfield (1845–1934) 140
   Julie English Early

Mary Gaunt (1861–1942) 147
   James Rigney

R. B. Cunninghame Graham
   (1852–1936) 154
   Ira Grushow

Beatrice Ethel Grimshaw
   (1871–1953) 163
   J. Allen Barksdale

H. Rider Haggard
   (1856–1925) 168
   John W. M. Hallock

Agnes Herbert (circa 1880–1960) 178
   Timothy S. Jones

W. H. Hudson (1841–1922) 181
   Joan Corwin

Sir Harry Johnston (1858–1927) 191
   Brian D. Reed

Mary Henrietta Kingsley
   (1862–1900) 201
   Catherine Barnes Stevenson

G. W. H. Knight-Bruce (1852–1896) 211
   James Rigney

Mrs. Aubrey Le Blond (1861–1934) 217
   Timothy S. Jones

Vernon Lee (1856–1935) 221
   Patricia O'Neill

Agnes Smith Lewis (1843–1926) and
   Margaret Dunlop Gibson
   (1843–1920) 232
   Barbara Brothers

Elinor Mordaunt (1872?–1942) 241
   Colleen Hobbs

Marianne North (1830-1890) 251
  Maura O'Connor

Frederick Courteney Selous (1851-1917) 256
  David C. Judkins

Robert Louis Stevenson (1850-1894) 268
  Gordon Hirsch

Ella C. Sykes (?-1939) 289
  Scott R. Christianson

Joseph Thomson (1858-1895) 294
  James Barszcz

Ethel Brilliana Tweedie (circa 1860-1940) 300
  Josephine A. McQuail

Travel Writing, 1876-1909 317
Books for Further Reading 327
Contributors 333
Cumulative Index 337

Plan of the Series

...Almost the most prodigious asset of a country, and perhaps its most precious possession, is its native literary product — when that product is fine and noble and enduring.

Mark Twain*

The advisory board, the editors, and the publisher of the *Dictionary of Literary Biography* are joined in endorsing Mark Twain's declaration. The literature of a nation provides an inexhaustible resource of permanent worth. We intend to make literature and its creators better understood and more accessible to students and the reading public, while satisfying the standards of teachers and scholars.

To meet these requirements, *literary biography* has been construed in terms of the author's achievement. The most important thing about a writer is his writing. Accordingly, the entries in *DLB* are career biographies, tracing the development of the author's canon and the evolution of his reputation.

The purpose of *DLB* is not only to provide reliable information in a convenient format but also to place the figures in the larger perspective of literary history and to offer appraisals of their accomplishments by qualified scholars.

The publication plan for *DLB* resulted from two years of preparation. The project was proposed to Bruccoli Clark by Frederick C. Ruffner, president of the Gale Research Company, in November 1975. After specimen entries were prepared and typeset, an advisory board was formed to refine the entry format and develop the series rationale. In meetings held during 1976, the publisher, series editors, and advisory board approved the scheme for a comprehensive biographical dictionary of persons who contributed to North American literature. Editorial work on the first volume began in January 1977, and it was published in 1978. In order to make *DLB* more than a reference tool and to compile volumes that individually have claim to status as literary history, it was decided to organize volumes by topic, period, or genre. Each of these freestanding volumes provides a biographical-bibliographical guide and overview for a particular area of literature. We are convinced that this organization — as opposed to a single alphabet method — constitutes a valuable innovation in the presentation of reference material. The volume plan necessarily requires many decisions for the placement and treatment of authors who might properly be included in two or three volumes. In some instances a major figure will be included in separate volumes, but with different entries emphasizing the aspect of his career appropriate to each volume. Ernest Hemingway, for example, is represented in *American Writers in Paris, 1920-1939* by an entry focusing on his expatriate apprenticeship; he is also in *American Novelists, 1910-1945* with an entry surveying his entire career. Each volume includes a cumulative index of the subject authors and articles. Comprehensive indexes to the entire series are planned.

With volume ten in 1982 it was decided to enlarge the scope of *DLB*. By the end of 1986 twenty-one volumes treating British literature had been published, and volumes for Commonwealth and Modern European literature were in progress. The series has been further augmented by the *DLB Yearbooks* (since 1981) which update published entries and add new entries to keep the *DLB* current with contemporary activity. There have also been *DLB Documentary Series* volumes which provide biographical and critical source materials for figures whose work is judged to have particular interest for students. One of these companion volumes is entirely devoted to Tennessee Williams.

We define literature as the *intellectual commerce of a nation:* not merely as belles lettres but as that ample and complex process by which ideas are generated, shaped, and transmitted. *DLB* entries are not limited to "creative writers" but extend to other figures who in their time and in their way influenced the mind of a people. Thus the series encompasses historians, journalists, publishers, and screenwriters. By this means readers of *DLB* may be aided to perceive literature not as cult scripture in the keeping of intellectual high priests but firmly positioned at the center of a nation's life.

*From an unpublished section of Mark Twain's autobiography, copyright by the Mark Twain Company

Plan of the Series

DLB includes the major writers appropriate to each volume and those standing in the ranks immediately behind them. Scholarly and critical counsel has been sought in deciding which minor figures to include and how full their entries should be. Wherever possible, useful references are made to figures who do not warrant separate entries.

Each *DLB* volume has a volume editor responsible for planning the volume, selecting the figures for inclusion, and assigning the entries. Volume editors are also responsible for preparing, where appropriate, appendices surveying the major periodicals and literary and intellectual movements for their volumes, as well as lists of further readings. Work on the series as a whole is coordinated at the Bruccoli Clark Layman editorial center in Columbia, South Carolina, where the editorial staff is responsible for accuracy of the published volumes.

One feature that distinguishes *DLB* is the illustration policy – its concern with the iconography of literature. Just as an author is influenced by his surroundings, so is the reader's understanding of the author enhanced by a knowledge of his environment. Therefore *DLB* volumes include not only drawings, paintings, and photographs of authors, often depicting them at various stages in their careers, but also illustrations of their families and places where they lived. Title pages are regularly reproduced in facsimile along with dust jackets for modern authors. The dust jackets are a special feature of *DLB* because they often document better than anything else the way in which an author's work was perceived in its own time. Specimens of the writers' manuscripts are included when feasible.

Samuel Johnson rightly decreed that "The chief glory of every people arises from its authors." The purpose of the *Dictionary of Literary Biography* is to compile literary history in the surest way available to us – by accurate and comprehensive treatment of the lives and work of those who contributed to it.

The *DLB* Advisory Board

Introduction

British Travel Writers, 1876–1909 covers a period that was tumultuous and contradictory, exciting and terrifying, modern and old-fashioned, nationalistic and self-critical, liberated and intensely repressed. Between 1876 and 1909 frontiers disappeared, as the curious and often rapacious Western eye had peered at least cursorily into the most remote corners of the world, and settlements covered much of North and South America. Railroads crisscrossed North America and Europe and extended substantially into Africa and Asia; ships propelled by steam mastered the oceans; and the first automobiles slowly and noisily chugged along streets intended for horse-drawn carriages.

The British Empire was at its zenith, and optimism was high. Scientists predicted the advancement of the "colored" peoples of the world under the guidance of the "superior" white man. Social engineers argued for quickly and "painlessly" killing "lesser" peoples, and Sigmund Freud uncovered the mysteries of dreams and infant sexuality. Man and machine seemed to be progressing to a golden age. Almost as quickly such a vision became tarnished by mounting expenses of governing the far-flung empire. The costs of conquering and ruling distant peoples included commitments of both money and men, as westerners clashed with one another over possession of prime territories such as South Africa. Meanwhile, the German states were rearranged for union and empire, and natives of czarist Russia moved inevitably toward revolution.

The writers covered in this volume share much with those in *DLB 166: British Travel Writers, 1837–1875*. Many of those early writers were still traveling and writing throughout this period, and their later works are among those listed in "Travel Writing, 1876–1909," the bibliography of travel books for this volume. For example, Isabella Lucy Bird, long-lived and sturdy despite her reputation as a sixty-year-old invalid, became increasingly adventuresome and traveled through some of the less accessible parts of the world – China, Mongolia, and India.

The introduction to the first volume of this series notes that many travel writers also wrote in different genres or held professions that took them to faraway places. In this volume such writers whose prominence was achieved in other literary fields include H. Rider Haggard, best known for his novels such as *King Solomon's Mines* (1885) and *She* (1886); Robert Louis Stevenson, famous now for *Treasure Island* (1883) and *The Strange Case of Dr. Jekyll and Mr. Hyde* (1886) but well known in his own time for travel writing; and Samuel Butler, whose travel works would most likely have gone unnoticed if not for the fame that he achieved in publishing the satiric romances *Erewhon* (1872) and *Erewhon Revisited* (1901) as well as his autobiographical novel, *The Way of All Flesh* (1903). Other writers such as Vernon Lee, Charles M. Doughty, Edith Durham, and W. H. Hudson were historians, illustrators, archaeologists, linguists, or scholars.

Some differences between the travel writers of the earlier volume and those in this one, however, are significant. Professionalism was increasingly becoming the norm: writers such as Durham and H. M. Tomlinson were professional journalists, editors, and correspondents accustomed to writing on contract and with deadlines. In addition, the style, shape, and content of travel writing changed in ways that reflected the various aims that motivated the writers to travel, the expanding canon of travel books to which writers contributed, and the diverse interests and backgrounds of readers. A writer such as Augustus J. C. Hare – whose books, like Durham's, are listed in "Travel Writers, 1876–1909" – wrote travel guides for John Murray and a series of books on Italian and French cities: *Walks in Rome* (1871), *Days near Paris* (1880), and *Cities of Southern Italy and Sicily* (1883). Biographies, histories, and other nonfiction books make up the canon of this highly prolific and professional writer who never wrote a novel and frequently illustrated his own works.

Scientific societies, particularly the Royal Geographical Society, sponsored much of the remaining discovery. Through their journals and meetings they sought to inform a growing number of members about strides in mapping Africa, and they encouraged the explorers whom they sponsored to write for the general public as well. In the 1860s Richard Francis Burton, John Hanning Speke, James Augustus Grant, David Livingstone, and Mr. and Mrs. John Petherick had produced books about Africa and the search for the source of the Nile

River, and many of them, like other explorers, wrote for *Blackwood's Magazine*.

Another change was in the attitude of British society toward missionary efforts. Although Victorians gradually acknowledged that missionary work failed, in most instances, to achieve its primary goal of converting "pagans," missionaries and other deeply religious travelers became prominent as travel writers. Such writers as Mildred Cable and G. W. H. Knight-Bruce traveled in the interests of religion but became embroiled in far more wide-ranging political and commercial issues.

Realities of Travel and Tourism

Many technical and economic innovations made travel easier, cheaper, and safer in the late nineteenth century. Railroad lines connected major cities throughout the Western world, including the United States, where a traveler could ride across the entire country. Railroads were also spreading rapidly outside the Western world, often financed by Western investors intent on opening further economic opportunities and thereby providing access into the interior for countless travelers other than those with business to conduct. By 1880 the United States had 87,800 miles of railway in operation; Great Britain, 17,900 miles; France, 16,400 miles; and Russia, 12,200 miles. Japan, China, India, Egypt, and South Africa also joined the race into the twentieth century with the development of their own railroads and steam shipping.

As the miles of track expanded, the comfort of the cars traveling the rails increased. Five months after the inauguration of the Orient Express in 1883, sleeping cars were introduced. Soon came the luxurious Calais-Mediterranean Express, which by 1897 ran three times weekly. An increasing number of pleasure seekers joined those who traveled for knowledge, adventure, or business. The telephone and the telegraph also improved communication and made travel easier and safer: Alexander Graham Bell invented the telephone in 1876, and London had its first telephone exchange in 1879. Few corners of the world were inaccessible.

Before the end of this period Henry Ford had built his first automobile, and in 1903, the same year in which taxis appeared in London, the Ford Motor Company began production and Orville and Wilbur Wright first flew their powered airplane. These advances in the means of transportation further democratized travel, just as the railway had done in the first half of the nineteenth century. Guidebooks in hand, the less wealthy and adventuresome found it possible to strike out on their own, or, through the services of a tourist company such as Thomas Cook's, which would provide tour guides to negotiate with dragomans and hotel keepers, these travelers could visit what had been remote corners of the world. Gertrude Margaret Lowthian Bell acknowledged in *The Desert and the Sown* (1907) that improved travel opportunities had changed travel writing, and her work included recommendations to help future travelers to areas that no Europeans had visited just a decade or two earlier.

The Mediterranean region was no longer viewed principally as a tourist attraction, the land of antiquity, the cradle of religions and civilizations. It also became the land of oil, as drilling began in Persia in 1901, and the Anglo-Persian Oil Company was formed by 1909. Bell, widely recognized as a travel writer on the Middle East, became a political adviser to kings, was appointed to the post of oriental secretary to the British high commissioner of Iraq and later to that of first director of antiquities by Iraq's King Faisal I, and was charged to found a national museum for the country.

The refinement of the tourism industry between 1876 and 1909 allowed thousands of neophyte tourists to visit previously exclusive regions. James Pemble, in *Mediterranean Passion: Victorians and Edwardians in the South* (1988), and James Buzard, in *The Beaten Track* (1993), note the rise of tourism and the changes wrought in the cultures and landscapes of the countries visited by tourists. This invasion provoked those visitors who considered themselves professional travelers. For the most part, travel writers in this volume describe their practices as contradictions of the popular tourist routines and rules, which prescribed where to go, what to see, and what to value — in the form of an increasingly intrusive commercial apparatus that was used to market travel to such tourists. As Buzard notes,

> Nervous jesting about the global "imperial" aspirations of the Thomas Cook Company, or about railways running to the top of the Himalayas, or about English or American dishes being offered at every table d'hote on earth all expressed some of the century's discomfiture concerning a cultural practice which, as it expanded, exercised a strange and unprecedented form of power over the character of places. Witnesses saw or suspected that tourism was capable of both physically remaking places (by introducing railways, hotels, restaurants, Thomas Cook offices, souvenir shops, crowds of tourists) and *re-presenting* them in a series of mnemonic stereotypes (symbols of Paris, Rome, Italy, the Rhine), and that it involved both material and "rhetorical" coercion.

Tourists were powerful colonialist forces: their existence altered economies, cultures, and political systems, and tourism gained momentum as the British Empire spread its military, political, and economic tendrils. For example, in *A Thousand Miles up the Nile* (1877) Amelia Anne Blandford Edwards describes how one of the gentlemen traveling in her party accidentally shot a child from a local village when he was aiming at a low-flying bird. The villagers rose up in anger but calmed down when they realized that the offender was British. A small bribe and some smooth talking sufficed to pay for the child's suffering. Edwards writes that residents of entire villages were killed or taken into custody for violating the freedom of British travelers and that British safety took precedence over justice or equity. While the disruptive effects of tourism were not generally so extreme, indigenous peoples were caught between their needs for tourist dollars and their desires to maintain their cultural values and ways of life. In seeking to accommodate to the needs and wants of visitors (and to do so as those visitors expected), natives often risked destroying their own cultural values and ways of life.

Despite claims to the contrary, the distinction between tourism and traveling that travel writers and commentators make is extremely unclear – even imaginary, in view of the realities of their traveling. For example, Cook's international corporation influenced tourists and travelers alike. At his urging, international banking and currency exchanges sprang up throughout the world to serve peripatetic westerners, and Cook's tours financed shipping lines, telegraphs, railroads, and hotels. Cook held the monopoly for carrying Egyptian government mails, and his fleet of pleasure steamers twice transported British government troops and stores up the Nile River: in 1884 in a failed effort to relieve Gen. Charles George Gordon at Khartoum and in 1896 to carry Gov. Gen. Horatio Herbert Kitchener to Atbara to avenge the earlier defeat of the British at the hands of the Arabs.

Cook's organization was therefore so ubiquitous that it served needs other than those of tourists, and his services expanded in the same way that Bell's had expanded. A list of those who attended the fiftieth-anniversary party of his company in 1891 attests to how varied and powerful his clients were. Such dignitaries included titled figures from Britain, Russia, Ireland, Saxony, and Germany. Kitchener, Cecil Rhodes, Haggard, Rudyard Kipling, and Prime Minister William Ewart Gladstone relied on Cook's. To travel nearly anywhere in the world, it was patently foolish to avoid Cook's organization. By 1900 Cook had offices in Cape Town, Bombay, Calcutta, Australia, New Zealand, the United States, Canada, India, and Egypt, and the company transported Indian princes and their entourages to England as well as Muslim pilgrims to Jidda. Whether one were a tourist or a traveler, Cook's agency was the organizer and facilitator of the journey.

Nor was the use of Cook's services the only similarity between tourist and traveler. Tourists and self-proclaimed travelers insisted on maintaining their worldviews and English comforts. Edwards, for example, sailed up the Nile with her writing desk and piano and enjoyed afternoon teas and formal dinners. Although American adventurer Henry M. Stanley lived off the land as he journeyed inland, he required English cooking – even of antelope – and he brought his rubber bath and English tea. Others read the landscape just as tourists did, seeing Egypt as the repository of ancient mysteries and unfortunately peopled by the detritus of centuries – a degraded society fallen from its glorious past. The differences between tourists and travelers are largely illusory and rhetorical.

Distinguishing clearly between a traveler and a tourist in this age is as impossible as it had been in earlier ages. Was a pilgrim a tourist or a traveler? Although most of the writers used the same facilities just as tourists did, travelers felt themselves to be different – and better. Traces of latent class prejudice appear in the works of such writers as Edwards, when she sniffs at Cook's steam tours of the Nile, or Durham, who studiously avoids territory familiar to British tourists. Victorian Britain was a class-conscious society, and the books of those considered to be the elite were as marketable to its reading audience as are the wares of the rich and famous to a twentieth-century American audience.

The Impact of Science

Scientific research greatly affected travel writers, even though Victorians were slow to make science a professional endeavor. Preferring to preserve a classical, literary focus in their curricula, universities ignored science as a formal course of study until late in the nineteenth century. The most famous Victorian scientists were talented and well-to-do amateurs who had the time and talent to dedicate to their avocations. For example, Charles Darwin, the renowned evolutionary theorist, and Havelock

Ellis, the British psychologist who was a contemporary of Freud and an advocate of social progress through science, had to defend their research from many in the academy who sought to squelch their scientific work. Vying with each other for coveted "royal" status that was thoughtfully granted by Prince Albert and then by Queen Victoria after his death, privately organized scientific societies provided the support and interest that academic organizations did not. The success of organizations such as the Royal Geographical Society (R.G.S.), founded in 1830, and the Royal Asiatic Society (R.A.S.), founded in 1822, encouraged professional amateurs, a few of whom proudly came to signify their membership in a select community of serious researchers by adding R.G.S. or R.A.S. after their names.

Although the Royal Geographical Society had its share of failures, it had also supported several notable travelers in the early Victorian period: Samuel White Baker, Livingstone, and Bird. In the late Victorian period the R.G.S. and a few other societies manipulated geographical research. A science-oriented traveler had no audience or support without the imprimatur of such an organization. Even token support, such as the lending of scientific apparatus, was sufficient to give credence to one's travel. Stanley proudly proclaimed his membership in the R.G.S., and Mary Henrietta Kingsley presented some of her most erudite essays at its meetings. Following in the footsteps of Stanley and Livingstone, Joseph Thomson traveled into Africa with the financial support of the R.G.S., and membership in the society became a hotly contested award that was valuable to the scientist, ethnographer, geographer, botanist, or mountain climber. Being a member of the R.G.S. ensured financial support, publishers, and scholarly and general reading audiences for one's work.

Though not yet incorporated into university curricula, many sciences took shape during the last decades of the nineteenth century. Victorians, famous for their craving for order and system, enthusiastically welcomed the developments of ethnography, anthropology, ecology, and psychology; they heralded the growth of such sciences as biology, botany, geography, chemistry, and entomology. This demand for order distinctly influenced travelers, both in regard to where they traveled and what they did once they had arrived at their destinations. Edwards, who was to found the first chair in Egyptology at University College, London, commented on Harriet Martineau's passion for pyramids by noting, "Miss Martineau tells how, in this part of the [Nile] river, she was scarcely satisfied to sit down to breakfast without having first explored a Temple; but I could have breakfasted, dined, and supped on Temples." Even though Edwards was not trained as a scientist, she helped to systematize Egyptology as a science, and her apparently romantic wanderings through ancient Egypt resulted in serious scholarly contributions.

Kingsley, another Victorian amateur scientist, had scientific motives for her travel: she wished to identify new fish species and collect native fetishes. She collected and pickled fish on the spot and enthusiastically observed native tribal behavior and traditions, even when her safety was threatened. She participated in the development of ethnography (a version of anthropology at the time), in which the researcher watches the subject and gathers information through taking notes, collecting artifacts, and studiously avoiding personal involvement. Despite what modern critics may believe, Victorian scientists were aware of the influence of white scientists on indigenous people, and these early researchers consciously tried to prevent their feelings and values from polluting their observations and conclusions. They also recognized that what they observed was tainted by their presence in any village or community and that it was essential to gather information from more than one informant. Kingsley worked hard at articulating the reality of native life without imposing her cultural prejudices or preconceptions.

Even such sporting activities as mountaineering assumed a scientific edge during this period. Mountaineering evolved from a part-time pursuit to a professional method of gathering information. Douglas W. Freshfield, member and president of the Alpine Club and a fellow of the R.G.S., was a pioneer of mountaineering, a purist who eschewed climbing apparatus such as grommets or crampons and preferred to free climb. As editor of the *Alpine Journal*, he encouraged a new generation of climbers to adopt the view that the goal of climbing was scientific, and he scoffed at climbing for the sheer sake of making a record. Mrs. Aubrey Le Blond, whose first husband had introduced her to alpine climbing, was to enjoy a long list of accomplishments and became the first president of the Ladies' Alpine Club, founded in 1907.

Travelers with a scientific intent eschewed political viewpoints or haggling and ignored or overlooked social problems that fell outside their scientific purviews. For example, Marianne North was monomaniacal about gathering and painting botanical samples. As a collector for the Royal Botanical

Society, she could – and did – remove rare varieties from their native soil for transport to England, where they would be grown in greenhouses if they survived the voyage. The tone of her commentary on natives is determined by how willingly they enable her to gather plants, how they treat or use their plants, and how well they feed, clothe, and protect her while she conducts her scientific endeavor. Doughty, an accomplished linguist, archaeologist, and geologist, focused on his archaeological discoveries in his travel writing. Such dedicated scientists and scholars frequently saw native residents of non-European lands as unworthy of their natural riches.

The Impact of Empire

In the period from 1876 to 1909 the British Empire was at its strongest, although it was burdened by outrageous expenses of maintaining armies and political and economic structures. England had colonies established throughout the world: India was the most well known, but the English also had outposts in Africa, the Bahamas, the Virgin Islands, Bermuda (which is still a British affiliate), and South New Guinea. In addition to maintaining colonies, England also accepted responsibilities for protecting Sarawak, Malaysia, Uganda, and Egypt and established trade agreements with China and Japan. Canada had gained independence from Britain in 1867 and Australia was to do so in 1901, but both countries kept close economic and political ties with England.

Any survey of the changing map of political-economic authority during the period is likely to be confusing. Nearly all parts of the world were under dispute, either by natives trying to reclaim their lands or by Europeans neglecting native peoples to fight for control with other nations seeking lands. For example, Korea became the object of a fierce battle between China and Japan, a dispute made even more complex when the British tried to impose economic pressure on the Japanese to force them to deal more reasonably with the Koreans. With the help of Stanley, who was disgusted with the unwillingness of both Britain and the United States to take control in Africa, King Leopold I of Belgium seized the Congo as his property in 1885. In the same year control of Khartoum shifted from Gordon to the Arabs, and this change caused reprisals that reverberated for many years. In 1888 Logbengula, king of Matabele, granted mining rights to Rhodes after having accepted British protection. By 1895 Rhodes's South Africa Company had secured control of the territory for the British and successfully blocked German, Dutch, or Portuguese attempts to wrest control from England.

Negotiating, bullying, mutinies, and aggression such as that which had secured the Rhodesian territory continued throughout other areas of the British Empire. In India, where control had passed to the British Crown from the British East India Company following the Indian Mutiny of 1857, the Indian National Congress accepted the benefits of British rule and economic ties until the beginning of the twentieth century, when the nationalist movement attracted support. The Belgians, French, Germans, and Americans vied with each other for profitable land around the world. Few of these aspiring colonial powers wanted political alliances with the native peoples of these countries; at most, westerners negotiated for the best possible economic and military positions within these countries – and then sought complete control as quickly as possible after having gained entrance.

Travelers followed the paths of imperial development, as their books readily attest. Stanley, whose brave and bold trip into the interior of Africa confirmed the source of the Nile River, describes in *The Dark Continent* (1899) the Western style of commerce in the oceanside cities that supplied food, guns, blankets, and maps to his army of three hundred bearers and guides. Without westernization his exploration would have been nearly impossible. Commerce opened new realms for scientists, adventurers, and explorers; political intervention secured their safety, even as it threatened the lives and livelihoods of native peoples. Margaret Brooke, Ranee of Sarawak, was perhaps one of the most peculiar remnants of an older form of imperialism. The sultan of Brunei had appointed her husband's uncle as raja of a small island community, and in 1868 her husband inherited this dominion, which in 1888 became a British colony. After producing the necessary sons for her businesslike husband, she returned to England to write of her experiences.

Without empire to familiarize the natives of foreign lands with British traditions and expectations, travelers would not have been able to travel so freely and safely. Following the paths of traders inland from coastal cities irrevocably changed by Western contact, travelers depended on European outposts for solace, supplies, and guidance. By being able to depend on such benefits that the empire conferred on them as travelers, few of the writers in this volume experienced real danger –

even when they were in the wildest, most remote areas.

Religious Imperialism

Travel and travel writing owe a debt to much-maligned missionary work, which continued throughout the nineteenth century, even though the discouraging results of such efforts eventually became known back in England. British evangelical groups continued to believe fervently that their mission was to convert the "savages" to Christianity, even if doing this meant risking British lives. They sought funding from their British followers, lobbied politicians for protection and aid, and recruited volunteers willing to be sent to far outposts to spread the Word.

To scoff at these earnest, dedicated people who left their homelands on missions to save the souls of their brothers and sisters is easy. Because most missionaries were unable to speak native languages and had little training to prepare them to pursue their aim in foreign cultural milieus, it is hard to believe that they could have been so naive as to believe that they could successfully convert to Christianity millions of devout Hindus, Buddhists, Muslims, and practitioners of various native African religions. By midcentury it was clear that they had failed. Livingstone, for example, did not convert one native in Africa. Missionaries in China and Japan were equally unsuccessful: missions were burned, and missionaries were beaten and killed in Chinese uprisings directed at ending European intervention in Chinese religious beliefs. Yet British religious bodies continued to send missionaries around the world on this patently fruitless and dangerous task, because they believed it was their duty to help their fellow men to find the right and true God.

Although modern critics have rejected the methods and goals of missionaries as racist and ethnocentric, it is important to remember that missionaries were motivated by earnest, if often misguided, motives: they wanted the best for their "dark brothers and sisters," and these religious leaders often antagonized British politicians, military leaders, and businessmen to defend what the missionaries regarded as the interests of their "flocks." Missionaries were genuinely trying to help, not hurt, the people whose needs they felt they had been called to fulfill. To their credit, missionaries established hospitals and schools, worked for better treatment of women and female babies, and advocated the humane treatment of the lowest and most miserable classes, such as the untouchables in India. Not all their efforts were characterized by intolerance and absurdity.

Many travelers in the nineteenth and twentieth centuries were missionaries or ordained ministers, and other travelers took advantage of missions throughout the world to facilitate their journeys. Cable and Francesca and Evangeline French, writers whose works are to be examined in a *Dictionary of Literary Biography* volume on British travel writing from 1910 to 1939, were longtime missionaries in China. Fluent in the requisite Chinese dialects, diplomatic enough to wear Chinese garb, and stubborn enough to stay in a country that patently did not welcome their overtures, they were far more educated and successful than most missionaries. Knight-Bruce was sent in 1886 to Africa to serve as bishop at Bloemfontein, where he promptly engaged in a heated conflict with Rhodes. Bird, whose travel writing is discussed in *DLB 166: British Travel Writers, 1837–1875,* in later years dedicated her travels to improving missionary work throughout the Far East and concentrated on China, where she financially supported hospitals and schools that she named after her deceased sister, Hennie, and her late husband, Dr. John Bishop.

The travel books written by missionaries most often are not religious tracts. Few publishers would have tolerated a completely Christianized travel book, if only because missionary work was so intensely controversial. Missionary books were published by religious tract societies and other specialized publishers, but missionaries demonstrated their abilities to write rousingly good travel books, with healthy doses of Christian rhetoric in the background. Missionaries such as Knight-Bruce, whose work was informed by sociopolitical as well as religious concerns, spoke confidently of coming changes in political and economic power and kept their religious missions in the background of their writings. Because missionaries so clearly stated their religious positions, readers today are suspicious of missionaries masquerading as travel writers, yet these missionaries were more well intentioned and generous than many European travelers. Their attempt to fit in in order to convert the pagans helped break down Western insularity.

Travel Writing

By adopting the reporting and recording habits of newly emerging sciences – geography, anthropology, archaeology, Egyptology, botany, linguistics, and others – travel writing became even more of a hybrid genre in this period. Many writers still claimed that they were unable to write well or carefully. Following their self-deprecating steps, Kingsley interrupted her narrative to note, "I must pause here to explain my reasons for giving extracts from my diary, being informed on excellent authority that publishing a diary is a form of literary crime." But this strategy was far less common than it had been: many travel writers of this period were professional writers who sought opportunities to write and were proud of their books as well-crafted, even artistic, pieces.

Professional journalists begin to proliferate. Stanley, perhaps the most famous – or most self-advertising – of the writers in this period, began as a journalist and discovered his passion for travel, adventure, and intrigue when he found Livingstone in Africa. Ethel Brilliana Tweedie was a journalist, editor, and travel writer. Living from 1860 through 1940, she enjoyed many changes in opportunities for both travel writers and women. Even travel writers who were not directly funded by newspapers often wrote articles or features for periodicals and thereby expanded the audience for their travel books.

Nor did travel writing become increasingly hybrid just through adopting the reporting habits of journalism and the sciences. Novelist and travel writer Mary Gaunt used travel so thoroughly in both her fiction and nonfiction that it is difficult to separate the two. As a naturalist, travel writer, and novelist, W. H. Hudson combined his keen eye for detail with a vivid imagination and a poetic style. In *Idle Days in Patagonia* (1893) he experimented by focusing on inactivity, an approach antithetical to typical travel books. W. H. Davies, author of the Super-Tramp series (beginning with *The Autobiography of a Super-Tramp,* 1908), reveals a subculture complete with its own hierarchy and rules: his books present a nontraditional travel account of the underworld of professional tramps, vagabonds, and petty thieves. Championed by influential sponsors, Davies returned to writing poetry after the success of his tramp books.

Readers often question the veracity of supposedly true travel narratives. Primarily a sensationalist fiction writer, Haggard was a novelist for whom travel and exotica were essential ingredients. In *A Winter Pilgrimage* (1901) he tried to inspire interest in travel to the Holy Land, but the book is a blend of truth and fiction. Most readers of travel literature are accustomed to querying the narrative voice and wondering just who, exactly, the traveler might be.

This curiosity is useful, because travel stories owe much to adventure tales, and a writer's concerns for arousing interest and excitement sometimes supersede those for truth or accuracy. R. B. Cunninghame Graham foregrounds this questioning in his brief sketches, impressionistic effects, elusive images, and experiments with various voices that systematically blur the lines between fact and fiction.

Late in the Victorian period the Aesthetic Movement was gaining momentum, and for the most part it hardly influenced those forthright and direct travel writers who had little patience with "art for art's sake" or amorality in art. Yet some travel writing shows signs of increased interest in creating an effect purely for the sake of art. More integrated than earlier forms of "word painting," the travel writing in Graham's poetic sketches incorporated detailed descriptions of nature that are full of vivid adjectives and too many adverbs. Description as an art shapes Stevenson's crisp, lively, and engaging characterizations and Vernon Lee's loose, impressionistic narrative frames. Rather than simply arousing readers' interest in travel or informing readers of strange facts, these writers crafted writing styles that inspired later generations to new stylistic heights. In fact, such a view of travel literature between 1876 and 1909 contributes to a redefinition of the aims of travel, which came to include that of contemplating beauty, of creating beauty, of finding oneself – aesthetic and personal aims that became crucial for writers and readers as the twentieth century progressed.

– *Barbara Brothers and Julia M. Gergits*

Acknowledgments

This book was produced by Bruccoli Clark Layman, Inc. Karen L. Rood is senior editor for the *Dictionary of Literary Biography* series. Denis Thomas and Samuel W. Bruce were the in-house editors.

Production manager is Samuel W. Bruce. Photography editors are Julie E. Frick and Margaret Meriwether. Photographic copy work was performed by Joseph M. Bruccoli. Layout and graphics supervisor is Emily Ruth Sharpe. Copyediting supervisor is Laurel M. Gladden Gillespie. Typesetting supervisor is Kathleen M. Flanagan. Systems manager is Chris Elmore. Laura Pleicones and L. Kay Webster are editorial associates. The production staff includes Phyllis A. Avant, Ann M. Cheschi, Melody W. Clegg, Patricia Coate, Joyce Fowler, Brenda A. Gillie, Stephanie C. Hatchell, Jeff

Miller, Kathy Lawler Merlette, Pamela D. Norton, Delores Plastow, William L. Thomas Jr., and Allison Trussell.

Walter W. Ross, Steven Gross, and Mark McEwan did library research. They were assisted by the following librarians at the Thomas Cooper Library of the University of South Carolina: Linda Holderfield and the interlibrary-loan staff; reference-department head Virginia Weathers; reference librarians Marilee Birchfield, Stefanie Buck, Stefanie DuBose, Rebecca Feind, Karen Joseph, Donna Lehman, Charlene Loope, Anthony McKissick, Jean Rhyne, Kwamine Simpson, and Virginia Weathers; circulation-department head Caroline Taylor; and acquisitions-searching supervisor David Haggard.

The publishers acknowledge the generous assistance of William R. Cagle, director of the Lilly Library, Indiana University, and his staff, who provided many of the illustrations in this volume. Their work represents the highest standards of librarianship and research.

Dictionary of Literary Biography® • Volume One Hundred Seventy-Four

British Travel Writers, 1876–1909

Dictionary of Literary Biography

Gertrude Margaret Lowthian Bell
(14 July 1868 - 12 July 1926)

Nancy V. Workman
Lewis University

BOOKS: *Safar Nameh – Persian Pictures: A Book of Travel,* anonymous (London: Bentley, 1894); republished as *Persian Pictures* (London: Benn, 1928; New York: Boni & Liveright, 1928);
Notes on a Journey through Cilicia and Lycaonia (Angers: Burdin, 1906);
The Desert and the Sown (London: Heinemann, 1907; New York: Dutton, 1907); republished as *Syria: The Desert & the Sown* (New York: Dutton, 1907; London: Heinemann, 1908);
The Thousand and One Churches, by Bell and Sir William M. Ramsay (London: Hodder & Stoughton, 1909);
Amurath to Amurath (London: Heinemann, 1911; New York: Dutton, 1911);
Palace and Mosque at Ukhaidir: A Study in Early Mohammadan Architecture (Oxford: Clarendon Press, 1914);
The Arab of Mesopotamia (Basra, Iraq: Government Press, 1917?);
Review of the Civil Administration of Mesopotamia (London: H. M. Stationery Office, 1920);
The Arab War: Confidential Information for General Headquarters from Gertrude Bell, Being Despatches Reprinted from the Secret "Arab Bulletin" (London: Golden Cockerel Press, 1940).

OTHER: Shirazi Hafiz, *Poems from the Divan of Hafiz,* translated by Bell (London: Heinemann, 1897);
"The Churches and Monasteries of the Tur Abdin," in *Amida,* by Max van Berchem (Heidelberg, 1910), pp. 224-262.

SELECTED PERIODICAL PUBLICATIONS – UNCOLLECTED: "Alps of Dauphine," *Nineteenth Century,* 47 (February 1900): 330-338;

Gertrude Margaret Lowthian Bell in 1923, from a drawing by John Singer Sargent, in The Letters of Gertrude Bell *(1927)*

"Turkish Rule East of Jordan," *Nineteenth Century,* 52 (August 1902): 226-238;

"Islam in India – A Study at Aligarth," *Nineteenth Century,* 60 (December 1906): 900-908;

"Palace in the Syrian Desert," *Quarterly Review,* 212 (April 1910): 339-368;

"Mount of the Servants of God," *Blackwood's Magazine,* 188 (September 1910): 354-355;

"Damascus," *Blackwood's Magazine,* 189 (April 1911): 539-549;

"Asiatic Turkey under the Constitution," *Blackwood's Magazine,* 190 (October 1911): 425-440;

"Post-Road through the Syrian Desert," *Living Age,* 280 (7 February 1914): 329-343;

"Post-Road through the Syrian Desert," *Living Age,* 280 (21 February 1914): 458-469.

In the introduction to *The Letters of Gertrude Bell* (1927) Lady Florence Bell enumerates the many achievements of her late stepdaughter: "Scholar, poet, historian, archeologist, art critic, mountaineer, explorer, gardener, naturalist, distinguished servant of the State, Gertrude was all of these, and was recognized by experts as an expert in them all." In fact, so accomplished was Gertrude Bell that another of her admirers, Vita Sackville-West, remarks on Bell's achievements in an introduction to an edition of *Persian Pictures* (1928), "Alpine guides remembered her as a skilled and fearless mountaineer; she could walk most men off their legs; she cared for gardening, fishing, archaeology, exploration, politics, poetry, friendship, and clothes. It is almost a relief to hear that she never could learn to cook or play the piano." Still others praise her political involvement with the newly emerging nations of the Middle East from 1914 until her death; some credit her diplomatic understanding as significant in establishing the nation of Iraq.

Bell was born in Durham, England, on 14 July 1868. As the daughter of Sir Hugh Bell and his first wife, Maria (Shedd), she was accordingly given all the advantages of her class and wealth. Born into a distinguished family of manufacturers and scientists, she became proficient at hunting, shooting, and riding at an early age. Even during her formative years she was an exceptionally bright young woman. She entered Queen's College, London, in 1885 and later transferred to Lady Margaret Hall, Oxford, where she won high honors in history in 1887. At a time when few women were enrolled at university, her achievements were noteworthy. Early letters, such as one written at age fourteen, record her extensive reading: "Monday (evening) Mrs Carlisle's Letters 1st volume; Tuesday Mrs C's letters finished, began Life of Macauley; Wednesday I Vol. Life of M finished, began II vol.; Thursday II vol. of Life of M finished. . . ."

When she had opportunities to travel, Bell did so. Among her first excursions were visits to Paris, Vienna, and Bucharest, places where the family's connections allowed her access to diplomats, politicians, scholars, and nobility: she sat beside ruling families at amusements, attended royal funerals and coronations, and was routinely introduced to foreign leaders. As H. V. F. Winstone notes, "Almost everyone she knew was in a position of some power or influence. Almost every branch of the family contained a noted politician, diplomat or scholar." In the years ahead these contacts provided her with the necessary letters of introduction for her own travels.

Bell displayed two early preoccupations – an interest in archaeological architecture and an abiding interest in politics. Perhaps prophetically, some of her early letters address the Irish independence movement, and some of her last letters discuss the independence of Middle Eastern nations. She first visited the Middle East in 1892, and many other journeys followed, including two world tours in which she visited the United States, India, Burma, and China. World War I forced her to abandon traveling, but, having mastered Arabic, Persian, and other languages, she ended her life working in the Middle East. Some of her mideastern journeys resulted in book-length studies, and her letters, journals, and field notes – particularly those dealing with the independence of Middle Eastern nations – continue to interest scholars. For the most part she did not write travel narratives of her visits to Europe or the United States, as she preferred to concentrate on those areas where she would eventually live.

Bell's private life, seldom revealed in her published travel narratives, appears in her correspondence and diary entries. Her mother died when Bell was three, and after her father's remarriage she spent her childhood with stepbrothers and stepsisters. She had a particularly close relationship with her stepmother, Florence Eveleen Eleanore Olliffe Bell, and maintained affection for her family throughout her life. So dutiful was Bell that even as a grown woman she continued to ask her parents' permission to visit foreign nations.

Her first travel narrative, *Safar Nameh – Persian Pictures: A Book of Travel* (1894), later republished as *Persian Pictures,* recounts her visit to Persia in 1892, a trip on which Bell and her traveling companions visited Germany and Austria en route to Constantinople, Tiflis, Baku, and the Caspian Sea. Taking the journey at a time when she was as yet unfamiliar with the language, she was studying with a

French tutor and using a dictionary to translate unfamiliar expressions and idioms when she wrote her notes. Communication between Bell and the tutor was difficult because the latter spoke English and French poorly and Bell had not yet mastered Persian or Arabic. Despite her reliance on a translator, she was thoroughly familiar with the history and politics of the regions she visited, and she frequently refers to past circumstances as she writes with what she calls "the silence of an extinct world still heavy upon us." She presents Persia as a richly textured landscape composed of remnants of the past that lurk beneath the sands, a palimpsest of culture.

Her letters provide much background information on her travels, although factual details and chronology are not apparent in her text. She wrote the manuscript for pleasure, and she published it anonymously because her family and several friends, including the publisher, asked her to do so. To comply with the standard for book-length studies, she was asked to add six additional chapters to her initial entries, and she did so, although she maintained that to do this was a "bore." Labeling the pieces "extraordinarily feeble," she eventually was swayed to publish the pieces by the pleading of her parents, although she announced, "I wish them not to be read."

Although Bell might not have wished to publish her work, she certainly had no difficulty imagining herself as a traveler. The conclusion to *Safar Nameh* discusses her motives for traveling. She scorns those who travel merely to re-create their lives in new settings; such people seek comfort in unfamiliar countries by carrying luncheon baskets and liquor flasks to lighten their burdens. She characterizes other travelers as intent on covering continents merely in order to be able to enumerate all the places they have visited rather than to transform themselves. By contrast, Bell finds the joy of travel in encountering traveling companions and in establishing transient friendships that reveal character "in as many different aspects as it would take ten years of the customary life to exhibit." By listening to the tales of such foreign characters, Bell can see into "the secret chambers" of remote worlds and landscapes; through the unfamiliar she can experience alternate realities of ancient cultures.

Ironically, however, Bell's protocol for travel does not accurately reflect her own practices. Winstone suggests that even in remote regions of the world Bell insisted on receiving daily copies of the London *Times,* if possible, and on eating with cutlery and fine linen. Thus, her remarks must be

Page from an undated letter by Bell to her stepmother, Florence (University Library, Newcastle upon Tyne)

seen as figurative, not literal. In addition, the contradictions between her statements and personal behavior reveal a Eurocentric attitude that cultural critics have been quick to address.

Bell went to Persia in the spring of 1892, when her uncle, Sir Frank Lascelles, was appointed minister to Tehran, and her visit was confined to that city and its environs. Traveling with Lady Lascelles and guides, Bell entered the region when cholera and typhoid were rampant, yet she seems to have had little fear that she would succumb. Indeed, the presence of the diseases comprises one of the many contrasts she sees between Middle Eastern and English life. Middle Eastern people greet potentially life-threatening illness with a fatalism that she finds confusing: "A wise philosophy bids men bear the inevitable evil without complaint, but we of the West are not content until we have discovered how far the coil is inevitable, and how far it may be modified by forethought and by a more complete knowledge of its antecedents." She goes on to praise Western notions of hygiene and sanitation that curb the spread of epidemics, and she questions why the East is less enlightened than the West in its practices.

Yet Bell frequently admits that it is impossible for an English person to understand the foreign culture:

> So in the wilderness, between high walls, the secret, mysterious life of the East flows on – a life which no European can penetrate, whose standards, whose canons, are so different from his own that the whole existence they rule seems to him misty and unreal, incomprehensible, at any rate unfathomable.

She also notes, "In vain you try to imagine yourself akin to these tented races . . . [for] the whole life is too strange, too far away. It is half vision and half nightmare; nor have you any place among dwellers in tents."

Bell's frequent references to the *Arabian Nights* and her fascination with gardens and desert landscapes in *Safar Nameh* reveal the visionary qualities of Persia. In this mode she dramatizes the many sheikhs she encounters as coming straight from the tales of Harun al-Rashid and his court, or as being actual *efreets* – imprisoned genies who had been banished by King Solomon. She describes meetings and dinners in tents, meals replete with highly sugared tea, witty conversation, and hospitable companionship amid mild breezes coming off the surrounding sands. In contrast, she reveals nightmare images in her descriptions of the severe living conditions, the constant presence of bugs and insects, and the unrelenting heat followed by unendurable cold and wind. Most threatening is the presence of disease and the fear of political insurrection, especially between warring sheikhs who massacre and ambush their enemies with no apparent provocation.

Bell also expresses her belief that native women, trapped in customs that deny them physical and psychological freedom, lead passive, meaningless lives. As she notes, "The sight of the andarun and its inhabitants knocks at the heart with a weary sense of discontent, of purposeless, vapid lives – a wailing, endless minor." Yet Bell rarely refers to women by name, and she generally interacts with the males of the culture, especially men of high rank and power. Forgetting that the many letters of introduction she carries have put her in a privileged position, Bell visits many restricted areas and is treated with respect and courtesy. Her guide refers to her as "excellency," and although she claims to sleep unprotected outdoors in the wild, her wealth and personal acquaintances clearly shield her.

Safar Nameh remains a dreamy narrative lacking the specific detail that characterizes her later books, which emphasize archaeological and historical facts. It is not surprising that *Safar Nameh* was not critically successful. Winstone includes an early critique of the work as evidence of its stylistic failings. Quoting Janet Hogarth, he writes that the book had "Charm but not actual achievement." It remains a preliminary exercise for her later travel narratives.

Bell's stay in Persia in 1892 also occasioned her meeting with Henry Cadogan, the first of two men whom her biographers identify as significant in her life. Ten years older than Bell and a minor diplomat whom she met in Tehran, Cadogan worked as a British legation secretary and was a sportsman who shared her interests in riding and in poetry, especially that of medieval Arabian poet Shirazi Hafiz, whose work she was to translate. However, her parents felt that Cadogan would be an inappropriate partner for their daughter and objected to his reputed indebtedness. Fearing also that his arrogant personality and his interest in gambling would discredit and dishonor their daughter, they recalled Bell to London, despite her entreaties. Her correspondence reveals her anxiety at this breakup with Cadogan, as she comments to friends about her lingering "fugitive sensations" associated with him. After leaving for home, Bell was never reunited with him: having fallen into a river during winter, chilling himself, and never recovering, Cadogan died of pneumonia six months later.

Although she was not fluent in Farsi when *Safar Nameh* had been written, Bell learned the language quickly and spent time translating the *Poems from the Divan of Hafiz* (1897) as her next undertaking. Although not a travel narrative, the translation offers insights into her understanding of Middle Eastern culture and her willingness to take on controversial subjects. Using library collections and volumes produced by Asiatic societies to acquire the background necessary for the work, Bell studied Sufism.

Hafiz, a fourteenth-century poet, is regarded as heretical for his proclamations on the nature of God and for his celebration of certain religious practices, especially the imbibing of liquor. In her introductory essay Bell presents Hafiz as a mixture of Sufi master and orthodox Muslim. For her, Hafiz lacks a concrete philosophical system, yet he embraces the more conventional aspects of Muslim ritual and deserves study as a poet. Indeed, she compares him to other medieval writers such as Dante, and she especially compliments his love poetry, which celebrates both the poet's union with God and the more sensual pleasures of the beloved. Her final evaluation is that Hafiz was "profoundly skeptical as to the infallibility of any creed . . . [but] an undercurrent of mysticism runs through the poems which it is impossible to explain away."

Bell concludes that notions of a "priestly" tavern keeper who uses wine as the "spirit of divine knowledge" originate in Sufi belief, not in the poet given to debauchery or licentiousness. She maintains that Hafiz sought fulfillment of desire in a final union with God, but he recognized that some compensations exist along the way. Whether or not Bell shared his point of view about drinking, she openly accepted his perspective and translated his poetry frankly, without concern for the proprieties of her day. Her translation was regarded as a literary achievement, for according to Winstone, a "distinguished Koranic scholar, A. J. Arberry," in 1947 called it the best of twenty translations with which he was familiar.

In accepting Hafiz's unorthodoxy, Bell also offers a continuing rationale for her lifelong atheism, which seems at odds with the extensive biblical allusions and references in her works. For Bell and for Hafiz, the sacredness of life was not confined to religious ritual or orthodox doctrine. Even as a young woman she had challenged Christian doctrine, with which she was well versed. Her biographers record frequent conversations and letters with her stepbrother Hugo, who eventually went on to be ordained and whose firm beliefs Bell challenged with her own healthy skepticism. Winstone maintains that Bell was an "unequivocal" atheist, even at university, and that her friends had expressed difficulty with her outspoken criticism of Christianity.

Bell's next major book, *The Desert and the Sown* (1907), is the most popular of her travel narratives. Written about her 1905 journey to Syria, it examines the area at a time when travel restrictions had eased and a new rush of tourism was beginning to influence how travel narratives were being written. Bell acknowledges this influence by remarking on the popularity of the Cook tours, and she inserts some travel "hints" that she has been asked to include. For instance, she advises travelers to carry sufficient currency to pay their debts, because acquiring it in the middle of the desert can be impossible. On this trip Bell was exploring rather remote parts of the region and often relying on inaccurate maps to navigate her caravan.

In her introduction to a 1985 edition of the work, editor Sarah Graham-Brown calls it the "liveliest account of her travel experiences outside her well-known letters," an accurate evaluation of the work. Bell's growing certainty about her authorial role eclipses the timidity she had shown in her previous volume. She openly acknowledges herself as an English representative who is asked to intervene in political disputes and to obtain political asylum; she is frequently asked to act as a mail courier and even to administer to the sick. Always conscious of her status and personal connections, she enters previously forbidden territory without much difficulty, although she acknowledges that she is being watched, that the Arabs are telegraphing her whereabouts to their superiors throughout her journey.

Bell's fascination with archaeology begins to appear in this work, although she remains more interested in describing the landscape of the desert and the customs of the people. Borrowing her title from a line of the *Rubáiyát of Omar Khayyám* (1859), she admits that the emptiness of the desert, the vastness of the land without cultivation or civilization, fascinates her. She anticipates the growth of the area and laments that eventually it will be overrun with villages and farms. For now, though, she is content to explore its wildness, to begin acquiring the skill to distinguish one spot from another in the seemingly unlimited desert. As do the Arabs, she prefers meeting the solitary shepherd and the roving herdsman, not the representatives of commerce and industry. She particularly welcomes the challenge of being alone for a long time: "you must go alone," she insists, and "the voice of the wind shall be heard instead of the persuasive voice of counsellors, the touch of the rain and the prick of the frost shall be spurs sharper than praise or blame."

Bell's journey was far from safe, and she frequently mentions tribal hostility and feuding. She finds wrong the notion that an "Arab nation" exists; at best, the term refers to a loose alliance of warring factions. She attributes much of the hostility among Arabs to poverty, because those fortunate enough to work are often paid starvation wages and are thus particularly subject to political and personal corruption in order to survive.

In emphasizing the rules of hospitality that govern the desert, Bell shows how even the poorest man will open his tent to welcome a traveler, even if doing so inconveniences the host and his tribe. For example, when she once needs a place of refuge, she finds "it was enough that I was cold and hungry and an Englishwoman" to be offered a place to clean her muddy clothes, discuss a recent book, and digest an "excellent dinner." Even when tired, Bell also responds with grace and courtesy: "I was wet through, but the obligations of good society had to be fulfilled, and they demanded that we should sit down on the divan and exchange polite phrases while I drank glasses of weak tea."

While Bell spends most of her time in the company of powerful Arab men, she sometimes ac-

Frontispiece and title page for the most popular of Bell's travel books

knowledges the presence of women, although the women are often unavailable to her. She accepts polygamy and its constraints, and she admits that when English women are married to Arabs, they must follow the rules of the household, no matter how restrictive. Visiting a harem, she criticizes its untidiness and wonders that its inhabitants, knowing of her coming visit, have not taken greater care. She remarks on the women's dirtiness, especially of their dress.

Overall the interests of *The Desert and the Sown* are more political than those of Bell's earlier work, and the book acknowledges an increase in the popularity of travel to the Middle East. She sees its readers as people who may one day trace her journey rather than remain armchair tourists. She frequently comments on the decline of the Turkish empire, and the book demonstrates her political acumen, her knowledge that eventually led to her success as a British diplomat. Robert Heussler calls the book's portrait of Bell's diplomatic skills an "enlightening, down-to-earth picture." Winstone maintains that the book established her fame as an explorer and that contemporary reviewers found the work equal to Charles M. Doughty's *Travels in Arabia Deserta* (1888). British nationals on duty in the East regarded it highly. Overlooked, however, in the praise it received were the photographs with which Bell illustrated her text. She had taken many of these photographs and apparently mastered the art of perspective and representation without formal training.

Bell's next travel narrative, *The Thousand and One Churches* (1909), was co-authored with Sir William M. Ramsay. Ramsay claimed that he wrote the preface and parts 1 and 4 (on the historical and religious background), but Dea Birkett and others believe that Bell actually wrote most of the text. That certainly is possible, as Winstone characterizes Ramsay as a "chaotic" traveler who, "if left to his own devices[,] would forget where his camp or his hotel was; on occasion he even lost his clothes."

The painstaking scholarship throughout the volume betokens authorial reliance on Bell's notes rather than Ramsay's, and the text certainly exhibits her style. Nonetheless, Ramsay undoubtedly provided some relevant information, even if Bell wrote more of the manuscript than parts 2 and 3, essentially the architectural discussions that are usually attributed to her. In addition to the detailed descriptions of the ruins and masonry that they examined, the authors supplied figure drawings, maps, and photographs and sometimes included earlier prints by German archaeologists who provided the bases for Bell and Ramsay's studies.

As Ramsay points out, the book's title is a misnomer for two reasons: first, they in fact examined only about twenty-eight Byzantine structures, presumably dating from the fifth to the eleventh centuries. However, sections of some of the buildings had been built much earlier, and evidence showed that the Christians had appropriated structures and modified them for their own uses. Because many of the ruins were badly damaged, it was sometimes difficult to determine actual boundaries and to know whether a ruin represented one building or two. Second, the ruins were not exclusively places of worship, as the title implies. Ramsay and Bell actually discuss mausoleums, fortresses, and water resources such as dams, cisterns, and aqueducts. Furthermore, many of the ancient sacred buildings had been used through the years in other ways, often as bakeries or mills. As a result, reconstructing their original appearances was difficult.

Both authors recognized the need for their detailed studies, because the area quickly was being destroyed, both by natural conditions such as the arid climate and by social erosion: people of the local community presumably had little knowledge of the archaeological importance of the area and were content to use it for their own mundane purposes. Ramsay remarked in 1909 that much of what they had originally seen in 1907 had already changed, and the discussion afforded by the book was already "the record of a vanished past" that existed only "in our photographs." Bell mentioned that she was working "for the future" so that modern archaeologists would have the benefit of her conclusions.

The site of their study, Maden Shehar, was a settlement consisting of an "Upper City" and a "Lower City" on two adjacent hills. Dating from the ancient Hittite civilization, followed by Roman and then Christian occupation, the area had always been sacred. Still later it became the site of Muslim communities. Thus, several monastic traditions had been represented, and the area showed how the Muslims had appropriated existing structures and converted some earlier Christian churches into mosques. Regardless of the ecclesiastical control, the area had continued to be a burial ground, and many church sites had memorials to revered leaders – gods and goddesses or saints and martyrs.

Bell provides both a narrative and elaborate architectural diagrams, especially of masonry and floor plans. Using her technical expertise, she corrects earlier archaeologists by pointing out that their models were not consistent with existing ruins. At other points she modestly allows that perhaps earlier archaeologists had been looking at different buildings from those that they had identified: "perhaps Smirnov's #20 may be our #28." Or she credits the discrepancies to the shallow excavations that Ramsay and she did. Limited by funds and time, they were unable to dig much deeper than the surface, and she speculates that deeper excavations would have unearthed different foundations. Despite the questions she raises about the accuracy of earlier findings, Bell generally accepts many ideas of those previous studies, which she mentions in the footnotes. Throughout her commentary, she uses the rhetoric of scientific discourse without attempting to personalize the material. Her particular interest involves the ornamentation of the ruins, especially features on doorways and walls. She maintains that the ornamentation reveals the "Asiatic character of the land," because it is distinct to the area. For her, "The East pursued its deep and silent way, assimilating what was brought to it and passing it out again marked with its own stamp," as the Roman system of government and transportation gave way to Middle Eastern successors.

Ramsay and Bell accepted the view that as religions evolve, they use existing structures that predate their own expressions. The more recent religions adapted from earlier religions those features of the existing sacred buildings that suited their rituals, or these new religions destroyed those structures that offered contrasting views. Thus, the mystery temple of the East gave way to the Christian basilica, and the altars originally devoted to goddesses were replaced with shrines for saints. However, differences between the Eastern and Western monastic traditions caused significant differences in the architecture: the West designed a medieval structure like Saint Galls, while the East held "largely to the somewhat inchoate scheme illustrated by the Rossicon on mount Athos."

Bell's letters offer a much more personal discussion of the circumstances of her work in Asia

Bell in front of her tent at Babylon, Iraq, in 1909

Minor than her published travel narratives. Unlike *The Thousand and One Churches,* which interests only scholars of antiquity, her letters are lively discussions of primitive working conditions – including the fleas and bugs in the ruins, as well as the intense labor needed to draw even the simplest diagrams. Often Bell's workday required rising at dawn, working steadily for twelve hours with only short breaks for breakfast and lunch, and resuming her drawing and writing well into the evening. Her correspondence also recounts emotional travails. During her visit to Asia, Bell felt guilty about missing family events, and she continually rationalized to her family why she chose to remain in the East rather than to return home. Uneasy at subordinating familial loyalty to personal interests, she compensated by remarking how satisfying her studies were, particularly because she was one of the first scholars seriously to study cuneiform churches.

Although the printed volume does not reflect it, the letters reveal that Bell was supervising a crew of sometimes thirty-one workers and that she was in danger from roving bandits. Even though she distrusted her guards, who were merely neighboring men with no skill at marksmanship and whom she found ill-equipped to handle real danger, she had to be guarded while asleep. Yet her letters also reveal the singular beauty of the area, especially her fascination with the butterflies that crowd the roads and the stillness of the landscape.

Another major difference between *A Thousand and One Churches* and the letters describing the work is in the humor of the latter. Her correspondence tells several funny stories about how difficult it was to convince the locals that, indeed, she really was interested in viewing ruins, not complete churches. Frequently when she asked to visit a site, she was told that it was not worth her time – yet as soon as she offered to pay for a guide, the local inhabitants would find many other ruins, often worthless modern structures instead of ancient ones. Bell concludes by acknowledging how the money she and Ramsay spent for wages and materials enormously affected the region: "I don't suppose so much money has passed hands in the Karadagh since the time of the Byzantine."

Following *The Thousand and One Churches* Bell published *Amurath to Amurath* (1911), an account of

Bell recording measurements of the palace at Ukhaidir, with the help of two Arab assistants in 1909

her journeys originating in Germany and Italy in 1909 and through Turkey and the Euphrates River basin via Beirut and Damascus. What the book also presents is a record of Bell's growing political awareness. As the preface indicates, the real subject of the book is the "months of suspense and terror" during her journey into a country experiencing "liberty" for the first time. Throughout the work Bell interjects her attitudes toward the governments and European involvement; she supports British intervention in Middle Eastern affairs, because such intervention supports the Turkish independence movement. In fact, she declares that she feels "glad" that "we were the first to hold out a helping hand." Unlike her other books, *Amurath to Amurath* is precisely dated so that readers can follow her progress day by day. As always, she illustrates her text with copious illustrations and photographs.

Bell finds that the new Turkish government, which came to power in 1908, was so remote from the people it governed that its citizens refused to take responsibility for what occurred. She characterizes the government as corrupt, as it relies on spies and local officials and ignores the increasing tension between civil authorities and religious factionalism. For example, she maintains that tension between Muslims and Christians obscures the real problem of poverty and that, as a result, attempts at reform are impossible. She also examines the failure of the government to pay military wages, the corruption of tax officials, and the people's open defiance of governmental proclamations on the quality of daily life. The work is far more a sociopolitical study than a standard travel narrative. It was valuable, however, because it came at a time when England desperately needed experts in the Middle East, and in conjunction with other events, its publication led to Bell's being approached to perform diplomatic missions and to write directly for the British government.

In addition to the current sociopolitical content Bell's book describes her interest in cities such as Damascus and Aleppo, and as she does in all her works, she connects the present with the preceding pagan and Roman times. She intersperses poetry throughout her discussion, and as always, she praises the desert and includes anecdotes about a culture that she cannot fully understand. For example, she fails to comprehend why the Bedouin will not eat fish, even though the rivers are full of them, and she recounts the arrival of new technology when she describes how the tribesmen confront an automobile for the first time.

Other anecdotes include Bell's entertainment by a performing monkey and her attempts to capture some thieves who have stolen supplies from impoverished natives. Despite her delight in exploration or adventure, *Amurath to Amurath* contains an undercurrent of both discomfort and fear. For the first time in her works Bell complains about the difficulty of traveling. She had written about this in her yet-unpublished letters, but her published works usually had maintained a stoicism that masked the realities of her undertakings. Yet in *Amurath to Amurath* she longs for lemons because she is always thirsty, and she admits, "for my part I longed for a table more than I could have thought possible. I was weary of sleeping on the stony face of the desert, of sitting in dust and eating my meals with a seasoning of sand." She also acknowledges having fears about her safety. Her reliance on being English – the shield that had afforded such reassurance through earlier travels – seems in *Amurath to Amurath* to have become quite the opposite: her nationality may actually make her a target for raiding nomads.

Winstone finds that the mixture of politics and scholarship in *Amurath to Amurath* was not altogether what the public wanted or expected, especially from the author of the popular *The Desert and the Sown*. Thus, "the public reaction to it was little more than polite. *Amurath to Amurath* was well received in some quarters and barely noticed in others." In one of the few extensive reviews of Bell's individual works Elizabeth Robins acknowledged that much of *Amurath to Amurath* is distinctly political and that "this part of the book seems to be offered as a contribution towards Western understanding of the unprecedented political crisis. . . . The motif of the new book is Freedom."

Robins notes that the theme of "social disorder" predominates and that Bell is particularly appalled at the waste she witnesses – at agricultural fields lying unused, at pasture land unused because shepherds fear raids on their livestock, and at the misuse of rivers that might provide potential irrigation. Robins praises Bell's ability to befriend locals who are willing to talk honestly to her about their apprehensions, but Robins is skeptical about Bell's presence: "I found myself wondering as I read her pages, what do these Turks, Arabs, Chaldeans, Devil-worshippers and the rest, what do they think of this fair-haired apparition out of the West, this woman equally concerned about current politics and Hittite inscriptions?" Robins concludes that "Miss Bell's record . . . is too honest to be all pleasant reading." Furthermore, Robins finds some of Bell's attitudes to be snobbish – especially in her display of aristocratic English "national pride." Robins argues that Bell defers to England especially when she "can take as simple-minded delight . . . a Moslem's compliment to her race."

Despite such critical disapproval the book was certainly useful to British authorities, for it seemed to express their sentiments about the need for political intervention in the Middle East. Bell had underscored the difficulty of defining "liberty" in a nation unfamiliar with the concept as most Europeans understood it. In *Amurath to Amurath* she recounts how one tribesman had "slashed the air defiantly with his tamarish switch as he proclaimed the liberties of the wilderness, the right of feud, the right of raid, the right of revenge – the only liberty which the desert knows." From her point of view political independence could not exist in a country that adhered to tribal laws and pledged alliance to ancient systems of resolution. Thus, outside assistance appeared to be necessary. Apparently British officials agreed, for by 1915 Bell was receiving secret government dispatches about Turkey and other Middle Eastern countries. She began writing her own bulletins and continued to do so for several years.

During her travels through Turkey, Bell met Lt. Col. Charles (Richard or Dick) Hotham Montagu Doughty-Wylie, the second significant man in her life. She met Doughty-Wylie, a vice-consul, while he was stationed in Konia, and Bell frequently saw him in the company of his wife, Lilian. As the nephew of Charles M. Doughty, the celebrated Arabist, he shared Bell's interests in Eastern archaeology; as a highly decorated soldier who had been wounded in several important battles, Doughty-Wylie was also her social equal. Many of their letters to each other were destroyed, but their remaining correspondence reveals their passionate devotion. The nature of their physical relationship is unclear; the letters of both writers directly acknowledge desires for sexual intimacy but also express frequent concerns for propriety, especially in references to Doughty-Wylie's marriage and the apparently warm friendship between Bell and his wife. For example, on 5 January 1914 Doughty-Wylie wrote, "I'm not going to write you a love letter. . . . Where are you? It's like writing to an idea, a dream." Such references may suggest that their relationship was idealized.

What is always apparent in their letters, however, is that the relationship, which began in 1912, sustained them both: they wrote daily as they traveled independently and sent letters by courier to each other in remote parts of the world. Bell must have been deeply in love with Doughty-Wylie, for she speaks of committing suicide if he were to abandon her. Yet she never had a chance to act on such a threat, for Doughty-Wylie died in the 1915 Battle of Gallipoli, where he became one of the most celebrated soldiers of his era.

Bell's final travel narrative, *Palace and Mosque at Ukhaidir: A Study in Early Mohammadan Architecture* (1914), was another book of historical archaeology. Focusing on the palace of Ukhaidir, located on the eastern side of the Syrian desert, it also examined neighboring ruins. According to Bell's preface, the palace was "practically unknown until the winter of 1908–9," and, indeed, opinions differed about whether the ruin was that of a palace, a temple, or a castle. However, Bell concluded that the structure had been a royal one because of the labor and resources that must have been used to build it.

At the same time that Bell was investigating the area, a Dr. Reuther and a team of German archaeologists were also excavating it. Rather than competing with her, they shared their findings with her, and she was to incorporate some parts of their discussion into her book. The preface to her book suggests that her study and manuscript preparation took four years to complete, but Bell wrote that this endeavor was her "chief delight" of the period. She adds that the undertaking brought her "amazement," "joy," and awe in seeing the palace's "solitary magnificence [that] rose out of the desert."

The book actually summarizes two separate journeys that Bell undertook in this area. The first started in Beirut in 1909, moved to Aleppo, and proceeded to the intersection of the Tigris and Euphrates Rivers, the ancient sites of Babylon and Sumeria. In 1911 Bell's second journey proved to be as dangerous as the first had been. The area was a "trysting place for raiding parties," and all her detailed work – drawing, measuring, and sketching the fortifications – had to be completed within four or five days, or her stay at the ruin would have provoked attack. As a result, she left the site of the palace and traveled around the perimeter of the area to visit other ruins. The featureless desert and lack of suitable water suggested to her that the earlier environment must have been very different from that which she was seeing in 1911.

Her inability to complete accurate measurements also impeded her work. The ruins were uneven, and the brickwork suggested the earlier presence of wooden structures that had decayed. Such complications made it difficult to imagine accurately how the rooms had been related to each other. Because few inscriptions remained, Bell could provide only guesses in trying to date sections of the site and sometimes had to allow her estimation range to encompass centuries. Furthermore, because the name *Ukhaidir* is "not mentioned by historians or geographers," she assumed that the name was a modern substitution. Attempting to associate the artifice she saw with earlier discussions was difficult, and she reluctantly concluded that, based on her analysis of its architectural features, the structure was probably from the eighth century.

Despite Bell's inability to verify historical information, Winstone praises this volume for comprising "Gertrude's most important exploratory journey," one of "actual geographical discovery" – but for the average reader, the volume is far too specialized to hold much interest. The discussions about foundations, room arrangements, and building materials are technical and complex. The book occasionally incorporates cultural discussions, particularly about the evolution of a nomadic culture into a settled one, but the scholarship is tedious. The many plates and photographs lack human content, and the text lacks the presence of an authorial personality, a voice delighted with wonder and awe. It is not surprising, therefore, that the work received little critical mention.

With the *Palace and Mosque at Ukhaidir* Bell ended her publishing of travel narratives. At the outbreak of World War I she became a diplomat writing for a specialized audience, one that sought her expertise in Arabic culture and her knowledge of Middle Eastern landscape. She was invited by the War Office to contribute reports, and in 1914 she joined the Arab Intelligence Bureau, an agency with which she assumed several positions during her tenure in Cairo. She also served as secretary to the civil commissioner in Baghdad and as a Middle Eastern adviser to Winston Churchill. For more than five years she wrote articles and weekly intelligence summaries for the British Colonial Office in a remarkable career of civil service that continued after the war.

The Arab of Mesopotamia (1917?) presents her understanding of the Ottoman Empire and emerging political events. Because it also describes several geographical issues, the collection can be regarded as a travel narrative, although not in the conventional sense. The book addresses themes mentioned in her earlier works, but, as one of the last public ex-

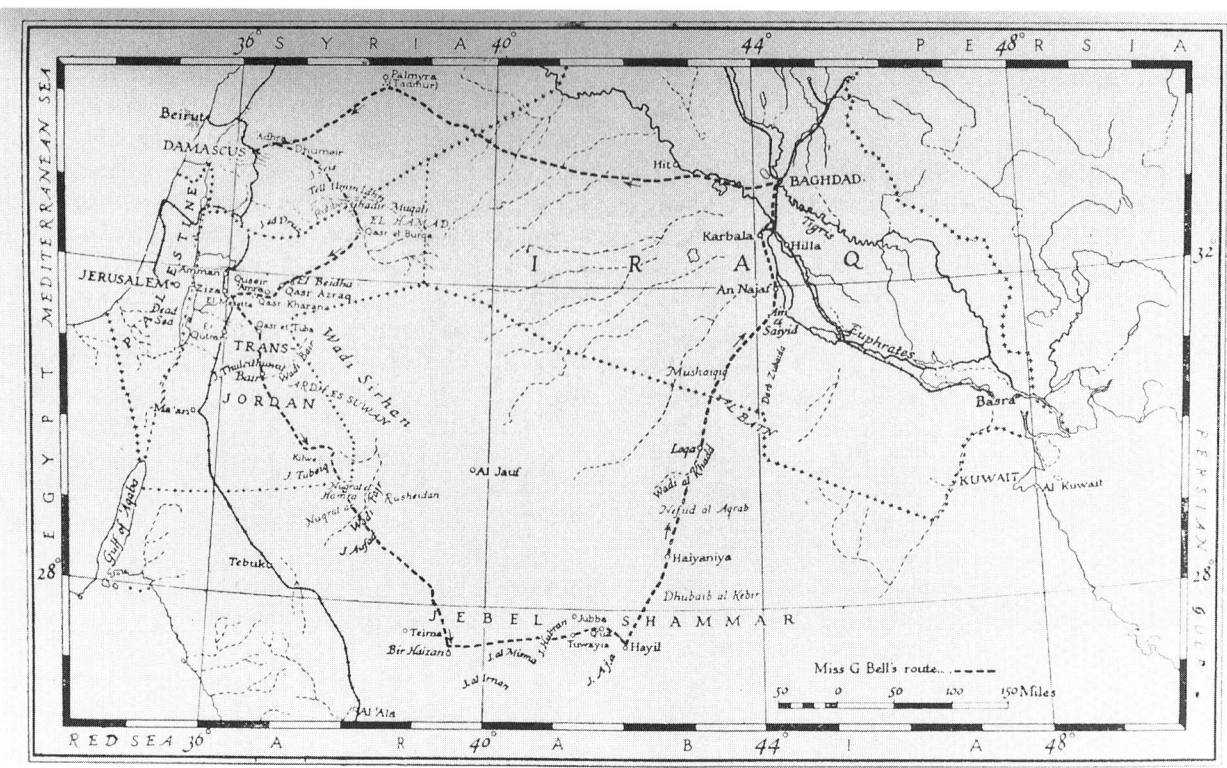

Bell's map of her central Arabian journey in 1913–1914 (from The Letters of Gertrude Bell, *1927)*

pressions of her attitude toward the Middle East, the volume does so far more directly. Privately printed by the Government Press, *The Arab of Mesopotamia* is a collection of essays. Only the last section of the small volume bears Bell's name, but the style and substance of the opening chapters bear her signature as well.

She announces that the focus of the essays is on the profound changes occurring in a nation moving from "a nomadic to a settled life" and that the work will examine tensions between various religious sects, particularly the Sunni and Shia sects. In Bell's estimation, one central difficulty for English readers in understanding the Ottoman culture was in the discrepancy between written rule and actual governance: the "unrecorded provisions of Government held sway," and actual power was vested in local headsmen, not elected officials. Because the country had adopted Napoleonic criminal codes and procedures, its system of jurisprudence was also difficult to evaluate by English standards. For these reasons – the "differing notions of statesmanship" – Europeans were confused about events occurring in the Middle East.

Indeed, Bell felt that most British politicians and investors had disregarded Turkey and other Middle Eastern nations until Germany, in its eagerness to appropriate territories, began its assaults on that area. Capital investing had seemed unattractive to the British, because they believed that the rate of return would not warrant the risks. Furthermore, nomadic people did not always observe territorial boundaries, but, on the positive side, Bell maintained that the culture was not litigious and relied on informal, rather than court-mandated, standards of rule. Because the desert economy was fragile and depended on long-established bonds between merchants and tribesmen rather than on agreements signed at bargaining tables in cities, traditional Western means of bargaining were futile. Bell supported British intervention in the Middle East because she believed that it would unify and strengthen the economy and political structure. Yet implicit in her commentary is a fear of the leadership in the area; she distrusted the local tribesmen and warring factions.

Another political work, *Review of the Civil Administration of Mesopotamia* (1920), was the last of Bell's books to appear during her lifetime. She continued to produce dispatches and other governmental works, but they were not intended for the general public or scholarly readers. Her tenure in Iraq has become regarded as controversial, especially because of her support for limited political intervention by Europe in Middle Eastern affairs between 1910 and 1920. Some critics accuse her of expressing pro-British attitudes during the fights over independence, but others see her support of indirect alliances between Britain and Iraq as a pragmatic solution to political circumstances.

In discussing the Iraqi political situation following World War I, Winstone shows how the intrigue and propaganda of the era make difficult any objective assessment of Bell's many roles during the period. He acknowledges, however, the many inconsistencies in her statements, and he concludes that even though sensitive political papers from the era "have become available, many a noted historian has tried to find his way through that labyrinth of opinions, myths and legends, and outright lies, and has had to admit in the end that he has been dealing not so much with fact as with romance." Similarly, Birkett maintains that Bell's pro-British attitudes and inconsistencies resulted from her belief that Europe offered a stable cultural and political alternative to the warfare characterizing newly emerging Middle Eastern countries.

As political alliances shifted during the 1920s, Bell no longer participated in important decision making. She increasingly turned her attention to the Department of Antiquities in Iraq and renewed her fascination with archaeological studies. When she was appointed curator of the Iraq Museum in Baghdad, she arranged for public exhibitions on Assyria, Babylonia, and Mesopotamia at a time when the discovery of King Tutankhamen's tomb in Egypt was overshadowing interest in these areas. She is credited with popularizing many monuments of the Middle East and with preserving relics that would have been otherwise destroyed.

Despite her hectic schedule, diary entries of this period reveal Bell's profound loneliness. She wrote enviously of her sisters, who had gone on to have their own families, and acknowledged that her adventurous life had brought despairing moments. Although she supervised the education of Faisal, the son of Iraqi king Faisal I and heir to the throne, she was isolated from important sources of power. Facing ill health and the loss of several close relatives, she took a fatal dose of sleeping pills and died 12 July 1926, at the age of fifty-seven. Her biographers disagree about whether or not her death was accidental, because the official inquiry did not publish its findings. In any event, Bell was buried with high honors in

Bell seated on her camel in Baghdad

the British cemetery in Baghdad, and eventually a room in the museum was dedicated to her. The memorial plaque there reads in part:

GERTRUDE BELL

Whose memory the Arabs will ever hold in
reverence and affection
Created this Museum in 1923
Being the Honorary Director of Antiquities for the Iraq
With wonderful knowledge and devotion
She assembled the most precious objects in it
And through the heat of the Summer
Worked on them until the day of her death.

Bell's family received accolades following her death. King George V wrote to her parents that he and the queen "are grieved to hear of the death of your distinguished and gifted daughter, whom we held in high regard." The London *Times* called her "distinguished," identified her as the only British woman "for whom the East has become a passion," and particularly praised her "English mettle" at hav-

Bell (seated on third camel from the left), Winston Churchill (on the camel immediately to Bell's right), and other participants in the Cairo Conference of 1921

ing endured the tribulations of recent political events.

Although Bell has been the subject of several biographies, little critical commentary exists on her travel narratives. Her canon, and especially *The Letters of Gertrude Bell,* have been frequently mentioned in the more generalized studies of the era and have been used extensively by critics of three schools. Colonialist critics such as Edward W. Said quote Bell to demonstrate certain methods of representation. Cultural scholars with particular interests in the Middle East, interests especially in women's institutions such as the harem, cite her comments on the roles of Middle Eastern women. Thus, critics such as Billie Melman and Judy Mabro examine Bell's descriptions of family life. Feminist scholars such as Sara Mills study Bell's work as representative of women's travel writing. For the most part, these scholars treat her work as a consistent whole, even though it spanned more than twenty years of travel writing. Rather than allow for changes in her attitudes or interests, critics present Bell as an exemplum embodying certain attitudes that they examine. References to her work are frequent but not extensive, and she is almost always studied along with other travelers, to whom she is compared.

The colonialist critics accuse Bell of embodying the worst of British imperialism. They align her with T. E. Lawrence and Charles M. Doughty, other notable travelers of the period, and maintain that these writers viewed Arab culture as "other" – that is, as a foreign entity, not a union of individual people with separate identities, but as a nation with a worldview different from and inferior to that of Europe.

Along with the colonialist critics, other scholars with a particular interest in the Middle East view Bell as continuing a literary tradition established by eighteenth-century travelers such as Lady Mary Wortley Montagu. Creating an "imaginary geography" of the East, those travel writers initially defined the East in terms of some exotic qualities – such as through poetic depictions of the desert landscape. Heavily influenced by the *Arabian Nights,* those writers of travel literature and fiction saw the East as sacred and as the origin of civilization and culture, but also as unchanging. They were particularly interested in unfamiliar Eastern institutions such as the harem.

During the eighteenth century writers were intrigued by the sexual possibilities of polygamy; in the nineteenth century other writers reduced harem culture to a Middle Eastern expression of separate spheres for men and women. Thus, in nineteenth-century writings the harem came to reflect domestic womanly virtues such as passivity, but also to reflect, in the minds of European travel writers, distinctly Arabic tendencies of indolence and uncleanliness. Feminist critics therefore suggest that Bell follows these nineteenth-century traditions – sometimes showing the harem as sexually permissive, but more often showing it as the site of sexual oppression, an enclosed area in which women are subjected to unfair limitations of their freedoms, particularly in education and marital choice. Such feminist critics find that, by presenting the harem through the lens of Victorian sensibility, Bell obscures its independent nature and the actual circumstances of the women's lives.

Other feminist critics see Bell's travel narratives as expressions of certain literary tropes and conventions, rather than as expressions of her personality. Citing her work, they see her as a model of a woman who used her writing as a transforming activity, one that allowed her to escape the restrictions of Victorian society. Mills, for example, sees what Bell omits from her narratives as evidence of the restrictions placed on female writers of the period: by failing to acknowledge the dangers to their own safety in pursuing their endeavors, women travel writers such as Bell inadvertently maintained the tradition of heroic adventure that male writers of the period used in order to applaud their contributions in discovery and exploration.

However, these critics also suggest that the prose of women travel writers such as Bell was rhetorically complex and that their travel narratives differ from those of male writers of the period. These critics conclude that the women travel writers, including Bell, constitute a separate literary tradition, one that is not part of the colonial structure. They cite Bell as an example of a writer who was more sensitive to Arab identity than were her male counterparts.

Because of the ways in which literary critics have used Bell's works, it is difficult to evaluate her contributions to the genre of travel writing. Without exception she is regarded as a major figure during the late nineteenth and early twentieth centuries, especially as her writing refers to the Middle East. Yet more often than not, critics use her writing to advance particular literary theses rather than to examine her works. As a result, Bell's individual works have been critically neglected, except by editors writing introductions or by biographers. Critics also often quote extensively from her letters rather than from her published texts, and this further blurs generic boundaries and the differences between commentary written for a private audience and that meant for public reflection. Just as Winstone suggests that it is difficult to evaluate Bell's political contributions, it is similarly difficult to assess her literary ones. She is rarely examined on her own terms.

Perhaps the most reasonable assessment of Bell's significance as a travel writer is included in Winstone's revised biography of Bell. In it he finds that Bell's work will remain of interest, especially "in light of changing world relations." In a new introduction to the revised biography, he notes that the Persian Gulf War of 1991 heightened interest in the history of the "creation of the Iraqi state" and that Bell's insights into that formation are enormously helpful. Her knowledge of the Middle East during the early twentieth century was crucial.

Yet except for *Safar Nameh* and *The Desert and the Sown* her travel narratives are today not likely to find receptive audiences because they are highly technical and specialized studies in archaeology and history. Of greater interest may be her letters, which reveal a complex personality in tracing her evolution from a youthful schoolgirl reading massive Victorian biographies to a mature woman commenting on the changing political milieu. As Winstone says,

> Her interests were those of the scholar, the historian and archaeologist; her approach to travel was philosophical, though it was accompanied by courage and endurance of the highest order. She did not seek to discover new places or to map unknown areas, and the efforts to portray her as the "Daughter of the Desert" as a kind of schoolgirl's "Lawrence of Arabia" do justice neither to her nor to her real achievements.

Instead, for him, Bell's willingness to use her privileged status for scholarship and writing ennobled her life. Comparing her to Charlotte Brontë's fictional Shirley Keeldar, he notes that Bell used the "gold given her by the industrial revolution to buy not privilege but the opportunity for noble performance."

Letters:

The Letters of Gertrude Bell, 2 volumes, edited by Lady Florence Bell (London: Benn, 1927; New York: Boni & Liveright, 1927);

The Earlier Letters of Gertrude Bell, edited by Elsa Richmond (London: Benn, 1937; New York: Liveright, 1937).

Biographies:

Ronald Courtenay Bodley, *Gertrude Bell* (New York: Macmillan, 1940);

Josephine Kahn, *Daughter of the Desert: The Story of Gertrude Bell* (London: Bodley Head, 1956);

Anne Tibble, *Gertrude Bell* (London: Blank, 1958);

Elizabeth Burgoyne, *Gertrude Bell, from Her Personal Papers,* 2 volumes (London: Benn, 1958–1961);

H. V. F. Winstone, *Gertrude Bell* (London: Cape, 1978; revised, London: Constable, 1993);

Susan Goodman, Gertrude Bell (Leamington Spa/Dover, U.K.: Berg, 1985);

Janet Wallach, *Desert Queen: The Extraordinary Life of Gertrude Bell, Adventurer, Adviser to Kings, Ally of Lawrence of Arabia* (New York: Doubleday, 1996).

References:

Dea Birkett, *Spinsters Abroad: Victorian Lady Explorers* (London: Blackwell, 1989);

Yvonne French, *Six Great Englishwomen* (London: Hamilton, 1953);

Sarah Graham-Brown, introduction to Bell's *The Desert and the Sown,* edited by Graham-Brown (Boston: Virago, 1985), pp. v–xviii;

Robert Heussler, "Imperial Lady: Gertrude Bell and the Middle East, 1889–1926," *British Studies Monitor,* 9 (Summer 1979): 3–22;

Stephen Hill, "Gertrude Bell," *Antiquity,* 50 (September–December 1976): 190–193;

Judy Mabro, ed., *Veiled Half-Truths: Western Travellers' Perceptions of Middle Eastern Women* (New York: I. B. Tauris, 1991);

Billie Melman, *Women's Orients: English Women and the Middle East, 1718–1918. Sexuality, Religion and Work* (Ann Arbor: University of Michigan Press, 1992);

Sara Mills, *Discourses of Difference: An Analysis of Women's Travel Writing and Colonialism* (New York: Routledge, 1991);

Elizabeth Robins, "A New Art of Travel: An Impression of Gertrude Lowthian Bell's Books, *The Desert and the Sown* and *Amurath to Amurath,*" *Fortnightly,* 95 (March 1911): 470–492;

Edward Said, *Orientalism* (London: Routledge & Kegan Paul, 1978);

Marion Tinling, "Gertrude Bell," in *Women into the Unknown: A Sourcebook on Women Explorers and Travelers,* edited by Tinling (New York: Greenwood Press, 1989), pp. 39–46;

H. V. F. Winstone, *The Illicit Adventure* (London: Cape, 1982).

Papers:

Bell's correspondence is spread among repositories such as the British Library, the Bodleian Library, the Cambridge University Library, and the Oriental Society of the Durham University Library. An extensive collection of more than two thousand items is located at the University Library of Newcastle upon Tyne. Additional field notes and papers are housed in the Royal Geographical Society, London. Official government documents related to the Middle East are housed in the Foreign and Commonwealth Office, the India Office, and the Public Record Office, London.

Hilaire Belloc
(27 July 1870 – 16 July 1953)

Brian D. Reed
Kent State – East Liverpool

See also the Belloc entries in *DLB 19: British Poets, 1880-1914; DLB 100: Modern British Essayists, Second Series;* and *DLB 141: British Children's Writers, 1880-1914.*

BOOKS: *Verses and Sonnets* (London: Ward & Downey, 1896);

The Bad Children's Book of Beasts (Oxford: Alden, 1896; London: Simpkin, Marshall, Hamilton, Kent, 1896; New York: Dutton, 1896);

Syllabus of a Course of Six Lectures on the French Revolution (Philadelphia: American Society for the Extension of University Teaching, 1896);

Syllabus of a Course of Six Lectures on the Crusades (Philadelphia: American Society for the Extension of University Teaching, 1896);

Syllabus of a Course of Six Lectures on Representative Frenchmen (Philadelphia: American Society for the Extension of University Teaching, 1896);

Syllabus of a Course of Six Lectures on Paris (Philadelphia: American Society for the Extension of University Teaching, 1897);

More Beasts (For Worse Children) (London & New York: Arnold, 1898; New York: Knopf, 1922);

The Modern Traveller (London: Arnold, 1898; New York: Knopf, 1922);

Danton: A Study (Oxford & London: Nisbet, 1899; New York: Scribners, 1899);

A Moral Alphabet (London: Arnold, 1899);

Lambkin's Remains, anonymous (Oxford: Proprietors of the J.R.C., 1900; New York: Mansfield, 1900);

Paris (London: Arnold, 1900; New York: Scribners, 1907);

Robespierre: A Study (London: Nisbet, 1901; New York: Scribners, 1901);

The Path to Rome (London: George Allen, 1902; New York: Longmans, Green, 1902);

Hilaire Belloc in Rome, 1901

The Aftermath; or, Gleanings from a Busy Life . . . Caliban's Guide to Letters (London: Duckworth, 1903; New York: Dutton, 1903);

The Great Inquiry, anonymous (London: Duckworth, 1903);

Why Eat? (London: Broadside, 1903);

Avril: Being Essays on the Poetry of the French Renaissance (London: Duckworth, 1904; New York: Dutton, 1904);

Emmanuel Burden, Merchant, of Thames St., in the City of London: A Record of His Lineage, Speculations, Last Days and Death (London: Methuen, 1904; New York: Scribners, 1904);

The Old Road (London: Constable, 1904; Philadelphia: Lippincott, 1905);

Esto Perpetua: Algerian Studies and Impressions (London: Duckworth, 1906; New York: McBride, 1925);

Sussex (London: Adam & Charles Black, 1906);

Hills and the Sea (London: Methuen, 1906; New York: Scribners, 1906);

The Historic Thames (London: Dent / New York: Dutton, 1907);

Cautionary Tales for Children, Designed for the Admonition of Children between the Ages of Eight and Fourteen Years (London: Nash, 1908; New York: Knopf, 1922);

The Catholic Church and Historical Truth (Preston: W. Watson, 1908);

An Examination of Socialism (London: Catholic Truth Society, 1908);

On Nothing and Kindred Subjects (London: Methuen, 1908; New York: Dutton, 1909);

Mr. Clutterbuck's Election (London: Nash, 1908);

The Eye-Witness: Being a Series of Descriptions and Sketches in Which It Is Attempted to Reproduce Certain Incidents and Periods in History, as from the Testimony of a Person Present at Each (London: Nash, 1908; New York: Dutton, 1924);

The Pyrenees (London: Methuen, 1909; New York: Knopf, 1923);

A Change in the Cabinet (London: Methuen, 1909);

Marie Antoinette (London: Methuen, 1909; New York: Doubleday, Page, 1909);

On Everything (London: Methuen, 1909; New York: Dutton, 1910);

The Church and Socialism (London: Catholic Truth Society, 1909);

The Ferrer Case (London: Catholic Truth Society, 1910);

On Anything (London: Constable, 1910; New York: Dutton, 1910);

Pongo and the Bull (London: Constable, 1910);

On Something (London: Methuen, 1910; New York: Dutton, 1911);

Verses (London: Duckworth, 1910; New York: Gomme, 1916);

The Party System, with Cecil Chesterton (London: Swift, 1911);

The French Revolution (London: Williams & Norgate, 1911; New York: Holt, 1911);

The Girondin (London & New York: Nelson, 1911; New York: Doubleday, Page, 1912);

More Peers: Verses (London: Swift, 1911; New York: Knopf, 1914);

Socialism and the Servile State: A Debate Between Messrs. Hilaire Belloc and Ramsay MacDonald (London: South West London Federation of the Independent Labour Party, 1911);

First and Last (London: Methuen, 1911; New York: Dutton, 1912);

The Battle of Blenheim (London: Swift, 1911; Philadelphia & London: Lippincott, 1936);

Malplaquet (London: Swift, 1911);

Waterloo (London: Swift, 1912; revised edition, London: Rees, 1915);

The Four Men: A Farrago (London & New York: Nelson, 1912; Indianapolis: Bobbs-Merrill, 1912);

The Green Overcoat (Bristol: Arrowsmith, 1912; New York: McBride, Nast, 1912);

Tourcoing (London: Swift, 1912);

Warfare in England (London: Williams & Norgate, 1912);

This and That and the Other (London: Methuen, 1912; New York: Dodd, Mead, 1912);

The Servile State (London & Edinburgh: Foulis, 1912; Boston: Phillips, 1913);

The River of London (London & Edinburgh: Foulis, 1912);

Crécy (London: Swift, 1912);

The Stane Street: A Monograph (London: Constable, 1913; New York: Dutton, 1913);

The Book of the Bayeux Tapestry, Presenting the Complete Work in a Series of Colour Facsimiles: The Introduction and Narrative by Hilaire Belloc (London: Chatto & Windus, 1913; New York: Putnam, 1914);

Poitiers (London: Rees, 1913);

Anti-Catholic History: How it is Written (London: Catholic Truth Society, 1914);

The History of England from the First Invasion by the Romans to the Accession of King George the Fifth (London & Edinburgh: Sands, 1915; New York: Catholic Publication Society of America, 1915);

A General Sketch of the European War, 2 volumes (London, Edinburgh, Paris & New York: Nelson, 1915, 1916); republished as *The Elements of the Great War,* 2 volumes (New York: Hearst's International Library, 1915, 1916);

High Lights of the French Revolution (New York: Century, 1915);

Land and Water Map of the War and How to Use It: Drawn under the Direction of Hilaire Belloc (London: Land & Water, 1915);

A Picked Company: Being from the Writing of H. Belloc (London: Methuen, 1915);

The Two Maps of Europe and Some Other Aspects of the Great War (London: Pearson, 1915);

At the Sign of the Lion, and Other Essays from the Books of Hilaire Belloc (Portland, Maine: Mosher, 1916);

The Second Year of the War (London: Burrup, Mathieson & Sprague, 1916);

The Last Days of the French Monarchy: With Many Illustrations from Paintings and Prints (London: Chapman & Hall, 1916);

Religion and Civil Liberty (London: Catholic Truth Society, 1918);

The Free Press (London: Allen & Unwin, 1918);

Europe and the Faith (London: Constable, 1920; New York: Paulist Press, 1920);

The House of Commons and Monarchy (London: Allen & Unwin, 1920; New York: Harcourt, Brace, 1922);

Pascal's "Provincial Letters" (London: Catholic Truth Society, 1921);

The Jews (London, Bombay & Sidney: Constable, 1922; Boston: Houghton Mifflin, 1922);

The Mercy of Allah (London: Chatto & Windus, 1922; New York: Appleton, 1922);

On (London: Methuen, 1923; New York: Doran, 1923);

The Road (Manchester: Hobson, 1923; New York: Harper, 1923);

Sonnets and Verse (London: Duckworth, 1923; New York: McBride, 1924; enlarged edition, London: Duckworth, 1938; New York: Sheed & Ward, 1939; enlarged again, London: Duckworth, 1954);

The Contrast (London: Arrowsmith, 1923; New York: McBride, 1924);

The Campaign of 1812 and the Retreat from Moscow (London & New York: Nelson, 1924); republished as *Napoleon's Campaign of 1812 and the Retreat from Moscow* (New York & London: Harper, 1926);

Economics for Helen (London: Arrowsmith, 1924; New York & London: Putnam, 1924); republished as *Economics for Young People: An Explanation of Capital, Labour, Wealth, Money, Production, Exchange, and Business, Domestic and International* (New York & London: Putnam, 1925);

The Political Effort (London: True Temperance Association, 1924);

The Cruise of the "Nona" (London: Constable, 1925; Boston & New York: Houghton Mifflin, 1925);

Mr. Petre: A Novel (London: Arrowsmith, 1925; New York: McBride, 1925);

Miniatures of French History (London, Edinburgh & New York: Nelson, 1925; New York & London: Harper, 1926);

A History of England, 4 volumes (London: Methuen, 1925-1931; New York & London: Putnam, 1925-1931);

The Highway and Its Vehicles, edited by Geoffry Holme (London: Studio, 1926);

Short Talks with the Dead and Others (Kensington: Cayme, 1926; New York: Harper, 1926);

Mrs. Markham's New History of England: Being an Introduction for Young People to the Current History and Institutions of Our Time (Kensington: Cayme, 1926);

The Emerald of Catherine the Great (London: Arrowsmith, 1926; New York & London: Harper, 1926);

A Companion to Mr. Wells's "Outline of History" (London: Sheed & Ward, 1926; San Francisco: Ecclesiastical Supply Association, 1927; revised edition, London: Sheed & Ward, 1929);

The Catholic Church and History (London: Burns, Oates & Washbourne, 1926; New York: Macmillan, 1926);

Mr. Belloc Still Objects to Mr. Well's "Outline of History" (London: Sheed & Ward, 1927; San Francisco: Ecclesiastical Supply Association, 1927);

The Haunted House (London: Arrowsmith, 1927; New York & London: Harper, 1928);

Oliver Cromwell (London: Benn, 1927);

Towns of Destiny (New York: McBride, 1927); republished as *Many Cities* (London: Constable, 1928);

James the Second (London: Faber & Gwyer, 1928; Philadelphia: Lippincott, 1928);

How the Reformation Happened (London: Cape, 1928; New York: Dodd, Mead, 1928);

But Soft – We Are Observed! (London: Arrowsmith, 1928); republished as *Shadowed!* (New York & London: Harper, 1929);

A Conversation with an Angel and Other Essays (London: Cape, 1928; New York & London: Harper, 1929);

Belinda: A Tale of Affection in Youth and Age (London: Constable, 1928; New York & London: Harper, 1929);

The Chanty of the Nona (London: Faber & Gwyer, 1928);

Do We Agree? A Debate between G. K. Chesterton and Bernard Shaw, with Hilaire Belloc in the Chair (Hartford, Conn.: Mitchell, 1928);

Survivals and New Arrivals (London: Sheed & Ward, 1929; New York: Macmillan, 1929);

Joan of Arc (London, Toronto, Melbourne & Sydney: Cassell, 1929; Boston: Little, Brown, 1929);

The Missing Masterpiece: A Novel (London: Arrowsmith, 1929; New York & London: Harper, 1929);

Richelieu: A Study (Philadelphia & London: Lippincott, 1929; London: Benn, 1930);

Wolsey (London, Toronto, Melbourne & Sydney: Cassell, 1930; Philadelphia & London: Lippincott, 1930);

The Man Who Made Gold (London: Arrowsmith, 1930; New York & London: Harper, 1931);

New Cautionary Tales: Verses (London: Duckworth, 1930; New York: Harper, 1931);

Why I Am and Why I Am Not a Catholic, by Belloc and others (New York: Macmillan, 1930);

A Conversation with a Cat and Others (London, Toronto, Melbourne & Sydney: Cassell, 1931; New York & London: Harper, 1931);

Essays of a Catholic Layman in England (London: Sheed & Ward, 1931); republished as *Essays of a Catholic* (New York: Macmillan, 1931);

Cranmer (London, Toronto, Melbourne & Sydney: Cassell, 1931); republished as *Cranmer, Archbishop of Canterbury, 1533–1556* (Philadelphia & London: Lippincott, 1931);

Nine Nines; or, Novenas from a Chinese Litany of Odd Numbers (Oxford: Blackwell, 1931);

On Translation (Oxford: Clarendon Press, 1931);

Six British Battles (Bristol: Arrowsmith, 1931);

Usury (London: Sheed & Ward, 1931);

An Heroic Poem in Praise of Wine (London: Davies, 1932);

Ladies and Gentlemen, for Adults Only and Mature at That (London: Duckworth, 1932);

The Question and the Answer (New York & Milwaukee: Bruce, 1932; London & New York: Longmans, Green, 1938);

Saulieu of the Morvan (New York: Ludowici-Celadon, 1932);

The Postmaster-General (London: Arrowsmith, 1932; Philadelphia: Lippincott, 1932);

Napoleon (London, Toronto, Melbourne & Sydney: Cassell, 1932; Philadelphia & London: Lippincott, 1932);

The Tactics and Strategy of the Great Duke of Marlborough (London: Arrowsmith, 1933);

Charles the First, King of England (London, Toronto, Melbourne & Sydney: Cassell, 1933; Philadelphia & London: Lippincott, 1933);

William the Conqueror (London: Davies, 1933; New York: Appleton-Century, 1934);

Cromwell (London, Toronto, Melbourne & Sydney: Cassell, 1934; Philadelphia & London: Lippincott, 1934);

A Shorter History of England (London: Harrap, 1934; New York: Macmillan, 1934);

Milton (London, Toronto, Melbourne & Sydney: Cassell, 1935; Philadelphia & London: Lippincott, 1935);

The Battle Ground (London, Toronto, Melbourne & Sydney: Cassell, 1936); republished as *The Battleground: Syria and Palestine, the Seedplot of Religion* (Philadelphia & London: Lippincott, 1936);

Characters of the Reformation (London: Sheed & Ward, 1936; New York: Sheed & Ward, 1936);

The Hedge and the Horse (London, Toronto, Melbourne & Sydney: Cassell, 1936);

The County of Sussex: With Six Maps in the Text (London, Toronto, Melbourne & Sydney: Cassell, 1936);

An Essay on the Restoration of Property (London: Distributist League, 1936); republished as *The Restoration of Property* (New York: Sheed & Ward, 1936);

Selected Essays, compiled by John Edward Dineen (Philadelphia & London: Lippincott, 1936);

The Crusade: The World's Debate (London, Toronto, Melbourne & Sydney: Cassell, 1937); republished as *The Crusades: The World's Debate* (Milwaukee: Bruce, 1937);

The Crises of Our Civilization (London, Toronto, Melbourne & Sydney: Cassell, 1937); republished as *The Crises of Civilization* (New York: Fordham University Press, 1937);

An Essay on the Nature of Contemporary England (London: Constable, 1937; New York: Sheed & Ward, 1937);

The Issue (New York & London: Sheed & Ward, 1937);

The Case of Dr. Coulton (London: Sheed & Ward, 1938);

Stories, Essays, and Poems (London: Dent, 1938);

The Great Heresies (London: Sheed & Ward, 1938; New York: Sheed & Ward, 1938);

Return to the Baltic (London: Constable, 1938);

Monarchy: A Study of Louis XIV (London, Toronto, Melbourne & Sydney: Cassell, 1938; New York & London: Harper, 1938);

Cautionary Verses: The Collected Humorous Poems (London: Duckworth, 1939); republished as *Hilaire Belloc's Cautionary Verses: Illustrated Album Edition with the Original Pictures* (New York: Knopf, 1941);

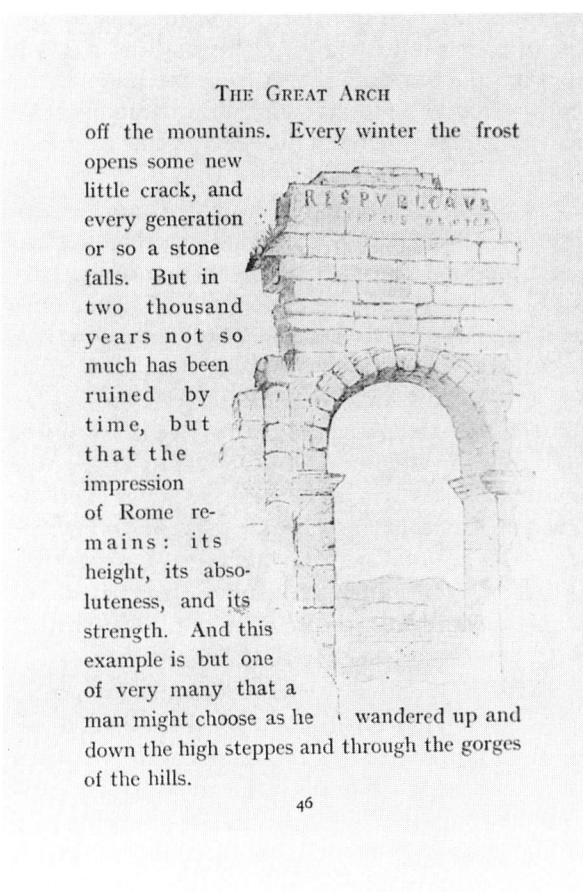

Page from Esto Perpetua: Algerian Studies and Impressions *(1906), with Belloc's drawing of one of the Roman ruins in North Africa*

On Sailing the Sea: A Collection of the Seagoing Writings of Hilaire Belloc, selected by W. N. Roughead (London: Methuen, 1939);

Charles II: The Last Rally (New York & London: Harper, 1939); republished as *The Last Rally: A Story of Charles II* (London, Toronto, Melbourne & Sydney: Cassell, 1940);

The Catholic and the War (London: Burns, Oates, 1940);

On the Place of Gilbert Chesterton in English Letters (London: Sheed & Ward, 1940; New York: Sheed & Ward, 1940);

The Silence of the Sea and Other Essays (New York: Sheed & Ward, 1941; London, Toronto, Melbourne & Sydney: Cassell, 1941);

Places (New York: Sheed & Ward, 1941; London, Toronto, Melbourne & Sydney: Cassell, 1942);

Elizabethan Commentary (London, Toronto, Melbourne & Sydney: Cassell, 1942); republished as *Elizabeth: Creature of Circumstance* (New York & London: Harper, 1942);

The Alternative: An Article Originally Written during Mr. Belloc's Parliamentary Days, for St. George's Review and Since Revised (London: Distributist Books, 1947);

Duncton Hill: Unaccompanied Part-Song for S. A. T. B., music by David Moule-Evans (London: Williams / Boston: B. F. Wood Music, 1947);

Selected Essays (London: Methuen, 1948);

Hilaire Belloc: An Anthology of His Prose and Verse, selected by Roughead (London: Hart-Davis, 1951; Philadelphia: Lippincott, 1951);

Songs of the South Country (London: Duckworth, 1951);

World Conflict (London: Catholic Truth Society, 1951);

The Verse of Hilaire Belloc, edited by Roughead (London: Nonesuch Press, 1954);

Essays, edited by Anthony Forster (London: Methuen, 1955);

One Thing and Another: A Miscellany from His Uncollected Essays, edited by Patrick Cahill (London: Hollis & Carter, 1955);

Collected Verses (Harmondsworth: Penguin, 1958).

OTHER: *Extracts from the Diaries and Letters of Hubert Howard with a Recollection by a Friend,* edited by Belloc (Oxford: Horace Hart, 1899);

Joseph Bédier, *Tristan and Iseult,* translated by Belloc (London: George Allen, 1903);

Alvin Langdon Coburn, *London,* introduction by Belloc (London: Duckworth / New York: Brentano's, 1909);

"The Place of a Peasantry in Modern Civilization," in *Co-operative Wholesale Societies Limited: Annual for 1910* (Manchester, 1910), pp. 279-298;

The Footpath Way: An Anthology for Walkers, introduction by Belloc (London: Sidgwick & Jackson, 1911);

Cecil Edward Chesterton, *The Perils of Peace,* introduction by Belloc (London: Laurie, 1916);

Ferdinand Foch, *The Principals of War,* translated by Belloc (New York: Holt, 1920);

Foch, *Precepts and Judgements: With a Sketch of the Military Career of Marshall Foch by Major A. Grasset,* translated by Belloc (New York: Holt, 1920);

"Mowing of a Field," in *Modern Essays,* edited by Christopher Morley (New York: Harcourt, Brace, 1921), pp. 113-127;

"On an Unknown Country," in *A Book of Modern Essays,* edited by Bruce Walker McCullough and Edwin Berry Burgum (New York & Chicago: Scribners, 1926), pp. 163-168;

Clare Leighton, *Woodcuts,* introduction by Belloc (London & New York: Longmans, Green, 1930).

Hilaire Belloc is chiefly remembered for his controversial political opinions, often belligerent character, and strong allegiance to the Catholic Church. His deep feelings about the important political and social issues of his day manifest themselves in most of his works. Although he has been criticized because of his tendency to sermonize, the sheer number of his published works, their versatility, and their stirring and often humorous style are impressive. Belloc was a historian, poet, essayist, biographer, and novelist, as well as an important and popular travel writer. His travel books were popular and critically respected works, especially *The Path to Rome* (1902) and *The Cruise of the "Nona"* (1925). Belloc traveled extensively in Europe and the United States, visited Moscow to do research, and traveled to Algeria to convalesce after an illness. Whenever he traveled, Belloc tried to write something about the people, customs, topography, and eating habits that he observed and that he could use in a narrative or collection of essays. In his impressive range of travel narratives, Belloc shows he is both a master of satire and allegory and a skillful observer of culture and landscape.

Joseph Hilaire Peter Belloc was born in a quiet suburb of Paris, Celle Saint-Cloud, on 27 July 1870. His parents were Louis and Bessie Belloc, née Parkes, who had met when Bessie, an English citizen, was vacationing in France. After the outbreak of the Franco-Prussian War later that year the French army began to install large artillery in the area because of its strategic position, and the Bellocs decided a move to Paris would be beneficial. When the Prussian army, under the direction of Otto von Bismarck, defeated the French on 4 September 1870, the Bellocs left for England.

At a young age Belloc foreshadowed his later career by writing poetry and drawing maps. His interest in verse and geography continued throughout his life and often appears in his travel narratives as interjections of off-the-cuff poetry and enlightening historical approaches to topography. When Belloc's father died, so did most French influences for the family, and Belloc became consummately English. When a house nurse began to call him "Hilary," even Belloc's name began to have an English pronunciation. In 1878 the family moved to Sussex, the locale of his travel story *The Four Men* (1912), and by age ten Belloc was sent to the Oratory School, run by John Henry Cardinal Newman. Belloc's Catholic education gave him a considerable knowledge of classical literature and history and a strong belief in religious doctrine. At seventeen he left the Oratory School without a clear direction. Since he was still a French citizen, it was difficult to decide where he should start a career.

Against the wishes of both his mother and Cardinal Newman, Belloc decided to return to Paris to join the French navy. His enrollment in the Parisian College Stanislas in October 1887 lasted only one term, supposedly because of his desire to wander the streets and gardens of Paris unescorted. The godlike way in which the students were to regard their schoolmaster and the oppressive atmosphere of the college were intolerable to Belloc. He left the college, lost his desire to join the navy, and returned to the farm life of Sussex. He soon became bored with the routine of agriculture, however, and left to spend the summer rowing on Irish rivers. At age nineteen he began work as a journalist, writing theater reviews for the *Pall Mall Gazette.*

Belloc's political ideologies had no place in theater reviews, so in 1889, with his friend A. H. Pollen, he founded the *Paternoster,* a monthly journal in which he could expound on his ideas. During this time Belloc met the love of his life, Elodie Hogan, a California native who was returning from a trip to Rome. While on a stopover in London, she attended the same tea party as Belloc, and he instantly fell in love with her. When she left to return to California, the adventurous and determined Belloc decided to follow her. By pawning his library and borrowing money, Belloc was able to afford passage to New York. When his money ran out, he traded drawings for necessities and walked along the railroad tracks toward San Francisco and the woman he adored. One month later, road weary and unkempt, he appeared on the Hogans' doorstep and presented himself to the family. His reception was cold, and Elodie Hogan refused his proposal of marriage. Belloc returned to England heartbroken.

Unbeknownst to Belloc, his sister, Marie, raised enough money to enroll him at Oxford. Balliol College allowed him to take the entrance exam because of his impressive lineage, and Belloc was matriculated in 1893. In a letter to Lady Juliet Duff dated 18 January 1919, Belloc reflects on his lifestyle as a young man: "When one is quite young then is the time to learn the world. One never learns it later. I have always been glad that I left school at 17, . . . went off to America from east to west, walked all over California and Colorado, went into the French Artillery and got into Balliol all before I was 22." Although he achieved First Class honors

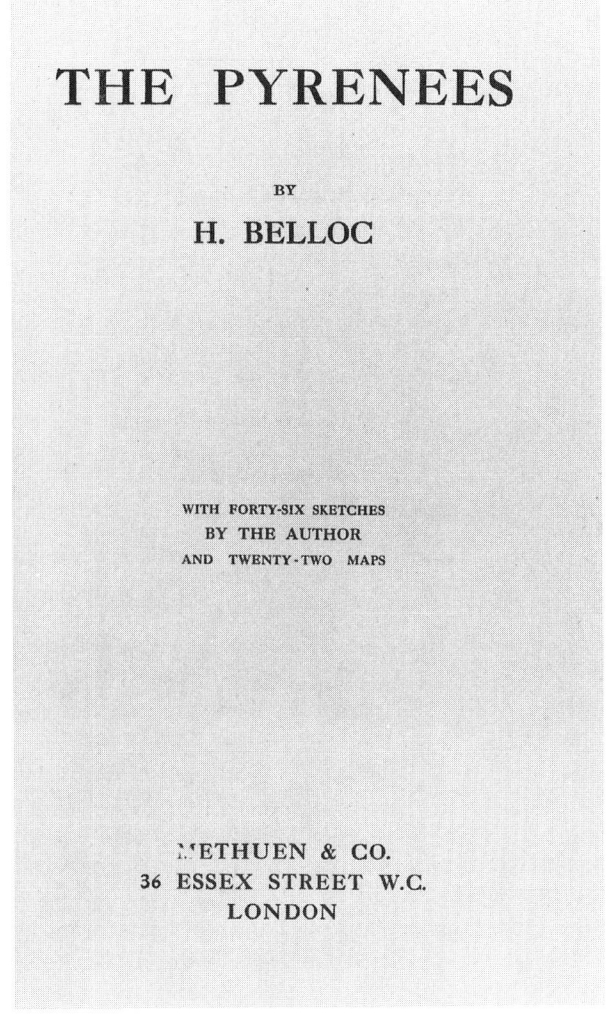

Frontispiece and title page for Belloc's 1909 guide to the European mountain chain that separates France from Spain

in 1895, Belloc believed he was not chosen for a fellowship at the university because of his strong Catholic beliefs, but the end of his academic career may have been more as a result of anti-Semitism on Belloc's part. Again he was disappointed and disheartened: twice he had been denied that which he desperately wanted.

The situation began to improve for Belloc in October 1895, when Hogan wrote to him of her unsuccessful stay at a Maryland convent. Belloc quickly took his mother and sailed for America. After lecturing across the country on his impressive knowledge of French history (which he would later publish in four volumes), Belloc arrived in San Francisco to find Hogan in ill health and mentally distraught. Upon her recovery, she consented to marry him, and on 16 June 1896 they were wed in California. Belloc wrote in a letter to his mother that their marriage "has brought me back from I cannot tell you what precipices of insanity and despair."

The newlyweds went back to England, and Belloc published his first two books: *The Bad Child's Book of Beasts* (1896), an illustrated collection of nonsense verse that was so successful its first edition sold out in four days, and *Verses and Sonnets* (1896), a book of serious poetry that received little notice. Belloc decided to pursue financial reward, so he continued writing satire and nonsense poetry and tutored students in history at Oxford. During this time he met G. K. Chesterton, and they quickly became notorious friends and allies. Their shared political and religious ideas were often unpopular and always controversial. During the South African Boer War (1899–1902), Belloc and Chesterton took the side of the Boers, whom they felt were rightly defending themselves against the imperialistic En-

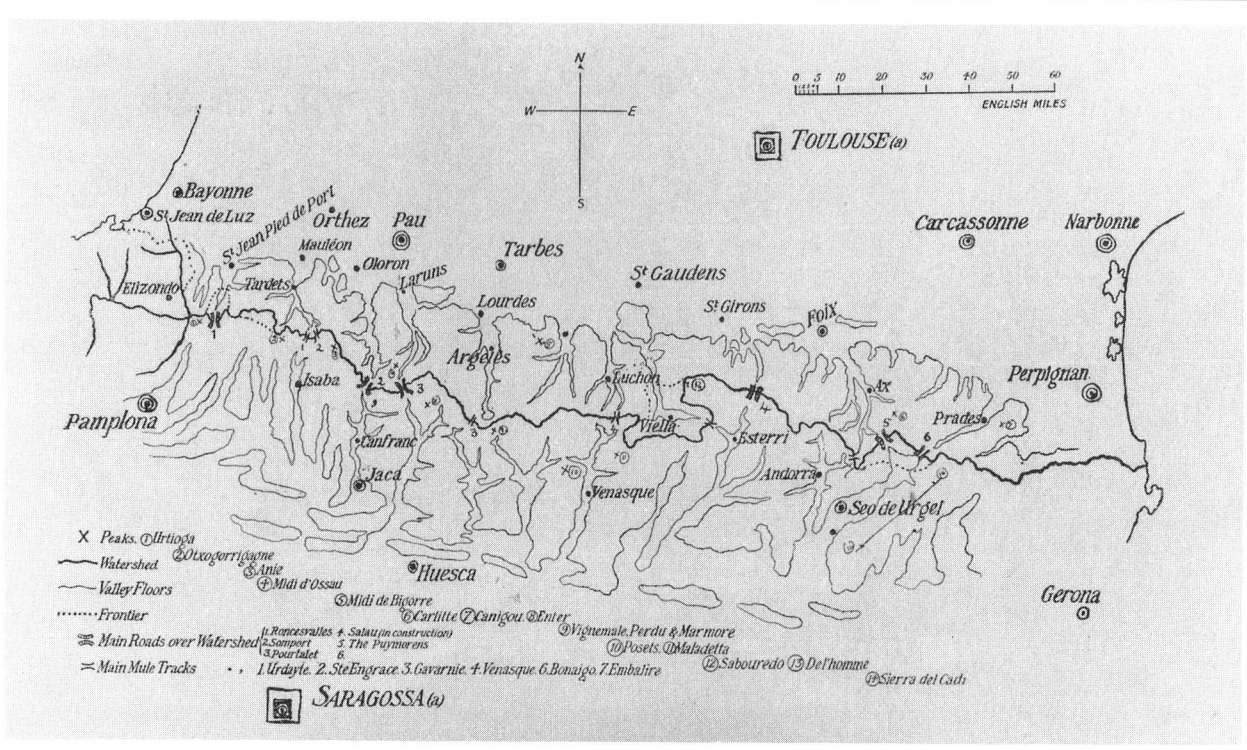

Map from Belloc's The Pyrenees

glish. Belloc's stance was considered by many to be unpatriotic and eccentric. He felt that his government was interested more in the precious gold mines that were under Boer control than the unpopular enslavement of the natives by the Afrikaners. Even amidst this controversy, Belloc was able to publish an important historical look at the origins of his earlier residence, *Paris* (1900), which was carefully researched and became a popular tour guide for English travelers.

Shortly thereafter he published *The Path to Rome,* the first of his travel books, considered by many to be his most important work. This eclectic and multifaceted book, replete with ideological musings, witty poetry, and personal sketches of towns and their terrains, recounts Belloc's journey on foot from Toul, France, across Switzerland to the ancient city and cultural center of Catholicism, Rome. The journey is narrated with the wit of Laurence Sterne and the spirituality of John Bunyan. When the book was published, Belloc was thirty-one years old, had two young children, and was known in literary circles as a somewhat eccentric yet important historian and aggressive journalist. The book would solidify Belloc's reputation as an important travel writer.

It is difficult to understand why Belloc so often left his friends and family to travel alone. Perhaps his mixed French and English heritage, his often unpopular religious and political treatises, and his ongoing desire for an individual spiritual epiphany may have led him to seek enlightenment continually on the road. Whatever his motivation, Belloc left Toul in early June 1901 with only a thin suit, one pair of heavy boots, and a meager amount of money. He slept in barns, fields, or an occasional country inn, and ate what was available to him, mostly ham, bread, and regional wine (he often extols their virtues at considerable length). His travel was slow and often perilous.

Belloc relates his failed attempt to cross directly over the Swiss Alps with a local guide: "the cold began to seize me in my thin clothes. My hands were numb, my face already gave me intolerable pain, and my legs suffered and felt heavy." As terror began to grip Belloc, his guide explained their situation in the midst of a snowstorm: "If you go back through it and lose your way, you are done for. If you halt in some shelter, it may go on for two or three days, and then there is an end of you." Belloc devised a different route, and after their descent he was even more determined to reach his goal. At the end of the book, he enters Rome with only twenty minutes to get to Saint Peter's to hear Mass. Oddly enough, he does not describe any part of Rome but instead uses his twenty minutes to enter a cafe and call "for bread, coffee, and brandy" and write down a travel song. The omission of the long-awaited description of Rome causes the Lector (a representation of the reader) to exclaim, "But this is a dogg –." Belloc's answer is "Not a word!"

The Path to Rome was one of Belloc's biggest successes (selling more than one hundred thousand copies) and helped establish him as an important literary figure. During the time Belloc was writing *The Path to Rome,* his literary output was tremendous. On his way to Toul he wrote *Robespierre* (1901), and throughout his thirties he published more than two books per year, all the while teaching, lecturing, writing reviews, and pursuing a political career.

Just eighteen months after he began his trek to Rome, Belloc started on another journey that would produce a published work. In December 1902 Belloc left Winchester with two friends and followed the Pilgrim's Way to Canterbury. His intention was to re-create the historic first pilgrimage to the cathedral, staying close to the original old road, and to enter the West Gate of Canterbury on the anniversary of Thomas à Becket's murder. Having just finished his spiritual journey to the center of Catholicism, Belloc wished to set out on a pilgrimage to the heart of Christianity in England. The book that came out of his experiences, *The Old Road* (1904), merges his skills as historian and traveler.

Belloc begins his book with an essay that outlines the central importance of roads throughout history, claiming that roads are "the most imperative and the first of our necessities." He carefully follows the antiquated "Old Road" that often runs through farmers' fields, across stone fences, and into difficult streams and marshes. At one point Belloc and his companion self-righteously steal a boat to cross a river. After they land safely on the other shore, Belloc relates: "we heard, from the further bank, a woman, the owner of the boat, protesting with great violence." This adventure did not deter the travelers on their journey, as Belloc explains: "We pleaded our grave necessity; put money in the boat, and then, turning, we followed the marshy path across the field." When Belloc finally enters Canterbury and sees the tower of the cathedral, he exclaims: "I had never known such a magic of great height and darkness." He entered the church and waited vainly for an epiphany or sight of the specter of Becket bemoaning his demise, but there was no such vision. The book closes ironically, with Belloc and his companions watching a woman dancing vigorously at a nearby inn.

In January 1905 Belloc decided to sail for Africa after his doctor recommended a period of convalescence to aid his recovery from pleuropneumonia. His journey took him to the old Roman stronghold of North Africa and resulted in *Esto Perpetua* (1906), another travel narrative in which history and topography are united. Following the pattern of his previous travel narratives, *Esto Perpetua* is as much about religion as it is about exploration. The idea for the North African trip came to Belloc when he saw a European craftsman making an Islamic crescent and a Christian cross to adorn the same building in Algiers. This contrast between faiths and the old pagan Roman ruins appealed to Belloc and gives the book its richness and variety.

Although Belloc could not understand or sympathize with Islam, he embraced classic Roman architecture and ingenuity and aligned himself with the ancients as a fellow European. As he journeys across a foliage-covered plateau and descends into the barren desert, he notes: "The Romans had once thoroughly possessed and tilled this land: the scrub had once been forests, the shifting soil ordered and bounded fields; but the Mohammedan sterility had sunk in so deeply that one could not believe that our people had ever been there." Belloc realized that both are a part of the same country just as two different theologies exist in the same space, but he resisted the concept of a universal church.

Belloc's refusal to immerse himself or accept Islamic culture may be most glaringly apparent during a meal that was prepared for him by an old Bedouin. His server brings him meat that he declares is mutton, but Belloc insists it is not, saying: "Good Lord! . . . it might be anything. There is no lack of beasts on God's earth." Belloc declares the meal "horrible," and, feeling God has revealed to him that it is camel meat, he refuses to eat it. When Belloc leaves the desert to return to the coast, he longs to reunite with his own people and "to get back to reasonable shrines." As he leaves port his thoughts are on Europe: "It was in this watch of the early morning that I called out to her *'Esto perpetua!'* which means in her undying language: 'You shall not die.'"

In December 1905, with the collapse of the Conservatives, Belloc began to concentrate on his political career. He ran for Parliament on a platform of Irish Home Rule, free trade, and reduction of British imperialism. When he was asked about his religious beliefs, Belloc took a rosary from his pocket and declared, "If you reject me on account of my religion, I shall thank God that He has spared

Belloc's boat, the Nona, *in which he sailed along the coast of England after his wife's death in 1914*

me the indignity of being your representative." Belloc's honesty won him the election.

During his controversial service to Parliament from 1906 to 1910, Belloc wrote both *The Historic Thames* (1907), a rather dry but colorfully illustrated book, which shows the river's historical significance from medieval times onward, and *The Pyrenees* (1909), which may still be the best guide to the European mountains. *The Pyrenees,* which Robert Speaight in his 1957 biography of Belloc calls another "historical topography," includes forty-six magnificent sketches made by Belloc during his trip through the mountains. While little time is spent describing vistas and precipices, the book presents a careful analysis of how to cross the mountains and what provisions to take. Its lighthearted philosophy of travel is illustrated with a step-by-step discussion about the best way to sip wine from a gourd (a goatskin flask) so that one may "drink as neatly and as cleanly as any Basque or Catalan" and a humorous evaluation of tents, which Belloc feels are "either very light – in which case [they are] always fairly useless – or [they are] heavy, in which case there is an end to your free going."

Shortly after the publication of *The Pyrenees* and during the general election, an outcry over electoral reform in the Parliament caused Belloc to reconsider running again. Belloc felt that the election was being called to save the party system he abhorred. In his final speech before the House of Commons, Belloc lambasted Parliament and later stated scathingly, "Perhaps they did not bribe me heavily enough, but in any case I am relieved to be quit of the dirtiest company it has ever been my misfortune to keep."

After quitting Parliament, Belloc began to spend more time at his Sussex home finishing works already started and beginning new histories and novels. During this period Belloc was putting the finishing touches on *The Four Men*, a travel allegory of a walking tour through his cherished county of Sussex that took him five years to write. Belloc was in his forties and realized how time had changed him and his Sussex landscape. The four men who undertake the journey ("Myself," the Poet, the Sailor, and Grizzlebeard) seem to be reflections of Belloc in his attempts to commune with Sussex before time and man changed its wholesomeness forever. The four characters meet by chance and form a pact to "walk through the whole county to the place we knew [home], and recover, while yet they could be recovered, the principal joys of the soul." As they walk across the English countryside, they discuss many esoteric issues, such as first loves and the mysteries of women, the best and worst things in the world, and the greatest or most sacred sounds. None of the issues are resolved with just one answer, but rather life remains a mystery to be pondered.

Throughout *The Four Men* the characters often break into song, most of which are sung like hymns. The November colors of the landscape are noted with a reverence that portends the approaching winter and the seasonal connection to man's mortality. Likewise, the story's ending is much different from the flippant and youthful final lines of *The Path to Rome,* to which it is often compared. Belloc concludes with serious considerations about the connections between humanity and the landscape. As he enters his valley alone, he composes a poem that transcends the metaphor of soil as death, and he exclaims: "Ah! but if a man is part of and is rooted in one steadfast piece of earth, which has nourished him and given him his being . . . he has outflanked Death in a way." Ironically, the meaning of life, the quest's goal, is revealed in the permanence of the ground the men have tread upon.

Belloc's next travel work, *The Stane Street* (1913), was not researched on foot but rather conducted from a motorcar, which Speaight claims was "liberally provided with cold chicken and champagne." The book, which like *Esto Perpetua* assumes that all of the West has a classical Roman foundation, retraces the old Roman military road that ran from London to the southwest port of Chichester. Complete with maps and topographical drawings by Belloc, *The Stane Street* attempts to discover the original alignments of the Roman road and theorizes as to the placement of the Roman camps along its route. Although many of his conclusions are merely postulations, Belloc feels that his cause is worthy because of the historical significance of a road that was designed "for military purposes alone": "No other road in Britain is so completely designed for the sole use of the army." Belloc concludes that more research must be done before any definitive pronouncement about the actual lay of the road can be made.

Years passed before the publication of Belloc's next important travel narrative. Much tragedy filled the gap. In 1914 Elodie died, and Belloc was distraught and full of guilt about his time away from home. His friends felt he was close to suicidal. World War I brought more tragedy for Belloc. In 1918 he lost his oldest son, Louis, who disappeared when his plane went down just weeks before the end of the war. Belloc withdrew from society and yearned for an earlier, better England.

The Cruise of the "Nona" describes a journey that began in 1914, just three months after Elodie's death. With his friend Phil Kershaw, Belloc sailed along the coast of England in his beloved boat. Although the book insinuates the journey was continuous, in actuality the cruise was often interrupted by business trips to London and an occasional docking in harbors that were more to the two men's liking. Full of opinions on religion, politics, and life, *The Cruise of the "Nona"* allows Belloc to reflect on the ills of England from her coastal waterways. Like a good dose of therapy, the cruise helped Belloc wrestle with his grief as he attempted to overcome it. While the tide is down and the winds are up, Belloc takes on the sea with his rugged boatmanship; when the water is calm, Belloc reposes with his pipe. The book is Belloc's most autumnal travel narrative, and while it is often full of faith and profundity, it is more often a book of despair and disillusionment. Belloc describes the work as having "only an accidental beginning," with no "real end, and nothing in particular between the two." At the end of his voyage over both calm and rough seas, Belloc connects the sea with immortality in the same way he connected the earth with the afterlife in *The Four Men*. His conclusion states, "The sea provides visions, darkness, revelations. . . . The sea has taken me to itself whenever I sought it and has given me relief from men. . . . It is the common sacrament of this world."

The Contrast (1923) came out of a group of exhaustive lectures Belloc gave in the United States about the differences between Americans and Europeans. While many of his assertions, such as his insistence that Americans are happier than Europeans, would be difficult to substantiate, Belloc finds differences between the two cultures that are genuinely fascinating. He observes that "wealth and opportunity in America connote the very opposites of what they do in Europe," claiming that rich Americans wish to surround themselves with noise and people while wealthy

Europeans desire "an extreme . . . seclusion" and have "a horror of noise." Later in the book, the mood changes. Belloc's bitter attack on European politics and his anti-Semitic statements about how America was handling the "Jewish Problem" may be partially responsible for the book's bad reviews and poor sales.

Belloc's final two travel books, *Towns of Destiny* (1927; published in England as *Many Cities,* 1928) and *Places* (1941), are collections of essays written while Belloc was exploring some of his favorite foreign places. Both books, which went into additional printings, display Belloc's skill as a keen observer and recall his earlier exuberance with travel. Until the end of his life Belloc continued to write, but after the 1920s he seldom traveled far from his Sussex home. In the late 1940s Belloc's memory and health began to fail. His eyes weakened, and he was unable to read as much as he desired. On 12 July 1953 Belloc's daughter Eleanor and his grandson Julian found him slightly smoldering after a fall from his fireside chair. They propped him up, and with a scorched and pale demeanor, he muttered, "This must be an awful bore for you." He died on 16 July 1953, as his family prayed and recited the rosary over him.

Belloc's entire life was made up of short and lengthy journeys, both physical and spiritual. His method of chronicling his travels by peppering them with political insights and theological ponderings makes his work in the genre uniquely entertaining. The landscape Belloc describes is most often the topography of humanity, and his destination is regularly spiritual. In old age, as he reflected on his well-known journey that took him to Rome, Belloc quietly stated his feelings about the trip that would lead him toward a path of future adventures: "I am very glad I went."

Bibliography:
Patrick Cahill, *The English First Editions of Hilaire Belloc* (London: Privately printed, 1953).

Biographies:
J. B. Morton, *Hilaire Belloc: A Memoir* (New York: Sheed & Ward, 1955);

Eleanor and Reginald Jebb, *Belloc, the Man* (Westminster, Md.: Newman Press, 1957);

Robert Speaight, *The Life of Hilaire Belloc* (London: Hollis & Carter, 1957);

A. N. Wilson, *Hilaire Belloc* (London: Hamilton, 1984).

References:
Jay P. Corrin, *G. K. Chesterton and Hilaire Belloc: The Battle against Modernity* (Athens: Ohio University Press, 1981);

Robert Hamilton, *Hilaire Belloc: An Introduction to His Spirit and Work* (London: Douglas Organ, 1945);

Marie Belloc Lowndes, *"I, Too, Have Lived in Arcadia"* (New York: Dodd, Mead, 1942);

Lowndes, *The Young Hilaire Belloc: Some Records of Youth and Middle Age* (New York: Kenedy, 1956);

John P. McCarthy, *Hilaire Belloc: Edwardian Radical* (Indianapolis: Liberty Press, 1973);

H. G. Wells, *Mr. Belloc Objects to "The Outline of History"* (New York: Doran, 1926);

Frederick Wilhelmsen, *Hilaire Belloc: No Alienated Man* (New York: Sheed & Ward, 1953);

Douglas Woodruff, ed., *For Hilaire Belloc: Essays in Honor of His 71st Birthday* (New York: Sheed & Ward, 1942).

Papers:
A Hilaire Belloc archive, with correspondence and literary manuscripts, is at the British Library.

James Theodore Bent
(30 March 1852 – 5 May 1897)

and

Mabel Virginia Anna Bent
(?–?)

Kelly Belanger
Youngstown State University

BOOKS: *A Freak of Freedom; or, The Republic of San Marino,* by Theodore Bent (London: Longmans, Green, 1879);

The Life of Giuseppe Garibaldi, by Theodore Bent (London: Longmans, Green, 1881);

Genoa: How the Republic Rose and Fell, by Theodore Bent (London: Kegan Paul, 1881);

The Cyclades; or, Life among the Insular Greeks, by Theodore Bent (London: Longmans, Green, 1885);

The Ruined Cities of Mashonaland: Being a Record of Excavation and Exploration in 1891, by Theodore Bent (London: Longmans, Green, 1892; London & New York: Longmans, Green, 1893);

The Sacred City of the Ethiopians: Being a Record of Travel and Research in Abyssinia in 1893, by Theodore Bent (London & New York: Longmans, Green, 1893);

Southern Arabia, by Theodore Bent and Mabel Bent (London: Smith, Elder, 1900);

A Patience Pocket Book, by Mabel Bent (Bristol: Arrowsmith, 1904);

Anglo-Saxons from Palestine; or, The Imperial Mystery of the Lost Tribes, by Mabel Bent (London: Sherratt & Hughes, 1908).

OTHER: Theodore Bent, "Modern Life and Thought among the Greeks," in *National Life and Thought of the Various Nations throughout the World,* by Eiríkr Magnússon (London: Unwin, 1891; New York: Stokes, 1891);

Thomas Dallam, *Early Voyages and Travels in the Levant,* edited by Theodore Bent (London: Hakluyt Society, 1893; New York: Franklin, n.d.);

Arthur W. C. Boevey, *The Garden Tomb, Golgotha, and the Garden of the Resurrection,* revised and enlarged by Mabel Bent and Miss Hussy (London: Committee of the Garden Tomb Maintenance Fund, 1931).

SELECTED PERIODICAL PUBLICATIONS – UNCOLLECTED: Theodore Bent, "Oliver Cromwell and Genoa," *Antiquary,* 4 (October 1881): 153–157;

Theodore Bent, "Researches among the Cyclades," *Journal of Hellenic Studies,* 5 (1884): 42–54;

Theodore Bent, "Recent Discoveries in Eastern Cilicia," *Journal of Hellenic Studies,* 11 (1890);

Theodore Bent, "A Journey in Cilicia Tracheia," *Journal of Hellenic Studies,* 12 (1891).

Theodore and Mabel Bent, like many well-to-do Victorians, traveled extensively in Africa, southern Arabia, and Mediterranean regions, and together they produced four travel books documenting their experiences. Although the two traveled as a team and many of their observations were recorded in Mabel's journal and photographs, Theodore wrote the first three books, and Mabel wrote the final one after her husband's death. Having written several historical works before trying his hand at travel writing, Theodore's extensive footnoting and meticulous descriptions of archaeological ruins made him an unusually scholarly travel writer. His publications in forums such as the *Journal of Hellenic Studies* indicate that, although he had no formal archaeological training, he was recognized as a dedicated and learned amateur. The sometimes dry, detailed ethnographic descriptions common in his work earned less admiration from tourists and general readers than from scholars, many of whom recognized his book on the Cyclades as a classic and considered him to be the "Fa-

James Theodore and Mabel Virginia Anna Bent

ther of Cycladic Prehistoric Studies," according to Al N. Oikonomides' preface to the 1965 edition of *The Cyclades; or, Life among the Insular Greeks* (1885). Mabel Bent, whose photographs and "Chronicles" (a journal she kept during all her travels with Theodore) were invaluable sources of ethnographic and archaeological data, merits similar recognition.

Theodore Bent, the son of James Bent, was born at Baidon House near Leeds, Yorkshire, on 30 March 1852. He received a classical education at Oxford University and graduated in 1875, two years before he married Mabel Virginia Anna Hall-Dare of Newtonbarry, County Wexford. John Pemble identifies Theodore Bent as an archaeologist and describes him as a figure only slightly less significant than Arthur Evans, a wealthy Welshman who excavated the city of Knossos in 1884, or Charles Newton, "the most distinguished archeologist of his generation." While Bent's work was of the highest scholarly quality, he "did not offer the explorer new Parthenons or majestic ruins of destroyed ancient cities such as Tetra and Pompeii." He traveled to excavate ruins, and his excavations in South Africa, for example, generated theories about the "authors" of the ruins and thereby stirred up controversy in the European scholarly community.

Mabel Bent's motives for travel are less clear. In fact, except through the travel books, little is recorded about her life, and she remains an obscure figure in the literature on travel writing. Her absence even from books such as Jane Robinson's *Wayward Women: A Guide to Women Travellers* (1990) is especially surprising and may result from the fact that she wrote collaboratively with her husband. Although in the prefaces to the Bents' first three travel books Theodore frequently acknowledges her vital role as a photographer, the importance of Mabel's contributions were revealed only when she wrote the fourth book.

Before 1883, when the Bents left for the Cyclades Islands, the subject of their first travel book, Theodore had already published three historical works on the Italian republics – one work being a biography of Italian liberator Giuseppe Garibaldi. These republics and the patriots who fought for their freedom fascinated Bent because, as he writes in his preface to *Genoa: How the Republic Rose and Fell* (1881), the republics had naval and commercial power disproportionate to their small size. As a result, Bent felt, they "must be regarded by the Anglo-

Saxon race with especial interest, as forming one of the steps in that ladder of progress by which [Great Britain has] succeeded in attaining such a pitch of commercial prosperity."

In conducting his research Bent made connections with officials in the Genoese government and gained access to historical manuscripts unavailable to foreigners. This ability to cultivate relationships with people of the countries he visited and researched served Bent well as a travel writer. His travel books invariably describe his remarkable good fortune in finding government officials, tribal leaders, and significant community members who were most accommodating in helping him gain access to homes, archaeological ruins, and cultural events such as weddings and funerals.

Bent met with such hospitality during a brief visit to the Italian republic of San Marino in 1877, a visit that became the basis for his first historical work, *A Freak of Freedom; or, The Republic of San Marino* (1879). While he was there Bent grew interested in the history of the small republic, a history that, as he noted, "mingles us in many of the most stirring scenes which have occurred on Italian soil." Upon leaving the republic, Bent was given a letter of thanks for his interest in the people and was made a citizen of San Marino. *A Freak of Freedom* presents Bent's understanding of how the inhabitants of San Marino came to feel so strongly about their freedom, and, from some of his "own sketches taken on the spot," it also includes illustrations of the private lives and customs of the citizens he met. These anecdotal sketches distinguish *A Freak of Freedom* from Bent's strictly historical work on the Republic of Genoa. They also represent a style Bent developed more fully when his work shifted from the genre of history to that of travel writing.

Bent's first travel publication, "Researches in the Cyclades," was a scholarly article in the *Journal of Hellenic Studies* and is not typical of his travel writing. Exclusively devoted to archaeology, it omits any anecdotal discussion of people and customs and offers detailed descriptions of the sort that typically appear in notes within his travel books. Bent focuses on findings from graveyards he excavated at the island of Antiparos: he diagrams bowls and collections of shells evidently intended as offerings to the dead, as well as marble versions of the human form in graves. Bent concludes by urging further study and more-extensive excavations.

Bent's first travel book, *The Cyclades; or, Life among the Insular Greeks,* includes archaeological data and also reveals ways in which his travels were enlivened and informed by the education he had received at Oxford. As Pemble points out, Bent and his classically educated countrymen felt a special kinship with Rome and Greece, because they could "see the landscapes of these places with the eyes of Homer, Thucydides, Herodotus, Virgil, Theocritus, or Livy." Evidence of the "light of recognition" pervades Bent's book on the Cyclades. He remarks, for example, how a particular story told by an island inhabitant reminds him of the adventures of Ulysses or how Herodotus is the earliest author to mention the eruption of a volcano on the island of Santorini. As the Bents awaited the arrival of their host on the island of Ios, Theodore writes that he "thought much about Homer, and wondered if he really did die here."

Although Bent's classical perspective makes his work significant, his ability to mingle with the people makes his study of the Greek islands even more notable. He writes in the preface that his work there had two focuses: Hellenic archaeology and the manners and customs of islanders. He focuses on the latter because, although the mainland of Greece had been "overrun by barbaric tribes" throughout the years, the political insignificance and unproductive soil of the Cyclades had enabled them to remain much as they had been in ancient times. In 1884 the Bents were among the first travelers to explore the region; in fact, Theodore comments on the islanders' surprise at seeing an English lady for the first time.

Perhaps because the islanders had not yet encountered many travelers and the exploitation that travelers sometimes bring, the Bents were able through their native servant to gain access to folklore that would have otherwise been impossible to study. Bent writes that "personal intercourse with the islanders in all grades of society, at their work and at their board, proved to us the most infallible methods of understanding their life and their superstitions as they exist today." Such close association with the islanders would not have been possible without the sincere respect that the Bents held for them.

Because the islanders accepted the Bents, the two were able to record a series of vignettes about people they visited. *The Cyclades* is divided into nineteen chapters, each chapter about a different island. In one, Theodore tells of a visit to a shoemaker recognized as the village bard. In the chapter on Antiparos, Bent tells of his trip to fish for octopus with an old man who shares a frightening story of how he was nearly murdered. On Sifnos the Bents spend the night with a captain and his wife and are surprised by one of the many pigs that roam the city

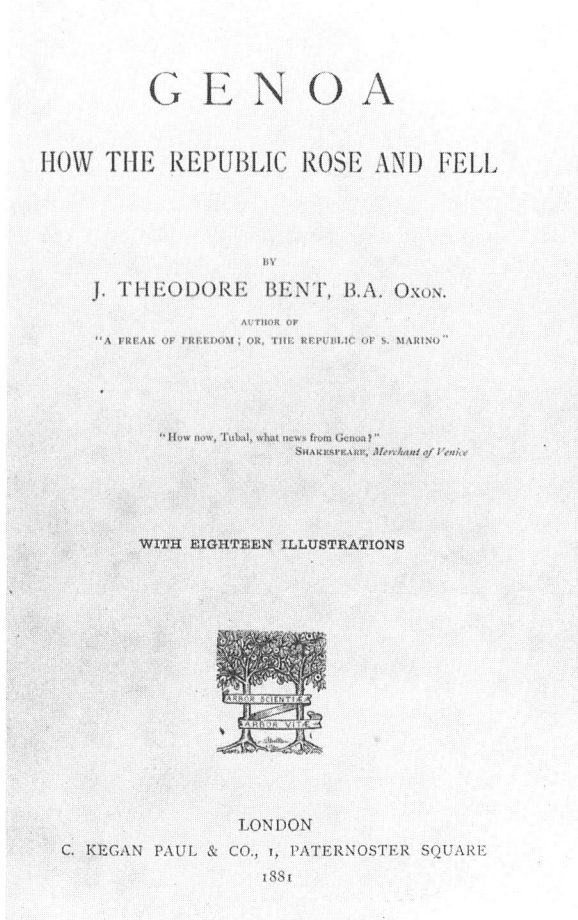

Frontispiece and title page for the last of James Theodore Bent's three books on the Italian republics

streets at night: the pig enters the house and pushes its way through the bedroom door. The Bents attend a ball on the island of Anafi, where they observe local dances and hear a song performed by a drunken shepherd who is considered the finest singer on the island. Bent remarks that he found attending the ball pleasant, for he and Mabel could "see the manners and the customs of [the islander's] private life to perfection." In fact, Bent was so successful at gaining access to cultural events that he sometimes found himself in uncomfortable situations. He tells, for instance, of a wedding that he attended on Santorini and in an insufferably hot cave with many female guests. In describing the dancing, he admits that "the ordeal for me was trying, for I became conscious that only women were admitted to this part of the ceremony, and I had seen more than I was intended to."

Some of Bent's questions also led him onto the "treacherous ground" of Greek politics, a subject that he eventually vowed to avoid. He was willing, however, to resort to an interviewing strategy that involved playing on the jealousy that Greek islanders felt toward inhabitants of other islands in the Cyclades. Bent writes that "this is always the way to see and hear anything in these islands. Tell everybody the beautiful things you have seen and heard in other islands, and you are sure to arrive at the best they have. Jealousy is wonderfully developed in these parts." When the Bents exclaimed to their hosts on Folegandros about the beauty of some embroideries that they had seen elsewhere, the husband said that he and his wife did not care about such things – but the wife displayed all the lace and embroidery that she could find. Through their understanding of the people, the Bents participated in various events – from the intensely private mourning of a family at a funeral to the festival of Tinos, one of the greatest events of Greek life, which drew forty-five thou-

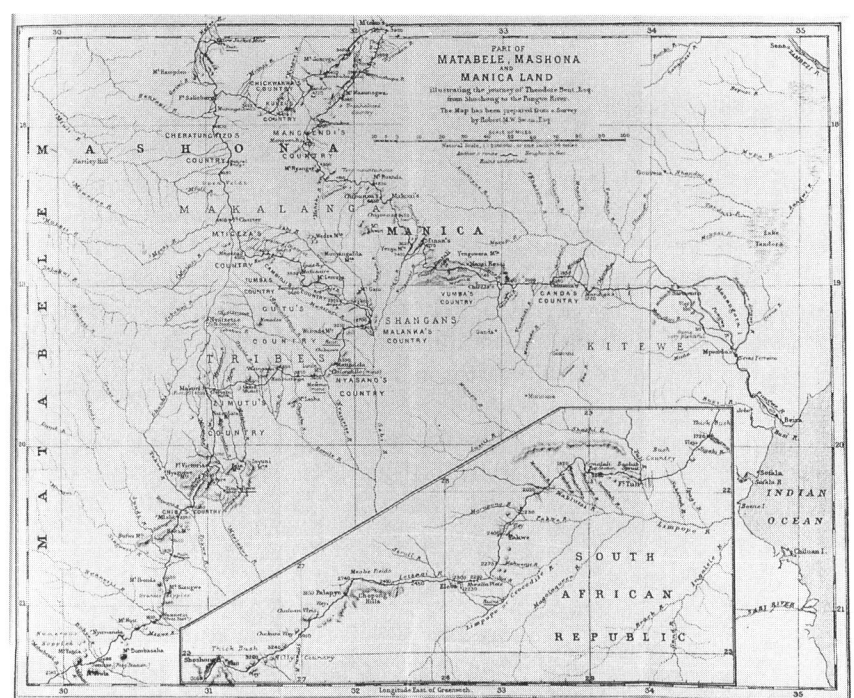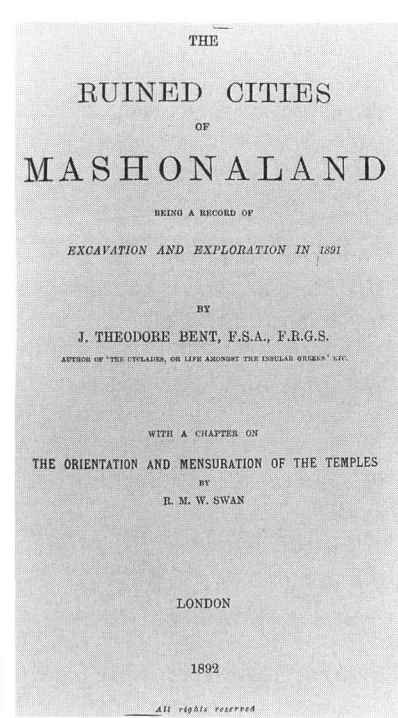

Foldout frontispiece and title page for the book based on the Bents' first trip to Africa

sand religious pilgrims to the island of Tinos each year.

Bent devotes an entire subchapter, one of his most lengthy descriptions of any event, to the festival. Aside from the sheer number of people involved in this event, the political importance of this festival made it particularly interesting to the Bents. Theodore explains that the Greeks' yearly excursion to Tinos was an opportunity for the oppressed Cretans to meet free people from Hellas, and thereby seeds of revolt against Turkish rule were sown. Because of his long interest in stories of how people sought freedom from oppressors, his focus on the festival of Tinos is not surprising. The chapter also demonstrates his ability to produce the ethnographer's "thick description," as he provides detailed descriptions and analyses of people, places, and events and even relates specific discussions. As they so often did, the Bents gained access through a friendly "informant" to the most private part of the festival. Bent writes that "with doors closed and windows bolted, to exclude the common herd, the miraculous [picture of the Madonna] was taken out of its box to be washed, and when it came out of its retirement I strained my eyes to see it."

The Bents' interests in archaeology and anthropology led them to focus on both archaeological ruins and people when they set out to investigate the ruined cities of Mashonaland in southern Africa. *The Ruined Cities of Mashonaland: Being a Record of Excavation and Exploration in 1891* (1892) consists of three chapters, the first of which focuses on the people and places they encountered on their journey to the ruins they excavated; the second describes the archaeology of the ruined cities; the final chapter describes both the people and ruins they encountered on their journey from the inner part of South Africa to the coast. The Bents had funded their trip to the Cyclades, but the Royal Geographical Society, the British Chartered Company of South Africa, and the British Association for the Advancement of Science funded in part this second major exploration. The support from these organizations suggests that the British government considered the Bents' work to be valuable. It added knowledge about the African continent, where Britain sought to stabilize its power.

Yet one consequence of accepting such funding was that the Bents surrendered some freedom to do as they pleased. Before the trip began, members of these societies questioned whether a lady ought to undertake such a rigorous journey. Theodore defends Mabel by insisting that he never questioned her abilities:

> fortified – by previous experience – we hardly gave these doubts more than a passing thought, and the event proved that they were wholly unnecessary. My wife was

the only one of our party who escaped fever, never having a day's illness during the whole year that we were away from home. She was able to take a good many photographs under circumstances of exceptional difficulty, and instead of being, as was prophesied, a burden to the expedition, she furthered its interests and contributed to its success in more ways than one.

In addition to Mabel, the third member of the exploration team was a cartographer, Robert McNair Wilson Swan. The presence of a cartographer and photographer doubtless accounts for the helpful maps, drawings, and diagrams that distinguish this text from the Bents' book on the Cyclades.

The three travelers left England in January 1891 and returned a year later. The first chapter of *The Ruined Cities of Mashonaland*, "The Kalahari Desert Route," narrates the travels of the group – accompanied by two wagons, thirty-six oxen, and "heaps" of tinned provisions – through the British colony of Bechuanaland to the Mashonaland ruins. The route the group chose led through the capitals of the most important tribal chiefs, the majority of whom, Bent reports, had little identity left to them and received monthly pensions from the British government. Aside from recounting an occasional potential danger or personal discomfort faced by the group, the narrative of the Bents' explorations depicts a trip that went quite smoothly. Although they had to inoculate their oxen against a fatal lung sickness then raging through the country and some of their animals succumbed to other diseases, the group never seemed to be in danger from either illness or hostile inhabitants. In fact, the inhabitants of the country were either friendly toward, or exceptionally fearful of, the British explorers, and Bent was pleasantly surprised by the performance of the natives he employed as excavators and interpreters. He writes that he felt it necessary to say "a few words in their favour, as it has been customary to abuse them and set their capabilities down as nought."

Finding some of the native peoples to be quick learners and excellent excavators, Bent began to question some of his stereotypical racist assumptions about the native peoples. His assumptions were further challenged when he met a chief who had a wide reputation for integrity and enlightenment. Before meeting him, Bent had been skeptical about the reputation of this chief, for a "negro [was] the case in point," and Bent expected to find him to be "a rascal and a hypocrite." However, after a week's stay with the chief, Bent concluded that "he is a veritable father of his people, a curious and unaccountable outcrop of mental power and integrity among a degraded and powerless race." Bent added that the chief might be the "only living negro whose biography would repay the writing."

Although some of Bent's assumptions about black people were changing, such statements reveal that he saw this man as an exception to the norm. Throughout this text Bent alludes to the process of "white men" appropriating the territory of the native people as a normal, even natural process. Generally throughout his travel writing Bent never questions the morality of colonialism or his right to conduct his research, as long as he does so with his customary respect for the native peoples.

Bent's assumption that Europeans can and should control Africa comes across as clearly in his third travel book, *The Sacred City of the Ethiopians: Being a Record of Travel and Research in Abyssinia in 1893* (1893), as it does in *The Ruined Cities of Mashonaland*. During the Bents' travel in Ethiopia (or Abyssinia) the Italians, not the British, had established some degree of control in the country, and from them the Bents sought support and guidance. Bent commented that "the sooner some European nation undertakes the government of Abyssinia, the better it will be for the country." His major concern was that poor peasants could not improve their conditions when the country was constantly torn by wars between their leaders. Because of such hostilities, the Bents had to cut short their explorations in Ethiopia, where they sought primarily to study Aksum, the center of Christianity in that country and its ancient capital. They were forced to stay for awhile in Asmara, the base of operations for the Italian Red Sea colony, before it was deemed safe to proceed to Aksum.

During the Bents' stay in Asmara they explored the city and went on expeditions to nearby monasteries. The chapter on Asmara in *The Sacred City of the Ethiopians* reveals both the advantages and disadvantages of having a female as a member of the party. In order to describe the dress of a grand Abyssinian lady, Theodore turns to Mabel for assistance: "A very grand lady we visited was dressed in a most magnificent overcloak, for the description of which – it is too complicated for a male – my wife's Chronicle is responsible." Later, however, Mabel was banned from entering the monastic grounds because she was female; as a result, the party failed to benefit from what Theodore calls her "detective camera."

Frustrated by being forced to wait miles from a monastery where her husband was exploring, Mabel designed a plan. Theodore writes that on a visit to another monastery she "dressed in a fashion

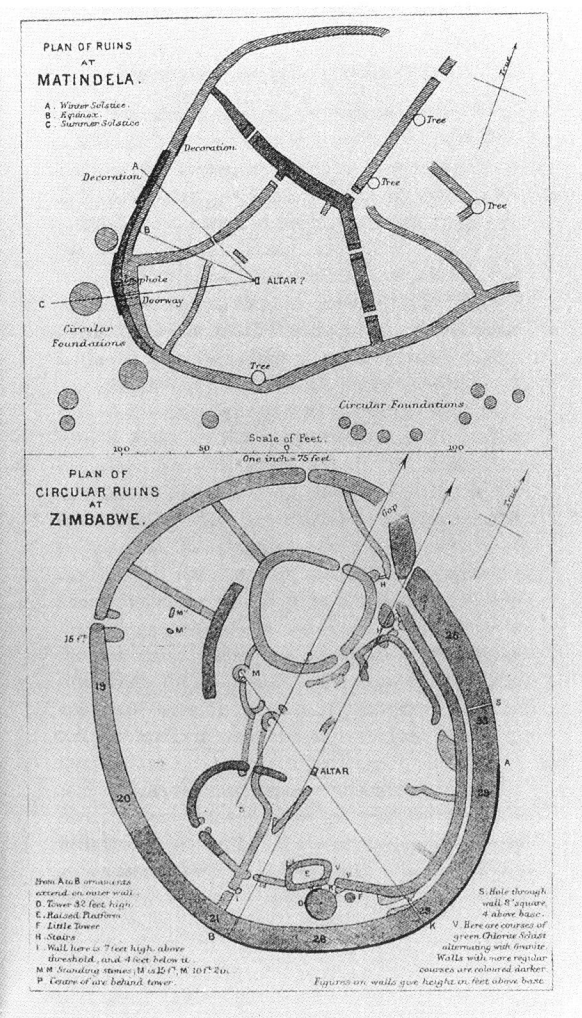

Diagrams by Robert McNair Wilson Swan for the Bents'
Ruined Cities of Mashonaland *(1892)*

that was calculated to deceive the most critical monk on the subject of her sex." Ironically, this disguise "was even more successful than she had hoped, for, owing to certain braids which adorned her jacket, the monks evidently mistook her for a general, and paid her the attention suited to her supposed rank." In the end "she was conducted into the church, shown the pictures and treasures, and treated in a manner that no female had ever before been treated in an Abyssinian monastery."

Finally being able to leave Asmara, the Bents traveled to Aksum. Two chapters of the book focus on this city: the first on Aksum as the Bents found it, the second on the antiquities of the city. Because of continued political unrest, the Bents unfortunately could stay only nine days, and they were prohibited from venturing more than ten miles outside the city. Once, when the Bents were "invited" to a military camp, Theodore used Mabel as an excuse for not reporting there: he wrote a letter to say he "regretted that the lady was too fatigued to undertake so arduous a journey." When they were finally forced to leave the city, political instability made it doubtful that they would safely reach Italian territory. Eventually they were met and, in effect, rescued by an Italian lieutenant with four hundred native troops. Tired from their various troubles and citing "excessive heat and unhealthiness of the climate" as reasons for not pursuing more-extensive archaeological work, they declared their journey terminated.

The Bents' final series of explorations occurred between 1889 and 1897, when they traveled to southern Arabia, the Sudan in Africa, and the island of Socotra. By the end of their journeys Mabel was ill with malaria and was carried aboard the homebound ship on a stretcher, and Theodore died of the same disease only four days after their return to England. As a result, using Theodore's notes, Mabel wrote their final book, *Southern Arabia* (1900). Compared to the previous works, this book is more dramatic, and its more suspenseful style may result from Mabel's authorship: she presents her feelings more frequently and vividly than her technically oriented husband had done in the preceding books. Even the parts focused on excavations are suspenseful, as Mabel describes their efforts to verify their hypothesis that the ruins are those of an ancient Phoenician empire. In any case, the style emphasizes how unusually filled with mishap were their travels in southern Arabia: Mabel describes the problems as having made the trip "quite the worst experience we had ever undergone in the course of any of our travels."

She lists four major sources for the book: several articles that her husband had published in *Nineteenth Century*, lectures that he had given to the Royal Geographical Society and the British Association, his notebook and her own "Chronicles," and various other books that she cites in the bibliography. Relying on these sources, Mabel wrote a work intended to help those "who, by following in our footsteps, will be able to get beyond them." The Bents traveled to seven distinct areas: the islands of Bahrain in the Persian Gulf (1889); the Masqat, south of Bahrain at the entrance to the Persian Gulf (1889 and 1895); the Hadhramaut, a broad valley running one hundred miles or so parallel to the coast of the Arabian Sea (1893-1894); the region across southern Arabia from Masqat to Aden (1894); the eastern Sudan in Africa (1895); the island of Socotra in the Indian Ocean (1897); and Yafei and Fadhli, countries of south Arabia (1897).

The most perilous of these journeys was the trip to the Hadhramaut, as "Harassed by Our Guides," Mabel's title for the chapter, indicates the great frustrations and anxieties that the Bents found while traveling there. This area had been previously explored by only one European because of the religious fanaticism of the inhabitants, yet the Bents were encouraged to travel there because a survey of it would be useful to the British government. In previously unsuccessful attempts to avoid problems explorers had attempted journeys to the Hadhramaut by disguising themselves as Arabs and traveling without trains of followers. The Bents, however, openly traveled and acknowledged the purposes for their travels. While such openness may have made their extensive journeys possible, it did not help them avoid seemingly endless problems – from dishonest employees to unfriendly Muslim tribes – that would have turned back most travelers.

The main sources of the Bents' frustrations were their guides, interpreters, servants, and excavators. At every stage of their journey and as often as several times a day, their guides would refuse to take them any farther unless the Bents immediately paid them money – presumably for unexpected food or transportation costs. The servants would become greatly alarmed if the Bents did not capitulate immediately to these attempts at extortion. Mabel describes such servant reactions in one instance: "The servants crowded round, scanning our faces, and in despair themselves, saying 'our lives are sacrificed,' and making great lamentations about their wives and families." To make matters worse, each time the Bents succumbed to the demands, they had to go through the lengthy process of unpacking their bags to get their money. Mabel writes, "plainly we were in their hands, and had to pay whatever Talib chose, as we might be hemmed in at any moment. We felt as if we were in a net." Unlike their earlier travel books, *Southern Arabia* is filled with phrases such as "abject distress" and "dismay reigned supreme among us" that describe the seriousness of their predicament.

However, when the Bents were not being plagued by their various problems, their lack of Arab disguises made it possible for them to help the native peoples by providing rudimentary medical care. Mabel's compassion for the people is evident, as she writes that "it grieved us sorely when poor souls came to us so hopeful and so confident of help, with a withered arm or an empty eye socket." The Bents also discovered new species of plant and animal life on their travels, and it pleased them to contribute to scientific knowledge.

Perhaps more notable than any record of the Bents' botanical or archaeological discoveries, however, is the extraordinary portrait the book provides of a nineteenth-century woman traveling in often unexplored, dangerous territories. Mabel repeatedly endured ridicule and condescension from native peoples who did not recognize her as a seasoned, capable explorer. She was actually one of Theodore Bent's "council of three" who were leaders of the exploration team and were responsible for making the party's major decisions. She also served, with her husband, as a supervisor of the men doing excavations.

Most often, the inhabitants of southern Arabia and the Sudan were simply amazed to see a woman with the exploration team, as they would crowd

James Theodore Bent with unidentified Arab visitors in Bahrain; illustration from the Bents' Southern Arabia
(1900)

around Mabel and shout that they wanted to see the "woman." When a group of women once asked to see her, Mabel sat in a chair as women surrounded her and implored her to remove her gloves, her hat, and her shoes. After doing all that they asked and even taking her hair down, her compliance ended when they wished to undress her completely.

In addition to describing herself as an object of such curiosity for the native peoples, Mabel also describes several incidents in which she served as a courageous and resourceful member of the exploration team. One incident occurred in the eastern Sudan, when the exploration party neared territory controlled by Muslim tribes who were enemies of the British and hostile to any Christian enterprise entering their secluded lands. Briefly separated from her husband, Mabel came face to face with a group of these enemies carrying spears and shields, some of whom were on camels, some on foot. Rather than panicking or running away, Mabel simply walked up to "the worst old man, grasped his hand, and wished him a happy day."

As the Muslim soldiers rushed toward her, she sat laughing at the man and said that she wanted to look at him. They stopped short and sat down nearby. When a member of the exploration team found her there, he assumed that she was the only survivor, until she asked how the British party could be so afraid of the soldiers when the soldiers were afraid of her. To shouts of "Peace, Peace," the "enemies" declared themselves her friends. In other, less dramatic situations Mabel's quick thinking allowed the group to take back to England valuable photographs that she had made and developed under trying conditions. One time, for instance, she devised a way to tie her camera to a rope and drop it into a downstairs room to which the Bents had been otherwise denied access.

Given her resourcefulness, courage, and remarkable stamina, it is not surprising that Mabel Bent survived, and even thrived upon, the many travels she undertook with her husband. Yet the sense of loss she expresses in the preface to *Southern Arabia* echoes in the last lines of the book, as she concludes in this simple, self-effacing way: "This is all I can write about this journey. It would have been better told, but that I only am left to tell it."

Although rarely acknowledged for her contributions to the Bents' travel writing, Mabel's publications after the death of her husband – *A Patience Pocket Book* (1904), *Anglo-Saxons from Palestine; or, The Imperial Mystery of the Lost Tribes* (1908), and the editing of a revised and enlarged publication of Arthur W. C. Boevey's *The Garden Tomb, Golgotha, and the Garden of the Resurrection* (1931) – demonstrate her ambition to write. Her continued publishing also indicates her importance to the travel writing published under her husband's name. Their collaboration benefited from their strengths – from Theodore's classical training and skill in archaeology and from Mabel's knowledge of photography and her persistence in keeping the detailed field notes that she called her chronicles.

Perhaps more than any other indicators, the governmental support that they received and the prompt publication of their works (usually within a year of their travels) attest to the value of their research and writing during their lifetimes. Publication of German scholar David H. Mueller's 1894 book on Theodore's impressions of Abyssinia suggests that the Bents' travels captured the interests of academics abroad. Mabel's significant role in the Bents' travel writing may gain for their work increased scholarly attention, as previously unrecognized contributions of women to travel literature continue to attract readers.

References:

David H. Mueller, *Epigraphische Denkmäler aus Abessinien nach Abklatschen von J. T. Bent* (Vienna: Kaiserliche Akademie der Wissenschaften, 1894);

Al N. Oikonomides, "Editor's Preface" and "Introduction to Cycladic Archaeology and Folklore," in Theodore Bent's *Aegean Islands; the Cyclades, or Life among the Insular Greeks* (Chicago: Argonaut, 1965), pp. v–ix, xvii–xxxvii;

John Pemble, *The Mediterranean Passion: Victorians and Edwardians in the South* (New York: Oxford University Press, 1988);

J. C. Rubright, "James Theodore Bent," in Theodore Bent's *The Cyclades, or Life among the Insular Greeks*, pp. xiii–xv.

Wilfrid Scawen Blunt
(17 August 1840 - 10 September 1922)

and

Lady Anne Isabella Noel Blunt
(22 September 1837 - 15 December 1917)

Syrine C. Hout
University of Maine at Fort Kent

See also the Wilfrid Scawen Blunt entry in *DLB 19: British Poets, 1880-1914.*

BOOKS: *Sonnets and Songs,* by Wilfrid Scawen Blunt, as Proteus (London: John Murray, 1875);

Proteus and Amadeus: A Correspondence, by Wilfrid Scawen Blunt (Proteus) and Charles Meynell (Amadeus), edited by Aubrey De Vere (London: Kegan Paul, 1878);

Bedouin Tribes of the Euphrates, by Lady Anne Blunt, edited, with a preface, by Wilfrid Scawen Blunt (2 volumes, London: John Murray, 1879; 1 volume, New York: Harper, 1879);

The Love Sonnets of Proteus, by Wilfrid Scawen Blunt, as Proteus (London: Kegan Paul, 1881; New York: Doxey, 1901);

A Pilgrimage to Nejd, the Cradle of the Arab Race: A Visit to the Court of the Arab Emir, and "Our Persian Campaign," by Lady Anne Blunt, edited by Wilfrid Scawen Blunt, 2 volumes (London: John Murray, 1881);

The Future of Islam, by Wilfrid Scawen Blunt (London: Kegan Paul, Trench, 1882);

The Wind and the Whirlwind, by Wilfrid Scawen Blunt (London: Kegan Paul, Trench, 1883; Boston: Tucker, 1884);

Ideas about India, by Wilfrid Scawen Blunt (London: Kegan Paul, Trench, 1885);

Justice and Liberty for Ireland: Extracts from the Speeches of W. S. Blunt to the Electors of Kidderminster, July 1886 (London: British Home Rule Association, 1886);

Mr. Wilfrid Blunt, Anti-Coercionist Candidate for Deptford: Who Is He? What Has He Done? Why Is He in Prison?, by Wilfrid Scawen Blunt (London: National Press Agency, 1888);

Wilfrid Scawen Blunt, circa 1860

In Vinculis, by Wilfrid Scawen Blunt (London: Kegan Paul, Trench, 1889);

A New Pilgrimage, and Other Poems, by Wilfrid Scawen Blunt (London: Kegan Paul, Trench, 1889);

The Love Lyrics & Songs of Proteus, by Wilfrid Scawen Blunt (London: Kelmscott, 1892);

Esther, Love Lyrics, and Natalia's Resurrection, by Wilfrid Scawen Blunt (London: Kegan Paul, Trench, Trübner, 1892); republished as *Esther: A Young Man's Tragedy, Together with the Love Sonnets of Proteus* (Boston: Copeland &

Day, 1895); republished as *Esther: A Young Man's Tragedy* (Portland, Maine: Bibelot, 1905);

Griselda: A Society Novel in Rhymed Verse, by Wilfrid Scawen Blunt (London: Kegan Paul, Trench, Trübner, 1893);

The Poetry of Wilfrid Blunt, edited by William Ernest Henley and George Wyndham (London: Heinemann, 1898);

Satan Absolved: A Victorian Mystery, by Wilfrid Scawen Blunt (London & New York: John Lane, 1899);

The Shame of the Nineteenth Century, by Wilfrid Scawen Blunt (London, 1900);

Love Poems of W. S. Blunt (London & New York: John Lane, 1902);

The Military Fox-Hunting Case at Cairo: Mr. Wilfrid Scawen Blunt to the Marquess of Lansdowne... A Supplement to the Blue Book Egypt 3, 1901 (London, 1902);

Fand of the Fair Cheek: A Three-Act Tragedy in Rhymed Verse, by Wilfrid Scawen Blunt (N.p.: Privately printed, 1904);

Atrocities of Justice under British Rule in Egypt, by Wilfrid Scawen Blunt (London: Unwin, 1906);

To the Rt. Honourable Sir Edward Grey, Bart., M.P., by Wilfrid Scawen Blunt (London: Chiswick, 1906);

Francis Thompson, by Wilfrid Scawen Blunt (London: Burns & Oates, 1907);

Secret History of the English Occupation of Egypt: Being a Personal Narrative of Events, by Wilfrid Scawen Blunt (London: Unwin, 1907; New York: Knopf, 1922);

Mr. Blunt and the "Times": A Memorandum as to the Attitude of the "Times" Newspaper in Egyptian Affairs, by Wilfrid Scawen Blunt (London: Published by the author, 1907);

The Bride of the Nile: A Political Extravaganza in Three Acts of Rhymed Verse, by Wilfrid Scawen Blunt (N.p.: Privately printed, 1907);

The New Situation in Egypt, by Wilfrid Scawen Blunt (London: Burns & Oates, 1908);

Denshawai Memorial School, by Wilfrid Scawen Blunt (London, 1908?);

India under Ripon: A Private Diary, by Wilfrid Scawen Blunt (London & Leipzig: Unwin, 1909);

Gordon at Khartoum: Being a Personal Narrative of Events, by Wilfrid Scawen Blunt (London: Swift, 1911; New York: Knopf, 1923);

The Italian Horror and How to End It, by Wilfrid Scawen Blunt (London: Bonner, 1911);

Lady Anne Blunt in Dublin, 1888 (Fitzwilliam Museum)

The Land War in Ireland: Being a Personal Narrative of Events, by Wilfrid Scawen Blunt (London: Swift, 1912);

The Poetical Works of Wilfrid Scawen Blunt: A Complete Edition, 2 volumes (London: Macmillan, 1914; Grosse Pointe, Mich.: Scholarly Press, 1968);

The Crabbet Arabian Stud, anonymous, by Wilfrid Scawen Blunt (London: Whittingham, 1915?);

History of the Crabbet Estate in Sussex, by Wilfrid Scawen Blunt (London: Chiswick, 1917);

My Diaries: Being a Personal Narrative of Events, 1888–1914, by Wilfrid Scawen Blunt (2 volumes, London: Secker, 1919–1920; New York: Knopf, 1921);

Poems, by Wilfrid Scawen Blunt, edited by Floyd Dell (New York: Knopf, 1923);

Desert Hawk: Abd el Kader and the French Conquest of Algeria, by Wilfrid Scawen Blunt (London: Methuen, 1947).

OTHER: Abu Zaid, *The Celebrated Romance of the Stealing of the Mare: An Arabic Epic of the Tenth*

Century, translated by Lady Anne Blunt, versified by Wilfrid Scawen Blunt (London: Reeves & Turner, 1892);

Wilfrid Scawen Blunt, "The Release of Arabi," *Nineteenth Century*, 32 (1892): 370;

Mu'Allakat, *The Seven Golden Odes of Pagan Arabia, Known Also as the Moallakat,* translated by Lady Anne Blunt, versified by Wilfrid Scawen Blunt (London: Chiswick, 1903);

Theodore Rothstein, *Egypt's Ruin: A Financial and Administrative Record,* introduction by Wilfrid Scawen Blunt (London: Fifield, 1910).

Wilfrid Scawen Blunt and Lady Anne Blunt, née Annabella Isabella Noel, later fifteenth Baroness Wentworth, were each descended from one of the families who had followed William the Conqueror from Normandy to England in 1066. Their marriage on 8 June 1869 at Saint George's Church in Hanover Square in London, after two months of engagement, placed Blunt, as he stated, "almost in the rank of the world's sublimities." As the granddaughter of George Gordon, Lord Byron, and Anna Milbanke, as well as the daughter of William King, Lord Lovelace, and Augusta Ada Byron, Lady Anne Blunt added £3,000 annually to Wilfrid Blunt's yearly income of £700 and also made it possible for him to share in literary fame.

Wilfrid Blunt had since his teenage years identified with Lord Byron. In looking back on his own fervent sexual and spiritual relationships with women before and after his marriage, Blunt was to describe his marital tie as a "practical romance." Indeed, the couple enjoyed adventurous lives, as they traveled more than two thousand miles from the Mediterranean Sea through the Arabian An Nafūd Desert to the Persian Gulf. On such travels they rode on Arabian stallions, which they preferred to camels, and their affinity for such horses led them to return to England with six mares – which were to be the bases for their Crabbet Arabian breed. Wilfrid Blunt's political activism and apprehensions about British imperialism merged with his interest in what these horses represented to him, as on 26 June 1880 he wrote in his diary: "If I can introduce a pure Arabian breed of horses into England and help to see Arabia free of the Turks, I shall not have quite lived in vain."

The subject of many critical studies and two biographies – Edith Finch's *Wilfrid Scawen Blunt, 1840–1922* (1938) and Elizabeth Longford's *A Pilgrimage of Passion* (1979) – Wilfrid Blunt's life has received more attention than that of Lady Anne, whose life has so far been written about only in connection with his. Not much is known about her before her marriage at the age of thirty-one, when her name was changed from Annabella to Anne. After her mother's death and when Annabella was fifteen, her father took her on holiday trips to the Continent, where she learned to speak French, German, Spanish, Italian, and Swiss patois – as well as Arabic, in which she later became fluent. She took lessons in violin from Joseph Joachim and in drawing from John Ruskin. She was humble, good-hearted, and well-bred, and Longford reports that her husband wrote how "she had never committed any act less than entirely conscientious and entirely honourable."

Blunt's biography, complete in its details of his private affairs, became possible only after the end of the fifty-year ban that he had put on the opening of his secret memoirs. Finally available to scholars in 1972 at the Fitzwilliam Museum in Cambridge, these diaries were essential for Longford's biography, in which she writes that these memoirs – covering the second half of Queen Victoria's reign, the entire Edwardian period, and part of George V's reign – reveal "the truth about eighty years of social history, seen through the eyes of a brilliant eccentric."

Wilfrid Blunt was born on 17 August 1840 at Petworth House, Sussex, and lost his father, Francis Scawen Blunt, owner of the two-thousand-acre Crabbet Park, when the boy was two years and four months old. His mother, Mary Chandler, converted to Catholicism in April 1851; she sent Wilfrid and his older brother, Francis, to Stonyhurst College, a Jesuit school in Lancashire, in 1852, and to St. Mary's College, Oscott, in 1855, the same year of her death. The three family members closest to Wilfrid – his mother, brother, and younger sister, Alice Mary – died from tuberculosis, and Blunt also contracted the disease during childhood and at age twenty-six lost one lung because of it. Following the deaths of his brother and sister Blunt inherited Crabbet Park along with the two other family estates, Newbuildings and Worth Forest, in 1872.

On 31 December 1858 the Foreign Office appointed young Blunt as an attaché to Athens, and during the ensuing years he was to serve in Frankfurt, Madrid, Paris, Lisbon, and Bern before retiring on 31 December 1869 to devote his time to writing. At age twenty-one he lost his faith in the Bible and started a lifelong search for what Longford calls a "Religion of Happiness" instead. Before embarking on his Middle Eastern journeys Wilfrid met two famous travelers: the forty-eight-year-old Sir Richard Burton in Buenos Aires in 1868, and Lady Anne's cousin, Robert Curzon, in West Sussex in 1870. Although Wilfrid was initially fascinated with Burton's personality, he denounced Burton's imperialistic attitude in 1906, for Burton, he felt, "showed little true sympathy with the Arabs he had come to know so well [and] would . . . have willingly betrayed them to further English, or his own professional interests."

Lady Blunt suffered miscarriages in 1870 and 1872 before giving birth to Judith Blunt in 1873. After the first of these mishaps Wilfrid took Anne for a Spanish tour that Longford describes as "perhaps the happiest" of many other journeys they undertook together. On 10 April 1873 they left for Turkey, and Longford writes that for Wilfrid "these times were [their] true times of marriage, more than in Europe, and they were happy times." To Anne the journey was "a new honeymoon," and during this time they bought their first eastern horse, Turkeycock. In the following year they traveled to Algeria, where Lady Blunt suffered yet another miscarriage. On 23 December 1875 they boarded a steamship to Cairo via the Suez Canal and arrived on 14 January 1876. During their three-and-a-half-month stay in that region the Blunts started to learn Arabic. Lady Anne's facility in learning Arabic, her passion for riding horses and camels, her sketching ability, and her habit of keeping a neat and detailed journal made her an ideal traveling companion. Leaving his wife to rest in Jaffa, Wilfrid went on to visit Jerusalem.

After discovering that the Royal Geographical Society's latest maps of central Arabia were based on Col. Francis Rawdon Chesney's surveys of 1837 and were therefore out of date, the Blunts were moved to explore that region and headed for Syria, then under Turkish rule, on 20 November 1877. Almost all the information about their two major trips to the Near East and the Arabian peninsula is reported in Lady Anne's travel narratives, *Bedouin Tribes of the Euphrates* (1879) and *A Pilgrimage to Nejd, the Cradle of the Arab Race: A Visit to the Court of the Arab Emir, and "Our Persian Campaign"* (1881), which are accompanied by her own maps, portraits, illustrations, and drawings. As editor of both works, Wilfrid wrote a preface and added a postscript and six chapters on various aspects of Bedouin life to the former; to the latter he added one chapter, appendixes in the form of notes on the history and geography of Arabia, and a memorandum on the Euphrates Valley Railway. His own accounts of these travels appeared later in his two-volume edition of *My Diaries: Being a Personal Narrative of Events, 1888–1914* (1919–1920).

In these four travel volumes – which Kathryn Tidrick calls "pedestrian" – Lady Anne distinguishes between notes taken on the spot, in form of dated entries, and rewritten material initially based on impressions and dialogues. She writes that notes "are our first impressions, and I write them down while they are fresh." She frames her narration throughout with locutions such as "now I return to my journal"; "now I am too busy to write more"; and "we have got a candle on the terrace, so that I can write." Similar references to writing – which she often did secretly, in addition to reproducing dialogue and stories that local people told her – are meant to authenticate her narration. She also refers to having to "condense" what she has written to keep her journal from becoming "a mere mass of notes, for the most part taken from conversations . . . with various interesting people here, and requir[ing] rewriting."

From Rhodes the Blunts sailed to Alexandretta and then traveled on mules for four days to Aleppo, where they met James Henry Skene, the British consul whom Longford identifies as a "fourteenth cousin of Anne." He provided them with an introduction letter that they could give to one of the four tribes of the Anazeh (the Fedaan, the Gomussa, the Roala, and the Sebaa) and that would enable them to join the tribe in riding southward. Skene proposed to travel with them as far as Deyr, a town west of the Euphrates River. In addition to their two riding mares, Hagar and Tamarisk, they bought Kars, a purebred Arabian stallion, for sixty-nine pounds.

Accompanied by a few Turkish officers and Skene, they left for Deyr on 9 January 1878. Skene, promising to meet them again in Deyr in March, left their company on 27 January. After crossing Mesopotamia and arriving in Baghdad, they were restless until they met an Indian exile, the previous king of Oude, who provided them with introduction letters to the sheikhs of Shammar, the enemies of the Anazeh. Excited by the prospect of a desert crossing, they purchased camels and were joined by their loyal servants, Hajji Mahmud from Alexandretta and Hanna from Aleppo. Their desert adventure peaked when they met and encamped with the handsome

Blunt on his Arabian stallion, Pharaoh; portrait by Lady Anne Blunt (Collection of Noel Anthony Scawen Lytton, fourth Earl of Lytton)

twenty-seven-year-old Faris, one of two rival leaders of the Shammars. Wilfrid swore blood brotherhood with Faris, whom Anne Blunt describes as a "*gentleman* of the desert" – someone whom Wilfrid perceived to be his Arabian counterpart.

Wilfrid Blunt admired the Bedouins for their primitive aristocracy, irreducible sense of independence, and simple belief in God. He also appreciated their hospitality, courage, protection of the weak, and pursuit of honor and pleasure. Anne seems to have had mixed feelings about them. On one hand, she makes sweeping generalizations about them – as in her statements, for example, that "truth is the exception, not the rule, among the Arabs" and that they were talkative, ungrateful, and meandering. However, she also extols certain features of Bedouin life that are not found in Europe, features such as their freedom of movement and ownership and their natural understanding of sketches and maps. She also depicts some Bedouins as paragons of beauty, talent, and strength such as one "could possibly find in any part of the world."

The Blunts' attitudes toward Bedouin women and the harem are also ambiguous. Lady Anne writes that, unlike the men, the women in Nineveh were "without ideas, good-natured, but quite uninteresting." Bedouin women in Hail were "satisfied to be bored" because of "the tyranny of fashion." Wilfrid gives some women more credit, as he notices that "in more than one Sheykh's tent, it is in the women's half of it that the politics of the tribe are settled." Anne writes that "Life in an Arab camp is terribly irksome," and impertinent women and children harass her so badly that she feels claustrophobic: in Palmyra, she writes, a "whole party of . . . tormentors . . . swarm like bees into the tent."

The son of the sheikh of Palmyra, Mohammed-Ibn-Aruk, was another Bedouin with whom Wilfrid swore an oath of brotherhood. Mohammed-Ibn-Aruk was convinced by a bribe and an appeal to Arab honor to take them to Jedaan, the Bedouin sheikh who unfortunately did not fulfill their romanticized image of him as an Anazeh hero. After meeting Medjuel, the sheikh of the Mesrab tribe, and Jane Digby, his English wife, in

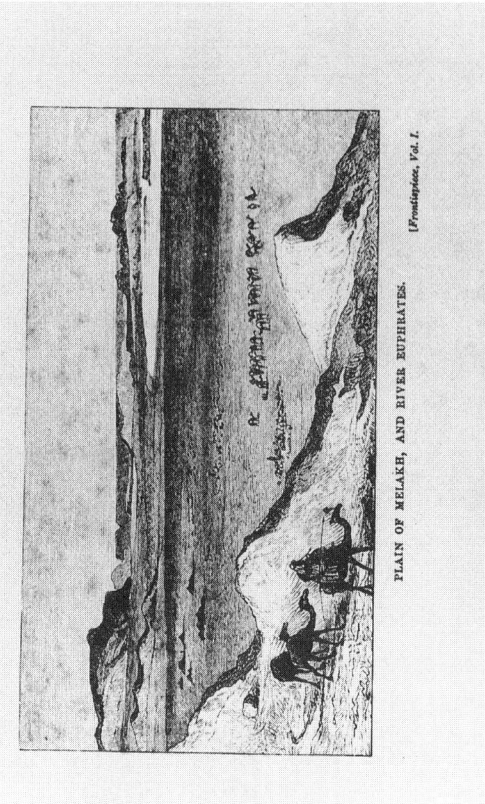

Frontispiece and title page for the first travel narrative by Lady Blunt and her husband

Damascus, the Blunts headed for Beirut and were back in civilization on 26 April.

Less than eight months later they were back in Syria on 6 December 1878 to meet their old friends and Abd el-Kader, the hero of the French war in Algiers. In the first lines of *A Pilgrimage to Nejd* Lady Anne writes that

> only yesterday [the Blunts] were still tossing on the sea of European thought, with its political anxieties, its social miseries and its restless aspirations.... The charm of the East is the absence of intellectual life.... Nobody here thinks of the past or the future, only of the present.

This journey to Nejd by the Blunts, who had promised Wilfrid's spiritual brother, Mohammed, to help him find a wife among his relatives there, was to be a pilgrimage – one in which their religion was romance.

Their caravan left Damascus for Jof, four hundred miles away, on 13 December. On 3 January 1879, after an arduous ride that had left Lady Anne with a sprained knee, they were attacked by twelve Roala horsemen, whose hostilities ceased when Anne surrendered in Arabic. To complete arrangements for Mohammed's marriage the following year, they paid fifty pounds for a girl named Muttra and then set out for the Nefud on 12 January. On the long trip to Jobba and Hail, Mohammed entertained them with thirst stories and other desert tales of horror. On 23 January they were welcomed in Hail by the emir, Mohammed Ibn Rashid, who showed them the latest toy he had acquired – a telephone. When Mohammed-Ibn-Aruk insisted on being recognized as the preeminent member of the traveling party, the emir became less hospitable to the Blunts, and they left Nejd.

As Longford remarks, the Blunts' claim that "no European nor Christian of any sort had penetrated *as such* before [them] to Jebel Shammar" is false, because Charles M. Doughty, author of *Travels in Arabia Deserta* (1888), had arrived in Hail just before them. After leaving Hail they met Ali Koli Khan, son of the khan of the Bactiari tribe of the northeastern Persian Gulf and the leader of the Persian pilgrimage. Accepting his invitation to visit him in Persia, the Blunts left with the pilgrims on 1 February, with Anne feeling uncomfortable among three thousand Muslim devotees and their five thousand camels. They arrived in Baghdad on 6 March, only to leave two weeks later for Bushire on the Gulf, in an attempt to help Capt. Verney Lovett Cameron, an African explorer, design plans for a railway between the Lebanese city of Tripoli and Karachi, Pakistan.

Much went wrong for the Blunts on this journey. Husseyn Koli Khan, whom they met on their way to Shustar, was traveling to Teheran and could not, therefore, invite them to his residence or give them mares. Things got worse when Wilfrid came down with dysentery on 8 April. Two events made Anne become a devout Catholic years later: Wilfrid's miraculous recovery and her vision of her three dead children on 12 April while the Blunts were on their way to Bushire, which they at last reached on 28 April.

Wilfrid Blunt pays tribute to the work of previous explorers of Arabia – William Gifford Palgrave, Francis Rawdon Chesney, Carsten Niebuhr, John Lewis Burckhardt, Heinrich Kiepert, Sir Austen Henry Layard, William Francis Ainsworth, Henry Baker Tristram, R. B. Cunninghame Graham, Georg August Wallin, and Carlo Claudio Camillo Guarmani – and he emphasizes what he and Lady Anne are contributing through correcting erroneous notions and providing new information about the region. For the populations of some towns, for example, Blunt presents more accurate figures than those given by Palgrave. He blames Palgrave's "contempt of all things Bedouin" for having misinformed readers about Arabian horses. Blunt updates, corrects, and redraws Chesney's maps of the desert and of the Euphrates and Tigris valleys, and he criticizes Niebuhr and Burckhardt for failing to discuss the topic of horses. The Blunts' major contributions as undisguised travelers include their ethnographic and geographic exploration, "with compass and barometer," of all of northern Arabia, the Nefud Desert, and Jebel Sham; their discoveries of inscriptions and of new flora and fauna; and their descriptions of the Bedouin political system.

Wilfrid Blunt also vehemently opposed the construction of a railway between Alexandretta, on the Mediterranean Sea, and Bushire, on the Persian Gulf: "If the recital of our passage through the uninhabited tracts . . . shall deter my countrymen from embarking their capital in an enterprise financially absurd," he wrote, "I feel that its publication will not have been in vain." The reasons for his opposition to such a railway were geographical, economic, and political – the hot weather of the Persian Gulf, the likelihood of contracting fever at the Syrian ports, the additional shipping cost and risk, and the dependence of travelers and shippers on the permission of whatever political power might prevail in Asia Minor. To corroborate his reasoning Blunt reiterated the opinions of English officers abroad and provided tables and figures. Instead of constructing such a railway, he proposed establishing a line of steamers on the Euphrates River. "My sympathy is with [the Bedouins]," he concluded, "and not with progress."

After Bushire they traveled by train to visit Edward Robert Bulwer, viceroy of India, and his wife in Simla. Blunt, not the staunch antiimperialist that he would later become, supported the second Afghan War that had begun in 1878, and he agreed to return to Hail on a secret imperial mission to extend British influence in central Arabia, but the plan never materialized. The Blunts' trip to India became the basis for his book *Ideas about India* (1885), which expresses the growing reservations he felt about the morality of British imperialism. In fact, Longford writes, by 3 July 1879 Wilfrid "left Simla with [his] faith in the British Empire and its ways in the East shaken to its foundations."

After some time in England the Blunts returned to the Middle East late in 1880, this time to Egypt. In the fall of 1881 Wilfrid met Ahmad Arabi, the Egyptian nationalist leader. At first he tried to mediate between the European interests of England and France, on one hand, and those of Arabi, on the other. After England and France issued their "Joint Note" backing the khedive Tewfik Pasha on 6 January 1882, Blunt fully supported the opposition of militant nationalist leader Arabi. He discussed with British prime minister William Gladstone on 22 March, and later with Edward, Prince of Wales, the prospect of establishing an Egyptian lobby in the English government.

Despite growing Egyptian opposition to British influence, Blunt purchased Sheykh Obeyd, near Heliopolis, as his new residence abroad, and two events made his support of oppressed nations a lifelong activity henceforward: the British bombardment of Alexandria on 11 July and the subsequent defeat of Arabi's forces on 13 September in Tel al-Kabir, where ten thousand casualties resulted. With Arabi captured and Cairo occupied by British troops, Wilfrid established an Arabi Defence Fund that accumulated £3,300 – to which Blunt had contributed £3,000 – and with

Lady Blunt with her horse, Kassida (Fitzwilliam Museum)

this Blunt succeeded in hiring a lawyer who managed to have Arabi banished to Ceylon instead of being hanged. Blunt's interests and experiences in the Egyptian struggle for independence inspired his publication of the political poem *The Wind and the Whirlwind* (1883).

Blunt's fascination with Islam, a result of his travels in Arab countries and conversations with Muslims, is evident in *The Future of Islam* (1882), a collection of essays that had been published in the *Fortnightly Review* during summer and autumn 1881. These were written for readers whom Blunt called "practical Englishmen," an audience that included politicians who might repair some of the damage that Europeans had generally inflicted on Islamic civilization. He explains Islamic doctrine and history by drawing analogies to Christianity, and he contends that Islamic scholarship became stagnant in the seventeenth and eighteenth centuries because of the unorthodoxy of Ottoman views. Blunt feels that two events can reconcile Islam with modernity and Christianity: not viewing the sheriat (the written code) and the Koran as unimpeachable authorities, and transferring the seat of political and intellectual power from Constantinople to Cairo or Mecca. Blunt often wore Bedouin dress in Sussex to express his Arabophilia.

In 1883 the marriage between Wilfrid and Lady Anne began to deteriorate, mainly because of his infidelity. Nevertheless, in autumn 1883 the Blunts went to Ceylon, where he met his exiled friend Arabi, and then the Blunts visited Madras, Hyderabad, and Calcutta, where they met the new viceroy of India, George Robinson, Marquess of Ripon, and attended the Indian national conference. Blunt recorded the details of this trip and expressed his full support of Indian self-determination in *Ideas about India* (1885) and *India under Ripon: A Private Diary* (1909).

Having returned to England in 1884, Blunt tried to enter politics officially by running as a candidate for Camberwell North in summer and fall 1885, after the Tories and Robert Gascoyne-Cecil, Marquess of Salisbury, had come to power when Gladstone's government had fallen in June. Although Charles Parnell and his Irish voters had promised to support Blunt, who advocated home rule for Ireland, Blunt was narrowly defeated. Between 23 March and 21 April 1886, when the "land war" was raging between British landlords and Irish tenants, Blunt made a trip to Ireland. After a second short trip there, between 15 and 31 May, Blunt again ran for Parliament, this time as a liberal Home Rule

Wilfrid Blunt's drawing of a dust storm in the desert (Collection of the Earl and Countess of Lytton, Lady Anne Lytton, Lady Winifrid Tryon, and Pearl Wheatley)

candidate at Kidderminster, and again he was defeated in a close vote.

Embittered by his failure to effect any political change, Blunt resolved to give up public service for contemplative life as a Roman Catholic intellectual. Attempting to achieve this metamorphosis, he and Lady Anne traveled in October to Rome, where he had an interview with Pope Leo XIII on 11 November 1886. That pursuit ended in January 1887, when Prime Minister Salisbury allowed the Blunts to return to their Egyptian estate, Sheykh Obeyd, from which they had been banished for nearly four years. Blunt was convinced that the British government had allowed them to do so in order to remove him from the continuing Irish political disturbances, and indeed, after five months in Cairo, Blunt returned to Ireland, where he attended antieviction meetings.

When he tried to give a speech on 23 October in Woodford, Galway County, in violation of the ban against antieviction gatherings, he was arrested, convicted, and sentenced to two months' imprisonment at hard labor in Galway and Kilmainhaim. In *The Land War in Ireland: Being a Personal Narrative of Events* (1912) Blunt wrote that this whole experience "deserves to be remembered in Irish history as being the first recorded instance, in all the four hundred years of English oppression, of an Englishman having taken the Celtic Irish side in any conflict, or suffered even the shortest imprisonment for Ireland's sake."

Yet he persisted: even before going to jail Blunt tried for the last time to run for Parliament in the by-election at Deptford, and the poll was declared on 29 February while he was already serving his sentence. His character having been attacked by a conservative opponent who had read at a meeting what he thought was one of Blunt's immoral poems from *The Love Sonnets of Proteus* (1881), Blunt lost again by a close vote. Released from prison on 6 March 1888, he realized that he had lost on both political and spiritual fronts: both Tories and Liberals had lost faith in him, and he had lost his faith in Roman Catholicism, as his poem *In Vinculis* (1889) was to reveal.

In autumn 1888 Blunt took Lady Anne and Judith to Greece to show them the country that had fascinated their ancestor, George Gordon, Lord Byron. The three spent the winter at Sheykh Obeyd, and he returned before they did to England in April 1889. Although he had been obliged to withdraw from political activity following his return to Egypt in 1887, he found indirect ways to continue to support Egyptian nationalism. He also found a way to combine travel with romance: he invited Mary Wyndham, Lady Elcho, a previous acquaintance who had become the wife of Hugo Elcho, to accompany him on a "desert honeymoon" that began in mid January 1895 at the Pink House at Sheykh Obeyd and lasted until 26 March.

Blunt's relationship with Lady Elcho resulted in a baby daughter, Mary, and officially ended in the spring of 1895. After visiting his cousin, Sir Philip Currie, Blunt returned to Egypt in October of that year. On 5 February 1897 Wilfrid, disguised in Arab dress and accompanied by two incompetent guides, left Cairo during the Ramadan holiday period for the Siwa oasis in the Western Desert. Weakened by hunger and thirst, they were attacked by two hundred armed men who battered and robbed Wilfrid. What saved him was a piece of advice that he had acquired from Doughty's narratives of his Arabian travels — that it is wise to remain passive and ask for the mercy of one's attackers. This forty-day episode in the wilderness altered his romantic notions about the Muslim faith and spirituality. On 24 March, one week after returning to Cairo, Blunt wrote in his diary, "The less religion in the world perhaps, after all, the better."

Two political events at the end of the nineteenth century made Blunt look back sadly on the period he preferred to call the "Eighteen Hundreds." The first was the scandal of the final battle with the Mahdists at Omdurman in 1898, when British commander Horatio Kitchener and Maj. William Gordon had massacred dervishes and desecrated the body of the Mahdi. Second, his concern for the fate of African blacks made him side with opponents of the Boer War. Blunt had responded to the first outrage by attacking Kitchener in a letter to the London *Daily News,* and continuing incidents in the second case prompted him to write the verse satire *Satan Absolved: A Victorian Mystery* (1899), assailing British pretensions in adopting what Rudyard Kipling called "the white man's burden" in Africa. Blunt summarized his great disappointment in his nation and countrymen in one sentence: "An abominable world it is, an abominable century, an abominable race."

Despite his low expectations for the new era, on 8 March Wilfrid and a party of Muslim pilgrims set out to visit Mount Sinai by boarding a steamship at Suez for Tor. News of the Boxer Rebellion in China was causing fear of the "Yellow Peril" in England, and Blunt, of course, sided with the oppressed Chinese "knocking the foreign vermin on the head," as he wrote in his diary. Publication of *The Shame of the Nineteenth Century* (1900), his article expressing his virulent responses to such events, was rejected by the London *Times* on Christmas Eve. On New Year's Eve he vented his frustration in another personal entry in his diary: "I bid goodbye to the old century, may it rest in peace as it has lived in war. Of the new century I prophesy nothing except that it will see the decline of the British Empire. Other worse Empires will rise perhaps in its place." The only happy event that year was the birth of the Blunts' first grandson, Anthony Scawen, on 7 April.

Wilfrid had reason to rejoice in 1902 when England and the Boer republics signed a peace agreement in 1902 and King Edward VII permitted the exiled Arabi to return to Egypt. Anne also felt happy and proud when she completed her only poem, an eight-line piece which had been occasioned by a foxhunting incident at Sheykh Obeyd and which she presented to Wilfrid on his sixty-first birthday. He congratulated her on having finally used her talent as Byron's granddaughter, and in her diary she responded, "Thirty-two years have I been waiting — I may say — for a compliment." The two had collaborated on three works — Lady Anne's two travel narratives of 1879 and 1881 as well as her translation of Abu Zaid's *The Celebrated Romance of the Stealing of the Mare: An Arabic Epic of the Tenth Century* (1892), which Wilfrid had set in verse. A final collaborative effort was to appear: *The Seven Golden Odes of Pagan Arabia, Known Also as the Moallakat* (1903) was also translated by Lady Anne and versified by her husband.

Wilfrid made his last desert journey, this time to Damascus, in March 1904. On 10 December of that year at Sheykh Obeyd he contracted malarial fever, which affected his spine for the next two years and may have occasioned his separation from Lady Anne — although he preferred the company of his nurse, Miss Elizabeth Lawrence, to that of his wife. On 28 November 1905 Lady Anne set off for Cairo to sell the outlying lands of Sheykh Obeyd, and she proved to be a shrewd businesswoman: she returned in mid June the following year with £35,000, which amounted to a profit of 4,000 percent.

Wilfrid never returned to his second home in Cairo, but from England he continued to scrutinize Egyptian politics. The main event was in May 1906 — the Denshawai Affair, when a riot had ensued over the hunting rights of British officers. One officer had died, and following his death many peasants were severely punished by public hanging, life imprisonment, or physical torture. Moved by this

Sheykh Obeyd, the Blunts' Egyptian estate; watercolor by Judith Blunt (Collection of the Earl and Countess of Lytton, Lady Anne Lytton, Lady Winifrid Tryon, and Pearl Wheatley)

tragedy, Wilfrid published *Atrocities of Justice under British Rule in Egypt* (1906), in which he listed all the injustices that British officials, especially Lord Evelyn Baring Cromer, had committed against the Egyptians since 1882. Lady Anne was particularly helpful in translating the Denshawai documents from Arabic into English for Wilfrid's book. Then in 1907 he published *Secret History of the English Occupation of Egypt: Being a Personal Narrative of Events* in addition to a second edition of *Atrocities of Justice*. Lady Anne also helped Wilfrid when he was finishing his two diary-histories, *Gordon at Khartoum: Being a Personal Narrative of Events* (1911) and *The Land War in Ireland: Being a Personal Narrative of Events*.

Wilfrid's relationships with both his wife and his daughter were in bad shape because of his entanglement with a young and devoted Dorothy Carleton, his last lover. He had met her during the summer of 1900 while he was visiting his relatives, Percy and Madeline Wyndham, in Wiltshire. Wilfrid vowed never to set foot again in Crabbet Park, which he had bequeathed to Judith. To justify Carleton's continued presence in his household, he adopted her as his niece in 1906, despite the consternation of his family.

Even in his old age Blunt was never abandoned by his many female friends. At the suggestion of Ezra Pound, who was at that time acting as secretary to William Butler Yeats, Lady Isabella Augusta Persse Gregory invited Blunt to a dinner where he was to be honored by a group of young poets. Blunt accepted the invitation, but at his wish Lady Gregory changed the location of the proposed affair from a London dinner to a luncheon on 18 January 1914 at his Newbuildings residence. Six young poets – Pound, Yeats, T. Sturge Moore, F. S. Flint, Richard Aldington, and Victor Plarr – read verse tributes to him and offered him a bas-relief by French sculptor Henri Gaudier-Brzeska. When Blunt was later to ask Lady Gregory, who had been a lifelong friend, to write a foreword to his two-volume memoirs, *My Diaries,* she immediately accepted.

At age seventy-four Blunt also received other distinguished guests – Winston Churchill, Lord Alfred Douglas, Francis Thompson, Hilaire Belloc, and Wilfrid and Alice Meynell – at Newbuildings, which remained his official residence until his death. Roger Casement, the Irish revolutionary, visited him on 13 May 1914, before British courts were to sentence Casement to be hanged for having participated in the Easter Rebellion of 1916. Blunt and Casement, who viewed each other as heroes and comrades, continued to exchange letters and sympathies during Casement's trial. Blunt was outraged by his execution and wrote that "Casement has accomplished everything I should have like[d] to do.... I was unworthy, and had the curse on me of my imperial English origin."

When World War I began in 1914, Blunt was disheartened at England's participation in the fight against Germany and lamented his country's lack of unarmed neutrality. On 31 December, Longford notes, he commented on the war in his diary: "It is not merely another year that has passed away but a whole cycle of years, a period in the history of the world, and the history of the human intellect." In the same month the two-volume complete edition of *The Poetical Works of Wilfrid Scawen Blunt* appeared, a collection of his works in different genres and poetic forms that received many positive reviews, especially one by poet Edward Thomas.

In 1915 Judith resolved to reconcile her parents and end their nine-year separation, so she arranged for Lady Anne to come to Wilfrid's residence at Worth Manor House on 12 May. Longford writes that, upon seeing Anne there and talking with her again, Blunt was to write: "I have never been more impressed with her intellectual superiority and the pleasantness of her conversation." An important business concern that was finally settled was the management of their Arabian stud. Anne was to own and supervise it until her death, after which Wilfrid was to do so – if he were still alive. After Wilfrid's farewell letter to her on 15 October, just before she left for Sheykh Obeyd, they never saw each other again. On 20 November 1915 Wilfrid's illegitimate daughter, Mary, was married to a British officer stationed in Cairo.

Suffering from dysentery and with her health generally failing, Lady Anne tried to finish both her fragmentary writings on the Arabian horse and her will. Wilfrid found that will "wicked," as it left nothing to him except the business papers on the stud. On 15 December 1917 Lady Anne died in the American hospital in Cairo, and she was buried in a nuns' cemetery at Jebel al Ahmar. Blunt designed a granite sarcophagus for her with the following inscription: "Here lies in the Egyptian desert which she loved LADY ANNE BLUNT." After World War II her grave was removed to a Catholic enclave in the Abbasieh Cemetery in Cairo.

Anne's will soured the relationship between Blunt and Judith, whose dispute over the disposal of the stud was resolved only through a lawsuit, which Judith won. The estrangement between Wilfrid and Judith ended in 1921, when he was persuaded to see again his three grandchildren – Anthony, now twenty-one years old; Anne, twenty years old; and Winnie, seventeen years old.

On his deathbed Blunt made one last attempt to achieve political justice: he asked his friend Winston Churchill to help Saad Zaghloul, an Egyptian national hero and leader of the 1919 rebellion against British imperialism, return from exile to Cairo. On the morning of 10 September 1922 Blunt died, and a letter of gratitude, announcing the release of a political prisoner whom he had helped, arrived from India two days later, too late for him to read.

Blunt fulfilled the Arabic proverb by which he had sought to live: "An hour of justice is worth seventy years of prayer." T. E. Lawrence called Blunt a prophet in whom there had been "a fire yet flickering over the ashes of old fury." On 12 September he was buried on his grounds without religious ceremony. A few people – including Dorothy Carleton, to whom he left Newbuildings, and his secretary, Sir Sydney Cockerell, to whose care Blunt had entrusted his secret memoirs – attended the funeral.

Wilfrid and Lady Anne Blunt were both pioneers. She was the first European woman to visit the Arabian peninsula, and he was the first Englishman to champion Arab nationalism. Notwithstanding their differences in temperament, they cooperated to contribute works that deepen the understanding and appreciation of non-European countries and cultures.

Biographies:

Edith Finch, *Wilfrid Scawen Blunt, 1840–1922* (London: Cape, 1938);

Elizabeth Longford, *A Pilgrimage of Passion: The Life of Wilfrid Scawen Blunt* (London: Weidenfeld & Nicolson, 1979; New York: Knopf, 1980).

References:

Thomas J. Assad, *Three Victorian Travellers: Burton, Blunt, Doughty* (London: Routledge & Kegan Paul, 1964);

Robert Davies, *Warriors and Gentlemen: The Occidental Context of the Arabian Travel Narratives of Burton, Blunt and Lawrence* (Loughborough: University of Technology, 1991);

Max Egremont, *The Cousins: The Friendship, Opinions and Activities of Wilfrid Scawen Blunt and George Wyndham* (London: Collins, 1977);

Frank Harris, "The Admirable Crichton: Wilfrid Scawen Blunt and the Right Hon. George

Wyndham," in his *Contemporary Portraits* (New York: Brentano's, 1923), pp. 17-33;

Reginald Hugh Kiernan, "Doughty and the Blunts," in his *The Unveiling of Arabia: The Story of Arabian Travel and Discovery* (London: Harrap, 1937; New York: AMS, 1975), pp. 268-288;

Shane Leslie, "Wilfrid Blunt, 1840-1922," in his *Men Were Different* (London: M. Joseph, 1937; New York: Books for Libraries, 1967), pp. 229-288;

Earl Noel Anthony Scawen Lytton, *Wilfrid Scawen Blunt: A Memoir of his Grandson* (London: Macdonald, 1961);

Mary Joan Reinehr, *The Writings of Wilfrid Scawen Blunt* (Milwaukee, Wis.: Marquette University Press, 1940);

James C. Simmons, "Sir Wilfrid Scawen Blunt and Lady Anne Blunt: Pilgrims to Najd," in his *Passionate Pilgrims: English Travelers to the World of the Desert Arabs* (New York: Morrow, 1987), pp. 255-290;

Kathryn Tidrick, "Wilfrid Scawen Blunt (1840-1922)," in her *Heart-Beguiling Araby* (Cambridge: Cambridge University Press, 1981), pp. 107-113;

Peter Upton, *Desert Heritage: An Artist's Collection of Blunt's Original Arab Horses* (London: Skilton & Shaw, 1980).

Papers:

Most of the Blunts' papers, including the secret diaries, are at the Fitzwilliam Museum, Cambridge, England.

Margaret Brooke, Ranee of Sarawak

(9 October 1849 – 1 December 1936)

Susan Morgan
Miami University

BOOKS: *My Life in Sarawak* (London: Methuen, 1913);

Impromptus (London: Arnold, 1923);

Good Morning and Good Night (London: Constable, 1934; New York: Hippocrene, 1984).

Lady Margaret Alice Lilly (de Windt) Brooke, Ranee of Sarawak, was part of a social set that included many other famous Victorians, from writers and painters to politicians and biologists. Brooke's renown, apart from her apparently vivid intelligence and charm, rested on the unusual circumstances of her life, which the popularity of her two books about her travels made well known in England. Apart from *Impromptus* (1923), a little-known collection of short stories set in England, she wrote only the two travel narratives, one twenty-one years after the other. These books, the first completely and the second partly, are memoirs of what Brooke regarded as the most exciting time of her life – her years as the ranee, or queen, in the southeast Asian state of Sarawak, a small area that is now a state in Malaysia, on the western coast of the island of Borneo. Brooke's travel memoirs constitute a vivid record of both Victorian society and a woman's view of her role in one of the most bizarre instances of nineteenth-century British imperialism. Her writing needs to be understood within the history of Sarawak.

In 1842 the sultan of Brunei officially appointed an Englishman, Sir James Brooke, to be governor, or raja, of Sarawak, one of the sultan's small states. In effect, Brooke owned Sarawak, and in 1868 his nephew, Sir Charles Brooke, inherited this privately owned dominion. The uncertainty of the legal status of Sarawak made it an unusual problem for the British government. The Brookes wanted Great Britain officially to recognize it as an independent nation. Throughout the century they tried to transform the private into the public, to turn their family kingdom into an internationally recognized country and themselves into recognized

Margaret Brooke, Ranee of Sarawak, portrait by Mrs. Alfred Sotheby (from Margaret Brooke, My Life in Sarawak, *1913)*

kings, by acquiring formal British recognition of their existence. Their struggle was waged in an extraordinary number of travel books about Sarawak published in England during the nineteenth and early twentieth centuries. The focus of their efforts was all in a name, and their argument a semantic contention – a matter of what the Brookes could call Sarawak and, consequently, what they could call themselves. The popular travel memoirs of life in Sarawak by Sir Charles's wife, Lady Margaret Brooke, were to be central voices in that public debate.

The story of Sarawak, retold repeatedly in the accounts of travelers who visited there, could well

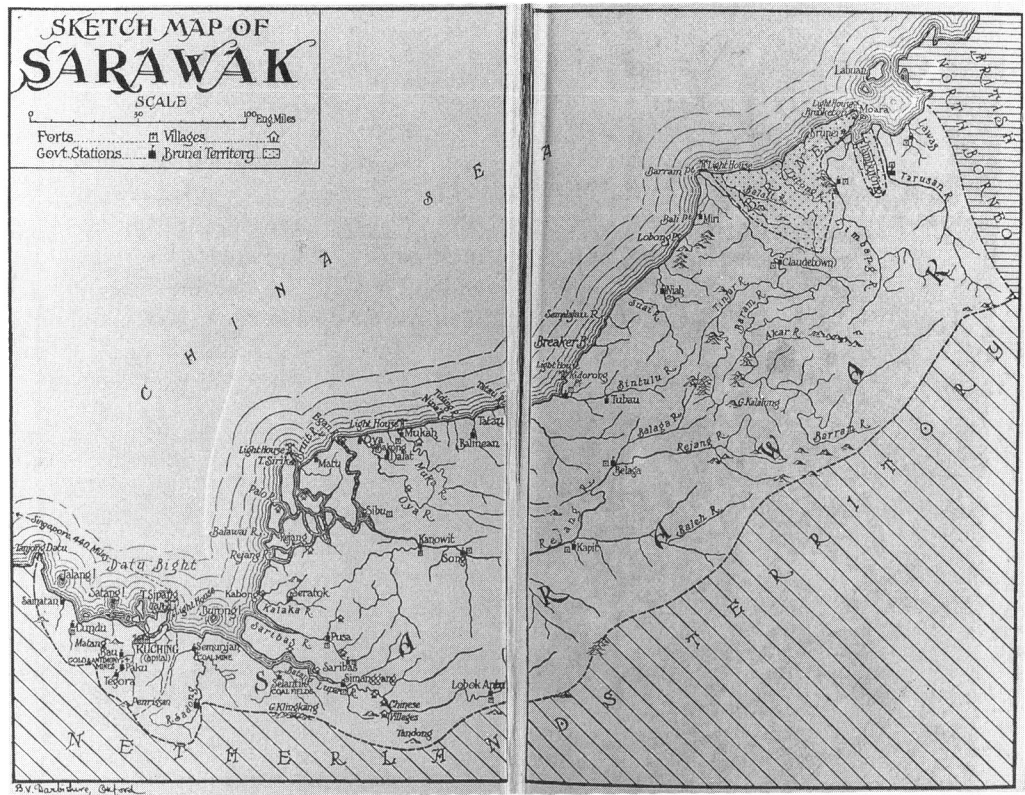

B. V. Darbishire's map of the Brookes' domain (from the front endpapers of My Life in Sarawak)

be called the great romance of British imperial history. The white rajas and the country they claimed to own attained a mythic significance in Victorian travel writing. The myth was created largely through the many books written about Sarawak by the actual participants – the Brookes and various people who were related to them or worked for them or both. The travel literature of Sarawak furnished the British public with a romance in which the white raja was the hero of that myth of isolated white men ruling over wild peoples in a tropical setting.

Audience fascination with travel narratives about the Brooke saga concerned the individualist promise at the heart of the imperial enterprise: that a single Englishman could go anywhere and become anything, could conquer pirates by sea and defeat jungle tribes by land, could get his very own country and become its king. That promise was a boy's adventure tale, for the Brookes' Sarawak offered a version of the life depicted on the island of Daniel Defoe's *The Life and Strange and Surprising Adventures of Robinson Crusoe* (1719), on Patusan in Joseph Conrad's *Lord Jim* (1900), in the India of Rudyard Kipling's *Kim* (1901), and in the Never Never Land of J. M. Barrie's *Peter and Wendy* (1911). Accounts of Sarawak displayed the familiar conventions of boys' adventure tales, one convention of which is the general absence of girls. Among the travel writings on Sarawak, Lady Margaret Brooke's two books hold special places as representations of the unusual role of a Victorian female placed in an imperial realm defined as quintessentially male.

At the age of forty Sir Charles Brooke, already having fathered a son by a Malay woman, went to England for the practical purpose of finding a British wife to provide him with a "proper" heir. He married nineteen-year-old Margaret de Windt in October 1869. By spring 1870 he was back in Sarawak with the bride, who was by then pregnant. After four years the couple had become parents of three surviving children (a fourth had been stillborn), and the Brookes left Sarawak on an ill-fated trip to England: all three children died on the ship.

The Brookes stayed two years in England, where Brooke gave birth to her fifth child, Vyner, in September 1874. In 1875 the Brookes returned to Sarawak, where a second son (Brooke's sixth child) was born in 1876. After her second four years in Sarawak, the family went again to England, where a third son was born in November 1879. Through ten years of marriage to a cold, indifferent man who did not love her, Brooke had borne seven children. Leaving their three sons safe in England, the Brooke parents went back to Sarawak in 1880. In 1882, having fathered three boys in twelve years of marriage, Sir Charles could feel secure in having an heir for his throne. He provided an allowance for Brooke and sent her permanently back to England, where she was to raise the children. The marriage was over, and from then on Brooke was not welcome in Sarawak.

In 1913 Brooke published *My Life in Sarawak,* an extremely popular travel memoir, just four years before Sir Charles, who by then had been raja for almost fifty years, died. In 1934 she published her final book and second Sarawak memoir, *Good Morning and Good Night,* which was also well received, although only about half of it recounts her travels in Sarawak. *Good Morning* offers more outspoken opinions and criticisms than does *My Life,* as by 1934 Raja Charles had been dead for almost twenty years. Brooke had lived in Paris, Italy, and London and was friendly with William Morris, Conrad, and Henry James, among others. One of her sons was dead, and another had been raja for almost twenty years. England had been through World War I, and literary conventions were changed. *Good Morning* constructs a self distinctly more outspoken, more given to taking verbal risks, than that of the careful narrator of the first book. Yet in spite of this change, both books present a figure who, to a greater degree in the second, breaks out of the feminine roles conventionally established in the imperial narrative.

However tamely the narrator of *My Life* represents herself and the Brooke presence in Sarawak, the lively narrative still refuses to offer a dutiful wifely version of his and hers accounts. Audiences read Brooke's memoir in part for its insider iconoclasm, for its revelations of what life was *really* like with the heroic white raja. The husband of both Sarawak books, first implicitly and, in the later book, explicitly, is unresponsive and silent. He offers a home life that his wife regards as bleak, and she characterizes him as completely lacking a sense of humor. She represents her marriage as a deep, often humiliating disappointment to her and a nuisance to her husband, whom she describes as being bored with matrimony. The two share no union or emotional connection, and their relationship involves only legal procreation, with Brooke as a baby-producing machine.

My Life begins with the journey to Sarawak, with the arrival at the port of Singapore. *Good Morning* begins much earlier, tracing Brooke's lineage and childhood. This portrait of her life before marriage emphasizes a key feature of her identity that the second narrative presents: Margaret de Windt was not, literally or emotionally, of British birth. Descended on her mother's side from prerevolutionary French aristocrats and the child of a half-French mother who had been raised in France on the ancestral estate restored to the family after the French Revolution, Brooke was also raised on that same estate. Her move to England was effectively an exile from her most beloved land.

In terms of social position, gender, and nationality, Brooke appears throughout her narratives as an outsider who is imaginatively as well as literally separated from the British in general and from her husband in particular. Her portrait of the Brooke domain is painted from the perspective of someone who considers herself not to be a British imperialist, and her position allows her to critique the ways of the British colonial community. That critique appeared in a small way in *My Life,* in Brooke's account of arguing against the imperial disdain for the locals that the travel writer Marianne North displays. North hates those whom she considers savages. But for Brooke, who defends the Dyak custom of headhunting, the Malays should not be considered savages but rather just people who are similar, possibly even identical, to the British – but for the difference of circumstances. *Good Morning* extends the motif in Brooke's often-stated dislike for the Englishwomen of Kuching, not only in their bickering about status but in their obsessions with things such as a half-ripe English strawberry while they are surrounded by the lush prodigality of the tropics.

The narrator of *Good Morning* frequently laughs at other colonial ladies for trying to be so British, while she is trying to be so Malay. Brooke's desire to be Malay appears in the long sections on her adopting Malay dress and making friends with the Malay women. The lonely young wife represents herself as finding her true place and appropriate identity in Sarawak. Brooke locates that identity, not in her imperial place as ranee, but in her role as a young woman who was never really from England and has been long ago displaced from France. In Sarawak, Brooke becomes part of a Malay woman's community and is delighted to be one of the women.

This narrator's sense of belonging to the Malay, rather than the white, world of Sarawak also appears in Brooke's first book. There she writes of

[22 Feb. 1924]

ASCOT HILL,
ASCOT, BERKS.

TELEPHONE: 314 ASCOT.

22—2—1924

Dear Mrs Conrad,

Mr Curle tells me, and can it really be true that you have very kindly asked me to luncheon with you on Friday the 29th at the Curzon Hotel. I accept with gratitude and pleasure! Such joy it will be to me to shake hands with your Great man, and inwardly to breathe my gratitude for all the pleasure I have had through his books!—

Thank you for being so kind! and it will be such a pleasure to meet you!

Yours very sincerely
Margaret
Sarawak

Letter from Brooke to Mrs. Joseph Conrad (Curle Ms. III; courtesy of the Lilly Library, Indiana University)

Brooke and her son, Capt. Harry Brooke, at Faversham to launch the ship Maimuna *in 1925 (Hulton Getty Picture Collection Ltd.)*

her sadness at not realizing what she had most hoped to find. She is particularly disappointed to be writing her book while living in London rather than Sarawak, where, surrounded by a loving husband, sons with wives and children, and family friends, she might have built a different life – if her husband had not coldly prevented it. This vision of a possible Sarawak self differs poignantly from that presented in the last chapters of *Good Morning*, a self belonging nowhere, moving among various houses in England, and living alone though surrounded by acquaintances.

Typical colonial exoticism underlay Brooke's narrative attempts to establish a Sarawak identity, but these were also responses to her authoritarian Victorian husband's oppressions and his refusal to allow her such an identity. Denied a place as a colonialist and existing only to produce children, she created in *Good Morning* a place where she could fit, as a part of a female colony in which women, whether their color be what she called white or brown, were all connected as members of the family of woman. Her identification with Malay women in Sarawak depended on the discomforting cultural similarities in roles enforced by imperial orders – roles in which all women, white and brown, ranees as well as Malays, were required to walk four paces behind the men.

Brooke's valuation of the importance of emotions leads her in *Good Morning* to question the principle of racial superiority, one of the foundations of European imperialist ideology. She wonders why white people have the vanity to think that they are somehow superior to the many different kinds of people whom they so ignorantly lump together as "black." When an Englishwoman asserts to Brooke that she, after all, would not like people to class her as "black," Brooke offers a twofold response. She first denies the reductively racist classification by pointing out that the indigenous peoples of Sarawak simply cannot all be labeled "black." She then places herself with those peoples by claiming that she would like to be considered a Malay. Her love for the people and the place is so strong, she says, that she wishes a passing angel would write *Sarawak* in golden letters across her heart when she dies.

The warmhearted young woman in a cold English marriage in a faraway world that she quickly comes to love is the dominating portrait of Brooke's two narratives of her life in Sarawak. It is hard to resist the charms of this amusing narrator who scorns the narrow colonial society, re-creates the delights of being among a tangled mass of Malay women hanging suspended in a large mat when the floor of a Dyak house collapses, and tells of crouching behind a piano all night while an anxious resident fearing possible Dyak attacks pushes plates of cold ham at her. The point is not to take the heroic pose of keeping a sense of humor amid the vicissitudes of life but rather to recognize that the vicissitudes are most often false alarms — and that is what is funny. Against the familiar masculine conventions of dangers and great adventures to be had, Brooke is unable to narrate anything more threatening in Sarawak than weak floorboards or the horrors of ham.

But despite having represented herself as standing outside the British colonial community and rejecting its myths of heroes and adventurers, Brooke also reveals herself in the two books to be a special member of that community. Because she is the ranee, as much as because of her own tastes, she does not mingle freely with the lesser administrators and their wives. Her claims to belong to the community of Malay women therefore do not comprise a rejection of imperialism, and this feminine voice often sounds remarkably like those of male travel writers about Sarawak, especially in their testimonials to the greatness of the Brooke reign over the peoples of Sarawak.

Her cheery accounts of adventurous trips in Borneo fail to mention that these were trips that Sir Charles Brooke made to appropriate more Malay lands and expand the borders of Sarawak. Both of Brooke's memoirs belong with the extensive myth-making travel literature about the white rajas of Sarawak. What the narratives conspicuously lack, in all their witty critiques of her life as Sir Charles's baby machine, is any doubt about her being needed to produce those babies. The unquestioned premise of *My Life* and *Good Morning*, the basis of Brooke's adult life and her two travel memoirs, is the imperial belief in the fundamental importance of providing Brooke heirs to the throne of Sarawak.

These two works are not private memoirs but works contributing to a stream of propaganda to shore up a political institution. Each having been published before a major war, both acknowledge the greatness of the Victorian imperial enterprise — even with its faults. The Brooke of the second half of *Good Morning* adopts her public identity as the ranee of Sarawak. Her presence and her book helped shape public opinion about Sarawak in the 1930s, and public opinion shaped answers to questions that determined the fate of Sarawak and the Brookes. Should England take over Sarawak as a colony? And, just as important to the Brookes, what compensations should England and/or the Sarawak treasury provide to the family who had created Sarawak as a nation and ruled it for a century as Britain had approved?

The last parts of *Good Morning* represent Brooke's royal identity as creating the kind of life she has, a life narrated as a series of ongoing vignettes about her encounters with other famous Victorians. Such vignettes include Sir Edward Burne-Jones showing her his pictures, A. C. Swinburne visiting her sick son, and Brooke falling off a ladder with Henry James. Because she was the ranee of Sarawak, Brooke's life was a public life, one set amid the rich, talented, and famous. Excluded by her husband from shaping an identity in Sarawak, Brooke adopted in England the imperial status that she had found in Sarawak to have been so privately unfulfilling.

References:

Sylvia Brooke, *The Three White Rajahs* (London: Cassell, 1939);

Susan Morgan, *Place Matters: Gendered Geography in Victorian Women's Travel Books about Southeast Asia* (New Brunswick, N.J.: Rutgers University Press, 1996);

Robert Pringle, *Rajahs and Rebels: The Ibans of Sarawak under Brooke Rule, 1841–1941* (Ithaca, N.Y.: Cornell University Press, 1970);

R. H. W. Reece, "A 'Suitable Population': Charles Brooke and Race-Mixing in Sarawak," *Itinerario*, 9, no. 1 (1985): 67–112.

Samuel Butler
(4 December 1835 – 18 June 1902)

Keith C. Odom
Texas Christian University

See also the Butler entries in *DLB 18: Victorian Novelists After 1885* and *DLB 57: Victorian Prose Writers After 1867.*

BOOKS: *A First Year in Canterbury Settlement* (London: Longmans, Green, 1863; New York: Dutton, 1915);

The Evidence for the Resurrection of Jesus Christ, as Given by the Four Evangelists, Critically Examined, anonymous (London: Williams & Norgate, 1865);

Erewhon; or, Over the Range, anonymous (London: Trübner, 1872; revised, 1872); republished under Butler's name (London: Trübner, 1873; revised again, London: Richards, 1901; New York: Dutton, 1910);

The Fair Haven: A Work in Defence of the Miraculous Element in Our Lord's Ministry upon Earth both as against Rationalistic Impugners and Certain Orthodox Defenders by the Late J. P. Owen, edited by W. B. Owen with a Memoir of the Author, anonymous (London: Trübner, 1873; New York: Kennerley, 1913);

Life and Habit: An Essay after a Completer View of Evolution (London: Trübner, 1877; New York: Dutton, 1910);

Evolution Old and New; or the Theories of Buffon, Dr. Erasmus Darwin, and Lamarck, as Compared with That of Mr. Charles Darwin (London: Hardwicke & Bogue, 1879; Salem, Mass.: Cassino, 1879; revised, London: Bogue, 1882; revised again, London: Fifield, 1911; New York: Dutton, 1914);

Unconscious Memory: A Comparison between the Theory of Dr. Ewald Hering, Professor of Physiology at Prague, and the Philosophy of the Unconscious of Dr. Edward von Hartmann; with Translations from these Authors (London: Bogue, 1880; New York: Dutton, 1910);

Alps and Sanctuaries of Piedmont and the Canton Ticino (London: Bogue, 1881; enlarged, London: Fifield, 1913; New York: Dutton, 1913);

Samuel Butler in 1862

Selections from Previous Works, with Remarks on Mr. G. J. Romanes' "Mental Evolution in Animals," and a Psalm of Montreal (London: Trübner, 1884);

Gavottes, Minuets, Fugues, and Other Short Pieces for the Piano, by Butler and Henry Festing Jones (London: Novello, Ewer, 1885);

Luck or Cunning as the Main Means of Organic Modification? An Attempt to Throw Additional Light upon the Late Mr. Charles Darwin's Theory of Natural Selection (London: Fifield, 1886);

Ex Voto: An Account of the Sacro Monte or New Jerusalem at Varallo-Sesia, with Some Notice of Tabachetti's Remaining Work at the Sanctuary of Crea (London: Trübner, 1888; revised and enlarged, 1889; London & New York: Longmans, Green, 1890);

Narcissus: A Dramatic Cantata in Vocal Score, With a Separate Accompaniment for the Piano-forte, by Butler and Jones (London: Weekes, 1888);

Shakespeare's Sonnets Reconsidered, and in Part Rearranged; with Introductory Chapters, Notes, and a Reprint of the Original 1609 Edition (London & New York: Longmans, Green, 1889);

A Lecture on the Humour of Homer, January 30th 1892; Reprinted with a Preface and Additional Matter from the "Eagle" (Cambridge: Metcalfe, 1892);

On the Trapanese Origin of the "Odyssey" (Cambridge: Metcalfe, 1893);

The Life and Letters of Dr. Samuel Butler, Headmaster of Shrewsbury School 1798–1836, and Afterwards Bishop of Lichfield, 2 volumes (London: John Murray, 1896; New York: Dutton, 1924);

The Authoress of the "Odyssey," Where and When She Wrote, Who She Was, the Use She Made of the "Iliad," and How the Poem Grew under Her Hands (London & New York: Longmans, Green, 1897; New York: Dutton, 1922);

Erewhon Revisited Twenty Years Later, both by the Original Discoverer of the Country and by His Son (London: Richards, 1901; New York: Dutton, 1910);

The Way of All Flesh, edited by R. A. Streatfeild (London: Richards, 1903; New York: Dutton, 1910);

Seven Sonnets and a Psalm of Montreal (Cambridge: Privately printed, 1904);

Ulysses: A Dramatic Oratorio in Vocal Score with Accompaniment for the Pianoforte, by Butler and Jones (London: Weekes, 1904; Chicago: Summy, 1904);

Essays on Life, Art, and Science, edited by Streatfeild (London: Richards, 1904);

God the Known and God the Unknown, edited by Streatfeild (London: Fifield, 1909; New Haven: Yale University Press, 1917);

The Note-Books of Samuel Butler: Selections, edited by Jones (London: Fifield, 1912; New York: Kennerley, 1913);

Butleriana, edited by A. T. Bartholomew (London: Nonesuch, 1932);

Further Extracts from the Note-Books of Samuel Butler, edited by Bartholomew (London: Cape, 1934);

Samuel Butler's Notebooks: Selections, edited by Geoffrey Keynes and Brian Hill (London: Cape, 1951; New York: Dutton, 1951);

The Note-Books of Samuel Butler, Volume I (1874–1883), edited by Hans-Peter Breuer (Lanham, Md.: University Press of America, 1984).

Editions: *The Shrewsbury Edition of the Works of Samuel Butler,* edited by Henry Festing Jones and A. T. Bartholomew, 20 volumes (London: Cape / New York: Dutton, 1923–1926);

Ernest Pontifex, or The Way of All Flesh, edited by Daniel F. Howard from Butler's manuscript (Boston: Houghton, Mifflin, 1964).

TRANSLATIONS: *The Iliad of Homer, Rendered into English Prose for the Use of Those Who Cannot Read the Original* (London & New York: Longmans, Green, 1898; New York: Dutton, 1921);

The Odyssey, Rendered into English Prose for the Use of Those Who Cannot Read the Original (London & New York: Longmans, Green, 1900; New York: Dutton, 1922).

Samuel Butler is not exactly well known today in English literature, though he may be better known and more widely read now than in his own time, the latter days of the Victorian period. He was generally considered a crackpot philosopher rather than a travel writer or novelist, and his books typically focused on his theories on matters involving religion (he doubted many orthodox beliefs), evolution (though a believer in evolution, he spent much of his time arguing with Charles Darwin), and psychology (he pioneered a neo-Lamarckian theory of heredity through unconscious memory). He found few takers for his opinions and published most of his books at his own expense. If his contemporaries knew any of his works at all, those works were *Erewhon; or, Over the Range* (1872), *Erewhon Revisited Twenty Years Later* (1901), and *The Way of All Flesh* (1903). They are still his most frequently read works, continuously in print in both hardback and paperback. In 1995, when *The New York Times Book Review* ran a series on literary rediscoveries, the American novelist William Gaddis wrote on Samuel Butler and *Erewhon,* although the main thrust of the article was not on Butler's quality as a writer but on the resemblances of the anti-Victorian utopia of *Erewhon* and Newt Gingrich's "Contract with America," signed by the Republicans in the 1995 U.S. Congress.

Samuel Butler was born on 4 December 1835 in the rectory at Langar, Nottinghamshire, the older surviving son of the Reverend Thomas Butler and

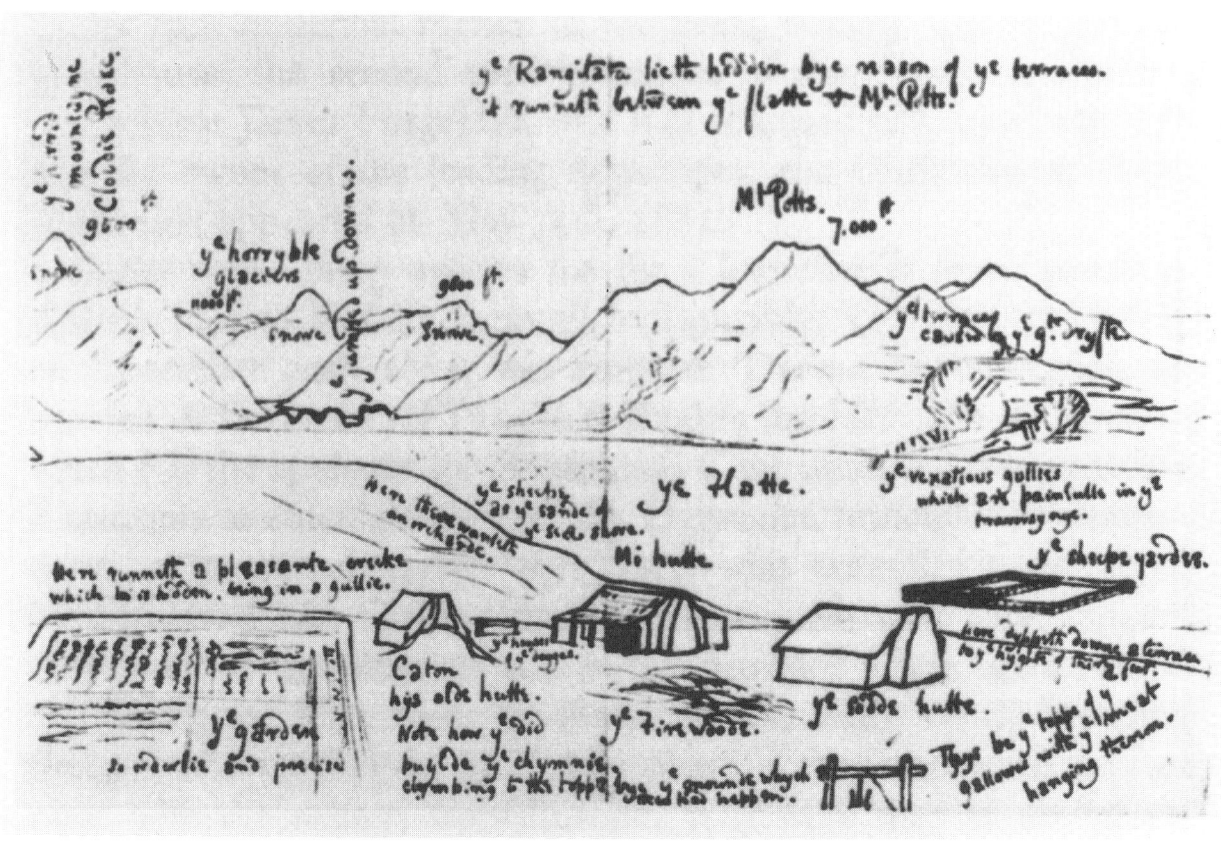

Drawing by Butler of the view from Mesopotamia, New Zealand, where he settled in the early 1860s (Canterbury Museum, Christchurch, New Zealand)

Fanny Worsley Butler, the daughter of a Bristol merchant. Reverend Butler was the son of Dr. Samuel Butler, the best-known headmaster of Shrewsbury School and later bishop of Lichfield. Samuel Butler had two sisters, Harriet and Mary, and a younger brother, Thomas, none of whom he ever got along with, just as he never quite secured the approval of his strict and religiously orthodox parents. As Butler wrote in his *Notebook* for March 1883, "If I break down prematurely and I often think I shall, it will have been due partly to the severity of my bringing up." Stemming from investments related to the Shrewsbury School, the family fortune was sufficient for physical comforts and travel. When he was only ten years old, Butler accompanied his family on his first trip to Italy. Like his father before him, he attended Shrewsbury School and St. John's College, Cambridge, where he finished the classical tripos with first class honors in 1858.

Destined by his family for the church, Butler refused ordination after working among the London poor for a year. Instead, he wanted to be an artist, but his father offered the alternative of passage to New Zealand, where Butler acquired a sheep station. He eventually accumulated a great enough fortune to sell his New Zealand holdings and return to London, where, taking rooms in Clifford's Inn, he settled down to live as he wished. Most of his friends were men, except for Elizabeth Mary Ann Savage, an intelligent but physically frail woman who was his best literary critic and adviser until her death on 22 February 1885. He never married.

Butler's life was difficult as he battled his disapproving family, who were agents of a repressive and hypocritical society, but he did have many pleasures, including art (he studied to be a painter), music (he was obsessed with George Frideric Handel, whose music he played and imitated in his own compositions), writing (he frequently worked at a desk in the Reading Room of the British Museum), and travel. Despite a constrained life in London, he traveled as far east as New Zealand and as far west as Canada and vacationed on the European continent virtually every year after he returned from the South Pacific. He believed strongly in the efficacy of travel and insisted on having his holidays even when his finances were at a low ebb. When he traveled in his later years with Henry Festing Jones, he always was in good humor, saying, according to Jones, that "A man's holiday is his garden." Butler's impulse to explore and discover dates back to his days as a young man in Canterbury Colony in New Zealand when he saw the Southern Alps in the distance and determined to cross them. During his years at Clifford's Inn in London, he took day trips to Margate, weekend trips to Gravesend or Boulogne, or, at least, accompanied by his manservant, Alfred, an evening at the pantomime. Even the books he published that are not clearly travel books are full of movement from one place to another. Butler's impulse to travel was of major importance in his life.

Lee E. Holt notes three travel articles by Butler in the *Eagle*, a publication of students at St. John's College, Cambridge, but they are most easily found now in volume one of the collected works of Butler under the title *A First Year in Canterbury Settlement* (1863). The earliest article is titled "Our Tour" (1859) and concerns a trip he and a friend took to Italy in June 1857. It is a daily account of places, hotels, local people, and scenery, showing an eye for detail and a youthful enthusiasm for adventure. At Dieppe he notes

> the novelty and freshness of everything you meet; whether it is the old bonnetless, short-petticoated women walking arm in arm with their grandsons . . . or . . . the plain evidence that meets one at every touch and turn, that one is among people who live out of doors very much more than ourselves.

Both of the other essays are titled "Our Emigrant" and signed "Cellarius," the first dated 1860 and the second dated 1861. They were incorporated into Butler's first book of travel, *A First Year in Canterbury Settlement*, comprising his letters home and the essays he sent to Cambridge, along with a preface written by his father. This collection was published by his family, and at first he liked the work, though later, in true Butlerian fashion, he grew to dislike it intensely. It is written clearly and with a fresh eye that reveals the pleasure Butler took in seeing a new country — and his relief at being away from England and his family. He relates how he landed in New Zealand at Port Lyttelton and as quickly as possible traveled to Christchurch. As soon as he could, he explored the territory west of Christchurch and found the place for his sheep station, called Mesopotamia. Later he would cross the Southern Alps and discover the pass that would become the gateway to Erewhon, his utopian land.

With a setting based on Butler's travels in New Zealand, one of Butler's earliest publications, *Erewhon,* was a utopian work that satirized many Victorian shibboleths and came so uncomfortably close to the truth that it attained fame if not popularity. It was read widely enough that when the author neared the end of his life and career, his *Erewhon Revisited* found for the first time a publisher willing to bear the expenses for the book's printing rather than demanding the cost from Butler. The sequel aroused the interest of the reading public as none of his books had before.

Following the publication of *Erewhon* Butler published works between 1873 and 1880 on evolution and on the theories of the unconscious advanced by Ewald Hering and Edward von Hartmann, as well as a defense of Christ's miracles. Butler was less comfortable with travel writing than with his work on unconscious memory and evolution. He believed, however, that travel was the cure for what ails a person suffering from a malady brought on by too much repression, too much family, and too much of the same society. His second travel book, *Alps and Sanctuaries of Piedmont and the Canton Ticino* (1881), was written about a journey he took with his ubiquitous traveling companion Jones, who became Butler's first major collaborator — the two men produced a few musical scores — and biographer. In the book Butler records his impression of the people he encountered in the Alps of Canton Ticino in southern Switzerland and in Piedmont and Lombardy in northern Italy.

Though Butler wrote copious notes on his travels, he spends more time in his travel writing discussing ideas than describing the country or the people. It is not that Butler did not appreciate the scenery or people's looks or customs — he always illustrated his ideas by referring to the physical world — however, he was more at home with concepts. In the first chapter, for instance, he discusses who had the greatest hold over Englishmen: Handel or William Shakespeare. Handel, says Butler, is "above Shakespeare and Shakespeare is above all others, except Handel himself." Then, beginning with the pass at Saint Gotthard, he describes the scenery by principles rather than by colors. "Mountain scenery, when one is staying right in the middle of it, or when one is on foot is one thing, and mountain scenery as seen from the top of a diligence very likely smothered in dust is another." Nonetheless, he does present pictures of the scene at last: "I shall never forget coming out of this tunnel one day late in November, and finding the whole Andermatt valley in brilliant sunshine, though from Fluelen up to

Frontispiece and title page for Butler's second travel book, about his journeys in the Alps of southern Switzerland and northern Italy

the Devil's Bridge the clouds had hung heavy and low. It was one of the most striking transportation scenes imaginable."

Critics of Butler have remarked on the effect his beloved alpine scene had on him: the release of tensions and the less formal style he displays, the appreciation of the time in which he lived as opposed to the past. The work is reminiscent of Robert Louis Stevenson's *Travels with a Donkey* (1879). Butler enjoyed the scenery, but the church sanctuaries pleased the artist in him. As he walked through the countryside from Faido to Fusio, however, his main focus was on art and architecture. Most striking were the chapels containing life-size figures posed in important scenes from the Bible. Others, however, do not find the artists and craftsmen so fine as he does. This concentration on the arts rather than the highways and byways caused his publisher, who suggested the book, to refuse the manuscript. Butler mentions little country inns and villages tucked away in the folds of hills but evidently did not do enough such description to satisfy the publisher. His drawings and illustrations should have made *Alps and Sanctuaries* a charming Baedeker-style guidebook. In spite of their dated look, they hold their charm today.

With the true loyalty of the aficionado, Butler tracked down one particular artist of the northern Italian chapels, Giovanni Tabachetti, and later wrote *Ex Voto: An Account of the Sacro Monte or New Jerusalem at Varallo-Sesia, with Some Notice of Tabachetti's Remaining Work at the Sanctuary of Crea* (1888), a book about the forty-five chapels on the sacred mountain of Varallo. Both Holt and Peter Raby comment on Butler's obsession with Tabachetti's portrayal of major events in the Bible, including the Fall of Adam and the life of Jesus. As in the case of *Alps and Sanctuaries*, he published *Ex Voto* at his own expense. He sold 117 copies of the second work, notably fewer than the first.

A late work that includes some travel writing is *The Authoress of the "Odyssey"* (1897), in which Butler exhaustively studies the evidence to support his intuition that the composer of the great epic was a young woman from Sicily. In particular, the landmarks of the poem were similar to those around Trapani and Mount Eryx in western Sicily. The proofs could be considered a tour of sorts of the Trapani in ancient times. There were some serious supporters of Butler's scholarship, but the book is not used as a work of travel per se.

Since *Erewhon* and *Erewhon Revisited* are stories of an imaginary country, they cannot strictly be called travel books, though they could well serve as guides to the terrain and natural life of New Zealand's South Island. Particularly, Butler's travelers explore the Rangitata River valley and the craggy mountain passes that lead into the Southern Alps. Obvious exceptions to the portrayal of actual geography are the enormous but fictional stone statues with holes in their heads that guard the entry to mythical Erewhon. The wind blowing through the holes makes such a terrifying moaning sound that possible interlopers are usually frightened away. One is reminded of the effect of gigantic organ pipes – a reasonable creation for Butler, who was himself an organist.

Butler's longtime habit of keeping notebooks to record thoughts and experiences provided material that has been edited and published and can be mined for further travel writing. Volume two of the Shrewsbury Edition of the *Notebooks,* for instance, includes "Material for a Projected Sequel to *Alps and Sanctuaries.*" The second paragraph is headed "Not to Be Omitted," and it contains a satisfying example of his sharp and humorous eye for people:

> The man who did not like parrots because they were too intelligent. And the man who told me that Handel's *Messiah* was "tres chic," and the smell of the cyclamens "stupendous." And the man who said it was hard to think the world was not more than 6000 years old, and we encouraged him by telling him we thought it must be even more than 7000. And the English lady who said of some one that "being an artist, you know, of course he had a great deal of poetical feeling." And the man who was sketching and said he had a very good eye for colour in the light, but would I be good enough to tell him what colour was best for the shadows.

Just before the notes that were "not to be omitted" are other notes typical of Butler's eye for little out-of-the-way hotels frequented by natives rather than tourists. At the Hotel de France in Montreuil-sur-Mer, he and Jones had a very good lunch and then met "the landlady, getting on towards eighty, with a hookish nose, pale blue eyes, and a Giovanni Bellini's Loredano Loredani kind of expression." Again he refers to art in describing the scene and the people.

The one book Butler wrote that is clearly a novel, *The Way of All Flesh,* was actually his first financial success, though he could never bring himself to publish it while he lived. His literary executors brought it out the year after his death. A semi-autobiographical narrative, it recognizably exposed Butler's own life and family to public scrutiny. Even his fictional surrogate, Ernest Pontifex ("bridge builder"), does not cast off his family without suffering a breakdown, the cure of which was more travel: Victorian doctors prescribed a change of scene whenever possible.

In Butler's autobiographical novel, which is so frank about his upbringing that he did not attempt to bring it out in his lifetime, Butler makes it clear that travel was the best medicine he could imagine. More than one travel writer began his or her career of traveling and writing through seeking better health – Stevenson, for example. Thus, while few of the books Butler published can be classified as travel books – even those being unconventional in their philosophical digressions – travel and writing about travel were important to him and helped shape many of his ideas. He has become a more important literary figure in succeeding ages than in his own.

Letters:

Samuel Butler and E. M. A. Savage, Letters 1871–1885, edited by Geoffrey Keynes and Brian Hill (London: Cape, 1935);

The Correspondence of Samuel Butler and His Sister May, edited by Daniel F. Howard (Berkeley: University of California Press, 1962);

The Family Letters of Samuel Butler, edited by Arnold Silver (Stanford, Cal.: Stanford University Press, 1962).

Bibliographies:

A. J. Hoppe, *A Bibliography of the Writings of Samuel Butler* (London: Bookman, 1925);

Carroll A. Wilson, *Catalogue of the Collection of Samuel Butler (of Erewhon) in the Chapin Library Williams College* (Portland, Maine: Southworth-Anthoensen, 1945);

Stanley B. Harkness, *The Career of Samuel Butler (1835–1902) A Bibliography* (New York: Burt Franklin, 1955);

Wayne G. Hammond, "Samuel Butler: A Checklist of Works and Criticism," 3 parts, *Butler News-*

letter, 3 (Summer 1980): 13–24; 3 (December 1980): 51–66; 4 (June 1981): 6–20.

Biographies:

Henry Festing Jones, *Samuel Butler: Author of "Erewhon" (1835–1902), a Memoir,* 2 volumes (London: Macmillan, 1919);

P. N. Furbank, *Samuel Butler (1835–1902)* (Cambridge: Cambridge University Press, 1948);

Philip Henderson, *Samuel Butler* (London: Cohen & West, 1953);

Peter Raby, *Samuel Butler: A Biography* (Iowa City: University of Iowa Press, 1991).

References:

William Gaddis, "Erewhon and the Contract with America: Wealth, Butler Knew, Was Virtue, Bad Luck a Crime," *New York Times Book Review,* 5 March 1995, pp. 6, 23;

R. S. Garnett, *Samuel and His Family Relations* (New York: Dutton, 1926);

Lee E. Holt, *Samuel Butler,* revised edition (New York: Twayne, 1989);

Thomas L. Jeffers, *Samuel Butler Revalued* (University Park: Pennsylvania State University Press, 1981);

Joseph Jones, *The Cradle of Erewhon* (Austin: University of Texas Press, 1959);

Malcolm Muggeridge, *Earnest Atheist: A Study of Samuel Butler* (London: Putnam, 1936);

Robert F. Rattray, *A Chronicle and an Introduction: Samuel Butler* (London: Duckworth, 1935);

Clara G. Stillman, *Samuel Butler: A Mid-Victorian Modern* (New York: Viking, 1932).

Papers:

The largest collection of Butler's papers is housed at the St. John's College Library, Cambridge. The Carroll A. Wilson Collection at the Chapin Library, Williams College, has many first editions and the original manuscript of the notebooks. Butler's correspondence is at the British Library.

Constance Gordon Cumming
(26 May 1837 – 4 September 1924)

Susan Schoenbauer Thurin
University of Wisconsin – Stout

BOOKS: *From the Hebrides to the Himalaya: A Sketch of Eighteen Months' Wanderings in Western Isles and Eastern Islands,* 2 volumes (London: Sampson Low, Marston, Searle & Rivington, 1876); revised in two parts as *In the Hebrides* (London: Chatto & Windus, 1883) and *In the Himalayas and on the Indian Plains* (London: Chatto & Windus, 1884);

At Home in Fiji (2 volumes, Edinburgh: Blackwood, 1881; 1 volume, New York: A. C. Armstrong, 1882);

A Lady's Cruise in a French Man-of-War (Edinburgh: Blackwood, 1882);

Fire Fountains: The Kingdom of Hawaii, Its Volcanoes, and the History of Its Missions, 2 volumes (Edinburgh & London: Blackwood, 1883);

Granite Crags (Edinburgh & London: Blackwood, 1884); republished as *Granite Crags of California* (Edinburgh: Blackwood, 1886);

Via Cornwall to Egypt (London: Chatto & Windus, 1885);

Wanderings in China, 2 volumes (Edinburgh: Blackwood, 1886);

Work for the Blind in China (London: Gilbert & Rivington, 1887);

"The Last Commandment." A Word to Every Christian, bound together with *Work for the Blind* (London: J. Nisbet, 1888);

Notes on China and Its Missions (London: Church Mission Society, 1889);

Two Happy Years in Ceylon, 2 volumes (Edinburgh: Blackwood, 1892; New York: Scribners, 1892);

The Inventor of the Numeral-Type for China, By the Use of Which Illiterate Chinese Both Blind and Sighted Can Very Quickly Be Taught to Read and Write Fluently (London: Downey, 1898);

Memories (Edinburgh: Blackwood, 1904).

SELECTED PERIODICAL PUBLICATIONS – UNCOLLECTED: "The Largest Extinct Volcano," *Scribner's Monthly,* 22 (June 1881): 272–275;

"Across the Yellow Sea," *Blackwood's,* 131 (May 1882): 623–634;

"The Last King of Tahiti," *Contemporary Review,* 41 (May 1882): 819–836;

"Early Spring in California," *Cornhill Magazine,* 48 (April 1883): 2115–2123;

"Oiling the Waves: a Safeguard in a Tempest," *Nineteenth Century,* 14 (April 1883): 704–716;

"De mortuis," Contemporary Review, 43 (June 1883): 858–869;

"The Temple of Heaven at Peking," *London Quarterly Review,* 64 (April 1885): 87–105;

"Some Eventful Voyages," *Blackwood's,* 147 (March 1890): 372–383;

"Easy Reading for Illiterate Chinese," *East and the West,* 2 (1904): 249–269.

Constance Frederica "Eka" Gordon Cumming was one of the most celebrated lady travelers in the last quarter of the nineteenth century. Essentially a tourist rather than an explorer, she was an astute observer in part because of her knowledge of botany and ethnology. Her social position piqued interest in her wanderings around the globe, while the warmth and human understanding suffusing her descriptions of little-known cultures made her travel books, according to W. C. Blackie in a July 1896 article in *Blackwood's,* "the most popular of our lady travellers." She produced hundreds of sketches and watercolor paintings, which illustrate her books and also were shown in several exhibitions. Late in life she became an advocate for missionary work among the blind in China, charity work she took up after spending six months in that country.

Cumming was born at Altyre, Morayshire, the twelfth child of Sir William Gordon Cumming and Lady Eliza Maria Campbell Cumming, who died in 1841 after the birth of her thirteenth child. Cumming was proud of her ancestry, writing in *Memories* (1904), her autobiography, that she was a daughter of the chief of clan Cumming, descendent of Charlemagne and of Red Comyn, who was slain by Bruce. Named for a brother and sister who died of scarlet fever just before she was born, Cumming spent her early years at Altyre and at age six went to live with a married sister at Cresswell Hall. When her father remarried, Cumming returned home as preparation for going to school in London. During her five years there, she spent holidays with relatives in England and summers back home in Altyre. After her father died in 1854 and until she began her travels in 1868, Cumming made her home with various relatives scattered about England and Scotland. She boasted of a large family that included fifty first cousins and a hundred second and third cousins. Her brother Roualeyn, whom in *Memories* she calls "the grandest and most beautiful human being I ever beheld," became well known as a South African lion hunter. His return, accompanied by his African companion and an enormous trophy collection, was one of the great events of her childhood. Because other relatives had ventured to India and Ceylon, Cumming grew up in an atmosphere redolent of travel and adventure.

Cumming enjoyed five years of crowded social life in the late 1850s and early 1860s, with parties and dancing until 4:00 A.M. These circumstances changed when five older brothers and sisters died within a few years of one another and the youngest half sister married. Cumming then embarked on a series of tours and long visits with friends and relatives that would culminate in twelve years of travel. Having the money and time to seize opportunity and follow her inclination, she returned home only twice: from 1870 to 1872 after visiting India and for several months in 1874 after having spent two years in Ceylon. Cumming has one of the longest and most varied itineraries of contemporary celebrated women travelers. Besides India and Ceylon she visited Fiji, Australia, New Zealand, Tonga, Samoa, Tahiti, California, Japan, China, and finally Hawaii before returning home.

Her first book, *From the Hebrides to the Himalaya* (1876), was an unwieldy work that later was revised and published as separate volumes, *In the Hebrides* (1883) and *In the Himalayas and on the Indian Plains* (1884). The volume about the Hebrides ostensibly recounts Cumming's wanderings from April to November 1868, but some experiences before and after the tour have been pieced in. The material is presented in the form of letters with some obvious additions of journal material, the overall effect being a loving but flawed account of these northern isles. Dorothy Middleton, in *Victorian Lady Travelers* (1965), dismisses Cumming as being almost unreadable for the mass of information crammed into her books; *From the Hebrides to the Himalaya* lends support to the accusation. Its unconscionable amount of history, legends, customs, and comparisons to other countries leave travel with a relatively minor role in this book, which nonetheless demonstrates Cumming's skill as a storyteller and eclectic collector of folklore.

The tour, like many of the author's travels, was unplanned. It began in Mull of Cantyre (Kintyre), where Cumming could sketch and her young half brother Fred could relax during his Easter vacation. After Fred returned to school, Cumming stayed on in Skye for four months, sketching, visiting, and making yachting excursions that included a short visit to the northernmost island in the Outer Hebrides. Cumming's love of the sea and her passion for sailing provide some of the most enjoyable passages in the book. She gives an especially appealing description of boating and sketching around the Quiraing, a fantastic array of towers and pinnacles composed of amygdaloidal traps (volcanic rock containing almond-shaped "bubbles" filled with secondary minerals such as calcite or quartz). Her tour of Fingal's Cave on Staffa produces a descriptive tone poem that serves as a fair companion to Felix Mendelssohn-Bartholdy's Overture in B Minor, op. 26, inspired by his visit to the cave in 1829.

The knowledge of botany Cumming displays in later travel books is confined in the first volume to a few enthusiastic comments on beautiful wildflowers and calling Mull a land of milk and honey. Ten years later she would be so stunned at the profusion of wildflowers and the bountiful agriculture in California that she would look back abashed at how grudgingly Scottish soil yields its harvest. On the other hand, her description of the Hebrides teems with sea life, especially the enormous colonies of sea lions on Skye and the waterfowl on Lewis with Harris, and includes a fascinating description of fishing with cormorants. A more mundane wildlife topic is the plague of mosquitoes on Skye, but here Cumming provides an interesting bit of popular culture: essence of lavender is an excellent insect repellent.

The culture and occupation of the inhabitants of the Hebrides are captured in vignettes that show Cumming's wide-ranging interests and benevolent heart. She commiserates with the kelp burners, whose production of the raw material for carbonate of soda and iodine had been undermined by cheaper, foreign products. A woman of her time and social class, however, Cumming regards the poverty of the peasantry more as a tourist attraction than as a social problem. The two-room bothies, nearly windowless and full of peat smoke with an outer room for cattle and an inner room for people, are objectified: either devoid of human presence or sanitized by the images of clean and neat churchgoers sallying forth. The human landscape is typically depicted in pastoral images as, for example, the Gaelic-speaking "picturesque lassies ... with bare feet and bright scarlet or white handkerchief on their glossy hair, half-hidden by the huge bundle of heather, which they would have to carry six or eight miles, that the men might rethatch their bothies."

Cumming is not only charmed by the "lassies," who on occasion take an interest in her sketching, but also delights in their allegiance to the unique Scottish costume. In this regard she sounds a note that echoes through all her later travel writings when she bemoans the loss of old customs and the adoption of a worldwide standard of dress. Cumming recognizes the advent of mass culture without understanding her own role as an instrument of change — that of the tourist-outsider introducing desirable elements of wealth and leisure from the industrialized world. In addition, her appreciation of cultural difference is dependent on the distinctions of class that the privileged would hardly desire to share with the underclass.

Cumming's antiquarian interests were stimulated by her tour of the Hebrides, and she uses the occasion of her sightseeing to relate the latest discoveries and theories. Her book is a virtual compendium of popular scientific information, some of which would be quite at home with notions of alien landing strips in the Andes. The antiquarian topic incidentally affords a good example of Cumming's convoluted organizational style. Her visit to a serpent-shaped mound near Oban, discovered by a Mr. Phené in 1871, leads Cumming to a long digression on serpent lore in ancient cultures. She discusses ancient Greece, Persia, and India, with excursions to places as disparate as Brittany and Wisconsin, ending eventually with a discussion of the Anglo-Saxons as the lost tribe of Israel. Later, in chapter 5, she abandons even the pretense of travel as a means of introducing folkloristic material and gives herself over to a comparison of customs, superstitions, and witchcraft around the world.

The visit to Iona is taken up almost entirely with legends about Saint Columba and the history of Christianity in the Hebrides, some of which is admittedly borrowed. Cumming's conventional religiosity, established in the descriptions of church services she attends on Skye, does not prevent her from taking a certain latitudinarian approach to fundamentalist activities. A Gaelic preacher who keeps his parishioners trapped for several hours raises her feminist hackles, which leads her to attack old church rules restricting women: "These law-givers, you perceive were *men*." Somewhat more respectful is her account of an outdoor, three-day revival meeting, interesting for its peripheral details (Cumming knew no Gaelic). The description also anticipates Cumming's accounts of missionary work in later books.

If Cumming is not deterred from making jokes at the expense of churchmen, neither is she reluctant to satirize the literary icon whose movements and musings James Boswell records in *A Tour to the Hebrides*. Cumming writes: "Speaking of rare guests, you will not travel far in Skye before hearing anecdotes of the pompous Dr. [Samuel] Johnson and his little toadying companion." Cumming portrays Johnson as a city slicker pathetically trying to rusticate. She repeats spurious, albeit humorous, stories of him wooing women with gifts of arithmetic books and testing "his drinking capacity against that of seasoned whisky-loving lairds." Popular authors such as Sir Walter Scott and Charles Kingsley are

 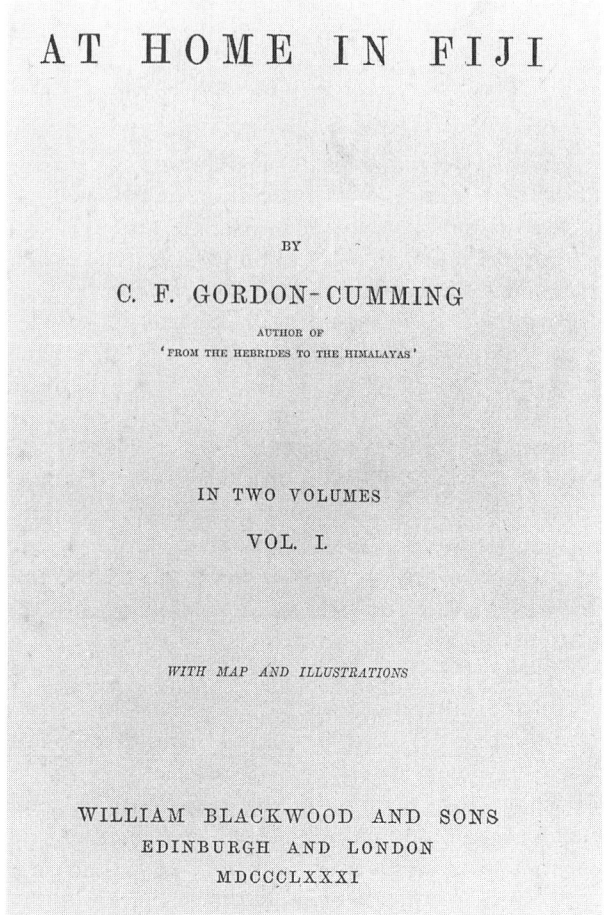

Frontispiece and title page for Constance Gordon Cumming's second book

more to Cumming's literary taste than Boswell and Johnson. She also honors Homer with an occasional allusion, such as describing dawn as "the first rosy flush."

Finally, *From the Hebrides* has a pilgrimage element. Deeply patriotic and proud of her Scots heritage, Cumming visits places associated with Prince Charlie and retells the Jacobite tale as found in James Browne's *A History of the Highlands and of the Highland Clans* (1840). In addition to a wealth of information about the Hebrides and Scotland in this book, the reader also meets an affable author who is willing to reveal her tastes and sensibilities.

Like several of Cumming's books, *From the Hebrides* ends abruptly. Once the tour is ended, with some extraneous family reminiscences included, Cumming gives a hasty summary of events that sent her abroad for the first time. Two weeks after the Hebrides tour ended, Cumming's young half sister Emilia, out in India with her husband, invited Cumming to visit them before they returned to England. Cumming says the proposal to see India as a *tourist* took her by surprise, but within a few hours of receiving the invitation she had begun preparing for the trip, and she set sail within a month.

The volume about the Indian journey begins as abruptly as the previous one ends, partly because the journey was so eventful as to detract from the larger story (it would later be developed into a separate book, *Via Cornwall to Egypt* [1885]). As a result, *In the Himalayas* opens without the arrival story found in many travel books, which produces a discomfiting sense of the story opening in medias res, one of several narrative problems arising from the author's almost complete reliance on letters and journals to compile her books.

In the Himalayas uses the chronology of the trip as organizing principle, beginning with Christmas in Calcutta and tracing the author's sojourn to Allahabad, Agra, Delhi, Simla, and the Himalayas and her return through Massourie, Hardwar, Benares, and Siva; the three-month trek in the Himalayas is the highlight of the narrative. Along the way Cumming stops frequently for sightseeing – and lecturing, one might say – for she again aims to be informative. To this end she does not trust merely her own observations but adds material gleaned from others, including residents and published experts on India. Aware that she is not covering new ground, she insists on sounding authoritative and thorough when describing Indian culture. At the same time, she reveals a sense of audience by maintaining a conversational style spiced with humor and human interest.

Cumming marks the inevitable rites of tourism with anecdotes about monkeys and magicians, snake tamers, yaks, elephant rides, varieties of local transportation, the natives, bazaars, sights off the beaten track, and a rope bridge crossing over a raging torrent. Broader topics of interest to Cumming include English life in India, colonial history, famous sights, religion, the caste system, and the flora and fauna of India.

The extent to which English life is replicated in India is a favorite theme. Cumming registers a certain condescension toward the luxury and comfort of colonial life, but when she comes to what she considers an outstation deep in the country, she is delighted to find an Englishman's home decorated in London fashion, complete with a piano. She disapproves of colonial manners and customs such as the "extreme simple law of chaperonage" – that is, it is essentially neglected – and the method of securing spouses by photograph and mail, a system to which she says she soon loses her English maid. Sounding a theme that recurs in many travel accounts by Victorian women, she complains most about the colonial version of English social habits, the accommodations to climate and conditions like midday calls, the evening drive, and "worst of all, the deadly-dull races."

While Cumming treats Anglo-Indians with disdain, she usually describes Indians with benevolence. Early in her visit she is bewildered and disgusted by colonial racism but also shows how quickly newcomers are indoctrinated. The ubiquity of male Indian servants soon teaches even Cumming's English maid to look on all men of color as if their reason for existence were to wait on white women. When an Englishman refuses to allow Cumming to sit next to an Indian lad, she unleashes a strong attack on English bigotry and is soon expressing embarrassment about the impoverished people who "minister so well to your luxuries," but who are paid so little that a day's wage for eight equals the price of a bottle of ale. A strong romantic element in her attitude toward the Indian is demonstrated in her equanimity regarding male nudity, which she excuses because of the Indian's lack of self-consciousness and his "beautiful silky brown colour." However, feminism rather than romanticism inspires her view of the low position of women in India. She shudders equally at old tales of the cruelties of the seraglio and the plight of present-day women relegated to small back rooms where "they live their dreary lives," unenlivened by an interest in the outer world.

Cumming went sightseeing to places of both English and Indian interest. The former allows her to recount items of colonial history in vivid detail, chiefly the Black Hole of Calcutta, the Well of Cawnpore, and events in the Mutiny of 1857, more than a decade earlier. Notable in her telling of these stories is a lack of vitriol. At times defending the Indians or even disputing tales of massacres, she appears to have a greater interest in sensationalism than in political correctness. Further evidence of this tendency lies in her frequent allusions to Charles Dickens and Sir Walter Scott, a family friend, and in the padding of her travel account with tales of wolf boys, human sacrifices, and similar events. Her descriptions of well-known sights such as the Taj Mahal and the Palace of Akbar often include not only tedious recitations of facts but also brilliant detail and entertaining stories such as that of the Indian queen reputed to have invented chess. Cumming, typical of informed tourists who chide their less informed comrades, complains that English colonials are uninterested in Indian culture while people in England lack knowledge of it.

Cumming is fascinated by Indian religious beliefs, and she takes pains to distinguish between major sects. As if to make clear that her own conventional religiosity is intact, however, she adds asides in which she calls legends about Krishna "grotesque mythological stories, such as the Oriental mind rejoices in." She singles out rituals and beliefs surrounding food for criticism, reflecting her own preoccupation with food. On the other hand, novel objects, such as the prayer wheels used by Tibetan Buddhists, catch the author's fancy. Fond of souvenir hunting, she describes with gusto her efforts to purchase prayer wheels from the devout, thus inadvertently betraying a naive insensitivity to local culture.

A subject that appeals to her typically Victorian interest in death and mourning is river burial in the sacred Ganges. This practice receives Cumming's intense interest both as a religious phenomenon and a health issue. She expresses sympathy for those too poor to afford cremation but horror at the results of their poverty: the carrion birds preying on

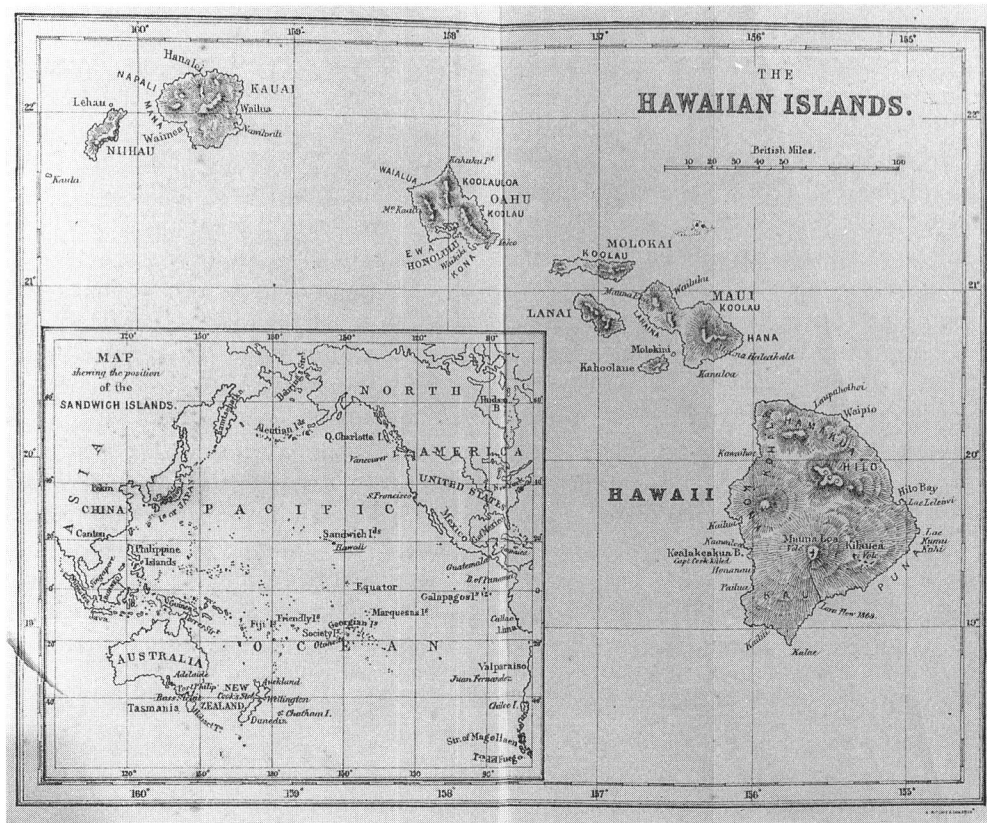

Map from Fire Fountains: The Kingdom of Hawaii, Its Volcanoes, and the History of Its Missions *(1883)*

the dead and the contamination of the water used for bathing and cooking. Because of her observations in India, Cumming became an advocate of cremation as a solution to England's overcrowded cemeteries. To this end, she includes in her book a long essay and notes arguing the value of cremation. Similar material is repeated in a later article as well in her book about China and in the Japan section of *Memories*.

Cleanliness, a motif in Victorian discourse, was essentially a class issue: an expression of middle-class fear of contamination by the lower classes. In the colonial world the desire for purity reflected racial as well as class prejudices. Cumming's preoccupation with the topic alternately supports and repudiates the views of her era regarding the poor and people of color, for she is disgusted by dirty, unkempt Yogis and fakirs and is afraid of vermin, but she also often remarks on the cleanliness of the poorest mud huts. Cumming shows her support for Indians by comparing them to Scottish peasants, and she finds no redeeming feature in the caste system. Caste and its association with religious sects such as Jainism compel her to applaud missionaries as bearers of enlightenment; she imagines that the Indians must be grateful for British rule that relieves them from oppressive custom.

After wending her way to Delhi, where her sister resided, Cumming spent several months visiting there and sightseeing in the Punjab. To escape the summer heat, Delhi matrons moved (as they do today) to Simla in the foothills of the Himalayas. With a strong aversion to heat, Cumming gladly formed part of the annual migration but felt frustrated with a repetition of Delhi social life, and she happily joined a small group for a three-month trek into the mountains. Her description of this episode offers a powerful comment on colonial life. In addition to thirty men – each was paid a mere six-pence a day to carry tents and supplies – the party was served by a cook, waiter, laundryman, water carrier, valet, gamekeeper, and groom. No provision appears to have been made for these servants, who slept on the ground around an open fire. Their means of existence "was a continuous mystery" to Cumming.

Making no apologies for her ample personage, Cumming, holding a white umbrella against sunstroke, insisted on being carried in a litter up steep and tortuous slopes. Her behavior seems to have been a source of difficulty, and some coolies mocked her as "little Miss Baby" and absconded rather than carry their burden. The other woman in the party walked every step of the way. That the Himalayan trek illustrates the limitations of Cumming's adventurousness was noted by contemporaries. An April 1876 *Blackwood's* review, "Mountaineering in the Himalayas," demonstrates a consciousness of Cumming's social rank by approaching her book with obsequious patronizing, but notes that her journey had been upstaged by longer, higher, and riskier journeys of both men and women. Cumming's petulance diminishes her admiration of the "Highlanders," especially the women in their colorful dress, as well as of the alternating forbidding and spectacular Himalayan scenery. However, she redeems herself by expressing concern about humanitarian and conservation issues: attacking English policy that supported opium poppy cultivation and asserting the need to preserve ancient forests.

The book ends almost as abruptly as it begins, with evidence of the traveler's weariness appearing in the final episodes. However, a warm English summer revived Cumming's spirits and enthusiasm for travel. A generation before Thomas Cook would organize guided tours to the Orient, Cumming proved that tourism of the sort the English had been enjoying in southern Europe for decades was possible in far-off colonies as well. That concept more than anything accounts for the appeal of her book on the Himalayas, which went through three editions.

During the next two years, Cumming paid many visits to relatives, including her first to Lady Middleton, a niece, at Applecross in the Highlands, where she met Isabella Lucy Bird, already well known for her book on travel in America. The two travelers formed a friendship that made it possible for Bird to read the proofs of *From the Hebrides to the Himalaya* while Cumming was in Fiji. First, though, Cumming went to Ceylon in 1872 at the invitation of old family friends. The book about that journey had to be postponed, since only four months after returning to Scotland in 1874 Cumming received an invitation to accompany Lady Gordon, the wife of the first governor of Fiji, on her journey to that island, which had just been made a crown colony. *At Home in Fiji* (1881) is about this third journey of 1875–1877. High-spirited, less crammed with information, and more personal than the earlier books, *At Home in Fiji* is an insider's picture of a newly formed colonial government that humanizes the institution, a fact that caused a reviewer in the *Nation* (1881) to fault the book, although it adds to the work's value today.

At Home in Fiji is a collection of letters to friends and relatives with all the problems of organization and benefits of style that such composition entails. Not only are there unnecessary repetitions, but political, economic, cultural, and personal material is presented piecemeal rather than in inclusive segments. In addition to these organizational problems, which plague her books from the first, Cumming had the misfortune of being frequently compared by contemporaries to the adventurous Bird, whose travels took her to some of the same places but whose books preceded and were more popular than Cumming's. An 1881 review in *Appleton's Journal*, for example, calls Cumming inferior to Bird both in matters of self-reliance and literary skill, then condescendingly recommends Cumming's book to the general reader as the best of four on Fiji published in 1881. The breadth and depth of Cumming's knowledge of Fiji along with the spontaneity of her writing seems to have appealed to readers, for the Fiji book went through multiple British and American editions.

The trip to Fiji began with a joke, for when asked about her next destination, Cumming had quipped, "Fiji," as being the most remote place imaginable. The journey, which included short stops in Paris, Marseilles, and Naples, ironically furnished Cumming with the only glimpse of the European continent she ever had, though she expresses no regret about that since the South Sea islands, rather than Europe, appealed to her sensibilities. It is not surprising then, that Australia, where the women and children in the governor's party stayed for several months while accommodations were arranged for them in Nasova, disappointed Cumming as being too English and devoid of Aborigines.

Once she was actually in Fiji, Cumming's good humor returned, nurtured by the esprit de corps often found among a small group of Europeans residing in a distant land. Although Fiji had been rapidly Christianized, with some nine hundred churches and fourteen hundred schools dotting the eighty inhabited islands in the archipelago, its colonial apparatus was just emerging. The capital consisted of little more than a single street with a strip of houses on one side. There were no carriages or horses, and Fijians resisted being turned into good English servants. After her first shock about the absence of European amenities, Cumming adapted to the relaxed atmosphere to the extent that she reports with remarkable equanimity the story of a Captain Olive, who had "gone native," and she gives a rather droll account at the expense of her countrymen on the preparation and use of *yangona*, the local stimulant drink.

Cumming makes good use of her enviable vantage point as a member of the governor's household. *At Home in Fiji* is punctuated with entertaining stories about the work of colonizing, such as training a local police force, planning a new capital, and waging military campaigns against the Fijians, who had sporadic returns to cannibalism. Cumming's discussions of trade, of the importation of foreign labor, and of land rights take an intelligent and sensitive approach to colonial politics. She is eloquent on the poverty and hardship endured by settlers, people often unprepared for either toil or the tropics. The Leefes, whom she visits on their island, Nananu, afford an example.

Cumming had romanticized a Mr. Leefe – who hailed from Cresswell, where Cumming spent several years of her childhood – as a sort of Robinson Crusoe. The life led by the Leefes disabuses Cumming of her fantasy. They labor incessantly raising Angora goats and coconuts, rarely leave their island, and have had but two visitors in ten years. Cumming depicts the isolation and dogged single-mindedness of the European settler as bereft of joy and ease. She cannot disguise her horror of this life. The island visit, like the trek into the Himalayas, brought out the worst in her: a disregard for the feelings of others and an obsession with her own comfort.

Cumming enjoyed a strong constitution that allowed her to nurse others who were ill, and she treats sudden illness and death, not infrequent visitors to the colonizers, with Dickensian sentimentality. This is especially the case in the death of the young wife of the attorney general, Mrs. Ricci, to whom Cumming had become attached. But the Fijians, too, became ill: Cumming reports with dismay a horrifying measles epidemic that killed one-third of the island's population. She lays blame for this catastrophe squarely on the British who bring the disease, but later she rejects Tahitian claims that Europeans introduced influenza to the island.

With characteristic energy Cumming availed herself of every opportunity to travel within Fiji, and as a result the largest portion of the book is devoted to the ethnological gleanings from these touristic activities. She depicts the Fijians as a fine, handsome people, and, despite the anomaly of their cannibal past, finds them naturally courteous and of gentle disposition. Some contemporary reviewers criticized Cumming for painting the Fijian character in too favorable a light, though they felt she gave an otherwise accurate picture of Fijian culture. Her enthusiastic descriptions of weddings and holiday festivities, and particularly of ceremonial dances, show her almost prepared to join in. Cumming's careful observations extend to Fijian topography, agriculture, and vegetation, and she often adds scientific detail in notes. Her Scottish childhood had acquainted her with amateur scientific activities such as fossil hunting, and these pursuits provided some preparation for her inquiries into forms of natural history in her travels.

As the title of her book indicates, Cumming adapted to Fiji well. Though her former fear of household pests had abated, she always traveled with her own mosquito net. She grew rather fond of the makeshift housing and staying overnight in Fijian homes, but she was less sanguine about the food. In her autobiography she writes that her school memories "seem to run largely to feeding!" and that in the Himalayas she was ecstatic at coming across a missionary couple who fed her freshly baked bread. In this light, her complaints about the difficulty of getting European food and a lack of trained cooks in Fiji are less those of a pampered aristocrat than of a gourmand.

At Home in Fiji abounds with so many cannibal stories and jokes that the reader wonders whether any Fijians looked on the plump Cumming with culinary interest. The cannibal motif in the book may be seen as an expression of sublimated sensuality. More than once Cumming delights in being alone with Fijians whom she imagines were probably cannibals only a year earlier. Such provocative flirtation with danger parallels the cannibal theme in some of H. Rider Haggard's novels, but Cumming, like fellow traveler Mary Kingsley, who lived among cannibals in West Africa, remained untouched, intact, pure. If Cumming's cannibal tales represent sensuality held at bay, her comments on clothing and bathing are candid expressions of indulgence. When the gentlemen tropicalize their attire, Cumming longed to do likewise, often noting the "so needlessly numerous" articles of dress European women wear. As with India, she never faults nudity or the undress of local people, but rather rhapsodizes on the beauty of glossy brown skin. Most telling, Cumming revels in her "enchanting" daily bath in

> a quiet pool on some exquisite stream, sometimes a clear babbling brook, just deep enough to lie down full

Frontispiece and title page for the book in which Cumming combined details of her 1868 trip from Scotland to India with episodes that occurred during later travels

length, beneath an overarching bower of great tree-ferns and young palm-fronds.... Of course we are not always burdened with bathing gowns.... sometimes I come on such irresistible pools when I am scrambling about alone, where the tall reedy grasses are so matted with large-leaved convolvuli, and not a sound is heard save the ripple of the stream.... I need scarcely say that our toilet on these expeditions is not very elaborate.

The sympathy for colonized peoples shown in Cumming's book on India becomes a regret over the destruction of culture and custom by the colonizers in Fiji. Proud of her own country, Cumming believes the Fijians are grateful to be delivered from misrule into the hands of an orderly British government, yet she sees the Europeanization of Fijians as a terrible violation. This dualistic view of colonialism is echoed in her comments on the double loss endured when works of traditional Fijian craftsmanship are sold to European collectors and then replaced by cheap imitations of British manufacture.

She then contributes to the problem, however, by joining in a competition with her friends to buy items of unique cultural value.

When it comes to the work of missionaries, Cumming's conventional religiosity and anthropological interests conflict strongly. Although she applauds the introduction of Western education and religion, she laments the destruction of local custom to the point of suggesting that Wesleyan missions go too far in forbidding traditional dances. At the same time she defends missionaries against criticisms leveled by other Europeans; in fact, her strong support of missionaries sets her apart from many Victorian travel writers.

Cumming's capacity for taking advantage of every opportunity to see new places is illustrated anew by her trip to New Zealand from Fiji. While Mrs. Gordon and the children took advantage of what urban comforts Auckland provided, Cumming set off on an energetic sightseeing tour to the mountains and volcanoes. New Zealand offered what she had missed in Sydney a year earlier: glimpses of indigenous people, the Maori, their customs and ceremonies. Typically, Cumming is precise and enthusiastic in describing all that she sees.

On 5 September 1877, only three weeks short of two years in Fiji, Cumming set off for Tonga, Samoa, and Tahiti, described in *A Lady's Cruise in a French Man-of-War* (1882). Cumming claims she hesitated to accept the unusual invitation to accompany Monsignor Elloi, a Roman Catholic bishop, on a tour of his Pacific island diocese, but the account she gives in *Memories* indicates that the scruples of Governor Gordon may have caused the hesitation. It was, perhaps, Cumming's most daring decision in all her travel, but the French bishop, the gentlemanly Captain Aube, the ship's officers, and the polyglot crew treated her royally. No doubt her good breeding and good French made her a congenial passenger. Cumming luxuriated in the three-thousand-mile cruise, pampered by an excellent chef, enjoying breakfasts that ended with a glass of chartreuse and a daily ration of good wine.

Like most of Cumming's books, *A Lady's Cruise* is in the form of journal letters to family and friends, with some historical and scientific information appended in notes. The travels of South Seas explorer Captain Cook are repeatedly invoked as a framework for and comparison with her own itinerary. Cumming's efforts at thoroughness and her willingness to use second-hand information even lead her to include long discussions of islands she did not visit, namely Fotuna and the Wallis Islands, which the bishop visited before his stop in Fiji; Easter Island; Java; and the Marquesas. Following her established pattern, the chronology of the trip is the organizing principle of the book so that facts, impressions, and the progress of the journey follow haphazardly – at times resulting in inadvertent comedy, as when a serious discussion of the archaeology of Polynesian pyramids is followed by a paragraph on pigeons.

Several themes appear insistently in these South Sea travels. One theme, not surprising given Cumming's religious bent and the fact that she was part of an inspection tour of missions, is the civilizing effect of missionaries on the "savages." Missionary tales often allow her to indulge her penchant for the grotesque and sensational in accounts of pre-Christian islanders and of missionaries' difficulties. A second theme is the corrupting influence of white traders whose irreligiosity and avarice undermine the efforts of the missionaries, a view that not all readers appreciated. The *London Quarterly Review* (1882), while giving *A Lady's Cruise* praise, cautions that Cumming's association with missionaries leads her to discredit Europeans unfairly. Local customs and the sensual detail of tropical delights compete with colonial politics as further themes in this travel book. The relatively short time Cumming spent on the island tours makes her reliant on secondhand stories without the moderating effect of familiarity with the peoples she describes; however, she does a service in collecting several myths and folktales from the peoples she visits.

Cumming's first impression of a new country is often negative, but as the length of her visit increases, so does her attachment to the place. Thus, her initial disappointment with Fiji changes to fondness during her two-year sojourn to the extent that Fiji becomes her touchstone for other Polynesian islands. Tonga, where she spends barely more than a week, seems flat and dull in comparison to Fiji, and Samoa, where she spends three weeks, seems disagreeably overrun by villainous Europeans. However, Alphonse Pinart, an archaeologist-scientist in the bishop's traveling party, moderates Cumming's negativism by encouraging her to take a more scientific approach to sightseeing. Thanks to his instructive presence, she comments knowledgeably on the curious geological and botanical features of the volcanic isles. Tongan archaeological sites inevitably lead her to compare them to similar sites, such as Stonehenge, while the distinct physical characteristics of the various Polynesian groups cause her to speculate on their origins, including advancing a bizarre theory about migration of peoples from North America.

Cumming's ire is incited against Europeans in Tonga for the over-Europeanization of dress and manners they had previously demanded of Tongans and in Samoa for the scramble for power among the Germans, French, and Americans. She sees the policies of the Tongan king George Tabou, who abolished laws enforcing European dress and established a Tongan parliament, as a rightful return to tradition. Cumming naively imagines that foreign disruption of a culture could somehow be controlled, that Polynesians could be Christianized yet continue to wear nothing but leaves, that they would receive Western education but be content with nothing but grass mats as furniture. The experience of Fiji misleads Cumming to believe in the possibility of Christianization without Europeanization, for custom remained strong in Fiji only because it had so recently adopted Christianity.

Without the slightest twinge of conscience, Cumming praises another Tongan policy, the use of prison labor for public works, so different from the pointless punishments she observed prisoners receive in Ceylon and so similar to the forced prison labor she observed in England. Perhaps in this respect, though, the Tongan motives were complicated by the fact that securing laborers for the industry and agriculture Europeans introduced was very difficult. A virtual slave trade, particularly of Chinese laborers, filled the need. In Tahiti Cumming was shocked by such practices, which she learned about because of the return of Gilbert Islanders after nine years as indentured laborers. Typically, Cumming responds with humanity and generosity to exploited people while neither surveying nor criticizing the larger issues from which specific instances of exploitation arise.

Most poignant in Cumming's Tongan visit is her account of the tiny French convent and its four ascetic nuns, so far and so long removed from their roots. The episode cries out for a candid discussion of the merits of self-sacrifice and missionary goals. Though Cumming cannot resist an occasional barb about Roman Catholic ceremonies, she displayed a broadmindedness about religion in her visits to the Catholic missions, often preferring to attend their Latin services rather than the Protestant ones in local languages. At times her religious views conflict with her ethnological interests, as when she ends a tale of mass conversion with regret that the result was a wholesale destruction of ancient idols and religious paraphernalia.

Bishop Elloi's Samoan stops, which included Pago Pago and Apia, gave Cumming an introduction to a culture that anthropologist Margaret Mead would explore in her *Coming of Age in Samoa* (1928) a half century later. It is tempting to compare these pre- and post-Freudian views of Samoa, Cumming's vaguely Rousseau-influenced approach with Mead's academic point of view. Yet Cumming in some ways sounds more modern, as when she characterizes Samoa as far more humane than pre-Christian Fiji, while Mead rather insistently uses "primitive" as a descriptor of Samoan culture. The shallowness of Cumming's expansive view of Dr. Peter Turner's South Seas College is exposed in Mead's pillorying of such training, but in at least one area Cumming anticipates Mead: a description of family structure with emphasis on the role of older siblings caring for younger ones.

Cumming attempts to make sense of the political situation in Samoa, the warring local factions and the intruding interests of various "utterly unprincipled whites." The scramble for dominion over Samoa that would continue for the rest of the century would also engage the interest of Robert Louis Stevenson, who died there in 1894. Cumming's sympathies with the Samoans and her eagerness to convey the villainy of the Germans and Americans results in entertaining though perhaps historically inaccurate sketches of, for example, a row between the British and Americans that drew in Captain Aube, who was called home in disgrace as a result. In contrast, the peaceful work of the London Mission Society leads Cumming to suggest that Samoa, like Fiji, would best be rescued by coming under British domination. Cumming's admission twice in her text that it is impossible to learn the truth about things in Samoa shows not only the limitations of the tourist but also the chaotic and perfidious nature of colonization. So disagreeable does she find the situation that she alters her travel plan and sails with the bishop to Tahiti rather than stay in Samoa to await a vessel by which she could return to Fiji.

Tahiti would entice Cumming to stay five months luxuriating in Elysium, as she puts it. She never faults Tahiti, a French protectorate since 1842 and a colony in 1880, for being far more Europeanized than Tonga or Samoa, perhaps because the well-established colonial life offers much comfort. Moreover, she was immediately embraced by the local aristocracy, the Salmon-Branders, to whom she discovered that she was distantly related. Becoming a personal guest of Mrs. Brander, who owned and personally oversaw a shipping fleet, Cumming was welcomed into the social circle of the popular governor and given a place in the entourage of the new Tahitian king and queen on a tour of their domain. Thus, family background and personal charm once again succeeded in giving Cumming a vantage point near the center of island life.

Cumming's artist's eye depicts the ease and beauty of life on Tahiti with images that correspond with those many other travelers recorded, from Louis-Antoine de Bougainville to Herman Melville to Paul Gauguin. Tahiti and Mooréa replaced Fiji as paradise in Cumming's imagination, to the extent that she often expresses regret at the prospect of returning to Britain. Tahiti is the only place that elicits this protest. Cumming admires the Tahitians even more than the Fijians for being gentle, affectionate, good-natured, and astonishingly beautiful. Their adoption of the sacque is perfect, and Cumming was soon walking barefoot and wearing the local costume. Climate and culture combined to relax her Protestant scruples so that she accepted

Tahitian-European alliances as a matter of course and adapted with good humor to loosely structured sleeping quarters. Her glorious alfresco baths in Fiji were supplanted by moonlight excursions to her personal pool. Food, always given a prominent place in Cumming's travels, is here described with Keatsian passion, especially fruits and roast suckling pig. Cumming's notes on the production of vanilla, breadfruit, and other valuable plants demonstrate attentive fieldwork. In short, she succeeds at making Tahiti thoroughly inviting.

Even Cumming's typical tributes to missionaries are mellowed by the Tahitian idyll, as she was lulled by the melodic *himenes,* part singing, she heard at every mission stop. Yet it is not quite paradise, as Cumming's distaste for the *Upa-upa,* the energetic, provocative national dance, signals. In spite of the warm hospitality she received from her French hosts, her English spleen was raised at French plans to dominate the Pacific by opening a canal across Panama (which the French never completed). Cumming also took a stern view of the French as colonists, finding them particularly wanting in the ability to establish a healthy island economy. Her brilliant description of the farming of the coral reef and an informative discussion of mother-of-pearl and bêche-de-mer, the sea slug deemed a delicacy in China for whose market it was gathered, display a good knowledge of colonial trade issues as well as an understanding of natural history. In sum, *A Lady's Cruise in a French Man-of-War* gives an attractive picture of Tahiti and its powerful appeal for the tourist.

After *A Lady's Cruise* Cumming dispensed with the chronology of her travel to publish her book about Hawaii. She sailed from Tahiti to California and did not visit Hawaii until the end of her years of travel, but because the books about Hawaii, Fiji, and the South Sea islands form a sort of trilogy with cross-references and comparisons, Cumming next published *Fire Fountains: The Kingdom of Hawaii* (1883). Her attention to thematic coherence is compromised, however, by the digressive introductory chapter, which chronicles an extraordinary event: a celebrated departure from Tokyo and rousing welcome in San Francisco in September 1879. The author was on the same ship as Ulysses S. Grant and his wife, who were returning from their triumphal world tour. Cumming basked in the reflected glory of her celebrated traveling companions and incidentally provides an admiring contemporary impression of Grant's fame and popularity in both Japan and California.

The rest of the first volume of *Fire Fountains* is devoted to the voyage to Hawaii, visiting in Honolulu and Hilo, and sightseeing among the volcanoes. The second volume begins with Cumming's departure and, echoing the ending of *From the Himalayas,* a caution about returning from warm places to Britain in winter. Other than that, the second volume has little personal experience and nothing original, being a recapitulation of Hawaiian history and mission history. Of note are Cumming's reprise of Captain Cook's attempt to subdue the Hawaiians, not without a certain macabre humor, and her horrific picture of the leper colony on Molokai, which she never saw. The book ends with an account by one of her missionary friends of an eruption of Mauna Loa in 1881, the year after she was in Hawaii.

Fire Fountains shows the toll of five years of continuous travel on the author. Cumming is often cranky and unenthusiastic. Although she had survived difficult, stormy voyages, the one between San Francisco and Honolulu is the only one about which she complains. Noisy children and a lack of facilities for the seasick result in the frequent use of the adjective *disgusting.* Travel between islands seems to have been equally infelicitous, and high winds and poor harbors make the island group seem barely approachable. Cumming's typical negative first impression of a new country was never stronger than in Hawaii, which she describes by repeatedly evoking images of hell. In the lava beds black desolation and fiery or dull red predominate. Mauna Kea's crater is a ghastly tableland; green mountain slopes lack form and interest; and vegetation is stunted and monotonous. Cumming is clearly unprepared for the untropical look of the leeward side of the islands. Although her descriptions gradually soften and change, Hawaii never approaches the beauties of Fiji and Tahiti. She even complains about the ubiquitous flower lei.

On the other hand, Cumming gives a vivid picture of the result of sixty years of colonial intrusion into the island, from the decimation of the original population to the development of Honolulu as an oasis in a parched land. For her, Hawaii represents the triumph of man over nature: what is nice is artificial, a result of reservoirs, irrigation, and artesian wells. A well-developed telephone system connecting all parts of the city is a further triumph over nature, as is the island economy that is based on plantations of nonindigenous fruits, rice, and sugarcane. Complementing her approval of human regulation of nature, Cumming portrays the Hawaiian past as a wanton disregard for nature and depicts natural man as subhuman. She is horrified at the massive slaughter of birds required to make the feather cloaks of ancient Hawaiian kings. The jocular tone with which she treated cannibalism in Fiji is here replaced by a view of it as demonic. In Tahiti, Cumming was indulgent about casual sexuality and relatively objective about infanticide, but she vehemently denounces such practices in precolonial Hawaii, declaring family life was nonexistent before the missionaries arrived. Even her old tolerance of nudity is muted.

The virtual elimination of all traces of original Hawaiian culture by missionaries, however, does not gain Cumming's approval. She bemoans the destruction of ethnological and antiquarian artifacts, even when she considers them, like the hula, repulsive. That such destruction is explained by the inability of one culture to value what is esteemed in another the author illustrates with both historical and personal examples, yet she does not apply these insights to contemporary Hawaii. In the 1870s white settlers represented about one-fourth of the population. Since Hawaiians, like the Fijians, were not inclined to become household servants or field laborers, plantation owners relied on imported labor, which in turn led to ethnic divisions within the newly constituted society. Siding with the white minority, Cumming reveals that her recent travel in China had not made her particularly sympathetic to its people, for in Hawaii she was contemptuous of Chinese who eschew field work for business or themselves become landowners.

There are, however, some things Cumming enjoyed in Hawaii. The American pioneer spirit as evidenced in the friendliness and independence of white settlers is a leitmotiv in her account. In fact, the only aspect of Hawaii that truly impressed Cumming was its pleasant, informal social life. Further, that white women did their own housework and girls saddled their own horses in the absence of servants charmed the aristocratic traveler. A Hawaiian activity Cumming honors with a five-page description full of fascination and admiration is surfing. Activities that invite comparison with other South Sea islands, such as poi preparation and "hog and dog" feasts, also receive much attention. The Hawaiian bathhouse with a clear stream running through it is, however, a feeble approximation of the alfresco bathing she enjoyed in Fiji and Tahiti.

Accurate and insightful on both the politics and production of sugar, Cumming foretells how the American economic colonization of Hawaii would lead to political hegemony. Far from criticizing this eventuality, Cumming portrays Hawaiian governmental institutions as absurd, with a bloated bureaucracy and a monarchy seemingly bent on self-satire. Cumming's comments on the declining native population and the short life span of the last five kings seem to predict the overthrow of the kingdom in 1893, an act for which the U.S. Senate formally apologized to native Hawaiians on 27 November 1993.

Almost against her will, Cumming left Oahu to see the craters and volcanoes on the big island, which ostensibly were the purpose of her travel. At first she behaved like a spoiled child, disappointed that Mauna Loa's head was in a cloud, and on a "seen one, seen them all" basis decided not to toil up to see the extinct crater of Mauna Kea. Rest, good food, and amiable companions altered her mood, however, and she made the difficult and dangerous trek to see the crater of Mauna Loa only to have an epiphany unlike any other noted in her travels. What at first seemed ugly, albeit colorful, lava spills gradually transformed into a vision of fomenting power. A crater quiescent one day suddenly burst into fiery rivers of molten lava the next. The rivers of fire were exciting, and Cumming dared to walk within ten feet of them. The strange chimneys and landscapes created by the action of the volcano captured the artist's imagination, and she settled down to sketch and paint, a dramatic illustration of the romantic belief in the power of nature as author of human creativity.

Once more Hawaii gave Cumming a stunning example of the power of nature when torrential rains turned a mountain stream into a seething cataract boiling madly into the sea a thousand feet below. So powerful a response did this evoke that the author did something she had never admitted to before: she forewent church to watch the river. Everything after the region of fire and flood is anticlimactic, though a later instance of tempestuous weather aroused the author's interest, and she advances a strange theory about controlling the sea, the subject of a separate journalistic piece titled "Oiling the Waves: A Safeguard in a Tempest" (1883). The first volume of *Fire Fountains* then dwindles into a series of unexceptional social visits to missions, settlers, and Hawaiian royalty. It ends with a chapter on such things as the sugar refining process, the introduction of the mongoose, and idiosyncrasies of the Hawaiian language that aims at providing information in an entertaining style.

Fire Fountains, like several of Cumming's books, had the misfortune of following the extremely popular account of Hawaii written by Isabella Lucy Bird. An *Athenaeum* review of Cumming's book (13 January 1883) begins by offering Bird extravagant praise, but it credits Cumming with more anthropological knowledge. Her book is judged temperate, judicious, and essentially "a good read." The Hawaii book was, however, doomed to a lesser place in Cumming's ouevre.

Fire Fountains was followed by *Granite Crags* (1884), Cumming's book about California, which she visited immediately after Tahiti. She had intended her visit to Tonga, Samoa, and Tahiti to end by joining Lady Gordon in Sydney at Christmas. Because of this plan she declined a cruise to the Marquesas, a decision she deeply regretted. Instead of returning to Fiji, she lingered in Tahiti until March 1878, when she took passage on a cargo ship laden with oranges for California, intending to disembark in Hawaii. Nature altered that plan. Unexpected winds and currents doubled the three-thousand-mile voyage, so instead of stopping at Honolulu, Cumming sailed to San Francisco, arriving at Easter. Hearing about the glories of the Yosemite Valley, she arranged a tour that turned into a four-month stay.

A compilation of letters, journals, and sketches Cumming did between April and August 1878, *Granite Crags* is a very uneven book, arguably the best example of both her strengths and weaknesses as a travel writer. The weaknesses are legion. Cumming was wholly unprepared for her experience in America – of course, she had intended to stop in Hawaii and came to California only by chance. Moreover, having just spent two and a half years in preindustrial Pacific regions, surrounded by a small, close circle of Europeans, San Francisco's cosmopolitanism as well as its frontier mentality took her by surprise. In addition, Cumming's typical credulity and fascination with sensationalism results in much disinformation in the book. The most glaring are outrageous statistics, such as her reports that California cabbages grow as high as a man, that pumpkins reach two hundred pounds even in a drought, and that a small farm is one of twenty thousand to forty thousand acres. Even if she did not suspect exaggeration and hyperbole, she should have used her own knowledge of botany and common sense to question the veracity of such data. Instead, smitten by the powerful images of bounty, she encourages Scotsmen to abandon their homeland and immigrate to California.

Nevertheless, *Granite Crags* offers a genial look at California one generation after the Gold Rush. Most interesting are the now almost forgotten details about the development of the Bay Area. Cumming's botanical interests were piqued by the method of turning barren sand dunes into the Golden Gate Park, then only six years old, while her warning about the danger in reclaiming parts of the bay from the sea seems destined to be repeated in discussions about potential earthquake catastrophe. Her descriptions of Oakland and other nearby cities, the ferry service across the bay, and American trains emphasize the rapid growth and metropolitan atmosphere of the area.

Cumming was highly impressed with American technology. In particular, San Francisco fire prevention and firefighting equipment receive glowing, detailed descriptions. A popular theme in the mid-Victorian period, the annihilation of distance as a result of new transportation and communication devices receives due attention from Cumming, who made her first use of a telephone on 26 April 1878, calling San Rafael. Among the marvels of modernity, however, some, such as the palacelike hotels with all their gadgetry, intimidated her, and the technology applied to funeral customs she describes as ludicrous. A century before Jessica Mitford's *The American Way of Death* (1963), Cumming writes a satiric set piece on American funeral practices: the expensive casket, embalming, and lifelike makeup of the corpse.

Native Americans provide material for Cumming's ethnological observations in *Granite Crags*, but with a marked difference from the way she treats indigenous peoples in the South Seas. In California she demonstrates the view she later attributes to a missionary in China: that it is easier to love the heathen in the abstract than in person. She romanticizes the solemn, picturesque red man in feathers and warpaint but gives a brutal description of the Digger Indians, in which dirt, the central image of Cumming's racial and class bias, figures prominently. The Indians' poverty repels her; their lack of English baffles her. On the other hand, she is sickened by the "pitiful history" of cruelties and reprisals exchanged by whites and Indians, her sympathies being with the Indians. Her oversimplified scenario of what should have been the alternative to white aggression fits with her religious views: the Indians, like the Fijians, should have been Christianized and thus pacified so that the land could have been taken over without bloodshed.

Cumming, who spends much of her time in America with a group of Englishmen, has little original to say about American manners and speech, but her newcomer's view yields a lively account of the culture – whether it is annoyance at being expected to carry her own bag, snobbery at the promiscuous use of the word *gentleman,* or being charmed by a youthful woman doctor. Her essay on

Illustration from In the Himalayas and on the Indian Plains *(1884)*

Middle English and Elizabethan roots of Americanisms, which is probably highly derivative, is entertaining and free of condescension about "correct" English.

On the topic of prejudice and frontier justice, she has a mixed understanding. Her criticism of the mistreatment of Chinese laborers gives an accurate and just account of that sordid element of Western expansionism and displays her usual sympathy for exploited peoples. On the other hand, she reveals her own prejudices in blaming "low Irish" demands for stopping Chinese immigration to California. More surprising, she praises the work of vigilante committees, reporting on tales of the rapid dispatching of criminals with great satisfaction.

By far the largest part of *Granite Crags* is devoted to trees and other natural wonders. Cumming, awed by the giant sequoias, repeatedly gives statistics on their size and age and is eloquent on the destruction of the forests by loggers, yet she takes the trouble to learn the particulars of logging to give balance to her views. Her utterances about the need for conservation, a theme to which she returns often, seem remarkably prescient. Clearly she was influenced by fellow Scotsman John Muir, whom she mentions frequently. So moved is she by the grandeur of Yosemite that she abandons her plan to sail to Hawaii in favor of spending the summer trekking in the mountains and sketching views of crags and falls. She carefully renders El Capitan, Bridal Veil Falls, the Sentinel, Cathedral Rock, Mirror Lake, Split Dome, and many other sights in word and picture. Accuracy was what she strived for in her art, so it is perhaps not surprising that she mentions Watkins's photographs of Yosemite with respect. Her book includes eight of the fifty drawings and watercolors she made of the area.

Before leaving the park she was gratified at being asked to have an exhibition of her work. She set up a makeshift gallery of borrowed bedsheets nailed to the outside of the house, pinning her sketches on the sheets. Locals as well as other tourists were enthusiastic about the show. Cumming's interest in the big trees of the Sierra Nevada was repaid with recognition of her as an artist of the American West. One of her paintings, *Father of the Grove* (Calaveras Grove), is in the Kahn Collection in the Oakland Museum.

Despite her "wilderness" experience, Cumming, like many other Victorian lady travelers, was at pains to cultivate a ladylike image – but with paradoxical results, often endangering rather than protecting her. In Cairo in 1868 she was thrilled, if embarrassed, to ride astride on an evening outing, alone with her Egyptian guide, but ten years later she insists on riding sidesaddle, even when warned about its treacherousness on steep mountains. Like a character in a novel of manners, she often complained that the least exertion was too fatiguing, and she was most disconcerted when, due to her man's hat and painting equipment, a rural woman insisted that she must be a man. Apparently some rough-and-ready frontiersmen were unimpressed with Cumming's stance, for more than one episode implies that she and her English gentleman companions in their tweeds and nailed boots were derided as greenhorns. At the same time, her contrariness about local custom appears an unconscious response to an admission she makes in her book about Hawaii, that as a child she was labeled a tomboy by her teachers.

Cumming seems to have enjoyed the rustic life she experienced in Yosemite Valley, but she was lonely. There are many passages about mail from home, and after her English friends left she took most of her meals alone. Perhaps that is why her conversations with Americans are treated as events to be recorded in detail. Fear also limited Cumming's experiences. She envied the many campers vacationing in the park, but the possibility of meeting a rattlesnake kept her in her cabin, and the only wild animal with which she became acquainted is the chipmunk. Stories about the gold rush indicate her love of vicarious adventure, though she appears happiest when contemplating a meadow of vibrant wildflowers. In sum, *Granite Crags* has a great deal of charm to which even its flaws contribute. Its effusive descriptions of Yosemite and giant sequoias, its active interest in a wide variety of people, and its revelation of a remarkable traveler all recommend it.

On 16 August 1878 Cumming sailed to Japan, reaching Yokohama on 6 September. After Japan was opened to foreigners in 1860, a flood of visitors and businessmen scoured the country in search of fame and fortune, and Japanism soon replaced chinoiserie as a Western fad. Because Cumming had been told that Japan as a subject was already hackneyed, she did not publish a book about this journey. However, her reticence may have been influenced by more-personal reasons. Her friend Isabella Bird, whom she chanced to meet in Tokyo, made a sensational trip to the North to see the aboriginal Ainu and published her account several years before Cumming began to write up her travels. Cumming's placid stay with wealthy friends could hardly compete with Bird's work, and Cumming waited until she wrote her autobiography in 1904 to give an account of her sojourn in Japan.

She did put her visit to good use in "*De mortuis*" (1883), an essay about cremation and Japanese funeral customs. As mentioned earlier, as a result of her travel in the East, Cumming became an outspoken supporter of cremation as a sanitary means of disposing of the dead. In *Memories* Cumming notes that the House of Commons debate on cremation in 1884 led to it being a recognized way of English burial by 1900.

Three more-lengthy books by Cumming would appear after *Granite Crags*. In the interim she published *Via Cornwall to Egypt,* ostensibly the story of her 1868 journey from Scotland to India, but experiences of subsequent trips along the same route are mingled with it. The book differs from Cumming's earlier ones in its relative compactness, in providing an index, and in including only one illustration, but it is similar in being a compilation of letters, journals, and magazine articles. The book draws together disparate material, from tales of shipwrecks to legends of King Arthur to sights in Gibraltar, Alexandria, Cairo, and Port Said. It includes a trip through the Suez Canal, which officially opened in 1869, to the Red Sea, as well as brief looks at Somalia and Aden. The initial trip to India carries the weight of the book both as a sensational sea story and as an immigrant tale replete with shipboard romance and the fears and idealism of those going out to the colonies for the first time.

The point of departure for the book is actually the end of the author's travels on 13 March 1880. In deep fog, only five miles from Anglesea, the boat in which Cumming was returning from many years abroad ran aground. The situation proved Cumming's mettle as few land experiences did, for she refused to get into the lifeboats, instead staying aboard to protect her paintings and journals amassed in distant lands and packing up the luggage of the passengers who abandoned the ship. The tale leads her to recount an astonishing number of sea disasters that she experienced, witnessed, or heard about. In each case human interest took precedence over the tempest, whether it be passenger behavior, coast guard incompetence, onshore looting of wrecks, or the frequency of loss of life. Incidentally, the chapter gives a glimpse into reckless shipbuilding in Cumming's era. But when the sea is calm and the vessel seaworthy, sailing is a magical link between past and future, the security of hearth and home and the adventure in exotic lands.

After the story of the shipwreck at the end of her travels, Cumming turns to her first journey abroad. The visit to Cornwall described in the second chapter of *Via Cornwall* was undertaken in 1868 to pass the time while the ship in which she had set out for India was damaged in high seas and had to return to Portsmouth for repairs. The unwonted delay afforded her the opportunity to see Portsmouth and Plymouth as well as the Cornwall associated with King Arthur. The reason for the untimely winter visit may explain the aura of unreality in Cumming's images of Cornwall. The attention to legend there, however, provides a thematic link to the later stages of the journey when Cumming recalls historic, mythic, and biblical associations with the part of the Mediterranean and Egypt through which she traveled. In this way too, Cumming's book gives a strong sense of travel as pilgrimage.

Cumming's tour through Egypt introduces several mid-Victorian issues, particularly the collision of vigorous economic and political expansionism with humanitarianism. She paints a macabre picture of the exploitation and greed of a rich nation taking what it needs from a poor nation in her account of the production of fertilizer – human remains in Egypt being composted to enrich Britain's agriculture. On other imperialist enterprises Cumming mixes pride in her nation's successes and horror at the excesses of military actions. On Gibraltar she takes a wry look at the tight military security of 1868, but she reports on the bombardment of Alexandria in 1882 with anguished excitement. She never questions the right of the British to protect the Suez Canal from falling under the control of the Khedive, is uninformed about the refusal of France and Italy to participate in the action, and gives a jumbled account of European military actions in Egypt and the Sudan. On the other hand, she sympathetically conveys the human dimension of these events.

Complementing the current-events material in this book is Cumming's studious attention to the past glory of the region. In describing its place as an ancient center of civilization and learning, Cumming displays familiarity with an impressive array of classical writers, many of whom she discusses at some length. In recounting Egypt's political history she focuses on several ancient water projects, such as the Mahmoudian Canal, while saying little about the building of the Suez Canal, just being completed at the time of her first journey through the region. Perhaps the engineering feat of the Frenchman Ferdinand de Lesseps rankles her, for she mentions him as a threat to British supremacy in the South Pacific in *A Lady's Cruise*. Cumming is more at home talking about religion, however, and Egypt presents the opportunity to delve into the history of Muhammadanism, the Coptic sect, and hagiographies.

Although Cumming is aware of the work of several Egyptologists and takes an interest in contemporary scientific discoveries, she can be mistaken. In particular, her theories about the Strait of Gibraltar being man-made and the pyramids being built as astronomical observatories for the magi are entertaining bits of Victorian ephemera, while her confusion about biblical geography – she places Abraham and Sarah in Egypt – reflects more piety than prudence. The area of knowledge more compatible to Cumming's abilities is ethnology, in which she effectively combines her skills of observation with humanitarian instincts. Her descriptions of fellahin and veiled women offer moving portraits of the particular oppression these people endured, and the account she gives of a dervish ceremony is a stunning picture of unusual religious devotion. Cumming's ethnological bent is also demonstrated in her broadmindedness about both race and religion.

Luxury Nile cruises would not become popular until the 1890s, but tourism in Cairo was well developed by the time Cumming visited. After describing the ritual donkey and camel rides to the pyramids, seeing churches and mosques, wandering in the bazaar, and souvenir shopping (she was forewarned about counterfeit artifacts), Cumming starts quoting William Makepeace Thackeray on hating sightseeing. She also commiserates with Muslim servants ministering to westerners insensitive to cultural differences and faults her countrymen who are unable to bend to the demands of being visitors, for "the rude British habit of despising everything foreign."

Via Cornwall to Egypt attracted little contemporary attention, but Cumming's next book, *Wanderings in China* (1886), was well received by both readers and reviewers. The book went through several editions and is her most significant work in terms of content and impact. Nearing the end of her travels, she had six years' experience in Asia to weigh against her observations in China. Beyond that, the mythic image of China in the West and England's involvement in the opium trade made the country a subject of enduring popular and political interest. Of all the places Cumming visited, the mysterious allure of China had the firmest hold on her psyche, a fact further demonstrated by her publication of five books about missionary work in that country.

Although the treaties ending the Opium Wars forced China to open its ports to trade, its cities to foreign residents, and its people to missionary activ-

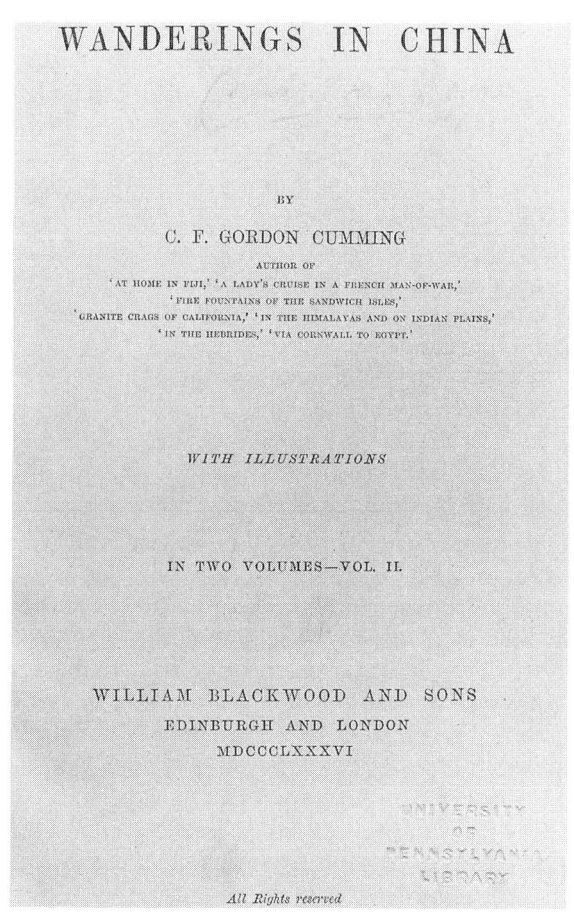

Frontispiece and title page for the book in which Cumming included discussions of the opium trade and missionaries in late-nineteenth-century China

ity, China's aura of impenetrability remained. In her six-month visit, Cumming got no nearer to authentic Chinese life than the secondhand tales of her various hosts. Moreover, her many letters of introduction meant that she rarely needed go beyond the environs of foreign enclaves. Even so, she compiled a wealth of information about China that gave her a good deal of credibility among her contemporaries. Just before her book came out, her old friend from Fiji, Henry Knollys, produced a book about China full of egregious errors and exaggerations, and Cumming's book was hailed by *The Athenaeum* (20 February 1886) as a welcome alternative.

Her accomplishment came in spite of the fact that she had not gone to China prepared for serious investigation; rather, she was seeking an escape from Japan's harsh winter. Finding the Shanghai of December 1878 cold and dirty, she sailed on to Hong Kong, arriving just in time to witness the Christmas fire that devastated a large part of the fledgling city of Victoria. She wrote a vivid account of the conflagration that typifies her privileged yet distant point of view in China. She spends the "livelong night . . . watching this appallingly magnificent scene" from the verandah of an Englishman's home on a hillside above the town. As fate would have it, Isabella Lucy Bird, Cumming's nemesis, also arrived that same night and managed to disembark from her ship and walk among the smoldering ruins, once again besting her friend and publishing her account in *The Golden Chersonese* (1883) several years before Cumming.

On the other hand, Cumming did not pretend to seek out spectacular or dangerous experience, being first and foremost a tourist and socialite who enjoyed parties and good food. Bird might have gone to Canton to inspect hospitals, but Cumming went to shop. The hubbub and the "Chineseness" of Canton enraptured her. The colonial social life she had found so dull in India years earlier she had come to enjoy, even coordinating her travel so as not to miss the horse races. She boasts about a languid houseboat trip on the River Min with her hostess Mrs. DeLano, supported by a staff of sixteen Chinese men. Like most Victorians in China, she relished an invitation to the home of a cultured mandarin, but when offered a chance to see ordinary Chinese homes, she judged it prudent to return to her English companions instead. In the company of missionary friends, she glimpsed the beautiful countryside and mountains but stayed close to the coastal cities with established European settlements – Amoy, Foo-chow, Shanghai, Ningpo, and Tientsin – then making a truly uncomfortable journey to Peking in a springless cart over the road apparently unrepaired for centuries.

The dismal picture Cumming gives of Shanghai, Tientsin, and Peking is understandable in that these cities have few natural beauties to recommend them. While the lack of drainage, the dust, and high walls may have lent little charm to Peking, it had its imperial monuments, lamaseries, and temples to interest the tourist. Cumming eventually tired of sightseeing, which, due to the communications of those times, often required a predawn departure. However, she made a coup by visiting the Temple of Heaven, then only occasionally opened to special tourists and never to women. As it happened, Ulysses S. Grant and his wife were in Peking, and Cumming and her missionary friend cunningly managed to be admitted to the temple in Grant's place. In spite of the confusion she caused, the Grants befriended her, and she later sailed with them to San Francisco from Japan.

Cumming's notes on the Temple of Heaven and the astronomical observatory on the city wall cover well-known information. She is less thorough about the ruins of Yuanmingyuan, the old summer palace. In 1860 the British and French, under the leadership of Lord James Elgin, sacked and burned the fabled palaces in the drive to open the interior of China and gain other concessions. The ruins elicit an expression of shame for the wanton destruction, though Cumming is silent on the political and economic motives behind it. Nor is she able to conjure up any sense of the humiliation the Chinese felt over the loss of this symbol of their cultural heritage. In fact, of all the people Cumming observes in her travel, the Chinese are the least substantial in her understanding yet ironically the only ones whom she attempts to help, which she does via missionaries.

Cumming's commentary perpetuates what had already become the stereotypical view of China – that infanticide was frequent, that necromancy and superstition pervaded daily life, that gambling was the national pastime, that ancestor worship was a prime cause of China's political stagnation, and that "everything in China is made to work by contraries." The difficulty Cumming had in valuing Chinese culture is exemplified by her seeing even its ancient achievements as evidence of contemporary decline as in this odd passage on the *Peking Gazette:* "This strange, stunted little gazette, which has thus survived seven centuries of dwarfed existence, is a characteristic example of many a Chinese institution, fairly commenced ere the rest of the world had emerged from barbarism, but then remaining spellbound, never developing."

Indulging the Victorian penchant for the grim and grotesque, Cumming reports tales of cannibalism, flood, and famine. She pays special attention to the condition of Chinese women and the painful practice of footbinding but betrays insensitivity by calling the "lotus foot" a "hoof." While Cumming's descriptions exoticize Chinese coiffure and dress, she frankly admits that the circumscribed life of wealthy women is suffocating; she prefers the company of men. She would like to identify with Chinese women, but her own independence makes her unable to do so.

On one of the most controversial topics of the time – the opium trade – Cumming speaks with force of conscience. She gives a thorough, albeit biased, summary of the debate over opium that waged throughout the nineteenth century, siding with missionaries who argued that the opium trade not only was immoral in itself but stymied their work. She vigorously excoriates as wrong the policy on which the opium trade, which was formally ended only in 1906, was based. However, her repugnance to the opium trade is compounded by the fact that it lies at the heart of the imperialist enterprise that she supports. To resolve this paradox, she decenters the opium trade, supports missionary work as productive of both religious and imperialist goals, and argues that China needs – perhaps deserves – to be thus conquered from within. This is accomplished by positing opium use as an us-and-them issue, first in allying it to degraded classes of people, then in fearing contamination – with America as the conduit – in Britain. Finally, after a lengthy discussion of the competition from new opium-producing countries in East Africa, Cumming seems to change sides, deeming the threat to trade the more serious of the problems opium presented.

Alongside the opium trade debate was a lower-keyed controversy over missionaries. In the effort to force foreign trade on China at any price, the Opium Wars had, in part, been rationalized in Britain as a means of Christianizing China. At home, the slow progress in gaining converts led to accusations of missionary lassitude, while in China the British connection to the opium trade complicated missionary work. Always supportive of missionaries, Cumming grew increasingly aware of the difficulties under which they labored, particularly of the physical dangers they faced. Her tales of missionary woe also reflect self-consciousness about numbers of converts, as in the case of the Baldwins, who labored for thirty years before having any measurable success. That the foreign mercantile community, as in Fiji, failed to support missionaries further exasperated Cumming. A stronger sense of irony might have enabled her to see the comic element in some of the squabbles between the Chinese, missionaries, and foreign merchants.

Her saccharine stories of missionaries and their converts often reveal a surprising credulity, but she was justifiably impressed with missionary work in orphanages and girls' schools, both neglected institutions in traditional Chinese culture. Medical missions also made an important contribution by introducing surgery and hygienic hospital care, though on this subject national pride more than good judgment influenced Cumming's views. Reflecting the attitude of her time, she rejected all Chinese medical practices as useless and later turned her interpretation of strange and disgusting medicines into a magazine article. While crediting opium rescue missions as a good cause, she barely hints at the ineffectiveness of the program they offer and never mentions the irony of the British vaunting a cure for an addiction they are largely responsible for causing.

Cumming's account of Rev. W. H. Murray's work among the blind is more satisfying. Murray combined his missionary zeal with a remarkable aptitude for language and the patient ingenuity of the Scottish workingman to invent a reading system for the blind. Cumming's conventional religiosity led her to believe the hand of Providence was involved in her meeting Reverend Murray, who had served as her guide in Peking. Once back in England, she began publicizing his work as a method of fund-raising for him, an effort redoubled after his presses were destroyed in an anti-Christian riot. She reproduces essentially the same biographical information in five books and pamphlets of different titles, and in *The Inventor of the Numeral-Type for China* (1898) she boasts of having raised £2,000 for Murray by 1887. Interestingly, her patronage of Murray's mission is the most frequently cited detail about her today. After publishing *Wanderings in China,* Cumming devoted much of the next five years to writing about Murray's mission. She lectured on missionary work and arranged several charitable exhibitions of her paintings and sketches.

Virtually all of Cumming's journalistic work, nearly fifty articles based on her travels, was published in the 1880s, the same period when most of her books were published. It was an extraordinarily productive decade for the author, who admitted that she published only out of sheer necessity — family resources having been lost to bad investments. Although she never left Britain after her return from abroad in 1880, she relied on her fame to give her an audience for one last book to produce the royalties she hoped would shore up her finances. She had not written about one of her journeys undertaken twenty years earlier, and this she took up in *Two Happy Years in Ceylon* (1892). The generous reviews of the book obscure the fact that it is her least substantial work.

What might have been an almost insuperable problem, the length of time between travel and publication, proved not to be one at all since Cumming simply resorted to her proven method of piecing together a book out of letters and journals written from the site. As a result, the spontaneity and fresh point of view are intact, and to compensate for time lag Cumming refers to correspondence with Anglo-Ceylonese friends over the years. The introductions and transitions written so long after the period of travel create both a sense of unreality and nostalgia in the book. The reader is constantly reminded of the contrast between contemporary times and the past. Ceylon "in those days" was paradise, a word appearing with great frequency, but the introduction of industry and political unrest alters the idyllic colony. The dramatic change in communications in the second half of the nineteenth century is another source of thematic contrast in the book. By reducing travel time and the exchange of letters to a matter of days or weeks rather than months, steamships and wire communications have destroyed the image of tropical fantasyland. The book's romantic images of brown-skinned men in outrigger canoes as much as the story of a leisurely two-year visit suggest an unrecoverable past.

The nostalgia of Cumming's Ceylon book stems both from the recognition of the end of an era and from fond memories. As a result, the book is something of a summation of her travel, for in retelling the story of her second sojourn abroad she constantly compares it with her later journeys. Cumming went to Ceylon in 1872 at the invitation of a family friend, Rev. Hugh W. Jermyn, who had been made bishop of Colombo, and she extended the visit by accepting invitations from new friends

Cumming in 1904 (photograph by Elliott & Fry)

in the colony. Both enjoyment of missionaries' hospitality and her own religious inclinations led Cumming to produce long, glowing accounts of church work in Ceylon. Typically, she seems to err on the side of generosity in this regard, making missionaries seem faultless bearers of enlightenment and civilization.

Conversely, Cumming portrays Buddhism as a foolish, benighted religion. Her first encounter with oriental religions in India had been educational, prompting her to study the philosophy and religious observances of several Hindu sects. In Ceylon, however, Cumming is more inclined to describe the external features of Buddhism, which leads almost inevitably to a comparison of Buddhism and Roman Catholicism, both of which she then unfavorably contrasts to Protestantism.

The author had sentimental connections to Ceylon since her brother John had attempted to develop a coconut plantation there before turning to professional hunting, dying two years before her visit. The twenty years of letters from John in Ceylon, William in India, and Roualeyn in southern Africa had incited Cumming's travel interest in general, but they also led to a paradoxical point of view. Her inclinations leaned toward conservation – wild flora and fauna were to be observed, enjoyed, sketched, and painted. The Ceylon book more than any of Cumming's other books abounds with enthusiastic, detailed descriptions of plants and animals; whole chapters are devoted to elephants and tropical flowers, for example. The family mythology, however, glorified the exploits of the three brothers, who made their mark as big-game hunters, and Cumming lovingly mentions the brothers who could kill thirty lions and leopards in a month. While never questioning hunting as a time-honored upper-class privilege, she does note that former great herds of animals were reduced almost to extinction in Ceylon and elsewhere.

On the whole, *Two Happy Years in Ceylon* is, as its title implies, light and cheerful, almost girlish in its effusive sensuousness when describing flowers and fruits, good food and beautiful scenery. Colombo, Batticaloa, and other cities and touristic sights are rendered with the verbal equivalency of a watercolor painting. Cumming climbed Adam's Peak, as she would later scale Fuji in Japan, proving that she was capable of the kind of physical exertion that she usually avoided. The trek and overnight stay were richly rewarded with feelings of natural piety. Contemporary reviewers took note of Cumming's book because of her fame rather than the intrinsic merit of the work. As to the content, Cumming would once again suffer an unflattering comparison to the daring exploits of Isabella Lucy Bird, whose *Journeys in Persia and Kurdistan* (1891) appeared the previous year.

Professional rivalry may have spurred Cumming to write her autobiography, *Memories,* in the year Bird died, though the book is as much a tribute to the Cumming clan as it is a personal story. It includes previously published material, a good deal of family history, and material omitted from the travel books but says little of her later life at Crieff, Scotland, where she spent her last forty-four years. *Memories* attracted little attention, for the lady traveler as a novelty represented a bygone era. After this last book Cumming retired into obscurity, dying on 4 September 1924.

That Cumming became a world traveler at all is in a sense because of chance, the series of invitations she received, rather than plans she conceived. However, once stirred into action as a result of opportunity others gave her, curiosity, a new global awareness, a thirst for new experience, and a genuine interest in other cultures spurred her on from country to country. These impulses intensified after she visited Fiji. For all that, she was not particularly bold as a traveler, keeping to already traveled routes and a gentlewoman's living standard to the degree possible in far-flung places.

The three hundred watercolors and hundreds of sketches Cumming produced during her travels show technical skill but unexceptional treatment of subject matter. A selection of these illustrate her books; others were exhibited as support for missionary work; and some were included in the Indian and Colonial Exhibition at South Kensington and subsequently in the Colonial Exhibition in Liverpool and Glasgow. Despite this recognition, in *Memories* Cumming reveals envy over the fame of another lady traveler-artist, Marianne North, whose vivid, realistic paintings of flowers are exhibited in Kew Gardens.

Cumming could, however, take heart in the recognition she received for her books. The most popular, such as *At Home in Fiji,* had dozens of reviews in Great Britain and the United States. Reviewers were almost universally kind though sometimes patronizing in saluting her panache and production of pleasing books for the general reader. They faulted her for unoriginality and lack of spunk. All the same, Cumming made a solid contribution to travel literature in her dispassionate, detailed representation of other cultures. She covers ethnographic subjects with professionalism. Her reliance on letters and journals creates organizational problems but gains by the spontaneity and enthusiasm of fresh experience. Her attitudes toward the imperialist effort in general and missionaries in particular reflect the spirit of her time, yet she consistently expresses nativist views. Finally, her views about travel for pleasure are thoroughly modern.

References:

W. H. D. Adams, *Celebrated Women Travellers* (London: Swan Sonnenschein, 1883), pp. 443-452;

Dea Birkett, *Spinsters Abroad: Victorian Lady Explorers* (Oxford & New York: Blackwell, 1982);

W. C. Blackie, "Lady Travellers," *Blackwood's,* 160 (July 1896): 49-66;

Frances Hays, *Women of the Day* (London: Chatto & Windus, 1885), p. 53;

Dorothy Middleton, *Victorian Lady Travellers* (London: Routledge & Kegan Paul, 1965).

Papers:

Several of Constance Gordon Cumming's letters and her diary of a visit to Fiji are in the National Library of Scotland. Correspondence of Cumming and her publisher, Chatto and Windus, are at the University of Reading.

W. H. Davies
(3 July 1871 – 26 September 1940)

Ira Grushow
Franklin and Marshall College

See also the Davies entry in *DLB 19: British Poets, 1880–1914.*

BOOKS: *The Soul's Destroyer and Other Poems* (London: Farmhouse, 1905; abridged edition, London: Alston Rivers, 1907);

New Poems (London: Elkin Mathews, 1907; revised edition, London: Cape, 1922; Boston: Humphreys, 1938);

Nature Poems and Others (London: Fifield, 1908; Boston: Humphreys, 1937?);

The Autobiography of a Super-Tramp (London: Fifield, 1908; New York: Knopf, 1917; enlarged edition, London: Fifield, 1920);

Beggars (London: Duckworth, 1909);

A Weak Woman (London: Duckworth, 1911);

The True Traveller (London: Duckworth, 1912);

Foliage (London: Elkin Mathews, 1913; Boston: Humphreys, 1938?; revised edition, London: Cape, 1922);

Collected Poems (London: Fifield, 1916; New York: Knopf, 1916);

A Poet's Pilgrimage (London: Melrose, 1918; New York: Cape & Smith, 1929);

Forty New Poems (London: Fifield, 1918);

Raptures (London: Beaumont, 1918);

True Travellers: A Tramp Opera (London: Cape, 1923; New York: Harcourt, Brace, 1923);

Collected Poems: Second Series (London: Cape, 1923; New York: Harper, 1923);

Later Days (London: Cape, 1925; New York: Doran, 1926);

The Adventures of Johnny Walker, Tramp (London: Cape, 1926);

The Song of Love (London: Cape, 1926);

Dancing Mad (London: Cape, 1927);

The Collected Poems of W. H. Davies (London: Cape, 1928; New York: Cape & Smith, 1929);

My Birds (London: Cape, 1933);

My Garden (London: Cape, 1933);

W. H. Davies (Hulton Getty Picture Collection Ltd.)

The Poems of W. H. Davies (London: Cape, 1934; New York: Oxford University Press, 1935);

Love Poems (London: Cape, 1935; New York: Oxford University Press, 1935);

The Loneliest Mountain and Other Poems (London: Cape, 1939);

The Poems of W. H. Davies (London: Cape, 1940);

The Essential W. H. Davies, edited by Brian Waters (London: Cape, 1951);

The Complete Poems of W. H. Davies (London: Cape, 1963; Middletown, Conn.: Wesleyan University Press, 1965);

Young Emma (London: Cape, 1980; New York: Braziller, 1981).

The reputation of W. H. Davies has undergone several revolutions since he first appeared on the literary scene at the beginning of the century. After several attempts at self-publication of his poems, Davies's writing was championed by Bernard Shaw, who contributed a preface and suggested the title for *The Autobiography of a Super-Tramp* (1908). Shaw's support brought Davies to public notice and thereby assisted him in achieving a less precarious livelihood and a wider audience for his poetry. As an artless and simple poet of commonplace nature, particularly coming after the decadent excesses of the fin de siècle, Davies was a welcome fixture of the Georgian literary scene. Proud of his naturalized citizenship in the republic of letters and with an almost awkward assumed gentility, Davies for the most part avoided any mention in his poetry of his colorful raffish adventures both in America and England as a tramp and a beggar. Yet even at the height of his poetic acclaim he continued to mine those experiences – grown dimmer from temporal distance and acquired domesticity – in his prose. The publication of *Young Emma* in 1980, long after Davies's death, reaffirmed his power to charm readers with his ingenuousness and to repel them with his vulgarity. The book presents a lightly fictionalized record of Davies's courtship of and marriage to Helen Payne, a young girl he had found on the streets of London.

William Henry Davies was born on 3 July 1871 in Newport, Monmouthshire, the fifth of the six children of Francis Davies and Mary Ann Evans, three of whom died in infancy; the fourth, his elder brother Francis, was feeble-minded. To compound the Dickensian misery of their childhood, the three survivors (William's sister Matilda was born in 1874) lost their father to tuberculosis when William was only three and saw their mother remarry within a year, leaving them to be raised by their paternal grandparents, who ran a pub in Newport. Growing up in a pub cannot have been too depressing for a spirited lad; in *The Autobiography of a Super-Tramp* Davies recalls his grandfather, a master mariner, as a colorful eccentric and his grandmother, though related to Henry Irving, as a somewhat puritanical woman with little use for reading and none for the stage. Among his childhood exploits Davies recounts being the leader of a gang of schoolboy shoplifters and being soundly caned when caught. The episode also brought his formal education to an end.

Davies had been introduced to the joys of literature by a friend and had conceived for himself a career as a writer such as George Gordon, Lord Byron, or Percy Bysshe Shelley. That the initial appeal may have been more to adventure than artistry may be deduced from his statement in *A Poet's Pilgrimage* (1918): "When I was between thirteen and fourteen I wanted to be a man of great literary genius; but when I was between seventeen and eighteen my ambition was to rob the rich by force and kill Indians for sport." Davies's attraction to literature, however, largely the product of self-education, never dissipated; in later life he continued to study even under the bleakest of circumstances, and his devotion to the world of letters manifested itself in such fanciful dreams as that of setting up a bookshop in London or of peddling his poems from door to door.

The death of his grandfather at the age of seventy-six in 1885 made it imperative to determine a more practical future for Davies, and he was bound as an apprentice to a picture framer and gilder. He was restive under his indenture, spending his evenings in writing or drinking, neither of which was conducive to a conscientious application to his trade. When his apprenticeship ended in 1891, he went on a spree in London that introduced him to a world of independence – and hunger. His grandmother's death in 1893 liberated him simultaneously from familial restraint and mindless dissipation: by the provision of her will she left each of her three grandchildren a weekly income of ten shillings to be distributed by her trustee. This providential stipulation assured that, whatever else might befall him, Davies was secure against absolute destitution.

After a brief attempt at living in his mother's home with her new family, Davies obtained an advance of fifteen pounds from the trustee and proceeded to Liverpool on his way to seek his fortune in America. In the *Autobiography* he confesses that

> I at once went to the steerage cabin and wrote a full description of the country, that very first evening aboard; telling of my arrival in America, and the difference between the old and the new world. This letter was given to the steward at Queenstown [i.e., Cobh, in Ireland], and was written to save me the trouble of writing on my arrival, so that I might have more time to enjoy myself. Several years elapsed before it occurred to me how foolish and thoughtless I had been. The postmark itself would prove that I had not landed in America, and they would also receive the letter several days before it would be due from those distant shores. I can certainly not boast a large amount of common sense.

The quotation is typical of Davies's autobiographical manner: after recording with a swagger an act of youthful bravado, he mitigates the heedlessness with a later sober reflection on its folly. Arriving in America in the midst of a depression, he had difficulty in finding steady employment, though there is little evidence that he sought it with much assiduity. He easily fell into the life of a tramp, particularly under the tutelage of a consummate beggar named Brum, who regarded any labor for pay or barter as beneath the dignity of his calling. Davies's wandering seems remarkably free of purpose, motivated neither by the spiritual quest of the knight-errant nor the nonconformity of the romantic picaresque rogue nor the outright rebelliousness of the beatnik storing up experiences for subsequent literary use. Like Brum, Davies came to regard "the travelling life" not as an unfortunate temporary expedient but as a profession. Tramping all over the eastern states, he occasionally had to serve time for vagrancy – a circumstance sometimes welcome, for he often designed his confinements to take place in winter at the most commodious jails.

Davies never quite arrived at Brum's perfection of the craft because he would temporarily stoop to labor as a fruit picker or canal worker, often dissipating what he saved from his wages over several weeks in a single evening's drinking binge. On several occasions he tramped his way back to Baltimore, which he regarded as his home port, and signed on as a worker on cattle boats. Such employment paid little more than his passage to Liverpool. After a short time in England he again grew restless and returned to America, where he found the life of a beggar far less demeaning and more rewarding. As he wrote in *Beggars* (1909):

> The American tramp sits comfortably at his camp fire, waiting the sound of a shrill whistle, or bell, which proclaims dinner-time. Five or ten minutes after hearing this sound, he dances out of camp, humming a tune, and goes begging as though he were going to a wedding, and he is often served with a hot dinner before the man of business can finish his own. But the poor English beggar makes funeral steps between meals, and asks for a mouthful of bread. The American beggar pulls the door bell, and makes himself heard the first time; but the English beggar timidly uses his knuckles on the back door, many times before he is heard. The American beggar rides on trains from town to town, but the English beggar tramps the hard roads. The English beggar explains his wants to the servants and children, but the American beggar asks to see the mistress. The American beggar, feeling himself a proud and free citizen, addresses himself familiarly to anyone; but the English beggar, feeling himself a despised outcast, will not speak except in want, or when he is first spoken to.

Back in America, Davies continued his wanderings along the Mississippi, his account of life in Arkansas being one of the highlights of the *Autobiography*. After a return to Europe in 1898 (a reason for his frequent returns was to collect on his legacy, which kept mounting during his absences) he conceived the idea of setting himself up as a secondhand bookseller, but this dream soon yielded to a more adventurous one – that of participating in the Klondike gold rush. In Ontario on his way west, Davies stumbled while trying to hop onto a moving train, an accident that crushed his foot, ultimately necessitating the amputation of his leg above the knee. Davies returned to Newport, his career both as a manual laborer and a tramp effectively terminated.

Davies thus found himself backed into a literary career, and for the next few years he lived in cheap lodging houses and Salvation Army shelters, peddling and begging during the day while writing poems and plays assiduously at night. After trying with little success to sell his poems from door to door, he managed in 1905 to solicit sales by mail in order to scrape together the nineteen pounds required to publish his first book of poems, *The Soul's Destroyer and Other Poems* (the title refers to alcohol), in an edition of 250 copies. A favorable notice in the *Daily Mail* stimulated fashionable purchases of the book, and Davies soon found himself championed by Edward Thomas and Edward Garnett. Their patronage enabled him to live a more settled life and to concentrate on his writing, but it was Shaw's promotional efforts on behalf of *The Autobiography of a Super-Tramp* that secured Davies's celebrity. Although Davies contended that fame in England "did not pay as well as begging in America," the income from his poetry, augmented by that from his sensational prose and supplemented by a civil list pension and support from the Royal Literary Fund, enabled him to spend the rest of his life in modest circumstances as a practicing poet. He died on 26 September 1940, two years after suffering a stroke, his constitution finally succumbing to the physical hardships he had undergone in his youth.

The reader who comes to Davies's travel books expecting a florid evocation of scenery or penetrating inquiry into the motivation of exotic characters is bound to be disappointed. Davies entertained no such lofty goals in recollecting his adventures as a vagabond, regarding his prose works largely as potboilers. For though the dream of becoming a poet was an enduring one for Davies, his exalted conception of literature made him chary of using his experiences as a tramp and a beggar as

Title page for Davies's best-known work

substance for composition. Understandably, therefore, the writings about his adventures are hardly deliberate reflections composed in tranquillity from journal notes and sketches of evanescent impressions. Rather they are anecdotal, arranged with some regard for chronology and thematic organization but with little apparent purpose beyond that of mere recollection. In the words of his biographer Richard J. Stonesifer,

> There is little in Davies's *Autobiography* that we can call thoughtful. He goes through a succession of adventures and he records them; he sees a variety of things and he reacts to them, simply, directly, often charmingly. But there is little evocation of background, little creation of mood. If he tells us things about the fields, if he takes us into the world of the doss-house to meet his colourful associates, he hardly ever tells us anything but the most obvious facts.

But by a broader conception of travel literature, the genre concerns itself less with geography or even cultural description than with reporting experiences of the unfamiliar. In Davies's prose the words *traveler* and *travel* do not convey what they do in that of Johann Wolfgang von Goethe or Henry James; rather they are euphemisms for *vagrant* and *vagrancy*. A vagrant may have an interesting tale to tell – often with the purpose of cadging money from his auditor – but it will rarely aspire to artistic merit. Davies's saving virtue as a compiler of autobiographical reminiscence lies precisely in his refusal to be genteel and to submit to the temptation of reading moral consequence into his adventures and anecdotes. His stories about himself or others are not intended to reveal character or to reconcile readers to his unconventional way of life as a young man. It is fairly difficult, after all, for the sophisticated reader to enter into any imaginative sympathy with professional vagabonds. Rather, Davies becomes the recorder – not the spokesman – of the subculture of beggars. The reader learns first and foremost that though men (and occasionally women) may have been led into a traveling life through unique and private circumstances, they share common traits and attitudes. While Davies does note some characteristics that distinguish beggars of different nationalities and observes that different locales provide different opportunities for them, he finds in his travels a pervasive similarity among them.

Perhaps the most surprising characteristic in the subculture of beggars to the lay reader is one that suffuses virtually all of Davies's reminiscences – namely, the pride that members feel in their profession. Beggary is not seen as a depressed condition, a reduced and straitened economic state, nor one to be described in negative terms, such as homelessness. The beggar enjoys sleeping in a bed as much as the respectable citizen; he merely does not insist that it be the same one each night. Moreover, the communal hobo campfire in summer is superior to the pathetic attempts of the bourgeoisie to approximate the outdoor life by camping, about which the only favorable thing that can be said is that they provide benefits to the tramps in the gear and food that are left behind. And in winter a comfortable jail cell, well heated and with ample meals, may be had at public expense.

Professional pride is the source of tension in many of the little dramas that Davies recollects, and the attainment of a geographical sobriquet, such as Cincinnati Slim or the Baltimore Kid, attests to the possessor's degree of proficiency. Around the campfire or in the lodging house each newcomer is queried and scrutinized, his daily performance sifted and status accorded on the basis of productivity and ingenuity. Davies records that on one occasion when he had the extraordinary good fortune early in the day to apply to a housewife who furnished him not only with an abundance of food but also of clothing, he rearranged the bounty into several parcels before returning to camp lest he appear to his mates as the recipient of random luck rather than as an assiduous and accomplished beggar. Though the beggar, particularly in England to avoid the penalties of vagrancy, may resort to peddling or even to singing in public places, protocol demands that the wares be virtually useless and the voice deliberately tuneless, mere pretexts for solicitation.

In much of his writing Davies takes great pains to discriminate the "downrighter," that is the true beggar, "a man that neither carries things to sell nor goes into a workhouse, but can beg the price of his bed every day," from the more occasional types: "navvies," "stiffs," and "narks." With a rather heavy-handed irony, he labels as enemies the workhouse stiff (who will stoop to entering that institution rather than begging the price of his bed at a lodging house) and the Salvation Army, which by sweeping through a residential area in its solicitation would take the bread out of the poor beggar's mouth.

Davies's recollections of his tramping life, therefore, are less a matter of geography or personal encounter than of general description of the economy of beggary – of the daily shifts and expedients demanded to attain success in the profession. Readers are told how the beggar's life is one of continuous foraging, his only recreation a drunken binge after a particularly successful day's "work." It is a solitary occupation, offering little opportunity for social activity. It is perilous as well. Subject to the debilitations of hunger, cold, and disease, the beggar enjoys no social safety net and can look forward to no retirement. Moreover, unlike any other practitioner of a trade or skill, he cannot confidently depend upon professional courtesy and must therefore be equally wary of the motives of his fellow travelers as well as of those whose charity he is seeking and of those who would impose it upon him. Lacking the systematic inquiry of a sociologist, Davies nevertheless is able to detail the etiquette of the doss house, where each man tends to his own meals and provides for his own security and where privileges may be had by bribing the nark, a species of resident tramp who serves as a virtual concierge and volunteer manager. No less impressive are his depictions of the hobo camp, the closest approximation available to beggars of a community, where potluck prevails, useful tips may be exchanged, and itinerant friendships are struck up.

Such society as exists among beggars is overwhelmingly male. Although Davies remarks the occasional presence in lodging houses of couples of men and women, the solitary nature of beggarly solicitation precludes working in pairs. The female equivalent that corresponds more truly to the male beggar is the prostitute. Although Davies is fairly reticent about his encounters with prostitutes in the *Autobiography*, he is much more expansive about the subject in the subsequent books that deal with his life as a beggar in England. Of course, he assumes no moral superiority to the prostitute; indeed, he recognizes a kinship in their status of tolerated disreputability. He is quick to note that the sporting house has often been for him a more efficient and bountiful responder to his needs for food and shelter than the loudly proclaimed charitable organizations, and he records some adventures with streetwalkers and courtesans, some of them at second hand, with affection and nostalgia even in instances where he has been set up or swindled. He recog-

Davies and his wife, Helen, in 1935

nizes that the span of the prostitute's working career is much shorter than that of the male beggar and that the hazards of her profession, both from clients and protectors, are far more dangerous than anything experienced by men.

However sympathetic Davies may be to members of a subculture cast out or voluntarily estranged from the prevailing bourgeois society, his toleration of otherness has distinct limits. In America he readily assimilated the attitudes of the dominant white majority: in his description of a lynching he witnessed in Tennessee he is an unmoved objective observer of an assembly of armed men who storm a jail and drag a black man from his cell with a rope around his neck. Any reservations he might have had about the proceedings are dissolved by the unmanly display of the wretch about to be hanged: "That this prisoner should have been so brutal and unfeeling in inflicting pain on another, and should now show so much cowardice in anticipation of receiving punishment inadequate to his offence, dried in me the milk of human kindness, and banished my first thoughts, which had been to escape this horrible scene without witnessing its end."

Davies stays to the end, his description of the event becoming virtually that of a sacred ritual. Though he is robbed in America by gangs of both whites and blacks, he characterizes black men as "certainly born thieves," adducing as evidence their nefarious proclivity to use the razor both as a tool of larceny and as a weapon of combat, all of which appears to vindicate in his mind the administration of rough justice in dealing with them. It may be some palliation of Davies's deplorable summary generalizations to observe that they seem to originate in personal encounter and that his travels were hardly conducive to mingling with the best elements of society. While, like most of his white contemporaries, he was easily persuaded of the racial inferiority of black people, he was certainly aware that environment, not just race or class status, played a considerable part in development. He finds, for example, that the backwardness of the inhabitants of Arkansas, "the last place in the world to recover from the Great Flood," is attributable in large measure to an isolated existence in swamps subsisting on a diet of cornbread and razorback hogs. Even the most presentable among them is tainted, as one anecdote reveals:

> Another time I met a native of this same state of Arkansas, who was well-dressed, and seemed to be more intelligent than others. In the course of conversation he asked me where I came from, and on being told that I came from England, he said, "You are a long way from home." The man certainly spoke with more culture than I had expected, and it filled me with astonishment when he requested me, in English better than my own, to say something to him in my own language for his amusement.

It is in his sketches of his fellow tramps, both the downrighters and down-and-outs, that Davies mainly excels, from the portrait of Brum, his American mentor, to that of Sullivan, who parlayed his lurid confession of sins at a mission house into marriage with a rich Christian widow. The London doss houses were particularly rich in eccentrics, such as the "finder" whose eyes were ever directed downward to the pavement and whose dodge was to return lost, abandoned, or stolen articles (on one occasion an empty envelope) to their owners for an expected gratuity. He had a fixed income so that he was not wholly dependent upon his scavenging for a livelihood; moreover, in a rare fit of confidence he revealed to Davies the contents of his locker, which included books, jewelry, handkerchiefs, and even the ferrules of umbrellas. The miscellaneous congregation of the doss house was a rich source of observation for Davies, who notes that it is not the absence of shelter but the absence of family that is the greater affliction for these inmates, one that is particularly felt as they grow sick and infirm: "All these poor invalids in common lodging-houses are under the impression that doctors, when they find that their patients have no friends, and cannot be thoroughly cured, kill them. That is why they are so stubborn, and fight till they cannot move, before they will enter a hospital." Davies found that even after he had abandoned the traveling life he could not wholly efface the memory of its peculiar conviviality. In *Beggars* he records a dream he had of a symposium of his erstwhile mendicant colleagues, whom he longs to join. In the dream the room changes into a cell, and he awakens to find himself in a cheap furnished cottage as a respectable but unhappy citizen.

The Autobiography of a Super-Tramp is Davies's best-known prose work, covering his life from childhood through his escapades as a tramp in America in the 1890s; his return to England only to be overcome by restlessness; his ill-fated attempt to join the Klondike gold rush, in which he lost his leg; his abandonment of the rugged tramping life for a beggary less itinerant and strenuous while attempting to establish a literary career for himself; and his final success in gaining recognition as a poet. It is an absorbing and fascinating memoir. It is less, however, an autobiography in any reflective or apologetic sense than a record of the colorful adventures of a vagabond arranged more or less chronologically. The artlessness of the narration and the utter absence of any progression other than that of time constitute at once the charm and the limitation of the work. The reader is enchanted, like a child listening to his grandfather tell of his youth, marveling that the kindly old man with the wooden leg could have had such an exciting life. There is little, however, to detain the more contemplative reader. Compared to George Orwell's *Down and Out in Paris and London* (1933), for example, it does not offer much by way of analysis or interpretation. As a testament to human endurance the *Autobiography* earns high marks, but the triumph is severely mitigated when we consider that the tribulations were largely and wantonly self-imposed. Moreover, the traveler seems so bent on getting from place to place that except in rare instances he neglects to impart impressions of the country he is passing through, an omission all the more striking in a poet of nature.

Those impressions, distilled many times over, became the substance of Davies's poetry. His prose confines itself to observations on men and manners, but chronological narrative was not the most effective matrix for the kind of insights Davies had into the subculture of beggars – the essay was. Clearly attempting to duplicate the commercial success of the *Autobiography*, Davies published two volumes of essays dealing with the life of itinerant mendicants both in America and England: *Beggars* and *The True Traveller* (1912). In these books Davies includes material that had not found its way into the earlier volume – condensed anecdotes and secondhand stories combined to illustrate some general truth, a lexicon of beggars' argot, little disquisitions on the reception of beggars in different communities or a diatribe against the institution of spring cleaning, and characterizations of groups of people such as lodging-house types and "boy desperadoes." This material is distinguished by being observed and commented upon, not necessarily participated in.

More detached and dry in tone than the *Autobiography*, these books never came close to repeating its success. They quickly went out of print, and copies are difficult to find today. Apparently Davies thought that their failure was owing to the absence of the first-person narrator, for he later conflated the two into *The Adventures of Johnny Walker, Tramp* (1926) as the middle volume in a trilogy of autobiographies concluding with *Later Days* (1925). As he explains in the foreword, "In doing this book, *Johnny Walker, Tramp*, I have used the experiences selected in *Beggars* and *The True Traveller*, but I have destroyed the essay-form, and made the book run as a story. This has been done successfully, I hope, without injury to the material contained in the two earlier books."

The destruction of the essay form is not as total as Davies wished – passages are lifted and transposed with virtually no alteration – and little narrative strength is gained by identifying the companion of *Beggars* as Brum in *Johnny Walker, Tramp* or for the sake of consistency transforming the Baltimore Kid to Boston Shorty. There is a distinct loss, however, in the disappearance of the chapters describing the lives of prostitutes, probably left out to preserve the reputation of an established poet, and in the omission of a few revealing anecdotes of Davies's childhood summoned up casually and associatively. In other words, Davies weakened the effectiveness of his material by his insistence on casting it into the form of personal adventure. He was no more successful in attempting dramatic form: *True Travellers: A Tramp Opera* (1923) is about as authentic a portrayal of underworld life as John Gay's *The Beggar's Opera* (1728), but it is mawkishly conceived and without a trace of the burlesque that enlivens Gay's masterpiece. Ironically, Davies's most successful narrative is the posthumously published *Young Emma* (the manuscript was kept under lock and key by his publisher for more than fifty years), his most candid work, in which the tension between his mudlark morality and genteel aspirations is most apparent.

Davies never regarded his prose writings as anything other than potboilers whose sensational content, in addition to providing him with substantial income, might secure a greater readership for his poetry. Although the *Autobiography* scarcely mentions his struggles as a creative writer beyond those of economics, for the 1920 edition of that popular work he appended five poems from *The Soul's Destroyer*. He was proud of the resilience he had shown in his youth but felt that there was little connection between his adventures as a tramp and his literary career. In this he was surely mistaken. His poetry exists today as a historical curiosity, a reminder of the taste of a former age. His prose, however, for all its ineptness, actually invokes that past age, not merely its posturing and aspiration. Davies is no modern François Villon, but his *Autobiography* will surely survive as a document of a time when romance still adhered to the life of the vagabond.

Biographies:

Osbert Sitwell, *Noble Essences* (Boston: Little, Brown, 1950);

Richard J. Stonesifer, *W. H. Davies: A Critical Biography* (London: Cape, 1963).

References:

William Cooke, "Alms and the Supertramp: Nineteen Unpublished Letters from W. H. Davies to Edward Thomas," *Anglo-Welsh Review*, 70 (1982): 34-59;

Thomas Moult, *W. H. Davies* (London: Butterworth, 1934).

Florence Douglas Dixie
(24 May 1857 – 7 November 1905)

Catherine Barnes Stevenson
University of Hartford

BOOKS: *Abel Avenged: A Dramatic Tragedy* (London: Moxon, 1877);

Across Patagonia (London: Bentley, 1880);

An Address to the Tenant Farmers and People of Ireland with Advice and Warning (Dublin, 1882);

A Defence of Zululand and Its King. Echoes from the Blue Books. With an Appendix Containing Correspondence on the Subject of the Release of Cetshwayo (London: Chatto & Windus, 1882);

In the Land of Misfortune (London: Bentley, 1882);

Ireland and Her Shadow (Dublin: Sealey, Bryers & Walker, 1882);

Waifs and Strays; or, the Pilgrimage of a Bohemian Abroad (London: Griffith, 1884);

Redeemed in Blood (London: Henry, 1889);

The Young Castaways; or, the Child Hunters of Patagonia (London: Shaw, 1889); republished as *The Two Castaways; or, Adventures in Patagonia* (New York: Dutton, 1890);

Aniwee; or, the Warrior Queen: A Tale of the Araucanian Indians and the Mythical Trauco People (London: Henry, 1890);

Gloriana; or, the Revolution of 1900 (London: Henry, 1890; New York: Standard, 1892);

Woman's Position, and the Objects of the Women's Franchise League (Dundee: Leng, 1891);

The Horror of Sport, Humanitarian League, Cruelties of Civilization, 2 (London: Reeves, 1895);

Little Cherie: Or, the Trainer's Daughter. A Racing and Social Novel (London: Treherne, 1901);

Songs of a Child and Other Poems by "Darling," 2 parts (London: Leadenhall Press, 1901-1902);

Isola; or, the Disinherited: A Revolt for Woman and All the Disinherited (London: Leadenhall Press, 1902);

The Story of Ijain, or the Evolution of a Mind (London: Leadenhall Press, 1903);

Toward Freedom; an Appeal to Thoughtful Men and Women (London: Watts, 1904);

Izra: A Child of Solitude (London: Long, 1906).

SELECTED PERIODICAL PUBLICATIONS – UNCOLLECTED: "Affums: A True Story," *Vanity Fair,* 26 (7 December 1881): 4-6;

"The Case of Ireland," *Vanity Fair,* 27 (27 May 1882): 301;

"Cetshwayo and Zululand," *Nineteenth Century,* 12 (August 1882): 303-312;

"Woman's Mission," *Vanity Fair,* 32 (16 August 1884): 114-115;

"Memories of a Great Lone Land," *Westminster Review,* 139 (March 1893): 247-256;

"The True Science of Living," *Westminster Review,* 150 (October 1898): 463-470;

"President Roosevelt's Gospel of Doom," *Weekly Times and Echoes,* 18 April 1903, pp. 1-10.

As a travel writer, novelist, and political and social activist, Florence Douglas Dixie used her writing to expose and to remedy the injustices in her world. Born into wealth and privilege, she rebelled against her lot by transforming herself into a passionate advocate of oppressed populations – women, the Irish, and the Zulu, as well as animals. The writer of her obituary in the London *Times* (8 November 1905) deemed her views "somewhat peculiar"; other contemporaries saw her as a courageous reformer and social visionary.

Florence Caroline Douglas and her twin brother were born on 24 May 1857, the youngest of six children born to Archibald William Douglas, the seventh Marquis of Queensbury, and Caroline Margaret Clayton Douglas, the daughter of Gen. Sir William Robert Clayton. An athletic child who joined her brothers in vigorous outdoor exercise, Douglas experienced turbulent family instability during her first seven years. Her father accidentally shot himself in 1860 while cleaning a gun; her eighteen-year-old brother died in 1865 in the first English attempt to scale the Matterhorn; and her mother was nearly deprived of custody of her children after she converted to Ca-

Frontispiece and title page for Florence Douglas Dixie's first travel book

tholicism in 1864. Caroline Douglas's flight to France with the twins is recorded in her daughter's autobiographical novel, *The Story of Ijain, or the Evolution of a Mind* (1903). With the exception of a short and traumatic stay in an English convent school, Douglas lived a nomadic existence in Europe from age seven to thirteen. Her childhood exposure to the legal disempowerment of women, foreign travel, and strong physical exercise laid the groundwork for the development of a hardy woman traveler and writer of strong feminist sympathies.

Between the ages of ten and seventeen Douglas began writing poems that were later published under the pseudonym "Darling" as *Songs of a Child and Other Poems* (1901–1902). Most of these are derivative of Romantic poetry – Wordsworthian meditations on nature and heroic ballads about the Douglas clan imitative of Sir Walter Scott – but some, such as "Poverty's Bitter Cry" and "Be Kind to the Erring Child," reveal a young mind challenging conventional pieties and looking for serious answers to important philosophical and social questions.

Rebellious independence of mind, Byronic self-glorification, and imitativeness of form mark Douglas's next two works. *Waifs and Strays; or, the Pilgrimage of a Bohemian Abroad* (1884), written in 1870 when she was thirteen, was clearly modeled on George Gordon, Lord Byron's *Childe Harold's Pilgrimage* (1812, 1816). A blank-verse drama Douglas wrote at fifteen, *Abel Avenged: A Dramatic Tragedy* (1877), depicts Cain as a heroic rebel against God, that "cruel tyrant" who ordained that mankind should be ignorant and sinful.

Despite her literary rebelliousness and her adolescent flouting of convention, Douglas did what was expected of women of her class: in 1875, at the age of eighteen, she wed the twenty-four-year-old Sir Alexander Beaumont Churchill Dixie, who was from a distinguished Leicestershire family. Two sons quickly followed: George Douglas Dixie (born in 1876) and Edward Wolston Beaumont Dixie (born in 1878). The couple had no other children. More than twenty years later, in an article titled "President Roosevelt's Gospel of Doom," which appeared in the *Weekly Times and Echoes* (18 April 1903), Dixie decried the "immorality of ruining women's health and making them untrue to themselves by forcing them to bear many children."

Clearly women's rights were beginning to occupy a central place in the mind of this young wife and mother. Soon after her marriage Dixie began writing a blank-verse tragedy, *Isola; or, the Disinherited. A Revolt for Woman and All the Disinherited* (1902). Isola, the play's intelligent, unconventional, and physically powerful heroine, rebels against her arranged marriage, demands equality with her husband, and then flees to join a band of insurgents fighting on behalf of her husband's illegitimate son. Her death ushers in a new era, which guarantees religious freedom, social justice, and equality, as well as women's control over their own bodies.

As *Isola* indicates, Dixie's restless, iconoclastic energy could not be contained by domestic obligations or released in conventional social interactions. She proclaims at the beginning of her first travel book, *Across Patagonia* (1880), that, "weary of the shallow artificiality of modern existence," she sought "a more vigorous emotion" than could be experienced in England. A six-month jaunt to Patagonia in the company of her husband, two of her brothers, and J. Beerbohm, a naturalist and travel writer, provided the panacea she sought. In this "outlandish" and desolate place she could adopt the stance of the adventurer, who explores "vast wilds, virgin as yet to the foot of man."

Across Patagonia is conventional in its organization: initial chapters present the motive for the journey, a catalogue of necessary equipment, emotions felt on parting, and an abbreviated account of shipboard life. Then the narrative unfolds in a strictly chronological fashion, tracing the travelers' progress from England to the "audacious" landscape of Rio, to the "desolate and dreary" stretches of the Patagonian desert, which seem "hardly of this world," and then to the wild and spectacular cordilleras of the Andes deep in Patagonia. The descriptions of stampedes, wildfires, hunts for exotic animals, and the extreme physical hardship of a six-hundred-mile trip on horseback make this narrative a lively and readable adventure.

Although readers are told little about the male travelers or their guides, they learn quite a bit about the unconventional, energetic, competent, physically hardy narrator, who shoots, rides, drinks with her male companions, sleeps in the open air with her saddle for a pillow, bathes in icy streams, and endures severe injuries with insouciance. After the group's packhorses stampede, leaving the party stranded without food and four days' walking distance from civilization, Dixie even acts the role of a heroine by single-handedly finding the horses and, riding bareback with her scarf serving as a bridle, driving them back to camp. She exemplifies what Mary Louise Pratt in *Imperial Eyes* calls "a sentimental hero whose travels assume the form of an epic series of trials, challenges, and encounters with the unpredictable."

In this hearty, buoyant book the voice of the defiant, socially concerned author of the early plays and poems is muted. Only in the poetic evocations of landscape, in the narrator's struggles with her feelings about hunting, and in her ambivalence about the natives are there intimations of her larger social and political views. At its best, Dixie's narrative captures the emotional tenor of the wild Patagonian landscape: the eeriness of its ancient forests, with charred giant trees lying like "weird and ghastly ... skeletons of a bygone age," or the unsettling effect created by shifting patterns of moonlight and heightened by the "mysterious wail of the grebe ... like the voice of an unquiet spirit."

The narrator's own spirit is unquiet about hunting: the excitement of the chase thrills her, yet, sensitive from childhood to the feelings of animals, Dixie is also repelled by it. Although she and her party kill chiefly for food and not for sport, she confesses to being haunted by the slaughter of innocent animals. The contest between her love for the aristocratic sport of hunting and her conscience was eventually resolved in favor of the latter: in the 1880s Dixie joined the Humanitarian League and published a pamphlet titled *The Horror of Sport* (1895), indicting hunters for inflicting physical and psychological pain on animals and asserting that hunting is barbarous.

Surprisingly, the author of *Isola* displays less interest in and sympathy for the people of Patagonia than she does for animals: the history, customs, and social conditions of the inhabitants of Rio or of the Patagonian desert evoke little comment. Initially she is content to make superficial judgments about "the natives," based solely on physical characteristics and rumor. Thus, she concludes that the inhabitants

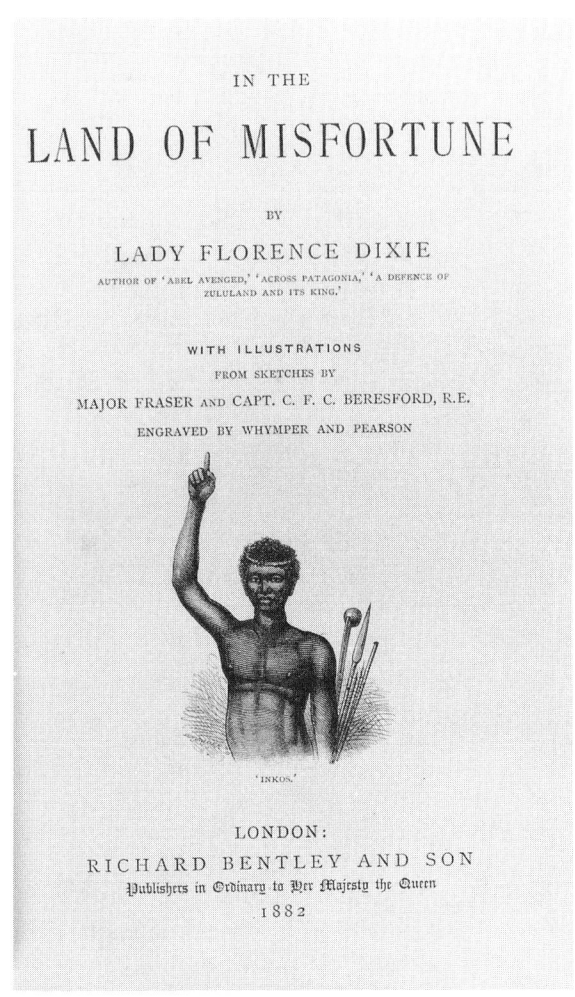

Frontispiece and title page for Dixie's book about South Africa, which includes her criticism of Britain's destruction of the Zulu nation

of Bahia de Todos los Santos must be happy because they are plump and chatter "like so many monkeys."

The natives of the desert interest her slightly more, however, and she records their physical characteristics, dress, and behavior. Although annoyed by their unceremonious intrusions into her camp and their pilfering, Dixie is outraged by the way the natives are cheated by white people, and she is horrified by their self-destructive behavior. The easygoing Tehuelch people are destroying themselves, she asserts, by a growing addiction to "aqua diente" (liquor). She predicts that "they will become nothing more than a pack of impoverished, dirty, thieving ragamuffins," yet she never enters sympathetically into the economic situations or colonial politics that might be producing this situation. In only a year's time, however, she was to become a public advocate of another oppressed population, the Zulus of South Africa.

Within a year of her return from Patagonia, Dixie was already planning another escape from England, but this time she longed for a more serious and professional outlet than that offered by travel for pleasure. History provided her with just the opportunity she was seeking. In 1880 the tensions between the British and the Afrikaner settlers in southern Africa finally erupted into a war. Envisioning herself as the Florence Nightingale of the First Anglo-Boer War, Dixie determined to go to South Africa; less conventionally, she also secured a commission from *The Morning Post* as its special correspondent. Arriving just as the war ended, Dixie never nursed a soldier, but she did travel for nine months, all the while writing columns for the newspaper about her experiences and her political opinions.

In November 1882 her second travel book, *In the Land of Misfortune,* appeared. As its title suggests, this work – which incorporates some of the material from her newspaper columns – is not simply an entertaining narrative of hard traveling in the colonies; it is also a political manifesto designed to educate its readers about two "misfortunes" for which Britain has been responsible in South Africa: the destruction of the Zulu nation and the surrender of the Transvaal to Afrikaner control as a result of the Pretoria Convention. Indeed, the narrative structure indicates the book's political intent: it begins and ends with vividly dramatic interviews with the deposed, imprisoned Zulu king, Cetshwayo, so as to make him seem like the presiding spirit of the whole journey. In fact, before her travel book even appeared Dixie had written a series of nine meticulously detailed articles attacking British treatment of the Zulus and urging the restoration of Cetshwayo to his throne. These were published first in *Vanity Fair* (10 December 1881–4 February 1882) and then collected and published as *A Defence of Zululand and Its King. Echoes from the Blue Books* (1882).

In the Land of Misfortune interweaves with its travel narrative commentary on that other misfortune of South Africa, the British conduct of the First Anglo-Boer War. The account of a journey from Cape Town to Natal, through the Transvaal, and then to Zululand includes several pilgrimages to battlefields. In these episodes the narrator engages in verbal landscape-painting while also celebrating British bravery, tragically wasted because of the government's hasty capitulation to the Boers. Dixie peppers her tale of arduous travels across the Transvaal with examples of the British settlers' disgust with and contempt for a government that seemed to have betrayed them. With no little British chauvinism, she also records her disgust at the physical appearance, manners, and racial politics of the Boers, the seeming victors in the conflict.

Though highly political, *In the Land of Misfortune* is also an exciting narrative of difficult travel, fraught with physical hardship and some danger, set against the backdrop of exotic and stirring scenery. The London *Athenaeum* of 25 November 1882 praised the book's immediacy and the "unconventional freedom" of its style, which enables a reader "to picture to himself more vividly than he could otherwise do what it is like to live in the veldt." For the narrator, the process of the journey and the imaginative associations that places evoke are more important than are scientific data about South Africa's climate, geology, flora, or fauna. When she looks at a glorious natural scene, she responds to its poetic effects. When she looks at a Zulu, she sees a man or woman who has been wrongly used, not the subject of scientific or scholarly study. Although committed to the reunification of the Zulu nation under Cetshwayo, Dixie provides little or no ethnographic data about these people. Equally unscientific is her approach to the exotic animals of Africa, which she sees with a hunter's, not a zoologist's, eyes. The other creatures she encounters on her travels – dogs, horses, trek oxen, mules – are lovingly and sympathetically described, and their rights are defended against human beings who would abuse them.

As in Dixie's earlier travel book it is the narrator of *In the Land of Misfortune* who compels interest: a vigorous, cool, competent woman who easily paces her male companions in their riding, hunting, or climbing and occasionally even bests them in dangerous situations. The woman who enjoyed challenging traditional ideas about her gender creates a comic subplot in her travel narrative in which her husband is depicted as frequently asleep and unwilling to be awakened to face the challenges of the moment. The reviewer in *Vanity Fair* (25 November 1882) praised the book and ironically joined in the domestic joke, calling the book "a lively account of personal adventures and discomforts, which there is some internal evidence to show were shared by Lady Florence's husband."

After the return from South Africa, the Dixies' travels ceased for medical and financial reasons. Florence turned her attention to writing and campaigning for causes such as the return of the Zulu king to his throne, Irish Home Rule, animal rights, and woman suffrage. She began producing a series of novels characterized by melodramatic plots, heroic action for noble causes, advanced social ideas, and one-dimensional characters. *Redeemed in Blood* (1889) is a sprawling story about two generations of Scottish aristocrats who end up in Patagonia for some highly contrived action sequences. *Gloriana; or, the Revolution of 1900* (1890) uses the form of a dream vision to depict a world in which sexual equality is achieved through the efforts of the protagonist, Gloriana, who, disguised as a man, becomes prime minister of England. It has achieved some present-day celebrity as an example of feminist utopian writing at the end of the century. Dixie's South American experiences laid the foundation for two other novels, adventure stories for children: *The Young Castaways; or, the Child Hunters of Patagonia* (1889) and *Aniwee; or, the Warrior Queen: A Tale of the*

Araucanian Indians and the Mythical Trauco People (1890).

Dixie's final autobiographical works, *The Story of Ijain, or The Evolution of a Mind* and *Izra: A Child of Solitude* (1906), recount her struggles from childhood onward against the narrow social, religious, and sexual codes of her time. These loosely structured episodic works encompass travel writing, utopian fantasy, and polemics on the oppression of women and animals and on contemporary politics. Dixie died suddenly of diphtheria on 7 November 1905 before any of her social visions had been realized.

Energetic and iconoclastic, Florence Douglas Dixie devoted her moderate literary talent to exposing injustice and to creating imaginary visions of a better world. In her writings about travels she developed an unconventional persona, eager to engage in adventure and to champion the cause of oppressed groups.

References:

Nan Bowman Albinski, "The Law of Justice, of Nature, and of Right: Victorian Feminist Utopias," in *Feminism, Utopia, and Narrative,* edited by Libby Falk Jones and Sarah Webster Goodwin (Knoxville: University of Tennessee Press, 1990), pp. 50-68;

Mary Louise Pratt, *Imperial Eyes: Travel Writing and Transculturation* (London & New York: Routledge, 1992);

Brian Roberts, *Ladies in the Veld* (London: John Murray, 1965);

Catherine Stevenson, *Victorian Women Travel Writers in Africa* (Boston: G. K. Hall, 1982), pp. 41-87.

Charles M. Doughty
(19 August 1843 - 20 January 1926)

Stephen E. Tabachnick
University of Oklahoma

See also the Doughty entries in *DLB 19: British Poets, 1880-1914* and *DLB 57: Victorian Prose Writers After 1867.*

BOOKS: *On the Jöstedal-Brae Glaciers in Norway* (London: Edward Stanford, 1866);

Documents épigraphiques recueillis dans le nord de l'Arabie, edited by Ernest Renan (Paris: Académie des Inscriptions et Belles Lettres, 1884);

Travels in Arabia Deserta, 2 volumes (Cambridge: Cambridge University Press, 1888; London & Boston: Cape & Warner, 1921); abridged by Edward Garnett as *Wanderings in Arabia,* 2 volumes (London: Duckworth, 1908; New York: Scribners, 1908);

Under Arms (Westminster: Constable, 1900);

The Dawn in Britain, 6 volumes (London: Duckworth, 1906);

Adam Cast Forth (Sacred Drama in Five Songs) (London: Duckworth, 1908);

The Cliffs (A Drama of the Time, in Five Parts) (London: Duckworth, 1909);

The Clouds (London: Duckworth, 1912);

The Titans (Subdued to the Service of Man) (London: Duckworth, 1916);

Mansoul; or, the Riddle of the World (London: Selwyn & Blount, 1920; revised, London: Cape, 1923);

Hogarth's "Arabia" (London: Chiswick, 1922).

SELECTED PERIODICAL PUBLICATIONS – UNCOLLECTED: "Die Sinai-Halbinsel," *Mittelungen der Kaiserlich und Königlichen Geographischen Gesellschaft in Wien,* 19 (1876): 268-272;

Report on His Arabian Explorations, *Globus,* 37 (1880): 201; 39 (1881): 7, 23; 40 (1881): 38; 41 (1882): 214, 249;

"Travels in North-Western Arabia and Nejd," *Proceedings of the Royal Geographical Society,* 6 (1884): 382-399;

Charles M. Doughty

Review of David G. Hogarth's *Travels in Arabia Deserta, Observer,* 19 March 1922, pp. 13-14.

Charles M. Doughty ranks as the most important British explorer of Arabia and one of the greatest travel writers of all time. His difficult solo journey in the northern Arabian peninsula from 10 November 1876 to 2 August 1878 is recorded in the classic *Travels in Arabia Deserta* (1888). This massive, encyclopedic work, some twelve hundred pages in length, remains not only the definitive Western account of Arabia in the nineteenth century but also a masterpiece of Victorian literary nonfiction, equal in quality to the writings of Thomas Carlyle, Walter Pater, and John Ruskin.

Travels in Arabia Deserta directly influenced another great book of Arabian travels, T. E. Lawrence's *Seven Pillars of Wisdom* (1926), and Lawrence has perhaps summed up Doughty's importance better than anyone else. In the introduction to the 1921 edition of *Arabia Deserta* (which Lawrence was instrumental in getting published) he wrote, "To have accomplished such a journey would have been achievement enough for the ordinary man. Mr. Doughty was not content till he had made the book justify the journey as much as the journey justified the book, and in the double power, to go and to write, he will not soon find his rival."

Doughty was an obsessive geologist, archaeologist, and linguistic antiquarian. He believed that understanding Arabia's topography and culture would help him understand the world and his place in it and that collecting large amounts of data might enable him to find the answers to many of the puzzles of life. Therefore, *Arabia Deserta* includes detailed information about northern Arabian towns and villages; the Wadi Hamd and Wadi er-Rumma watercourses; geology; Arabian antiquities; Arabic linguistic mannerisms; the customs and economics of Bedouin life; the biography and methods of governance of Mohammed Ibn Rashid, the ruler of northern Arabia; attitudes toward the West; religious beliefs and practices; jinni tales and superstitions; horse and camel husbandry; and even the postal system. Doughty's poetic talent and interest in language enabled him to present vast amounts of information and the story of his interaction with the Arabs with utmost authenticity and vividness in a style uniquely fashioned for that purpose.

At first glance Doughty the poet, antiquarian, devout Christian, and nationalistic Englishman seems an unlikely Arabian traveler, but closer inspection of his life reveals how, without having Arabian exploration as an early purpose, he developed exactly the mental qualities and education that would ideally suit him for this task. His most important personal trait was a consistent ability to succeed in spite or even because of personal problems and disabilities. Charles Montagu Doughty was born on 19 August 1843 at Theberton Hall, Sussex, England, to the Reverend Charles Montagu Doughty, the descendant of a conservative landowning family, and his wife, Frederica (née Hotham), of an adventurous family that favored naval careers. As a baby he was so frail that his father had him baptized soon after birth, although later in life he was "tall and strongly, though not heavily built, with aquiline features and a thick beard which was reddish in early and middle life," according to W. D. Hogarth, the son of Doughty's biographer, David G. Hogarth.

Doughty's mother died a few months after his birth, his father when he was six. Doughty and his older brother grew up in the care of an uncle, Frederick Goodwin Doughty. In the Doughty family there was an atmosphere of respect for religion, which clearly influenced Doughty's later interest in biblical lands and in Bedouin life. Moreover, Doughty's position as an orphan undoubtedly contributed to his self-sufficient nature, as well as to his serious attitude. At school he was shy and led a solitary existence but was too good a fighter to be attacked with impunity by the other boys.

Doughty was greatly disappointed in 1856 when at the age of thirteen he failed a physical examination and was denied admittance to the Royal Navy because of a slight speech defect. He was, therefore, unable to carry on the naval tradition of his mother's family. Instead of feeling self-pity, however, Charles resolved to write an epic of old Britain in which he could express the love of country and adventure denied expression in his lost naval career. His interest in oral as well as written expression may have owed something to his speech problem.

From 1861 to 1863 Doughty attended the conservative Gonville and Caius College, Cambridge, but in 1863 he transferred to Downing College, Cambridge, which was more hospitable to the new science of geology, in which he was strongly interested. This choice of subject was probably motivated by Doughty's interest in the Bible as much as it was by an interest in science. At that time geology was at the center of the controversy between Christianity and science: the work of Charles Lyell, one of the first important geologists, had sometimes contradicted biblical statements about the age of the earth. Moreover, the Darwinian controversy concerning the origin of human beings was raging and was felt in geological circles, since geology was concerned with the fossil record. Doughty may have thought that if he could use the new methods of science to understand the history of the earth, he might possibly be able to solve religious questions, such as the origin of life and the universe. Unlike some of his contemporaries, Doughty felt that science would ultimately support religious belief rather than undercut it.

He was drawn to the bare places of the earth where he could see the history of the planet and its inhabitants in the clearest form. While still an undergraduate at Downing College in 1863-1864, he lived among isolated farmers and hunters for nine months studying a glacier in Norway, about which he published a paper in 1866. The search was personal as well as scientific. Doughty had the capacity for solitary contemplation, and in such a bare environment he could probe his essential nature stripped of the accretions of civilization.

Although he was a diligent student and had conducted original research, Doughty received a "second" rather than a "first" in his final B.A. natural science examinations. The reason given by his examiner was that Doughty could not seem to choose which data to present but instead displayed all of his knowledge unselectively. While this trait may have been a fault in regard to the examination, Doughty's obsessive notice of and compulsion to record almost every detail of his journey in Arabia would later prove to be one of the great strengths of *Arabia Deserta*.

After graduation in 1866 Doughty was able to study and travel at leisure, even though around this time his family suffered a financial setback that rendered him needy for most of his life. For four years he studied older English literature in London and at the Bodleian Library, Oxford, in preparation for the great epic of England that he wanted to write, and then from 1871 to 1875 he traveled as a studious "world's wanderer," as he calls himself in *Arabia Deserta*, in Europe and North Africa.

In Sicily in 1872 he witnessed the eruption of Vesuvius, which he recalls in *Arabia Deserta,* and he frequently returned in later life to Italy, which he loved. From Sicily he went to North Africa, where he became interested in the Semitic world. In the summer of 1873 he went to Italy again and then to Greece, drawn ever eastward. In Greece in 1874 he decided to plunge into the Levant by visiting the sites holiest to Christianity, proceeding immediately from Acre in what is now Israel to Mount Carmel, Nazareth, Samaria, and Nablus on his way to Jerusalem. He then traveled to Bethlehem and Hebron before going on to Galilee, Baniyas, and Damascus, from whence he returned to Jerusalem by way of Baalbek and Jaffa.

In the fall of 1874 Doughty left Palestine via Al-'Arish and spent a short time in Egypt before embarking on a three-month camel journey through the Sinai desert in 1875. This sojourn is recorded in a brief report to the Royal Geographical Society of Austria, in *Arabia Deserta* retrospectively, and almost half a century later in the poem "Mansoul," in which he stresses the difficulty of the topography. He left Sinai and arrived at Ma'an and Nabataean Petra in what is now Jordan in May 1875. In that area he heard of the inscriptions of Medain Salih and nearby Hejr, and he decided to return later to make "squeezes," or impressions, of them, as he relates in the opening section of *Arabia Deserta*. Doughty opted to live in Damascus for one year in order to familiarize himself with the Arabic language and culture as preparation for his archaeological journey. During that time he tried without success to secure funding from the Royal Geographical Society for the proposed trip. He did not even receive passport documents from the British consul in Damascus, who felt that it was too dangerous. Finally, against the advice of his British and Arab friends and with very meager resources, he embarked on his journey to the Nabataean monuments on the northern edge of Arabia.

The forty chapters of *Arabia Deserta* comprise nine plot sections, detailing the fantastic journey that began when Doughty, clothed as a poor Syrian, left Damascus on 10 November 1876, and which ended on 2 August 1878 when he appeared in the Arabian seaport of Jidda, far to the south. In addition to some medicines to sell to the Arabs, Doughty's meager equipment included some pages from Speght's 1687 edition of Geoffrey Chaucer's *Canterbury Tales* (circa 1387), a King James Bible, and two books in German about previous Arabian travels. These books indicate Doughty's interest in the English language and past ages, in religion, and in scientific exploration.

In the first section (volume one, chapters 1-3) of *Arabia Deserta*, Doughty describes the colorful preparation and departure of the hajj, or pilgrimage caravan, and his own exit from Damascus two days later, clothed as a poor Syrian seller of medicines. He further describes the interaction among the various groups of Muslim pilgrims and between them and himself as an overt Christian. Throughout his journey Doughty never made any attempt to hide his religion, refusing even to appear to convert to Islam to smooth his way. He dwells on the biblical account of the area's history and on the topography of the regions through which the caravan passed on its way to Medain Salih and the holy cities of Mecca and Medina. Despite some perhaps retrospective anger at the way he was treated as a Christian in Arabia, Doughty's writing here is lighthearted and enthusiastic.

Chapters 4-7 describe Doughty's experience of life in the *kella,* or caravan way station, at Medain Salih. Old Haj Nejm, the gatekeeper, and the irascible Mohammed Aly, the tower keeper, emerge as fully rounded characters, as does Doughty. As if de-

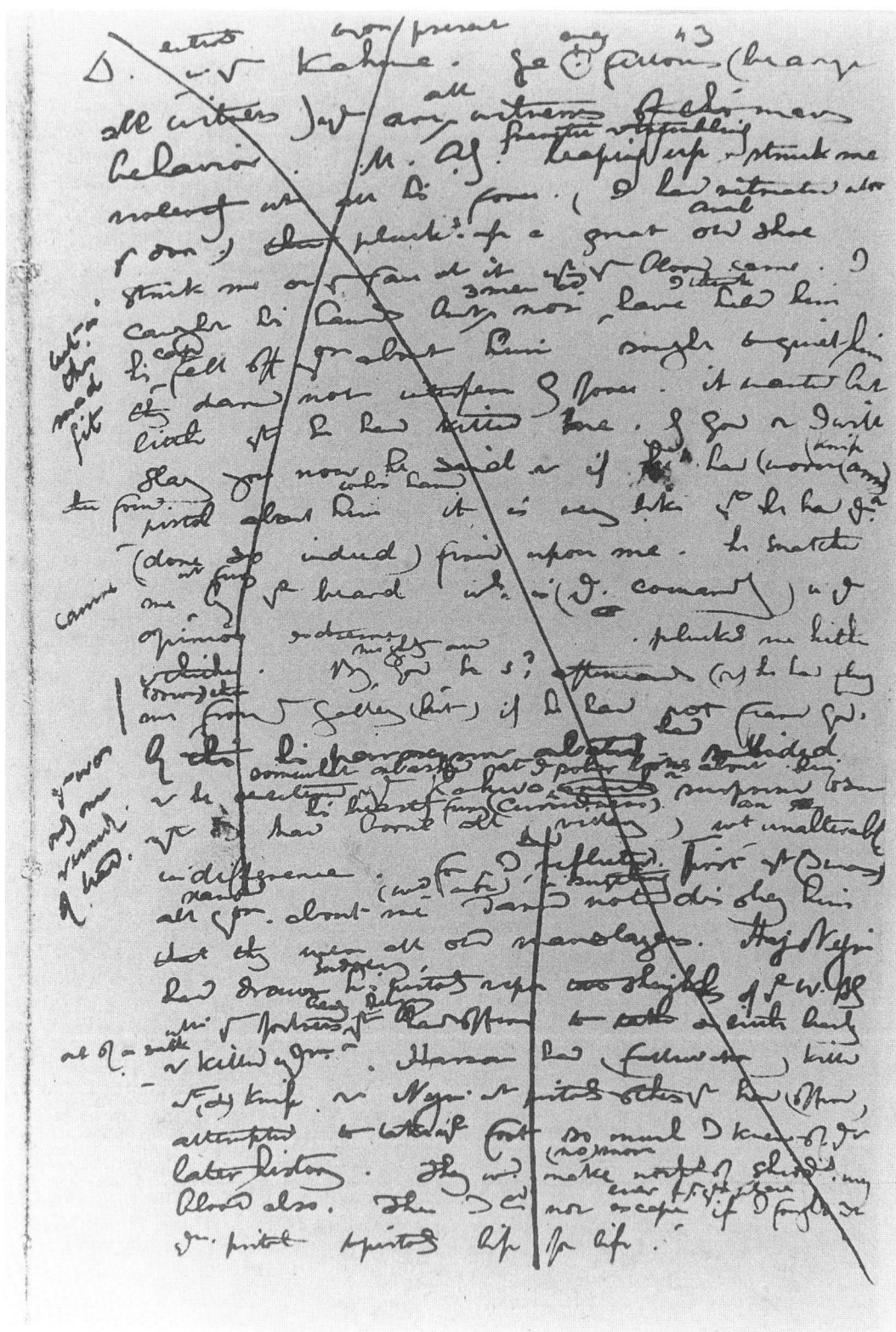

Page from Doughty's Arabian notebooks recording an angry encounter with Mohammed Aly, the tower keeper of the kella, or caravan way station, at Medain Salih (Fitzwilliam Museum, Cambridge)

scribing a scene from *The Arabian Nights,* Doughty relates how he controlled Mohammed Aly's furious and apparently inexplicable outburst of rage by means of Christian passive resistance. This dangerous encounter provides a lively literary counterpoint to Doughty's disappointing exploration of Nabataean sepulchres, which he ultimately describes as ugly "rat holes."

Chapters 8–13 of volume one tell of Doughty's stay with Sheikh Zeyd of the Fejir or Fukara Bedouin. At the end of chapter 7, Doughty sends his pressings of the monuments back to Damascus with the returning caravan but decides to remain behind to study Bedouin ways, despite warnings. Like many European travelers to Arabia before him, Doughty views Bedouin freedom and austerity favorably. But unlike most European travel writers, he records in candid detail his arguments and fallings-out with the tribesmen, as well as their failings. In the ambiance of light domestic dramatic comedy, Doughty describes his role as a go-between between Zeyd and his disaffected wife Hirfa. Here as elsewhere in *Arabia Deserta,* he does not present the Bedouin and other Arabs in terms of favorable or unfavorable stereotypes but as complex human beings. Rejecting previous European falsifications of Arabian experience and romanticization of the Bedouin in particular, he comments in *Arabia Deserta* that there is "so little (or nothing) of *'Orientalism'*" in Arabia. In his determination to record the truth of his experience, he rejects "tales rather of an European Orientalism than with much resemblance to the common experience."

By the first week of May 1877 Doughty was ready to leave Arabia, having stayed longer than he had planned. But owing to illness, he sojourned for four additional months with the Moahib and then with some Fukara and Wady Aly Bedouin who were camped on the great *harra,* or lava field, near Medain Salih. Despite near-starvation conditions, his enjoyment of Bedouin life increased. The story of that sojourn and of his trip to Hayil, the capital of Mohammed Ibn Rashid, ruler of northern Arabia, fills chapters 14–21 of volume one.

The last part of chapter 21 and all of chapter 22 of the first volume, and chapters 1 and 2 (and the first part of chapter 3 of the second volume) tell of Doughty's stay at the Ibn Rashid court. As if writing a Shakespearean tragedy, Doughty recounts the many murders surrounding Mohammed Ibn Rashid's rule, but he concludes that Mohammed is a worthy ruler. Mohammed and his relative and adviser, Hamud, query Doughty about the potential value of petroleum and ask how electricity works. This is one of the earliest discussions of Arabian oil on record. Doughty cures Hamud's child and wins Ibn Rashid's respect, but Doughty's stubborn failure to pay obeisance to the Emir Mohammed and the restiveness of a xenophobic population (made worse by the Russo-Turkish war that was raging at the time and that was interpreted in Arabia as a Christian-Muslim confrontation) resulted in his expulsion from Hayil on 20 November 1877.

Chapters 3–8 of volume two detail Doughty's difficult journey to the northern Arabian city of Kheybar, the site of legends about an ancient Jewish population but in Doughty's time a poor, superstitious, and disease-infested village ruled by a brutal and corrupt Turkish commandant named Abdullah, who keeps Doughty virtually imprisoned from the end of November until the middle of March 1878. Doughty's difficult stay in Kheybar is redeemed, however, by his friendship with Mohammed en-Nejumy, who offers him comfort and aid. Finally, a letter from Abdullah's enlightened commander, the pasha of Medina, results in Doughty's being sent back to Hayil.

With the true spirit of discovery and the desire to contribute to science that motivated him throughout his journey, Doughty wanted only to go to Bahrain and then to India, thus accomplishing what would have been the west-east crossing of Arabia, but this was not to be. When he arrived at Hayil a second time, Ibn Rashid's chamberlain, Aneybar, sent him back to Kheybar instead of onward to Bahrain. The tale of Doughty's return to Hayil, expulsion, trouble with guides who desert him, and arrival at the town of Boreyda is told in chapters 9–11 of volume two. At Boreyda he barely escaped attack by an aroused population and precipitately departed for the town of Aneyza.

Doughty's troubles increased the deeper he went into Arabia, and the tone of his tale gets more ironic as he remembers – with the help of notes he took during the journey – the difficulties and humiliations he faced. Chapters 12–16 of volume two tell of Doughty's positive first reception by Aneyza's ruler Zamil and by two of its leading merchants, Abdullah el-Kenneyny and Abdullah el-Bessam, including good food and cultured discourse. But the people turned against him owing to incitement by local preachers. He was driven out, recalled, and finally departed for Mecca and the coast with a merchant caravan.

As a Christian, Doughty could not enter Mecca, but even to pass near it was dangerous enough. In an episode worthy of H. Rider Haggard's *King Solomon's Mines* (1886), Doughty was attacked as an infidel by the half-mad Sheikh Salem and his associate Fheyd, who were also interested in stealing Doughty's few goods. Chapters 16–18 of volume two tell of Doughty's amazing showdown with Fheyd, in which Doughty offers Fheyd his own pistol butt first, showing Fheyd that he does not fear him. Although Fheyd did not shoot Doughty, he hit him with a heavy camel stick, leaving him weak and dizzy. Owing largely to the intervention of Maabub, a servant of the sharif of Mecca, who was worried about British reaction should Doughty be harmed, the much-tormented Doughty arrived safely at Taif, where the sharif punished the two miscreants.

Doughty still wished to explore more Arabian geography, including the Wadi Duwasir and the forbidding Empty Quarter (not crossed by a European until Bertram Thomas did it in 1931), but sickness prevented it. A triumphant Doughty in aristocratic Arab dress, a true prince of the spirit, descended to the port of Jidda and to the welcome of the British officialdom that had denied him proper papers at the start of his journey, thus causing many of his troubles. This epic tale seems incredible, but Doughty affirms that all of it is true when he writes in his preface to the second edition that "the haps that befell me are narrated in these volumes: wherein I have set down, that which I saw with my eyes, and heard with my ears and thought in my heart, neither more nor less." His account has been confirmed in many details by later travelers, and his book was relied upon by British intelligence in both world wars, so accurate was its information.

After he left Arabia Doughty went to India, where he was hospitalized with exhaustion, ophthalmia, and bilharzia. He also read a paper about his trip to the Royal Asiatic Society branch in Bombay. After he got back to England in late 1878, he discovered that the Royal Geographical Society was not particularly interested in his results, so he published his first articles on his explorations in the German journal *Globus* in 1880–1882. He did not address the Royal Geographical Society until 1883 and received its award only in 1912.

Doughty labored on the draft of *Arabia Deserta* from 1879 until 1883 and continued to revise it until it was published in 1888. He worked from notes taken, sometimes with great difficulty, during his trip. During the time he was writing *Arabia Deserta*, he also gathered "word notes": small slips of paper on which he would jot down a word or phrase and many synonyms for it while he worked, walked, or talked. From this self-composed thesaurus he would select the one word that was exactly right in a given context. A parcel of several word notes, collectively labeled "Desert," includes the stylistically and spiritually evocative "I am miswent, infinite, illimitable, solitudines, inhuman, forlorn, lorn, lone, wild waste moors, desolate, forlaetynsse, dry uplands," and "it is as a pass through the nether world," clear indications of Doughty's feelings as revealed in *Arabia Deserta*.

Doughty's painstaking process of collecting and selecting words reveals the same love of language that led to the compilation of the *Oxford English Dictionary* by James A. H. Murray during the same period. In writing *Arabia Deserta*, Doughty drew many words and expressions from Chaucer and Edmund Spenser, his favorite authors, and from other older English literature. He did not hesitate to use Arabic words in his sentences when he felt they were appropriate for the feeling he was trying to convey. Doughty's ultimate goal was to revitalize Victorian English, which he asserted was stale and empty. When H. W. Bates, the assistant secretary of the Royal Geographical Society, wanted Doughty to render his lecture into more usual English for publication in the society's proceedings in 1883, Doughty wrote that "as an English Scholar I will never submit to have my language of the best times turned into the misery of today – that were unworthy of me."

Because of its unique style and in spite of its obvious value, *Arabia Deserta* was rejected by four commercial publishers before Cambridge University Press accepted it, and even that press tried to get Doughty to agree to have it revised by professor Robertson Smith, a leading Semitic expert and advocate of Doughty's manuscript. His stubborn refusal of this request, so similar to his refusal to hide his religion or his opinions during his Arabian journey, fortunately prevailed, and the result has become a literary classic.

The celebrated first sentence of *Arabia Deserta*, which has had at least one article devoted entirely to it, offers the reader an appreciation of the difficulty that publishers found in his style:

> A new voice hailed me of an old friend when, first returned from the Peninsula, I paced again in that long street of Damascus which is called Straight; and suddenly taking me wondering by the hand "Tell me (said he), since thou art here again in the peace and assurance of Ullah, and whilst we walk, as in the former years, toward the new blossoming orchards, full of the sweet spring as the garden of God, what moved thee, or how couldst thou take such journeys into the fanatic Arabia?"

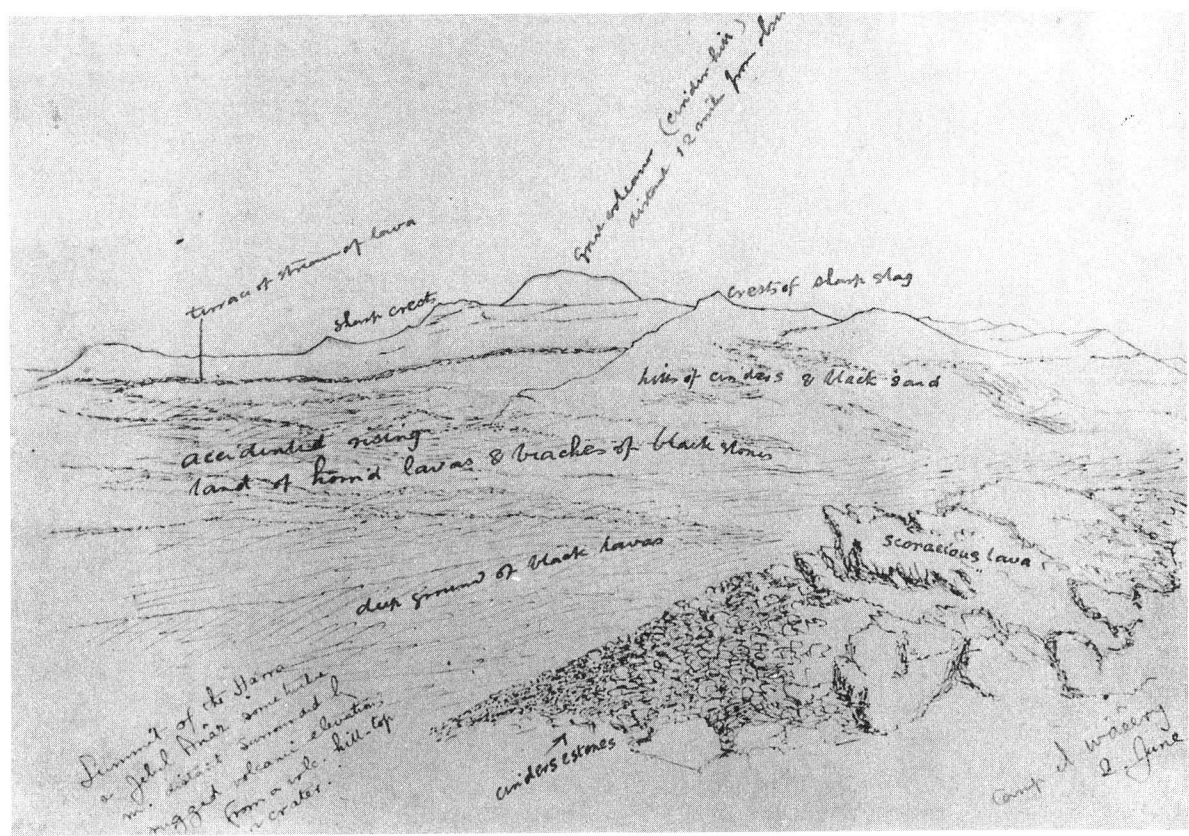

Doughty's geological sketch of the Aueyrid harra, *or lava field, 2 June 1877 (from David G. Hogarth,* The Life of Charles M. Doughty, *1928)*

Doughty spends the rest of the book's two volumes answering this question (if in an indirect way). First and most obviously, Doughty wanted to add to the store of Western knowledge about the then little-known Arabian peninsula. Second, Doughty went into "the fanatic Arabia" alone despite all danger because he was a man who best found out who he was by means of opposition – in this case, opposition from a harsh natural environment and from the followers of another religion. By opposing them, he discovers his own hardiness and beliefs. Third, like a religious hermit or saint, he welcomed austerity and found himself stripped to essentials when alone and practically naked in the desert. By suffering, Doughty may have felt that he was doing penance for man's original sin in the Garden of Eden, which the first sentence of the book recalls. Moreover, by turning the other cheek and surviving, he may have felt that he was proving the truth of Christianity. Fourth, Doughty went to Arabia because of his interest in the keepers of an ancient oral tradition, the Bedouin. At least one reason he wrote *Arabia Deserta* was to see to what degree he could replicate ancient oral utterance in English, thus creating in some sense another bible. Fifth, Doughty was proving his ability to live among an old, hardy people and survive, thus becoming for a time like the early British tribesmen about whom he would write in *The Dawn in Britain* (1906). There are other reasons that each reader may discover; in terms of Doughty's motivation, *Arabia Deserta* is ultimately as inscrutable as the sands of the desert.

Although some early reviewers expressed consternation at Doughty's style, others approved and understood that they were dealing with something unique and wonderful. The *Cambridge Review* for 15 February 1888 opined: "The orientalist, the archaeologist, the student of folk-lore, will peruse the book with interest," but "it is the lover of English literature who may be expected to derive the most pleasure and profit from Mr. Doughty's performance. His style is such as would impart a freshness to the driest matter." Fellow Arabian traveler Richard Francis Burton, writing in *The Academy* for 28 July 1888, expressed some pique that Doughty had not consulted Burton's *Personal Narrative of a Pilgrimage to Al-Madinah and Meccah* (1855–1856) and took issue with some aspects of Doughty's knowledge of Arabic and with his Christian forbearance in the face of Muslim opposition. But Burton concluded that "the geographer, the epigraphist, and the student of Arabic" would "attach the highest importance" to Doughty's narrative, which was also "right well told" and "pleasant for the reminiscences of days when English was not vulgarised and Americanised." During the same period, poet Robert Bridges and artist Edward Burne-Jones expressed their high esteem for *Arabia Deserta,* and it was one of William Morris's favorite books toward the end of his life.

Despite the praise of the few able to appreciate it, the book went out of print until a 330,000-word abridgment by Edward Garnett, *Wanderings in the Desert,* appeared in 1908. In 1921, owing to the advocacy of T. E. Lawrence, *Arabia Deserta* was republished in its original form. As a descriptive bibliography by Philip M. O'Brien (published in *Explorations in Doughty's "Arabia Deserta,"* 1987) makes clear, *Arabia Deserta* has been in print in one edition or another with few gaps ever since; it has also been translated into French, German, Hebrew, and Swedish. Ironically, Doughty regarded the work now considered his masterwork as an aberration in his poetic career and burned the manuscript of the book in his garden, claiming that it was of no interest.

Outside of Middle Eastern scholars, Doughty is not as well known to readers as he might be, largely because of the initial difficulty of his style. In 1923 the noted critic Leonard Woolf, writing in *The Nation & Athenaeum* for 27 October, addressed this problem directly when he wrote: "The book is a work of the highest art.... No doubt there are certain obstacles in the way of a wide popularity for it, but they are not, I think, insuperable. The greatest obstacle is Mr. Doughty's style. To most people, when they begin to read 'Arabia Deserta,' the style appears to be inconceivably contorted, crabbed, archaic. The sentences twist themselves into a curious kind of gnarled Elizabethanism which is unlike anything which you may have previously met with between the covers of a book." Yet Woolf ultimately found the style, "which began by being a weariness to you, is in the end, if you will persist, the real source of intense pleasure. The highest triumph of the writer is to evolve a style which is exactly fitted to express the subject matter of his book."

Travels in Arabia Deserta is perhaps most clearly understood stylistically as one of the few truly bilingual works in English literature. Doughty was a linguistic scholar and antiquarian, and he felt that the English of the Victorian period was inadequate. He inserted Arabic words and rhythms into his English when he thought that would help the reader get closer to the Arabian reality that he had experienced. Although any skilled reader of modern English can follow and enjoy Doughty's masterpiece, readers must ideally have an interest in the Middle East to motivate their efforts, and knowledge of both a Semitic language and Middle English helps even more. Obviously, this limits Doughty's readership, but the simplification of Doughty's style by H. L. MacRitchie in a 1989 edition, while changing the impact of the original, may have some success in attracting a wider audience for the book.

Many of the most important literary artists of the twentieth century have concurred with Woolf that Doughty's work is strong and exciting. William Butler Yeats read *Arabia Deserta* three times, and it is likely that Yeats based most of his desert poetry, including the well-known "Second Coming" (1921), on it. The novelist Wyndham Lewis praised the book's "archaic but delightful jargon," while T. S. Eliot, who justly attacked Doughty's poetry for lack of concreteness, called *Arabia Deserta* a "great work" and said that it, along with the novels of Henry James and Joseph Conrad, might indicate the future direction of English prose. D. H. Lawrence read it twice, and Aldous Huxley described *Arabia Deserta* as the work of a "most splendid, archaic stylist." John Bickerson "Binx" Bolling, the hero of Mississippi novelist Walker Percy's *The Moviegoer* (1961), has a one-book library consisting of *Arabia Deserta.* Canadian critic Douglas Grant has named it "one of the greatest travel books in English."

Present-day literary scholars have come to understand that *Arabia Deserta* occupies a central place in the important Victorian prose tradition. Thus, in Stephen E. Tabachnick's collection of essays about *Arabia Deserta,* William N. Rogers II writes: "One should not allow a perception of Doughty's initially perceived exoticism of subject matter and concern with style to blind one to the fact that in his deep and central concerns he is a Victorian in a sense much pervaded by Carlylean values, creating a work that contributes much in the scientific spirit of the age, while simultaneously asserting a spiritual and moral purpose to counteract a merely physical, mechanistic view of the universe." Canadian literary scholar Richard Bevis finds that "not only is the book aging well; it actually seems to be growing with time, rising out of the rubble of modern prose." The linguist Edward Levenston of the Hebrew University, Jerusalem, writes that "though occasionally disconcerted, those readers who acquire a taste for Doughty read him mainly with awestruck

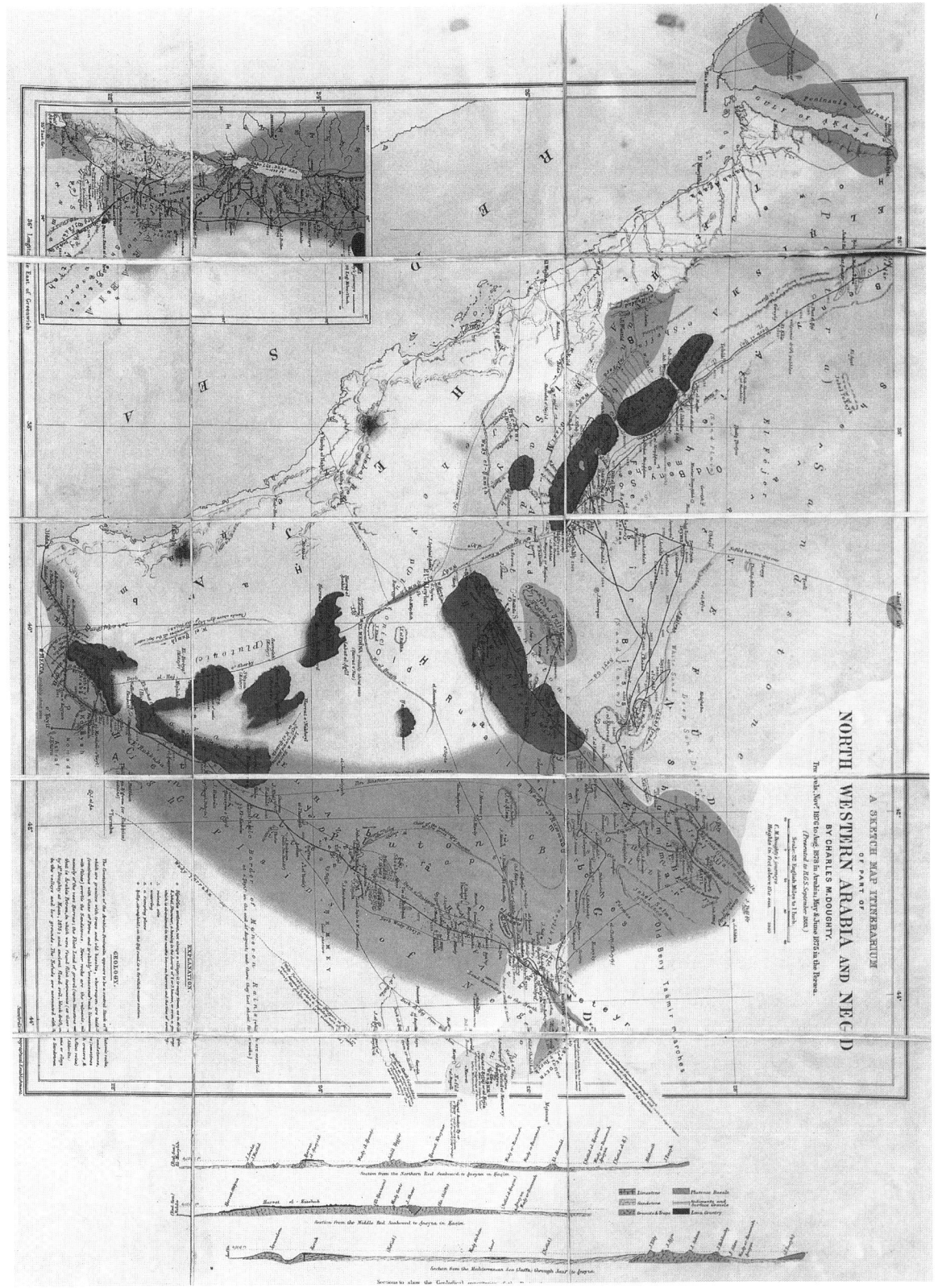

Map from Travels in Arabia Deserta *(1888; courtesy of the Lilly Library, Indiana University)*

delight: awe at his versatility and delight in his mastery of his resources." Annette M. McCormick has pointed out the virtuosity of Doughty's use of Hebrew parallelism that informs his English, so that his book constantly echoes the Bible.

It is not only recent literary scholars, but recent scientists who have recognized the importance of *Arabia Deserta*. In Tabachnick's collection the British geographer J. M. Wagstaff finds that "the diversity and richness of *Travels in Arabia Deserta* make it an immense quarry for geographers." Phillip Hammond of the University of Utah declares that modern archaeologists have learned more about the Nabataeans (the desert dwellers of circa 200 B.C.-A.D. 200), who created the city of Petra from Doughty than from all other writers. Geologists Reginald Shagam and Carol Faul, while deciding that Doughty's geological descriptions lack depth, praise his mapping effort. The historian Bayly Winder says that "no single work paints an overall picture of nineteenth-century Arabian society that tells us more than does *Travels in Arabia Deserta*." Middle Eastern anthropologist Robert Fernea states that "after many years of trying to communicate in various dialects of Arabic," he finds Doughty's language in *Arabia Deserta* "unmistakably authentic" and an amazingly successful attempt to escape the problems of distortion that occur when one simply translates one language into another. Some native speakers of Arabic have concurred in finding Doughty's style authentic, and Arabs have written several doctoral dissertations on *Arabia Deserta*. Issam Safady, the author of a 1968 University of Kentucky doctoral dissertation about Doughty's book, finds that it is "not about the land he visited, but *of* it."

Doughty's book is especially important today when, owing to modern technology, there is more interaction between different cultures than ever before, for it represents an early attempt to replicate cross-cultural experience. Not content to describe another culture from the outside, Doughty attempted to bring the reader into its linguistic and life rhythms, however great an effort that might require of the reader. In his preface to the first edition he wrote: "The book is not milk for babes: it might be likened to a mirror, wherein is set forth faithfully some parcel of the soil of Arabia."

Doughty was able to turn his frequently adversarial relationship with the Arabs into the stuff of great literature and history by accurately recording his own prejudices (and unexpected sympathies) and theirs. As a traveler Doughty did not try to disguise himself as a Muslim or conceal his feelings, including his often irrational dislike of certain aspects of Islam, but instead directly confronted the Arabs with them. That they tolerated him for the most part and even liked him despite his bluntly stated criticism of Islam, polygamy, and slavery is remarkable, given that he had no passport and little money and often had to depend entirely on their hospitality. That a Russo-Turkish war was being fought during the period of his travels – leaving him open to the charge of spying for Christendom – added to his difficulties. Doughty has conscientiously recorded not only his own often tactless statements but also the Arabs' varied responses to them, ranging from bemused politeness and intellectual debate to fierce and prejudiced hostility. This authentic and often contentious cross-cultural dialogue makes *Arabia Deserta* a unique and uniquely honest record of great historical value. As a writer, Doughty frequently shows himself to disadvantage in confrontations with the Arabs. Thus, when the Bedouin defend polygamy on the ground that it takes time and experience to find a good partner for life, Doughty is unable to provide a good counterargument. When Hamud innocently asks about the Christians' eating of pork (forbidden to Muslims), Doughty loses his temper for no apparent reason. In these cases and others Doughty reveals his own prejudices no less than those of the Arabs and lets the reader decide who is right. Where else in nineteenth-century British literature can one find such a genuine, honest, and equal confrontation between two different systems of value and two different religions and cultures?

Another aspect of the richness of *Arabia Deserta* is that Doughty's opinions about Islam and other features of Arabian life cannot easily be summed up. On one hand Doughty will write with obvious prejudice that "the nations of Islam, of a barbarous fox-like understanding, and persuaded in their religion, that 'knowledge is only of the koran,' cannot now come upon any way that is good." On the other he writes of the Arabs that

> Mohammed's sweet-blooded faith has redeemed them from the superfluous study of the World, from the sour-breathing inhospitable wine; and has purified their bodies from nearly every excess of living. . . . Marriage is easy from every man's youth; and there are no such rusty bonds in their wedlock, that any must bear an heavy countenance. The Moslem's breast is enlarged; he finds few wild branches to prune of his life's vine, – a plant supine and rich in spirit, like the Arabic language. There is a nobility of the religious virtue among them.

A thorough reading of *Arabia Deserta* reveals that one of Doughty's major characteristics is self-contradiction and that he is as likely to praise the Arabs in one place as he is to criticize them in another.

Taking Doughty to task for the overt religious and national prejudices that were common in the nineteenth century (among Arabs as well as Europeans) is to state the narrowly obvious while evading the genuine cultural complexity of *Arabia Deserta*. After a careful study of this issue, Janice Deledalle-Rhodes (in an article in the Tabachnick collection) concludes that Doughty's willingness to debate with the Arabs is a mark of respect and that "among nineteenth-century travellers to the Middle East Doughty alone, to my knowledge . . . treated the Arabs as men and equals." Doughty sums up his relationship with the Arabs when he writes in a preface to the second edition that he did "not greatly err, when I trusted my existence . . . amongst an unlettered and reputed lawless tribesfolk." For "there passed over me, amongst the thinly scattered, generally hostile and suspicious inhabitants of that Land of wilderness, nearly two long and partly weary years; but not without happy turns, in the not seldom finding, as I went forth, of human friendship amongst Arabians and even of some very true and helpful friendships; which, from this long distance of years, I vividly recall and shall, whilst life lasts, continue to esteem with grateful mind." It is also important to remember that Doughty says no less than three times in his book that the Bedouin way of life is the best in the world.

The attitude of many recent scholars may be summed up in the words of the French critic François Pouillon, who states his surprise that it should be precisely a writer deeply prejudiced against Islam who has left one of the best and most sympathetic portraits of nineteenth-century Arabia. The reason for this is that Doughty's contradictory personal sympathies and prejudices constitute only one small part of *Arabia Deserta* and that the complex, well-rounded Arab characters in the book constantly provide responses to Doughty's claims.

Travels in Arabia Deserta proves once again that great art follows no rules or ideology and that superb language, subtle characterization, a unique vision, fascinating facts, and an amazing story will render a work of art timeless in spite of its author's human foibles, even when they appear in the work itself. Khalil, as Doughty called himself in Arabia and in *Arabia Deserta,* with his blunt and often tactless honesty, his self-effacement, and his unshakable conviction that "truth may walk through the world unarmed," ultimately strikes the reader as a com-

Doughty in Arab garb (from Travels in Arabia Deserta*)*

plex human being. Readers and writers have been impressed not only with the majesty of Doughty's epic tale of personal suffering and triumph, with the sometimes painful honesty of his reportage of the human foibles of himself and others, and with his ability to recapture the feel of another culture, but also with the brooding philosophical questing that penetrates the entire work.

The reader senses that Doughty is trying to find not so much "the Story of the Earth, Her manifold living creatures, the human generations and Her ancient rocks," as he writes in the preface to his second edition, but rather something that lies entirely in the spiritual realm, particularly in the discipline of austerity and trial. When Doughty emerges from his tribulations at Taif, the reader feels that, whatever Doughty's faults, he has suffered and learned: "The tunic was rent on my back, my mantle was old and torn; the hair was grown down under my kerchief to the shoulders, and the beard fallen and unkempt; I had bloodshot eyes, half blinded, and the scorched skin was cracked to the quick upon my face." In 1886 Doughty wrote, "I am by nature self-willed, headstrong, and fierce with opponents, but my better reason and suffering in the world have bridled these faults and in part extinguished them." It is precisely this process of Khalil's painful growth into greater humanity and deeper knowledge of self and world that will continue to attract readers to *Arabia Deserta.*

Doughty's life after Arabia was relatively uneventful and devoted to poetry that has not stood the test of time. He published his Nabataean work, which was well received, in Paris in 1884 under the title *Documents épigraphiques recueillis dans le nord de l'Arabie.* His poetry is marred by excessive nationalism, an emphasis on war, and a lack of concreteness. In the poetry he did not have the Arabs to oppose his views. In 1900 he published *Under Arms,* a thin volume of hackneyed, chauvinistic verse praising the British role in the Boer War. In 1906 *The Dawn in Britain,* a patriotic epic that he had been working toward since graduation, appeared. Doughty goes back to the earliest times in Britain but is unable to create characters who have the authenticity of the genuine people who appear in *Arabia Deserta.* Doughty could work only from a basis of fact and lacked the ability to create striking imagery from imagination alone. Only in his slim verse drama *Adam Cast Forth (Sacred Drama in Five Songs)* (1908), which recreates to some degree his Arabian travails, does the reader feel the authenticity that comes from Doughty's genuine experience.

The Cliffs (A Drama of the Time, in Five Parts) (1909) and *The Clouds* (1912) represent an early form of science fiction, in which Doughty correctly predicted World War I and its use of the submarine, mine, airship, and torpedo, but they have few other redeeming features. *The Titans (Subdued to the Service of Man)* (1916) is antiquarian in its attempt to go back beyond Adam and Eve to the formation of the earth and the giants who supposedly lived on it then, but the poem never comes alive. *Mansoul; or, the Riddle of the World* (1920) is based on the medieval dream-vision format but is ruined by clichéd, low-level philosophical speculation and outbursts against the kaiser.

Doughty was married in 1886 to Caroline Amelia McMurdo, the daughter of a general, and with her he had two daughters, Dorothy (born in 1892) and Freda (born in 1894). The family remained in financial need until near the end of Doughty's life, when he received a small inheritance. They were forced to move about frequently, as the suitcase in the Gonville and Caius Library, used to hold Doughty's word notes, sadly testifies. Doughty and his family often spent time in Italy, but in 1899 they settled permanently in England, living in Tunbridge Wells, Eastbourne, and finally after 1923 in Sissinghurst.

Doughty died on 20 January 1926, having been recognized for his travels by the Royal Geographical Society (1912), with honorary degrees from Oxford (1908) and Cambridge (1920), and with warm admiration from T. E. Lawrence, who consulted him before taking his first Middle Eastern walking trip in 1909. Later, Lawrence would make his reputation by destroying the Hejaz Railroad that Doughty had envisaged in *Arabia Deserta.* The man who wrote *Travels in Arabia Deserta* was cremated at Golders Green, but the self-effacing yet staunch hero Khalil will live on forever in its pages.

Bibliography:

Philip M. O'Brien, "Charles M. Doughty's *Travels in Arabia Deserta* and Its Abridgements: A Descriptive Bibliography," in *Explorations in Doughty's "Arabia Deserta,"* edited by Stephen E. Tabachnick (Athens: University of Georgia Press, 1987), pp. 223-253.

Biography:

David G. Hogarth, *The Life of Charles M. Doughty* (Oxford: Oxford University Press, 1928).

References:

Thomas J. Assad, *Three Victorian Travellers: Burton, Blunt, Doughty* (London: Routledge, 1964);

Richard Bevis, "Spiritual Geology: C. M. Doughty and the Land of the Arabs," *Victorian Studies,* 16 (December 1972): 163-181;

Jonathan Bishop, "The Heroic Ideal in Doughty's *Arabia Deserta,*" *Modern Language Quarterly,* 21 (March 1960): 59-68;

Peter Brent, *Far Arabia: Explorers of the Myth* (London: Quartet, 1979), pp. 135-145;

Richard Francis Burton, "Mr. Doughty's Travels in Arabia," *Academy,* 34 (28 July 1888): 47-48;

Janice Deledalle-Rhodes, "The Image of Islam in Nineteenth Century English Travel Literature," *International Jounal of Moral and Social Studies,* 1 (Autumn 1986): 265-280;

T. S. Eliot, "Contemporary English Prose," *Vanity Fair,* 20 (July 1923): 51-98;

Barker Fairley, *Charles M. Doughty: A Critical Study* (London: Cape, 1927);

Edward Garnett, "Books Too Little Known," *Academy and Literature* (24 January 1903): 86-87;

Douglas Grant, "Barker Fairley on Charles Doughty," *University of Toronto Quarterly,* 36 (1967): 220-228;

David G. Hogarth, *The Penetration of Arabia: A Record of the Development of Western Knowledge Concerning the Arabian Peninsula* (New York: Stokes, 1904);

R. H. Kiernan, *The Unveiling of Arabia: The Story of Arabian Travel and Discovery* (London: Harrap, 1937), pp. 268-276;

T. E. Lawrence, Introduction to Doughty's *Travels in Arabia Deserta* (London: Cape, 1921), pp. xxv-xxxv;

Annette M. McCormick, "An Elizabethan-Victorian Travel Book: Doughty's *Travels in Arabia Deserta*," in *Essays in Honor of Esmond Linworth Marilla,* edited by Thomas Kirby and William Olive (Baton Rouge: Louisiana State University Press, 1970), pp. 230-242;

McCormick, "Hebrew Parallelism in Doughty's *Travels in Arabia Deserta*," in *Studies in Comparative Literature, Humanities Series,* 11, edited by Waldo F. McNeir (Baton Rouge: Louisiana State University Press, 1962), pp. 29-46;

Sari Nasir, *The Arabs and the English* (London: Longman, 1979), pp. 83-89;

François Pouillon, Critical afterword to Doughty's *Travels in Arabia Deserta,* selected passages chosen by Garnett, translated by Jacques Marty (Paris: Editions Payot, 1990), pp. 331-368;

Edward Said, *Orientalism* (New York: Pantheon, 1978);

Sir Ronald Storrs, "The Spell of Arabia: Charles Doughty and T. E. Lawrence," *Listener,* 38 (25 December 1947): 1093-1094;

Stephen E. Tabachnick, "Adam Cast Forth: The First Sentence of Doughty's *Arabia Deserta*," *Pre-Raphaelite Review,* 1 (May 1978): 49-63;

Tabachnick, "Burton's Review of Doughty's *Arabia Deserta*," in *In Search of Sir Richard Burton,* edited by Alan H. Jutzi (San Marino, Cal.: Huntington Library, 1993), pp. 47-59;

Tabachnick, *Charles Doughty* (Boston: Twayne, 1981);

Tabachnick, "Two 'Arabian' Romantics: Charles Doughty and T. E. Lawrence," *English Literature in Transition: 1880-1920,* 16 (1973): 11-25;

Tabachnick, ed., *Explorations in Doughty's "Arabia Deserta"* (Athens: University of Georgia Press, 1987);

Walt Taylor, *Doughty's English,* S.P.E. tract no. 51 (Oxford: Clarendon Press, 1939);

Kathryn Tidrick, *Heart-Beguiling Araby* (Cambridge: Cambridge University Press, 1981), pp. 136-156;

Anne Treneer, *Charles M. Doughty: A Study of His Prose and Verse* (London: Cape, 1935).

Papers:

Charles M. Doughty's notebooks for *Arabia Deserta* are in the Fitzwilliam Museum, Cambridge; his "word notes" and some letters and manuscripts are in the library of Gonville and Caius College, Cambridge; and the manuscript for "The Utmost Isle," an early version of *The Dawn in Britain,* is in the British Library.

Amelia Anne Blandford Edwards

(7 June 1831 – 15 April 1892)

Patricia O'Neill
Hamilton College

BOOKS: *My Brother's Wife: A Life History* (London: Routledge, 1855; New York: Harper, 1856);

The Ladder of Life: A Heart History (London: Routledge, 1856; New York: Harper, 1864);

A Summary of English History: From the Roman Conquest to the Present Time, with Observations on the Progress of Art, Science, and Civilization, and Questions Adapted to Each Paragraph (London & New York: Routledge, 1856); revised as *Outlines of English History* (Boston: Hickling, Swan & Brewer, 1857);

The Young Marquis; or, a Story from a Reign (London, 1857);

The History of France: From the Conquest of Gaul by the Romans to the Peace of 1856 (London & New York: Routledge, 1858);

Hand and Glove: A Tale (London: Darton, 1858);

Sights and Stories: Being Some Account of a Holiday Tour through the North of Belgium (London: Emily Faithfull, 1862);

Ballads (New York: Carleton, 1862; London: Tinsley, 1865);

The Story of Cervantes; Who Was a Scholar, Poet, a Soldier, a Slave among the Moors, and the Author of "Don Quixote" (London: Routledge, Warne & Routledge, 1863);

Barbara's History, 3 volumes (London: Hurst & Blackett, 1864; New York: Harper, 1864);

Half a Million of Money, 2 volumes (Leipzig: Tauchnitz, 1865; New York: Harper, 1866; revised edition, London: Tinsley, 1866);

Miss Carew: a Novel (London: Hurst & Blackett, 1865; New York: Harper, 1865);

The Four-Fifteen Express (New York: Happy Hour Library, 1867);

Debenham's Vow (London: Hurst & Blackett, 1870; New York: Harper, 1870);

In the Days of My Youth (London: Hurst & Blackett, 1873; Philadelphia: Porter & Coates, 1874);

Monsieur Maurice, and Other Tales, 3 volumes (London: Chapman & Hall, 1873);

Untrodden Peaks and Unfrequented Valleys: A Midsummer Ramble in the Dolomites (London: Longmans, Green, 1873);

A Night on the Borders of the Black Forest (Leipzig: Tauchnitz, 1874; New York: Stokes, 1890);

A Thousand Miles up the Nile (London: Longmans, Green, 1877; New York: Caldwell, 1877; revised edition, London: Routledge, 1889);

Lord Brackenbury: A Novel (London: Hurst & Blackett, 1880; New York: Harper, 1880);

On the Dispersion of Egyptian Antiquities in Connection with Certain Recent Discoveries of Ancient Cemeteries in Upper Egypt (Leide: Brill, 1884);

On a Fragment of Mummy-Case Containing Part of a Royal Cartouche (Leide: Brill, 1884);

Pharaohs, Fellahs, and Explorers (London: Osgood & McIlvaine, 1891; New York: Harper, 1891).

OTHER: Fanny Loviot, *A Lady's Captivity among Chinese Pirates* translated by Edwards (London: Routledge, 1858);

A Poetry-Book of Elder Poets, Consisting of Songs & Sonnets, Odes and Lyrics, Selected and Arranged with Notes from the Works of the Elder English Poets, by A. B. E., edited by Edwards (London: Longmans, 1879);

A Poetry-Book of Modern Poets, Consisting of Songs and Sonnets, Odes and Lyrics, Selected and Arranged with Notes from the Works of the Modern English and American Poets . . . by A. B. E., edited by Edwards (London: Longmans, 1879);

"Mummy," in *Encyclopaedia Brittanica*, ninth edition (Edinburgh: Black, 1885);

G. Maspero, *Egyptian Archaeology*, translated by Edwards (London: H. Grevel, 1887);

Edouard Henri Naville, *The Season's Work at Ahnas and Beni Hasan*, with an introductory essay by Edwards (London: Egypt Exploration Fund, 1891).

SELECTED PERIODICAL PUBLICATIONS – UNCOLLECTED: "Lying in State in Cairo," *Harper's New Monthly Magazine*, 65 (July 1882): 185–204;

"Excavations at Sãn," *Academy*, 26 (12 July 1884): 34–35, 66–67;

"The Story of Tanis," *Harper's New Monthly Magazine*, 73 (October 1886): 710–738;

"Nature of the Egyptian Ka," *Academy*, 35 (5 January 1889): 12–13;

"My Home Life," *Arena*, 4 (August 1891): 299–310;

"Art of the Novelist," *Contemporary Review*, 66 (March 1894): 225–242.

Amelia Anne Blandford Edwards was an accomplished woman of letters when she decided to turn her summer expeditions of mountain climbing in the Dolomites and sailing up the Nile into popular travel books. Like other Victorian adventuresses she was middle-class, middle-aged, and unmarried. In her writing she combined the expertise of a geologist, archaeologist, and art critic with the style of a witty storyteller. The stories told in both *Untrodden Peaks and Unfrequented Valleys* (1873) and *A Thousand Miles up the Nile* (1877) reveal the independence of an early feminist and the contradictory manners and assumptions of British travelers in the age of empire. A self-made woman in several genres, Edwards eventually applied her literary talents to popular accounts of the expeditions and discoveries of British Egyptologists. Travel writing for Edwards bridged the gap between literature and science and initiated her career as a public intellectual with wide-ranging concerns and influence.

Amelia Anne Blandford Edwards was the only child of a retired army officer who worked for the city branch of London and Westminster Bank and a lively Irish woman, a descendant of the Irish barrister Robert Walpole. Born in London on 7 June 1831, Edwards spent many of her summers and holidays with country relatives, including her cousin Mathilda Betham-Edwards, who also became a novelist and journalist. Edwards's mother provided her early education and supported her precocity by engaging a tutor – usually reserved for boys preparing for college. According to her cousin, Edwards's mother forbore giving her daughter domestic training: "The first woman Egyptologist never threaded a needle or made a cup of tea in her life." Instead Edwards studied music seriously for several years. She also received art instruction from her mother and later took classes in Rome. Despite the family's limited means, mother and daughter also frequented the theater. At age seven Edwards had a poem, "The Knights of Old," printed in a weekly journal, and at age nine she won a prize for a temperance story. Her "The Story of a Clock" was published in 1843, and Edwards attracted the attention of editor and illustrator George Cruikshank with the sketches for a story she submitted to Cruikshank's *Omnibus* in 1845. By the age of twenty-one she was an accomplished artist, musician, singer, and actress, but when her father's bank failed, a check from *Chambers's Journal* for a story she had written decided her upon a literary career.

Edwards supported her family by writing stories for such magazines as *Household Words* and *All the Year Round*. She also contributed reviews of books and music and criticism of drama and art for journals such as the *Saturday Review*, the *Graphic*, the *Illustrated London News*, and the *Morning Post*. Throughout her career Edwards's facility with language and her broad knowledge of culture impressed her readers and colleagues. Her editor at the *Academy* claimed that she was "a model contributor – never declining a request, punctual to her promises, writing in a clear, bold hand, and considerate of printer as well as editor." Her unrelenting

THE SASSO BIANCO FROM VAL CORDEVOLE. [*Frontispiece.*

The summit of Monte Pezza in Italy, which Edwards ascended in 1872; frontispiece for Untrodden Peaks and Unfrequented Valleys *(1873)*

sense of duty is captured as well in "My Home Life," an autobiographical essay written the year before she died: "For at least the last twenty-five years, I have rarely put out my lamp before two or three in the morning. Occasionally when work presses and a manuscript has to be despatched by the earliest morning mail, I remain at my desk the whole night through, and I can with certainty say that the last chapter of every book I have written has been finished at early morning." Talented and disciplined, Edwards published the first of her eight novels, *My Brother's Wife: A Life History* (1855), at age twenty-four.

In 1860 her parents died, and Edwards, unmarried and almost thirty years old, accepted the hospitality of a widow, Mrs. Ellen Braysher. Unlike other young women who relied on the patronage of wealthy dowagers, Edwards felt much affection for this friend who allowed her sufficient independence to pursue her own interests. Edwards continued to write prolifically. In addition to fiction she published textbook histories of England and of France, a translation of a French novel, and poetry.

Her first travelogue is really a children's book. *Sights and Stories: Being Some Account of a Holiday Tour through the North of Belgium* (1862) is a collection of fictional episodes about Mr. Butler, the head usher at a boys' school, and the five students he takes with him on a walking tour through Belgium during midsummer holiday. Each chapter recounts the adventures of the boys and includes instruction on art, architecture, history, and geography. Through dialogue and description, Mr. Butler relates the places and people they meet to other historical events around the world. *Sights and Stories* is full of biographies of famous people and anecdotes on a variety of contemporary subjects. At each stopping place in the journey, Mr. Butler introduces the boys to local points of interest, and they experience some aspect of the culture, which Mr. Butler interprets as part of their general education. For instance, a farmer's wife serves the boys hot sugar-cakes, which she names *Suiker-kookjes*. Mr. Butler adds, "I remember to have met with an account of these cakes in a work of Southey's – Southey, you know, was the author of that beautiful sonnet which I read to you the other day on leaving Bruges. He says, if I recall his meaning rightly, that these *Suiker-kookjes* are peculiar to Belgium; so there are nationalities, you see, even in the matter of cakes." Edwards's ability to synthesize different kinds of knowledge in brief and amusing episodes such as this one served her well as a writer of both fiction and nonfiction. As in all her travelogues, Edwards includes her own watercolor drawings and sketches.

During the 1860s Edwards pursued her literary career through stories and sketches for popular journals. In 1864 she published what was to be her most popular novel, *Barbara's History,* which was translated into German, French, and Italian. Of this novel Katharine Sarah Godsden Macquoid says the chapters are "so full of local history and association that one thinks it might be well to have the books for companions when visiting the places described." Beginning in the rural England of Edwards's childhood, the novel's protagonist travels to Europe and marries an inveterate traveler whose adventures have included a liaison with a woman from Capri. Barbara's character is precocious. Her early life with a curmudgeonly aunt who keeps a hundred pigs as a measure of her wealth and status is amusing, while her infatuation with an older man is presented touchingly and unsentimentally from a child's point of view.

Before marrying, Barbara attends an art school in Germany, which offers Edwards a chance to discuss education. Barbara's travels alone and with her husband, Hugh, allow Edwards to comment on the art and life of Europe. Hugh's observations, in particular, contrast the exotic styles of foreign places to conventional social life in England. Consequently, when Barbara learns that her husband has been supporting a mistress, she leaves him to live as an artist on the Continent. After her child is born, she finds work and a place to live through the help of an elderly professor. Apparently self-sufficient, Barbara and her former classmate and best friend, Ida, live within the art community in Rome until she learns of her husband's illness. The ending is melodramatic, but the intelligence of Edwards's characters and her authoritative discussions of painting and architecture show how much the author's life and interests contributed to the story.

Despite the seemingly autobiographical aspects of the book, Edwards's cousin claimed that "she fell in and out of love with persons as well as with pursuits." Like other Victorian women travelers, Edwards's wanderlust had little to do with romance. Perhaps for that reason, her novels are more notable for their settings than their plots. Though most of the local details in her early works come from travel books and other secondary sources, her novels are consistent in their expression of what

Phyllis Grosskurth has called "Edwards's craving for adventure, and to an even greater degree, her curiosity about far-off places." Grosskurth quotes a character from *In the Days of My Youth* (1873): "Next to a military life I think that of a traveller, who turns his back upon railroads and guides – must be the most exciting and the most enviable under heaven. Since reading these books, I dream of the jungle and the desert, and fancy that a buffalo-hunt must be almost as fine sport as a charge of cavalry." Except for occasional visits to Germany, Greece, and Italy, however, Edwards had not yet had the opportunity to explore the "untrodden" paths for which she and her characters longed.

In the spring of 1872 Edwards was able to finance a trip to the Dolomites – limestone mountains in the southeastern area of Tirol more or less inaccessible to casual visitors and challenging even to the members of the Alpine Club. *Untrodden Peaks and Unfrequented Valleys: A Midsummer Ramble in the Dolomites* established Edwards's reputation as a Victorian adventuress. Traveling unchaperoned with a woman friend, identified simply as "L," was bold enough, but she also pursued routes beyond those traveled by fashionable English society. Edwards proudly notes in her preface that they met only three parties of English during the whole expedition. Like other travel writers, Edwards sees the attraction of such a remote area in terms of its freedom from "the flood of Cook's tourists" and its possibilities for "those who love sketching and botany, mountain-climbing and mountain air, and who desire when they travel to leave London and Paris behind them, the Dolomites offer a 'playground' far more attractive than the Alps."

The Dolomites, as Edwards notes, derive their name from Dieudonné de Gratet de Dolomieu, an eighteenth-century French natural philosopher who brought attention to their interesting geological character. Edwards includes an outline of the coral reef theory of a Baron von Richthofen, which she adapted from an essay by George Churchill, a prominent mountain climber and contemporary writer. Originating in prehistoric ages as coral reefs, the mountains' abrupt and fantastic shapes gave Edwards many opportunities for indulging her pleasure in sketching and painting.

Like other Victorian women travelers who explored regions remote from England, Edwards emphasizes the physical hazards of her journey. She declares the passes too difficult for ladies on foot. Their first adventure concerns the acquisition of the only two sidesaddles to be had in the region. Yet Edwards also takes the role of cultural guide. She visits the birthplace of Titian and describes the local artists and art of the region. To mitigate criticism of her unwomanly exploits in the mountains, Edwards takes a self-deprecating tone, reporting wittily on the incongruities between their civilized habits and the "wild" circumstances of their journey.

On the one hand, Edwards describes the special needs and comforts of two Englishwomen and one lady's maid. In addition to the sidesaddles, they bring along tea and chocolate, two bottles of cognac and four of Marsala, as well as an etna stove. On the other hand, their actual journey through the Dolomites and their original ascent of Sasso Bianco, the rocky summit of Monte Pezza, prove Edwards's courage and stamina. In the choice of local guides over a cosmopolitan courier, her sense of practicality overrides convention: only men native to the region could take them through the passes and help them negotiate lodgings and provisions in the many small and isolated villages they visit. But if, as Philippa Levine suggests, Edwards "sensed the absurdity of such social nicety" as a male chaperon, she also relished her own sense of decorum and custom. Despite poor roads and generally rough conditions, Edwards and L rejoiced in their ability to make a meal out of tea and hard-boiled eggs. The juxtaposition of concern for propriety and acts of boldness in Edwards's narrative makes her appear amiable rather than heroic. In this way Victorian women made the travel narrative a genre suitable for women writers and readers.

For Edwards, part of the pleasure of travel was in spending part of the day sketching the scene of either the local populace or the landscape. As Levine notes, "this meshing of the visual and the verbal was one of Amelia's most marked talents." Edwards's illustrations and her descriptions of the landscape, although rarely effusive, occasionally express a Romantic tribute to nature's beauty and sublimity. Her anecdotes about people she meets also tend to the dramatic rather than the sentimental: individual characters are transformed into types of humanity. Thus Edwards developed a literary prose style and format between the informal and subjective approach of autobiography and the more formal and impersonal approach of a scientific or scholarly report.

In describing the various Dolomite peaks Edwards often uses the language of art, giving a sense of the composition of the scene. She describes Monte Civita near Caprile as a "sheer, magnificent wall of upright precipice, seamed from crown to foot with thousands of vertical fissures, and rising in a mighty arch towards the centre." But she is also capable of adding a poetical touch by describing the effect of the sun and mists on the color and mood of the peaks. Before breakfast she sees the Civita, "like a beautiful ghost, draped in haze against a background of light. I thought it then, for simple breadth and height, for symmetry of outline, for unity of effect, the most ideal and majestic-looking mountain I had ever seen; and I think so still." Much of the landscape of piled rock suggests to Edwards scenes of catastrophe and ruin. The effects of ancient bergfalls, or volcanic action, create the impression of desolation. One of the most interesting spots Edwards visited was the lake of Alleghe, where a bergfall in 1771 caused two whole villages to be submerged. Told by the local people that the walls and roofs of the villages could still be seen, Edwards picked a clear day to go out on the lake where indeed she was able to discern what looked like the substructures of several houses. According to Victorian biographer Sarah Knowles Bolton, "no tourist will ever go over the route taken by Miss Edwards without looking for the buried villages."

Although many Victorian travelers took colored pencils and watercolors with them when they traveled, Edwards's sketches of the peaks seemed a particularly strange pastime to the villagers of the Dolomite region. One old woman, not satisfied when Edwards explained that her picture would enhance her memory of the mountains and allow her to share her experience of them with friends, was even more surprised that Edwards had traveled so far to see such scenery. "Have you, then, no mountains and no trees in England?" she asked as she went on her way.

In Ampezzo Edwards is delighted to find the local people willing to be sketched. One young woman stands motionless for more than half an hour so that Edwards can capture the details of her native costume. In response to Edwards's question, she explains that she is not married because young men do not marry poor girls. Despite her curiosity, Edwards refrains from asking her any more personal questions but remarks of the woman's "natural dignity and reserve." When Edwards's sketching is finished, the woman leaves, "going home over the mountain, alone." Although Edwards offers no opinion about the woman question, which was much in the news at home, her travelogues present such episodes to focus her readers' attention on the views and conditions of women outside Victorian England.

The subject of marriage comes up again later in their journey in the village of Selva, where Edwards and her friend are confronted by four women

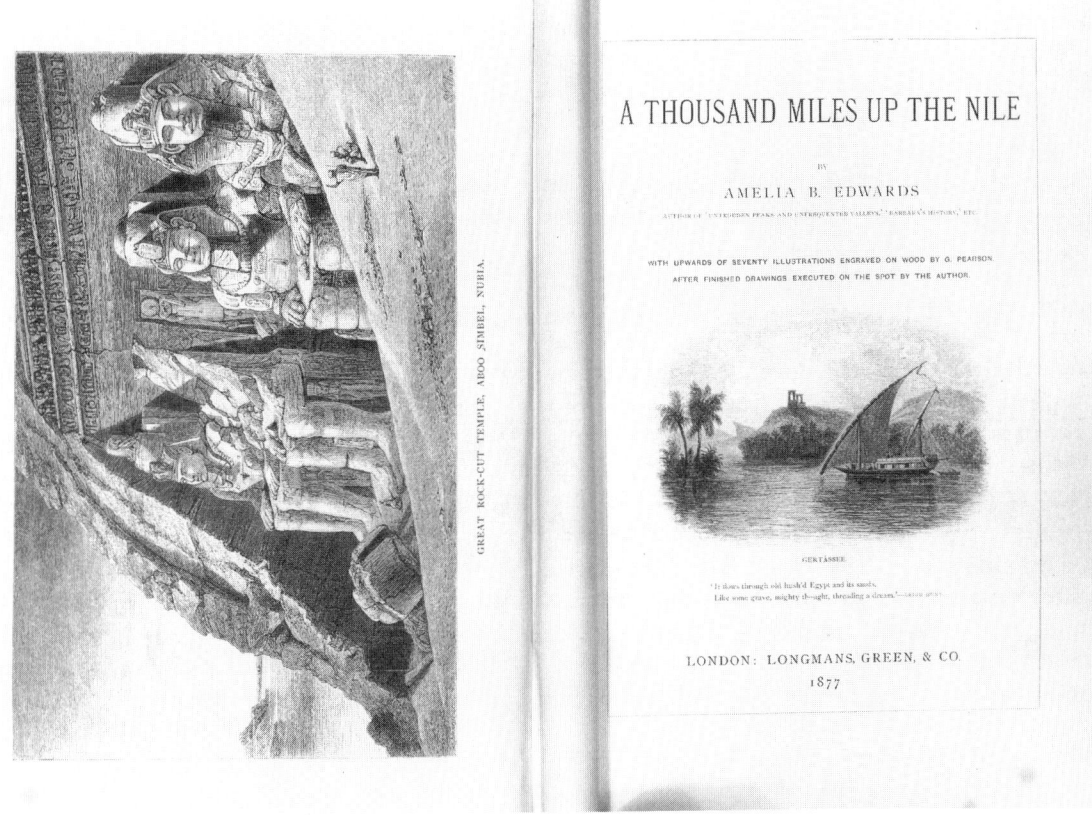

Frontispiece and title page for Edwards's popular account of her travels in Egypt (courtesy of the Lilly Library, Indiana University)

of the house, who are astonished by their clothes and equipment. When they admit to having come from England and that they are not sisters nor married, the women cry " 'poverine' in chorus, with an air of such genuine concern and compassion that we are almost ashamed of the irrepressible laughter with which we cannot help receiving their condolences." Without further commentary, Edwards manages to convey her independence and avoids giving offense to her more conventional readers.

Edwards explains that one of the purposes of the trip is to visit Cadore, the birthplace of Titian. They tour the small house where Titian spent his childhood and view an early painting of the Madonna and child with an angel (supposed to be a self-portrait of the artist), kneeling on one knee and holding out a small picture. The picture had been painted over. Edwards's background in painting and her reliance on earlier accounts allowed her to distinguish among the many reputed works of Titian they are invited to see in the area. She willingly visited and recorded any relic or cultural treasure her local guides offer.

In many of her descriptions of places, Edwards also describes the local economy. She condemns Auronzo for its dirt and poverty, but she also notes that the commune is in debt and overpopulated. Similarly, Edwards contrasts the benefits of prosperity with the cost of picturesqueness in her account of Predazzo, a town of foundries and timber yards. Her character as a traveler is most apparent, however, in her recommendation of Tirolean hotel keepers. Although she reports condescendingly of "that perfume of antique Republicanism that seems yet to linger in the air of all that was once Venice," she records that these innkeepers are usually from old families and do not extend hospitality to strangers for the money merely. Her respect for their hereditary nobility appears in her criticism of the hauteur and troublesomeness of some English tourists. Such bad manners, she claims, will cause these independent innkeepers to be replaced by "a class of extortionate speculators, probably Swiss," which will end all communication between the native population and the traveler, and "the simplicity, the poetry, the homely charm of the Dolomite district will be gone for ever."

The poetry of the place notwithstanding, Edwards sometimes describes her hosts as if they were natives living remote from the industrial advances of Europe. Their surprise at the wonders of Edwards's etna stove or binoculars is compared to the "open-mouthed astonishment of a savage" or to the response of people "as ignorant and uncivilised as aboriginal Australians." Edwards's personal sense of dignity and superiority as an Englishwoman is only momentarily threatened when a German girl appears with her knapsack, apparently traveling alone and scoffing at the idea of using a guide. Edwards calls her a "Phenomenon" and reports that she supped with the men and regaled them with stories of her adventures. Such a character underscores Edwards's own cultivated position between the natives' savage ignorance and the potentially unfeminine or vulgar behavior of the modern traveler.

Despite such digressions and anecdotes, Edwards's main concern in *Untrodden Peaks* remains the mountains. Each chapter leads the reader over new terrain to identify individual peaks and make the mountain passes familiar. Edwards's spirit of adventure is fulfilled through the scientifically and socially recognized achievement of being the first recorded climber of Sasso Bianco in the Monte Pezza.

The climax of most travelogues occurs when the traveler either gains sight of a particular destination or discovers something unexpected along the journey. According to Edwards, the "words 'prima ascenzione' are Cabalistic, and haunt the memory strangely. They invest the Monte Pezza with a special and peculiar interest; so that it is no longer as other mountains are, but seems henceforth to have a halo round its summit." Yet the climb itself is neither eventful nor particularly revelatory. A mountain whose height makes it of the second class, Sasso Bianco has the advantage of standing "in the very centre of the Dolomites, like the middle ball upon a Solitaire board, surrounded on all sides by the giants of the district." Edwards and her companion choose a fine day and pass some idyllic spots outside Caprile, making their way through fields, forests, and pastures, until the mules must be left behind and the actual climb begin. Amid anxiety that the mist coming up from the valley will obscure the panoramic view they anticipate, Edwards, L, and their guides mount the rocks "like going up the steps of the Great Pyramid." At the summit they break for lunch and wave to L's maid, who has been stationed to verify their ascent. Edwards gives an account of the various peaks that can be seen from their vantage point. In all they identify fifty, including the most famous of the Dolomites. Edwards regrets being unable to sketch an outline of the panorama. Because the horizon is never completely clear from mist, they also gain no southern view of the valleys that lead to Venice and the Adriatic. Edwards uses the mountaineer's technique of determining the height of the mountain by taking the temperature at the summit and estimates it as about eight thousand feet. After two hours they descend and are greeted appreciatively by the locals who "took the expedition as an indirect compliment paid to themselves." According to Edwards, from no other point could one see all of the Dolomite giants at once. The rest of the book describes their return to their starting point at Caprile and their final excursion through the area on the way to Botzen.

The reviewer for the *Athenaeum* placed Edwards's book in the tradition of other Dolomite travelogues including the early nineteenth-century writings of Sir Humphry Davy and the guidebooks of Churchill and Gilbert, to which Edwards refers for background information. Predictably, the reviewer noted Edwards's style and imagination without granting her the authority of a male mountain climber or literary personage. According to the reviewer, Edwards's narrative had "much that is new told in a light and lively style, but with a suspicion of exaggeration here and there." While treating some of the details of Edwards's narrative skeptically because they do not match his own experience of the Dolomite region, the reviewer admits that, despite some inaccuracies and "sundry slips of grammar and spelling, Miss Edwards's book may be commended to all who love mountain travel, either in theory or practice. She has a quick eye for scenery and costume, can discuss a picture critically, and set forth the merits of native artists." He also compliments her engravings.

A second edition published in 1890 is dedicated to "My American Friends in all parts of the World." Levine, who wrote the introduction to Virago's 1986 edition of *Untrodden Peaks and Unfrequented Valleys,* recommends Edwards's travelogue for the "unquenchable vitality" that fills the pages of "this funny and endearing account of her travels." Edwards's travelogue proved her competence not only as an independent woman but also as a writer of nonfiction. Although she remained interested in literature, she wrote only one novel in the remaining twenty years of her life.

The year following their trip to the Dolomites, while on a sketching holiday in France, Edwards and L decided to leave the rainy weather behind and go to Egypt. They arrived in Cairo on 29 November 1873, hired a passenger boat called a *dahabeeyah* with accommodations for themselves and three other passengers, and headed up the Nile with a crew of twenty men. *A Thousand Miles up the Nile* describes their journey from Cairo to Abu Simbel near the southern border of present-day Egypt. Practicing Arabic with the merchants of Cairo and studying hieroglyphics and Egyptian history along the way, Edwards combines again serious instruction with personal anecdotes about her adventures sailing and exploring the Nile regions.

Two years in the writing, Edwards's book is an epic within the genre of travel literature. Unlike Lucie Duff Gordon's *Letters from Egypt* (1865), which concerns the lives and manners of the people and the cruelty of the government, Edwards's travelogue avoids overt commentary on the contemporary political situation, concentrating instead on the history of ancient Egypt in the art and architecture of its monuments. The wealth of historical and archaeological information, given in a prose style at once pictorial and lucid, as well as the more than eighty illustrations reproducing Edwards's drawings, demonstrate Edwards's full scope as a writer, artist, and scholar. Although it was praised as a guidebook for travelers to the region, by its second edition in 1888 Edwards disclaimed knowledge of the current conditions for travel. Instead she added notes and updated archaeological information gathered from British and French excavators whose work she knew from firsthand reports. What began as a travel adventure story like *Untrodden Peaks* became for Edwards and her Victorian readers a valuable introduction to Egyptology.

Edwards's preface again prepares her readers with an overview of the kind of region she explored and provides clues to her purposes as a travel writer. First, she contends that "the physique and life of the modern Fellah is almost identical with the physique and life of that ancient Egyptian laborer whom we know so well in the wall paintings of the tombs." What she enumerates as similarities refer to social manners and the food and dress of the people. Unlike some Victorians, Edwards does not characterize Egyptians, past or present, as primitive Europeans. Instead they represent a separate history as timeless as the pyramids. Despite her admiration for their prodigious past, Edwards's informal generalizations about the character of modern Arabs bespeak a European sense of superiority. She describes the crowd in Cairo gathered for a procession as "a harmless, unsavoury, good-humoured, inoffensive throng" and the Muhammadan on holiday as a "child" who "loves noise and movement for the mere sake of noise and movement." The variety of people and sights in Cairo provide an entertaining prelude to the journey proper, but the city and its wares are less interesting to Edwards than relic hunting through the debris of Saqqâra.

As the voyage begins Edwards describes the individual members of her heterogeneous crew, which includes Egyptians from different cities as well as Syrians and Nubians. As a group they represent to Edwards's eyes both living vestiges of the ancient Egyptians and the noble simplicity of trustful children in an otherwise sophisticated modern world. When the wind dies and their only hope of progress requires the crew to drag the boat upstream, Edwards registers the strangeness of her situation: "the sight of the trackers jarred, somehow, with the placid beauty of the picture. We got used to it, as one gets used to everything in time; but it looked like slaves' work, and shocked our English notions disagreeably." Edwards's attitude toward the Egyptians vacillates throughout the journey between an active respect for the inhabitants of a strange land and the detached condescension of an imperial guest.

In the first part of her journey Edwards visits some of the local villages. The poverty of the children reminds her of what she had seen in Irish villages, and later she reports sympathetically the plight of a poor woman whose son had not returned or sent word home. Her almost maternal concern for their suffering remains personal and anecdotal. She makes no attempt to understand the political or social context in which such hardships occur. When she meets with Egyptian officials, her attention is focused on the differences between European and oriental customs of hospitality. Nevertheless, unlike some travelers to Africa, Edwards never suggests that the native culture is morally degraded. Wisely, she emphasizes the need for diplomacy and a certain amount of protection before wandering into remote parts of the region. Religious and social practices she describes deferentially, if a bit incredulously. In relating the problems caused by an insensitive fellow traveler who could not resist the temptation to shoot wherever game might appear, Edwards remains a neutral observer of cultural differences, with only a hint of ridicule for both crocodile hunters and the incongruities of local forms of justice.

Her purpose in journeying up the Nile has less to do with nineteenth-century travelers and Egyptian society, however, than with the manners and customs of the ancient Egyptians. After her initial curiosity is satisfied, Edwards avoids visiting local villages. The sight of the suffering children and the

The partially excavated Sphinx; illustration from A Thousand Miles up the Nile *(1877)*

squalid conditions of life seem too horrible, and there is nothing she can do for them. To her crew, she shows more consistent concern. When their boat is stalled for several days outside Philae and the crew's rations are depleted, Edwards, against the advice of her cook, raids her own supplies for eggs, biscuits, chocolates, and tobacco to keep up the spirits as well as the strength of her boatmen. If there is little of the missionary or philanthropist in Edwards, she nevertheless exhibits a hardy sense of team spirit.

The crew responds loyally to the respect and honor that Edwards shows them. As a precaution against misunderstanding or hostility, Edwards never goes ashore without being accompanied by one of the crew, usually a man named Salame. After giving a quick character sketch of the man, she demonstrates his "good breeding" by relating that he fasted all day long rather than eat while she was engrossed in painting. Because of their isolation from Europe the people of the Nile required more study than can be garnered from the standard guidebooks. Edwards claims that her daily contact with the crew gives her occasion to know their character intimately. If, according to Edwards, the "Fellah is half a savage," on the whole the Egyptians' good points outnumber the bad. With characteristic urbanity, Edwards asks, "what man or nation need hope for a much better character?"

Despite her generalizations about character and her pseudoscientific categorization of the Nubians as people belonging to "a lower ethnological type," Edwards catches herself in the act of reducing the people and the place to mere artifacts for her own use and entertainment. After describing the interesting sights and scenery around the first cataract, she writes, "It is all so picturesque, indeed, so biblical, so poetical, that one is almost in danger of forgetting that the places are something more than beautiful backgrounds, and that the people are not merely appropriate figures placed there for the delight of sketchers, but are made of living flesh and blood, and moved by hopes, and fears, and sorrows, like our own." Such moments suggest the self-consciously literary aspects of the travelogue. They also suggest the advantages of the travelogue for dramatizing what is otherwise merely factual information. The people of the Nile function in Edwards's narrative as more than the exotic background for the conventional feelings of a fictional heroine and more human than in accounts by statistically minded government agents. By making the place into the main character, Edwards's narrative explores its affects while preserving its mystery.

The physical types and social customs of Edwards's crew provide a comforting sense of continuity with the ancient Egyptians, represented on the walls and in the statuary of the monuments of the Nile. In contrast, the dimensions and age of the monuments suggest the dizzying effects of time and cultural difference. To realize the duration of six or seven thousand years is, as Edwards notes, no easy task. In her detailed descriptions of each major temple and ruin along the Nile, Edwards conveys an enormous breadth of learning about the past. She instructs her readers several times on the importance of knowing something about the vast history that is embodied by the pyramids and temples. To that end, she clarifies a common misconception about hieroglyphic writing. Jean-François Champollion's discovery of the alphabetic and syllabic basis of the signs enables even the self-taught tourist to decipher many of the inscriptions. According to Edwards, scholarly advances in translating Egyptian literature as it appears on papyrus, stone, wood, and other materials excavated from the temples have unraveled even the secret of the Sphinx – a pictorial rendition of which Edwards uses as an epigraph for her book. Unlocking the mysteries of ancient Egyptian hieroglyphs holds the same fascination for Edwards as her childhood reading of *The Manners and Customs of the Ancient Egyptians* and *1,001 Arabian Nights*. Later, through her work in support of further excavation of Egyptian temples, Edwards's skill in reading and accurately drawing Egyptian characters helped popularize interest in Egyptian literature and culture.

Her most important advice to travelers is to see the Nile region in the order of its construction, "for the history of ancient Egypt goes against the stream. The earliest monuments lie between Cairo and Siout, while the latest temples to the old gods are chiefly found in Nubia." Thus she and L made brief visits to the Great Pyramid in Cairo and to Thebes on their way up the Nile, but she reserves commentary until their return. Before reaching the great temples at Abu Simbel, Edwards dedicates a whole chapter to the history of Ramses the Great and his wives. According to Edwards, the traveler is ill-equipped who goes through Egypt without something more than a mere guidebook knowledge of this military leader, builder of great temples, and pharaoh at the time of Moses. One of the most interesting aspects of Edwards's writing is her attention to the domestic details of the various dynasties, including the love of Ramses and Nefertiti.

In describing the temples of Egyptian pharaohs, Edwards refers to the works of earlier explorers, such as Mariette Bey, but she also combines discussion of the dimensions of the temple or the history depicted on its walls with an artist's perspective on the color or excellence of the portraiture. On the whole, Edwards engages her Victorian readers by applying a Ruskinian view of the relations of art and social history. The mosque of Sultan Hassan is praised for its beauty because it "was built at just that happy moment when Arabian art in Egypt, having ceased merely to appropriate or imitate, had at length evolved an original architectural style out of the heterogeneous elements of Roman and early Christian edifices." Likewise, Edwards describes the wall sculpture of Amanda as from the period of the Egyptian Renaissance that, like Ruskin's view of Renaissance Italian art, combines tradition with a measure of individual liberty. The men who worked the beauty out of the rock at Abu Simbel are called the "Michael Angelos of their age."

Edwards's best writing appears, however, in her unobtrusive yet evocative descriptions of the daily scene, the vegetation that marks the shoreline, the texture of the sand, the color of the sky, and the beauty of the night or the sunrise as they depart or arrive at a new site. Her occupation most days is to sketch, sometimes the whole exterior and sometimes detailed corners of the interior of the temples. Her verbal descriptions of what she draws lead to accounts of the scene's religious or historical significance. Encounters with the local population develop into dramatic vignettes that give variety to the narrative without diverting attention from the author's main interest.

If, along the way, Edwards points out the natural beauty of the landscape of the Nile, her most intense moments of awe and self-revelation occur on the sites of Abu Simbel and Philae. In the former Edwards wanders alone among the halls and rooms in the late afternoon, "like a shade among shadows." The influence of the place is so weird and awful that one afternoon she experiences a sudden panic and can hardly move to escape. Edwards also makes her debut as an excavator. One of her company discovers a buried entrance, which they hope will lead to a tomb. After a day of digging they descend into a chamber that seems a pylon to the Great Temple. Despite its lack of important relics or treasures, the excitement of their discovery prompts them to make it official by sending a telegram to the London *Times*. According to her American colleague William Copley Winslow, "as an incipient Egyptologist in 1874, she 'wriggled in' through an aperture about a foot and a half square." Her party also celebrates their find by carving their initials on a blank wall over the inside of the door. In the second edition of her work, the more experienced Edwards adds a note of concern for the destruction of Egyptian monuments by treasure hunters and tourists:

> There is no one to prevent it; there is no one to discourage it. Every day, more inscriptions are mutilated – more tombs are rifled – more paintings and sculptures defaced. The Louvre contains a full-length portrait of Seti I, cut out bodily from the walls of his sepulchre in the Valley of the Tombs of the Kings. The Museums of Berlin, of Turin, of Florence are rich in spoils which tell their own lamentable tale. When science leads the way, is it wonderful that ignorance should follow?

The incipient Egyptologist who could carve her initials and sift through the rubble of graves for relics also sets her crew to using gallons of coffee to restore the color on the face of a Colossus that had been disfigured by a cast taken for the British Museum. Like the restoration work of the Gothic revival in England, Edwards's attempts to preserve Egyptian monuments were sometimes unorthodox, but her spirit of reverence for the ancient culture deepens as the journey progresses.

In terms of beauty both of landscape and temple, Edwards seems most enthusiastic about Philae, which she compares to Athens: "perfect grace, exquisite proportions, most varied and capricious grouping, here take the place of massiveness; so lending to Egyptian forms an irregularity of treatment that is almost Gothic, and a lightness that is almost Greek." On their return to Cairo, Edwards spends a few moments at sunset eulogizing the silence and melancholy beauty of Philae, appreciating not only the solemnity of the painted columns and the mystic chamber of Osiris, but the river, the palms, a woman crooning to her baby on a neighboring island, and the sounds of nature. In literary terms, this is the climax of the narrative, for here Edwards focuses on her subjective response to the scene: "Lingering till it is all but dark, I at last bid them farewell, fearing lest I may behold them no more."

The rest of the journey proceeds more quickly because, as Edwards realized from her traveling companions, not everyone is as enthusiastic for temples as she is. Earlier in her narrative Edwards complains of the difficulty of painting a landscape that appears all yellow, but in Thebes she notes the difficulty of finding a language to describe the temples, comparing the task to that of describing mountains: "no two are alike, yet all sound so much alike when described that it is scarcely possible to write about them without becoming monotonous." Unlike mountains, however, temples provide stories that compare with the edifices and stories of medieval and Renaissance Italy. The Temple of Gournah, for example, is as distinctively a family memorial as the Medici Chapel, and one of its profile-portraitures of the Ramses dynasties reminds Edwards of Dante in his elder days.

Related to Edwards's interest in excavation and preservation of the temples are her observations about the black market in antiquities and her praise for the Boulak Museum in Cairo. Attempts by local merchants to sell Edwards fake scarabs and the apparent European obsession with mummies are described ironically. The wholesale pillage of Egyptian antiquities had been condoned by earlier viceroys of Egypt, but Edwards praises the work of Ismail Pasha, who was making such exports illegal. Instead, the government supported a national museum in Cairo, where Edwards encounters an impressive display of portrait statues made of private individuals.

The final day of their journey is spent in Ghîzeh to see the pyramids and the Sphinx. From the top of the Great Pyramid, Edwards describes the various sites and recognizes their function as great cemeteries despite ingenious theories about their use as observatories. The desert landscape allows Edwards to note monuments and landmarks in all directions, and she remarks in her last sentence the direction of Thebes, Philae, and Abu Simbel, the places she has left behind. Unlike *Untrodden Peaks,* this narrative avoids any sense of closure, perhaps because Edwards saw Egypt's history as larger than any human life.

It is no wonder that Edwards's journey launched a new career for her as an Egyptologist. As a polymath who had already written textbooks on European history and knew the technical aspects of art, music, and literature, Edwards found much in Egypt's past to absorb her attention. In the three years between her return to England and the publication of *A Thousand Miles up the Nile,* she familiarized herself with the works of French and English scholars and archaeologists. According to William Copley Winslow, she was interested in a variety of topics, from philological and ethnical questions to the comparative study of contemporaneous arts and sciences of the region. Reviewers praised her travelogue for its accuracy and its helpful tips for travelers. According to J. S. Cotton, the editor of the *Academy,* the cheaper edition became "as indispensable as Murray or Baedeker." More to the point was the praise of John Addington Symonds, a respected man of letters and Edwards's editor at the *Academy:* "It is this enthusiasm for old Egypt, running powerful and deep throughout the volume as an undercurrent to its many other interests, that gives its real charm to the work." Indeed, even more than her earlier travelogue, Edwards's work on Egypt focused on issues of scholarship rather than tales of domestic relations or adventure.

In the ensuing years, Edwards's interest in Egyptology dominated her time and energy. She wrote one more novel, published in 1880, *Lord Brackenbury,* which like her others is highly descriptive of place and character and was popular enough to be translated into French, German, and Russian. But her literary work turned to book reviews and popular essays on the discoveries of explorers in Egypt. In addition to letters to the London *Times,* Edwards made almost weekly contributions to such journals as the *Academy* in London and frequent contributions to *Harper's* in the United States.

In 1883, at the age of fifty-two, Edwards became honorary secretary for the Egypt Exploration Fund, which she cofounded with Sir Eramus Wilson, an enthusiast of Egyptian history, and professor Reginald Stuart Poole, the keeper of the British Museum coins department. The fund was a more practical organization than the scientific societies: In its first year it sent French explorer Edouard Henri Naville on an expedition that uncovered the sites of Pithom and Goshen. These expeditions provided new treasures for museums and increased knowledge of Egyptian history, which Edwards interpreted for readers and contributors. According to Poole, "On her fell the duty of maintaining the subscriptions to the Fund in England, and of corresponding with the explorers and editing the Memoirs – a labour on which she spared no pains and made many lasting friends and not a single enemy." Generous intellectual and financial support came from the United States, especially from the efforts of William Copley Winslow of the Boston Museum.

In addition to her responsibilities as fundraiser and publicist, Edwards presented scholarly papers: "On a Fragment of a Mummy-Case" at Leiden in 1884 and "The Dispersion of Antiquities" at the Orientalist Congress in Vienna in 1885. These addresses sought to elicit support for cataloguing the private collections of wealthy amateurs in Europe and America and thus to recover some of the lost chapters of Egyptian history. Her correspondence with one of the fund's explorers, Flinders Petrie, led to her being the first to offer positive identification of the Cypriot, Phoenician, and other characters on the potsherds found at Fayûm. These inscriptions suggested new dates for the introduction of the Greco-Phoenician alphabet in Egypt and involved Edwards in controversies about the movements and interactions of ancient Mediterranean cultures. She also contributed an article on Egyptian mummies to the ninth edition of the *Encyclopaedia Britannica.*

According to Winslow, Edwards was in advance of the most contemporary discoveries. Her expertise meant that whatever was uncovered would be properly appreciated, and her ability to explain the explorer's finds helped create a receptive audience for further excavations. Despite her expertise, Edwards never claimed to be a scholar. Instead she enjoyed publicizing and explaining the importance of the discoveries made by explorers in the field. American friends and supporters, however, conveyed a more profound appreciation of Edwards's contributions. Columbia College distinguished her in 1887 with a degree of doctor of letters, and Smith College's award of a doctor of laws was the first ever accorded a woman in the United States.

In 1889, at the request of twenty-five college presidents and literary notables such as Robert Lowell, Edward Whittier, and Oliver Wendell Holmes, Edwards went to the United States to deliver a total of 120 public lectures. Those collected in *Pharaohs, Fellahs, and Explorers* (1891) exemplify the popular style developed by Victorian scientists such as Thomas Henry Huxley. Detailed and authoritative, these essays make the remote periods of Egyptian history accessible through analogies to modern culture. By focusing on simple or interesting figures to represent the complexities of the entire field of Egyptology, Edwards inspired support and interest in biblical archaeology, art history, and Egyptian religion, literature, and science. One of the essays is on Queen Hatasu, whom Edwards introduces as the Queen Elizabeth of Egyptian history. After describing the political climate in which Hatasu ascended the throne, Edwards records her achievements as one of the great "builder-sovereigns of Egypt." Edwards's accounts not only summarize the points of view of other scholars but also make original arguments by interpreting hieroglyphs,

Illustration from Pharaohs, Fellahs, and Explorers *(1891), a collection of public lectures Edwards delivered in the United States in 1889*

sculptures, portraits, and personal effects assembled from a variety of locations. Her interest in Egyptian queens and goddesses was particularly suited to encourage women in a field that had been dominated by male explorers and scholars.

Lecturing was a particularly successful way for Edwards to communicate with her audience. According to one auditor, "the picturesqueness of her style, the interest of her facts, and the sympathetic charm of her delivery have evoked unwonted enthusiasm.... Herself a skillful artist, she can, in an instant, deftly illustrate with chalk some hieroglypic puzzle or curious relationship between Egyptian and Greek arts." Edwards attributed her success as a lecturer to early training in voice and to acting in amateur theatricals in her teens.

Interest in Edwards as a personality led to an interesting autobiographical piece for the American journal the *Arena,* titled "My Home Life." The essay begins as a travelogue. Edwards describes the landscape and architecture of the village of Westbury-on-Trym where she lived for more than twenty-five years. She includes a historical review of some of the great houses in the area before she turns modestly to her own home, The Larches. After detailing the structure and grounds, she turns to the interior: "As soon as the front door is opened, the incoming visitor finds himself in the midst of modern Egypt, the walls of the hall being lined with Damascus tiles and Cairene woodwork, the spoils of some of those Meshrabeeyeh windows which are so fast disappearing both in Alexandria and Cairo." More treasures are on display as the visitor is led to the library, but many more, enough to furnish a small museum, according to Edwards, are hidden in drawers and boxes.

Before describing them, Edwards reveals the contents of an extensive collection of books, pictures, and artifacts. Shelves line every wall with books three deep representing different stages in her literary and scholarly career. She notes her possession of an 1863 folio edition of Gustave Doré's "Don Quichotte"; John Addington Symonds's volume of poetry, *Many Moods* (1878), which was dedicated to her; and Robert Browning's inscribed first volume of *The Ring and the Book* (1869). The library bespeaks the cultivated sensibilities of an independent scholar.

The reader then learns more of her treasured collection of Egyptian antiquities, including mummified hands and other more gruesome "fragments of spiced and bituminized humanity." Finally Edwards describes herself in ethnographical terms as "essentially a worker" who regularly stays at her tasks until two or three in the morning. She closes with a romantic anecdote of hearing the song of a nightingale from the pear tree outside her window as she finished the writing of her Dolomites travelogue.

According to Macquoid, Edwards was actively involved in public movements of the day. She was a member of the Biblical Archaeological Society and the Society for the Promotion of Hellenic Literature. She was also vice president of a society for promoting woman suffrage and an honorary member of the Anti-Vivisection Society. In 1889, during her lecture tour in the United States, Edwards broke her left arm but continued to meet her engagements. Two years later she took ill after a visit to the London docks, where she was in charge of authorizing the distribution of antiquities to various museums in England and America. She died on 15 April 1892, at the age of sixty. She endowed a chair of Egyptology at University College, London. Her library was given to Somerville Hall, then a college for women at Oxford.

Although no official biography has been written, Amelia Edwards has been recognized for her travel writing and her contributions to Egyptology. As a self-taught scholar and a gifted writer, Edwards not only brought remote and inaccessible places to those who were less intrepid but also showed that travel could be a productive and liberating experience for women. While the novel provided financial independence for many Victorian women writers, travel writing provided a new genre with fewer conventions. For Edwards, travel writing provided the impetus for intellectual as well as literary achievement. Traveling to remote regions allowed her the rare opportunity to contribute observations and information to the otherwise male-dominated scientific circles and organizations back home. Edwards's success in writing travelogues and in organizing the Egyptian Exploration Fund proved that women could participate in fieldwork and scholarship.

Although Edwards remained more a journalistic than an academic commentator, her writing and lectures educated a large and supportive audience in the history of Egypt and the wonders of archaeological research. As late as the 1960s, David Crownover invoked Edwards's name and drawings to support UNESCO's plan to salvage the monuments of Nubia threatened by the Aswan High Dam. Edwards's travel writing, then, exemplifies one way in which the past may be recovered and preserved for future readers and explorers. In combining scientific learning with wit and personal experience, Edwards's work epitomizes the interdisciplinary nature of Victorian culture and the possibilities for Victorian women to go beyond the confines of domestic duties and the representation of romantic heroines.

References:

Mathilda Betham-Edwards, "Amelia Blandford Edwards," in her *Friendly Faces of Three Nationalities* (London: Chapman & Hall, 1911), pp. 43–70;

Sarah Knowles Bolton, "Amelia Blandford Edwards," in her *Famous Types of Womanhood* (New York: Crowell, 1892), pp. 327–350;

David Crownover, "Amelia Edwards and the New Aswan Dam," *Expedition,* 4 (Spring 1962): 24–27;

Maria Frawley, *A Wider Range: Travel Writing by Women in Victorian England* (Cranbury, N.J.: Associated University Presses, 1994);

Phyllis Grosskurth, "Amelia Edwards: A Redoubtable Victorian Female," *A Review of English Literature,* 6 (1965): 80–92;

Philippa Levine, Introduction to *Untrodden Peaks and Unfrequented Valleys,* by Edwards (London: Virago, 1986), pp. xv–xviii;

Katharine Sarah Godsden Macquoid, "Julia Kavanagh, Amelia Blandford Edwards," in *Women Novelists of Queen Victoria's Reign: A Book of Appreciations* (London: Hurst & Blackett, 1897), pp. 251–274;

William Copley Winslow, "The Queen of Egyptology," *American Antiquarian and Oriental Journal,* 14 (November 1892): 305–315.

Papers:

Amelia Anne Blandford Edwards bequeathed her Egyptological library and collections to University College, London, and her personal library to Somerville Hall, Oxford.

Matilda Barbara Betham-Edwards

(4 March 1836 – 4 January 1919)

Cynthia Ellen Patton
Mesa State College

BOOKS: *The White House by the Sea: A Love Story,* anonymous, 2 volumes (London: Smith, Elder, 1857);

Charlie and Ernest; or, Play and Work: A Story of Hazlehurst School (Edinburgh: Edmonston & Douglas, 1859);

Now or Never: A Novel (Edinburgh: Edmonston & Douglas, 1859);

Ally and Her Schoolfellow: A Tale for the Young (London, 1861);

Holidays among the Mountains; or, Scenes and Stories of Wales (London: Griffith & Farran, 1861; London: Griffith & Farran / New York: Dutton, 1880?);

Little Bird Red and Little Bird Blue: A Tale of the Woods (London: Low, 1861; New York: Gregory, 1862?);

John and I, anonymous, 3 volumes (London: Hurst & Blackett, 1862); as Betham-Edwards, 1 volume (London: Chapman & Hall, 1876);

Snow-Flakes, and the Stories They Told the Children (London: Low, 1862; London & New York: Routledge, 1884?);

Scenes and Stories of the Rhine (London, 1863);

Doctor Jacob (3 volumes, London: Hurst & Blackett, 1864; 1 volume, London: Bradbury, Evans, 1868; Boston: Roberts, 1869);

Lisabee's Love Story, 3 volumes (London: Hurst & Blackett, 1865);

The Primrose Pilgrimage: A Woodland Story (London: Griffith & Farran, 1865);

The Wild Flower of Ravensworth (London, 1866);

A Winter with the Swallows (London: Hurst & Blackett, 1867);

Dr. Campany's Courtship, and Other Tales, anonymous (London: Bradbury & Evans, 1868);

Through Spain to the Sahara (London: Hurst & Blackett, 1868);

Kitty (3 volumes, London: Hurst & Blackett, 1869; 1 volume, New York: Harper, 1870);

The Sylvestres, 3 volumes (London: Hurst & Blackett, 1871); republished as *The Sylvestres; or, The Outcasts* (Philadelphia: Lippincott, 1892);

Holiday Letters from Athens, Cairo, and Weimar (London: Strahan, 1873);

Mademoiselle Josephine's Fridays, and Other Stories (London, 1874);

Felicia (3 volumes, London: Hurst & Blackett, 1875; 1 volume, London: Chatto & Windus, 1879);

Walking with the World (Philadelphia, 1875);

Minna's Holiday; or, Country Cousins, and Other Stories (London, 1876);

Bridget, 3 volumes (London: Hurst & Blackett, 1877);

A Year in Western France (London: Longmans, Green, 1877);

Brother Gabriel, 3 volumes (London: Hurst & Blackett, 1878);

Holidays in Eastern France (London: Hurst & Blackett, 1879; New York: Harper, 1879);

Friends over the Water: A Series of Sketches of French Life (London: Houlston, 1879);

Forestalled; or, The Life-Quest, 2 volumes (London: Hurst & Blackett, 1880; New York: Lovell, 1880);

The Famous Women Library, 6 volumes (London: Griffith & Farran, 1880); republished as *Six Life Studies of Famous Women* (London: Griffith & Farran / New York: Dutton, 1880);

The Starry Blossom, and Other Stories for the Young (London: Japp, 1881);

Exchange No Robbery: A Novel (New York: Harper, 1882); republished as *Exchange No Robbery; or, Fated by a Jest: A Novel* (New York: Munro, 1882); republished as *Exchange No Robbery, and Other Novelettes,* 2 volumes (London: Hurst & Blackett, 1883);

Pearla, 3 volumes (London: Hurst & Blackett, 1883); republished as *Pearla; or, The World after an Island: A Novel,* 1 volume (New York: Harper, 1883);

Disarmed, anonymous, 2 volumes (London: Bentley, 1883); republished as *"Disarmed!" A Novel* (New York: Harper, 1883); revised as *Disarmed* (London: Methuen, 1891);

Love and Mirage; or, The Waiting on an Island: An Out-of-Door Romance, anonymous (New York: Harper, 1884); republished as *Love and Mirage; or, The Waiting on an Island, and Other Tales,* 2 volumes (London: Hurst & Blackett, 1885);

Poems (London: Kegan Paul, Trench, 1884; revised and enlarged edition, London: Nutt, 1907);

Half-Way: An Anglo-French Romance (1 volume, New York: Harper, 1885; 2 volumes, London: Low, Marston, Searle & Rivington, 1886; revised edition, London: Sampson Low, 1889);

The Flower of Doom; or, The Conspirator, and Other Stories (London: Ward & Downey, 1885); republished as *The Flower of Doom, and Other Stories* (New York: Harper, 1885);

Next of Kin Wanted: A Novel, 2 volumes (London: Bentley, 1887); republished as *Next of Kin – Wanted: A Novel* (New York: Harper, 1887);

The Parting of the Ways: A Novel (3 volumes, London: Bentley, 1888; 1 volume, New York: Lovell, 1890);

The Roof of France; or, The Causses of the Lozère (London: Bentley, 1889);

For One and the World: A Novel, 2 volumes (London: Ward & Downey, 1889; New York: Lovell, 1890);

A Romance of the Wire (New York: Lovell, 1890; London: Blackett, 1891);

A Dream of Millions, and Other Tales (London: Sampson Low, 1891);

Two Aunts and a Nephew (London: Henry, 1891);

A North-country Comedy (London: Henry, 1891); republished as *A North Country Comedy* (Philadelphia: Lippincott, 1892);

The Romance of a French Parsonage (New York: Lovell, 1891); republished as *The Romance of a French Parsonage; or, The Double Sacrifice,* anonymous, 2 volumes (London: Chapman & Hall, 1892);

France of To-Day: A Survey Comparative and Retrospective, 2 volumes (London: Percival, 1892–1894); republished as *France of To-Day: A Survey, Comparative and Retrospective* (New York: Lovell, Coryell, 1892);

The Curb of Honour (London: Black, 1893); republished as *The Curb of Honor* (New York: Anglo-American Publishing, 1893);

A Romance of Dijon (London: Black, 1894; New York: Macmillan, 1894);

The Golden Bee, and Other Recitations (London: Dean, 1895);

The Dream-Charlotte: A Story of Echoes (London: Black, 1896; New York: Macmillan, 1896);

Reminiscences (London: Redway, 1898; revised edition, London: Unit Library, 1903);

A Storm-Rent Sky: Scenes of Love and Revolution (London: Hurst & Blackett, 1898);

The Lord of the Harvest (London: Hurst & Blackett, 1899);

A Suffolk Courtship (London: Hurst & Blackett, 1900);

Anglo-French Reminiscences, 1875–1899 (London: Chapman & Hall, 1900);

Mock Beggars' Hall: A Story (London: Hurst & Blackett, 1902);

East of Paris: Sketches in the Gâtinais, Bourbonnais, and Champagne (London: Hurst & Blackett, 1902; New York: Dutton, 1902);

Barham Brocklebank, M.D. (London: Hurst & Blackett, 1903);

A Humble Lover (London: Hurst & Blackett, 1903);

Home Life in France (London: Methuen, 1905; Chicago: McClurg, 1905);

Martha Rose, Teacher (London: Long, 1906);

A Close Ring; or, Episodes in the Life of a French Family (Bristol: Arrowsmith, 1907);

Literary Rambles in France (London: Constable, 1907; Chicago: McClurg, 1907);

French Vignettes: A Series of Dramatic Episodes, 1787–1871 (London: Chapman & Hall, 1909; New York: Brentano's, 1909);

Unfrequented France, by River and Mead and Town (London: Chapman & Hall, 1910; New York: Stokes, 1910);

French Men, Women and Books: A Series of Nineteenth-Century Studies (London: Chapman & Hall, 1910; Chicago: McClurg, 1911);

Friendly Faces of Three Nationalities (London: Chapman & Hall, 1911);

In the Heart of the Vosges and Other Sketches by a "Devious Traveller" (London: Chapman & Hall, 1911; Chicago: McClurg, 1912);

In French-Africa: Scenes and Memories (London: Chapman & Hall, 1912; Chicago: McClurg / London: Chapman & Hall, 1913);

From an Islington Window: Pages of Reminiscent Romance (London: Smith, Elder, 1914);

Under the German Ban in Alsace and Lorraine (London: Dent, 1914);

Hearts of Alsace: A Story of Our Time (London: Smith, 1916);

Twentieth-Century France: Social, Intellectual, Territorial (London: Chapman & Hall, 1917);

War Poems (Bristol: Arrowsmith, 1917);

Mid-Victorian Memories (London: John Murray, 1919; New York: Macmillan, 1919).

OTHER: Arthur Young, *Travels in France during the Years 1787, 1788, 1789,* introduction by Betham-Edwards (London: Bell & Daldy, 1889); republished as *Arthur Young's Travels in France during the Years 1787, 1788, 1789,* edited by Betham-Edwards (London: Bell, 1892); republished, with introduction, biographical sketch, and notes by Betham-Edwards (London: Bell, 1905);

Edward Robert Bulwer-Lytton, *Poems of Owen Meredith (the Earl of Lytton),* edited, with an introduction, by Betham-Edwards (London & New York: Scott, 1890);

Young, *The Autobiography of Arthur Young, with Selections from His Correspondence,* edited by Betham-Edwards (London: Smith, Elder, 1898);

French Fireside Poetry, with Metrical Translations and an Introduction, translated by Betham-Edwards (London: Allen & Unwin, 1919; Boston: Small, Maynard, 1921).

SELECTED PERIODICAL PUBLICATIONS – UNCOLLECTED: "Three Experiments in Co-Operative Agriculture," *Fraser's Magazine,* 11 (April 1875): 529–539;

"The International Working Men's Association," *Fraser's Magazine,* 12 (July, August & September 1875): 72–87, 181–194, 300–311;

"Progress of Colonisation in Algeria," *Fraser's Magazine,* 16 (October 1877): 422–441;

"A Chapter on French Geography," *Macmillan's,* 50 (May 1884): 47–52;

"The Decadence of French Cookery," *Longman's,* 7 (March 1886): 509–517.

The announcement of Matilda Barbara Betham-Edwards's death in the London *Times* was subtitled "An Interpreter of France" – a fair summation of her career as most readers remembered it and as she wanted it to be remembered. In her sixty-two years as an active writer Betham-Edwards produced novels, short stories, poems, articles on contemporary events, and volumes recording her travels in Europe (especially France) and North Africa. In 1913, toward the end of her long life, she was delighted by the selection of her first novel, *The White House by the Sea* (1857), as one of the volumes to be published in the New World's Classics series. Before her death she was granted the even greater honor of a civil list pension by the British government. She was a friend of such literary luminaries as George Eliot, George Henry Lewes, Coventry Patmore, and Sarah Grand; Barbara Leigh Smith Bodichon, the pioneer of women's higher education in England, was her traveling companion as well as one of her closest friends. But the crowning achievement of Betham-Edwards's life came in 1891, when France awarded her the title of Officier de l'Instruction Publique de France. Grand writes that the "bonny-looking, little elderly *gentlewoman*" delighted in featuring that title prominently at the head of her books and articles, and in wearing and being photographed wearing the purple button to which the distinction entitled her. It was the most gratifying public proof that Betham-Edwards had succeeded in the central work of her life: introducing English readers to what she called "the most splendid civilization in the world."

Matilda Barbara Edwards was born on 4 March 1836 at Westerfield, Suffolk. Her father, Edward Edwards, was a farmer prosperous enough to live in an Elizabethan manor house with his large family; her mother, Barbara Betham (whose maiden name the writer later adopted as part of her own), came from a distinguished, part-French literary family and, as a child, had received a letter from Mary Lamb. In later life Betham-Edwards deliberately obscured her early biographical details, for, as she wrote in her revised edition of *Reminiscences* (1903),

Frontispiece and title page for Betham-Edwards's first travel book, based on her visit with friends near Algiers

she believed that "family circumstances are of no general interest, family sorrows . . . too sacred for printers' copy."

Reminiscences reveals that she had two sisters who were close to her own age and were "the younger children of a numerous family." Her formal schooling was limited; Betham-Edwards considered her earliest teachers to be the Bible and the works of William Shakespeare and John Milton, and by the age of twelve she had read and reread the family's small library of English classics. At ten years old she was sent to a day school at nearby Ipswich, and in the two years she was there the director of the school, an enthusiast of France and the French language, passed both her enthusiasm and her knowledge to young Betham-Edwards. This formal education ended abruptly when Betham-Edwards's mother died and the combination of household duties and her father's indulgence left the book-loving twelve-year-old to her own intellectual devices.

At about age eighteen Betham-Edwards made the disastrous experiment of entering Mimosa House, a boarding school at Peckham, as a "governess-pupil," exchanging her services as a teacher of the junior girls for lessons in music, drawing, and dancing. The arrangement lasted only six months; the moral and intellectual atmospheres of the school were coarse, and only a growing friendship with her cousin Amelia Ann Blandford Edwards (later to be a distinguished Egyptologist) made the six months bearable for Betham-Edwards. At the end of that time she escaped to Suffolk, where she reveled in outdoor activities, the peace of the countryside, and the familiar rituals of the local farmers' year. Those scenes remained one of the main sources of her fiction throughout her life, and they spurred her to begin her first novel soon after she returned home.

This novel was *The White House by the Sea,* which she finished just as she reached her twentieth birthday in 1856. In those days before parcel post the family grocer arranged to carry the manuscript to London, and to everyone's surprise, including Betham-Edwards's, the well-known publishing house of Smith, Elder promptly accepted it in 1857. Her only payment was twenty-five copies of new novels by other writers, but Betham-Edwards recalled that *The White House by the Sea* was "well printed, well bound, well advertised, and presented to the world in excellent company," and the young novelist's career was launched in earnest.

Soon after the publication of her first novel Betham-Edwards began to travel. In order to improve her knowledge of languages, she went with her cousin Amelia to the Continent, and she spent the next seven years in Germany, Austria, and France. In Europe the studious country girl's life became even more eventful than the publication of her first novel had promised to make it. During a brief stop at Munich she encountered an acquaintance from an earlier part of her journey, a Hungarian patriot in exile, and she declined on the spot his fervent proposal of marriage. While staying at Heidelberg a little later, she declined a different proposal – that of being adopted by a wealthy Englishwoman. Neither incident seems to have loomed large in the young writer's imagination, at least not in comparison with two other, less obviously thrilling, events.

While living in Frankfurt, Betham-Edwards studied German with a schoolmistress, Fräulein Fink, and heard about the scandal of an English clergyman who had embezzled a large sum of money donated by the people of Frankfurt for his projected crusade to convert the Jews of Jerusalem. These events provided the seeds of her successful novel *Doctor Jacob* (1864). In *Reminiscences* Betham-Edwards writes that the false crusader's story "with its picturesque surroundings might well have proved the genesis of a novelist as well; who could have helped putting it upon paper?" Later in her career the novel based on personal – but not too personal – experience of colorful people in colorful places would become as characteristic of Betham-Edwards's work as the nonfictional accounts of those experiences. Even in early works such as *The White House by the Sea* and *Doctor Jacob* it is easy to see how much her love of distinctive places and the real lives of people in them shaped her writing.

This first travel experience was ended by the death of Edward Edwards in 1864. Betham-Edwards returned to Suffolk to manage her father's lands in partnership with her only unmarried sister until that sister also died in 1865. This severed Betham-Edwards's last tie to the countryside of her childhood. Her short period as a "girl farmer" remained vivid in her memory; when writing of her life and of farm life in France, she refers to it repeatedly. But, as she acknowledges in *Reminiscences,* "the climate of my native place was rude, the intellectual resources within reach were *nil,* [and] I naturally turned my eyes elsewhere." "Elsewhere" meant London and "the life of Letters, Art, and Science."

In London, Betham-Edwards lived in Kensington, and during her first three years there she made acquaintances among leading writers, scientists, and social thinkers: Richard Monckton Milnes, Lord Houghton; Herbert Spencer; Charles Bradlaugh; and, most important of all to the young writer, Barbara Bodichon and her husband, Dr. Eugène Bodichon. Almost immediately the two women became fast friends, and toward the end of 1865 the Bodichons invited Betham-Edwards for a long visit to their winter home outside Algiers. The sights and, even more, the people of Algeria stimulated Betham-Edwards's imagination in a new and fruitful way. *A Winter with the Swallows* (1867) is her first actual volume of travel writing, though she had already experimented with fiction that imitated the form of travelers' stories in *Holidays among the Mountains* (1861).

In a conversational manner *A Winter with the Swallows* mixes Betham-Edwards's personal experiences and observations with general information. She blends the traveler's enthusiasm with the Suffolk farmer's shrewdness, and she is disarmingly frank about her prejudices as well as her pleasures. When she disembarks from her ship at Algiers, for example, she records that "a splendid young Arab took possession of me and my baggage, and whilst he was very likely making a sum after this fashion, – so much feminine greenness, so many francs, – I was admiring his beautiful oval face, and his perfectly proportioned limbs that seemed cast in bronze." Later, describing a visit to a Berber village, she wishes that she "could lecture an assemblage of Kabyle farmers upon iron ploughs, and artificial manure, and team threshing-machines."

In this early stage of her career as a travel writer Betham-Edwards follows a conventional pattern, both in her strictly chronological organization and in many of her observations. The feminist influence of Barbara Bodichon can be detected in Betham-Edwards's remarks on the degraded condition of women in Algeria. The influence of guidebook writing is evident in Betham-Edwards's closing section, with its practical notes for readers who might wish to follow the routes of her journeys to and through North Africa. Three qualities of *A Winter with the Swallows* foreshadow Betham-Edwards's more mature travel writing: her fascination with people, rather than landscape; her emphasis on accepting and adapting to as many local ways as possible; and her confidence in her own judgment, no matter what "experts" might believe.

These same qualities distinguish her second book about her travels with Barbara Bodichon, *Through Spain to the Sahara* (1868). During the winter of 1866-1867 Bodichon and Betham-Edwards made what the latter described years later as "a somewhat venturesome and perilous journey" back to Algiers by way of France, Spain, Gibraltar, and Oran. As in *A Winter with the Swallows,* the perils in *Through Spain to the Sahara* seem more attractive than risky, and again Betham-Edwards stylistically invites the reader to join in the experience and to trust her judgment rather than common wisdom. "I don't at all understand the usual tone taken by tourists in describing Spanish inns," she writes. "We never found anything to complain of.... Carry with you a tolerable supply of patience and gold pieces, you will do very well." Along the way Betham-Edwards also shows the interest in ordinary people and their fleeting, half-noticed lives that was to typify her work, both as a novelist and as an observant traveler:

> You see faces that tell their own story, and in a moment they have vanished. You are let into little domestic scenes touching, or comic, or painful, or passionate, as the case may be. You cannot stop five minutes at a village station, or linger five minutes in a village waiting-room, without being moved to smiles or tears.

After her second winter in Algeria, Betham-Edwards returned to England, to use that habit of observation in her fiction writing and to begin the series of journeys and extended stays in France that would eventually make her reputation as a travel writer. Back in England in 1868, she was introduced by Barbara Bodichon to two leading literary figures of the day, George Eliot and George Henry Lewes. Betham-Edwards describes them vividly in *Reminiscences* and in her posthumously published volume of memoirs, *Mid-Victorian Memories* (1919). During the nine years following her return to England she wrote four novels, many short stories, articles for periodicals such as *Fraser's Magazine* and *Cornhill Magazine,* and *Holiday Letters from Athens, Cairo, and Weimar* (1873), a short book of travels.

In her "letters home" in this last work the practical traveler again shrewdly notes the foolishness of those who complain that modern travel is unromantic: "rapturous impressions do not wear off the sooner because you sleep upon them in a comfortable bed." Also as in her two earlier travel books, Betham-Edwards does not hesitate to borrow from other writers when her invention flags or the subject fails to engage her – but when she borrows, in this book as in all her books, she is scrupulous about naming and praising her sources. One friend, Grand, wrote after Betham-Edwards's death that "she remained efficient to the last by resolutely economising her strength." Although *Holiday Letters* suggests that its author was economizing her strength for her real interest, the travels in France that she had already begun at the urging of Eugène Bodichon, it is well organized and presents interesting accounts of her acquaintance with the Goethe family and with Franz Liszt.

Betham-Edwards's first extended stay in France began in August 1875. It was meant to be a visit of twelve days, but it turned into a yearlong succession of visits and journeys through much of the western French countryside. She began in Nantes at the house of Dr. Ange Guépin and his wife – friends of Dr. Bodichon and, like him, fervent supporters of the French Republic, anticlerical freethinkers, philanthropists, and social reformers. Throughout the next forty years, while continuing her prolific career as a novelist and periodical contributor, Betham-Edwards studied in minute detail the political sympathies, the economic conditions, and above all the regional character of nearly every part of France.

The series of thirteen books that she wrote as a result of these studies begins with *A Year in Western France* (1877) and ends less than two years before her death, with *Twentieth-Century France: Social, Intellectual, Territorial* (1917). The form of these books that she published changed throughout the years – from straightforward travel narratives to collections of miscellaneous essays on political geography, striking personalities, and national traits – but the distinguishing qualities of Betham-Edwards's earliest travel writing still remained. The most important feature, the novelist's emphasis on ordinary characters within a setting rather than on exotic incidents or on the setting alone, is acknowledged near the beginning of *Anglo-French Reminiscences, 1875-1899* (1900): "These reminiscences are of men and women, not of sites and scenes; they form a record of social intercourse, no collection of travellers' tales."

A Year in Western France and its companion volume, *Holidays in Eastern France* (1879), open the series by stressing the value of "travel" rather than "tourism." Betham-Edwards came to know France – the real France, beyond Paris and beyond the most

 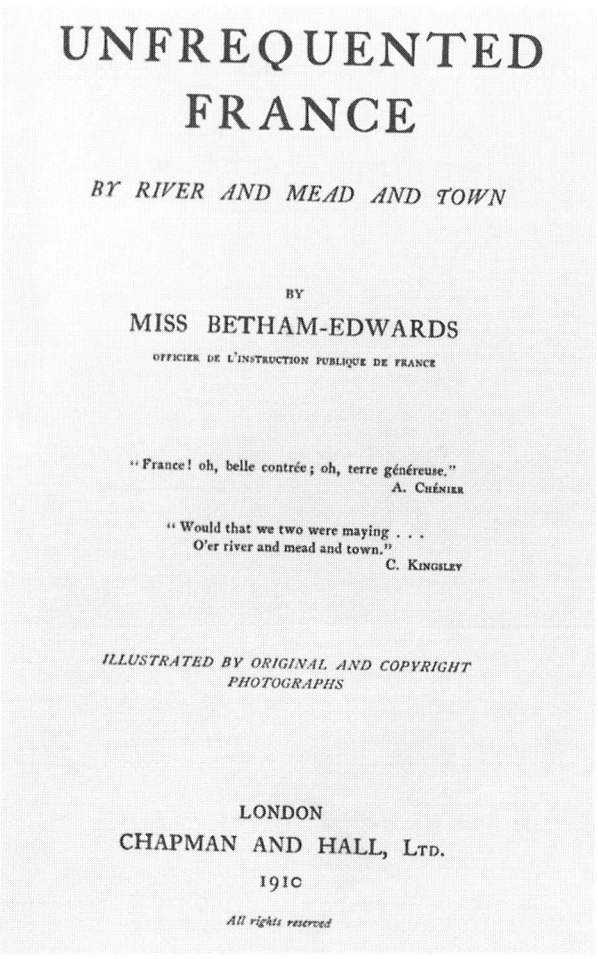

Frontispiece and title page for a collection of travel sketches previously published in Betham-Edwards's Holidays in Eastern France *(1879) and* The Roof of France; or, The Causses of the Lozère *(1889)*

famous towns and cathedrals in the provinces — through the help of her French acquaintances and friends. For her, "the real France" soon came to mean working France: as she writes in the preface to *Holidays in Eastern France,* "I saw, not only places, but people, and not only one class, but all.... Wherever I went, moreover, I felt that I was breaking new ground, the most interesting country I visited being wholly unfamiliar to the general run of tourists."

Not everything in that interesting country was instantly appealing to her, however; in both of these early books on France, Betham-Edwards struggles, even with the guidance of her French friends, to exercise her judgment free of standard English prejudice. In *A Year in Western France* she complains about hotels and reports having scolded noisy servants as a typical tourist might have done, and she does not hesitate to repeat the conventional English judgment on French national character: "*Il faut s'amuser*

is the guiding maxim of existence." At the same time, though, her analytical habit of mind forces her to admit that, despite the shameful love that Frenchwomen have for fine clothing, they are admirable mothers and housewives and that the national character of France is, in fact, "fundamentally democratic."

Only two years later, in *Holidays in Eastern France,* Betham-Edwards moves closer to her mature appreciation for French life and manners without abandoning the light, conversational style of *A Winter with the Swallows*. In Brittany she notes that women's finery is laid aside on working days and that "nowhere in France do people work harder than here." In *Holidays in Eastern France* even terrible hotels are not disasters; though Betham-Edwards occasionally warns "fastidious travellers" away from certain areas of rural France, she defends one of "the worst in the matter of accommodation" by praising the honesty and willingness of the local people to oblige the traveler. By 1879 the conversion of her identity from that of an Englishwoman to that of an adopted Frenchwoman was well under way.

The next decade was a busy one for Betham-Edwards the novelist as well as for Betham-Edwards the traveler and travel writer. *The Dictionary of National Biography* remarks that between 1857 and 1917 there were only eight years when she failed to publish either a new book or a new edition of one of her older books. She continued to visit France, to stay as long and travel as widely as possible while she was there, and to make notes for future works. *Friends over the Water: A Series of Sketches of French Life* (1879) and *The Roof of France; or, The Causses of the Lozère* (1889) focus, as do her previous travel books, on people in their native places; these volumes also display Betham-Edwards's staunch anticlericalism, her growing admiration for the thriftiness and family loyalty she finds among all classes of French society, and her affection for the French language and landscape.

The book that Betham-Edwards seems to have valued most was *France of To-Day: A Survey Comparative and Retrospective* (1892–1894), a two-volume political, social, and economic survey of the country. *France of To-Day* is organized in Betham-Edwards's characteristic fashion, as a series of journeys that any other traveler could undertake. She is careful to describe only what she has observed and, as always, to concentrate on the lives of French people rather than on descriptions of scenery and monuments that any standard guidebook could provide. She sets out her guiding principle in an introductory statement that might characterize any of her books on France: "A year of honest investigation, a few months of sympathetic intercourse, will better serve us than accumulated hours of laborious study at home; but a glimpse, a hint, a perception, will not suffice."

Her theme, repeated in much the same wording in every chapter, is that France embodies "the joy of civilization" more than any other nation. In developing this theme, she describes lending libraries for working people in Nevers, the improved treatment of animals in Clermont-Ferrand, the importance of museums in country towns all over France, and the "social, moral, and material ascendency [sic]" of the French peasant landowner. Without ignoring the faults that she sees – the lack of spirituality among peasants, the "immoral" influence of Catholic theology among Frenchwomen, and the distrust between the French middle class and the peasants – Betham-Edwards insists on the veracity of her generalization drawn from hundreds of closely packed particular observations: France shows the highest level of democracy and civilization combined in the modern world.

In *France of To-Day* Betham-Edwards for the first time resorts to a technique that appears repeatedly in her later travel books: she borrows extensively from her previously published works. As was a common practice in Victorian publishing, her earlier volumes of travels collect articles from periodicals. *A Year in Western France* includes six chapters that had appeared (in virtually identical form) in *Fraser's Magazine* in 1875. *France of To-Day,* moving beyond this convention, incorporates sections of all Betham-Edwards's earlier (and, by 1892, out-of-print) books on France – a method of filling out her surveys of foreign life that is even more obvious in *Anglo-French Reminiscences, 1875–1899, Unfrequented France, by River and Mead and Town* (1910), *Under the German Ban in Alsace and Lorraine* (1914), and her "return" to Algeria, *In French-Africa: Scenes and Memories* (1912).

After 1884 Betham-Edwards lived quietly at her house in Hastings, seldom traveling outside England and relying on her memories, published and unpublished, to create new novels and travel books. When she repeats material in her travel writing, her descriptions of people, places, and events change very little, for the most part, from one printing to another. What does change is the connecting material – the general statements that link particular episodes and the descriptions of some incidents that she left out of earlier books – and therefore the overall effects of her experiences. She remembers events less joyfully than before; disappointments and discomforts are more prominent in the later accounts of her experiences. As Betham-Edwards grew older, her style became more formal and "literary" as well.

In place of the conversational tone of *A Winter with the Swallows,* for example, *In French-Africa* offers such descriptive phrases as "the laziest loon imaginable" and "the simple canticles among the wild African hills." Both Betham-Edwards's friends and her critics noticed the change, and Grand most gently describes it in the personal sketch of her friend that introduces Betham-Edwards's posthumous *Mid-Victorian Memories:* beginning with her early works, Grand notes, Betham-Edwards "spoke as she wrote," although "her later style in writing was sprinkled with preciosity and became somewhat archaic." Betham-Edwards's first "teachers" – Shakespeare, Milton, and the Bible – returned forcefully to her mind and her writing after she reached old age.

Photograph of Darby and Joan Vendée published in Home Life in France *(1905) to illustrate Betham-Edwards's study of the French national character*

The great exception to this stiffness in Betham-Edwards's style after middle age is a book often praised by early-twentieth-century readers and often mentioned by other writers after her death: *Home Life in France* (1905). Rather than taking the form of a series of journeys or examining France region by region, this book presents the country to English readers through a group of witty character sketches. The nation is its people: Betham-Edwards had been making that argument about various countries for nearly forty years, and in *Home Life in France* she makes it again through the form of her book. Chapters include every stage of the French national character and its idiosyncrasies, beginning with babyhood, covering both sexes and a range of occupations for both, and ending with an overview of the state of Anglo-French relations.

Betham-Edwards had seldom attacked the problem so directly, and never in such an extended work. Rather than waiting for her English readers to abandon their stereotypes and draw the correct, unflattering comparisons between themselves and the French people about whom she had written for years, she organizes the book around the stereotypes and piles detail upon detail to undermine them. In a sense, *Home Life in France* marks a full circle to Betham-Edwards's career as a writer about French society. *A Year in Western France* had begun with stereotypes that she gradually abandoned as she traveled farther and learned more; twenty-eight years later she returned to her starting point, ready to use her knowledge to lead readers down the same road that she had traveled to a personal entente cordiale.

For the last twenty-five years of her life Betham-Edwards lived in retirement at her "Villa Julia" in Hastings, with the company of Emily Morgan, her faithful maid, and such friends and neighbors as Grand, Patmore, and the positivist thinker Frederic Harrison. She enjoyed visiting and being visited by her friends, who both dreaded and loved her exacting demands on their hospitality, but the outside world was never allowed to interrupt her quiet life. Throughout her career she had remained carefully detached from public life and public issues: she had refused to support Barbara Bodichon's work on behalf of woman's rights as adamantly as she had refused to serve on charitable boards and committees. She maintained the same detachment even after the outbreak of World War I; the war hardly affected her, either emotionally or physically.

In 1916, almost sixty years after publishing her first novel, Betham-Edwards wrote her last novel, *Hearts of Alsace,* and she complained half seriously to her friends that the year "won't, I fear, bring me my deserts, viz., the title of *Baroness* as accorded to Miss [Angela] Burdett-Coutts," the banker and philanthropist. Despite – or perhaps because of – the honors she had received from her adopted country of France, Betham-Edwards never gained more public recognition from the British government than a civil list pension. Toward the end of 1918 she was eagerly awaiting publication of a collection of memoirs, personal sketches of her famous friends. She had begun a new novel to be called "Bitter Sweet" and was planning yet another novel, the title of which was too exciting to divulge. Friends thought that she had recovered from a fall outside her house in March, but she suffered a stroke on 8 December and died on 4 January 1919.

The London *Times* obituary about her on 7 January 1919 took little notice of her many novels. Matilda Barbara Betham-Edwards was to be most remembered as the interpreter of her adopted nation to the nation of her birth. In an era when travel writers were often expected to thrill their readers with accounts of exotic locales, not to mention feats of physical or moral courage, Betham-Edwards's quiet crossings and recrossings of France may have lacked obvious audience appeal. *The Dictionary of National Biography* summed up her work, travel writing and novels alike, concisely and accurately: "She made repeated use, always with freshness, of a comparatively narrow range of material – recollections of Suffolk, strong anti-clerical and other prejudices, lasting enthusiasm for certain persons and certain books, and above all an intense attachment to France." She praised friendship in everything she wrote that was based on her travels: friendship between individuals, between families, between nations. In her travel writing she tried to cement those friendships for the reading public, as she had cemented them in her own life.

References:

Helen C. Black, *Notable Women Authors of the Day* (Glasgow: Bryce, 1893);

Sarah Grand, "Personal Sketch: Matilda Barbara Betham-Edwards, 1837[*sic*] – 1919," in Betham-Edwards's *Mid-Victorian Memories* (London: John Murray, 1919), pp. vii–lxvi.

Douglas W. Freshfield
(27 April 1845 – 9 February 1934)

Julie English Early
University of Alabama in Huntsville

BOOKS: *Across Country from Thonon to Trent: Rambles and Scrambles in Switzerland and the Tyrol* (London: Spottiswoode, 1865);

Travels in the Central Caucasus and Bashan: Including Visits to Ararat and Tabreez and Ascents of Kazbek and Elbruz (London: Longmans, Green, 1869);

Italian Alps: Sketches in the Mountains of Ticino, Lombardy, the Trentino, and Venetia (London: Longmans, Green, 1875);

A Tramp's Wallet of Alpine and Roadside Rhymes, anonymous (London: Privately printed by Elkin Matthews, 1890);

Memorials of Henry Douglas Freshfield, by Freshfield and Augusta Charlotte Freshfield (London, 1892);

The Exploration of the Caucasus, 2 volumes (London & New York: Arnold, 1896);

Round Kangchenjunga: A Narrative of Mountain Travel and Exploration (London: Arnold, 1903); reprinted, with an introduction by Harka B. Gurung (Katmandu, Nepal: Ratna Pustak Bhandar, 1979);

Hannibal Once More (London: Arnold, 1914);

Unto the Hills (London: Arnold, 1914);

The Life of Horace Benedict de Saussure, by Freshfield and Henry F. Montagnier (London: Arnold, 1920);

Below the Snow Line (New York: Dutton, 1922; London: Constable, 1923);

Quips for Cranks and Other Trifles (London: Privately printed by W. Clowes & Sons, 1923).

OTHER: *Alpine Journal,* 6-9, edited by Freshfield (1872-1880);

The Caucasus, in *Illustrated Travels: A Record of Discovery, Geography, and Adventure,* volume 2, edited by H. W. Bates (London & New York: Cassell, Petter & Galpin, 1875);

A Handbook for Travellers in Switzerland, the Alps of Savoy and Piedmont, the Italian Lakes, and Part of Dauphiné, sixteenth edition, revised, edited by Freshfield (London: John Murray, 1879); seventeenth edition, revised (London: John Murray, 1886);

Hints to Travellers, Scientific and General, fifth edition, enlarged, edited for the Council of the Royal Geographical Society by Freshfield, H. H. Godwin-Austen, and J. K. Laughton (London: Royal Geographical Society, 1883); sixth edition, revised and enlarged, edited by Freshfield and Capt. W. J. L. Wharton (London: Royal Geographical Society, 1889); seventh edition, revised and enlarged, edited by Freshfield and Wharton (London: Royal Geographical Society, 1893);

Clinton Thomas Dent, *Mountaineering beyond the Alps*, contributions by Freshfield (London: Longmans, Green, 1892; Boston: Little, Brown, 1892);

"On Mountains and Mankind," in *Smithsonian Institution: Annual Report, 1904* (Washington, D.C.: Government Printing Office, 1905), pp. 337–354;

Pope Pius XI, *Climbs on Alpine Peaks,* translated by J. E. C. Eaton, foreword by Freshfield (London: Unwin, 1923).

In his eight books on mountain travel and mountaineering, his extensive writing for and editing of mountaineering periodicals, and his leadership of societies dedicated to mountain exploration, Douglas W. Freshfield represents the learned gentleman of independent means, zealously dedicated to his avocation. "With the passing of the great pioneer," the editor of the *Alpine Journal* remarked at Freshfield's death in 1934, "comes the close of an epoch in Mountaineering History; Freshfield was a survivor – probably the sole – of the so-called Golden Age." Freshfield's perspectives were those of the well-placed Victorian, yet much of the information in his works on the Caucasus Mountains and the Himalayas would remain significant into the mid twentieth century. His later writings, in particular, would realize the ideal of what Freshfield called, in his article "On Mountains and Mankind" (1905), the merger of "a man of science and a man of letters."

Douglas William Freshfield was born in Hampstead on 27 April 1845, the only son of Henry Ray Freshfield, a solicitor for the Bank of England, and Jane Quentin Crawford Freshfield, the daughter of William Crawford, member of Parliament for the City of London. In 1854 Freshfield began ten years of annual Alpine travel with his parents during his summer holidays from Eton and, later, from University College, Oxford. His mother's *Alpine Byways; or, Light Leaves Gathered in 1859 and 1860* (1861), by "A Lady," and her *A Summer Tour in the Grisons and Italian Valleys of the Bernina* (1862) recount the family's experiences in less frequented regions of the Swiss and Italian Alps, including Freshfield's 1861 climb of Monte Nero in the Bernina – the first of his twenty-seven first ascents.

While at Oxford, Freshfield began traveling in the mountains with friends; his journal of these trips was published as his first book, *Across Country from Thonon to Trent: Rambles and Scrambles in Switzerland and the Tyrol* (1865). He graduated from Oxford with honors in law and history in 1868. That year he ventured beyond the Alps to the mountains of the Near East, a journey that resulted in his *Travels in the Central Caucasus and Bashan: Including Visits to Ararat and Tabreez and Ascents of Kazbek and Elbruz* (1869). In 1869 he married Augusta Charlotte Ritchie, daughter of William Ritchie, advocate-general of Calcutta. Freshfield was called to the bar in 1870, but, having private means, including substantial acreage in Sussex, he never practiced. He and his wife would have five children; they would pay tribute to their only son, who died in childhood, in *Memorials of Henry Douglas Freshfield* (1892).

Freshfield became a member of the Alpine Club in 1864; from 1869 to 1871 he served on its governing committee; from 1872 to 1880 he edited its *Alpine Journal;* he served as vice president from 1878 to 1880 and as president from 1893 to 1895; and he was made an honorary member in 1924. He contributed at least 176 long, signed articles to the *Alpine Journal* and, particularly during his editorship, many more unsigned pieces. His work also appeared in general-interest periodicals such as *Fraser's, The Academy,* the *Edinburgh Review, Saturday Review,* and *Popular Science Monthly,* and he was a frequent contributor to the *Proceedings of the Royal Geographical Society* and the *Geographical Journal.* He became a fellow of the Royal Geographical Society in 1869, served as honorary secretary from 1881 to 1894, received the Founder's Medal in 1903, served as president from 1914 to 1917, and was appointed trustee in 1924.

As editor, he shaped the concerns of the *Alpine Journal* to emphasize the latter part of its subtitle, *A Record of Mountain Adventure and Scientific Observation;* the previous editor, Leslie Stephen, had given preference to adventure. In both societies he worked tirelessly to develop geography as a discipline and to create a broader audience for geographical knowledge. In the Royal Geographical Society he was the principal advocate for establishing and subsidizing a chair of geography at Oxford in 1888 and at Cambridge in 1903. In 1893 he battled George Nathaniel, first Baron and first Marquis Curzon of Kedleston, over the issue of admitting women to the society. He indefatigably urged improvements in ordnance mapping and in the collecting and cataloguing of photographs. Against some fellows' objections to popularizing methods, he urged public lantern-slide lectures and a more attractive, illustrated *Geographical Journal* to replace the society's stuffy *Proceedings.* He long championed medal recog-

Title page for the work that was the standard source of information on the Caucasus Mountains for at least half a century

nition of mountaineers, and he was a member of the committee to develop Everest expeditions. His commitment to professionalization transformed the Royal Geographical Society's *Hints to Travellers* from collected tips to substantial essays on the methodologies of geography and ethnography. Under his editorship, the fifth (1883), sixth (1889), and seventh (1893) editions emphasize scientific observation and knowledge.

Believing that mountain travel and exploration should be conducted for the sake of acquiring knowledge, Freshfield scorned those who sought only daring conquests of peaks. A prudent mountaineer, he had little use for dangerous feats and deplored the advent late in the century of mechanical aids such as crampons, swivels, and pitons that he thought fostered such feats. His sympathy for mountaineers who died on slopes was, in many cases, tempered by anger at the self-induced "slaughter of Incompetents and Innocents." Considerably after judicious use of additional gear had been widely accepted, Freshfield still held that nailed boots, ice axes, and rope were the only appropriate adjuncts to the climber's own developed balance, rhythm, and stamina. In his contribution to Freshfield's obituary in the *Alpine Journal,* E. J. Garwood recalled that Freshfield rarely carried a rucksack and climbed in a gray cutaway tailcoat and trousers rather than in a special mountaineering outfit.

Freshfield's early books are the works of a young man with few doubts. *Across Country from Thonon to Trent* and *Travels in the Central Caucasus and Bashan* each detail a single trip; although climbing is at the heart of both books, they are broadly focused accounts of cultural encounters and observations.

In its prefatory account of Freshfield's journey from Cairo through Syria, *Travels in the Central Caucasus and Bashan* shows the influence of Alexander William Kinglake's wildly successful *Eöthen* (1844). As a Near Eastern tourist Freshfield, like Kinglake, enacts the amused and only semitolerant young English gentleman, superior to all he sees and alert to the cringing obsequiousness, shirking laziness, wily greed, and insolence that he expects to be endemic to the "oriental mind." Less in the spirit of *Eöthen*, however, *Travels in the Central Caucasus and Bashan* is burdened by complaints, apologetically acknowledged as "almost too trifling to record," that demonstrate "the difference between travel in a country organised for pleasure-visitors, and one entirely, so to speak, in a state of nature." The paradox of Freshfield's early books is their combination of a desire for the unspoiled "state of nature" on a mountaintop or in an un-Westernized village, on the one hand, with a demand for accommodation to Western habits. In Syria and the Caucasus he seeks authenticity off the beaten path, yet regrets that in only a few places "are English tastes well understood." Mountain peoples are primitive barbarians, with their Christianity or "Mohammedanism" only a thin veneer over superstition. Russian officialdom is nearly equally uncivilized. Freshfield's ethnographic comments summarize the type, the "usual" characteristics, but seldom render the individual.

The discussion of the traveler's trials and the swift classification of "the natives" is far less successful than the spirit Freshfield brings to scholarly debate and to portraying the experience of mountain climbing. In Syria he explores the remote "Giant Cities" of Bashan that Joseph Porter had argued in 1874 were the cities of the ancient giant people of Og. Freshfield dismisses the claim, finding a misreading of the evidence that would be acceptable only to those with "a preconceived theory to support." His narrative gains vigor and loses its ethnocentric posturing when he writes of the mountains. His first climb in the Near East, Mount Ararat, is a disappointment. Freshfield succumbs to mountain sickness and waits on a lower rock shelf while the leader cuts more than a thousand steps in a sheer ice face, but ultimately exhaustion prevents ascent of the last thousand feet. Chastened by the failure of their eighteen-hour climb, the party prepares for the goal of the journey, the two highest peaks of the Caucasus Mountains. They soon achieve a first ascent of the 16,558-foot Kazbek, but the careful observation of the glaciers, passes, and lesser peaks separating it from the 18,481-foot El'brus forms the most lasting contribution of the exploration. For more than two months they traverse the 120 miles of the central Caucasus Mountains that separate Kazbek and El'brus, crossing eleven passes between eight thousand and twelve thousand feet and exploring the sources of eight rivers and the flanks of the main chain. The climax of the narrative is the first ascent of El'brus, Europe's highest peak. Throughout, Freshfield delineates the character of the Caucasus Mountains by comparing the chain to the Alps, a region with which most of his readers would have been familiar.

Freshfield's third book, *Italian Alps: Sketches in the Mountains of Ticino, Lombardy, the Trentino, and Venetia* (1875), published midway through his term as editor of the *Alpine Journal*, reflects his greater experience both as writer and mountaineer. Rather than conveying the adventure and achievement of a single journey, Freshfield synthesizes multiple visits; drawing on previously published articles and "a patchwork from the journals of seven summers," he addresses the problems produced by the mountaineering mania and the rush to the Alps of the 1850s and 1860s. In 1865 the deaths of climbers in the party of the preeminent mountaineer Edward Whymper on the return from the first ascent of the Matterhorn prompted the *Times* of London to question the sense, even the morality, of mountaineering. Freshfield saw, even apart from the controversy over the Whymper party's errors, that increased Alpine travel, diminishing the exclusiveness of the experience, prompted ill-advised ascents on ever more hazardous routes under unwise conditions simply to secure distinction. *Travels in the Central Caucasus and Bashan* represented Freshfield's own commitment to untrammeled regions and significant first ascents, but in *Italian Alps* he argues for the less dramatic and less frequented southern Alpine slopes on which "an eclectic wanderer," as he characterizes himself, would find compelling interest and a broad range of pleasures both below and above the snow line. In his preface he addresses the paradox that mountain-travel writing celebrates singular adventure yet cultivates increasing numbers of devotees. Seeing it as "inevitable that these mountains should be brought before the world," he justifies his own contribution: "It seemed better that they should be introduced by one who had with them a friendship of some years standing rather than by a new acquaintance." He followed this principle in preparing two editions of the John Murray publishing firm's *A Handbook for Travellers in Switzerland, the Alps of Savoy and Piedmont, the Italian Lakes, and Part of Dauphiné* (1879, 1886).

By 1896 Freshfield had completed three major journeys in the Caucasus Mountains – in 1868,

Freshfield and two members of his 1899 expedition to climb Kangchenjunga, the Himalayan mountain in the background

1887, and 1889 – and had become steeped in English and Continental mapping and literature of the region since his own contribution more than twenty-five years earlier. His study of the chain increased his respect for its dangers even for prudent climbers; his somber third trip had sought the fate of his friends W. F. Donkin and H. Fox, who had disappeared in 1888. In *The Exploration of the Caucasus* (1896) Freshfield writes for an audience of his peers who are versed in the scientific debates on glaciation and the geologic formation of mountains and who have broad knowledge of history and literature. Nonetheless, *The Exploration of the Caucasus* is a highly readable, eclectic, and successful merger of geographic and cultural exposition, scholarly scientific debate, detailed observation, and personal narrative. Freshfield's impatience with folly is directed at the overspecialized "scientists" (he places the word in quotation marks) who theorize without benefit of direct experience, and he corrects their misinterpretations by pointing out features that would be "obvious . . . to every mountain traveller who uses his eyes." The two-volume work, illustrated with outstanding photographs by Vittorio Sella, was regarded as standard for several decades after Freshfield's death.

Freshfield returned, with greater maturity, to the narrative of a single expedition in *Round Kangchenjunga: A Narrative of Mountain Travel and Exploration* (1903) to describe his 1899 exploration in the Himalayas of the world's third-highest peak. In seven weeks the party made the difficult circuit of the 28,208-foot massif, reconnoitering its approaches and delineating its five main glaciers and many of its minor ones. Breaching closed Tibetan borders, they were the first westerners to view Kangchenjunga from other than its southern, Sikkimese face. Freshfield found the extraordinary "variety of expression in the vast landscape . . . an almost unbelievable vision," and he describes it with evocative precision:

> At dawn the lower mountains shone . . . through a veil of amethystine vapour; the damp air converted the naturally green hills of the middle distance into blocks of blue or mauve colour, recalling the background of an

old Venetian picture. Soon the sun's shafts, darting into the thousand hollows . . . drew the warm, moist air out of their depths in transparent waves. These, as they rose into a colder atmosphere, were condensed and formed into tall, luminous, sharp-edged columns of clouds, which rose vertically until, caught by some upper current, they bent and broadened at the top, and finally broke up into detached fragments, which floated away slowly northwards to lodge in the hollows of the snows during the hours of noontide heat.

In Freshfield's view, "there are few mountains . . . of more formidable aspect"; the peak would not be achieved until 1955. Although his party was hampered by unusually heavy snows even at low altitudes, Freshfield's book remains a classic of Himalayan literature, warranting its republication in Nepal in 1979 as a work that, according to the editor, "epitomises an accomplished scholar's appreciation of nature and mountain exploration."

At fifty-four, Freshfield reflects in *Round Kangchenjunga* that the lure of mountains for him has been spiritual, physical, and intellectual: "The human imagination, as much as the human body, needs a playground, and at the same time the human intellect resents a void. . . . the mountain [stands] as a symbol of the existence of something beyond and above the common ken of the dwellers at its feet." He acknowledges that he has "travelled and climbed for scenery first, and science afterwards." He recognizes as "preposterous" vanity the mountaineer's "serene conviction that he alone can properly understand and appreciate the divine architecture," but he also believes that technical knowledge lifts "him for the moment to this upper level of intelligence. He knows by experience what the few discover by intuition, and the many never discover at all."

Freshfield's last significant exploration of territory new to European mountaineers was his 1905 journey to Africa's Mountains of the Moon, or Ruwenzori, a group of six peaks (the tallest 16,763 feet) long postulated as the source of the Nile. Plagued by incessant rain, the expedition stalled at twelve thousand feet. Freshfield's reflections on the journey were drawn together with accounts of his travels to Corsica, Greece, Japan, Morocco, the Alps, and the Apennines, most revised from previous publication in the *Alpine Journal*, for his last book of mountain travel, *Below the Snow Line* (1922). With this "bundle of memories of walks and climbs among the lesser ranges," Freshfield reiterated the commitment expressed fifty years earlier in *Italian Alps* to the challenges and pleasures of rough mountain walking. Reviewers recognized mastery both of material and style, admiring the work's genial and experienced authority. The 22 March 1923 *Times* review termed the collection "a classic"; the *Saturday Review* (21 April 1923) admired Freshfield's "gift for apt and exact description" and was confident that "no guide could be found more nicely fastidious, more acutely discriminating, more sure and catholic in taste. One is quite certain that what he approves deserves approbation."

With his most vigorous climbs behind him, Freshfield pursued mountain scholarship in *Hannibal Once More* (1914), his contribution to the debate over the site of Hannibal's crossing of the Alps, and in *The Life of Horace Benedict de Saussure* (1920), his biography of the eighteenth-century Genevese "father of mountain science." Freshfield's characterization of de Saussure as "far more of a mountain traveler and a scientific observer, a geological student, than a climber" is a good summary of Freshfield's own career, as well. In another personal echo, he is reminded by de Saussure's purple silk frock coat of "the difference between the man who climbed Mont Blanc in that garment and the modern gymnast who thinks himself par excellence the mountaineer."

Freshfield's acerbic view of others' pretensions, failings, and intellectual follies surfaces in all his books but is given fullest play in two privately printed volumes of poetry, *A Tramp's Wallet of Alpine and Roadside Rhymes* (1890) and *Quips for Cranks and Other Trifles* (1923). His commercially published book of verse on his mountain experiences, *Unto the Hills* (1914), more soberly renders the spiritual aspects of his vocation.

Freshfield synthesized the broad range of perspectives he brought to his writing in his 1904 address to Section E (Geography) of the British Association, revised and printed the following year in the *Smithsonian Institution Annual Report* as "On Mountains and Mankind." Freshfield prefaces his agenda for future scientific inquiry into mountain geology, botany, and medicine by reminding his specialized audience that "geography is concerned with the interaction between Man and nature in the widest sense." He calls on biblical, classical, and medieval literature to place the nineteenth-century fascination with mountains on a continuum that suggests "a healthy, primitive, and almost universal instinct" appealing to "the artistic and scientific mind." Significant in religion, history, science, nature, and art, mountains are "the best companions in the vicissitudes of life."

The eclectic interests and trained intelligence that informed Freshfield's books were represented in commitments to associations apart from those concerned with mountain exploration: he was an active conservationist in Sussex, an officer of the Hellenic and Roman Societies, and chairman of the committee of the Society of Authors. He was awarded honorary degrees by the University of Geneva and his own University College, Oxford.

Of Freshfield's books, only *Round Kangchenjunga* has been reprinted. Limited attention has been given to his substantial career, and that only in histories of the societies in which he assumed a prominent role.

References:

"In Memoriam: Douglas William Freshfield (1845-1934)," *Alpine Journal*, 46 (1934): 166-176;

Thomas George Longstaff, "Douglas Freshfield, 1845-1934," *Geographical Journal*, 83 (April 1934): 257-262;

Sir Arnold Henry Moore Lunn, *A Century of Mountaineering, 1857-1957* (London: Allen & Unwin, 1957);

Carlo Mariani, *L'Ombrello di Freshfield: Relazioni di Viaggo e Storia dell'Esplorazione nelle Alpi Apuane, 1865-1905* (Pisa: Giardini, 1986);

Hugh Robert Mill, *The Record of the Royal Geographical Society, 1830-1930* (London: Royal Geographical Society, 1930);

Arnold Louis Mumm, *The Alpine Club Register, 1864-1876* (London: Arnold, 1923);

Vittorio Sella and Edmund J. Garwood, *Catalog of a Collection of Photographs . . . Taken during the Tour of Kanchinjinga Made in 1899 by Mr. D. W. Freshfield* (London: Spottiswoode, 1900).

Papers:

The Royal Geographical Society has a collection of Douglas W. Freshfield's correspondence from 1881 to 1934; the Alpine Club's collection encompasses the years 1912 to 1923. Smaller collections include letters to Oscar Browning at Cambridge's King's College Modern Archive Center and to W. M. Conway at Cambridge's Department of Manuscripts and University Archives. London University holds correspondence with Violet Maram.

Mary Gaunt

(20 February 1861 – 19 January 1942)

James Rigney
Roehampton Institute, London

BOOKS: *Dave's Sweetheart,* 2 volumes (Melbourne: Mullen, 1894; London: Arnold, 1894);

The Moving Finger (London: Methuen, 1895);

Kirkham's Find (London: Methuen, 1897);

Deadman's (London: Methuen, 1898; New York: New Amsterdam Book Company, 1899);

The Arm of the Leopard: A West African Story, by Gaunt and John Ridgewell Essex (London: Richards, 1904);

Fools Rush In, by Gaunt and Essex (London: Heinemann, 1906);

The Silent Ones, by Gaunt and Essex (London: Laurie, 1909);

The Mummy Moves (London: Laurie, 1910; New York: Clode, 1925);

The Uncounted Cost (London: Laurie, 1910; New York: Clode, 1910);

Alone in West Africa (London: Laurie, 1911; New York: Scribners, 1912);

Every Man's Desire (London: Laurie, 1913);

A Woman in China (London: Laurie, 1914; Philadelphia: Lippincott, 1914);

The Ends of the Earth: Stories (London: Laurie, 1915);

A Broken Journey: Wanderings from the Hoang-Ho to the Island of Saghalien and the Upper Reaches of the Amur River (London: Laurie, 1919);

A Wind from the Wilderness (London: Laurie, 1919);

The Surrender, and Other Happenings (London: Laurie, 1920);

Where the Twain Meet (London: John Murray, 1922; New York: Dutton, 1922);

As the Whirlwind Passeth (London: John Murray, 1923);

The Forbidden Town (London: Unwin, 1926; New York: Clode, 1926);

Saul's Daughter (London: Unwin, 1927);

George Washington and the Men Who Made the American Revolution (London: Black, 1929);

The Lawless Frontier (London: Benn, 1929);

Joan of the Pilchard (London: Benn, 1930);

Reflection – in Jamaica (London: Benn, 1932);

Harmony: A Tale of the Old Slave Days in Jamaica (London: Benn, 1933);

Worlds Away (London: Hutchinson, 1934).

Mary Gaunt's adventurous solo journeys to West Africa and China provided material for, and a degree of realism to, the novels and short stories with which she supported herself. Her writing was also a means of asserting her independence in her relationships with her brothers Guy and Ernest, both of whom rose to the rank of admiral in the Royal Navy. Ernest Gaunt served in China during the Boxer Rebellion in 1898–1899, was severely wounded in Somalia in 1903 while leading a landing party to avenge the death of an Italian naval officer, and during World War I commanded the First Battle Squadron of the British fleet at the Battle of Jutland. Guy Gaunt served in the Philippines and China, took command of the British consulate at Apia in Samoa during a rebel attack, served on the battleship *Vengeance* in China during the Russo-Japanese War, and was British naval attaché in Washington, D.C., during World War I. This background accounts for some of Mary Gaunt's conventional loyalty to the British Empire, the frequency with which sailors feature in her works of fiction, and the attraction for her of China and the African coast.

Gaunt borrowed settings and incidents for her tales from the stories her brothers told of their adventures – she thanks one of her brothers for correcting the manuscript for her short story "The Mate's Salvage," published in her 1915 collection *The Ends of the Earth* – but her travels enabled her to add her own experiences as a woman from the margins of the empire encountering the colonial heritage and practices of Europe. The purposeful vigor with which she sought out both the exotic and the mundane in the countries she visited and the tension generated by her need to compete with her brothers account for all that is best in her travel writing. In her later years, when she denied herself the stimulation of demanding journeys, her travel writing lost the qualities that made it so admired in her day and so interesting to later readers.

Mary Eliza Bakewell Gaunt was born on 20 February 1861 in the settlement of Indigo in colonial Victoria, Australia. She was the first of six children of William Henry Gaunt, an English-born civil servant who served as a magistrate in the Australian goldfields, and Elizabeth Mary Gaunt. In *Alone in West Africa* (1911) Gaunt writes of a hereditary predisposition to travel that is typified by her family's presence in Australia; she says that she cannot "remember when any one of us would not have gone anywhere in the world at a moment's notice." Among her father's responsibilities was the protection of Chinese miners, who were often targets of violence by whites. These early experiences gave her an insight into racial animosity that she would bring to her later writings.

Gaunt was educated at Grenville College in Ballarat and in 1881 became one of the first women to matriculate at the University of Melbourne. She discontinued her studies that year, however, and turned to writing, contributing reviews and articles to Australian newspapers such as the *Age*, the *Argus*, the *Australasian*, and the *Sydney Mail*. Her earnings from journalism and from her contributions to popular reference works, including *Cassell's Picturesque Australasia* (1887), enabled her to travel to England and India in 1890–1891.

For her first novel, however, published in 1894 in Melbourne and London, Gaunt chose a setting from her childhood. *Dave's Sweetheart,* a romance set in the Victoria goldfields, initiates her use of the theme of two men of opposed moral qualities fighting for the love of a strong-willed woman. Another theme – the ways in which isolation in pioneer settlements brings out strengths and weaknesses in various individuals – recurs throughout her fiction and provides a regular point of interest in her travel writing.

In 1894 Gaunt married Hubert Lindsay Miller, a physician from Warnambool, Victoria. Miller encouraged her to continue writing under her maiden name. Over the next four years she wrote several romances with an Australian setting: *The Moving Finger* (1895), *Kirkham's Find* (1897), and *Deadman's* (1898). Miller died in 1900, leaving his widow with an income of only thirty pounds a year. Rather than return to live with her parents in rural Victoria, Gaunt launched her career as a travel writer for much the same financial reasons and in much the same personal circumstances as her predecessor, Louisa Stuart Costello, the first female professional travel writer.

Dependent, like many Australian writers of the period, on the London literary market, Gaunt moved to London after Miller's death. She found both the climate and the society of her new home uncongenial. The poverty she found in Victorian England would color her perception of deprivation and suffering in her travels in West Africa and China. In her Jamaican travel book, *Where the Twain Meet* (1922), she says that she saw the English poor as slaves in a bitter, cold, and cheerless country. She goes on to say that since she had a living to earn, she had no time to investigate this situation. Her living depended, she soon discovered, on the evocation of distant lands; she found that she had no taste or talent for stories of English life.

Gaunt collaborated with an English author, John Ridgewell Essex, on two novels set in West Africa: *The Arm of the Leopard* (1904) and *Fools Rush In* (1906). The commercial success of these books enabled her to make her first voyage to West Africa. On her return from the trip in 1908 she and Essex wrote *The Silent Ones* (1909), the tale of an Ashanti, James Craven, who, having studied at Cambridge and in Germany, no longer fits in when he returns

Porters carrying Gaunt across the Tano River during her West African travels

home to Africa. Gaunt returned to the character of the educated black the following year in *The Uncounted Cost.*

The Silent Ones and *The Uncounted Cost,* as well as Gaunt's novel *The Mummy Moves* (1910), were published by T. Werner Laurie. That firm commissioned her first travel book, *Alone in West Africa.* In addition to providing an account of her travels, *Alone in West Africa* is a perceptive study of the difficulty Europeans had in adapting to the continent and the undesirability of African adaptation to Western ways. Interracial marriage and the adoption by Africans of Western styles of dress strike her as particularly lamentable. Disapproval and ostracism are the only responses to mixed marriage that Gaunt can envisage. In her 1913 novel, *Every Man's Desire,* the heroine discovers that her husband, a colonial administrator in Africa, has gone through a form of marriage with a native woman; Gaunt lays some of the blame for such situations on white women who, unlike herself, will not risk going to the tropics. During her travels along the West African coast and into the interior Gaunt did not feel safe around half-civilized black men; she regretted the loss of dignity Africans suffered when they put on Western clothes and adopted Western pretensions. She was particularly scornful of Liberia, where these practices had been encouraged.

Although Gaunt writes encouragingly in *Alone in West Africa* of the economic prospects of the Gold Coast and exhorts her English readers to make the most of them, she was unimpressed with the group of English traders with whom she dined: they were unshaven, and their shirts needed washing. The civility of the French and the efficiency of the Germans, on the other hand, earned her admiration. She recommends the use of a competitive examination, similar to that used in the Indian Civil Service, to select administrators in Africa. Above all, she thinks, "Women are the crying need; quiet, brave, sensible women who are not daunted because the black cook spoils the soup, or the black laundryman ruins the tablecloth, who will take an intelligent view of life, and will make what is so much needed – a home for their husbands." Gaunt's 1926 novel, *The Forbidden Town,* tells the story of just such a woman, who goes to Africa to manage an estate on the West Coast and combines commerce and philanthropy in a way that traders and missionaries cannot.

Gaunt's having been brought up in the wilds of Australia accounts for some of her impatience with the English she meets in Africa. She is particularly disdainful of women who refuse to experience foreign travel or who allow themselves to be prevented from doing so by their husbands. When an English official tells her that his wife could not possibly be brought to Africa because she has such a delicate complexion that she must wash her face in distilled water, Gaunt comments that such a woman bought her complexion at a heavy cost if it deprived her of the joy of seeing new countries. She also says that there is no poverty in Africa to compare with the "unutterable woe" that can be found in an English city.

While most travel writing of the time emphasized the hardship and danger endured by the author and the marvels and curiosities he or she observed, Gaunt's style is low-key and matter-of-fact. The dangers of travel only surface in her fiction, where they take the form – in works such as "The Last White Woman" in *The Ends of the Earth* and her novel *The Lawless Frontier* (1929) – of the plot device of the woman lost in the wilderness. Among Gaunt's distinguishing qualities as a travel writer is her ability to laugh at herself. Having made a difficult crossing of the Eveto Ranges, she settles for the night at a store operated by the Swanzey trading company:

> There are difficulties connected with lodging in a cocoa store, especially when you are surrounded by a population who have never seen a white woman before. I needed a bath, but how to get it I hardly knew, with eyes all over the place, so at last I put out the lights and had it in the dark, and went to bed in the dark, and as I was going to sleep I heard the audience dispersing, discussing the show at the top of their voices. As I did not understand what they said I did not know whether they found it satisfactory. At least it was cheap, unless Swanzey's agent charged them.

Soon after returning from her second trip to Africa in 1910, Gaunt set out for China – a place to which, like Africa, she claims to have been attracted since childhood. Gaunt's brother Lance was related by marriage to George Ernest Morrison, an Australian who was the London *Times* correspondent in Beijing and a political adviser to the government of China, and Morrison provided Gaunt with a base in the Chinese capital. (Morrison's own career as a travel writer could have served as a model for Gaunt's: while still a student he had walked 750 miles from Melbourne to Adelaide, often through unexplored country, and sold an account of his journey to the *Leader;* soon afterward he had traveled 3,000 miles from China to Burma, much of the way on foot, and published his account of the trip as *An Australian in China, Being the Narrative of a Quiet Journey Across China to British Burma,* 1895.)

In the book that resulted from her trip, *A Woman in China* (1914), Gaunt says that she went to China to gain material for a novel rather than to travel as a tourist. The work is a description of a more conventional journey than *Alone in West Africa,* yet the trip was not without its dangers and hardships. Gaunt expresses an uncharacteristic degree of anxiety for her safety while traveling in China: considering the low value that seemed to be placed on women in China, she felt a general sense of threat and insecurity. She depicts the women of China, who must suffer the agony and indignity of footbinding, as the crippled products of an effete civilization. Photographing a court eunuch, sheepishly looking at the camera while seated on a stone "that had been a seat perhaps when Kublai Khan built the palace," she sees him as "the representative of the old cruel past that pressed men and women alike into the service of the great."

Though she admires the majesty of parts of Beijing and the remnants of ancient China, Gaunt is sensitive to the squalor that underlies them and to the incongruities exacerbated by the modernizing and Westernizing trends of the postrevolutionary government. The tatters beneath the gown that she notices at the funeral of the dowager empress are always metaphorically visible to Gaunt in China. Of the imperial pleasure palace at Jehol she comments: "Always it was the same, desolation and dirt and ruin, and the young man who was showing us everything made as if he wanted to impress on us that it did not matter. He belonged to the modern world and these were past and gone."

Although she regrets some of the influence of modernization, she looks on traders more approvingly in China than she did in Africa. She is at first shocked by the incongruous presence of billboards advertising the American Tobacco Company in Inner Mongolia, but then she is filled with admiration for the way the company is bringing China into the modern world. Missionaries in China do not receive her approval; the best that can be said of them is that they are pioneers for trade. This lack of sympathy for missionaries is carried over from *Alone in West Africa,* yet in *A Woman in China* she recounts moving tales of their sufferings during the Boxer Rebellion when she visits the sites of massacres and atrocities. Gaunt also writes with sensitivity and perception of how galling and insulting the Chinese must find the presence of foreign legations in their

Frontispiece and title page for Gaunt's narrative of her trip home from China, during which threats of attacks by outlaws cut short the journey she had planned to make

cities and the restrictions that are placed on Chinese traffic near them.

In the walled town of Tsung Hua Chou in northern China a serving woman pretends to offer her a child to frighten the other children with the possibility that they, too, will be given away to the white woman if they do not behave. This incident takes Gaunt back to her own childhood, when white nurses would pretend to give away children to Chinese vegetable sellers; she is struck by the way she has taken the place of the "smiling, bland John Chinaman of my childhood." Although she cannot completely shed her Western preconceptions, Gaunt's willingness to try to enter into the minds of the people she encounters – a willingness that seems to diminish in her later books – is part of the appeal of her works.

Gaunt's travels through western China and Inner Mongolia were undertaken in conditions of great difficulty, by unsprung cart and flatboat. She writes, however, with her usual self-deprecating humor about how she was carried backward in a chair down from the Nine Dragons Temple and how the chief magistrate of Pa Kou provided her with what he described as cavalry protection: a soldier armed with a fly-whisk. In 1914, at the end of her travels, she rented a temple in the hills west of Beijing and settled down to compose *A Woman in China*. A few months later she set out to travel once more to the Chinese interior but was forced to abandon this journey because of the threat of banditry along the way. She decided, instead, to make her way back to England via Siberia.

The trip home became the subject of her third travel book, *A Broken Journey* (1919), and also provided material for her short-story collection *The Surrender, and Other Happenings* (1920). *A Broken Journey* is a slow-moving and disengaged book that lacks the vigor of her earlier works. World War I impinged on her journey when the ship on which she was traveling back to England was intercepted by a German raider and the British sailors traveling on board were taken prisoner. Her main concern through most of the closing chapters of the book is her efforts to get the dog who accompanied her through China past Swedish quarantine.

Gaunt concludes the book by quoting from a letter from her brother Cecil, a soldier serving in the Mesopotamian campaign. The letter quotes the lines: "Salt with desire of travel / Are my lips: and the winds wild singing / Lifts my heart to the Ocean, / And the sight of the great ships swinging." Gaunt comments: "And my heart echoed 'And I too! And I too!' " These arduous journeys, to places women rarely went (and where, indeed, male travelers were few), were undertaken in the cause of independence. At one level the independence is financial: the journeys were commercial exercises related to her writing. At the same time, they asserted her right to behave in a way permitted to her brothers but deemed, as she recalls, "improper for a young lady" and "likely to inhibit my chances of marrying." In *Alone in West Africa* she sees her youth in the Australian bush as the source of her independence, while at the same time it showed her the limitations to which women were subject: "Only for us girls there was no prospect. Our world was bounded by our father's lawns and the young men who came to see us and made up picnic parties to the wild bush around for our amusement.... It was not bad, but it was not as good a life as the boys of the family were having." Gaunt's travel writing shows her living the life she wished.

For the next decade Gaunt lived in Bordighera, Italy, with a female companion and concentrated on mining her travel experiences for material for her fiction. She published several novels, including *A Wind from the Wilderness* (1919), whose heroine is an American missionary doctor in China, and *As the Whirlwind Passeth* (1923) and *Joan of the Pilchard* (1930), both of which are set in colonial Australia. The collection *The Surrender, and Other Happenings* includes stories about the bandits of the Chinese interior and the U-boats of the North Sea.

Gaunt produced two more travel books, both about Jamaica: *Where the Twain Meet* and *Reflection – in Jamaica* (1932). Both are less satisfactory than her earlier works; Jamaica was a less exotic location than West Africa or China, and there are no adventures for her to relate. The itinerary for the journey recorded in *Reflection – in Jamaica* was planned by the editor of a local newspaper. What she produces in each case is a travel book seen mainly from the verandas of hotels and plantation houses and, unlike *Alone in West Africa* and *A Woman in China*, designed for the ordinary tourist, who could replicate Gaunt's itinerary. Although she does not consider herself a tourist, she concedes that tourists serve an important function in Jamaica's development and "I, who love the wilds must not complain." *Harmony: A Tale of the Old Slave Days in Jamaica* (1933), the novel that derived from the trip, is similarly two-dimensional.

Where the Twain Meet is a sympathetic account of the history of Jamaica from the time of Christopher Columbus. The book draws heavily on historical documents and the work of earlier historians (notably Lady Mary Nugent's *Journal of Residence in Jamaica from 1801 to 1805,* first published in 1839 and republished in 1907). The historical passages are interspersed with Gaunt's contemporary observations. Gaunt tries to link her visit to Jamaica to her earlier trip to Africa by suggesting that her journey re-creates that of slaves traveling from Africa to the West Indies.

Her visits to plantations and reading of historical records make her "dimly understand" what the poor and the slaves suffered in earlier days, but the sight of "clothed and ardent church-goers who a little while before had been naked savages" is offered as a sign of the shared progress of blacks and whites in Jamaica. Her conclusion is that the descendants of slaves have attained a standard of living that they never could have known had their ancestors not been sold into slavery.

Gaunt is struck by Jamaica's combination of a primeval landscape with a high degree of civilization; it is possible to observe the countryside without being inconvenienced by its rugged features. This quality leads Gaunt to a more sedate and sedentary approach than she has shown before. She talks more about comfort in *Reflection – In Jamaica* than in any other book and depicts herself as seeking it out more than before. The change is understandable, considering that she was in her sixties by this time; but along with it comes a lessening of the interests and sympathies that marked her earlier works. She confesses that it is difficult to enter another's mind and that the point of view of her servants in Jamaica remains a mystery to her. This insight into local minds was something she sought and managed to evoke (no matter how incompletely) during her journeys in Africa and China; she is now content to offer a twentieth-century reading of Lady Nugent's journal and a picture of the past and survivals of planter society in Jamaica.

Gaunt was forced to abandon her home in Fascist Italy and flee to the south of France in 1940; she had to leave behind most of her possessions, including the mementos of her travels. Her health had never been robust; but in her younger days she had traveled strenuously, and she had evangelized both in her travel books and in her fiction for fresh air as a preventative for disease. She died of respiratory illness in Cannes on 19 January 1942.

Arriving in Beijing, Gaunt had wondered aloud how she would get along in China. A young man was sitting opposite her in the railway carriage: " 'Oh, you'll be all right,' said he. 'The Chinese'll like you because you're fat and o – ' and then he checked himself, seeing, I suppose, the dawning wrath in my eyes. The Chinese admire fat people and they respect the old, but I had not been accustomed to looking upon myself as old yet, though I had certainly seen more years than he had, and as for fat – well I had fondly hoped my friends looked upon it as pleasing. With these chastening remarks sinking into my soul we rolled into the railway station." Stout and indomitable, Gaunt imposed her will on porters, on officials, and on conventions as to what a woman may do. She recounts in *A Broken Journey* that she showed her earlier book on China to a male friend, who objected to the "fuss" she made about the condition of women in China and warned that unless it was modified people would take her for a suffragette. Gaunt replied that she was an ardent suffragist who believed that a woman was most valuable neither as an angel nor as a slave but as a useful citizen. She concludes that her friend not only knew nothing about the women of China who swarmed around him but probably knew nothing about the women of his own race. Her career as a travel writer represents Gaunt's lifelong program of asserting her right to be counted as a useful citizen in the literary world.

Bibliography:

Ian McLaren, *Mary Gaunt: A Cosmopolitan Australian* (Parkville: University of Melbourne Library, 1986).

References:

Margaret Bradstock, "Mary Gaunt in China," *Southerly*, 50, no. 4 (1993): 151–160;

Sue Martin, " 'Sad Sometimes, Lonely Often . . . Dull Never': Mary Gaunt, Traveller and Novelist," in *A Bright and Fiery Troop: Australian Women Writers of the Nineteenth Century,* edited by Debra Adelaide (Ringwood, Victoria, Australia: Penguin, 1988), pp. 183–197.

R. B. Cunninghame Graham
(24 May 1852 – 20 March 1936)

Ira Grushow
Franklin and Marshall College

See also the Graham entries in *DLB 98: Modern British Essayists, First Series* and *DLB 135: British Short-Fiction Writers, 1880–1914: The Realist Tradition.*

BOOKS: *The Nail and Chainmakers,* by Graham, J. L. Mahon, and C. A. Conybeare (London: London Press Agency, circa 1889);

Economic Evolution (Aberdeen: Leatham / London: Reeves, 1891);

Notes on the District of Menteith, for Tourists and Others (London: Black, 1895);

Father Archangel of Scotland and Other Essays, by Graham and Gabriela Marie Cunninghame (London: Black, 1896);

The Imperial Kailyard: Being a Biting Satire on English Colonisation (London: Twentieth Century Press, 1896);

Aurora La Cujini, a Realistic Sketch in Seville (London: Smithers, 1898);

Mogreb-el-Acksa: A Journey in Morocco (London: Heinemann, 1898; revised edition, London: Duckworth, 1921; New York: Viking, 1930);

The Ipané (London: Unwin, 1899; New York: Boni, 1925);

Thirteen Stories (London: Heinemann, 1900; Harmondsworth & New York: Penguin, 1942);

A Vanished Arcadia: Being Some Account of the Jesuits in Paraguay, 1607 to 1767 (London: Heinemann, 1901; New York: Macmillan, 1901);

Success (London: Duckworth, 1902);

Hernando de Soto: Together with an Account of One of His Captains, Gonçalo Silvestre (London: Heinemann, 1903; revised, 1912; New York: Dial, 1924);

Progress and Other Sketches (London: Duckworth, 1905);

His People (London: Duckworth, 1906);

Faith (London: Duckworth, 1909);

Hope (London: Duckworth, 1910);

Charity (London: Duckworth, 1912);

Success and Other Sketches ((London: Duckworth, 1912);

A Hatchment (London: Duckworth, 1913);

Scottish Stories (London: Duckworth, 1914);

Bernal Díaz del Castillo: Being Some Account of Him, Taken from His True History of the Conquest of New Spain (London: Nash, 1915; New York: Dodd, Mead, 1915);

Brought Forward (London: Duckworth, 1916; New York: Stokes, 1917);

A Brazilian Mystic: Being the Life and Miracles of Antonio Conselheiro (London: Heinemann, 1920; New York: Dodd, Mead, 1920);

Cartagena and the Banks of the Sinú (London: Heinemann, 1920; New York: Doran, 1920);

The Conquest of New Granada: Being the Life of Gonzalo Jimenez de Quesada (London: Heinemann, 1922; Boston: Houghton Mifflin, 1922);

The Dream of the Magi (London: Heinemann, 1923);

The Conquest of the River Plate (London: Heinemann, 1924; Garden City, N.Y.: Doubleday, Page, 1924);

Inveni Portam: Joseph Conrad (Cleveland: Rowfant Club, 1924);

Doughty Deeds: An Account of the Life of Robert Graham of Gartmore, Poet and Politician, 1735–1797, Drawn from His Letterbooks and Correspondence (London: Heinemann, 1925; New York: MacVeagh, 1925);

Pedro de Valdivia, Conqueror of Chile (London: Heinemann, 1926; New York & London: Harper, 1927);

Redeemed, and Other Sketches (London: Heinemann, 1927);

Bibi (London: Heinemann, 1929);

José Antonio Páez (London: Heinemann, 1929; Philadelphia: Macrae-Smith, 1929);

The Horses of the Conquest (London: Heinemann, 1930);

Writ in Sand (London: Heinemann, 1932);

Portrait of a Dictator: Francisco Solano Lopez (Paraguay, 1865–1870) (London: Heinemann, 1933);

With the North-west Wind, by R. B. Cunninghame Graham; and a Tribute by Edward Carpenter (Berkeley Heights, N.J.: Oriole Press, 1934);

Mirages (London: Heinemann, 1936);

R. B. Cunninghame Graham His Last Letter to Herbert Faulkner West (Tempe, Ariz.: Edwin B. Hill, 1948);

Two Letters on an Albatross, by Graham and W. H. Hudson (Hanover, N.H.: Westholm, 1955);

Three Fugitive Pieces (Hanover, N.H.: Westholm, 1960).

Editions: *Thirty Tales and Sketches,* edited by Edward Garnett (London: Duckworth, 1929; New York: Viking, 1929);

Rodeo: A Collection of the Tales and Sketches of R. B. Cunninghame Graham, edited by Aimé Felix Tschiffely (London & Toronto: Heinemann, 1936; Garden City, N.Y.: Doubleday, Doran, 1936);

The Horses of the Conquest, edited by Robert Moorman Denhardt (Norman: University of Oklahoma Press, 1949);

The Essential R. B. Cunninghame Graham, edited by Paul Bloomfield (London: Cape, 1952);

Selected Short Stories, edited by Clover Pertíñez (Madrid: Alhambra, 1959);

The South American Sketches of R. B. Cunninghame Graham, edited by John Walker (Norman: University of Oklahoma Press, 1978);

Selected Writings of Cunninghame Graham, edited by Cedric Watts (Rutherford, N.J.: Fairleigh Dickinson University Press, 1981; London: Associated University Press, 1981);

The Scottish Sketches of R. B. Cunninghame Graham, edited by Walker (Edinburgh: Scottish Academic Press, 1982);

The North American Sketches of R. B. Cunninghame Graham, edited by Walker (Tuscaloosa: University of Alabama Press, 1986).

OTHER: William Stirling, *The Canon: An Exposition of the Pagan Mystery Perpetuated in the Cabala as the Rule of All the Arts,* preface by Graham (London: Matthews, 1897);

Santiago Pérez-Triana, *Down the Orinoco in a Canoe,* introduction by Graham (London: Heinemann, 1902);

Moussa Aflalo, *The Truth about Morocco: An Indictment of the Policy of the British Foreign Office with Regard to the Anglo-French Agreement,* preface by Graham (London & New York: John Lane, 1904);

Ida A. Taylor, *Revolutionary Types,* introduction by Graham (London: Duckworth, 1904);

Charles Rudy, *Companions in the Sierra,* introduction by Graham (London: John Lane, 1907);

Gabriela Cunninghame Graham, *Santa Teresa: Being Some Account of Her Life and Times. Together with Some Pages from the History of the Last Great Reform in the Religious Orders,* 2 volumes, preface by R. B. Cunninghame Graham (London: Black, 1907);

Gabriela Cunninghame Graham, *The Christ of Toro, and Other Stories,* preface by R. B. Cunninghame Graham (London: Nash, 1908);

Martin Andrew Sharp Hume, *True Stories of the Past,* preface by Graham (London: Nash, 1910);

Emily, Sharifah of Wazan, *My Life Story,* edited by S. L. Bensusan, preface by Graham (London: Arnold, 1911);

Walter Shaw Sparrow, *John Lavery and His Work,* preface by Graham (London: Kegan Paul, Trench, Trübner, 1911);

William Henry Koebel, *In Jesuit Land: The Jesuit Missions of Paraguay,* introduction by Graham (London: Paul, 1912);

Seventeen-year-old R. B. Cunninghame Graham dressed as a gaucho

Marguerite Radclyffe Hall, *Songs of Three Counties, and Other Poems,* introduction by Graham (London: Chapman & Hall, 1913);

Arthur C. Rickett, *William Morris: A Study in Personality,* introduction by Graham (London: Jenkins, 1913);

Charles Rosher, *Light for John Bull on the Moroccan Question, with a Note on Tripoli,* preface by Graham (London: Henderson's, 1913);

Miguel de Cervantes Saavedra, *Rinconete and Cortadillo,* translated by Mariano J. Lorente, preface by Graham (Boston: Four Seas, 1917);

Fritz W. Up de Graff, *Head-Hunters of the Amazon: Seven Years of Exploration and Adventure,* introduction by Graham (London: Jenkins, 1923);

Gustavo Barroso, *Maripunga,* translated, with an explanatory preface, by Graham (London: Heinemann, 1924);

Charles Simpson, *El Rodeo: One Hundred Sketches,* introduction by Graham (London: John Lane, 1925);

Joseph Conrad, *Tales of Hearsay,* preface by Graham (London: Unwin, 1925);

Agube Gudsow, *The Princess Biaslantt: A Caucasian Story,* introduction by Graham (London: Heinemann, 1926);

Guy de Maupassant, *Tales from Maupassant,* edited by James Eveleigh Nash, preface by Graham (London: Nash & Grayson, 1926);

Joseph Morewood Dowsett, *The Spanish Bull Ring,* preface by Graham (London: Bale, 1928);

Cecil Herbert Prodgers, *Adventures in Bolivia,* preface by Graham (London: John Lane, 1928);

William Henry Hudson, *Far Away and Long Ago, a History of My Early Life,* introduction by Graham (London & Toronto: Dent / New York: Dutton, 1931);

Tate Gallery, *Exhibition of Paintings of Scenes of Gaucho Life in the Province of Entre Rios by C. Bernaldo de Quirós. January, 1931,* introduction by Graham (London, 1931);

John Ressich, *Gallop!* (London: Benn, 1932);

Robert Kirk, *The Secret Commonwealth of Elves, Fauns and Fairies,* introduction by Graham (Stirling: Mackay, 1933).

In his own day R. B. Cunninghame Graham was seen as a colorful personality and was often likened to Don Quixote as a passionate idealist in a cynical age. As a writer and scholar he was an unremitting critic of what he considered an effete and materialistic Western civilization; he sought to record for a comfortable bourgeoisie his remembered impressions of the more elemental life he had encountered in Latin America, the southwestern United States, North Africa, and Scotland. His principal literary genre was the sketch – an elusive form, usually reminiscent but sometimes narrative, sometimes essayistic, sometimes wholly fictional, and sometimes unabashedly autobiographical. The common feature of his sketches is their brevity: most are fewer than ten pages in length. He was regarded by friends and fellow writers such as W. H. Hudson and Joseph Conrad as a writer's writer; if he is neglected today, it may be because the impressionistic effects he aimed for are now sought in poetry and the novel.

The eldest of three sons of William Cunninghame Bontine, a major in the Scots Greys, and Anne Elizabeth Elphinstone Fleeming, Robert Cunninghame Bontine – by the terms of an entail he bore the surname Bontine during his father's lifetime — was born in London on 24 May 1852. On his father's side he could claim descent from the Scottish kings, but his appearance was influenced by his mother's Spanish and Italian ancestry. He was schooled at Hill House, Leamington, and from 1865 to 1867 at Harrow. In 1866 his father became insane and was confined under a doctor's supervision for the rest of his life. Robert Bontine's education continued in Brussels.

Responsive to tales of European adventures in the New World, Bontine disdained the military careers pursued by his father and maternal uncles and set sail for Argentina in 1870 to learn ranching. He returned home after two years an accomplished gaucho, having experienced civil war and typhus. He returned to South America in 1873; there he cultivated maté in Paraguay for export and witnessed the devastating aftermath of that country's war against Brazil, Argentina, and Uruguay. In addition to many of his sketches, this sojourn would inspire two books: *A Vanished Arcadia* (1901), a description of the experimental colony founded by the Jesuits in Paraguay in the sixteenth century, and *Portrait of a Dictator* (1933), about Francisco Solano López, whose policies decimated the population of Paraguay in the nineteenth century. In the mid 1870s he shuttled back and forth between the hemispheres, spending time in Britain, France, and Spain when not engaged in ranching pursuits in Argentina and Texas. In Paris he met Gabriela de la Balmondière, the nineteen-year-old daughter of a Chilean merchant of French extraction; they eloped to London and were married at the Strand Register Office on 24 October 1878. He was biding his time until he could assume his inheritance on the death of his father, but he was also stocking his memory. In the words of his biographers Cedric Watts and Laurence Davies, "In the 1870s he saw the scenes and had the adventures; from the 1890s onwards he wrote them down."

Succeeding to his father's extensive Scottish estates in 1883, he changed his name to Robert Bontine Cunninghame Graham. He was defeated when he stood for Parliament from North-West Lanark in 1885 but was elected the following year. Although he was elected as a Liberal, he was in actuality the first socialist member of Parliament, and he quickly established a reputation as an outspoken firebrand and defender of the underdog. On 13 September 1887 he was suspended from the House of Commons for offensive language; when asked to withdraw his utterance, he responded, "I never withdraw," a declaration that became so well known that Bernard Shaw was able to encapsulate the character of Sergius Saranoff in *Arms and the Man* (1894) merely by having him repeat it. Graham was sentenced to prison for six weeks for his part in the Bloody Sunday demonstration in Trafalgar Square on 13 November 1887. He became president of the Scottish Labour Party in 1888, a champion of such "lost" causes as the eight-hour workday, Irish Home Rule, woman suffrage, and the nationalization of mines, many of which were realized long after he retired from Parliament in 1892; he declined to stand again partly because of the expense but largely because he believed that the working class should be represented by one of its own.

He had found his legacy saddled with debt, but his taste for travel for pleasure, research, and financial opportunity had not diminished. Its scope, however, was diminished: if he no longer undertook to raise livestock in South America, he sought gold in Spain, using Pliny's *Historia Naturalis* as a guide. Moreover, his wife's health, never robust (she would die childless in 1906), could not endure the cold, damp Scottish winters. Reluctantly, he sold his estate, Gartmore, in 1898, and he and his wife moved to London.

The sale of Gartmore liberated Graham from financial insecurity; once unburdened of that anxiety, he turned to authorship with a passion. He had already achieved some notoriety as a polemical political journalist, and he easily transferred the vigor of that mode of composition to his more creative efforts. Beginning serious writing only in middle age, Graham became extraordinarily prolific, publishing a book virtually every other year from 1895 until his death in 1936. Writing as an amateur, he could be indifferent to the pressures and prejudices of the marketplace; the relative unfamiliarity of his subject — the European presence in South America — brought him an inquisitive readership that had few models with which to compare him. This freedom, however, was not an unalloyed good. Graham's writing, while capable of great poignancy, often seems underedited and without direction. Elaborate rhetorical structures arise that serve no apparent purpose, and his digressions often strike the reader as self-indulgent. It is impossible to chart phases of his literary career or to assert that he matured significantly as a writer: to the end he comes across more as a man of action than of words.

His books can be divided into three main classes: collections of sketches, histories of the Spanish Conquest, and biographies. The material of the last two groups is drawn largely from chronicles, diaries, and other written documents; these works, which deal with the past, cannot be considered

Graham disguised as an Arab sheikh in 1897, during the adventures he recounted in Mogreb-el-Acksa: A Journey in Morocco *(1898)*

travel literature. The sketches, however, for which Graham is best known today, are another matter: though not all of them recount actual journeys, travel is at their core. The problem with the sketches is that while Graham assumes various personae, experimenting with multiple narrators and oblique points of view, he is never quite able to submerge his own personality into those of his characters. As a result the characters' motives are not always clear, and the plots are often confusing.

The most effective of the travel sketches are those that are least crowded with incident and most contemplative. "Animula Vagula" (the title, based on a poem by the Roman emperor Hadrian, means "elusive little soul"), included in the collection *Redeemed, and Other Sketches* (1927), is a Conradian yarn told by an orchid hunter to a group of other expatriates aboard a steamer on the Magdalena River in Colombia. The story is of the summary inquest into the death of an unidentified Englishman or American whose body is brought out of the forest to the river settlement by Indians. The narrative of the proceedings is full of insightful details that reveal the attitudes toward one another of the Indi-

ans, the Colombian official, and the European, but at the center of the sketch is the narrator's imagined tale of the nameless wanderer's passage, by stages, from the security of civilization to the indifference of the jungle. As in Conrad's *Heart of Darkness* (1902), this journey comes to be seen as emblematic of the human condition. In "Calvary," included in *Thirteen Stories* (1900), the direction of the voyage is reversed: the reader is given a vision of the agony of horses from their roundup on the Argentine pampas through the squalor of their overcrowded passage across the Atlantic to their senseless deaths as overworked jades drawing London cabs.

Not all the sketches are so tragic, but few are lighthearted. Most of them examine the essential nature of men and women stripped of their sophisticated trappings. "El Tango Argentino," included in *Brought Forward* (1916), ironically juxtaposes two occasions on which the narrator watched the dance: the first is a stylized rendition executed by two cosmopolites in a Buenos Aires ballroom; the second is a memory, revived in Paris, of a violent confrontation in a remote Argentine *pulpería* (tavern). Other sketches probe the clash and transformation of cultures in the New World, the antagonism and fusion of Spanish and Indian, the more pernicious encounter between gringo and Indian, the persistence of Spanish-Portuguese rivalry in incidents along the border between Uruguay and Brazil, and the atavistic rites and memories of *el gaelico* (the Gaelic) of the Scots-Argentines.

Although Graham's sketches are filled with local color and italicized foreign terms and abound with explanations of curious customs, of the differences between the tack employed by gauchos and Moors, and of peculiar features of the landscape, these details are invoked not for their own sake but to authenticate the point of his fictions. One generally expects from travel literature less artifice and more self-revelation, more seemingly spontaneous self-discovery and less pronounced moral or aesthetic purpose. This tendency is manifested to different degrees in three of Graham's works that cannot be subsumed under the three main categories. The slightest of these is his second published book, *Notes on the District of Menteith, for Tourists and Others* (1895). This small octavo of eighty-five pages, replete with tipped-in photographs of lake views and landmarks of his ancestral turf, has the look and heft of a tour guide like many others produced by the publishing house of Adam and Charles Black, and one half-expects to find hotel advertisements in the back. Though the book begins with chapters headed "Descriptive" and "Historical," it is soon ob-

"The Old and The Young Self," Max Beerbohm's caricature of Graham (from Beerbohm, Observations, *1925)*

vious that the reader's guide bears less resemblance to Karl Baedeker than to Laurence Sterne. The work is a whimsical and sometimes melancholic lament for the passage of ancient times with their noble monuments, curious traditions, and colorful personages. Speaking in his own voice, Graham establishes himself as a quirky critic of the emotionally and imaginatively depressed quality of modern life – a posture, indeed, of virtually all of Graham's writing, but one often muted in his sketches by his assumption of other personae.

It is not only the presence of an actual journey, which gives structure to the work, that contributes to the excellence of the second of Graham's travel books, *Mogreb-el-Acksa: A Journey in Morocco* (1898), an account of Graham's failed expedition to Tarudant, a Moroccan city forbidden to infidels. In addition, there is an incisive and engaging narrator, no longer the agreeable eccentric of *Notes on the District of Menteith* but an observer and commentator whose experiences and reflections lead the reader to question the presumed superiority of Western civilization. In the years immediately before the Algeciras Conference and the Agadir Incident, Morocco was virtually the last country in Africa to resist European political and economic exploitation. In the opening pages of the book, as his ship approaches port, Graham strikes the keynote of an indigenous culture in peril:

> The city [Larache, El Areish], painted red, white, and blue, and the whole scale of tints from brown to Naples yellow, stands on a hill, and is today the haunt of consuls of all nations, who have replaced the pirates of the times gone by. Consuls of France and Spain, of Portugal, of Montenegro, Muscat, Costa Rica, Brazil, United States, their flag-staffs rear aloft from almost every house-top, and their great flags, large in proportion to the smallness of the state they represent, flap in the breeze; the caps of liberty, the rising suns, and other trade-marks of the various states seeming to wink and to encourage one another in the attempt to be the first to show the glories of the commercial system to the benighted Moors.

Amid the machinations of missionaries, military adventurers, and commercial intriguers, Graham's subterfuges – he attempts to disguise himself

Graham's gravestone, which bears his old Argentine cattle brand

first as a Turkish doctor and then as a Moorish sheikh – seem altogether innocent; they are motivated by intellectual curiosity rather than by gain. Though the greatest of Graham's tribulations is his detention by the Kaid el Kintafi – nominally for his own protection – the narrative is enlivened by the descriptions of his surroundings; his interviews with his captor, his fellow prisoners, and other travelers; and his frequent digressions. Edward Garnett, whose editorial prodding imposed a measure of literary discipline on Graham, is surely right in his observation in the introduction to the American edition (1930) that "what often makes for a weakness in [Graham's] Sketches, namely the breaking of the scene's atmospheric illusion by the irruption of his asides and comments, becomes in his travel book an artistic strength."

That strength derives from the naturalness of the narrator, who seems wholly at ease in discoursing at length, for example, on Moorish saddlery and horsemanship or, though his own Arabic is quite limited, on the peculiarities of the local language. There is some humor in his attempts to make good his disguise as a doctor and in his playing Gulliver to the Kaid's King of Brobdingnag. But the humor of the work is more sardonic than lighthearted. Even in his informative footnotes Graham is not above a gratuitous jab at national hypocrisy, as when he defines *jihad:* "religious war – generally applied to a war entered into from self-interest, as that of the United States against Spain." As in his writings about South America, where the cruelty and excesses of the conquistadores are palliated by comparing them to the subtler torments executed on native populations by modern imperialists, Graham extenuates the enormities of the Moors as at least taking place within a coherent social framework:

> I found that, though the Kaid oppressed and plundered all the district, his oppression was in a measure balanced by his charity, for he fed all the poor of the valley, and dispensed his hospitality to all and sundry who passed his gates. So that, take it for all in all, his tyranny was only different in degree from that of the manufacturer in the manufacturing towns of England, who lives upon the toil of several thousand workmen, discharges no one useful function to the State, his works being run by paid officials, and he himself doing nothing but sign his letters, whilst he uses the money wrung from his workmen to engage in foreign speculations, to swindle the inhabitants of distant lands; and for all charity subscribes to missions to convert the Jews, or to send meddling praters to insult good Catholics in Spain.

In his encounters with the Arabs, Berbers, and Jews of Morocco, virtue and probity are measured by resistance to the contagion of European ways. *Mogreb-el-Acksa* is a tribute to a way of life in which even "the domestic animals and man understand one another better than they do in any other country I have seen" – a way of life doomed to be swept away by technology and materialism.

Graham's last true travel book (as opposed to the fragmented and fictionalized recollections of his sketches) is the product of a journey undertaken twenty years after the one to Morocco. Although he opposed entry into World War I on the ground that it did not serve the interest of the common people (he had similarly characterized the Boer War as a conflict between two burglars), once war was declared Graham, then in his sixties, volunteered for service and was entrusted with missions commensurate with his expertise. He made two trips to South America, the first to Argentina to procure horses for the battlefields, the second to Colombia to investigate the feasibility of establishing a packinghouse for the shipment of beef across the Atlantic. Out of the second trip arose *Cartagena and the Banks of the Sinú* (1920) – an uneven work that is part guidebook, part history, and part journal. It is not until halfway through the book that the reader learns why the traveler has come to Colombia, and then only obliquely. The summaries of the chronicles of the conquest and the celebration of life on the *estancia* (ranch) have a faded glory about them, as though the harsh reality of modern warfare, which has extended even to this remote backwater, has rendered them romantic irrelevancies.

Though his writing tended increasingly toward the historical and biographical, Graham had not wholly given up on the contemporary world. He stood again for Parliament in 1918 as a Liberal and came in third; afterward his political activities were devoted to the cause of Scottish nationalism. His long and colorful life came to an end, fittingly, in a land he had memorialized. In January 1936 he made another trip to South America; shortly after arriving in Buenos Aires he contracted a bronchial infection that developed into pneumonia, from which he died on 20 March.

The major theme of Graham's writing – the impact of a "superior" civilization on a supposedly more primitive one – is still relevant. Moreover, as one who traveled widely, Graham is able to apply a comparative perspective to the situations he describes, characterizing, for example, the flight of a party of Mescalero Indians as a hegira or contrasting the "quite Arcadian" brigands of Morocco to their government-subsidized counterparts in Mexico. The form in which he cast most of his work, however – the sketch – seems rather obsolescent, both commercially and psychologically. One longs for more substance; Graham's experiences serve him as little more than isolated memories, a quarry from which he mines pebbles rather than blocks of stone to fashion into statues. He seemed to be aware of this limitation. In *Inveni Portam* (1924), an obituary tribute to Conrad, he praises *Nostromo* (1904), a novel to which Graham himself had contributed a wealth of details, for "its immortal picture of the old follower of Garibaldi, its keen analysis of character, and the local colour that [Conrad] divined rather than knew by actual experience, its subtle humour, and the completeness of it all, forming an epic, as it were, of South America, written by one who saw it to the core, by intuition." Graham had neither the talent nor the inclination to write an epic. But in the less demanding form of the travel book, he created a minor classic in *Mogreb-el-Acksa*.

Letters:

Joseph Conrad's Letters to R. B. Cunninghame Graham, edited by Cedric Watts (Cambridge: Cambridge University Press, 1969).

Bibliographies:

John Walker, "A Chronological Bibliography of Works on R. B. Cunninghame Graham (1852-1936)," *Bibliotheck,* 9 (1978): 47-64;

Walker, "R. B. Cunninghame Graham: An Annotated Bibliography of Writings about Him," *English Literature in Transition,* 22 (1978): 78-156;

Walker, *Cunninghame Graham and Scotland: An Annotated Bibliography* (Dollar: Mack, 1980).

Biographies:

Herbert F. West, *A Modern Conquistador: Robert Bontine Cunninghame Graham: His Life and Works* (London: Cranley & Day, 1932);

A. F. Tschiffely, *Don Roberto* (London: Heinemann, 1937);

Hugh McDiarmid, *Cunninghame Graham: A Centenary Study* (Glasgow: Caledonian Press, 1952);

Richard E. Haymaker, *Prince-Errant and Evocator of Horizons* (Kingsport, Tenn.: Kingsport Press, 1967);

Cedric Watts and Laurence Davies, *Cunninghame Graham: A Critical Biography* (Cambridge: Cambridge University Press, 1979).

References:

Laurence Davies, "Cunninghame Graham's South American Sketches," *Comparative Literature Studies,* 9 (September 1972): 253–265;

Davies, "R. B. Cunninghame Graham: The Kailyard and After," *Studies in Scottish Literature,* 11 (1974): 156–177;

Frank MacShane, "R. B. Cunninghame Graham," *South Atlantic Quarterly,* 68 (1969): 198–207;

Jeffrey Meyers, "Robert Bontine Cunninghame Graham: The Genius of Failure," *London Magazine,* 15 (October–November 1975): 54–73;

James Steel Smith, "R. B. Cunninghame Graham as a Writer of Short Fiction," *English Literature in Transition,* 12 (1969): 61–75;

R. W. Stallman, "Robert Cunninghame Graham's South American Sketches," *Hispania,* 28 (1945): 69–75;

John Walker, "Cunninghame Graham and the Critics: A Reappraisal," *Studies in Scottish Literature,* 19 (1984): 106–114;

Walker, "The Scottish Writings of R. B. Cunninghame Graham," *Scottish Tradition,* 13 (1984–1985): 25–33;

Cedric Watts, *R. B. Cunninghame Graham* (Boston: Twayne, 1983).

Beatrice Ethel Grimshaw

(1871 – June 1953)

J. Allen Barksdale
Bowling Green State University

Frontispiece

BOOKS: *Broken Away* (London & New York: John Lane, 1897);

Vaiti of the Islands (London: A. P. Watt, 1906; New York: A. Wessels, 1908);

From Fiji to the Cannibal Islands (London: Eveleigh Nash/G. Bell, 1907); republished as *Fiji and Its Possibilities* (New York: Doubleday, Page, 1907);

In the Strange South Seas (London: Hutchinson, 1907; Philadelphia: Lippincott, 1908);

The New New Guinea (London: Hutchinson, 1910; Philadelphia: Lippincott, 1911);

When the Red Gods Call (London: Mills & Boon, 1911; New York: Moffatt, Yard, 1911);

Guinea Gold (New York: Moffatt, Yard, 1912; London: Mills & Boon, 1912);

Three Wonderful Nations: Tonga, Samoa, Fiji (Dunedin: Union Steamship Company of New Zealand, 1913);

The Sorcerer's Stone (London: Hodder & Stoughton, 1914; Philadelphia: Winston, 1914);

Red Bob of the Bismarks (London: Hurst & Blackett, 1915); republished as *My Lady of the Island* (Chicago: McClurg, 1916);

A Coral Queen (New York, 1917; London: Newnes, 1925);

Nobody's Island (London: Hurst & Blackett, 1917; Garden City, N.Y.: Doubleday, Page, 1923);

Kris-Girl (London: Mills & Boon, 1917);

The Terrible Island (New York: Ridgeway, 1919; London: Hurst & Blackett, 1920);

The Coral Palace: 'Twixt Capricorn and Cancer (London: Mills & Boon, 1920);

The Little Red Speck and Other South Sea Stories (London: Hurst & Blackett, 1921);

My South Sea Sweetheart (London: Hurst & Blackett, 1921; New York: Macmillan, 1921);

Conn of the Coral Seas (New York: Macmillan, 1922; London: Hurst & Blackett, 1922);

The Valley of Never-Come-Back, and Other Stories (London: Hurst & Blackett, 1922);

The Sands of Oro (Garden City, N.Y.: Doubleday, Page, 1924; London: Hurst & Blackett, 1924);

The Candles of Katara (London: Hurst & Blackett, 1925);

The Wreck of the Redwing (New York: Holt, 1927; London: Hurst & Blackett, 1927);

Black Sheeps Gold (London: Hurst & Blackett, 1927; New York: Holt, 1927);

Eyes in the Corner and Other Stories (London: Hurst & Blackett, 1927);

The Paradise Poachers (London: Hurst & Blackett, 1928);

My Lady Far Away (London: Cassell, 1929);

The Star in the Dust (London: Cassell, 1930);

Isles of Adventure: Experiences in Papua and Neighboring Islands (London: Jenkins, 1930); republished as *Isles of Adventure; From Java to New Caledonia but Principally Papua* (Boston & New York: Houghton Mifflin, 1931);

The Mystery of Tumbling Reef (London: Cassell, 1932; Boston & New York: Houghton Mifflin, 1932);

The Long Beaches and Other South Seas Stories (London: Cassell, 1933);

Victorian Family Robinson (London: Cassell, 1934; New York: Longmans, Green, 1935);

Rita Regina (London: Jenkins, 1939; New York: Arcadia House, 1940);

South Sea Sarah. Murder in Paradise. Two Complete Novels (Sydney: New Century, 1940).

Perhaps best known to contemporaries as the self-proclaimed first white woman to set foot in the cannibal lands of Borneo and New Guinea, Beatrice Ethel Grimshaw enjoyed a lengthy career as an author of travel books and romance novels set in exotic locales. Equipped with an early training in journalism, Grimshaw possessed not only valuable knowledge of botany, zoology, and anthropology but also the rare ability to balance the factual and the fantastic successfully. Attentiveness to detail matched with a knack for storytelling provided an ideal voice for Grimshaw's intrepid curiosity.

Information about her early life is minimal. Born in 1871 in Cloona, County Antrim, Ireland, Grimshaw was educated at Caen, Normandy; Victoria College, Belfast; and Bedford College, London. Possessing an unusual amount of formal education for a woman of her era, Grimshaw rejected domestication at an early age. Upon graduation, she returned to live with her family in Dublin and embarked on a career in journalism.

In her book *Isles of Adventure: Experiences in Papua and Neighboring Islands* (1930) Grimshaw reflects on this period of her life: "Being young and rather brazen and full of the 'beans' that go with a good muscular system, I started to teach other and older people their jobs." Indeed, her work at this time, subeditor of a sporting paper followed by the position as editor of the *Society Journal*, seems noticeably pedestrian compared to her subsequent projects. Unsatisfied by life in Dublin, she moved to London, hoping for expanded opportunities. In 1897 Grimshaw's first novel was published, a romance titled *Broken Away*.

The mummy of a Malekulan chief of New Hebrides; photograph by Grimshaw for From Fiji to the Cannibal Islands *(1907)*

However, Grimshaw's true passion was not a comfortable career in journalism but rather a desire to see the world – particularly the more remote areas of the South Sea Islands. As she would explain in the autobiographical first chapter of *Isles of Adventure,* Grimshaw deduced that the most practical way of achieving her goals was to persuade a shipping company to grant her free passage. In return, she would provide them with high-quality written publicity to attract tourists, entrepreneurs, and businesses to the generally undeveloped lands.

With transportation provided by the Union Steamship Company of New Zealand, Grimshaw undertook her first major expedition in early 1904. Embarking from San Francisco, she sailed first to Tahiti, followed by a four-month voyage through the South Pacific and an additional two months on the island of Niue. During this trip Grimshaw visited Tonga, Samoa, Fiji, Rarotonga, and some of the Cook Islands. After traveling thousands of miles and interacting with indigenous populations,

Grimshaw returned to London with enough material to publish two lengthy travelogues in 1907.

Unlike many travelogues, which pay clear attention to dates and itineraries, Grimshaw's writings are largely absent of such chronology. She seldom makes more than an offhand reference to the year of her travels or how long they last. Furthermore, several of her books are actually comprised of anecdotes from journeys several years apart; consequently, two separate books may be a continuation of a single trip. It is impossible to say whether Grimshaw felt such dates and itineraries were unimportant or if she used the omission as a stylistic device to draw the reader into the carefree idealistic lifestyle of the islands.

In the Strange South Seas (1907) and *From Fiji to the Cannibal Islands* (1907; published in America as *Fiji and Its Possibilities*, 1907) recount Grimshaw's wanderings among the islands. The style of her later works, characterized by a thorough attention to details and an often politely humorous tone, is already fully developed here. When applicable, she begins the descriptions of each area with its colonial history, making certain to highlight the achievements of white settlements. Grimshaw provides an exhaustive picture of a region's fauna and wildlife. As she recounts her adventures, she provides narratives depicting the customs and lifestyles of the native populations. Early in *In the Strange South Seas* Grimshaw proudly endorses British colonialism, carefully equating the aspiration of conquest with her vagabond leanings: "In the heart of every Briton a wanderer once has lived. If this were not so, the greatest empire of the world would never have been."

Grimshaw's epics of tropical adventure reveal her twin intentions as an author. While highly competent in *National Geographic*–style reporting, Grimshaw is also aware of the reader's desire for suspense, danger, and excitement. Both books contain ample discussions of cannibalism, head-hunting, poisoning, and tribal magic. Of course, this material is carefully spread throughout the texts because the central motivation for Grimshaw's work was to encourage tourism and development – not to frighten people away. She skillfully mixes tales of exotic jungle adventure with romantic descriptions of the South Sea allure.

Her work's suspense comes from Grimshaw's practice of creating the illusion that she is alone among the savages. Sometimes this was indeed the case, but when it was not, the author is careful to present guides, escorts, and ship captains as comic characters offering minimal protection. Generally she insisted on arriving and remaining at her desired destinations without help from guides.

Reviews of these two volumes were often negative. *Fiji and Its Possibilities* was dismissed by an unattributed critic for the *Annals of the American Academy of Political and Social Science* as "diffuse and unsatisfactory," while a reviewer in *The Nation* complained that "she tires the reader by repetitions, long digressions, feebly humorous passages, frequent allusions to cannibalism, and prolix accounts of trivial and uninteresting incidents." *In the Strange South Seas* elicited criticisms from the *Times Literary Supplement;* specifically, Grimshaw was reprimanded for relying "on the loose talk of traders" for information and charged with creating a frequently inaccurate work. Despite such criticisms, the publication of these books on both sides of the Atlantic as well as subsequent reprintings indicate that Grimshaw's work enjoyed some popularity.

Grimshaw also used her knowledge of the islands to provide background for a series of romance novels. *Vaiti of the Islands* (1906), a story concerning the half-caste daughter of a Maori princess and the drunken English captain of a trading schooner, is the first of Grimshaw's "island romances." Interestingly, this work elicited more favorable reviews than the author's travel narratives. A review for the 28 November 1908 issue of *The New York Times* congratulated the author for "creating something new in heroines" and recommended the novel to all "wandering seekers of excitement."

Now a professional traveler, Grimshaw secured passage on the steamer *Merrie England* and departed in November 1907 for New Guinea. This voyage, coupled with a second trip in 1909, would be detailed in *The New New Guinea* (1910). Curiously, her interest in the region was sparked by the suicide of the English chief judicial officer several years earlier. Although still a British colony, New Guinea was far from settled. As late as 1903 there had been violent massacres of settlers by the native population. In anticipating her voyage Grimshaw explained that "the books told of cannibals and crocodiles, fevers and snakes and swamps, unexplored rivers, unknown mountains. It sounded interesting, but calculated to give the unescorted woman wanderer food for rather serious thought."

More than her previous works, *The New New Guinea* is partially intended as a resource book for potential visitors to the island. In addition to hints spread throughout the text, Grimshaw provides appendixes containing information on travel arrangements as well as medical precautions for visitors. Despite this nod to practicality, Grimshaw relishes

Skulls with clay portrait masks – headhunters' trophies on the Sepik River, New Guinea; photograph from Grimshaw's Isles of Adventure *(1930)*

the often-real danger attached to her surroundings. Her desire to redefine the position of the woman traveler is best illustrated by her insistence upon sampling deep-sea diving on the Torres Strait. Although Grimshaw braved the shark-infested waters, she subsequently recalled upon entering the water: "At this point my fiction broke up, and I realized that I was extremely afraid. The sobering truth, I think, is that a woman is always afraid of doing dangerous things." Rather than conceding a docile role for women in society, however, Grimshaw is actually arguing that this fear of a legitimate danger indicates a degree of common sense that make women far better equipped to function outside the home than men think. *In the Strange South Seas* explicitly states a similar idea: "We are not as clever as men – but neither are we stupid." *The New New Guinea* received more favorable reviews than Grimshaw's volumes on Fiji. Particularly laudatory was Forbes Lindsay's article in *The New York Times* praising "the lifelike pictures of the native tribes, and charming descriptions of the country."

Grimshaw followed the story of her New Guinea adventures with another romance novel, *When the Red Gods Call* (1911). Two years later Grimshaw repaid her debt to the Union Steamship Company by authoring a thirty-two-page pamphlet, *Three Wonderful Nations: Tonga, Samoa, Fiji* (1913), a brief history and description of the areas. She does not discuss her personal experiences but merely depicts the islands as ideal vacation spots. For the next seventeen years Grimshaw would devote her writing entirely to fiction, generally further variations on the tropical romance motif.

During her second trip to New Guinea in 1909, Grimshaw visited the Solomon Islands. Incidents from this trip were combined with writings from additional South Seas excursions in 1912 and 1928 for Grimshaw's final travel book, *Isles of Adventure*. In some ways this work may be her most successful, both as a limited attempt at autobiography and for the wide range of topics she discusses. Within this volume Grimshaw recounts her adventures in Java, New Caledonia, Ile Nov, and the infamous Sepik River Valley – the location of practicing headhunters at the time of her visit.

As always, Grimshaw is strangely ambiguous about the people she observes. Explaining that the "Black Peril" renders it essential for all white women to carry guns – "Brown Men are violent and unrestrained of feeling, like all savages" – the author later concedes of the same natives: "Brains these folks have in abundance. Strange though it may seem they have much natural goodness of character too."

Following the publication of this work Grimshaw continued to travel among the islands, finally settling in Bathurst, Australia. She turned her attention to novels and short stories, generally romances set in exotic locales. Completing nearly thirty such works, she retired in 1940 after the publication of *Rita Regina* (1939). Grimshaw died in June 1953 at the age of eighty-two.

For modern scholars the most germane study of Grimshaw arises from her role as a woman brazenly dismissing the prescribed domesticity of her age as well as her identity as a firsthand observer (some would argue participant) of the colonial process. Although the breadth of her writing would suggest that she is an ideal subject for such inquiry, Grimshaw is actually far more enigmatic than a casual survey would indicate. She was highly aware of violating the standard role model of early-twentieth-century women and addresses this issue repeatedly in her books — the reader may occasionally feel that many of her adventures would have come across as far less perilous if the subject were a man. She was fascinated by the gender customs and role of women among the societies she encountered but was rarely permitted any real contact with the women she sought to study. A nearly unanimous practice among the indigenous peoples was to hide all females at times of danger or uncertainty. As a result, the natives generally cloistered the women far away as soon as Grimshaw's expedition was spotted. Furthermore, her interactions with and observations of other women were generally limited or at best closely monitored by men.

Grimshaw's position in the hierarchy of colonialism is also difficult to determine. To be certain, she was quick to glorify the achievements of British paternalism and all too often succumbs to racism and cultural bias when discussing indigenous peoples. However, to dismiss her as a crass, myopic Anglophile is unfair. Although Grimshaw never failed to locate faults, she overwhelmingly felt an attraction to the preindustrial cultures she examined, finding that way of life more often than not to be infinitely preferable to the one she left at home. In *Isles of Adventure* Grimshaw unknowingly provides a flattering but accurate epitaph: "I have written as a traveller, a wanderer, to whom new and strange things are the chief happiness of my life."

H. Rider Haggard

(22 June 1856 – 14 May 1925)

John W. M. Hallock
Temple University

See also the Haggard entries in *DLB 70: British Mystery Writers, 1860–1919* and *DLB 156: British Short-Fiction Writers, 1880–1914: The Romantic Tradition.*

BOOKS: *Cetywayo and His White Neighbours; or, Remarks on Recent Events in Zululand, Natal, and the Transvaal* (London: Trübner, 1882);

Dawn (3 volumes, London: Hurst & Blackett, 1884; 1 volume, New York: Hurst, 1886?);

The Witch's Head (3 volumes, London: Hurst & Blackett, 1885; 1 volume, New York: Appleton, 1885); republished as *The Witch's Head: A Novel* (New York: Munro, 1885);

King Solomon's Mines (London, Paris, New York & Melbourne: Cassell, 1885); republished as *King Solomon's Mines: A Novel* (New York: Harper, 1886; revised edition, London: Cassell, 1905; abridged edition, London & New York: Longmans, 1930);

She: A History of Adventure (New York: Harper, 1886; London: Longmans, Green, 1887); republished as *She; or, Adventures in the Caves of Kor* (Philadelphia: Franklin News, 1887);

Jess (London: Smith, Elder, 1887); republished as *Jess: A Novel* (New York: Harper, 1887); revised as *Jess* (London: Smith, Elder, 1898);

Allan Quatermain: Being an Account of His Further Adventures and Discoveries in Company with Sir Henry Curtis, Bart., Commander John Good, R. N., and One Umslopogaas (London: Longmans, Green, 1887; New York: Harper, 1887);

A Tale of Three Lions (New York: Lovell, 1887); republished as *Allan the Hunter: A Tale of Three Lions* (Boston: Lothrop, Lee & Shepherd, 1898);

Maiwa's Revenge; or, The War of the Little Hand (London & New York: Longmans, Green, 1888; New York: Harper, 1888); republished as *Maiwa's Revenge: A Tale of Adventure* (Chicago: Universal, 1888); republished as *Maiwa's Revenge* (New York: Lovell, 1888);

H. Rider Haggard

Mr. Meeson's Will (London: Blackett, 1888); republished as *Mr. Meeson's Will: A Novel* (New York: Harper, 1888); republished as *Mr. Meeson's Will: A Story of Adventure* (New York & Chicago: Ogilvie, 1888);

My Fellow Laborer (New York: Munro, 1888);

Colonel Quaritch, V. C.: A Tale of Country Life (3 volumes, London: Longmans, Green, 1888; 1 volume, New York: Lovell, 1888);

Cleopatra: Being an Account of the Fall and Vengeance of Harmachis, the Royal Egyptian, as Set Forth by His Own Hand (London: Longmans, Green, 1889; New York: Munro, 1889);

Allan's Wife and Other Tales (London: Blackett, 1889; New York: Lovell, 1889);

Beatrice: A Novel (London: Longmans, Green, 1890; New York: Munro, 1890);

The World's Desire, by Haggard and Andrew Lang (London: Longmans, Green, 1890; New York: Harper, 1890);

Eric Brighteyes (London: Longmans, Green, 1891; New York: Lovell, 1891);

Nada the Lily (London & New York: Longmans, Green, 1892);

Montezuma's Daughter (London: Longmans, Green, 1893; New York: Longmans, Green, 1893); republished as *Montezuma's Daughter: A Romance* (New York: McKinlay, Stone & Mackenzie, 1909);

An Heroic Effort (Frome & London: Butler & Tanner, 1893);

The People of the Mist (London & New York: Longmans, Green, 1894); republished as *The People of the Mist: A Novel of Adventure* (New York: Review of Reviews, 1894);

Church and State (New Style): An Appeal to the Laity (London: McCorquodale, 1895);

East Norfolk Representation (Norwich: Norfolk Weekly Standard and Argus, 1895);

Lord Kimberly in Norfolk (Great Yarmouth: East Norfolk Printing, 1895);

Speeches of the Earl of Iddlesleigh and Mr. Rider Haggard (London: Published for the National Society for the Prevention of Cruelty to Children, 1895);

Joan Haste (London & New York: Longmans, Green, 1895);

Heart of the World (New York: Longmans, Green, 1895; London, New York & Bombay: Longmans, Green, 1896);

The Wizard (Bristol: Arrowsmith / London: Simpkin, Marshall, Hamilton, Kent, 1896; New York, London & Bombay: Longmans, Green, 1896);

Doctor Therne (London, New York & Bombay: Longmans, Green, 1898);

Swallow: A Tale of the Great Trek (New York, London & Bombay: Longmans, Green, 1899);

A Farmer's Year: Being His Commonplace Book for 1898 (London, New York & Bombay: Longmans, Green, 1899);

The Last Boer War (London: Kegan Paul, Trench, Trübner, 1899); republished as *A History of the Transvaal* (New York: New Amsterdam Book Company / London: Kegan Paul, Trench, Trübner, 1899);

The New South Africa (New York: Pearson, 1900);

Black Heart and White Heart, and Other Stories (London, New York & Bombay: Longmans, Green, 1900); republished in part as *Elissa: The Doom of Zimbabwe; Black Heart and White Heart: A Zulu Idyll* (New York, London & Bombay: Longmans, Green, 1900);

Lysbeth: A Tale of the Dutch (London, New York & Bombay: Longmans, Green, 1901);

A Winter Pilgrimage: Being an Account of Travels through Palestine, Italy, and the Island of Cyprus, Accomplished in the Year 1900 (London, New York & Bombay: Longmans, Green, 1901);

Rural England: Being an Account of Agricultural and Social Researches Carried Out in the Years 1901 & 1902, 2 volumes (London, New York & Bombay: Longmans, Green, 1902);

Pearl-Maiden: A Tale of the Fall of Jerusalem (London, New York & Bombay: Longmans, Green, 1903);

Stella Fregelius: A Tale of Three Destinies (London, New York & Bombay: Longmans, Green, 1903; New York, London & Bombay: Longmans, Green, 1903);

The Brethren (London, Paris, New York & Melbourne: Cassell, 1904; New York: McClure, Phillips, 1904);

A Gardener's Year (London, New York & Bombay: Longmans, Green, 1905);

Report on the Salvation Army Colonies in the United States and at Hadleigh, England, with Scheme of National Land Settlement (London: Printed for His Majesty's Stationers Office, 1905); enlarged as *The Poor and the Land: Being a Report on the Salvation Army Colonies in the United States and at Hadleigh, England, with a Scheme of National Settlement and an Introduction* (London, New York & Bombay: Longmans, Green, 1905);

Ayesha: The Return of She (London: Ward, Lock, 1905; New York: Doubleday, Page, 1905);

The Way of the Spirit (London: Hutchinson, 1906);

Benita: An African Romance (London, Paris, New York & Melbourne: Cassell, 1906); republished as *The Spirit of Bambatse. A Romance* (New York: Longmans, Green, 1906);

Fair Margaret (London: Hutchinson, 1907); republished as *Margaret* (New York: Longmans, Green, 1907);

Reports of the Royal Commission on Coast Erosion, 3 volumes, by Haggard and others (London, 1907–1911);

The Real Wealth of England (London: Ward, Lock, 1908);

The Ghost Kings (London, Paris, New York, Toronto & Melbourne: Cassell, 1908); republished as *The Lady of the Heavens* (New York: Lovell, 1909);

The Yellow God (New York: Cupples & Leon, 1908); republished as *The Yellow God: An Idol of Africa* (London, New York, Toronto & Melbourne: Cassell, 1909);

The Lady of Blossholme (London: Hodder & Stoughton, 1909);

Morning Star (London, New York, Toronto & Melbourne: Cassell, 1910; New York, London, Bombay & Calcutta: Longmans, Green, 1910);

Queen Sheba's Ring (London: Nash, 1910; New York: Doubleday, Page, 1910);

Regeneration: Being an Account of the Social Work of the Salvation Army in Great Britain (London: Longmans, Green, 1910; New York & London: Longmans, Green, 1910);

Rural Denmark and Its Lessons (London, New York, Bombay & Calcutta: Longmans, Green, 1911);

Red Eve (London, New York & Toronto: Hodder & Stoughton, 1911; Garden City, N.Y.: Doubleday, Page, 1911);

The Mahatma and the Hare: A Dream Story (London, New York, Bombay & Calcutta: Longmans, Green, 1911; New York: Holt, 1911);

Reports of the Dominions Royal Commission on the Natural Resources, Trade, and Legislation of Certain Portions of His Majesty's Dominions, 24 volumes, by Haggard and others (London: His Majesty's Stationers Office, 1912–1917);

Marie (London, New York, Toronto & Melbourne: Cassell, 1912); republished as *Marie: An Episode in the Life of the Late Allan Quatermain* (New York: Longmans, Green, 1912);

Child of Storm (London, New York, Toronto & Melbourne: Cassell, 1913; New York: Longmans, Green, 1913);

The Wanderer's Necklace (London, New York, Toronto & Melbourne: Cassell, 1914; New York: Longmans, Green, 1914);

A Call to Arms to the Men of East Anglia (London: Clay, 1914);

The Holy Flower (London, Melbourne & Toronto: Ward, Lock, 1915); republished as *Allan and the Holy Flower* (New York: Longmans, Green, 1915);

The Ivory Child (London, New York, Toronto & Melbourne: Cassell, 1916; New York: Longmans, Green, 1916);

The After-War Settlement & Employment of Ex-Service Men in the Overseas Dominions: Report to Royal Colonial Institute (London: Published for the Royal Colonial Institute by the Saint Catherine Press, 1916);

Finished (London, Melbourne & Toronto: Ward, Lock, 1917; New York: Longmans, Green, 1917);

Love Eternal (London, New York, Toronto & Melbourne: Cassell, 1918; New York: Longmans, Green, 1918);

Moon of Israel: A Tale of the Exodus (London: John Murray, 1918; New York: Longmans, Green, 1918);

When the World Shook: Being an Account of the Great Adventure of Bastin, Bickley and Arbuthnot (London, New York, Toronto & Melbourne: Cassell, 1919; New York: Longmans, Green, 1919);

The Ancient Allan (London, New York, Toronto & Melbourne: Cassell, 1920; New York: Longmans, Green, 1920);

The Missionary and the Witch Doctor (New York: Paget Literary, 1920);

Smith and the Pharaohs, and Other Tales (Bristol & London: Arrowsmith/Simpkin, Marshall, Hamilton, Kent, 1920; New York: Longmans, Green, 1921);

She and Allan (London: Hutchinson, 1921; New York: Longmans, Green, 1921);

The Virgin of the Sun (London, New York, Toronto & Melbourne: Cassell, 1922; Garden City, N.Y.: Doubleday, Page, 1922);

Wisdom's Daughter: The Life and Love Story of She-Who-Must-Be-Obeyed (London: Hutchinson, 1923; Garden City, N.Y.: Doubleday, Page, 1923);

Heu-Heu; or, The Monster (London: Hutchinson, 1924; Garden City, N.Y.: Doubleday, Page, 1924);

Queen of the Dawn: A Love Tale of Old Egypt (London: Hutchinson, 1925; Garden City, N.Y.: Doubleday, Page, 1925);

The Treasure of the Lake (London: Hutchinson, 1926; Garden City, N.Y.: Doubleday, Page, 1926);

The Days of My Life: An Autobiography, edited by Charles James Longman, 2 volumes (London, New York, Toronto, Bombay, Calcutta & Madras: Longmans, Green, 1926); republished in part as *A Note on Religion* (London, New York, Toronto, Bombay, Calcutta & Madras: Longmans, Green, 1927);

Allan and the Ice-Gods: A Tale of Beginnings (London: Hutchinson, 1927; Garden City, N.Y.: Doubleday, Page, 1927);

Mary of Marion Isle (London: Hutchinson, 1929); republished as *Marion Isle* (Garden City, N.Y.: Doubleday, Doran, 1929);

Belshazzar (London: Stanley Paul, 1930; Garden City, N.Y.: Doubleday, Doran, 1930);

The Private Diaries of Sir H. Rider Haggard, 1914-1925, edited by D. S. Higgins (London: Cassell, 1980; New York: Stein & Day, 1980).

OTHER: Ella Haggard, *Life and Its Author: An Essay in Verse,* with a memoir by H. Rider Haggard (London: Longmans, Green, 1890);

Bertha Jebb, *A Strange Career: Life and Adventures of John Gladwyn Jebb,* introduction by Haggard (Edinburgh & London: Blackwood, 1894);

Alexander Wilmot, *Monomotapa, Rhodesia: Its Monuments and Its History,* with a preface by Haggard (London: Unwin, 1896);

William A. Dutt, *The King's Homeland: Sandringham and North-West Norfolk,* introduction by Haggard (London: Black, 1904);

Thomas Adams, *Garden City and Agriculture: How to Solve the Problem of Rural Depopulation,* introduction by Haggard (Hitchen: Garden City, 1905);

Home Countries [*J. W. Robertson Scott*], *The Case for the Goat,* introduction by Haggard (London: Routledge, 1908).

SELECTED PERIODICAL PUBLICATIONS – UNCOLLECTED: "The Transvaal," *Macmillan's Magazine,* 36 (May 1877): 71-79;

"About Fiction," *Contemporary Review,* 51 (February 1887): 172-180;

"Suggested Prologue to a Dramatised Version of *She*," *Longman's Magazine,* 11 (March 1888): 492-497;

"An Incident of African History," *Windsor Magazine,* 13 (December 1900): 112-119;

"Kaffir Telegraphy," *Spectator,* 3887 (27 December 1902): 1026;

"Lost on the Veld," *Windsor Magazine,* 17 (December 1902-May 1903): 185-194;

"The Chinese in South Africa," *Spectator,* 4032 (7 October 1905): 522;

"A Literary Coincidence," *Spectator,* 4138 (19 October 1907): 565.

H. Rider Haggard brought more of the world into Victorian homes than almost any other author, yet Haggard was technically not a travel writer. Instead, he fused his extensive travels with fiction and philosophical inquiry. Like the cumbersome souvenirs that decorated his home and the artifacts that he used as fashion trinkets, Haggard seemed out of place in Great Britain. His sympathies were global. While colonists were busy exporting British culture, Haggard became a medium by which a great knowledge of other realms from Peru to Denmark was imported into the collective English imagination.

Haggard in 1881, soon after his return from Africa

Nearly named Sylvanus by his parents, Henry Rider Haggard was the sixth son born into Ella and William Haggard's family of ten children. The Norfolk family boasted wealth and ancient genealogy. Reared in India, Ella Haggard opened her son's eyes to "the blessed kingdom of Romance," according to Haggard's dedication for *The Brethren* (1904). Haggard developed jaundice after his birth, and a large Lowestoft china bowl that happened to be conveniently located was used to perform a makeshift christening of the infant. Though the incident was minor, it proved prophetic. Substituting the Eastern vessel for the orthodox basin seemed to anoint the boy with unconventional powers of imagination and an unquenchable desire to travel.

Nicknamed "Nosey" by his schoolmates, young Haggard was to be christened a third time by his devoted African servant, Mazooku, who appears as a character in *The Witch's Head* (1885). Mazooku was to call Haggard "Lundanda," Zulu for "the tall and pleasant-natured one." Another figure from Haggard's past – that of an old button-eyed, woolly-haired doll that he had called "She-Who-

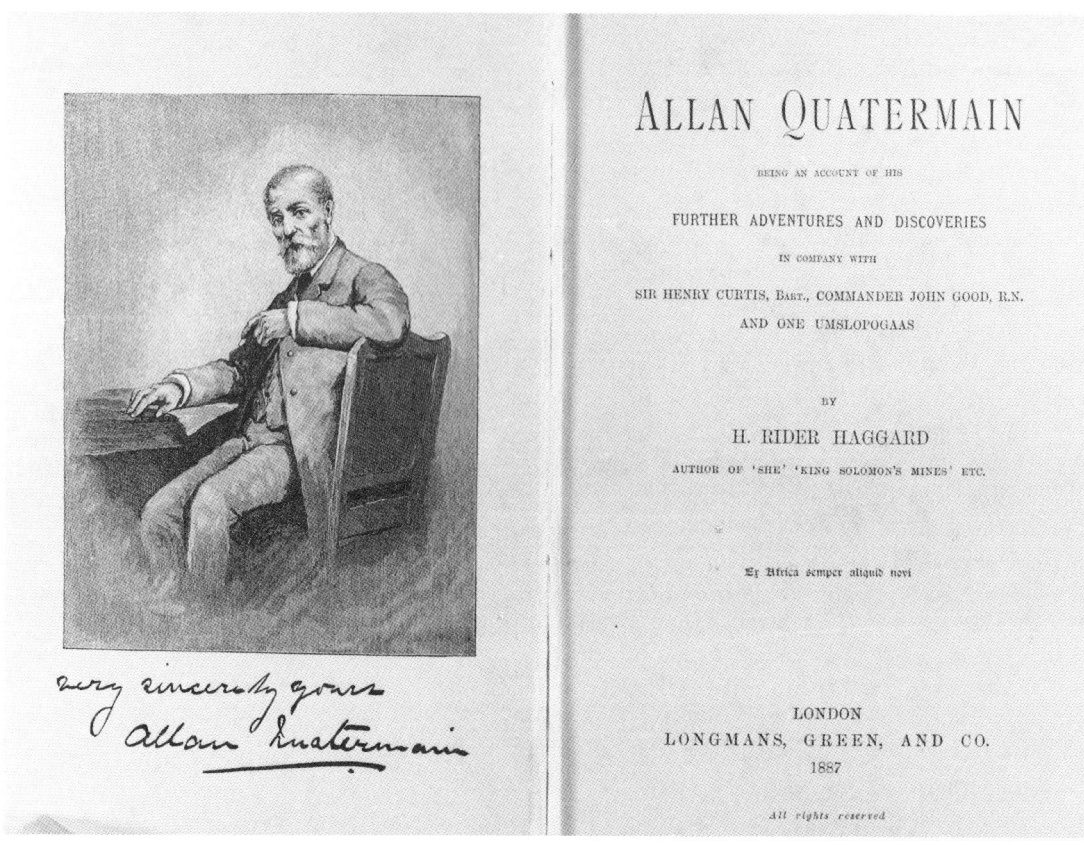

Frontispiece and title page for one of the novels in which Haggard drew on his experiences in Africa

Must-Be-Obeyed" – was also to appear in his fiction. He was to transfer this title to Ayesha, his most famous fictional femme fatale. Haggard's own most enduring title was created by the British press, which would come to hail him as "King Romance."

Haggard was to write fifty-eight adventure novels, based largely on his travels and other autobiographical experiences, although his early reading evidently influenced his career. The young Haggard obsessively read Daniel Defoe's *The Life and Strange and Surprising Adventures of Robinson Crusoe* (1719), and he was delighted by *The Arabian Nights' Entertainments,* Alexandre Dumas's *The Three Musketeers* (1844–1845), and Edgar Allan Poe's poems. While few of his works were set in England, he would name his most famous character, Allan Quatermain, after a kindly neighboring farmer. Haggard was to claim that Quatermain was only himself fictionalized in various settings.

Haggard's curiosity about other worlds was apparent from his childhood fear of ghosts and the omens of death he reported at Ipswich Grammar School. After Haggard failed his army entrance examination, his father sent him to London to apply to the Foreign Office in 1873. Haggard became a frequent guest of Lady Anne Paulet, a Spiritual Athenaeum member who hosted séances that would appear often in Haggard novels, such as *Love Eternal* (1918). He also fell in love with Lilly Jackson, the daughter of a prosperous Yorkshire farmer, but his feelings were subordinated to the will of his father, who arranged for the young man to pursue his study of French in Tours. Perhaps because he could only love from afar, the relationship of flesh to spirit became a theme in many of Haggard's works.

When an unexpected appointment under the lieutenant governor of Natal sent Haggard to South Africa in 1875, he was thrown into a heated triangle of conflict among blacks, Boers, and Britons that led to annexation of the Transvaal on 12 April 1877. Haggard himself raised the British flag. At the age of twenty-one he soon boasted of being the youngest head of a department in South Africa. His "A Zulu War Dance" appeared in *Gentleman's Magazine* (July 1877), and his articles on the Boer Transvaal began to appear in *Cornhill Magazine*. As a correspondent reporter, Haggard discovered that London society functioned quite well without him; Lilly Jackson had married another man.

Haggard rebuilt Hilldrop, a bungalow that was nicknamed "Palatial" and then still later known as "Mooifontein," which he shared with Arthur Cochrane in Pretoria. Here Haggard resigned from the government and began an ostrich farm. Pretoria had to be defended in 1879, and Haggard recounted the battle in four works: *The Witch's Head;* "The Tale of Isandhlwana and Rorkes Drift," which appeared in *The True Story Book,* edited by Andrew Lang in 1893; the title story of *Black Heart and White Heart, and Other Stories* (1900); and *Finished* (1917). In 1879 he left South Africa for England, where he became engaged to Louisa Margitson, heiress of the Ditchingham estate. Her uncle took legal action to prevent the financially uneven match, but the two married in 1880 and returned to Africa, where Arthur John ("Jock") Rider was born amid the chaos and danger of war. Fearing for the safety of his wife and infant, Haggard returned to England and applied his political experience toward the study of law.

Cetywayo and His White Neighbours; or, Remarks on Recent Events in Zululand, Natal, and the Transvaal (1882) publicly criticized the popular policies of British prime minister William Gladstone and was well reviewed, although it was a financial failure. In 1883 Haggard's second child, Agnes Angela Rider, shared her name with the first heroine of Haggard's fiction, Angela Caresfoot of *Dawn* (1884). Set in Norfolk in Brantham Abbey, this novel was also highly autobiographical, as Haggard's family home was Brandenham Hall. Although *Dawn* was transformed into a popular play in 1887, critics were bored with the novel. Another rural tale, *The Witch's Head,* was also autobiographical in being centered on a failed affair and an advantageous marriage. Only the section that was set in South Africa and recalled Haggard's encounter with hyenas received praise. Haggard passed his law examinations at about the time his second daughter was born.

Practicing law amid England's growing agricultural crisis, Haggard sought escape in reading Robert Louis Stevenson's *Treasure Island* (1883). When his brother wagered that Haggard could not create a story as good as Stevenson's, Haggard began writing feverishly. *King Solomon's Mines* (1885) won the gentleman's bet Haggard had with his brother, and Stevenson wrote a letter to Haggard to warn him against writing with reckless speed. *King Solomon's Mines* was based on a visit to the caves at Wonderfontein, which provided the stalactites that Haggard's creativity used to entomb royal corpses in the novel's icy "Place of Death." Haggard also incorporated the name of Umslopogaas, that of a

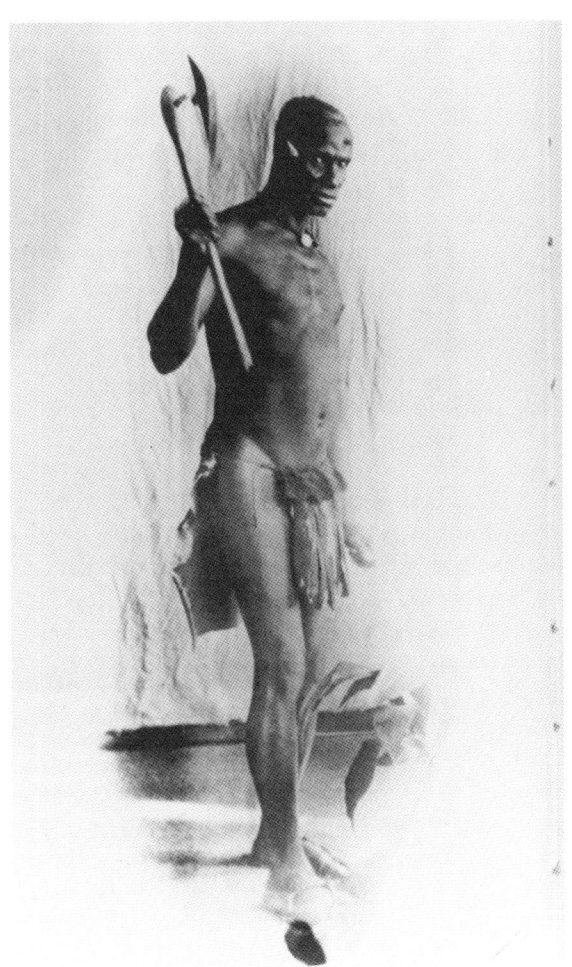

Umslopogaas, a Zulu prince on whom Haggard based a character in several of his popular novels

Zulu prince of Swaziland, who appears in several novels. Haggard's work tantalizes readers by erasing boundaries between domestic England and foreign culture. In Kukuanaland, the primordial Eden of *King Solomon's Mines,* the heroes are saved from witches by scientific calculations of an eclipse. Despite the fantastic features of his story, Haggard's use of his real travels convinced some readers that the legendary treasure actually existed. By September 1885 30,000 copies of the novel had been sold in England; more than 650,000 copies were printed in Haggard's lifetime.

By having English scenes provide the framework for an African adventure, *She: A History of Adventure* (1886) provides a fantastic account of matriarchal rule among the Amahagger people. Necrophilia, embalming, and a curious mixture of the modern and the magical illustrate how Haggard meshes the familiar with uncharted realms. The narrator in *She* places a perfectly preserved foot in a

Gladstone bag, and a native goddess is compared with Mary, Queen of Scots. A secret pit leading to a fire of rejuvenation is measured by the dome of Saint Paul's Cathedral; the main thoroughfare of the Temple of Truth is the width of the embankment of the Thames River.

Akin to the true-life setting that inspired the "Place of Death" in *King Solomon's Mines,* the setting of Kor in *She* was based on myths that Haggard had heard about a strange, prehistoric city in Zimbabwe. The lore of the Lovedu Tribe provided the origins of Haggard's white female magician. Even the deism of *She* is swayed by faithless civilizations, and the novel concludes that "morality [is] an affair of latitude." *She* was soon parodied by Andrew Lang's *He* and was presented as a drama at the Gaiety Theatre in the Strand on 6 September 1888. In these two novels Haggard finally capitalized on exotic settings. As his greatest novels, both of these have never gone out of print and have been the bases for motion pictures.

In 1887 Haggard stopped practicing law. *Allan Quatermain* (1887), which was a sequel to *King Solomon's Mines,* and *Jess* (1887) were being serially published in *Longman's Magazine* and *Cornhill Magazine,* respectively. Riding a tide of confidence in his literary abilities, Haggard wrote a presumptuous essay, "About Fiction," for the *Contemporary Review.* Although critics lambasted his argument against realism, public approval of his fiction was demonstrated by increased sales of books. *Maiwa's Revenge; or, The War of the Little Hand* (1888) sold twenty thousand copies on the day it was published, and Haggard's readership showed no signs of becoming travel-sick.

Trip after trip resulted in new characters in new lands. Haggard's visit to Egypt in 1888 became a basis for *Cleopatra: Being an Account of the Fall and Vengeance of Harmachis, the Royal Egyptian, as Set Forth by His Own Hand* (1889). His Icelandic voyage in the second half of 1888 preceded his writing of *Eric Brighteyes* (1891), and his Mexican tour of 1890 similarly informed *Montezuma's Daughter* (1893). While searching for Aztec treasure, Haggard received the shocking news that his only son had died of peritonitis.

In 1892 Haggard brought two women into the world: the fictitious heroine of *Nada the Lily* (1892), a Zulu story; and his third daughter, Lilias Rider, who would become his first biographer. The novel elaborated the battle-ax tale that Umslopogaas had once reported to Haggard. Haggard used his British associations and literary success to enter politics, as he became a Unionist candidate and ran for Parliament in 1895. Defeated in this aim, he turned to journalism and, with William A. Wills, became co-director of *The African Review of Mining, Finance and Commerce. A Farmer's Year: Being His Commonplace Book for 1898* (1899). This post led Haggard into a position as an agricultural reformer, and after an official tour of twenty-seven counties in 1901, he published a two-volume work, *Rural England: Being an Account of Agricultural and Social Researches Carried Out in the Years 1901 & 1902* (1902).

A Winter Pilgrimage: Being an Account of Travels through Palestine, Italy, and the Island of Cyprus, Accomplished in the Year 1900 (1901) is Haggard's only actual "travel book." Even in this work mystic impressions dominate. If the site under discussion in a particular part of the narrative is not utterly spectacular, Haggard adds a suspiciously fictitious detail – as in the somber caption beneath his photograph from Palestine: "Shepherd carrying a lost sheep." By the time Haggard reaches the pools of Solomon and Solomon's quarries, his fictions are animating the tour more than the actual journey is inspiring his text.

Haggard's aims in writing this travel narrative and in visiting the Holy Land were to revive tourism in Cyprus, a possession of the British Empire, and to demonstrate to the "jaded reader" that "on this side [of] the grave there is no new thing." *A Winter Pilgrimage* begins with a majestic description of the seedy Charing Cross Station, "for it is bold to break away from the accepted formula of books of travel consecrated by decades of publication." From his description of European hotels as models for hell to that of "a hideous apology for a road" to Jaffa, Haggard's tone as a "railway Jonah" brings refreshing humor to this travelogue. The narrative also frequently recounts previous expeditions he has made.

As the Tuscan wines of Florence remind Haggard of his Pretoria vineyards, the reader is always reminded of the public figure behind the words. Having left the tour guide in order to read the ancient graffiti of Pompeii for himself, Haggard vows not to challenge Edward Bulwer-Lytton's masterpiece, *The Last Days of Pompeii* (1834), with "the pen of fiction." Keeping himself before the reader, Haggard also mentions the Cyprus antiques that he has submitted to the British Museum as well as his proposal to make the island a military training ground for troops en route to India. He often observes problems, such as the deterioration of orange trees, and he seldom fails to offer solutions. The self-assured prose of *A Winter Pilgrimage* is seasoned with

Frontispiece and title page for Haggard's only travel book, written to arouse readers' interests in touring the Holy Land (courtesy of the Lilly Library, Indiana University)

foreign terms and traces holy pilgrimages to A.D. 333.

At the time of Haggard's pilgrimage, Cyprus, the fabled abode of Venus, still observed "Whitsunday," an annual nude-bathing holiday for both sexes. The book demonstrates his knowledge about the Phoenicians, who are described as "those English of the ancient world," and traces historic turns in Cyprus by Quakers and Jews. The mule ride to Colossi recalls Haggard's Mexican adventures, and he compares a Cypriot wedding to an Icelandic funeral. He even compares methods of "tombing" that he observes to those of Egyptians. Only after discussing controversial techniques of capital punishment does Haggard conclude that "this matter is too large to enter upon in the pages of a book of travel."

Haggard contradicts his usually imperialistic tone by simultaneously criticizing Turkish domination and the British "plethora of authority" over natives in Cyprus and other colonies. His histories, often rich in national propaganda, include that of "Beyrout," which he characterizes as overly influenced by the French and overpopulated by Americans. He embraces Christian perspective, and biblical signposts mark the long horse ride to Galilee and Nazareth. The narrative reports on the virgins of Saint Claire, who cut off their noses to avoid being raped by the Saracens. "This is no fiction of the romancer," Haggard assures the cautious reader.

Often Haggard's voice remains cynical about the authenticity and condition of Holy Land sites. His dramatic renditions of apocalyptic literature mesh with a dire realism that conveys the filth and stench of alleys encircling the temple in Jerusalem. Along with a nephew who accompanies him, Haggard instructs the reader in the arts of hoodwinking. He blames the proliferation of tourist traps on the naive visitor and the "enterprising" guide. Disgusted at the arrant nonsense sold to the gullible, Haggard is good-natured enough to confess, "I purchased salvation to the value of ninepence in small change."

The real curiosities at the Church of the Sepulchre, the alleged burial site of Christ, are the faithful – who worship what is geographically improbable and who view the skeptical author "as an emissary of anti-christ." The fanaticism of Christians is mirrored by a Muslim family that throws stones at

Haggard on his last visit to Egypt in 1924, a year before his death

Haggard when he accidentally steps on a Muhammadan tomb. Haggard ironically deduces that the British commander Charles Gordon's tomb is the actual tomb of Christ, of which "the guidebooks take no notice." At the same time the narrative's irreverence toward Jericho, "a horribly foul village," contrasts with its admiration for the zealous piety of Russian pilgrims.

A Winter Pilgrimage is, as Haggard writes, "one more of life's turned leaves," and his insertion of recollections and proposals creates a meandering style somehow appropriate to the genre of travel writing. As Britain's foremost adventure novelist, Haggard used his travelogue not only to explore this world but also to reach the boundaries of nonfiction through the imaginative embellishments of his account.

Haggard's interest in the travelogue did not compete with his love of fiction. His pilgrimage of 1900 provided settings for *Pearl-Maiden: A Tale of the Fall of Jerusalem* (1903) and *The Brethren*, while a trip to Egypt in 1904 inspired *The Way of the Spirit* (1906). Returning home from this latter trip via Spain provided a basis for the Spanish setting of *Fair Margaret* (1907). During a tour of the United States in 1905 to report on the Salvation Army's labor colonies, Haggard befriended President Theodore Roosevelt and expanded an earlier parliamentary report for wider readership. This work, *Report on the Salvation Army Colonies in the United States and at Hadleigh, England, with Scheme of National Land Settlement* (1905), received much attention, as did *Ayesha: The Return of She* (1905), a novel that reincarnated Ayesha and relocated her in central Asia.

In 1906 Haggard joined the Royal Commission on Coast Erosion and Aforestation and in 1912 was made a knight bachelor for public service. Haggard also completed his autobiography in 1912, which he gave to his publisher, Charles Longman, who was to publish it after Haggard's death. These memoirs proved premature, however, for in the same year that they were written he was appointed to report on trade and agriculture for the British Empire – a post that required that he tour sites such as India, Canada, Newfoundland, Australia, and New Zealand. He also continued to write fiction: *Marie* (1912), the first work of a Zulu trilogy, was followed by *Child of Storm* (1913) and *Finished*, a novel that re-created Haggard's historic role in the annexation.

Haggard returned to South Africa in 1914 and visited the Zimbabwe excavations as well as his old Zulu servant, Mazooku. In Zululand he publicly advocated native rights and deplored the position of the Boers, who were rumored to be dealing secretly in slavery even after it was to have been abolished at the Sand River Convention in 1852. In 1916 Haggard returned to South Africa and then went to Australia, New Zealand, and Canada to research British military settlements in these dominions. He continued to use travel as a springboard for fiction, and a year after having published *Moon of Israel: A Tale of the Exodus* (1918), Haggard was made knight commander of the British Empire. He returned to Egypt in 1924 and died the following year.

Haggard used travel as a metaphor of the human spiritual journey. The pilgrimages of his fiction and nonfiction reveal the heart of Victorian crisis: lost faith in ancient truths and speculation about the new humanism that would replace them. While Haggard's geographical settings were new to his audience, the aim of his travel writing was conventional. He used travel to search for ideals and for means of resolving political tensions and preserving ideologies in a rapidly changing era.

Bibliographies:

James Edward Scott, *A Bibliography of the Works of Sir Henry Rider Haggard, 1856-1925* (London: Mathews, 1947);

Denys Edwin Whatmore, *H. Rider Haggard: A Bibliography* (Westport, Conn.: Mecklar / London: Mansell, 1987);

Lloyd Siemens, "The Critical Reception of Sir Henry Rider Haggard: An Annotated Bibliography, 1882-1991," *English Literature in Transition, 1880-1920,* special series, no. 5 (1991).

Biographies:

Lilias Rider Haggard, *The Cloak That I Left: A Biography of Rider Haggard* (London: Hodder & Stoughton, 1951);

Morton Cohen, *Rider Haggard: His Life and Works* (London: Hutchinson, 1960);

Peter Berresford Ellis, *H. Rider Haggard: A Voice from the Infinite* (London & Henley: Routledge & Kegan Paul, 1978);

D. S. Higgins, *Rider Haggard: The Great Storyteller* (London: Cassell, 1981);

Tom Pocock, *Rider Haggard and the Lost Empire* (London: Weidenfeld & Nicolson, 1993);

Victoria Manthorpe, *Children of the Empire: The Victorian Haggards* (London: Gollancz, 1996).

References:

Norman Etherington, *Rider Haggard* (Boston: Twayne, 1984);

Wendy R. Katz, *Rider Haggard and the Fiction of Empire: A Critical Study of British Imperial Fiction* (New York: Cambridge University Press, 1987).

Papers:

Although Haggard's materials are scattered, the Norfolk Record Office has a collection, as does the Huntington Library in San Marino, California; the Lockwood Memorial Library at the State University of New York in Buffalo, New York; and the Columbia University Library, New York City.

Agnes Herbert
(circa 1880 – 7 February 1960)

Timothy S. Jones
Augustana College

BOOKS: *Two Dianas in Somaliland: The Record of a Shooting Trip* (London & New York: John Lane, 1907);

Two Dianas in Alaska (London & New York: John Lane, 1908);

The Isle of Man (London & New York: John Lane, 1909);

The Life Story of a Lion (London: A. & C. Black, 1911);

Casuals in the Caucasus: The Diary of a Sporting Holiday (London & New York: John Lane, 1912);

The Moose (London: A. & C. Black, 1913);

The Elephant (London: Hutchinson, 1916; New York: Stokes, 1917);

Northumberland (London: A. & C. Black, 1923);

Korea (London: A. & C. Black, 1924).

OTHER: *Black's Guide to Scotland, South-East*, revised by Herbert (London: A. & C. Black, 1925).

Agnes Herbert begins her first book, *Two Dianas in Somaliland: The Record of a Shooting Trip* (1907), with the explanation: "It is not that I imagine the world is panting for another tale about a shoot. Simply – I want to write." Write she did – more than one thousand pages in three books of sporting adventures in Somalia, Alaska, and the Caucasus. Accompanied by her cousin Cecily, native guides, and occasionally by her cousin Ralph Windus and another man known only as "the leader of the opposition," Herbert collected an awesome assortment of big-game trophies and travel tales.

Agnes Elsie Diana Thorpe was born around 1880, the daughter of James Bateman Thorpe, and spent her youth on the Isle of Man. The name Herbert was that of her first husband, who died prior to her hunting expeditions, and she continued to use it professionally, even after marrying Comdr. Archibald Thomas Stewart in 1913. She appears to have taken advantage of her interim widowhood to travel and indulge her sporting interests. In *Casuals in the*

Agnes Herbert in native parka dress; frontispiece for Two Dianas in Alaska *(1908)*

Caucasus: The Diary of a Sporting Holiday (1912) she describes an encounter with a woman ardently searching for a second husband: "She didn't seem to realize one bit that the widow who marries a second time does not deserve to be one." By the time she remarried, Herbert was in her mid thirties and had composed all three of her hunting books.

Herbert's robust sense of humor is often directed at men and members of society. Commenting in *Two Dianas in Somaliland* on male desire for the

Frontispiece and title page for the last of Herbert's hunting books

forbidden, she suggests that Eve plucked the fruit in Eden "at the express wish of Adam who wanted it badly, and had not the moral courage to take it for himself." In *Casuals in the Caucasus* she describes a hostess in Gibraltar as having "the graceful contemplative hauteur of a woman who has been called beautiful by the Society papers so often that she had come to believe it to be a fact and not merely an optical illusion on the part of the reporter." The tone might become arrogant were Herbert not willing to poke fun at herself as well. On several occasions she apologizes for digressing with the query, "Did you ever know a woman who could stick to the point?" The irony is that these digressions contribute to the unique voice of Herbert's texts, in which the author distinguishes herself from the traditional male hunting monologue and its devotion to the business of shooting.

Many contemporary reviewers found that Herbert's style took some getting used to. *The Spectator* review of *Two Dianas in Somaliland* complained, "The reader never quite knows when the author will harass him with some piece of doubtful taste or doubtful humour" – possibly referring to the author's flippancy toward a missionary and Holy Scripture. On the boat to Aden, a priest was raising money for a painting of Moses to decorate his chapel. Herbert drove him off by suggesting that Moses was the illegitimate son of the Egyptian princess and that she pretended to find him among the bulrushes in order to diffuse a potential scandal. In the end, however, Herbert's nerve and courage during the hunt won this and other reviewers over.

There are moments of danger and suspense in *Two Dianas in Somaliland* – a native companion is killed by a rhinoceros, and Herbert's arm is torn open by an oryx – but more interesting to the modern reader is Herbert's commentary on all aspects of the traditionally male pastime of big-game hunting. From selecting guns to buying supplies to dealing with porters to pulling the trigger, she consciously contrasts her experience with that of the men who have hunted and written before her. When she considers an opportunity to give advice on marksmanship, she thinks the better of it and concludes: "All these tomes of wisdom were written for man by man. I tried to follow out their often entirely opposite advice, but after a while, being a woman and therefore contrary, I 'chucked' all systems and manufactured rules for myself. I don't close either eye when I shoot. I shoot with both eyes open."

About the same time, Herbert's publisher asked her to write the text for a color picture book about the Isle of Man. The result is a description that outshines the watercolor illustrations by Donald Maxwell. *The Isle of Man* (1909) discusses history, archaeology, government, and language. The reader is taken along on a herring-fishing trip and tours of the island by land and sea. Herbert occasionally includes experiences and recollections from her childhood and concludes the book in the manner of William Shakespeare's comedies with the description of a wedding celebration. W. Ralph Hall Caine, writing his own volume on the island, credits Herbert's book with an important position in Manx literature, but he also echoes the common complaint, "Why will she not be more serious?"

A second criticism introduces *Two Dianas in Alaska* (1908): a friend who had received a complimentary copy of the Somalia book returned it to the author with the complaint that it contained too much killing. Shooting wild game, Herbert responds, is no different from butchering domestic sheep or cattle. She suspects that the friend meant that the killing is unfeminine, but Herbert ultimately is not interested in making apologies: "We went to Alaska to shoot, and – we shot."

Again in the company of Cecily, she set out by steamer for New York and then by train for Butte, Montana. There they met up with "the opposition" from the Somalia expedition, Henry Windus, and "the leader." Joining forces, they chartered a boat in Vancouver and sailed up the Inland Passage to Alaska, where they shot bear, moose, caribou, and Dall sheep. Several of these chapters are written by "the leader," including an account of how Herbert's keen reflexes and sure aim saved him from an enraged bear. The interplay between these two characters and their narratives proves to be a driving force in the book. In the end it is Cecily and Henry who marry and sail off for a honeymoon in San Francisco, but Herbert gives a clear sense that her relationship with "the leader" is more than casual.

The last of Herbert's hunting volumes, *Casuals in the Caucasus,* contains the least narration of the hunt. Early in the book Herbert complains of the tedious descriptions of "getting there" that clutter so many travel books. As a reader she prefers to pass over lengthy explanations and "skip on to where, with large margins and many spaces, the people talk." As a writer she fills the journey with garrulous people who open their mouths to prove themselves fools. A single, unusually picturesque description of Constantinople serves largely to set up the rudely comic response of her cousin Kenneth: " 'Ripping!' he said, 'I call it simply ripping!' "

Fully a third of this book passes before the hunting party fires a single shot. The women sailed from Gibraltar to the Black Sea port of Batoum (Batumi) and then proceeded by train to Tiflis (Tbilisi) and horseback to Signakh (Signachi). Finally, they ascended a river valley into Dagestan to hunt tûr, a Caucasian ibex, and gray bear. Stocked with these trophies, the two women returned to Tiflis and made a second trek across the Caucasus Range to Vladikavkaz and then to the Karabada region near Mount El'brus, where they became the guests of a local nobleman they had met on the train from Batoum and hunted ollen, a variety of deer, more ibex, and a variety of fowl. As autumn arrived, it became clear that the nobleman had his own predatory designs set on Cecily, and the two Dianas made their escape to Tiflis, and thence to England.

After hanging up her rifle and trophies, Herbert continued to write, composing guidebooks to Northumberland (1923), Korea (1924), and southeastern Scotland (1925). She also served the Society of Women Journalists as vice chair from 1929 to 1933 and as vice president beginning in 1939, and for many years she edited the *Writer's and Artist's Yearbook*.

Herbert's contributions to English journalism were significant and varied, and her hunting narratives offer an active and engaging account of the travel experiences newly opened to women in the early twentieth century. Her book provides both sharp observations about the people and places that she encountered and distinctive and humorous commentary on the traditionally male practice of big-game hunting.

Reference:

Jane Robinson, *Wayward Women* (Oxford: Oxford University Press, 1990), pp. 69–70.

W. H. Hudson

(4 August 1841 – 18 August 1922)

Joan Corwin

See also the Hudson entries in *DLB 98: Modern British Essayists, First Series* and *DLB 153: Late-Victorian and Edwardian British Novelists, First Series.*

BOOKS: *The Purple Land That England Lost: Travels and Adventures in the Banda Oriental, South America,* 2 volumes (London: Sampson Low, Marston, Searle & Rivington, 1885); revised as *The Purple Land: Being the Narrative of One Richard Lamb's Adventures in the Banda Oriental, in South America, as Told by Himself* (London: Duckworth, 1904; New York: Grosset, 1904);

A Crystal Age (London: Unwin, 1887; revised, 1906; New York: Dutton, 1906);

Argentine Ornithology: A Descriptive Catalogue of the Birds of the Argentine Republic, 2 volumes, by Hudson and Philip Lutley Sclater (London: Porter, 1888); Hudson's notes revised as *Birds of La Plata,* 2 volumes (London: Dent / New York: Dutton, 1920);

The Naturalist in La Plata (London: Chapman & Hall, 1892; New York: Appleton, 1892);

Fan, the Story of a Young Girl's Life, 3 volumes, as Henry Harford (London: Chapman & Hall, 1892; New York: Dutton, 1892); republished, by Hudson (London: Dent / Toronto & New York: Dutton, 1923);

Idle Days in Patagonia (London: Chapman & Hall, 1893; New York, 1898);

Birds in a Village (London: Chapman & Hall, 1893; Philadelphia: Lippincott, 1893); revised and enlarged as *Birds in Town and Village* (London: Dent / New York: Dutton, 1919);

Lost British Birds (London: Chapman & Hall, 1894); revised and enlarged as *Rare, Vanishing & Lost British Birds, Compiled from Notes by W. H. Hudson,* edited by Linda Gardiner (London: Dent / New York: Dutton, 1923);

British Birds (London & New York: Longmans, Green, 1895);

Birds in London (London, New York & Bombay: Longmans, Green, 1898);

W. H. Hudson

Nature in Downland (London & New York: Longmans, Green, 1900; New York: Duckworth, 1928);

Birds and Man (London & New York: Longmans, Green, 1901; revised, London: Duckworth, 1915; New York: Knopf, 1916);

El Ombú (London: Duckworth, 1902; enlarged edition, London & Toronto: Dent / New York: Dutton, 1923); republished as *South American Sketches* (London: Duckworth, 1909); revised as *3 Tales of the Pampas* (New York: Knopf, 1916);

Hampshire Days (London & New York: Longmans, Green, 1903);

Green Mansions: A Romance of the Tropical Forest (London: Duckworth, 1904; New York: Putnam, 1904);

A Linnet for Sixpence! (London: Royal Society for the Protection of Birds, 1904);

A Little Boy Lost (London: Duckworth, 1905; New York: Knopf, 1918);

The Land's End: A Naturalist's Impressions in West Cornwall (London: Hutchinson, 1908; New York: Appleton, 1908);

Afoot in England (London: Hutchinson, 1909; New York: Knopf, 1922);

A Shepherd's Life: Impressions of the South Wiltshire Downs (London: Methuen, 1910; New York: Dutton, 1910);

Adventures Among Birds (London: Hutchinson, 1913; New York: Kennerley, 1915);

Far Away and Long Ago: A History of My Early Life (London: Dent / New York: Dutton, 1918);

Roff and a Linnett: Chain and Cage (London: Humanitarian League, 1918);

The Book of a Naturalist (London & New York: Hodder & Stoughton, 1919);

Dead Man's Plack and An Old Thorn (London & Toronto: Dent / New York: Dutton, 1920);

A Traveller in Little Things (London & Toronto: Dent, 1921; New York: Dutton, 1921);

A Hind in Richmond Park (London & Toronto: Dent, 1922; New York: Dutton, 1923);

Seagulls of London (London: Shorter, 1922);

Ralph Herne (New York: Knopf, 1923);

Tales of the Gauchos: Stories (New York: Knopf, 1946);

William Henry Hudson's Diary Concerning His Voyage from Buenos Aires to Southampton on the Ebro (Hanover, N.H.: Westholm, 1958).

Collections: *The Collected Works of W. H. Hudson*, 24 volumes (London: Dent / New York: Dutton, 1922–1923);

A Hudson Anthology, edited by Edward Garnett (London & Toronto: Dent / New York: Dutton, 1924);

The Best of W. H. Hudson, edited by Odell Shepard (New York: Dutton, 1949).

W. H. Hudson is remembered almost exclusively for the novel *Green Mansions: A Romance of the Tropical Forest* (1904), which became his first substantial financial success after a writing career of more than four decades. Hudson, however, was much more than a romance writer. His best writing, in fact, combined the nature essay with the travel essay, resulting in works that have made him an apt successor to Henry David Thoreau, whom he admired. In addition to essays and romances Hudson produced short stories and a nostalgic account of his early years in the Argentine pampas, an account still considered an especially fine work of autobiography.

The fourth child and third son of New England parents who had immigrated to South America, William Henry Hudson was born on 4 August 1841 ten miles from Buenos Aires at Quilmes, in the La Plata region of the Argentine pampas. He spent the first five years of his childhood at a small *estancia,* or cattle ranch, named the Twenty-Five Ombus for the trees that grew there. Just before his fifth birthday his family moved to the Acacias, a larger property that included a general store, gardens, an orchard, and a plantation. The many trees at the Acacias attracted exotic birds, inspiring in Hudson a fascination with bird life that would last his lifetime.

The wild plains of Hudson's childhood offered ample opportunity for both active adventure and dreamy contemplation. Argentina was the country of the romantic and brutal gauchos, or South American cowboys, and colorful, eccentric neighbors, the closest of whom lived two miles away. In such a setting, and in spite of his parents' attempts to have their children tutored, Hudson enjoyed a largely unrestricted childhood spent mostly outdoors in pursuits such as riding and hunting. As a result, he was exposed to a wealth of diverse and splendid forms of wildlife, and he began to evince an almost trancelike absorption in them, his first expression of an appreciation of nature that would characterize his mature life and writings.

Some months before Hudson's fifteenth birthday, he caught typhus while on a trip to Buenos Aires. The formerly active youth was forced to endure a long convalescence from this illness and from a subsequent rheumatic fever, which permanently impaired his heart and left him subject to attacks and palpitations. Although he lived to be eighty-one, he suffered from a constant fear of heart failure, a fear that added piquancy to his love of life. The illness initiated a lifetime of bad health and instilled in Hudson a fear of death, but it also introduced him to books, which he had scorned until then as a poor substitute for experience; during this period his love of reading became firmly established. He read history and natural science and was strongly impressed by Gilbert White's *The Natural History and Antiquities of Selborne* (1789). A few years later, his brother would introduce him to Charles

Darwin's *On the Origin of Species* (1859), which would shake his already weakening religious faith. Although he was always careful to insist on specific points of difference with Darwin, Hudson eventually became an evolutionist.

The years of Hudson's young manhood did not begin auspiciously: in 1857 his father, Daniel, a gentle and naive soul, lost the Acacias as a result of his trust in the good faith of mankind: he had neglected to obtain the title for the property, which he had restored at his own expense, and it was awarded to a subsequent claim. Then in 1859, two months after Hudson's eighteenth birthday, his mother died; Caroline Kimble Hudson had been the only person who shared his intense appreciation of nature, a bond that had made them especially close and had meant a great deal to the dreamy youth.

Hudson spent his early twenties in the saddle, working on his father's land, visiting old gaucho friends at their ranchos, and attending celebrations and cattle markings. He traveled widely, living on the hospitality of others or sleeping in the open, often encountering dangers, natural and human. In those years he was creating the habits of a lifetime, wandering, at once detached from and alert to human society, mentally recording every fine detail of landscape and fauna that filled him with his particular enjoyment of the natural world.

In 1865 Hudson was first introduced to the idea of becoming a scientific bird collector, a vocation that was attractive to him because it involved being surrounded by nature in a pursuit he enjoyed. He began hunting birds, preparing them as specimens, and sending them, often with letters describing their habits, to Prof. Spencer Fullerton Baird of the Smithsonian Institution in the United States and to Dr. Philip Lutley Sclater of the Zoological Society in London. After his father's death in 1868, Hudson continued to travel and collect, covering much of Argentina and as far away as Brazil, Uruguay, and the Rio Negro. Sclater published the letters he received from Hudson in the Zoological Society's proceedings from 1869 to 1873, Hudson's first published work. As a result, before he turned thirty Hudson had already achieved a name for himself as a naturalist and writer among scientific circles in both countries. With this success, he began to hope that he might support himself on his earnings from such specimens, but he found that the cost of traveling and of shooting, preserving, and shipping birds made his vocation impractical.

Hudson was discouraged by the failure of his collecting career, and he was disturbed by the advance of civilization on the great natural wilderness of his homeland. Unlike his siblings, he had always felt a strong tie with England through Daniel Hudson, whose father had come from Devon. This attachment was strengthened by the flattering reception Daniel Hudson's specimens and writings had received in London, so in 1874 he decided to immigrate to what he called the land of his desire.

Hudson docked in Southampton, and after a brief tour of the countryside he continued on his way to London to seek the help of professionals such as Sclater and the ornithologist John Gould in making a career as a naturalist. Their reception of him was chilly, and the experience made him bitter against such scientists for the rest of his life. Now in his mid thirties, Hudson turned primarily to writing as a career, living in London at a boardinghouse run by Emily Wingrave, a former singer who may have been as much as fifteen years Hudson's senior; they were married in 1876.

From the late 1870s to the mid 1880s the Hudsons ran a succession of boardinghouses, each of which failed, and the couple were left supporting themselves on what Emily made teaching music lessons and on the meager income Hudson received from magazines for his short stories and essays. The Hudsons were so poor that one week they were forced to live on a tin of cocoa and milk. They were delivered from this level of destitution in 1886 when Emily inherited Tower House on St. Luke's Road, Westbourne, where they lived and let rooms. Although Hudson remained quite poor in property throughout the early decades of his writing career, he was becoming rich in friends, among them the writers George Gissing; Morley Roberts, who left a long and moving memoir of Hudson; and the flamboyant R. B. Cunninghame Graham, who shared with Hudson a romantic nostalgia for the South American pampas and their gauchos.

At the center of all of Hudson's themes, in both his fiction and his nonfiction, is the conflict between nature and civilization. His intense appreciation of nature was a result of a remarkably acute receptivity to sensory stimuli combined with a contemplative temperament. His response to nature he called "animism," which he defines in *Idle Days in Patagonia* (1893) as "the mind's projection of itself into nature, its attribution of its own sentient life and intelligence to all things." As a result his writing about animals, while wonderfully precise, has a touch of humanity and mysticism that raises it above the level of most naturalists' accounts. Hudson's attachment to nature made him extremely nostalgic about his life in Argentina, where the wildlife was not only more abundant but more colorful

Frontispiece and title page for Hudson's account of an 1870 trip to Rio Negro (courtesy of the Lilly Library, Indiana University)

and varied. Not only did he feel a strong attachment to his own past, but he felt a connection with the past that predated his civilization's existence, a past, he believed, in which primitive man was in harmony with nature. It is not surprising that the writer who felt such a reverence for nature and such a bond with the preindustrial past expressed strong support for the conservation of wildlife.

Almost all of Hudson's writing, his fiction included, has some connection with travel. Even his bird essays are often organized around the occasion of travel. The two primary geographical sources for his material were the English countryside and the Argentine pampas.

Early in his London writing career, Hudson flirted with poetry, decided he was not cut out for it, and destroyed his unpublished efforts. From the appearance of his letters to Sclater in the *Proceedings of the Zoological Society of London* until the end of his life, Hudson published 221 essays, reviews, stories, and letters to the editor in more than thirty different journals, magazines, and newspapers. Almost all of these were collected into books during his lifetime. His first primarily literary effort, an article titled

"Wanted – A Lullaby," which he wrote pseudonymously as Maud Merryweather, was published in *Cassell's Family Magazine* in March 1875. His first piece of fiction appeared eight years later; it was the short story "Pelino Viera's Conversion," in the *Cornhill Magazine* (1883), a recounting of a gaucho legend about a mysterious bird woman.

Hudson's next literary milestone was the publication in 1885 of the unsuccessful historical romance *The Purple Land That England Lost*. *The Purple Land* is less a coherent novel than a string of stories held together by the career of a picaresque hero named Richard Lamb, an Englishman resident in South America. Two years later Hudson produced another work of fiction, a utopian fantasy titled *A Crystal Age* (1887), in which a young Englishman named Smith suffers a fall in a hunting accident and awakens thousands of years later to find an idyllic pastoral society that has reestablished man's lost ties with nature. *Ralph Herne,* Hudson's only fiction published in serial form, appeared in the magazine *Youth* in 1888. The story concerns a young Englishman who immigrates to Buenos Aires to become a physician. It reappeared in book form in 1923, a year after Hudson's death.

In 1888 Hudson turned to a kind of writing more clearly suited to his style and temperament – the descriptions of bird life, appearance, song, and habits that he contributed to Sclater's *Argentine Ornithology* (1888). The writing is the best of its kind and makes up the bulk of the volume. Hudson felt that Sclater was stingy with praise and credit, and the experience of the collaboration became for him part of a bitter lesson he was learning about professional natural scientists in England. Hudson later extracted his part of *Argentine Ornithology*, revised it, and in 1920 published it as *Birds of La Plata*.

Ironically, 1892 was the year that Hudson published his worst piece of fiction, a cloyingly sentimental novel titled *Fan, the Story of a Young Girl's Life*. He wrote this, purely to take advantage of the market for such melodrama, in the same year that he published *The Naturalist in La Plata*, the work that would make his name as a writer. Hudson's first book-length treatment of his native land's natural life was also his first complete volume of writings in a genre that would produce his greatest works, the travel/nature essay. The book was a success: combining description of exotic wildlife with a substantial amount of evolutionary debate, it brought Hudson to the front of public consciousness. Praised for both its style and its natural history, it is distinguished from Hudson's other writings by the large amount of space he devotes to scientific polemic, particularly his arguments against Darwin's contention that sexual selection is responsible for birdsong and play in animals. Among those impressed by Hudson's work was Alfred Russel Wallace, one of the leading naturalists of the day.

The following year Hudson produced another moderate popular success and one of his finest travel books, *Idle Days in Patagonia*, an account of a trip he took twenty-three years earlier to the Rio Negro to collect animal specimens for museums in England and America. A series of essays and episodes in the style of the earlier *Naturalist in La Plata*, *Idle Days* is innovative for a work of travel. For one thing, it concerns "idle," or inactive, days rather than the kind of busy chronological itinerary that had previously made up the structure of most travel books. Hudson's idleness was the result of a wound he inflicted on his knee when a gun he was examining unexpectedly discharged. Coming at the beginning of his tour of the Rio Negro, this accident meant that his initial observations had to be made from a hospital bed. As a result of this inactivity, his explorations are aimed as much inward as outward.

The book lingers over details of animal life and terrain by indulging in associative contemplation, both reflective and speculative. It is a record of Hudson's change of heart: from energetic collector of dead specimens to patient observer of animal life and of the magic that animates the natural world. Also, more intensely than are Hudson's earlier writings, *Idle Days* is concerned with the past – the recovery of the past through the senses and the recovery of a prehistoric identity, a oneness with the primitive peoples of the past who had lived as a harmonious part of nature.

The book's best-known essays are "Snow, and the Color of Whiteness" and "The Plains of Patagonia." The first concerns the effect on one's mood of the color white in certain animals and landscapes and is the essay in which Hudson defines his idea of animism; the second is a remarkable chapter on the ability of vast wilderness solitudes to send the visitor "back to a mental condition we have outgrown," a primitive alertness. Another important essay is the closing chapter, "The Perfume of an Evening Primrose," in which Hudson deals with involuntary memory activated by smells, which produce the phenomenon of temporarily restoring the past: "I am actually a boy again."

Before the publication of *The Naturalist in La Plata* and *Idle Days in Patagonia*, Hudson had been making a new kind of friend: those who were interested, as he was, in bird conservation. At that time in England, private collectors and women's fashions, which included the wearing of wild birds on ladies' hats, were decimating the wild bird population. In 1891 Hudson joined the recently formed Society for the Protection of Birds, and in 1893 he turned from describing South American bird life to devoting whole books to British birds. In that year appeared *Birds in a Village*, the longest and title essay of which describes the wild birds of the village of Cookham Dean. Hudson later updated *Birds in a Village*, and in 1919 it was republished with added material under the title *Birds in Town and Village*. *British Birds*, virtually a reference book covering the bird species of the British Isles, followed in 1895. *Birds in London*, which appeared in 1898, is a catalogue like *British Birds*, organized geographically as a tour of London's gathering places for birds, such as its parks and cemeteries; it includes an indictment of collectors and is the first full-length written expression of Hudson's conservationism. These increasingly popular books on birds and the many pamphlets on bird conservationism that Hudson wrote for the Society, such as *Lost British Birds* in 1894, made him an extremely valuable member.

Hudson's English bird books were the first fruits of his travels in his new environment. The little he had earned during his years of poverty had enabled him to make short journeys, often with Emily, to the rural areas immediately surrounding London for walking tours. As his income increased his travels became more extensive. Often he would find shelter in the homes of country folk, for whom he grew to have great affection and respect. These longer journeys provided the material for what would become the popular, wide-ranging books of nature essays he produced in his mature writing period, from 1900 to his death in 1922. They also became essential to Hudson's emotional well-being as he grew more and more dissatisfied with London and the modern industrial world.

From 1889 to 1911 Hudson traveled extensively, though still modestly, throughout the English countryside, gathering material for what many readers consider his finest books. The first of these, *Nature in Downland* (1900), describes the villages and towns of Sussex, including the house in Goring where Richard Jefferies, an earlier renowned countryside writer, had died. Like most of Hudson's mature work, the book incorporates references to his wide reading. It is also the first in a series of his mature works that focus on human beings rather than birds and the significance of their experiences of life.

Nature in Downland appeared in a year that saw Hudson's naturalization as a British citizen and brought a much-needed new source of income. Edward Grey, later Viscount Grey of Fallodon, had been impressed as early as 1893 by *The Naturalist in La Plata* and *Idle Days in Patagonia*. After the publication of *Nature in Downland,* he recommended Hudson for a civil list pension of £150 a year, and in 1901 Hudson was notified of this stipend, which would considerably ease his straitened financial condition.

Birds and Man (1901) was the first of Hudson's books of bird essays to exhibit his maturity as a stylist. In it Hudson claims that the full appreciation of all natural life requires the emotional, as distinct from sentimental, participation of the observer, something impossible when one is confronted with a dead, stuffed specimen in a museum or drawing room. He emphasizes the unity of all living things and the humanlike characteristics of birds and other animals. Hudson also put together his finest collection of short stories, *El Ombú* (1902); the best of the collection are the title story and another titled "Marta Riquelme." The book was admired by the editor Edward Garnett, who became Hudson's great friend.

Another record of Hudson's rambles, *Hampshire Days,* appeared in 1903; it concentrates on the human inhabitants of that district. The book touches on several Hudsonian themes: an ancient barrow inspires in Hudson the illusion of the primitive inhabitants of England; a clergy/lepidopterist who despairs of a poor harvest of hummingbird moths ignites the author's contempt for collectors; and a poverty-stricken old couple who live off the produce of their garden provide an object for his compassion, as well as an opportunity to demonstrate his sensitive ear for local country dialect.

In 1904 Hudson published the book that would make his fortune in his old age. *Green Mansions: A Romance of the Tropical Forest* concerns a young Venezuelan political refugee who flees his country to live with a remote and savage Indian tribe. In the wilderness he encounters and falls in love with Rima, a mystical, birdlike young woman whom the Indians later kill. Ironically, the lush and romantic description of the South American rain forest in which the story takes place was based not on Hudson's own experience – he had never been to the upper Amazon River region – but on the works of other naturalists, particularly Henry Walter Bates's travel classic *The Naturalist on the River Amazons* (1863). Only moderately successful at first, *Green Mansions* experienced a best-selling revival in America twelve years later. In 1905 *Green Mansions* was followed by *Little Boy Lost,* a piece of fiction intended primarily for children.

For the rest of his life, with the exception of his autobiography, his revised bird books, and a collection of short stories, Hudson devoted himself primarily to essays based on material gathered during his rambles throughout England. *The Land's End: A Naturalist's Impressions in West Cornwall* (1908) is a largely unflattering description of the area and its people. In spite of some rich chapters on the seascape, the Cornish wren, and the furze shrub, the book is marked by Hudson's disappointment in the region. It was followed by *Afoot in England* (1909), a collection of twenty-five essays, many of them previously published. Because *Afoot in England* is devoted to several regions, it lacks the coherence of the other rambles volumes; but it is notable for its discussion of a wide variety of authors and contains many amusing anecdotes that reveal human nature.

Then in 1910 Hudson's masterpiece in the genre appeared: *A Shepherd's Life: Impressions of the South Wiltshire Downs.* Limited to the Salisbury Plain region of Wiltshire and centering on the figure of Caleb Bawcombe, the fictional name of a shepherd Hudson had befriended, it is one of the most tightly

Frontispiece and title page for Hudson's book based on his experiences in the Salisbury Plain region of Wiltshire

structured of his books, even though it purports to digress as Hudson allows one idea to trigger another – dogs, cats, birds, foxes, poaching, the past, or the human characters who people the area – from chapter to chapter. The figure of Caleb centralizes these ideas; his more than fifty reminiscences act not only as inspiration, but also as margins and a foundation for the book. It is a moving record of a period and lifestyle that was disappearing from the English countryside, a fact that gave Hudson's writing a quiet poignancy. The book includes some of his most important themes: nostalgia for a past (recent and ancient), communion with nature, and the resourcefulness of the human spirit.

In 1911 Emily Hudson became ill and required Hudson's almost constant attendance. Nevertheless, he was able to leave her for a visit to Wells-Next-the-Sea in Norfolk, where he completed *Adventures Among Birds* (1913), a book of varied essays based on his experiences in several different rural locales, including the coast of Norfolk with its large flocks of migratory wild geese. The book presents enlarged versions of some essays that had earlier appeared in periodicals.

Emily became bedridden in 1914. Hudson himself fell seriously ill two years later and decided to take a rest cure at the Convent of the Sister of the Cross in Cornwall, a district he had grown to love in spite of his first impression years before. "On the second day of my illness," he writes in *Far Away and Long Ago: A History of My Early Life* (1918) "during an interval of comparative ease, I fell into recollections of my childhood, and at once I had that far, that forgotten past with me again as I had never previously had it." The result was *Far Away and Long Ago,* a moving and much-praised record of Hudson's childhood and adolescence in South America. The book is a testimony to how much the author valued memory and to his ability to find meaning in past experience. Its evocation of a lost time and place instantly made it valued by armchair travelers.

In 1919 and 1920, respectively, Hudson published his revision of *Birds in Town and Village* and *Birds of La Plata,* the latter drawing on the same talent that makes his autobiography so vivid, his astonishingly accurate recollection of concrete sensory detail. During this period his books of essays became even more far-ranging in subject and tone than before. *The Book of a Naturalist* (1919) includes several essays on snakes, to which he at one time had hoped to devote an entire volume. In 1920 he also produced his second and last volume of fiction, *Dead Man's Plack and An Old Thorn,* which comprises the two stories of its title.

The title of the essay collection *A Traveller in Little Things* (1921) refers to a traveling salesman's description of Hudson. The drummer had taken Hudson for a salesman, too, but one who dealt in small wares. Hudson is pleased with the description and implies that he did indeed travel to experience, and to relate his experience of, all the bright diverse little things of the natural world. Of these, the old samphire gatherer and a village full of apple blossoms inspire vivid essays. Less profound than his earlier collections of essays, and even superficial in places, *A Traveller in Little Things* is, nevertheless, entertaining and often touching.

In the same year that *A Traveller in Little Things* was published, Emily Hudson died. Hudson's own health failed soon after. For some time before this decline, he had been unable to sustain the strenuous exercise that his walking and cycling tours required and was sometimes forced to travel by motorcar, a conveyance he had held in great disdain. He was at home in Tower House, working on the first draft of his final book of essays, when he died in his sleep on 18 August 1922.

The book, *A Hind in Richmond Park,* which was tidied up for publication the same year by Hudson's devoted friend Morley Roberts, is uncharacteristically focused on a single central theme – the senses in animals and man – and more than usually liquid and conversational (Roberts described it as "near his talk"). The hind of the title is a deer Hudson was watching in the great enclosed park of Richmond Manor; his observation of the animal inspires a series of fluidly connected thoughts on nature and mankind. As in his earlier writings, he is concerned with the appreciation of beauty felt by all living things. He discusses the forms of human expression such appreciation takes – such as music and the plastic arts – and reveals his deep love of poetry, especially that of Geoffrey Chaucer. He concludes, however, that a full and satisfactory expression of the experience of beauty has yet to be evolved.

Although W. J. Keith has, with some justice, called Hudson's writing "stylistically fastidious," most of Hudson's critics have found his style uniformly simple and concrete. He is considered a master of the essay form largely because he eschewed the kind of ponderous language of which its practitioners are often guilty, relying more on details of concrete evidence and observation to support his generalizations than on elaborate philosophical constructions. He often treated narrative, related simply and unsentimentally, as just such evidence, and both his essays and his stories are profoundly moving in their quiet and careful rendering of experience. Many contemporary writers wished to emulate what they considered his transparent style, which captured the essence of the thing he described. His confidential and often humorous tone gave his writings the appearance of facility, but in fact this illusion of ease was won through hard work and revision. Readers have also admired Hudson's musical ear, remarking especially on the rhythms in his prose. His rendering of dialect was accurate and lyrical, and he may well have come closest of any English writer to capturing in print that elusive music birdsong.

The structures of Hudson's books reflect his preoccupation with the interconnectedness of all living things and with the workings of thought and memory. He worked from close observation of the particular to evolve broad and far-reaching generalizations. His individual essays are often formed by the fluid movement of the mind touching on what it perceives and what it associates with these perceptions, while the books consist of loosely related collections of such essays, whose variety reveals the author's conviction that the most diverse forms of life are connected.

In spite of his many friendships, Hudson was a private man, extremely reserved and independent. Some saw him as a magnificent wild creature, awkwardly restrained by the civilized life he endured. As he aged, he could become irascible and contemptuous, though never violent, choosing for his targets particularly those despoilers of bird life and those proponents of industrialism whom he felt responsible for destroying the beauty of the natural world. His friends, however, attested to a warmth of manner and a gentleness that surpassed that of anyone they knew. He was especially fond of children, believing them to be closer to nature than their jaded elders. In later life he could count many accomplished novelists and travel writers among his friends: Joseph Conrad, John Galsworthy, T. E. Lawrence, Wilfrid Scawen Blunt, H. J. Massingham, W. H. Davies, and Edward Thomas, a young writer whose death in World War I soured Hudson on the war effort.

Page from Hudson's manuscript for Far Away and Long Ago: A History of My Early Life *(1918; Pierpont Morgan Library)*

Hudson was buried in Worthing beside his wife. His grave site carries the inscription "He loved birds and green places and the wind on the heath, and saw the brightness of the skirts of God." Several friends met a few months later to devise a memorial for him. The result was a bird sanctuary in Hyde Park, the memorial stone for which was sculpted by Jacob Epstein and depicts the bird-woman Rima. It was dedicated by Prime Minister Stanley Baldwin.

Hudson's influence on travel and nature writing, and perhaps on writing in general, is not currently widely recognized. His style, however, impressed such notables among his contemporaries as Conrad and Galsworthy, the latter contributing a foreword to the republication of *Green Mansions*. His *Idle Days in Patagonia* may come to be seen as a significant link in the evolution of travel writing, one which transformed the travel book from an exterior journey into a journey of the soul.

Letters:

Letters from W. H. Hudson, 1901-1922, edited by Edward Garnett (New York: Dutton, 1923);

Men, Books, and Birds. Letters to a Friend (London: Nash & Grayson, 1925);

Letters from W. H. Hudson to Garnett, edited by Garnett (London: Dent, 1925);

W. H. Hudson's Letters to R. B. Cunninghame Graham, with a Few to Cunninghame Graham's Mother, Mrs. Bontine, edited by Richard Curle (London: Golden Cockerel, 1941);

"W. H. Hudson's Lost Years," *Saturday Review of Literature,* 30 (12 April 1947): 15-17;

Letters on the Ornithology of Buenos Ayres, edited by David R. Dewar (Ithaca, N.Y.: Cornell University Press, 1951);

Two Letters on an Albatross, by Hudson and R. B. Cunninghame Graham, and edited by Herbert Faulkner West (Hanover, N.H.: Westholm, 1955);

Birds of a Feather: Unpublished Letters of W. H. Hudson, edited by Dennis Shrubsall (Bradford-on-Avon, Wiltshire: Moonraker, 1981);

Landscapes and Literati: Unpublished Letters of W. H. Hudson and George Gissing, edited by Shrubsall and Pierre Coustillas (Salisbury: Michael Russell, 1985).

Bibliographies:

G. F. Wilson, *A Bibliography of the Writings of W. H. Hudson* (London: Bookman's Journal, 1922);

John R. Payne, *W. H. Hudson. A Bibliography* (Folkestone, Kent: Dawson / Hamden, Conn.: Archon, 1977);

Dennis Shrubsall, "Updating Hudson's Bibliography, I," *English Literature in Transition 1880-1920,* 31, no. 2 (1988): 186-188;

Shrubsall, "Updating Hudson's Bibliography, II: Correspondence and Books," *English Literature in Transition 1880-1920,* 31, no. 4 (1988): 437-444.

Biographies:

Morley Roberts, *W. H. Hudson. A Portrait* (London: Nash & Grayson, 1924; New York: Dutton, 1924);

Dennis Shrubsall, *W. H. Hudson, Writer and Naturalist* (Tisbury: Compton, 1978);

Ruth Tomalin, *W. H. Hudson: A Biography* (London: Faber & Faber, 1981);

Amy D. Ronner, *W. H. Hudson: The Man, the Novelist, the Naturalist* (New York: AMS, 1986).

References:

Ford Madox Ford, "W. H. Hudson," in *Portraits from Life* (New York: Houghton Mifflin, 1937), pp. 38-56;

John T. Frederick, *William Henry Hudson* (New York: Twayne, 1972);

Robert Hamilton, *W. H. Hudson: The Vision of Earth* (London: Dent, 1946);

Richard E. Haymaker, *From Pampas to Hedgerows and Downs. A Study of W. H. Hudson* (New York: Bookman Associates, 1954);

W. J. Keith, *The Rural Tradition: A Study of the Non-Fiction Prose Writers of the English Countryside* (Toronto & Buffalo: University of Toronto Press, 1974);

H. J. Massingham, *Untrodden Ways* (London: Unwin, 1923).

Papers:

Hudson destroyed his personal papers and correspondence, leaving instructions for his executors to do the same. Nevertheless, almost thirteen hundred letters by Hudson are held by the Royal Society for the Preservation of Birds Library, London; the City of Manchester Central Library; the Lockwood Memorial Library of the State University of New York at Buffalo; and the Harry Ransom Humanities Research Center, University of Texas at Austin. Hudson's early letters to Spencer Fullerton Baird are in the Smithsonian Institution, Washington, D.C.

Sir Harry Johnston
(12 June 1858 – 31 July 1927)

Brian D. Reed
Kent State University – East Liverpool

BOOKS: *Report on the Natural History of Mossâmedes and District, and of South-Western Africa Generally: With Reference to the Proposed Expedition of the Earl of Mayo* (London: Clowes, 1882);

A Visit to Mr. Stanley's Stations on the River Congo (London: Privately printed, 1883);

The River Congo, from Its Mouth to Bólóbó; with a General Description of the Natural History and Anthropology of Its Western Basin (London: Low, Marston, Searle & Rivington, 1884; revised, London: Low, Marston, 1895);

The Kilima-Njaro Expedition: A Record of Scientific Exploration in Eastern Equatorial Africa (London: Kegan Paul, 1886);

Report on a Journey up the Cameroons River from Bell Town to Wuri and Budiman (London: Her Majesty's Stationers' Office, 1886);

The History of a Slave (London: Kegan Paul, Trench, 1889; New York: Munro, 1889);

Livingstone and the Explorations of Central Africa (London: Philip, 1891); republished as *Livingstone and the Exploration of Central Africa* (London: Philip, 1912);

Report of the First Three Years' Administration of the Eastern Portion of British Central Africa, March 31, 1894 (London: Harrison, 1894);

Report on the Relation between Malarial Fever among Her Majesty's White Troops at Port Louis, Mauritius, and the Meteorological Elements of Temperature, Rainfall, and Relative Humidity for the Year 1889; with a Preliminary Sketch of the Medical Topography of the Island and the Epidemic of Malarial Fever in 1866–67 (Edinburgh, 1894);

British Central Africa: An Attempt to Give Some Account of a Portion of the Territories under British Influence North of the Zambezi (London: Methuen, 1897; New York: Arnold, 1897);

France: Report on the Regency of Tunis during the French Protectorate (London: Harrison, 1898);

A History of the Colonization of Africa by Alien Races (Cambridge: Cambridge University Press,

Sir Harry Johnston in 1922

1899; revised and enlarged, Cambridge: Cambridge University Press, 1913);

Preliminary Report by Her Majesty's Special Commissioner on the Protectorate of Uganda (London: Her Majesty's Stationers' Office, 1900);

General Report by Sir H. Johnston on the Uganda Protectorate (London: Her Majesty's Stationers' Office, 1901);

The Uganda Protectorate, 2 volumes (London: Hutchinson, 1902; New York: Dodd, Mead, 1902);

British Mammals (London: Hutchinson, 1903);

The Nile Quest: A Record of the Exploration of the Nile and Its Basin (London: Lawrence & Bullen, 1903; New York: Stokes, 1903);

Handbook to British Central Africa (Shire Highlands and Nyasaland) (London: British Central Africa Company, 1904?);

Liberia, 2 volumes (London: Hutchinson, 1906; New York: Dodd, Mead, 1906);

The Language of Liberia (London: Hutchinson, 1906);

George Grenfell and the Congo, 2 volumes (London: Hutchinson, 1908; New York: Appleton, 1910);

A History and Description of the British Empire in Africa (London: National Society's Depository, 1909); republished as *Britain across the Seas: Africa; a History and Description of the British Empire in Africa* (London: National Society's Depository, 1910);

The Negro in the New World (London: Methuen, 1910; New York: Macmillan, 1910);

The Negro and Religion (London, 1910);

The Opening Up of Africa (London: Williams & Norgate, 1911; New York: Holt, 1911);

Pioneers in West Africa (London: Blackie, 1911);

Views and Reviews from the Outlook of an Anthropologist (London: Williams & Norgate, 1912);

Pioneers in Australasia (London: Blackie, 1912);

Pioneers in Canada (London: Blackie, 1912);

Phonetic Spelling: A Proposed Universal Alphabet for the Rendering of English, French, German and All Other Forms of Speech (Cambridge: Cambridge University Press, 1913);

Common Sense in Foreign Policy (London: Smith, Elder, 1913; New York: Dutton, 1913);

Pioneers in India (London: Blackie, 1913);

A Survey of the Ethnography of Africa: And the Former Racial and Tribal Migrations in That Continent (London: Royal Anthropological Institute, 1913);

Pioneers in South Africa (London: Blackie, 1914);

Pioneers in Tropical America (London: Blackie, 1914; New York: Dodge, 1914);

A Gallery of Heroes and Heroines (London: Gardner, Darton, 1915);

Political Geography of Africa before and after the War (London: Royal Geographical Society, 1915);

Are We a Logical People? Native Races and the Great War: A Letter to You from Sir Harry Johnston (London, 1915?);

The Truth about the War: Lest We Forget (London: Review of Reviews, 1916);

The Black Man's Part in the War (London: Simpkin, Marshall, Hamilton, Kent, 1917);

David Livingstone (London: Kelly, 1917);

On the Urgent Need for Reform in Our National and Class Education; Delivered at South Place Institute on May 30, 1918 (London: Watts, 1918);

The Gay-Dombeys: A Novel (London: Chatto & Windus, 1919; New York: Macmillan, 1920);

A Comparative Study of the Bantu and Semi-Bantu Languages, 2 volumes (Oxford: Clarendon Press, 1919–1922);

The Backward Peoples and Our Relations with Them (London: Oxford University Press, 1920);

Mrs. Warren's Daughter: A Story of the Woman's Movement (London: Chatto & Windus, 1920; New York: Macmillan, 1920);

The Man Who Did the Right Thing: A Romance of East Africa (London: Chatto & Windus, 1921; New York: Macmillan, 1921);

The Veneerings: A Novel (London: Chatto & Windus, 1922; New York: Macmillan, 1922);

Little Life Stories (London: Chatto & Windus, 1923; New York: Macmillan, 1923);

The Story of My Life (London: Chatto & Windus, 1923; Indianapolis: Bobbs-Merrill, 1923);

The Natives of Nyasaland: A Short Sketch of Their History and Characteristics (London, 1924);

Relations (London: Chatto & Windus, 1925; New York & London: Harper, 1926);

Handbook to British Central Africa (Blantyre, Malawi: Rotary Club of Blantyre, 1985).

OTHER: David Kerr Cross, *Health in Africa,* introduction by Johnston (London: Nisbet, 1897);

The Living Animals of the World, edited by Johnston and Charles John Cornish (New York: Dodd, Mead, 1901);

Ham Mukasa, *Uganda's Katikiro in England,* introduction by Johnston (London: Hutchinson, 1904; Freeport, N.Y: Books for Libraries Press, 1971);

John Cathcart Wason, *East Africa and Uganda: Or, Our Last Land,* preface by Johnston (London: Griffiths, 1905);

Ellen C. Parsons, *Christus Liberator: An Outline Study of Africa,* introduction by Johnston (New York: Macmillan, 1905);

James F. Cunningham, *Uganda and Its Peoples,* preface by Johnston (London: Hutchinson, 1905; New York: Negro Universities Press, 1969);

Karl Georg Schillings, *With Flashlight and Rifle,* introduction by Johnston (London: Hutchinson, 1906);

Edmund Dene Morel, *Red Rubber: The Story of the Rubber Slave Trade Flourishing on the Congo in the Year of Grace 1906,* introduction by Johnston (London: Gresham, 1906; New York: Unwin, 1906);

Alfred J. Swann, *Fighting the Slave-Hunters in Central Africa,* introduction by Johnston (London: Seeley, 1910; Philadelphia: Lippincott, 1910);

Sir Hanns Vischer, *Across the Sahara from Tripoli to Bornu,* foreword by Johnston (London: Arnold, 1910);

Allan J. Macdonald, *Trade, Politics, and Christianity in Africa and the East,* introduction by Johnston (London: Longmans, Green, 1916; New York: Negro Universities Press, 1969);

Carlo Paladini, *Impero e libertà nelle colonie inglesi,* preface by Johnston (Florence: Bemporad & Figlio, 1916);

Ernest Frederick Spanton, *In German Gaols: A Narrative of Two Years' Captivity in German East Africa,* prefatory note by Johnston (London: Society for Promoting Christian Knowledge, 1917);

Thomas Alexander Barns, *The Wonderland of the Eastern Congo,* introduction by Johnston (London & New York: Putnam, 1922);

Douglas Elliot Charles Romaine Stirke, *Barotseland: Eight Years among the Barotse,* introductory chapter by Johnston (London: Bale & Danielsson, 1922; New York: Negro Universities Press, 1969);

The Outline of the World To-Day, 3 volumes, edited by Johnston and Leslie Haden Guest (London: Newnes, 1923-1924); republished as *The World of To-Day: The Marvels of Nature and the Creations of Man,* 4 volumes (New York & London: Putnam, 1924-1925);

Cuthbert Christy, *Big Game and Pygmies,* introductory chapter by Johnston (London: Macmillan, 1924);

The Old Navigators, edited by Johnston (London: Collins Clear-Type Press, 1926?);

Adventurers and Discoverers, edited by Johnston (London: Collins Clear-Type Press, n.d.);

A Book of Empire Heroes, edited by Johnston (London: Collins Clear-Type Press, n.d.);

The Conquest of the Sea, edited by Johnston (London: Collins Clear-Type Press, n.d.);

The Great Travellers, edited by Johnston (London: Collins Clear-Type Press, n.d.);

Modern Travellers, edited by Johnston (London & Glasgow: Collins Clear-Type Press, n.d.).

SELECTED PERIODICAL PUBLICATIONS – UNCOLLECTED: "The Okapi: The Newly Discovered Beast Living in Central Africa," *McClure's Magazine,* 17 (September 1901): 497-501;

"The Pygmies of the Great Congo Forest," *McClure's Magazine,* 18 (February 1902): 349-353;

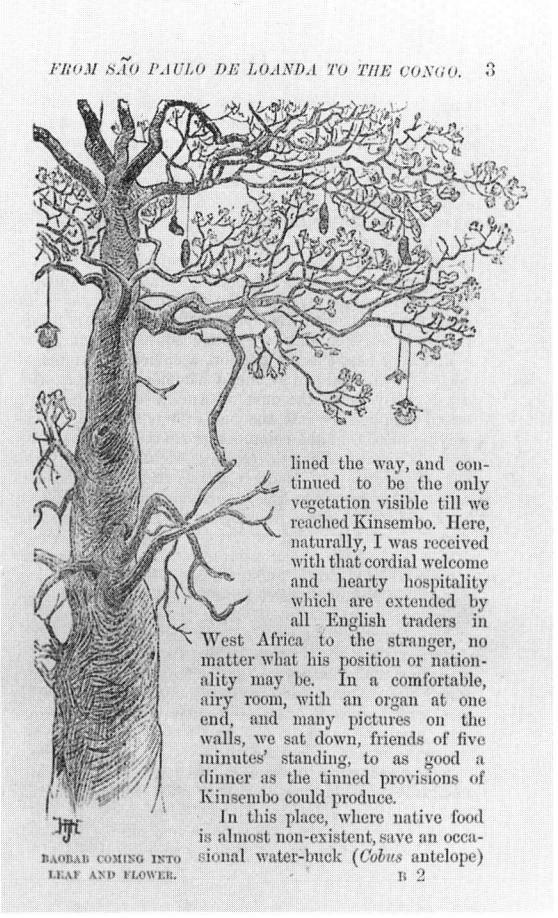

Page with one of Johnston's illustrations in The River Congo, from Its Mouth to Bólóbó *(revised, 1895)*

Johnston and Ernest Lyon, "The Black Republic – Liberia," *National Geographic,* 18 (May 1907): 334-343;

"Where Roosevelt Will Hunt," *National Geographic,* 20 (March 1909): 207-256;

"Haiti, the Home of Twin Republics," *National Geographic,* 38 (December 1920): 483-496;

"Race Problems in the New Africa," *Foreign Affairs,* 2 (June 1924): 598-612.

Sir Harry Johnston was known throughout Europe as an explorer, naturalist, painter, and author. He traveled extensively to gather geographic information and botanical samples as well as to record the languages and customs of native peoples. He is best remembered for his imperialistic British interests and for his ethnocentric attempts to make sense of the complex African continent.

Henry "Harry" Hamilton Johnston, the eldest of twelve children, was born in South London on 12 June 1858 to Esther and John Johnston. John John-

Johnston's oil painting of the okapi, an animal he discovered while exploring the forests of the Congo (from Roland Oliver, Sir Harry Johnston & the Scramble for Africa, *1957)*

ston, who also had two children from a previous marriage, was a wealthy director of a London insurance company, and his work often took him throughout western Europe and Scandinavia and occasionally to remote parts of Russia and South Africa. He studied the writings of the early African explorers and held a particular enthusiasm for the work of Dr. David Livingstone, a passion that his son Harry later shared.

Having begun painting at an early age, young Johnston earned commissions for his work from both the London Zoological Society and the Royal College of Surgeons. By age sixteen he was attending classes in French, Italian, Spanish, and Portuguese at King's College and taking courses in painting at the Royal Academy. He was a precocious child who was dedicated to his academic studies and found various ways to develop his talents and interests. To test his linguistic and artistic studies, he persuaded his father to finance a trip to Spain and the Balearic Islands in the summer of 1876. These travels took him to Paris, Marseilles, Barce-

lona, Palma, and Sóller. When he returned to London later that year, he renewed his determination to continue painting, and he also promised himself that he would travel again and more ambitiously. Yet the next three years were filled with domestic tragedy and periods of depression.

In November 1877, Johnston later wrote, he became "too stupefied with grief even to go on with his work" when his mother died during childbirth. One year after her death Johnston's stepbrother, George, having contracted typhoid fever, also died. During this period Johnston began to abandon his Christian faith, as he acknowledges in his autobiography, *The Story of My Life* (1923), when he writes of the despair he felt at these deaths in his family:

> There are millions of similar deaths occurring every year, showing ... that in this incredibly vast universe the overruling "intelligent" power – if there be one – has no conception of what agonies we poor human ants on this tiny planet can suffer.

In November 1879, after a bout of ill health that most believed to be hypochondria, Johnston returned to traveling. He chose to go to Tunis, as he explains in his diary, not only because he wished to convalesce but because he was "so anxious to be great." He felt that either his art or his wits could bring him fame and fortune through an adventure in an exotic, unknown region. In Tunis, where he hoped to establish himself as an important modern painter, he attempted an artistic project that was far more ambitious than anything he had done: for seven months he worked on a colossal canvas depicting the Olive Tree Mosque with its cosmopolitan crowd of worshipers.

At the time of Johnston's visit to Tunis the Moors, Bedouin, Jews, and French inhabited the city, which was under the jurisdiction of Turkey, and the politics of Tunisia fascinated him as much as the spectacle of the Olive Tree Mosque did. When France began to appropriate the land to build railways and telegraph lines, British and Italian interests in the area grew. When Johnston was commissioned to write a monthly article for the London *Globe,* he learned much about European expansion in Africa. Soon he realized, as he wrote in a letter to the permanent undersecretary of state for foreign affairs in 1881, that "the whole tenor of my life has been changed" through this experience and this developing interest in the British Empire, and he resolved to pursue that interest as a writer by focusing on English interests abroad.

On his return to England, Johnston put aside his artistic career and began to study the Bantu languages seriously. In hopes of someday gaining a commission to Africa, he spent hours reading in the British Museum about the complications of classifying the Bantu tongue. When an opportunity arose in 1882 for him to accompany amateur zoologist Dermot Robert Wyndham Bourke, the earl of Mayo and an enthusiastic explorer, on a hunting expedition through Portuguese Angola, Johnston jumped at the chance to use his knowledge.

During late 1882 and early 1883 he developed his skills of observation in preparing reports and notes – first as he accompanied Bourke through Angola and later as Johnston set off down the Congo River with only African guides. Near the mouth of the river his desires for exploration were further whetted when he met Henry M. Stanley, the famous American explorer of Africa who was then employed by King Leopold II of Belgium. Stanley was undoubtedly a major influence on Johnston's still growing Africanist ideals. In *The River Congo, from Its Mouth to Bólóbó; with a General Description of the Natural History and Anthropology of Its Western Basin* (1884) Johnston extensively details his eight-month trip along the Congo basin. Loaded with Johnston's excellent botanical and anthropological illustrations, *The River Congo* was republished in several translations (including a popular French edition) and became the seminal early work about the Congo region. With this book Johnston established himself as a serious explorer and African authority.

Various British organizations that had vested interests in Africa paid close attention to Johnston's detailed reports. Upon his return to England, he gave popular lectures that provided information deemed important by investors. In 1884 he was summoned to the British Foreign Office to discuss what the government felt were perplexing activities in the Congo by King Leopold II. This discussion led the British government and businessmen to provide the financial backing and resources necessary for Johnston to be able to continue his explorations in a costly expedition through the mountains of East Africa.

In 1884 Johnston set out for Mount Kilimanjaro with the sponsorship of the British Association and the Royal Society and with the equipment of the Royal Geographical Society. He was charged to bring back unclassified vegetation from the heights of the East African mountains and to gather information about the region. The beautifully illustrated *The Kilima-Njaro Expedition: A Record of Scientific Exploration in Eastern Equatorial Africa* (1886) was the result. Johnston's artistic background enabled him to represent the landscape in splendid detail. As in many of his works that were to follow, *The Kilima-Njaro Expedition* is divided into two parts: the first being a travel narrative, which in this case recounts his exploration of the mountain, and the second comprising zoological, anthropological, and botanical studies. Both his love for describing nature and his concern for the success of the expedition are evident in his almost poetical first impressions of the mountain:

> Kilima-Njaro was weird in the early flush of dawn, with its snowy crater faintly pink against the sky of deep blue-grey, wherein the pale and faded moon was sinking, and the stars were just discernible; but as the stronger light prevailed, and the clouds which concealed the base of the mountain disappeared, its appearance was disappointing.

Missing the comforts of home, Johnston – with the help of twenty-eight Swahili natives – constructed a long, three-roomed cottage with surrounding gardens and a poultry yard from which he could base his explorations. Mandara, the chief of the local Moshi tribe, developed close relations with Johnston through an exchange of gifts and occasional assistance, such as that which Johnston offered when the neighboring Chagga states attacked the Moshi. Historians are unsure about the nature of Johnston's participation, but much of his recounting of "guns firing and flags flying" as the reinforcements marched under a "magnificent flight of rockets" are mainly Johnston's egotistic re-creations.

Yet whatever the real circumstances were, Johnston overstepped his instructions and turned the Kilimanjaro expedition into a political undertaking. He negotiated treaties and felt that he was an unofficial representative of the British government. In his autobiography Johnston admits, "I had become far too much interested in the political future of Equatorial Africa." Johnston's strong feelings about Great Britain's role in East Africa soon led to his being appointed to key government posts in the Foreign Office and transformed him from an explorer to an exploiter, a major agent in the European race to colonize Africa.

In 1885 he became the vice-consul in Cameroon and the Niger Delta, where he honed his varied talents while writing his first novel, *The History of a Slave* (1889). Soon afterward Johnston created quite a stir in London when he tricked a troublesome oil merchant, King Ja Ja of the Niger Delta, and had him forcibly expelled. This led the prime minister to

Pioneers in the British colonial development of south central Africa, at Kimberley, South Africa, in 1890: (seated) Rochfort Maguire, Johnston, Cecil Rhodes, and Archibald Colquhoun; (standing) J. A. Grant, John Moir, and Joseph Thomson

appoint Johnston as commissioner and consul general for Mozambique and the Nyasa districts.

In support of the "Caucasian effort" Johnston soon subdued and stabilized the Africans at his new post. He formed treaties, fought a war with Arab slave traders, encouraged white settlements and white-operated plantations, and fought off the Portuguese, who also had interests in the Nyasa region. Johnston left for London in 1896 to recuperate from blackwater fever and reflect on his accomplishments, and he was knighted by Queen Victoria on his return. With hopes of attaining a permanent imperial post as high commissioner to South Africa, he began writing accounts of his first proconsulship in *British Central Africa: An Attempt to Give Some Account of a Portion of the Territories under British Influence North of the Zambezi* (1897). But the post he desired was awarded to a more politically influential nobleman, and Johnston became frustrated and bitter.

In 1896 Johnston's moods were partially checked by his marriage to Winifred Irby. While honeymooning in Italy, Johnston was offered a consul generalship in Tunis, and he accepted this position somewhat reluctantly. The two years he spent in Tunis were dedicated mostly to scholarly pursuits.

His repose was happily interrupted in 1899, when he was offered the position of special commissioner to Uganda. Johnston states in his autobiography: "I was to be given the military rank of Commander-in-Chief in the War Office List, so as to make it conclusive that I could dominate the British general officers who had not managed very well to suppress the mutiny of the Sudanese soldiers [at Khartoum in 1885]." In addition to increasing his salary, power, and prestige, Johnston was gratified with the renewed opportunity that his acceptance of this post afforded him to continue research on the Bantu languages. "I do not think for a moment I contemplated refusing," he wrote of this decision. Yet this was to be Johnston's last tour of duty in Africa.

Johnston faced difficult problems in Uganda, where Catholics, Protestants, and Muslims were fighting for control of the region, and no formal

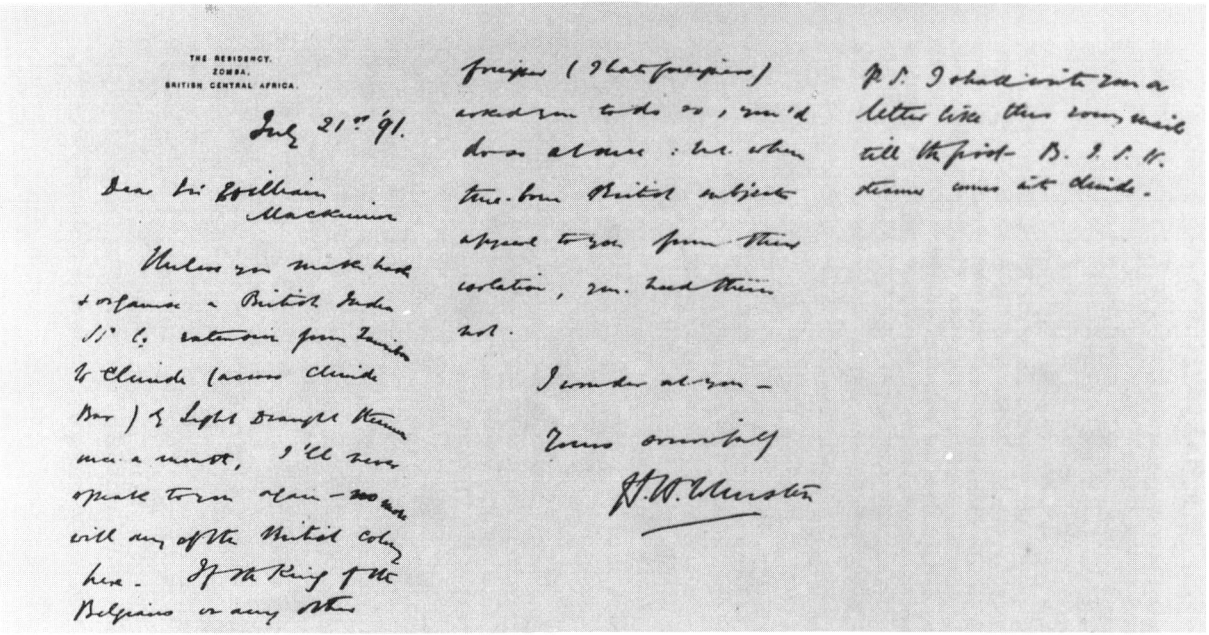

Letter from Johnston to shipping entrepreneur Sir William Mackinnon (Mackinnon Papers, School of Oriental and African Studies, London)

government was in place following the downfall of the British East Africa Company. But this was the kind of challenge in which Johnston reveled. He soon established a civil government that replaced the military regime, encouraged the expansion of the Uganda railway, and forged the important Uganda Agreement. This agreement may have been the most liberal British-African treaty at the time, because it protected both the ruling class and the commoners while retaining much of the traditional political system.

He also continued his research on the Bantu languages, collected flora and fauna of the area, and began composing his two-volume account of Ugandan life and landscape, *The Uganda Protectorate* (1902). Full of watercolor paintings of indigenous plants and birds and detailed descriptions of the native people, Johnston's book about Uganda displays his love for African beauty. In the section of the book that comprises his travel narrative Johnston describes the economic and physical allure of the region:

> Banana plantations grow everywhere in splendid luxuriance. Whatever is not cultivated field is tropical forest of grand appearance. The grey parrot of West Africa ... fills the air with its shriekings and whistlings ... and large horn-bills with enormous white casques people these lofty trees, from which indiarubber lianas sway like ropes to be used in a moving stage scenery.

When Johnston returned to England after his two-year tour, he received a hero's welcome, but while he was entertained and honored by dignitaries and politicians, he was carefully being eased out of the Foreign Office by officials who found him unorthodox. By early 1902, after repeated rebuffs, Johnston began to think about pursuing other careers. He hoped to supplement his meager pension by concentrating on literary and artistic ventures, and the bulk of his important writing followed.

The Nile Quest: A Record of the Exploration of the Nile and Its Basin (1903), an extremely ethnocentric yet enjoyable account of the search for the sources of the Nile River, was written at top speed and published during the first year of his official retirement. In it Johnston exclaims, "only the Caucasian race ... has cared for geography in the past," and "the Negro of Papua and Africa has never cared to ascertain whence rivers flowed and whither." He adds that "Fate has ordained that the entire basin of this [Nile] river and its tributaries ... should come under the political control of Great Britain." Aside from these prejudices, *The Nile Quest* is one of Johnston's best books and an important and accurate account of the scramble to solve one of Africa's great mysteries – the source of the Nile.

In the years before the outbreak of World War I the amount of work Johnston published was tremendous. Besides numerous articles and book reviews he published sixteen books, including the

Frontispiece portrait of John Hanning Speke and title page for Johnston's The Nile Quest (1903)

two-volume *Liberia* (1906), which presents much information about the history of the Liberian republic and the anthropology and flora of the region. Liberia had begun to interest Johnston while he had been touring with Bourke in 1882, and he planned a two-month excursion through the Liberian forests in 1904 to record data for his book. As a result of his trip, he returned home impressed by "the splendid timber, the abundance of rubber producing trees and lianas, and the extraordinary variety of other forms of vegetable wealth."

But *Liberia* is not just a prospectus for the exploration of the country. Much of the book recounts Johnston's coming to terms with the cannibalistic natives. In a section of the book that describes his travels on foot, Johnston specifically notes the industriousness of the natives and justifies their "lust for human flesh" by hypothesizing that they need the salt of human blood because of their isolation from the natural salt of the coast. Possibly the mixed reviews for this massive work stemmed from what was deemed an overly sympathetic view of the natives.

Another two-volume work of this period, *George Grenfell and the Congo* (1908), provided a history and description of the Congo regions. Commissioned by the Baptist Missionary Society, this work presents information about missionary George Grenfell, presents his reports involving the Congo and its people, and recounts his undertakings there. Although reviews of the book were bad, the first volume was fairly well received by the public. In highlighting the evils that white civilization brought to an undeveloped country, it tells the story of a missionary who fought diseases, mapped out the landscape, and formed governments in an unknown land.

In response to the prodding of U.S. president Theodore Roosevelt during the winter of 1908–1909, Johnston toured the West Indies and the southern United States to gather information for his next important work, *The Negro in the New World* (1910). His trip through the southern United States

Johnston's photograph of villagers in Pangwa, near Liberia

in 1908 was significant in that it allowed Johnston to meet distinguished African Americans such as Booker T. Washington and the botanist George Washington Carver, the latter of whom taught him about the distribution of plants throughout North and South America. Most certainly Johnston's experiences with these men should have shaped his perspectives. Johnston also visited the homes of black families, and he concluded that there was no difference "in surroundings, in culture, [or] in decorum, between the lives of these absolute negroes . . . and the lives of Anglo-Saxon white people earning the same wages." His later works, *The Opening Up of Africa* (1911) and *Common Sense in Foreign Policy* (1913), present an overview of British colonization and propose a "healthy imperialism through humanity." But in his later book, *The Backward Peoples and Our Relations with Them* (1920), Johnston discusses the blacks and other races as if they are lesser citizens than whites, a decidedly racist end to his studies of race. Yet even with such a disturbing characterization of these races as "sub-species," Johnston's books are well worth reading for his observations about the plight of blacks in the United States and in the West Indies.

With the coming of World War I, the focus of Johnston's literary production shifted. He patrioti-cally spent his time lecturing troops and gathering information for future volumes, and on his lecture tour Johnston interviewed Senegalese troops, from whom he gathered additional information about a previously unknown form of the Bantu tongue. He spent most of the war years working on his important scholarly piece, *A Comparative Study of the Bantu and Semi-Bantu Languages* (1919), a work that was supplemented with a second volume in 1922.

By the end of the war Johnston's poor health, possibly exaggerated by his hypochondria, caused him to stay close to home. He began to demand privacy and spent his quiet hours reading, researching, and writing a series of popular yet simplistic novels based on the works of Charles Dickens. His autobiography, *The Story of My Life,* shows the astonishing versatility and energy of a man who had played vital roles in colonizing Africa. Yet during his final years Johnston became frail and bedridden, and in 1925 he suffered a serious stroke from which he never fully recovered. He died in obscurity in Nottinghamshire on 31 July 1927 while on a journey to visit his sister.

Sir Harry Johnston was a man of many talents — most of which he used, for better or worse, as he pioneered through much of the African continent. His knowledge of Africa was unrivaled in his own day. Hailed as the seminal authority on Bantu languages and noted for his careful botonical research, Johnston remains an important figure for study by scientists, linguists, naturalists, and historians. His artfully illustrated books on African customs and geography are unparalleled in their detailed descriptions. His works are valuable resources for scholars and offer pleasurable reading for travel enthusiasts.

Bibliography:

James A. Casada, *Sir Harry Hamilton Johnston: A Bio-Bibliographical Study* (Basel, Switzerland: Basler Afrika Bibliographien, 1977).

Biographies:

Alex Johnston, *The Life and Letters of Sir Harry Johnston* (London: Chatto & Windus, 1929);

John Vernon Wild, *The Story of the Uganda Agreement* (London: Macmillan, 1955);

Roland Oliver, *Sir Harry Johnston & the Scramble for Africa* (London: Chatto & Windus, 1957).

References:

Charles Chaillé-Long, *My Life in Four Continents* (London: Hutchinson, 1912);

Richard James Hammond, *Portugal and Africa 1815–1910: A Study in Uneconomic Imperialism* (Stanford, Cal.: Stanford University Press, 1966);

Alexander John Hanna, *The Story of the Rhodesias and Nyasaland* (London: Faber & Faber, 1960);

Donald Anthony Low and Robert Cranford Pratt, *Buganda and British Overrule* (London: Oxford University Press, 1960).

Papers:

A large collection of Johnston's manuscript materials, letters, and sketches is at the National Archives of Rhodesia in Salisbury. Additional papers are housed at the Public Records Office and the Royal Geographical Society Archives in London.

Mary Henrietta Kingsley
(13 October 1862 – 3 June 1900)

Catherine Barnes Stevenson
University of Hartford

BOOKS: *The Ascent of Cameroons Peak and Travels in French Congo* (Liverpool: Journal of Commerce Printing, 1896);
Travels in West Africa, Congo Français, Corisco and Cameroons (London: Macmillan, 1897; London & New York: Macmillan, 1897; abridged edition, London & New York: Macmillan, 1897);
West African Studies (London & New York: Macmillan, 1899; expanded edition, London & New York: Macmillan, 1901);
The Story of West Africa (London: Marshall, 1899).

OTHER: Richard Edward Dennett, *Notes on the Folklore of the Fjort – French Congo,* introduction by Kingsley (London: Folk-Lore Society, 1898);
George Kingsley, *Notes on Sport and Travel,* edited, with a memoir, by Mary Kingsley (London: Macmillan, 1900).

SELECTED PERIODICAL PUBLICATIONS – UNCOLLECTED: "The Development of Dodos," *National Review,* 27 (March 1896): 66–79;
"Travels on the Western Coast of Equatorial Africa," *Scottish Geographical Magazine,* 12 (March 1896): 113–124;
"The Throne of Thunder," *National Review,* 27 (May 1896): 357–374;
"Black Ghosts," *Cornhill Magazine,* 74 (July 1896): 79–92;
"Two African Days' Entertainment," *Cornhill Magazine,* 75 (March 1897): 354–359;
"Fishing in West Africa," *National Review,* 29 (April 1897): 213–227;
"The Native Populations of Africa," *Spectator,* 83 (15 May 1897): 695–696;
"The Fetish View of the Human Soul," *Folklore,* 8 (June 1897): 138;
"African Religion and Law," *National Review,* 29 (September 1897): 122–139;
"A Parrot Story," *Cornhill Magazine,* 83 (September 1897): 389–391;

"The Hut-Tax in Africa," *Spectator,* 84 (19 March 1898): 407–408;
"The Liquor Traffic with West Africa," *Fortnightly Review,* 69 (April 1898): 537–560.

Mary Henrietta Kingsley – traveler, writer, and political activist – discovered no new territory, but she brought West African culture and politics to the attention of late-nineteenth-century British readers. John Flint credits her with having "revolutionize[d] the attitude toward Africa of British officials and the informed public." Her paradoxical, ultimately tragic personality and her entertaining yet scholarly travel books and articles have fascinated several generations of readers.

Mary was the first child of George Kingsley and Mary Bailey, Kingsley's housekeeper until their marriage four days before Mary Kingsley was born. Her father, a member of the well-known literary family that included the writers Charles and Henry Kingsley, trained as a doctor but spent most of his life traveling with wealthy patrons. A student of ethnography, he produced several books and articles, including the travel narrative *South Sea Bubbles* (1872), which he wrote with G. R. C. Herbert, the earl of Pembroke. Kingsley's father often left her mother and younger brother, Charles (born in 1864), alone to handle the economic and psychological consequences of his long absences.

Kingsley's memoir to her father's *Notes on Sport and Travel* (1900) recalls that her mother endured "an unbroken strain of nervous anxiety" about her husband that led her literally to shut herself away from the world. In this lonely environment young Mary, as her mother's "handyman" and "chief officer" about the house, assumed a heavy burden of adult responsibility. Books and the newspaper the *English Mechanic*, an invaluable guide to household repair, were the chief companions of this intelligent young girl who, receiving no formal education, tried to teach herself, sometimes using hopelessly outdated books. In 1899 she angrily told her publisher, Macmillan, that "being allowed to learn German was all the paid-for education I ever had – 2000 pounds was spent on my brother." Her father allowed her to pursue those German lessons chiefly so that she could assist him in the scholarly projects he undertook when he retired from travel in 1879. Her experience in collecting travelers' accounts and reading German authors on sacrificial rites provided her with training in the rudiments of ethnography and the techniques of research that proved to be invaluable in her own travels.

Kingsley had to subordinate scholarly research, however, to domestic responsibilities – particularly after 1888, when both her parents required nursing. In an autobiographical sketch on 20 May 1899 she was to describe those years as ones of "work and watching and anxiety," all of which ended suddenly in 1892, when both parents died within six weeks. Her unconventional, lonely childhood in the company of demanding and eccentric parents had trained her in self-reliance and caretaking. It had also left her feeling alienated in terms of social class (she dropped her *h*s), appropriate norms for female conduct, and personal identity. In a letter to her acquaintance Maj. Matthew Nathan, on 12 March 1899 she confessed, "I am no more a human being than a gust of wind is. I have never had a human individual life. I have always been the doer of odd-jobs and I lived in the joys, sorrows, and worries of other people." But that family life had also whetted her appetite for travel, particularly to places where she could continue her scholarly research. As she added in her autobiographical sketch, "When there were no more odd jobs . . . to do at home, I, out of my life in books, found something for which I had been taught German[:] . . . the study of early religion and law, and for it I had to go to West Africa." To increase the scientific value of this trip, Kingsley agreed also to collect fish specimens for the British Museum.

To travel to the west coast of Africa was a daring exploit for a single woman with no previous travel experience. At this time West Africa was a particularly dangerous place to visit because of disease, the reputed savagery of the native peoples, and rivalries between European powers engaged in a scramble for control of the territory. Nonetheless, Kingsley made two voyages there. The first, a five-month trip in 1893, took her to ports along the west coast, then inland and north through French- and German-controlled territories, and ending in the British Niger Coast Protectorate, from which she sailed home. Katherine Frank remarks that Kingsley wrote "Bights of Benin," a travelogue based on this voyage, and submitted it to Macmillan before leaving again for West Africa in December 1894.

For this second trip she traveled down the coast to Calabar in the Niger Coast Protectorate, where she spent four months, and then to the port of Glass, from which she took a solitary trip up the Ogooué River to collect fish specimens. Accompanied only by African porters and trading with local peoples in order to sustain her journey, she crossed through areas of West Africa that Europeans had never before traversed. She concluded her odyssey with a mountaineering exploit that made her the first white woman to climb Mount Cameroon and the third person to ascend its difficult southeastern face. She returned home ten months later in November 1895 with sixty-five species of fish, three of which were entirely new and were subsequently named after her; eighteen species of reptiles; and enough material for two books and many articles.

Because of her lack of formal education, Kingsley was understandably self-conscious about her literary ability. On 18 December 1894 she jokingly referred to her writings as "well-intentioned word-swamps" in a letter to Macmillan. Her lectures and articles of 1896 and 1897 not only tested public response to her ideas but also helped her overcome some misgivings about her ability to write clear, grammatical prose. While keeping her name in the public eye, they also educated the public about the African continent and cultures. The first article to appear was "The Development of

Dodos," in March 1896, which uses an ironic parable to attack popular notions about the childishness of Africans and their "arrested" psychological, intellectual, and moral development. Kingsley's agenda becomes clear when she urges readers to "put yourself in [the African's] place, and you'll understand his attitude." "Black Ghosts," an article that anticipates the style of Kingsley's later travel books by integrating scholarly observation with amusing personal narrative, appeared four months later in *Cornhill Magazine*.

As she was working on the manuscript of her first major book in 1897, Kingsley published five articles, all of which present ethnological information through sprightly narration. "Fishing in West Africa" hilariously describes the mishaps of the narrator as she searches for specimens. "Two African Days' Entertainment" recounts Kingsley's "entertaining" ministrations to a dog thought to be hydrophobic and a man suffering from brain fever. An equally amusing persona in "A Parrot Story" narrates a digressive African folktale that illustrates the religion and the family relationships prevalent in Africa while also revealing the narrator's fascination with the area: "One of the charms of studying in West Africa is the keen human interest that comes off every mortal subject like the scent off a flower."

More scholarly, but nonetheless entertaining, are "African Religion and Law," an analysis of the religious basis of many African legal institutions, and "The Fetish View of the Human Soul," an examination of the animistic religious beliefs in West Africa. These articles show Kingsley developing the paradoxical but engaging persona that shines through her travel books – that of a self-mocking woman both coolly competent and hilariously inept, a serious scientist who is also a clown.

In writing her first book, Kingsley had to overcome many difficulties related to style, cultural expectations about women authors, and literary technique. Sara Mills has shown that disbelief and charges of exaggeration were common responses to the narratives of women travel writers. Thus, Kingsley felt a special need to vouch for the factual accuracy of her work and to dispel any prejudices about her scientific objectivity. As a result, she not only read widely and corresponded extensively with experts but also omitted from her discussion aspects of West Africa – such as encounters with wild animals – that she felt might generate disbelief or that she felt her personal experience left her unable to discuss. In addition, as a woman who was eager not to seem "unladylike" by shocking her audience, Kingsley confessed to Macmillan in November 1896 that she had removed "all the poetry and bad language except for the native legal oaths" as well as some details of her experience in nursing men with yellow fever and sick or drunken traders. Aware of the multiple audiences of her work and the differing readers she was addressing in writing a popular yet scientific travel book, Kingsley created a series of different narrative personae in her writing: the serious scholar, the rakish male adventurer, the slapstick comic, and the prim Victorian lady.

Furthermore, as a self-taught and unsystematic writer Kingsley found it difficult to organize readable narrative from her chaotic journal entries about the multifariousness of West Africa. In *Travels in West Africa, Congo Français, Corisco and Cameroons* (1897) she wittily directs the attention of her reader to her frustrated attempts to construct a travel narrative from her "bush diary." "My notes for a day," her narrator announces, "will contain facts relating to the kraw-kraw, price of onions, size and number of fish caught, cooking recipes, genealogies, oaths (native form of), law cases and market prices." The diversity of her materials and the different audiences whom she wished to reach led Kingsley to consider producing two works: a light, popular travel narrative to which, in a letter to Macmillan on 11 December 1895, she jokingly referred as "the log of a lighthearted lunatic" as well as a scientific work for a scholarly audience at home and for friends in West Africa. Yet she ultimately decided not to publish a pure travelogue but to try in one narrative to meld scientific writing and personal narrative.

That narrative, *Travels in West Africa,* derives its structure from the shape of Kingsley's second journey and humorously recounts many of her adventures; however, it also includes ethnological data and an extensive scholarly appendix on West African trade, labor, disease, reptiles, and fish. In *Imperial Eyes: Travel Writing and Transculturation* (1992) Mary Louise Pratt identifies two distinct but complementary kinds of travel writing: the objective, scientific mode – which can in fact mask imperialistic designs on the landscape – and the sentimental mode, which dramatizes the protagonist's responses to experiences.

Kingsley's writing encompasses both modes. Her journey into the heart of Africa's uncharted lands is in many ways a psychological odyssey that recalls the experiences described by late-nineteenth-century novelists. The journey into what she calls "the great, grim twilight regions of the forest" seems a spiritual progress toward an "enlightened" state.

Kingsley dressed in mourning following the deaths of her parents in 1892

When the traveler first embarks, she is virtually blind to what surrounds her; she is an illiterate in a vast, rich library. In this preliminary stage she realizes how dependent she is on other people or on familiar surroundings, and this realization forces her to discover in herself the resources she needs to meet the demands of this new milieu. Gradually the voyager learns to see anew: she learns to see "inextricable tangles" of vegetation with almost preternatural clearness, and "a whole world grows up out of the gloom." In this "illumined" state the traveler feels "a sense of growing power" that climaxes in her conviction of invincibility: "Put me where you like in an African forest, and as far as the forest goes, starve me or kill me if you can." If Joseph Conrad's Marlow or H. Rider Haggard's Alan Quartermain finds the bestiality of the primal self in the hinterlands of Africa, Mary Kingsley discovers something different: the power and self-sufficiency of the African "natural man" — qualities that were available to a white woman, too, if she left behind the restraints and dependencies of the "civilized" world.

Although Kingsley admits that, by itself, the journal of an isolated white person in the interior of Africa would prove "exceedingly interesting" for "psychological study," the focus of her work lies instead "on the state of things in general in West Africa." Because journal entries distort experience by leaving the reader with a vivid impression of relatively minor events but little sense of the meaning of the whole, Kingsley restricts her use of these immediate records of her experiences in order to fulfill the broader aims of the narrative — to educate and entertain. She has therefore produced a remarkable, multifaceted book: a scientific tract with a psychological valence and an engaging plotline, a travel book with the weight of a scholarly treatise, and a political document that is disarmingly like a picaresque novel.

Each of Kingsley's books begins with an avowal that she is not "a literary man" but an untutored recorder of facts and experiences. She was in fact a gifted stylist whose pose of "plain speaking" masked considerable artistry. She valued her style enough to resist the attempts of certain friends to edit her writing into more concise scientific prose; in fact, Stephen Gwynn reports that she observed, "I'm popular because I am natural." Rhythmic, powerfully descriptive, allusive, and metaphorical, Kingsley's prose captures the color, detail, and ambiance of a landscape without falling into cloyingly studied or bombastic prose. In addition, the chatty, digressive, witty narrative voice of *Travels in West Africa* has its antecedents in the travel writers and novelists of the eighteenth century.

The chapter headings, for example, employ a tongue-in-cheek irony reminiscent of Henry Fielding's chapter headings to *The History of Tom Jones* (1749) or *The History of the Adventures of Joseph Andrews, and of his Friend Mr. Abraham Adams* (1742). Like the picaro, Kingsley is an outsider who fits in no particular social context but moves easily through diverse situations — from the homes of colonial government officials to the outposts of missions or the rugged, all-male bush stations of traders; the villages populated by cannibals; the lone safaris through hostile territory. Her adventures are presented in a digressive, episodic structure and narrated in baroque sentences. The narrative tries to present the spontaneity of the traveler in the process of writing:

> I must forthwith stop writing . . . or I shall go on telling you stories and wasting your time, not to mention the danger of letting out those which would damage the nerves of the cultured of temperate climes, such as . . . the moving story of three leeches and two gentlemen; . . . and the reason why you should not eat pork along here because all the natives have either got the guineaworm,

or kraw-kraw or ulcers; and then the pigs go and – dear me! it was a near thing that time. I'll leave off at once."

Immediacy is a hallmark of the narrative. When the action is exciting or funny, the narrative shifts from the past tense to the present-tense format of a diary. For example, a local government official having trouble with his Turkish trousers

> writes furiously – blotting paper mislaid – frantic flurry round – pantaloons won't stand it – grab just saves them – something wanted the other side of the room – headlong flight towards it – "now's our chance," think the pantaloons, and make off – recaptured.

On other occasions the narrator directly addresses her audience, forestalls their objections, and tries to carry her point by a joke or an apt analogy.

Kingsley's literary sophistication is evident in her use of allusion and in the range of her humor. Literary allusions throughout the text to authors such as Charles Dickens, Robert Louis Stevenson, Rudyard Kipling, Mark Twain, Samuel Johnson, and Johann Wolfgang von Goethe and to the Bible help to bring Africa into an understandable frame of reference for the reader. Kingsley similarly makes alien, unimaginable African scenery accessible to Victorian readers by comparing it to scenes in paintings by Joseph Turner, George Frederic Watts, or the Pre-Raphaelites. Her humor encompasses slapstick, parody, verbal irony, mock-heroic satire, and gallows humor. On 6 June 1900 the London *Evening News* praised her writing for being rich "in that quality in which the writings of women seldom excel: the divine quality of humor."

Parody of the style and content of male travel writing about Africa enlivens *Travels in West Africa* and warns readers not to expect a predictable narrative of dramatic encounters and heroic postures. Occasionally the narrator sets the stage for such a familiar scene, only to undercut her reader's expectations. Kingsley, who held that it was "utter idiocy" to flourish a revolver and threaten to shoot in Africa, sometimes aimed at the language with which the white hunter in Africa conventionally portrayed himself:

> I know exactly how I ought to have behaved. I should have felt my favourite rifle fly to my shoulder, and then, carefully sighting for the finest specimen, have fired. The noble beast should have stumbled forward, recovered itself, and shedding its life blood behind it have crashed away into the forest. I should then have tracked it, and either with one well-directed shot have given it its quietus, or have got charged by it, the elephant passing completely over my prostrate body; either termination is good form.

She also pokes fun at the myth of the "daring rescue." When one of her bearers gets in trouble in a local village, she wryly observes, "I dare say I ought to have rushed at him and cut his bonds, and killed people in a general way with a revolver, and then flown with my band to the bush; only my band evidently had no flying in them." As this narrator suggests, canny calculation of risks and a knowledge of psychology are more germane to the situation than schoolboy histrionics. In her account of an escape from an armed party of cannibals who are chasing her canoe, the voice of the reliable scientist parenthetically deflates the romantic, exaggerated prose of the adventure story: "Regardless of danger, I grasped the helm, and sent our gallant craft flying before the breeze down the bosom of the great wild river (that's the proper way to put it, but in the interests of science it may be translated into crawling towards the middle)." In *Imperial Eyes* Pratt observes that Kingsley's irony both "mocks the self-importance and possessiveness of her male counterparts" and establishes "her own form of mastery, deployed in a swampy world of her own that the explorer men have not seen or do not want."

As she learned from her tutors, the trading men on the coast, irony is also an effective form of mastering fear, a defensive strategy of dusting over a rotting corpse "with jokes" so that "it would hardly show at a distance." She warns any newcomer to the coast to "take the most cheerful view of these statements, . . . take every care short of getting frightened, which is as deadly as taking no care at all." Yet her own encounters with danger are comic interludes. With mock indignation she writes:

> I hate holes, and especially do I hate these African ones, for I am frequently falling, more or less, into them, and they will be my end. . . . All . . . sorts I have tried, having pitched by day and night into those, from three to twelve feet deep . . . and also into those from twenty to thirty feet deep with pointed stakes at the bottom, artfully disposed to impale the elephant. . . . But my worst fall was into a disused Portuguese well of unknown depth. . . . Just as I was convinced that my fate was an inglorious and inverted case of Elijah . . . I was being carried off, alive.

The matter-of-fact tone of the narrator's scientific interest in the precise depth of the holes and the "artful" arrangement of the deadly spikes directs attention from the suffering and danger. Emotions such as fear, pity, or anger are thus restrained, her courage minimized, and life-threatening situations detoxified.

This conscious and persistent self-mockery in much of Kingsley's writing serves both psychological and literary ends. It not only suggests her insecurity and negative self-image but also serves as an amusing, surprisingly complex literary device. Although such self-deprecation should engender a distrust in the authority and reliability of the narrator, in practice it wins the reader's sympathy and liking for the entertaining – albeit inept and struggling – narrator who always comically needs hairpins and is plagued by an "intolerable habit of getting into water" and transporting reeking "abominations full of ants," her specimens. Sometimes as a hilariously awkward clown she falls through roofs, tumbles into game pits, or steers her canoe into conservatories and the sides of hospitals.

Mills has interpreted this self-mocking humor, one source of the continuing popularity of *Travels in West Africa,* as a destabilizing force that subverts both the "feminine discourse" expected of a Victorian woman and the discourse of imperialism practiced by male travel writers. Kingsley's humor is indeed subversive in attacking the jargon of contemporary psychology, newspaper reporting, and religion, especially when public rhetoric toward Africa masks social and political attitudes that she regards as wrong or dangerous. For instance, she strongly disagrees with those who argue that Africans are innocent children, mystical dreamers, cruel savages, or "undeveloped" white men who should be judged by European standards; instead, she contends that they are human beings who possess "a remarkable mental acuteness and a large share of common sense" and who have a rich, distinctive cultural heritage. Her humor often startles readers from complacently assuming that they are culturally superior and forces them to acknowledge the legitimacy of African cultural institutions.

She concludes her narrative of a tense encounter between her bearer, Kiva, and one of his creditors in a remote Fang village by drawing a startling cultural parallel. The creditor plans to collect his debt by seizing and eating Kiva, and Kingsley speculates that "evidently this was a trace of an early form of the Bankruptcy Court; the court which clears a man of his debt, being here represented by the knife and the cooking pot.... There is always some fragment of sound sense underlying African institutions." With a masterful control of tone, Kingsley evokes not only laughter but also some serious thought about cultural comparison. Frequently her satire is directed against people in England who, knowing nothing about Africans, want nonetheless to "civilize" them. A discussion of the Mother Hubbard, the garment imposed on African women by missionaries, becomes a satiric outburst about European attitudes:

> Forgive me, but I must break out on the subject of Hubbards; I will promise to keep clear of bad language, let the effort cost me what it may.... I think these things are one of the factors producing the well-known torpidity of the mission-trained girl; and they should be suppressed in her interest, apart from their appearance, which is enough to constitute a hanging matter.... What idea the pious ladies in England, Germany, Scotland, and France can have of the African figure I cannot think, but evidently part of their opinion is that it is very like a tub.

The final line, worthy of Jonathan Swift, cuts to the central issue: European philanthropists regard the "native" not as a flesh-and-blood reality with a sophisticated culture but as a slightly dangerous curiosity to be tailored to fit their conceptions.

Kingsley frequently uses comic reversals to force her readers to see things as she believes an African would. Through these reversals she aims to create a fresh climate of opinion in which new verities can replace old prejudices. For example, she ironically describes the way in which Europeans "inform" themselves about African matters:

> Now polygamy is . . . a difficult thing to form an opinion on, if . . . you go and make a study of the facts.... It is therefore advisable to follow the usual method employed by the majority of people. Just take a prejudice of your own, and fix it up with the so-called opinions of people who go in for that sort of prejudice too.

But she also uses her wit more subtly to humor readers into suspending judgment about African practices that were likely to shock Western sensibilities and evoke hostile responses. In discussing the practice of cannibalism among the Fang tribe she observes, with carefully modulated understatement, that it presents no danger to white travelers but is "a bother" because the traveler must labor to keep her "black companions from getting eaten." She then shifts her discussion to adopt the African point of view:

> The Fan is not a cannibal from sacrificial motives.... He does it in his common sense way. Man's flesh, he says, is good to eat, very good, and he wishes you would try it. Oh dear no, he never eats it himself, but the next door town does.... He does not buy slaves and fatten them up for his table as some of the Middle

Kingsley seated between Sir Claude MacDonald, commissioner and consul general of the Niger Coast Protectorate, and his wife, Ethel, among other members of the colonial community at Calabar, May 1895

Congo tribes I know of do. He has no slaves, no prisoners of war, no cemeteries, so you must draw your own conclusions. No, my friend, I will not tell you any cannibal stories.

Although some contemporary critics of *Travels in West Africa* accused Kingsley of being flippant about cannibalism, it is possible to see her humor as an attempt to forestall any immediate, stereotypical response from a European reader. The amusing recreation of the cannibal's self-justifying voice in the preceding passage encourages the reader to listen sympathetically and to avoid a culturally conditioned response. Throughout her travel writing Kingsley eschews sensationalism and cultivates a slightly amused objectivity about controversial topics such as African polygamy, human sacrifice, the murder of twins, and trial by ordeal. For instance, when she finds "a human hand, three big toes, four eyes, two ears, and other portions of the human frame" in a bag hanging in a hut where she is spending the night, Kingsley calmly examines and then replaces these repugnant relics; speculates about their origin (they are perhaps "mementoes" of earlier meals); and comments wryly that, though it is "touching" that cannibals keep such remembrances, "it's an unpleasant practice when they hang the remains in the bedroom you occupy." The narrator of *Travels in West Africa* and of the later *West African Studies* (1899) clearly uses humor as a political and educational tool, simultaneously demolishing misconceptions about, and shaping new attitudes toward, Africans.

Far from being disturbed by cannibalism, Kingsley felt akin to the least westernized Africans with whom she believed she shared an essentially spiritual outlook. Gwynn quotes her as declaring, "I am a firm African." Work in postcolonial theory has made late-twentieth-century readers acutely conscious of the representational problems and distortions to be found in texts purporting to tell "the truth" about a colonized people. In addition, critics such as Mills and Alison Blount foreground the complexities of the process by which a white woman such as Kingsley, constrained by the gender norms of mid-Victorian culture and condoning imperial power, represents and identifies with Africans. Late-twentieth-century readers are aware that Kingsley's empathy with Africans was grounded in problematic assumptions about race and gender as well as in some deep-seated psychological need. For

Illustration in Kingsley's Travels in West Africa *(1897). Kingsley is believed to have taken this photograph.*

example, she believed in a division of races that paralleled the division of genders: she identified the "Teutonic" and "Hebrew" races as masculine, while the "Negro" and "Melanesian" were feminine. The feminine races were destined to remain inferior to the masculine, just as "a great woman, either mentally or physically, will excel an indifferent man, but no woman ever equals a really great man."

The difference between men and women travelers aroused the interests of reviewers of *Travels in West Africa,* many of whom focused on the unconventional personality and original style of the writer rather than on the scientific and ethnographic materials. On 6 February 1897 the *Athenaeum,* however, wisely observed that despite the "comical" approach that Kingsley adopts, her work "must undoubtedly be taken seriously, for she displays keen powers of observation, far keener than those of most men who visit the coast." The reviews were largely positive, and the book was a popular success that went into its fifth edition six months after publication. In the next few years Kingsley proved to be an extremely popular and adept lecturer as well as a polemicist for a set of attitudes and policies that the "Liverpool" faction, the commercial interests in the area, favored toward West Africa. At the same time, she began work on a more scholarly production incorporating ethnological material that either had been "crowded out" of the first book or, according to her preface to this second book, had required "further investigation and comparison."

This second book, *West African Studies,* does incorporate some material deliberately excised from her first account – for example, the raucous narrative of life on a west-coast steamer that Kingsley's friends had found "too racy." This description of the bugs that "used to come in round the hanging lamp at dinner time" at Cabinda, for example, might have unsettled a squeamish reader: "Ever and anon a big beetle with a terrible boom on would sweep in, go two or three times round the room and then flop into the soup plate, out of that, shake himself like a retriever and bang into some one's face, then flop on the floor."

But despite narrative pieces about her travels, *West African Studies* is not essentially a travel book but a six-hundred-page compendium of personal narrative, scholarship, and polemic about West Africa. It includes an extensive appendix written by John Harford on early voyages to the Oil Rivers, another written by Count C. N. de Cardi on the natives of the Niger Coast Protectorate, and one on seventeenth-century trade goods. The book extensively discusses fetish, imperialism, trade, and commerce in West Africa, as well as African ideas about property, law, and religion. But this is no dry tome. Kingsley often becomes impassioned in attacking the imperial imagination that has conjured a "fiend-child" of the African and shaped policies on the basis of this illusion. Her text aims to acquaint the English with "the real African that does exist and suffers" from mistaken colonial policy.

West African Studies also contains what Kingsley called "the most difficult thing I have ever had to do," a one-hundred-page critique of the crown colony system of governing West Africa and a proposal for restructuring that system of imperial control. As a supporter of trading interests centered in Liverpool and as an admirer of African culture as it existed, Kingsley wanted to replace the centralized bureaucratic government of British colonies in West Africa with a council of merchants who would rule indirectly through existing social organizations in West Africa. Pratt has commented that Kingsley was "an imperialist but a passionate anti-colonialist" who lobbied for the view that "expansionism and frontier relations were best left in the hands of traders."

During this period Kingsley was also writing periodical articles that advocated her political views. She also produced two more books: *The Story of West Africa* (1899), a short narrative history of the region, and an edited volume of her father's writings, *Notes on Sport and Travel*. These years of political struggle, writing, and lecturing were physically and emotionally exhausting for Kingsley. She endured a battle with typhoid fever, lost a close friend, had a frustrated (possibly romantic) relationship with Maj. Matthew Nathan, and tried to keep house for her peripatetic brother.

In March 1900 she decided to travel to South Africa to collect fish from the Orange River. Since South Africa was then in the throes of the Second Boer War, Kingsley was in a potentially fatal spot. Gwynn's biography quotes one of Kingsley's letters to Sir Arthur Lyall, to whom Kingsley confesses that she feels that her "home" is "in the valley of the shadow of death" and she must "reconcile . . . to it, build my Shimbec there and settle down." One can only speculate about how consciously she was seeking death.

As soon as she arrived in South Africa, Kingsley volunteered to nurse Boer prisoners of war at Simonstown. Conditions were appalling; planning and scientific administration were lacking. In a letter published in the London *Times* (8 August 1900) she described and criticized what she saw to her friend Alice Stopford Green. Exhausted and weakened after two months of "the stench, the washing, the enemas, the bed pans, the blood," she contracted enteric fever and died on 3 June 1900. She was buried at sea off the Cape of Good Hope. The tributes that poured in after her death stressed her courage, her humanity, and her scholarly achievements; they often expressed wonder that a woman could show such intelligence, stamina, and political good sense while still remaining feminine. Such tributes ironically reinforced the same gender dichotomies that had been so central to Kingsley's life and so constraining to her sense of herself. As a tribute to her scholarly contributions, the African Society (now the Royal African Society) was founded in her memory and carried her picture on the title page of its journal until the 1923–1924 issue.

Mary Kingsley was a paradoxical figure: a woman who sometimes vehemently insisted on her femininity and sometimes referred to herself as a man; a daring and independent female traveler who opposed woman suffrage and the admission of women into the Royal Geographical Society; a supporter of British imperialism and also a harsh critic of certain aspects of imperial policy; a serious, objective, scientific observer of African phenomena and a rakish, self-mocking clown. Scholars have celebrated her as a student of African affairs, as a tragically divided woman, as a feminist heroine, and more recently as a writer in whose works the conflicting discourses of imperialism, gender, class, and race destabilize each other. Common to these different readings of Mary Kingsley's life and works is a recognition of the importance of her writings to an understanding of late-nineteenth-century West Africa, of colonial politics at the turn of the century, and of the achievements of Victorian women travel writers.

Bibliography:

Dea Birkett, *Mary Kingsley (1862–1900): A Biographical Bibliography* (Bristol: Bristol University Press, 1993).

Biographies:

Stephen Gwynn, *The Life of Mary Kingsley* (London: Macmillan, 1933);

Kathleen Wallace, *This is Your Home: A Portrait of Mary Kingsley* (London: Heinemann, 1956);

Olwen Campbell, *Mary Kingsley: A Victorian in the Jungle* (London: Methuen, 1957);

Katharine Frank, *A Voyager Out: The Life of Mary Kingsley* (Boston: Houghton Mifflin, 1986);

Robert D. Pearce, *Mary Kinglsey: Light at the Heart of Darkness* (Oxford: Kensal, 1990);

Dea Birkett, *Mary Kingsley: Imperial Adventuress* (New York & Basingstoke: Macmillan, 1992).

References:

George Baber, *Journey among Cannibals: The Story of Mary Kingsley* (London: Lutterworth, 1959);

Dea Birkett, *Spinsters Abroad: Victorian Lady Explorers* (Oxford & New York: Blackwell, 1989);

Alison Blount, *Travel, Gender, and Imperialism: Mary Kingsley and West Africa* (New York & London: Guilford, 1994);

Colin Clair, *Mary Kingsley: African Explorer* (Watford: Bruce & Gawthorn, 1963);

John Flint, Introduction to *West African Studies,* by Kingsley (London & New York: Cass, Barnes & Noble, 1964);

I. M. Holmes, *In Africa's Service: The Story of Mary Kingsley* (London: Saturn, 1949);

Cecil Howard, *Mary Kingsley* (London: Hutchinson, 1957);

Sara Mills, *Discourses of Difference: An Analysis of Women's Travel Writing and Colonialism* (London & New York: Routledge, 1991);

Mary Louise Pratt, *Imperial Eyes: Travel Writing and Transculturation* (London: Routledge, 1992);

Catherine Stevenson, *Victorian Women Travel Writers in Africa* (Boston: Twayne, 1982).

Papers:

Kingsley's important correspondence with Macmillan is part of the Macmillan Papers at the British Library; her letters to Alice Stopford Green are at the National Library of Ireland, as are her letters written to Stephen Gwynn. Her letters to Maj. Matthew Nathan are in the Rhodes House Library, Oxford University.

G. W. H. Knight-Bruce

(23 March 1852 – 16 December 1896)

James Rigney
Roehampton Institute, London

BOOKS: *Journals of the Mashonaland Mission, 1888 to 1892*, edited by Louisa Knight-Bruce (London: Society for the Propagation of the Gospel in Foreign Parts, 1892);

Memories of Mashonaland (London & New York: Arnold, 1895);

Gold and the Gospel in Mashonaland, 1888: Being the Journals of: 1. The Mashonaland Mission of Bishop Knight-Bruce 2. The Concession Journey of Charles Dunell Rudd, edited by Constance E. Fripp and V. W. Hiller (London: Chatto & Windus, 1949).

SELECTED PERIODICAL PUBLICATIONS –
UNCOLLECTED: "From Bloemfontein to the Zambesi," *Mission Field*, 34 (July–December 1889): 263–270, 298–307, 342–350, 378–391, 417–425, 457–466;

"Notes on a Journey through Mashonaland in 1889," *Proceedings of the Royal Geographical Society*, 6 (June 1890): 346–352;

"The Country East of the Junction of the Sabi and the Odzi Rivers," *Geographical Journal*, 1 (April 1893): 344–345.

G. W. H. Knight-Bruce's travels through southern Africa in the 1880s and 1890s coincided with a period of intense economic development and imperial expansion. The missionary motives that brought Knight-Bruce to southern Africa operated in an often tense relation to the exploitative colonialism that attracted most popular attention. J. P. Fitzpatrick's *Through Mashonaland with Pick and Pen* (1892), the anonymous pamphlet *With Mr Rhodes Through Mashonaland* (1892), and Rose Blennerhasset and Lucy Sleeman's *Adventures in Mashonaland by Two Hospital Nurses* (1893) fed public interest in the region. Although Mashonaland, which formed part of Rhodesia (modern-day Zimbabwe), was created as a colony by the businessman Cecil Rhodes, it was formed as a colony by the efforts of missionaries. The discovery of gold increased the level of infiltration by hunters and prospectors, and Rhodes's eventual success lay in the way in which he channeled these isolated attempts at penetration. Bishop Knight-Bruce had affinities with both types of early white travelers – the transient hunter and the settler missionary. Missions were often seen as an impediment to commercial development: Knight-Bruce's energetic travels in the region, undertaken before those of Fitzpatrick, Blennerhasset, and Sleeman, took the missions into parts of southern Africa that had yet to be penetrated by investment and sought to secure for the church a position from which the power of commerce could be mediated and modified.

George Wyndham Hamilton Knight-Bruce was born on 23 March 1852 at Heathfield House, Keston, Kent, the eldest child of James Lewis Bruce Knight-Bruce, a judge, and his wife, Caroline Newte Knight-Bruce. George went to Eton in 1864 but remained only until 1867. His family later accounted for this short stay as the result of a riding accident. Five years later, in 1872, he matriculated to Merton College, Oxford, from which he graduated in 1876 with a fourth-class degree in history. He was ordained a deacon in the same year and a priest in the year following, both in the Diocese of Gloucester. In 1878 he was appointed to a curacy in Cornwall, beginning a long association with the west of England. In the same year Knight-Bruce made an advantageous marriage to Louisa Torr, the daughter of John Torr, member of Parliament for Liverpool, whose money had made a significant contribution to the establishment of the Diocese of Liverpool. Two of Louisa's brothers became clergymen; both served on the council of Knight-Bruce's Mashonaland Mission Association, and one, the Reverend T. J. Torr, contributed £350 to the cost of his brother-in-law's expeditions. Louisa Knight-Bruce accompanied her husband on many expeditions: her account of the Ngwato chief, Kgama III, whom she and Knight-

G. W. H. Knight-Bruce's journal entry for 30 July 1888, describing his trip down the Zambezi River (from Gold and the Gospel in Mashonaland, 1888, *edited by Constance E. Fripp and V. W. Hiller, 1949)*

Bruce visited in Bechuanaland, published as *The Story of an African Chief: Being the Life of Khama* (1893), was enormously popular, running to six editions in seven years.

In 1883 Knight-Bruce went to London to serve in the East End Mission of Bishop Walsham How, becoming a curate in the parish of Bethnal Green from 1884 to 1886. Knight-Bruce was like many other young men of his background who experienced a call to missionary work either in the deprived areas of England or in foreign countries, but Knight-Bruce was distinctive in the energy with which he committed himself to the rigors of missionary travel once circumstances brought him to Africa.

During 1885 Bishop How was approached by representatives of the clergy in southern Africa to suggest a candidate for the post of bishop of Bloemfontein. The job required qualifications above those normally sought in a bishop – principally that the candidate had to be young, fit, and wealthy. The size of the diocese and the difficulties involved in traveling through it lay behind the first two criteria, while the debts that the diocese currently bore meant that the bishop's stipend was extremely small and would need to be supplemented from the private means of the incumbent. Knight-Bruce met all the qualifications, and in October 1885 he accepted the offer. On 25 March 1886, having previously been awarded the degree of doctor of divinity by Oxford University, he was consecrated bishop of Bloemfontein. Knight-Bruce, his wife, and his chaplain sailed from Dartmouth on 23 July 1886 and arrived in Cape Town on 12 August.

During his time as bishop of Bloemfontein, Knight-Bruce paid off the debt in the diocese by economies (including taking no salary for himself) and by raising money from English well-wishers. Like most missionaries his continuing good relations with supporters in England was essential to the success of his work, and it was for this audience in particular that Knight-Bruce wrote and published the various narratives of his later travels.

One condition imposed by Knight-Bruce's principal sponsor, the Society for the Propagation of the Gospel, was that he keep a journal during his travels. From these three one-hundred-page notebooks Knight-Bruce composed what is in effect one travel narrative covering both his 1888 journey to the Zambezi River and his expedition in Mashonaland in 1891. The record exists, however, in several forms, each designed to serve different functions in relation to different audiences, and each revealing a different facet of Knight-Bruce. From the notes in his diary of his first expedition, Knight-Bruce published an article titled "Notes on a Journey through Mashonaland in 1889" in the *Proceedings of the Royal Geographical Society* in 1890; a version had also appeared in *The Mission Field* in 1889. This earliest version is not entirely accurate since passages of editorial précis have been included and editorial notes occasionally interrupt the narrative. Much of the same material is found in *Journals of the Mashonaland Mission, 1888 to 1892* (1892), edited (with some editorial intervention) by Louisa Knight-Bruce, and in *Memories of Mashonaland* (1895). The record of Knight-Bruce's journey was published in a complete version for the first time in 1949 as *Gold and the Gospel in Mashonaland, 1888*. The various versions of Knight-Bruce's record can be accounted for as ways of meeting the needs of different audiences, from those interested in African travel and exploration to those specifically concerned with missionary record. The different versions also reveal successive aspects of Knight-Bruce, giving him a depth, complexity, and dignity that is sometimes missing in some of the more self-effacing versions.

Knight-Bruce's time as bishop of Bloemfontein involved him in a two-year cycle of preaching, consecrations, and social events. In the spring of 1886 he undertook preliminary travels into the wilder parts of his diocese with a long ride into the Malato Mountains: as far as his health permitted, Knight-Bruce was an outdoorsman, a good horseman and shot who enjoyed the opportunities for these activities found in southern Africa. By 1888 Knight-Bruce had extended his vision of his diocese beyond its existing limits, and after consultation with travelers such as the legendary Frederick Courteney Selous about routes and equipment, he set out to explore the north of the diocese around the Zambezi River with the aim of developing missionary opportunities in Mashonaland.

After a month of traveling through the northern reaches of his diocese with his wife, Knight-Bruce reached Bulawayo, the capital and the home of Lobengula, the chieftain of the Matabele, who controlled access to the land of the Mashona. Lobengula was initially reluctant to assist Knight-Bruce because of his suspicion that missionaries would curtail his annual raids against the Mashona to gain captives and to "blood" his young warriors. However, after much negotiation with the chief Knight-Bruce finally got permission to proceed into Mashonaland. From Bulawayo he traveled northeast to Gwelo and then due north to Gatooma, striking northeast again to the Hunyani River and the site of Fort Salisbury. At the Hunyani River he was forced to abandon his wagons and most of his animals because of tsetse flies and proceeded the rest of the way on foot. He pushed as far as Spolilo, near the border with Portuguese territory. He then traveled seventy miles downriver before walking back to Chilimanzi and then returning to Bulawayo. The round-trip from Bulawayo covered 535 miles in forty days. Knight-Bruce covered between twenty-five hundred and three thousand miles of territory between April and December 1888, much of it hitherto unexplored by Europeans.

Knight-Bruce encountered Mashonaland at a special moment in its history. The spread of gold fever from the Transvaal filled the territory with prospectors and speculators, lured by reports in works such as E. P. Mathers's *Zambesi, England's El Dorado* (1891). When he entered Mashonaland the first thing that struck Knight-Bruce was a disquieting solitude: a recent raid by the Matabele had killed or driven off most of the population, and the English traveler whom Knight-Bruce expected to meet at Zumbo, Richard Foster, was presumed murdered by the roadside. Making his way through a wide sea of grass and across exposed hillsides, suffering from malaria and the pain from his recently broken shoulder, Knight-Bruce was conscious of his pioneering role in the region and writes eloquently, especially in *Memories of Mashonaland,* of the silence of the country. When he returned, some years later, to the village of Sipiro, which he first visited in 1888, the village chief said to him: "You and Selous are the only two people known in this country." This conjunction with Selous is a testimony to Knight-Bruce's achievement in traveling in such little-known territory. When he looks back on this journey in *Memories of Mashonaland,* he expresses some regret for the way in which the untouched wilds and the stillness have vanished: "The whole face of the country seems different. The old order has changed, and the romance is gone; but perhaps it is some comfort to feel that, rightly or wrongly, the change must have come sooner or later in Mashonaland."

The forces of change were close behind Knight-Bruce as he journeyed through Mashonaland. Cecil Rhodes, the prime minister of the Cape Colony, once he had gained a mining concession in the region from the Matabele rulers, established the British South African Company in 1889 and sent in troops to secure the concession. He recognized the need for a missionary presence to support his operations. When the African bishops decided to constitute Mashonaland as a missionary diocese and invited Knight-Bruce to accept the see, the £1,000 funding by the Society for the Propagation of the Gospel was matched by £500 from Rhodes.

Whereas Knight-Bruce had taken only Africans with him on his 1888 expedition, the journey to establish the Mashonaland diocese began with his sending two white chaplains ahead of him into the territory that he had already opened on his earlier trip. Knight-Bruce soon passed them, covering 120 miles of the journey from Umtali to Fort Salisbury in just five days, planning mission centers, negotiating with local chiefs, and treading a fine line between the needs of the missions and his relations with Rhodes's operation. In a speech in Vryburg in December 1888, shortly after returning from the Zambezi journey, he had protested against the pressure that Rhodes was bringing to bear on Lobengula, and he opposed the extension of Rhodes's commercial concession and was one of those who petitioned the British government without success for the extension of the Crown protectorate as far as the Zambezi River. Rhodes's complex mix of piety and profiteering made life difficult for Knight-Bruce: for example, his progress was frequently impeded by the Portuguese, who felt impinged upon by the Chartered Company and were understandably suspicious of Knight-Bruce's relations with Rhodes. In March 1891, leaving a Portuguese settlement in which he had been hospitably welcomed, Knight-Bruce met Eustace Wickham-Fiennes, an officer of Rhodes's company, about to attack the camp. Knight-Bruce carried a message for Wickham-Fiennes to abort the attack in light of a recent agreement between the company and the Portuguese, and he recalled: "as I had just left, laden with their hospitality and kindness, it would have been painful to know that they were being attacked by an English police force."

Knight-Bruce felt obliged to accompany Rhodes's armed forces during the Matabele Wars, stressing that he went as bishop of the territory in which the war took place, rather than as chaplain to the company's forces. Knight-Bruce's bravery on this occasion is striking: during the battle at the Shangani River he assisted both English and Mashona casualties, rescuing each under fire. He volunteered to negotiate with the Matabele king, Lobengula, who by that stage had, in desperation, burned his own camp at Bulawayo and fled to the North, but was dissuaded by the English administrator Dr. Leander Jameson, who feared that Knight-Bruce would be killed on the way. Immediately after a church parade, held on 12 November 1893 at the conclusion of the Matabele Wars, Knight-Bruce gave Holy Communion to some of the men:

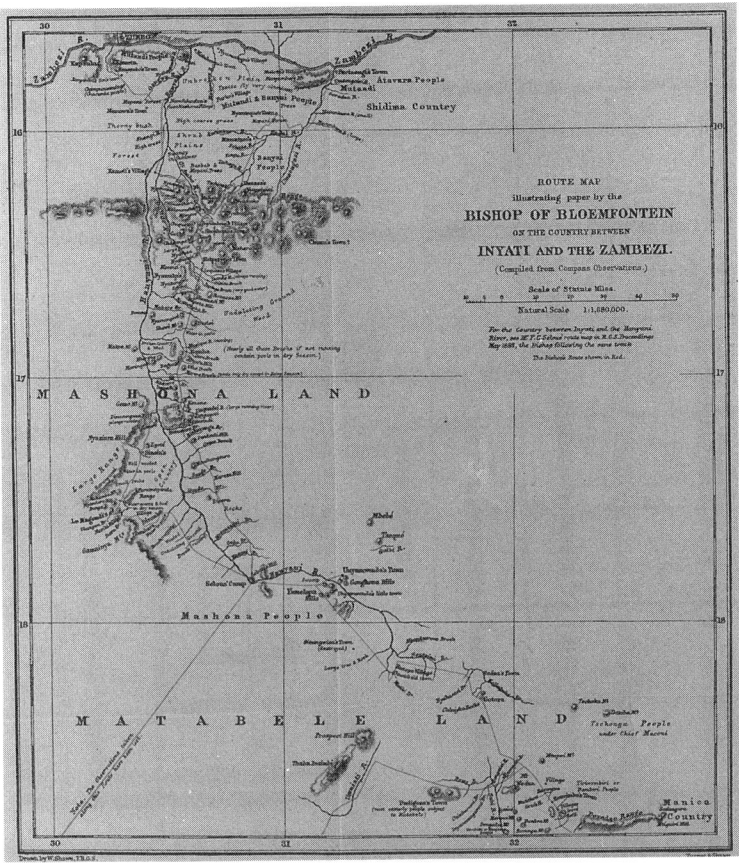

Map from Knight-Bruce's "Notes on a Journey through Mashonaland in 1889," published in the Proceedings of the Royal Geographical Society *in June 1890*

I think it was one of the most beautiful services I have ever known – the perfect peace after the life of fighting and noise, and dust and heat; the looking back into the plunge into the unknown, that had been made by the men; the strange end to the long series of unexpected acts that only culminated here – all affected us very strongly. I have never seen a more utterly-reverential body of men gathered before their God.

In the midst of his exposure to the worst side of imperial expansion, Knight-Bruce found comfort in the transfiguring quality of religion, expressing himself in this passage, for almost the only time in his narratives, in a way that transcends the often conventional terms in which he discusses his religious experience in Africa.

In *Memories of Mashonaland* Knight-Bruce stresses the importance for anyone planning to travel on foot in Africa to make a close study of the native character "for a year or so before starting." Knight-Bruce's attitude toward the natives with whom he came into contact was, like that of many of his contemporaries, frequently impatient and patronizing, yet he also shows a sensitivity to what he sees as the relativism of the native viewpoint. He records in *Memories of Mashonaland* that he tried to be led by the example of Selous and David Livingstone, both of whom treated natives well and whose memoirs record "none of the abuse of the 'nigger' which seems to delight some writers on the subject." As part of his intention to proselytize for the missions, he is eager to stress how the presence of missionaries improves the natives not merely in moral terms but as resources: a Christian native makes a far better bearer than one who is not. In *Journals of the Mashonaland Mission* he expresses the view that "No one who has not had dealing with the really heathen native can credit what degradation of humanity they are. To live somewhat intimately among them is the best refutation of the belief that heathen natives are better than Christian, and is the strongest argument for the necessity of raising them." Yet Knight-Bruce was also aware that the effects on natives of contact with Europeans was not always beneficial. On traveling from Fort Salisbury to the southern tribes of Mashonaland he remarks: "It is extraordinary how far nicer they are when they have not had to do with a low kind of Europeans. If only these countries could be colonised by a high-minded set of Englishmen, what an unspeakable difference it would make in the education and in the whole future existence of the natives!"

Knight-Bruce's journeys were remarkable achievements; the anonymous editor who prepared the journals for publication in *The Mission Field* said that they were "an admirable instance of Christian Missionary enterprise, and not inferior to any other achievement in South African travel." When Knight-Bruce met the nurses who had walked 140 miles from the coast to join him at the site of his mission hospital, he remarked that their journey was an achievement that would perhaps not be immediately appreciated. He comments: "In Africa there seems to be too strong a tendency to self-advertisement to allow any unadvertised work to be much recognised."

Knight-Bruce's own reticence is best illustrated by his tendency in many of the versions of his travels to edit out references to his illness or to refer to them only slightly in passing. In *Journals of the Mashonaland Mission,* for example, he comments: "Certainly sitting on a horse in a blazing sun with a bad head for nine hours seems to make one conscious of one's own weakness and the length of a day." Yet it is clear from a reading of all the versions of his journal that he was always handicapped by poor health but pushed his body constantly. Knight-Bruce was an unrelenting man, in many respects like a hero in a John Buchan novel. Knight-Bruce's willingness to travel in dangerous conditions, on more frequent and longer journeys when weak and sick, without transport for stores, and with poor food suggests that exhaustion was an end in itself. He frequently notes in *Memories of Mashonaland* that he missed much of beauty along the way because he was too tired to care about what he saw.

Eventually Knight-Bruce's health collapsed under the strain of his travels: during January and February 1894 he suffered a severe attack of blackwater fever that was expected to prove fatal. On his recovery his doctors insisted that he leave Mashonaland for England, which he did, settling in early 1894 in Bovey Tracey in Devonshire, where he worked as curate in the parish. When the vicar of Bovey Tracey died in October of that year, Knight-Bruce succeeded him and carried on an active ministry in the parish; to this work he added the responsibilities of assistant bishop of Exeter in February 1895. During the same time Knight-Bruce regularly traveled to London to attend meetings of the Society for the Propagation of the Gospel and the Mashonaland Mission Association. On 10 and 11 December 1895, in cold and wet weather, Knight-Bruce visited the sick around Bovey Tracey and caught a chill: the chill developed into pleurisy and pneumonia, and he died on 16 December.

In her edition of her husband's journal Louisa Knight-Bruce remarks that "England raised a million in a few weeks to work Mashonaland gold, and sent out a number of men for every available post; how much will England give in prayer and help and dedicated life to work for the Golden Harvest that Angel hands will garner?" G. W. H. Knight-Bruce gave generously of a dedicated life during eight years of missionary zeal and pioneering spirit. His character and his writing lack the complexity of his predecessor, David Livingstone, but the tensions in Knight-Bruce's situation, his pioneering role in southern African travel, and the juxtaposition of his journeys with the commercial and imperial penetration of the region make for interesting narratives and demonstrate their author's heroism.

Reference:

Constance E. Fripp, *George Wyndham Hamilton Knight Bruce,* Gray Centenary Pamphlets, second series, no. 2 (Cape Town: South African Church Publications, 1947).

Papers:

The manuscript of G. W. H. Knight-Bruce's journal is held in the former Central African Archive, now part of the National Archives of Zimbabwe in Harare. Other papers, including letters, are in the Bishop Bousfield papers within the church in the Province of South Africa Archives, held at the University of Witwatersrand in Johannesburg.

Mrs. Aubrey Le Blond
(1861 – 27 July 1934)

Timothy S. Jones
Augustana College

BOOKS: *The High Alps in Winter; or, Mountaineering in Search of Health,* as Mrs. Fred Burnaby (London: Sampson Low, Marston, Searle & Rivington, 1883);

High Life and Towers of Silence, as Elizabeth Main (London: Sampson Low, Marston, Searle & Rivington, 1886);

My Home in the Alps, as Mrs. Main (London: Sampson Low, Marston, 1892);

Hints on Snow Photography, as Mrs. Main (London: Sampson Low, Marston, 1894);

Cities and Sights of Spain: A Handbook for Tourists (London: Bell, 1899; revised, London: Bell, 1904);

True Tales of Mountain Adventure for Non-Climbers Young and Old (London: Unwin, 1903; New York: Dutton, 1903);

Adventures on the Roof of the World (New York: Dutton, 1903; London: Unwin, 1904);

The Story of an Alpine Winter (London: Bell, 1907);

Mountaineering in the Land of the Midnight Sun (London: Unwin, 1908; Philadelphia: Lippincott, 1908);

Charlotte Sophie, Countess Bentinck: Her Life and Times, 1715–1800, 2 volumes (London: Hutchinson, 1912);

The Old Gardens of Italy: How to Visit Them (London & New York: John Lane, 1912; revised, London: John Lane, 1926; New York: Dodd, Mead, 1926);

Day In, Day Out (London: John Lane, 1928).

OTHER: Henry Inigo Triggs, *The Art of Garden Design in Italy,* twenty-eight photographs by Le Blond (London: Longmans Green, 1906);

E. F. Benson, *Winter Sports in Switzerland,* reproductions from forty-seven photographs by Le Blond (London: George Allen, 1913);

Charlotte Amélie Aldenburg, *The Autobiography of Charlotte Amélie, Princess of Aldenburg, Née Princess de La Trémoïlle, 1652–1732,* translated and edited by Le Blond (London: Nash, 1913; New York: McBride, Nast, 1914);

Mrs. Aubrey Le Blond (portrait by Mary Macleod; from Mrs. Aubrey Le Blond, Day In, Day Out, *1928)*

Otto Victor Maeckel, *The Dunkelgraf Mystery,* collaboration with Le Blond (London: Hutchinson, 1929);

Louis Hubert Gonzalve Lyautey, *Intimate Letters from Tonquin,* translated by Le Blond (London: John Lane, 1932).

On a fall day in 1881 Elizabeth Burnaby (later Elizabeth Le Blond) and a party of other young people set out from the Chamonix valley for a hike up a mountain to Pierce Pointue. Most of the hikers were satisfied upon reaching their destination early in the day, but Burnaby and another woman decided to continue to the Grand Mulets. When darkness fell they were forced to spend the night with their guides in a cabin, and the next morning, lacking the help of a maid, Burnaby had to put on her boots without assistance for the first time in her life. Returning to Chamonix, they were greeted by anxious

Le Blond during the years of her first marriage – to Col. Fred Burnaby, who was also a travel writer

friends and reproachful relatives. Burnaby, however, had discovered a new experience on the borders of Victorian society and was to become one of the leading alpinists of her time.

The daughter of Baronet St. Vincent Hawkins-Whitshed and Anne Alicia Handcock of Killincarrick, Wicklow County, Elizabeth Le Blond belonged to a family of noble lineage. Throughout her life she enjoyed the resources to travel and live abroad, and, as a member of this social class, she engaged in many typical philanthropic activities – assisting the wounded at Dieppe, heading the appeals department of the British Ambulance Committee during World War I, and later leading efforts to rebuild the cathedral at Rheims. She also published a biography of her ancestor, Charlotte Sophie, Countess Bentinck, and translated the letters of Louis Hubert Gonzalve Lyautey.

Her love of adventure was apparent in her first marriage. At the age of seventeen and in her first season in London she met and married Col. Fred Burnaby, the noted Victorian soldier, correspondent, and author of *A Ride to Khiva: Travels and Adventures in Central Asia* (1876) and *On Horseback through Asia Minor* (1877). He was killed at Abu Klea in northern Sudan in 1885, and Le Blond was to devote fifty pages of her own memoirs, *Day In, Day Out* (1928), to an account of his life and adventures.

In the winter following her marriage Burnaby became ill when a snowstorm stopped her train to Marseilles. She spent the spring recovering in Algiers and gave birth to a son in May. When she became ill again that summer, she followed a doctor's advice to convalesce at Chamonix, and during that time she made her climb to the Grand Mulets. One week later she abandoned an attempt to climb Mont Blanc, but during the summer of 1882 another attempt was successful. She followed this success with ascents of the Col du Tacul, the Grand Jorasses, the Aiquille du Midi, the Great Saint Bernard, and several other peaks. Her climb of the Aiquille du Midi was the first winter ascent of that peak, and the *Alpine Journal* (May 1883) called her string of successes "one of the most brilliant chapters in the history of winter mountaineering."

Approval of her accomplishments, however, was not universal. After a look at Burnaby's sunburned face, Lady Renira Bentinck, her great aunt, wrote to her mother: "Stop her climbing mountains! She is scandalizing all London and looks like a red Indian!" But she not only continued to climb but also began to write about her adventures. Her first book, *The High Alps in Winter; or, Mountaineering in Search of Health* (1883), details her introduction to mountaineering and her winter climbs of 1882 and 1883. She describes her trips by train and carriage through Switzerland and gives the winter climber tips on subjects such as how to keep wine from freezing.

Although the *Alpine Journal* had applauded her exploits, a reviewer for the journal was disappointed with her book and called it "a collection of slight and hasty sketches filled out into a volume by the use of the largest type" and "a volume which is probably the flimsiest and most trivial that has ever been offered to the alpine public." In fact, this reviewer was unimpressed with her successful climbs, which he attributed to the skill of her guides, although he allowed her some credit for "perseverance." He was clearly concerned about the decline in the reputation of some alpine peaks if women could so easily climb them.

Another remark in this review – that "winter mountaineering is hardly the peculiar and personal revelation the authoress seems to fancy it" – is also revealing. Alpine climbing was an adventure that enthusiasts enjoyed while contributing to the traditions that were developing around it: writing about an ascent meant becoming a part of the history of events involving a mountain and its human climbers, not merely chronicling one's own exploit. In writing her memoirs years later, Le Blond was clearly aware of this and acknowledged this early reviewer's criticism of the "personal" orientation of her narrative: "I might have written the whole history of winter climbing up to that time, and have produced a book of real value. But I climbed like a child, ardently, engrossingly, thinking not at all of onlookers and indeed unaware that such even existed." Although she referred to a few previous climbers in *The High Alps in Winter,* she also gave Mark Twain attention for his description of the Gorner Grat glacier. Her only experience with the literature of mountaineering, she admitted, had been as a child listening to her mother read from Edward Whymper's *Scrambles amongst the Alps in the Years 1860–69* (1871).

The haste with which the first book was published following the news of her achievements may also account for some of what appeared to be its deficiencies. At least, this is what a writer for the *Alpine Journal* expressed in a review of her second book, *High Life and Towers of Silence* (1886), which was published under the name of Elizabeth Main after she had married J. F. Main, a physician. The reviewer claims that this work is "a vast deal better" than the first and "a pleasantly written volume which will agreeably while away an hour or two on a winter's afternoon, though it does not contain the tale of any hairbreadth escapes, and is not likely to rank among the Alpine classics." Recounting perilous incidents was a staple of most alpine literature, and Main's narration as a travel writer rather than as an adventurer stands out. In the opening chapters she assumes the point of view of an average English visitor to the Alps and describes the sights, characters, and customs that one might encounter. The bulk of the book describes ascents of the Bieshorn, Matterhorn, Dent du Géant, Riffelhorn, and Wieshort, and Main says that the purpose of such descriptions is to "give the very least amusement to even one of my readers."

This traveler's sensibility also informs her third book, *My Home in the Alps* (1892), published after the Mains had settled in Saint-Moritz. Conceived as a guide for English visitors, *My Home in the Alps* includes many short anecdotes and summarizes the natural history of the Alps. As an aid to amateur alpinists, the book introduces readers to the charac-

Le Blond's photograph of her two Swiss guides crossing a snow slope in Norway

ter and skills of alpine guides. Main also recounts some of her own adventures, including thrilling avalanche stories, in order to introduce inexperienced climbers to some of the difficulties they might face.

Adventure is the chief subject of two anthologies of tales – *True Tales of Mountain Adventure for Non-Climbers Young and Old* (1903) and *Adventures on the Roof of the World* (1903) – that she later compiled and published under the name of Elizabeth Le Blond, following the death of Dr. Main in 1892 and her marriage to E. B. Aubrey Le Blond in 1900. In these books Elizabeth Le Blond aimed "to seek out tales of adventure easily intelligible to the nonclimber, to edit them in popular form, to point out the lessons that most adventures can teach to those who may climb themselves one day." She gathered these stories largely from remote sources and provided introductory comments. In *Adventures on the Roof of the World* Le Blond includes only a single narrative of her own experience, an encounter with an avalanche on the Hohberghorn. Yet the format of these volumes pleased a reviewer for *The Nation* (14 May 1903), who noted that the narratives in *True Tales of Mountain Adventure* were preferable to mere records of the whole experience of climbing a mountain in that "the exciting moments are singled out and compressed together" in Le Blond's volume. Another reviewer for *The Athenæum* (14 February 1903) suggested that "mountain misadventures" might be a better choice for the title.

Sometime near the date of her marriage to Aubrey Le Blond, Roman, the son of her favorite guide, Josef Imboden, was killed in a climbing accident, and she left the Alps and began climbing in Norway. These expeditions resulted in many first ascents in the Lyngenfjord region, and Le Blond was to write of these in *Mountaineering in the Land of the Midnight Sun* (1908). By this time her two anthologies had attracted readers and had earned reviews in major periodicals, and *Mountaineering in the Land of the Midnight Sun* was published in both England and the United States and reviewed in the London *Times Literary Supplement*, *The Nation*, and *Nature*. This readership appreciated her eye for her surroundings and her attention to the aesthetic experience that the mountains provided. While paying scant attention to the business of mountain climbing, the reviewer for *The Nation* comments that "the book shows, as no other within our knowledge does, how delightfully a summer can be passed in camp or in one of the little settlements in this region" and cites several examples of Le Blond's descriptive prose. Le Blond, while more adept than she had been at technically recording the rigors of mountaineering in her previous works, continues to exhibit a broad interest in alpine regions. The last third of *Mountaineering in the Land of the Midnight Sun* is devoted to the activities of three summers spent camping, fishing, hiking, boating, and photographing in arctic Norway.

Le Blond was also to write two guidebooks for tourists – *Cities and Sights of Spain: A Handbook for Tourists* (1899) and *The Old Gardens of Italy; How to Visit Them* (1912) – as well as a novel, *The Story of an Alpine Winter* (1907), and a biography, *Charlotte Sophie, Countess Bentinck: Her Life and Times, 1715–1800* (1912). After settling in Switzerland following the death of her second husband, she had become adept at winter photography and had written a manual, *Hints on Snow Photography* (1894). With her photography skills she was to illustrate Henry Inigo Triggs's *The Art of Garden Design in Italy* (1906) and E. F. Benson's *Winter Sports in Switzerland* (1913). In 1907 she helped to found and was then elected as first president of the Ladies' Alpine Club.

In her memoirs, *Day In, Day Out,* Le Blond augments her record of travel and adventure with tales of bicycling in the Alps. She also gives a limited account of her journey to Asia with her husband in 1912 and 1913; she describes only significant landmarks in China, Korea, Japan, and Russia and comments briefly on political tensions. In contrast, her account of a journey from Switzerland to England in August 1914 is detailed, and she describes the tension of her service in France during World War I and her tour of the battlefields afterward. The memoirs conclude by recounting several journeys to the United States to visit her son in Los Angeles; this narrative returns to a travel guide format in being written for the British tourist exploring North America. It details train routes across the continent and major attractions such as the Grand Canyon, Yellowstone Park, and Hollywood.

Following Le Blond's death the London *Times* (28 July 1934) credited her as an alpine pioneer but stated that "She climbed in a skirt which hardly reached her knees, though she in her turn, at all events at first, equally disapproved of women who climbed with no skirt at all." Although the term is now out of fashion, Le Blond was a *lady alpinist,* one who believed that breeding and character were invaluable on the summit of a mountain and that a woman might possess these virtues as well as a man. She was also honest about her motivations, as she had written in *The High Alps in Winter* that "to do something that no one else has done is pleasant, for the gratification of our vanity has a large share in the proportion of happiness allotted to humanity." It is hard to imagine one of Le Blond's contemporary male mountaineers making such a public confession. Such attitudes were not in accord with the male tradition of alpine adventure, and their expression added a new voice to the literature of mountaineering.

References:

"Mrs. Aubrey Le Blond, Mountaineering for Women" *Times* (London), 28 July 1934, p. 14;

E. L. Stutts, "Mrs. Aubrey Le Blond (1861-1934)," *Alpine Journal,* 46 (November 1934): 382-384.

Vernon Lee
(Violet Paget)

(14 October 1856 – 13 February 1935)

Patricia O'Neill
Hamilton College

See also the Lee entries in *DLB 57: Victorian Prose Writers After 1867; DLB 153: Late-Victorian and Edwardian British Novelists, First Series;* and *DLB 156: British Short-Fiction Writers, 1880–1914: The Romantic Tradition.*

BOOKS: *Studies of the Eighteenth Century in Italy* (London: Satchell, 1880; Chicago: McClurg, 1908);

Tuscan Fairy Tales, Taken Down from the Mouths of the People, anonymous (London: Satchell, 1880);

Belcaro: Being Essays on Sundry Aesthetical Questions (London: Satchell, 1881);

The Prince of the Hundred Soups: A Puppet-Show in Narrative (London: Unwin, 1883; New York: Lovell, 1886);

Ottilie: An Eighteenth-Century Idyll (London: Unwin, 1883): republished with *The Prince of the Hundred Soups* (New York: Harper, 1886);

The Countess of Albany (London: W. H. Allen, 1884; Boston: Roberts, 1884);

Euphorion: Being Studies of the Antique and the Medieval in the Renaissance (2 volumes, London: Unwin, 1884; Boston: Roberts, 1884; revised, 1 volume, London: Unwin, 1885; Boston: Roberts, n.d.);

Miss Brown: A Novel (3 volumes, Edinburgh & London: Blackwood, 1884; 1 volume, New York: Harper, 1885);

A Phantom Lover: A Fantastic Story (Edinburgh: Blackwood, 1886; Boston: Roberts, 1886);

Baldwin: Being Dialogues on Views and Aspirations (London: Unwin, 1886; Boston: Roberts, 1886);

Juvenilia: Being a Second Series of Essays on Sundry Aesthetical Questions (2 volumes, London: Unwin, 1887; 1 volume, Boston: Roberts, 1887);

Hauntings: Fantastic Stories (London: Heinemann, 1890; New York: Lovell, 1890);

Vanitas: Polite Stories (London: Heinemann, 1892; New York: Lovell, Coryell, 1892); republished and expanded as *Vanitas: Polite Stories, Including the*

Vernon Lee (Violet Paget)

Hitherto Unpublished Story Entitled "A Frivolous Conversation" (London: John Lane, Bodley Head, 1911; New York: John Lane, 1911);

Althea: A Second Book of Dialogues on Aspirations and Duties (London: Osgood, McIlvaine, 1894); republished as *Althea: Dialogues in Aspirations and Duties* (London & New York: John Lane, 1910);

Renaissance Fancies and Studies: A Sequel to "Euphorion" (London: Smith, Elder, 1895; New York: Putnam / London: Smith, Elder, 1896);

Art and Life (East Aurora, N.Y.: Roycroft, 1896);

Limbo and Other Essays (London: Richards, 1897); republished and enlarged as *Limbo and Other Es-*

says, to Which Is Now Added "Ariadne in Mantua" (London: John Lane, Bodley Head / New York: John Lane, 1908);

Genius Loci: Notes on Places (London: Richards, 1899; London: John Lane, Bodley Head / New York: John Lane, 1908);

The Child in the Vatican (Portland, Maine: Mosher, 1900);

Chapelmaster Kriesler: A Study of Musical Romanticists (Portland, Maine: Mosher, 1901);

In Umbria (Portland, Maine: Mosher, 1901);

Ariadne in Mantua: A Romance in Five Acts (Oxford: Blackwell, 1903; Portland, Maine: Mosher, 1906);

Penelope Brandling: A Tale of the Welsh Coast in the Eighteenth Century (London: Unwin, 1903);

Hortus Vitae: Essays on the Gardening of Life (London: John Lane, Bodley Head / New York: John Lane, 1903);

Pope Jacynth and Other Fantastic Tales (London: Richards, 1904; London: John Lane, Bodley Head / New York: John Lane, 1907);

The Enchanted Woods, and Other Essays on the Genius of Places (London: John Lane, Bodley Head / New York: John Lane, 1905);

Sister Benvenuta and the Christ Child: An Eighteenth Century Legend (New York: Kennerley, 1905; London: Richards, 1906);

The Spirit of Rome: Leaves from a Diary (London: John Lane, Bodley Head / New York: John Lane, 1906);

The Sentimental Traveller: Notes on Places (London: John Lane, Bodley Head / New York: John Lane, 1908);

Gospels of Anarchy and Other Contemporary Studies (London & Leipzig: Unwin, 1908; New York: Brentano's / London: Unwin, 1909);

Laurus Nobilis: Chapters on Art and Life (London: John Lane, Bodley Head / New York: John Lane, 1909);

Beauty and Ugliness and Other Studies in Psychological Aesthetics, by Lee and Clementina Anstruther-Thomson (London: John Lane, Bodley Head / New York: John Lane, 1912);

Vital Lies: Studies of Some Varieties of Recent Obscurantism, 2 volumes (London: John Lane, Bodley Head / New York: John Lane / Toronto: Bell & Cockburn, 1912);

The Beautiful: An Introduction to Psychological Aesthetics (Cambridge: Cambridge University Press / New York: Putnam, 1913);

The Tower of the Mirrors, and Other Essays on the Spirit of Places (London: Lane, Bodley Head / New York: Lane / Toronto: Bell & Cockburn, 1914);

Louis Norbert: A Two-Fold Romance (London: John Lane, Bodley Head / New York: John Lane, 1914);

The Ballet of the Nations: A Present-day Morality (London: Chatto & Windus, 1915; New York: Putnam, 1915);

Peace with Honour: Controversial Notes on the Settlement (London: Union of Democratic Control, 1915);

Satan the Waster: A Philosophic War Trilogy (London: John Lane, Bodley Head / New York: John Lane, 1920);

The Handling of Words and Other Studies in Literary Psychology (London: John Lane, Bodley Head, 1923; New York: Dodd, Mead, 1923);

The Golden Keys and Other Essays on the Genius Loci (London: John Lane, Bodley Head, 1925; New York: Dodd, Mead, 1925);

Proteus; or, the Future of Intelligence (London: Kegan Paul, Trench, Trübner / New York: Dutton, 1925);

The Poet's Eye: Notes on Some Differences between Verse and Prose (London: Hogarth Press, 1926; Folcroft, Pa.: Folcroft Library, 1974);

For Maurice: Five Unlikely Stories (London: John Lane, Bodley Head, 1927; New York: Arno, 1976);

Music and Its Lovers: An Empirical Study of Emotional and Imaginative Responses to Music (London: Allen & Unwin, 1932; New York: Dutton, 1933).

Collections: *A Vernon Lee Anthology: Selections from the Earlier Works Made by Irene Cooper Willis,* with an explanatory note by Lee (London: John Lane, Bodley Head, 1929; Folcroft, Pa.: Folcroft Library, 1977);

Etudes et Réflexions sur l'Art, edited by Berthe Noufflard (Paris: Editions Corrêa, 1938);

The Snake Lady, and Other Stories, edited by Horace Gregory (New York: Grove, 1954);

Supernatural Tales: Excursions into Fantasy, introduction by Irene Cooper Willis (London: Owen, 1955); republished as *The Virgin of the Seven Daggers* (London: Transworld, 1962);

Pope Jacynth and More Supernatural Tales: Excursions into Fantasy (London: Owen, 1956).

OTHER: C. P. Stetson, *La donna et l'economia sociale,* introduction by Lee (Florence: Barbèra, 1902);

Saint Mary Magdalen, *The Life of Saint Mary Magdalen,* introduction by Lee (London & New York: John Lane, 1904);

"Notes on Arthur Lemon," in *Memorial Exhibition of Works by the Late Arthur Lemon* (London: Goupil Gallery, 1913), pp. 5–9;

"The Democratic Principle and International Relations," in *Towards a Lasting Settlement,* edited by Charles R. Buxton (London: Allen & Unwin, 1915; New York: Macmillan, 1916), pp. 201–216;

Richard Wolfgang Semon, *Mnemic Psychology,* introduction by Lee (London: Allen & Unwin, 1923);

Clementina Anstruther-Thomson, *Art and Man: Essays and Fragments,* edited, with an introduction, by Lee (London: John Lane, Bodley Head, 1924; New York: Dutton, 1924);

An Exhibition of Drawings by Aubrey Waterfield, introduction by Lee (Aulla, Italy, 1927);

"J. S .S. In Memoriam," in *John Sargent,* by Evan Charteris (London: Heinemann, 1927), pp. 233–255;

Irene Forbes-Mosse, *Don Juan's Daughters, Together with Dream Children and The Burden,* preface by Lee (London: John Lane, 1930).

SELECTED PERIODICAL PUBLICATIONS – UNCOLLECTED: "Les Aventures d'une pièce de monnaie," *La Famille* (Lausanne), no. 10 (May 1870): 233–237; no. 12 (June 1870): 268–271; no. 14 (July 1870): 327–334;

"Vivisection: An Evolutionist to Evolutionists," *Contemporary Review,* 41 (May 1882): 788–811;

"Sketches in Tangier," *New Review,* 2 (March 1890): 221–228;

"A Worldly Woman (Part I)," *Contemporary Review,* 58 (October 1890): 520–541;

"A Worldly Woman (Part II)," *Contemporary Review,* 58 (November 1890): 693–711;

"On Modern Travelling," *Macmillan's Magazine,* 69 (February 1894): 306–311;

Review of *Florentine Painters of the Renaissance,* by Bernard Berenson, *Mind,* new series 2 (1896): 270–272;

"An English Writer's Notes on England," *Atlantic Monthly,* 84 (July 1899): 99–104;

"The Art and the Country: Tuscan Notes," *Contemporary Review,* 77 (April 1900): 541–555;

"An English Writer's Notes on England," *Atlantic Monthly,* 88 (October 1901): 511–519;

"In the Tuscan Maremma," *Harper's Magazine,* 106 (January 1903): 237–244;

"Psychologie d'un écrivain sur l'art (observation personnelle)," *Revue Philosophique,* 56 (September 1903): 225–254;

"Essais d'esthétique empirique: l'individu devant l'oeuvre d'art," *Revue Philosophique,* 9 (January–February 1905): 43–60, 133–146;

Violet Paget in 1871, the year after her first historical sketches were published in La Famille, *a Swiss newspaper*

"Warwickshire," *Planet,* 23 January 1909, pp. 23–24;

"The Sense of Nationality," *Nation* (London), 12 (12 October 1912): 96–98;

"An English Writer's Notes on England: Things of the Past," *Scribner's Magazine,* 54 (August 1913): 177–194;

"An English Writer's Notes on England: Things of the Present," *Scribner's Magazine,* 54 (November 1913): 609–619;

"An English Writer's Notes on England: The Celtic West (Cornwall, Wales, Ireland)," *Scribner's Magazine,* 54 (December 1913): 712–724;

"The Effect of Italy," *Literary Digest,* 49 (11 July 1914): 65;

"Ghosts in a Roman Photograph Album," *English Review,* 30 (May 1920): 431–438;

"Dark, Many-Towered Bologna," *North American,* 216 (July 1922): 83–88.

Because she wanted recognition for her work on aesthetics and politics, Vernon Lee sometimes complained of the popularity of her travel essays. Yet her seven collections of travel writing composed between 1897 and 1925 represent not only the excellence of Lee's writing but the transformation of the travel essay. Unlike the pseudoscientific journalism of Victorian travelers to foreign lands and the narrowly focused guides to art and architecture of Europe, Lee's essays helped make travel literature into an artistic genre. Like other literary artists of the late nineteenth century, Lee developed an impressionistic manner of description as a point of departure for self-reflection and intellectual inquiry. In addition, Lee offered an expatriate's and internationalist's perspective on Europe before World War I. Her descriptions of days spent exploring odd corners of mostly well-known towns in western Europe reflect her love of nature and her understanding of the bonds that underlie apparent differences between the histories and cultures of various nations.

Born Violet Paget to Henry Ferguson Paget and his wife, Matilda, expatriate British parents living in France, Lee's childhood was spent moving from one place to another on the Continent, where middle-class accommodations were more affordable than in England. For the first ten years of her life Lee and her family spent most of their time in Germany. She received instruction from a series of governesses and tutors as well as from her mother and half brother, Eugene Lee-Hamilton. In Nice in the winter of 1866–1867 the Pagets met the Sargent family. Lee became close friends with future painter John Singer Sargent and his sister, Emily, and at Mrs. Sargent's instigation, the two families agreed to spend the next winter together in Rome. According to Lee, Mrs. Sargent taught her to pay homage to the "Spirit of Localities," something Lee was later to call the *Genius Loci*. Mrs. Sargent's enthusiasm for sightseeing, her excursions in and around Rome with her own children and Lee, and her vivid stories about other places sparked Lee's imagination. Lee and the Sargent children learned about Italian history and culture by collecting ancient coins and bits of marble. These antiquarian interests and her early decision to pursue a literary career culminated in a short story, "Les Aventures d'une pièce de monnaie," published just before her fourteenth birthday.

Encouraged by her brother and mother, Lee sought out English and Italian publishers for her essays on Italian music and literature. These essays, published in 1880 as *Studies of the Eighteenth Century in Italy,* established Lee as a writer and scholar. She began regular visits to London and became close friends with Walter Pater, his sister, and many other important artists and writers in the Aesthetic Movement. A prolific writer and a strong personality, Lee apparently intimidated and offended people as readily as she made friends and associates among English, French, German, and Italian intellectuals.

She wrote in many genres, including the novel, and impressed her contemporaries with works such as *Baldwin: Being Dialogues on Views and Aspirations* (1886) and *Althea: A Second Book of Dialogues on Aspirations and Duties* (1894), which considered aesthetic and ethical problems in the form of philosophical dialogues. Although her literary style owes much to Pater and John Addington Symonds, her interests extended to political issues, including questions of international relations and women's rights. In the late 1890s, in addition to reviews and essays on aesthetics and ethics, she began writing weekly travel essays for the *Westminster Gazette*. Her first collection, *Limbo and Other Essays* (1897), introduces her philosophical approach to travel writing.

Less cosmopolitan than Henry James in her love of travel, Lee introduces her collection with an allusion to Dante's early fourteenth-century *Divina Commedia*. Limbo, she recalls, is a place for worthy people who have missed a necessary rite or have been born before the time of Christ. The traveler is likewise a worthy but excluded outsider, and traveling provides a kind of limbo for those who have not found peace or kindliness or justice in the world. Limbo thus affords a refuge as well as a place of exile from the immediate and fashionable world.

The first essay, "In Praise of Old Houses," received special attention from reviewers because of its perceptive commentary on why the past appeals to the modern traveler. According to Lee, nature does more for man than humankind does, and yet the old towns and houses of human social life can become natural objects: "Time gives them infinitely more variety and charm," as the crumbling walls of old buildings remind one of childhood or of the common joy and suffering of humanity. Lee's passion for the past is not nostalgic, however. People do not make the past charming; instead, the past makes people charming. Such subtly ironic observations undermine the superficial effusions of ordinary travel writing for tourists.

In addition to celebrating a sense of the past, Lee posits an impersonal, or formalist, attitude toward landscape. In concentrating on the "tilt of a roof," "the bend in a road," or the "undulation of a field," Lee emphasizes the subjective claims of the physical texture and the psychological effects of the landscape. In painting as well as in traveling, Lee prefers "topographical charm" to the light and atmosphere preferred by the Impressionists. Like a latter-day Romantic, Lee values the influence of the imagination more than that of the perceiving eye. The mind's power of association connects mountains, valleys, and rivers to sentiment and fancy, and many of her essays explore the relations of landscape and sentiment.

For instance, Lee contends in "On Modern Traveling" that she prefers "obscure places . . . and people who provide the sentimental traveler with droll and melancholy and perhaps chivalrous stories." Thus, she appreciates Robert Browning's choice of Asolo for the setting of "Pippa Passes," for his poem gives the place an otherwise unwarranted interest. According to Lee, "the cosmopolitan abroad desists from flannel shirts because he is always at home," but those who are attracted to a place by the spirit that pervades it – by the shadows of its history and legends rather than its present – those who are willing to live "prosaically" in inns and boardinghouses in order to see an event – a procession, a street fight, or a pilgrimage – are those who are apt to discover what is "particularly characteristic" about the place.

The specificity that makes Lee's impressions valuable as observations depends on her knowledge of social history as well as of art and literature. In "Old Italian Gardens" she notices how both nature and social organization have contributed to the peculiar beauty of those gardens. The heat of the intense sun and the wealth of the Pope's nephews created the conditions for Italian potted flowers, fountains, and garden sculptures such as those of Gian Bernini:

> mistaken as indoor decoration, as free statuary in the sense of the antique, this sculpture has after all given us the only works which are thoroughly right in the open air, among the waving trees, the made vegetation which sprouts under the moist, warm Roman sky, from every inch of masonry and travertine.

Many of the old gardens, Lee knows, have been abandoned – their gates, like the world they represented, have been left leading nowhere.

Lee's historical interests grow philosophical in "About Leisure." Meditating on a photograph of "St. Jerome in his Study," Lee understands the importance not of "leisure time," but of "time in which we can feel at leisure." She argues for a way of life in which not only the material, or even intellectual, but also "the special, mysterious commodity called charm" sustains people. If most people cultivate themselves as vegetables are cultivated for market,

STUDIES

OF THE

EIGHTEENTH CENTURY IN ITALY,

BY

VERNON LEE.

LONDON:
W. SATCHELL AND CO.,
12, TAVISTOCK STREET, COVENT GARDEN.

1880.

Title page for Lee's first book

Lee insists on leisure to cultivate the soul, for "the type of all leisure is art." She thus suggests how her methods and ends are universal. Unlike the travel writer whose graceful descriptions of places provide a kind of elitist guidebook for the leisured classes, Lee conveys the ethical value of even simple forms of aesthetic experience.

Some have accused Lee of being an aesthete interested only in art for art's sake in her early writings and of being a tiresome moralist in her later writings; however, her travel essays demonstrate a fine balance of Paterian rhetoric and a new woman's independence of thought. Critics of *Limbo and Other Essays* were generally appreciative. One reviewer described her style as "buoyant, lucid, redolent of warm airs," and another complimented its refined thought and beautiful prose. F. B. R. Hellems in *Dial* declared that Lee is "master of an easy, almost conversational style, but is more than the unusually clever woman she seems – she is clairvoyant and sympathetic," but he also speculated that those who did not share "her good fortune in leisure and travel" would not enjoy her writing. Yet Lee's travel essays remained popular, perhaps because her quest was as much spiritual as geographical. To emphasize the intangible quality of her travels, Lee's next volume invoked a divinity, the Genius

Loci, who served as Lee's muse and guide in all her subsequent travels.

Genius Loci: Notes on Places (1899) suggests how travel writing must have appealed to or at least consoled Lee for the often uneasy relationships she had with her family and closest friends. She compares the friendship of places to friendship with people: "Charming us, . . . raising our spirits, . . . [and] subduing our feelings into serenity and happiness . . . are the highest gifts of our human affection; and surely we receive them equally, nay, sometimes even better, from the impersonal reality whom I call, for want of a better name . . . the *Genius Loci*." This "impersonal reality" is no personification but a "spiritual reality" imbuing the peculiar features of a landscape or surroundings.

In *Genius Loci* Lee's travels include Germany and France as well as Italy. She emphasizes the national characteristics of each country and gives each a personality. Her Germany, for instance, is "not the one which colonises or makes cheap goods, or frightens the rest of the world in various ways; but the Germany which invented Christmas trees, and [Jacob and Wilhelm] Grimm's Fairy Tales, and [Johann Sebastian] Bach, and [Wolfgang Amadeus] Mozart, and which seems to be vouched for in a good many works of classic literature." In fact, almost all of Lee's journeys in Germany highlight its romantic character. Because she idealized German culture, her anguish grew as tensions increased among the European powers before World War I. Her pacifism was ultimately grounded in her deep feeling of friendship for all the places that she had known and written about.

If Germany evoked associations with literary works and musicians rather than commerce, France — especially southern France — presented contradictory impressions. On the one hand appeared the French genius "for turning into a kind of poetry the peaceful sensual needs of life," and on the other, the history of French intolerance and insensitivity to the sufferings of other people. Italy, Lee's home for most of her life, provided the most consistently welcome and interesting places to explore. *Genius Loci* introduces the reader to Siena through the medieval works of Simone Martini and to the statuary of Michelangelo through a description of a marble quarry in Tuscany. Lee's itinerary — including Germany, France, Italy, and later Switzerland and England — never varied. Despite having made a brief journey to Greece and having offers to go to more-exotic places, Lee wrote only on the Europe that was familiar to her. Her familiarity with these places allowed her to experience their differences from each other only as differences in temperament. As she says in the last essay of *Genius Loci,* "one wants, if one really cares for places . . . to feel what the life of that particular place has been striving after through the uneasy centuries — what has been, to put it pedantically, the formula of its evolutions."

Lee's commitment to her general impression of places may have been, as Burdett Gardner has claimed, part of her compulsive efforts to distance and sublimate her passion for real people. Her father and mother died in 1894 and 1896, respectively. The brother whom she had nursed for many years recovered, only to leave for the United States; Clementina Anstruther-Thomson, the friend who had supported Lee through her nervous depressions for twelve years, apparently tired of the relationship, and after 1900 she visited and corresponded with Lee only occasionally. Such personal losses must have encouraged Lee's investment in the unchanging familiarity of certain settings.

The Enchanted Woods, and Other Essays on the Genius of Places (1905) makes a case for enriching life "not by making far-fetched plans, not by the seeking of change and gain; but by the faithful putting to profit of what is within our grasp." What sounds like resignation or muted ambition in this preface, however, is also Lee's response to imperialism in the Edwardian years. According to Peter Gunn, she outspokenly criticized Italian political adventures in East Africa, as even earlier she had opposed British policy in the Boer War. Her celebration of "putting to profit what is within our grasp" presents an isolationist's view of travel rather than the expansionist views of other travel writers.

The essays in *The Enchanted Woods* elaborate the themes of Lee's earlier collections. Although the importance of place continues to supersede that of people, Lee shows that the Genius Loci also grants "very human favors." The kindness of the people of a Swiss village more than compensates Lee for the absence of her friend, and Lee praises Italy for being "the land of friends, real, living, incomparable . . . revealed in their everyday talk." An exhibition in Paris, however, offends her sensibilities. She criticizes the "disregard to all sense of geography implied in the gazebos of the Rue des Nations (let alone disregard to all other decent feelings to bringing over savages to stare at)." Since only the immediate and particular experience of physical and cultural conditions matter to Lee's sense of place, the exhibition seems to her a "stupid, wicked carnival sacrilege towards the *Genius Loci*."

In contrast to the simulated travels presented by the Paris exhibition, Lee's travels deepen her understanding of her own culture. Thus, a house in Savoy more perfectly accords with Lee's sense of the romantic past than her rereading of Jean-Jacques Rousseau. The dreamy atmosphere surrounding a castle in the Euganean Hills seems more important than the fact that she knows the castle owners. According to Lee, "among the many pleasant things of travel, me thinks we should include, as so much to the good, that which our fancy adds to places."

Yet the material means of travel, such as the motorcar, affect the fancy. The motorcar, as a new mode of travel reserved for the wealthy, convinces Lee "of the real unreality of things, of their becoming only because we happen to see them." As motorcars slow for wagons and dogcarts, Lee feels a sense of "enlarged brotherhood," and when they begin moving faster she is exhilarated by their "triumph over steepness." Nevertheless, she mourns the loss of the "whole real shape of the earth's surface and the complexities of its ways." Such travel turns the worship of the Genius Loci into a spiritual snapping of a photograph, another new technological achievement that provides an apt analogy for the unusual sensations of early motorcar travel. In a different essay, on Maremma, Lee finds that even riding in a saddle seems somehow to compromise the spirit of travel. Only with the feet can one possess a country. Thus, in Viterbo "it was the consistence of the earth which told me, even before the shapes of the hills and houses, that I was back on Roman territory."

Though Lee insists that an experience of place is necessarily subjective, a scene also must provide something tangible to stimulate one's response. In Tuscany the smell of burning olive twigs allows her to create her vision of Greece and Sicily, "the real South which I shall never go to." The celebration of what is at hand, of the everyday and humble, in these essays becomes the occasion, and perhaps the necessary ground, for Lee to express her vision of the world.

In 1903 Edith Wharton asked for Lee's help with her book on Italian villas and gardens. Percy Lubbock described Lee's style of observation as the two women toured a villa near Siena:

> There was no hurrying of Vernon Lee while her inexhaustible mind was at work, absorbing and straining and philosophizing an impression of beauty, were it the twentieth time she stood in that garden and searched its appeal; no experience, be it ever so familiar, was to be dismissed as finished and settled in the past, its history was always in the making.

The place that aroused these qualities of Lee's personality and style was Rome. Her next volume of travel essays, *The Spirit of Rome: Leaves from a Diary* (1906), reflects her love of the city where Mrs. Sargent had first introduced her to the Genius Loci. In her preface Lee acknowledges her long personal associations with Rome, and her essays describe what Rome has come to represent to her after so many years. Although only a page or two long, most of the entries have the compressed intensity of a poem by John Keats. Describing the wall of a villa, Lee writes of its "fragments of delicate dolphined frieze."

Time provides the theme of *The Spirit of Rome*. The churches are "inhabited" by generations; the cypresses of the graveyard "do not seem to unite folks with the terrible unity Death, so much as with the everlasting life of the centuries"; and the city is "a gigantic stage, splendidly impressive to eye and fancy, where Time has strutted and ranted, and ever will continue." The continuity of the spirit of Rome comes in part from its literary associations. The "dreary, squalid" Palazzo Cenci with its sordid story of Renaissance tyranny and patricide is enveloped with the "power of romance" through the poetry of Percy Bysshe Shelley, as is the site of Browning's poem on Renaissance decadence, "The Bishop Orders His Tomb at Saint Praxed's Church."

If the timelessness of Rome persists through its art and atmosphere, Lee recognizes that the narrowly historical-minded reader does not share her interest in "the metaphysical essence of the past." She criticizes a friend's desire to see the site where Julius Caesar fell: "I cannot say how deeply, though vaguely, I felt the meaningless tragic triviality of these successive generations of reality, in the face of that solemn, meaningful abstraction which we call history, which we call humanity, the centuries, Rome." Lee's confidence in meaningful abstractions remained, and so did the tragedy of political assassinations.

Already holding an increasingly marginal position in post-Victorian literature, Lee published *The Sentimental Traveller: Notes on Places* (1908), a work that Virginia Woolf reviewed negatively. Lee's title recalls the persona of Henry James in his travel writings of the 1880s and 1890s. Both James's and Lee's essays express subjective states: "Sentiment remains, but like wonder and belief, changes its object." Uninterested in what the tourist sees, the roads Lee travels return her to her "innermost self, the self which never changes.... These places, these hills and rivers and houses, possess me as before, their power unchanged." Although modernist writers were also to explore the inner experience of

A sketch of Lee by John Singer Sargent (Ashmolean Museum, Oxford University)

their characters, they were more skeptical of the "meaningful abstractions" that Lee cultivated.

In private Woolf worried that her own writing was turning vaporous in the style of Vernon Lee, a style that Woolf's review of *The Sentimental Traveller* characterizes as being egotistic and impressionistic, and Woolf enjoins Lee to pursue the greater value of concentrating on "real people, for their characters are more profoundly realised." Woolf's criticisms of Lee's writing in *The Sentimental Traveller* have some truth: both geographically and thematically Lee's wanderings have little focus. Nevertheless, given Lee's priorities as a travel writer interested in the continuities of history and humanity, the volume contradicts the idea that personal (or national) identity limits consciousness.

In dedicating *The Sentimental Traveller* to Irene Forbes-Mosse, Lee's neighbor and traveling companion, Lee notes the value of having friends of different nationalities. Just as books add meaning to the visually perceived landscape, friendships add meaning to books. As the essays in the volume at-

test, Lee enjoyed visiting towns associated with Forbes-Mosse's ancestors such as Bettina von Aemin, whose correspondence with Johann Wolfgang von Goethe provided Lee with many anecdotes. But the significance of these associations between places and books and between books and friends of different nationalities also allows Lee to criticize what she describes in the preface as "our times of dog-in-manger and bluster-to-keep-up-your-courage nationalism." She traces the legendary associations of the river Narni with the Sybil, or she uses the myth of Icarus to suggest the spirit of the Euganean Hills. Roads are as "rivers made by human hands, down which language, art, customs, civilization flow." By idealizing the spirit of place, Lee is intentionally sentimental in order to recall the cultural richness of the European past to her readers.

If her journeys do not lead her to the untrodden paths of earlier travel writers in foreign lands, her fancy sometimes leads to danger and adventure. In Paris, where she visits the graveyards of members of Forbes-Mosse's family who had been guillotined, Lee empathizes with them in their fate by imagining – ironically, given her interest in socialism at the time – her own death at the hands of workmen or philanthropic friends. She also pays tribute to a heroic life in her portrait of a deceased friend: Madame Blanc-Bentzon combined the "vanished past" of an eighteenth-century salon intellectual with "a future still existing only in dreams."

For Lee, the act of traveling carries penalties: "The soul suffers from all this parting." Yet in paying respect to the places she has known, Lee invites the reader to sympathize with her feelings of "happiness tinged with respect and gratitude." Reading travel literature in this way should prepare readers to "journey through the world, or through their lives, less as bagmen [traveling salesmen] than as Sentimental Travellers."

During these years Lee was also publishing essays on aesthetics, a play, two novels, and many short stories, and her interest in the art of writing and in the lives of artists is apparent in the many allusions and associations that inform her travel essays – especially those of her next collection, *The Tower of the Mirrors, and Other Essays on the Spirit of Places* (1914). In this work Lee uses literary and intellectual figures from the past not only to enhance the atmosphere of the place but also to gauge how far the modern world has fallen. Lee had spent most of the six years following her publication of *The Sentimental Traveller* at home in Tuscany, where she was active in charitable endeavors and extensive correspondence. Her international standing had given her grounds for hoping that she could influence those who were anticipating war. In the context of her protests against the Italian-Turkish conflict and the buildups of arms in France, Germany, and England, *The Tower of the Mirrors* presents a renewed plea for wider sympathies and multiple perspectives.

The title is taken from the poet Virgil, but Lee presents the image of a tower of mirrors in terms of "the universal mirrorings by which all outside things exist as we know them only in the reflecting and refracting mirrors of our memory and our emotions." Such a perspective requires "the trouble to ascend into towers, and go to the expense of furnishing our soul with as many mirrors as possible, and a steady well-oiled winch wherewith to turn them in some directions and avert them from others." Lee's inward eye depends on a varied, complex view of the world, and the essays of the collection reveal how her perspective is refracted by the inadequacies of modern life.

The city of Poitiers exemplifies the evils of "half-modern France" in the "choking and deviating of the country's life for the sole benefit of Paris, and the loss of tradition, of vital organization in this sterilizing process." The condition of Switzerland seems better than that of bigger, busier countries, because its people are "still peasants at bottom" and the country benefits from its combined rusticity and civilization. The house of Swiss theologian and mystic Johann Lavater in Zurich provides the occasion to "rest a moment in the thought of those dear narrow times when trees were planted and houses built, and poems composed, in leisure and for all eternity." She praises Wetzlar, in western Germany, only because it provided the settings for romances by Goethe and Browning. But the gardens of Jena in eastern Germany fall short of their association with intellectual history: the works of science that were undertaken at such a place should be romantic and should serve the purposes of peace.

In this volume even Italy appears in a negative light. While recovering from influenza in Pisa, Lee idealizes the "sea magic" of the place about which Shelley and Byron had written, yet she wonders what will redeem her generation in the eyes of posterity so that modern remains may appear "exquisitely lichened and nobly overgrown." Bocca D'Arno has been "desecrated by mankind's sadly brutal love of air and war," and only the dockyards of Viareggio retain their moral charm through the making and mending of its boats. Such a place undermines the "prostitution of Italy to idle strangers," for "one can think of the real and the present instead of blinking them." Lee ends the volume by describing a fellow traveler on the train from Rome to Pisa who sleeps during the day and reads his newspaper at night: "He strikes me as vaguely allegorical of so much traveling through the world, and also, even more through life." While he tries to read news of events that will be altered on the morrow, he misses the sunlight "along the Etruscan sea, among the asphodels of the pasture and the myrtle thickets of the hillsides."

Lee's outspoken pacifism during the war offended many old friends. Staying in the English countryside or in rooms in Chelsea, she collaborated with intellectuals such as Bertrand Russell to oppose the war. A few sympathetic journals were willing to publish her work, and she concentrated most of her literary energies in writing a satire, *Satan the Waster: A Philosophic War Trilogy* (1920). Gunn has praised this work as a "brilliant analysis of the psychological effects of war on individuals" and "perhaps the most complete expression of the many sides of [Lee's] complex personality." After the war she returned to Italy and slowly mended fences with those who respected her sincerity. She also acted as a mediator between friends from different sides of the conflict. She worried about the ill effects of the defeat of Germany and about the continued hostility of the Allies toward their former enemies. The rise of Benito Mussolini and Adolf Hitler was to confirm her mistrust of political leaders and movements.

Although deafness kept Lee from enjoying much social life, she continued writing. A collection of her essays on aesthetics, *The Handling of Words and Other Studies in Literary Psychology* (1923), drew commentary from leading academics such as I. A. Richards and, later, René Wellek. In 1925 Lee published *The Golden Keys and Other Essays on the Genius Loci*, her last collection of travel essays. Although most of the pieces in this volume were written before the war, their tone is elegiac. As the preface and the final essay clearly show, Lee had sensed the tremendous consequences that the war would have not only for the places she had visited but also for the survival of the Genius Loci. In the preface she writes:

> The *Genius Loci* is that portion of nations and civilizations which, while it speaks aloud in their philosophy and poetry and music, and is written clearly in the shapes of their buildings, addresses itself to the initiate mind in their humbler habits, kindly and gracious, sometimes childish and funny.

Letter from Lee to Joseph Conrad (courtesy of the Lilly Library, Indiana University)

But even the humbler aspects of life in Europe had turned melancholy for Lee. If an old house in France reminded her of another old house that she had loved in Italy twenty-five years earlier, remembrance was no longer an art but "a grace" needed to replenish the "shrinking present and future ... from the heaped-up past." Memories, personal and cultural, had always been a way for Lee to connect individual places or national cultures with the transcendent needs and desires of humanity. Her essay on "The Old Bologna Road" reminds her readers of the beauty and importance of the road that carried not only Napoleon Bonaparte and Giuseppe Garibaldi but also Johann Winckelmann and Goethe, Byron, Shelley, and Browning. In the same way, a Roman road in England, though utterly English in its utilitarian perfection, recalls the past unity of diverse countries such as England and Italy.

One of the last essays in *The Golden Keys* signals the irreversible changes that war had wrought for disciples of the Genius Loci. Lee recalls an inscription that she had seen at the villa of Tiberius and that notes the seconds required for a stone to drop to the sea from the spot where Tiberius, the second emperor of Rome, had cast his slaves. The remembrance makes her ponder how history serves tourism: "Will *our* horrors also, so immeasureably greater and more scientific than those of poor artless Antiquity, amuse the leisure of peaceful future generations? Such at least seems the only durable result of wars and massacres: sale of *souvenirs* and motor trips to Marathon and Waterloo." Unable to sanction or to reverse such a rewriting of Europe's past, in her last essay Lee acknowledges the loss of the world that she had known and written about. She imagines her generation, like the people of Messina returning after the earthquake of 28 December 1908, clambering up and down Europe after the war, "unable to guess under which mound of plaster and of rags there lies our treasure and so much of our heart."

Lee had almost enough notes about places to write another book, but these notes remained fragmentary and unpublished. They indicate clearly that even after the war she found those places she had visited still to be haunted by associations with the past. Several notes mention Ravenna, both for its surviving medieval castle and for its debased postwar condition. Longer notes on a French-style cottage that she passed while living in England express her sense of having been something of a refugee in her own country. Although Lee could occasionally revive her interest in the spirit of places, it is clear that the pleasure of travel and of writing about her travels was exhausted. Such writing belonged to an earlier Vernon Lee and an earlier epoch. In her last years Lee finished work on *Music and Its Lovers: An Empirical Study of Emotional and Imaginative Responses to Music* (1932) and a few collections of short stories. In February 1935 she died in her home at Tuscany.

Aldous Huxley wrote appreciatively of *The Golden Keys,* "I tend to like best the papers about the places I know myself.... They recall old delights and make one observe details which I had not noticed or insufficiently appreciated." He also regarded highly Lee's comments on the Genius Loci and its influence in the life of ordinary people. In comparing Lee's work to that of William Wordsworth, Huxley noted: "Of the theory and practice of the Natural Pieties [Lee's] books are a most delicate and beautiful exposition." Lee's travel writings remain, like half-forgotten landmarks, essential for the recuperation of a literary genre and a cultural landscape. They contribute to an understanding of the interplay of nature and culture, and they testify eloquently to a "disinterested interest in Men and Things."

Letters:

[Irene Cooper Willis, ed.], *Vernon Lee's Letters* (London: Privately printed, 1937).

Bibliographies:

Phyllis F. Mannocchi, "'Vernon Lee': A Reintroduction and Primary Bibliography," *English Literature in Transition 1880-1920,* 26, no. 4 (1983): 231-267;

Carl Markgraf, "'Vernon Lee': A Commentary and Annotated Bibliography of Writings about Her," *English Literature in Transition 1880-1920,* 26, no. 4 (1983): 268-312.

Biography:

Peter Gunn, *Vernon Lee: Violet Paget 1856-1935* (London: Oxford University Press, 1964).

References:

Richard Cary, "Aldous Huxley, Vernon Lee and the *Genius Loci,*" *Colby Library Quarterly,* 5 (June 1960): 128-140;

Burdett Gardner, *The Lesbian Imagination (Victorian Style): A Psychological and Critical Study of "Vernon Lee"* (New York: Garland, 1987);

Percy Lubbock, *Portrait of Edith Wharton* (London: Cape, 1947; New York: Appleton, 1947);

Virginia Woolf, "The Sentimental Traveller," in *The Essays of Virginia Woolf,* edited by Andrew McNeillie (New York: Harcourt Brace Jovanovich, 1986), I: 157-158.

Papers:

The Vernon Lee Collection is at Colby College, Waterville, Maine.

Agnes Smith Lewis
(11 January 1843 – 23 March 1926)

and

Margaret Dunlop Gibson
(11 January 1843 – 11 January 1920)

Barbara Brothers
Youngstown State University

BOOKS: *Eastern Pilgrims: The Travels of Three Ladies,* by Lewis, illustrated by Gibson (London: Hurst & Blackett, 1870);

Effie Maxwell, 3 volumes, by Lewis (London: Hurst & Blackett, 1876);

Glenmavis, 3 volumes, by Lewis (London: Hurst & Blackett, 1879);

The Brides of Ardmore: A Story of Irish Life, by Lewis (London: Stock, 1880);

Glimpses of Greek Life and Scenery, by Lewis (London: Hurst & Blackett, 1884);

Through Cyprus, by Lewis (London: Hurst & Blackett, 1887);

Life of the Rev. Samuel Savage Lewis, by Lewis (Cambridge: Macmillan & Bowes, 1892);

How the Codex Was Found: A Narrative of Two Visits to Sinai, from Mrs. Lewis's Journals, 1892–1893, edited, with an introduction, by Gibson (Cambridge: Macmillan & Bowes, 1893);

Two Unpublished Letters, by Lewis (Cambridge: Privately printed, 1893);

A Lady's Impressions of Cyprus in 1893, by Lewis (London: Remington, 1894);

In the Shadow of Sinai: A Story of Travel and Research from 1895 to 1897, by Lewis (Cambridge: Macmillan & Bowes, 1898);

Light on the Four Gospels from the Sinai Palimpsest, by Lewis (London: Williams & Norgate, 1913);

Margaret Atheling, and Other Poems, by Lewis (London: Williams & Norgate, 1917).

OTHER: Panagiotes G. Kastromenos, *The Monuments of Athens: An Historical and Archaeological Description,* translated by Lewis (London: Stanford, 1884);

Euripides, *The Alcestis of Euripides,* translated by Gibson (London: Williams & Norgate, 1886);

Margaret Dunlop Gibson

James Young Gibson, *The Cid Ballads and Other Poems and Translations from Spanish and German,* 2 volumes, edited by Gibson, with a memoir of James Gibson by Lewis (London: Kegan Paul, Trench, Trübner, 1887);

A Translation of the Four Gospels from the Syriac of the Sinaitic Palimpsest, translated by Lewis (London: Macmillan, 1894);

The Four Gospels in Syriac, Transcribed from the Sinaitic Palimpsest, translated and edited by Robert Lubbock Bensly, J. Rendel Harris, and F. Crawford Burkitt, with an introduction by

Lewis (Cambridge: Cambridge University Press, 1894);
Catalogue of the Syriac Manuscripts in the Convent of S. Catherine on Mount Sinai, Studia Sinaitica, no. 1, edited by Lewis and Harris (London: Clay, 1894);
An Arabic Version of the Epistles of St. Paul to the Romans, Corinthians, Galatians . . . , Studia Sinaitica, no. 2, edited by Gibson (London: Clay, 1894);
Catalogue of the Arabic Manuscripts in the Convent of S. Catherine on Mount Sinai, Studia Sinaitica, no. 3, compiled by Gibson (London: Clay, 1894);
Some Pages of the Four Gospels Re-transcribed from the Sinaitic Palimpsest, Studia Sinaitica, no. 4, by Lewis (London: Clay, 1896);
Apocrypha Sinaitica, Studia Sinaitica, no. 5, edited and translated by Gibson (London: Clay, 1896);
A Palestinian Syriac Lectionary, Containing Lessons from the Pentateuch, Job, Proverbs, Prophets, Acts, and Epistles, Studia Sinaitica, no. 6, edited by Lewis, with a glossary by Gibson and notes by Eberhard Nestle (London: Clay, 1897);
The Story of Ahikar from the Aramaic, Syriac, Arabic, Armenian, Ethiopic, Old Turkish, Greek and Slavonic Versions, edited by Lewis, Frederick C. Conybeare, and Harris (London: Clay, 1898; revised and enlarged, 1913);
The Palestinian Syriac Lectionary of the Gospels, Re-edited from Two Sinai Mss. and from Pre de Lagarde's edition of the Evangeliarium Heirosolymitanum, by Lewis and Gibson (London: Kegan Paul, Trench, Trübner, 1899);
An Arabic Version of the Acts of the Apostles and the Seven Catholic Epistles, Studia Sinaitica, no. 7, edited by Gibson (London: Clay, 1899);
Select Narratives of Holy Women, from the Syro-Antiochene or Sinai Palimpsest, as Written above the Old Syriac Gospels, Studia Sinaitica, nos. 9–10, 2 volumes, edited and translated by Lewis and Francis Burkitt (London: Clay, 1900);
Palestinian Syriac Texts from Palimpsest Fragments in the Taylor-Schechter Collection, edited by Lewis and Gibson (London: Clay, 1900);
Apocrypha Arabica, Studia Sinaitica, no. 8, edited and translated by Gibson (London: Clay, 1901);
Apocrypha Syriaca: The Protevengelium Jacobi and Transitus Mariae, Studia Sinaitica, no. 11, edited and translated by Lewis (London: Clay, 1902);
The Didascalia Apostolorum in Syriac . . . , Horae Semiticae, no. 1, edited by Gibson (London: Clay, 1903);
The Didascalia Apostolorum in English, Horae Semiticae, no. 2, translated by Gibson (London: Clay, 1903);

Agnes Smith Lewis

The Mythological Acts of the Apostles, translated and edited by Lewis (London: Clay, 1904);
Supplement to the Palestinian Syriac Lectionary, by Lewis (Cambridge: Cambridge University Press, 1906);
Forty-one Facsimiles of Dated Christian Arabic Manuscripts, Studia Sinaitica, no. 12, edited and translated by Lewis and Gibson, with an introduction by David S. Margoliouth (Cambridge: Cambridge University Press, 1907);
Codex Climaci Rescriptus, Fragments of Sixth Century Palestinian Syriac Texts of the Gospels, Horae Semiticae, no. 8, edited by Lewis and Nestle (Cambridge: Cambridge University Press, 1909);
The Old Syriac Gospels, or Evangelion Da-Mepharreshê, Being the Text of the Sinai or Syro-Antiochene Palimpsest, edited by Lewis (London: Williams & Norgate, 1910);
The Commentaries of Isho'dad of Merv, Bishop of Hadatha (c. 850 A.D.) in Syriac and English, Horae Semiticae, nos. 5–7, 10, and 11, edited and trans-

lated by Gibson and Harris (Cambridge: Cambridge University Press, 1911-1916);

The Forty Martyrs of the Sinai Desert, and The Story of Eulogios; from a Palestinian Syriac and Arabic Palimpsest, Horae Semiticae, no. 9, edited by Lewis (Cambridge: Cambridge University Press, 1912);

Leaves from Three Ancient Qurâns, Possibly Pre-Othmanic, with a List of Their Variants, edited by Lewis and Alphonse Mingana (Cambridge: Cambridge University Press, 1914).

Before 1892, the year in which they discovered the Mount Sinai Palimpsest (or Lewis Codex), Agnes Smith Lewis and Margaret Dunlop Gibson, her twin sister had traveled to Greece, Egypt, and Cyprus. Their discovery was the event that began their careers as biblical scholars, and their journeys provided bases for the three well-written travel books they published about their experiences. The people among whom Lewis traveled regarded one of her books, *Glimpses of Greek Life and Scenery* (1884), so highly that it was translated into Greek, and her command of that language was so expert that she translated Panagiotes G. Kastromenos's *The Monuments of Athens: An Historical and Archaeological Description* (1884).

Though Lewis gave her sister a pseudonym in the narratives, Gibson was more than Lewis's traveling companion. Gibson contributed the sketches that illustrate their first book, *Eastern Pilgrims: The Travels of Three Ladies* (1870), and she later edited *How the Codex Was Found: A Narrative of Two Visits to Sinai, from Mrs. Lewis's Journals, 1892-1893* (1893), Lewis's diaries, and added an introduction. Gibson was also the photographer of the palimpsests they examined and of their travels throughout the Middle East, and she edited and translated many other works both on her own and with her sister. Living together even during their brief marriages, the two sisters shared life and work to such a degree that, as Whigham A. Price writes, the fellows of the Cambridge colleges referred to them by a single name – the Giblews. Their production of biblical scholarship and translations was amazing, and for these accomplishments the two earned many honorary degrees. Except for the first two of these, an honorary doctorate of philosophy degree and a master of liberal arts degree that Lewis received from the University of Halle in 1899, the sisters shared in this recognition: they received degrees as doctors of law from Saint Andrews University in 1901, doctors of divinity from Heidelberg University in 1904, and doctors of letters from Trinity College, Dublin, in 1911.

Agnes and Margaret Smith were born 11 January 1843 in Irvine, Scotland. Their mother, Margaret, died three weeks after they were born, and they were reared by their father, John Smith, who introduced them to the Latin, French, and German languages. Dr. William B. Robertson, John Smith's close friend and a man highly educated and fluent in many European languages, became another important person in the young girls' lives. Robertson wrote to the twins while they were attending boarding school at Irvine Academy, Birkenhead, England, and he sometimes accompanied the family on holidays abroad. The Smith girls were included in the adult male world of their father and Robertson, and that environment contributed much more to the development of their intellectual independence, faithfulness to the Presbyterian Church, fluency in language, and confidence and knowledge as travelers than the formal schooling they received at Irvine or later at the Kensington finishing school in London. Before they were sent to London in 1861 to complete their formal educations, John Smith inherited £200,000 from a distant relative. The learning and means of the two sisters made marriage unlikely: education was not an asset for women, and their wealth placed them above their social class.

Eighteen months after their father's death in 1866 the sisters decided to travel to the Near East with Grace Blyth, a young teacher with whom they had grown quite friendly while at school in Kensington. Thomas Cook was organizing trips to the Middle East, and although upper-middle-class women might subscribe to these by 1868, Lewis wrote that she and her sister decided that traveling with the services of an escort would prevent them from "gathering the information" they sought. Thus, provided with "the best of all auxiliaries – viz: a knowledge of the French, German, and Italian languages" – and accompanied by Blyth, a woman old enough to provide the respectability the party needed, the Smiths set off on the first of their six trips to the Orient, despite "the consternation expressed by our friends at the idea of three ladies venturing on so lengthened a pilgrimage alone. 'Do you think they will ever come back?' They are going amongst Mohammedans and barbarians...."

The title of their narrative of this first journey, *Eastern Pilgrims: The Travels of Three Ladies,* reflects their religious motives – "to know more of our Father's universe ... [in order to] rise to broader conceptions of Himself" – as well as their destination. Intellectual curiosity; a determination that

Frontispiece and title page for Lewis's book on Greece

their gender should not prevent them from enjoying a full, adventurous life outside the domestic walls that confined their Victorian sisters; and deep religious devotion characterized the Smiths throughout their lives. These qualities are evident in the narrative of their journey.

In the chronological record of discussions and events recounted in *Eastern Pilgrims* Lewis refers to herself in the third person as Agnes and assigns pseudonyms to her sister and their friend. Margaret becomes Violet and Blyth becomes Edith – Lewis continues to use these pseudonyms in subsequent travel books. Somewhat in the manner of E. M. Delafield describing her travels in *Straw without Bricks: I Visit Soviet Russia* (1937), Lewis records as much about the travelers, their idiosyncrasies and discussions, as about the places and people that the three visit. On the trains and boats they take in traveling first through Europe – Luxembourg, Triev, Ems, Frankfurt, Nuremberg, Vienna, Munich, Innsbruck, Salzburg, Pesth – and then to the Orient – Constantinople, the Bosporus, Stamboul, Smyrna, Patmos, Rhodes, Cyprus, Jaffa, Port Said, Alexandria, Jerusalem, Damascus – the three travelers talk with passengers.

For the first leg of their journey their companions in their train compartment are Germans. A husband declares that women should not be given the franchise, but when he dozes, his formerly silent wife expresses a contrary view. With other Germans the three British travelers discuss English cooking, which some characterize as wretched and others as healthy. On another train an elderly Italian woman talks with them about the "necessity of education amongst the working classes of Italy and England." With still other travelers they discuss religion and colonial governments – questions about whether the people of Gibraltar and the Ionian Islands, Corfu in particular, have prospered more under the English or the Greeks, and whether the

English or the French are better at governing. Lewis notes how "stiff" the English people are, how indisposed they are to learn other languages, and how Americans are so concerned with commerce, "so engrossed, they have no time to cultivate literature or poetry."

Though she remarks on the sights, the religions, and the differences in the practices and beliefs of people from various countries, Lewis does not provide the amateur sociological, anthropological, or scientific records that are typical of many other travel books, including her own later works. Instead, *Eastern Pilgrims* is the story of a journey that begins with the preparations: the purchase of luggage – a trunk made by Edward Cave, a basket "covered with strong tarpaulin," and a "small leather pormanteau called 'The Gladstone'" – and of mackintosh sheets, sidesaddles, and riding costumes. The sisters resisted taking a portable bath, something other travelers to the Orient and Africa frequently purchased for journeys to such places viewed as uncivilized. Though the sisters carried a Murray tourist guidebook and a Bradshaw railway guidebook, they do not always follow the advice of such books. For example, in the Bosporus, Lewis and Gibson chose to visit Büyükdere, which the Murray guidebook labeled as a place where "ladies have no business," and they sought the advice of fellow travelers about places to stay and to eat.

Shortly after their arrival in Cairo, Lewis comments, "I have heard a great deal about the poetry of the East," but "I have not yet found out where it is." She questions whether there is

> any poetry in condemning women to walk with a dirty piece of clothe before their mouths? Is there any poetry in the bleeding sides of the donkeys? Is there any poetry in being half suffocated with dust every time you go out, and seeing so many people blind of an eye? Is there any poetry in the mosquitoes? Is there not great sameness in the blue green of the palm-tree and the prickly pear, and in the dull green of the olive? Is there not great sameness in gazing at a landscape of sandhills? Commit me to Northern climes for beautiful scenery, and for poetry too! . . . I am the more thankful for having been born in Britain, every day I am abroad.

She rejects the sights of Egypt, but not merely from chauvinism or homesickness.

After two weeks in Cairo the three hire a dragoman to take a boat up the Nile River on 15 December. Having suffered from cold and snow penetrating sizable holes in their railway car on the train to Varna, a fall off a camel in Cyprus, a missed train to Ephesus, and various attempts of boatmen and others to fleece them of money, the three take in stride yet another unpleasant experience – an infestation of rats on the riverboat. That Lewis can meet any challenge is perhaps best evidenced by her handling of the dragoman's attempts to prevent the three travelers from sailing to Wadi Halfa, the destination of their trip on the Nile. He claims at first that they cannot pass the cataract and then that they must be responsible for damages to the boat. Checking their contract and the Murray guidebook, Lewis declares that they are not responsible for damages to the boat and that the dragoman must take them to their destination. After various delays that the dragoman causes, they reach Wadi Halfa precisely when Edward, Prince of Wales, is making a royal visit.

Though the three have managed quite well, Edith (Blyth) comments that someone ought to set up an office in Cairo to assist travelers and make arrangements for them. (Thomas Cook is in fact already doing so, although Edith does not know this.) In her view, travelers would be well advised to pay a commission for such services because the dragomen are "generally very clever rogues." Even the extremely independent Lewis doubted the wisdom of sending people with tuberculosis to Egypt because of the treatment they would receive at the hands of the necessary but undependable dragomen.

On their return from Cairo the three visited Jerusalem by horseback and then traveled to sites in Greece and Italy before arriving home in June 1869. Giving Hamilfield, the family home, to their father's junior partner, the Smiths then settled near Blyth in London. After publishing her account of their journey, Lewis completed a poem on the life of Queen Margaret of Scotland, one that she had begun as a young girl but would not publish until 1917. Between 1876 and 1880 the sisters continued to travel, and Lewis wrote and published three novels: *Effie Maxwell* (1876), a fictionalized autobiography; *Glenmavis* (1879); and *The Brides of Ardmore: A Story of Irish Life* (1880). While the novels are more satisfactory than the verse (some of which was to appear in Lewis's travel books, where it provides proof that she was no poet), she wisely began to pursue a scholarly life that combined the religious devotion, intellectual acumen, and love for learning, travel, languages, and antiquity that she and her sister shared.

To prepare for a visit to Greece, in 1878 the sisters undertook the study of ancient Greek with a Mr. Vice of King's College. With Blyth again as a companion, the sisters set sail on the Orient liner *Iberia* in January 1883. As Lewis indicates in the opening of *Glimpses of Greek Life and Scenery,* their interests in the people and geography of Greece had been aroused during a brief stopover in Athens on their return journey from the Holy Land in 1869, and these interests were the motives of this trip. As in her first travel book, Lewis uses pseudonyms for her sister and their friend, opens with a statement of their motives for the trip, and relates significant details of the journey. One event is that of a storm at sea on their trip from Gravesend to Naples, a storm that Lewis faces with her usual bravery, although it is so severe that even members of the crew fear for their lives. Later, on the boat from Naples to Athens the sisters report being heartened that their fellow passengers think that their Greek pronunciation and grammar are fine, although the sisters' archaic words sometimes evoked laughter.

Even a century later Lewis's description of Athens is so precise – blending streets, ruins, people, and history – that her reader can imaginatively wander the streets with the unchaperoned women and compare the sights to those of Athens today. Lewis notes that Karl Baedeker, publisher of one of the most popular travel guidebooks of the period, had warned the traveler "never to wander from the chief thoroughfares," and one passenger on the boat from Naples to Athens had warned the sisters and Blyth that women should not "step round the corner unaccompanied." Yet the three find the city to be the safest in Europe, and they walk, ride, and take boats to explore not just the mainland but also the Greek islands. Lewis describes what she sees and also depicts the scenes through the words and stories of those who have visited, studied, and lived in Greece. What makes her travel book distinct from those by some predecessors is that her allusions to authorities include Murray and Baedeker among the previous literary travelers (William Wordsworth and George Gordon, Lord Byron, for instance), writers of ancient Greece, and the historians whom she acknowledges as travel experts.

To twentieth-century readers Lewis's frequent references to religious beliefs and practices, such as the sermon she recounts and a narrative of the sisters' Bible study on a particular Sabbath during their stay in Greece, are likely to appear excessive and outdated. Some of the poetry she quotes is also juvenile, dated, or otherwise offensive to the literary tastes of such readers. Yet many of her discourses read well: on climbing the mountains of Greece (nothing, not even snow, deters Lewis), on finding spring water to drink and avoiding fevers, on experiencing a slight earthquake, on traveling by horseback through Boeotia, and on assisting the Greeks "to fit together the shattered stones of [the nation's] unique temple" – especially since the British and Germans have so pillaged the treasures of the Parthenon. Lewis blends the past and the present, the stories and the sites, making apparent her breadth of knowledge:

> Betwixt these two seas, bordered by the mountains of northern Greece and of the Peloponnesus, stretched the little plain.... Turning towards the south, we saw the glittering, snowy pile of Cyllene, with the ruins of Sicyon, once a flourishing city, which an earthquake made empty of men.... Range after range rises here in succession, and we at once saw how hard it must have been for the Turks to subdue the inhabitants. We had soon the pleasure of drinking at the well of Pirene, famed for its excellent water. Pirene is supposed to have become a spring, instead of a woman, through her tears at the death of her son Cenchreus, whom Artemis had unwittingly killed.

In fact, Lewis's knowledge and the succinctness with which she sets forth her observations make her books still worthwhile and pleasurable to read. It is easy to understand why the book was translated into Greek and was effective in encouraging tourists.

Sketches accompany the text, and Lewis discusses the language and character of contemporary Greeks in a separate chapter. Readers will also enjoy sharing history in the making, as the trio observe the digging of the Peloponnesian canal on their way to Corinth and often visit with Heinrich Schliemann and his wife. Schliemann was conducting an island dig for the cave of Circe. Lewis's descriptions of travel provisions and arrangements are particularly valuable to the travel historian. The women bring with them travel beds – "strong iron bedsteads, cork mattresses, with large mackintosh valances, and waterproof portmanteaus to hold them. The whole apparatus cost seven pounds, and was furnished by Messrs. Maynard and Harris, of Leadenhall Street." Though they find less discomfort than they had been led to expect from reading *A Tour in Greece, 1880* (1882) by Edward, Lord Windsor, and Richard Ridley Farrer, Lewis notes that Farrer's advice to "sleep in flannel . . . saved us from many a cold [because] Greek country houses are somewhat too well ventilated," and "always roll[ing] and unroll[ing] the mattresses . . . taking care that neither they nor the blankets should touch the floor" helps them to avoid "the usual plagues of Greek houses." On their way home from Greece, Margaret married James Young Gibson, Blyth's

Frontispiece and title page for Lewis and Gibson's book about their most significant scholarly discovery

cousin, whom Margaret had known since 1871 and to whom she had once been engaged.

After a year of marriage James Gibson died, and in 1886 the two sisters, this time unaccompanied by Blyth, returned to the Mediterranean. Lewis describes their travels to Port Said; the Suez Canal, where Agnes has her first camel ride on a side trip to "Ain Mousa" and contemplates going to Sinai on their next trip; and Cairo before beginning their extended stay on Cyprus. In *Through Cyprus* (1887) Lewis refers to the works of such travel writers as Samuel White Baker and Isabella Lucy Bird, whose accurate descriptions of Cyprus and Japan other travelers praise. While in Cairo, Lewis visited Mary Louisa Whately's school, stayed in Shepheard's Hotel, and studied Arabic. She describes the sights of Cairo that they revisited to "refresh our memories" as well as those they were seeing for the first time, and she finds Cairo much improved since their 1868 visit: its streets are cleaner, and the khedive's band now plays every afternoon in a beautiful public garden.

They also attended English polo games, toured the Arab Museum and the Boulak Museum to gaze on the mummy of Ramses II, took excursions to the tombs of the caliphs and the Sphinx, and climbed the Great Pyramid. Lewis visited and dined with a native congregation and witnessed the marriage ceremony of some Copts, something that other English travelers were unlikely to do: the British generally did not know the language, and the "etiquette of our insular manners" called for "ignoring the presence of anyone else in the room," a "haughtiness of the passive kind" that made British etiquette "differ widely from that of every other civilized nation." She offers other criticisms of the English attitude of moral superiority and of the claim that the British are the civilizers of the world.

From Cairo the sisters traveled to Alexandria and Beirut, where they took an excursion to Nahr-

Engraving of the convent of Saint Thekla, Ma'lula, a mountain village that Lewis and Gibson found to be "one of the most picturesque towns we have ever seen" (from Lewis, In the Shadow of Sinai, 1898)

el-Kelb before going on to Cyprus. The book includes a map of their journey around the island. Adventuresome as always, they rode mules and slept in tents in order to visit all of Cyprus. Lewis describes the various sights of the island, its monasteries and villages, its people and priests. The book closes with a history of Cyprus and a statement of its potential for successful English colonization because of its fertility and climate; temperature and rainfall charts are included.

In December 1887, following the publication of the Cyprus travel book, Agnes married Samuel Lewis, the librarian of Corpus Christi College, Cambridge, and Margaret accompanied the two on their honeymoon tour of Greece. The Lewises rented a home in Cambridge, next door to the John Maynard Keynes family, before they moved into their new house, Castlebrae, in March 1890. Agnes, however, proved to be no more fortunate than Margaret had been, as Samuel died at Easter 1891.

Once again one of the sisters used travel to assuage the sorrow of the other, and in January 1892 Lewis and Gibson left for the first of six trips to Mount Sinai and Saint Catherine's Convent, where they discovered the Syriac palimpsest, more important to biblical scholarship than the discovery of the Dead Sea Scrolls has been in the twentieth century. Aside from Lewis's publication of her husband's biography, the remainder of the sisters' publications are travel accounts related to their trips to visit and revisit the monastery and its library and many works of scholarship and translation that their discovery made possible.

Gibson published the popular account, *How the Codex Was Found: A Narrative of Two Visits to Sinai, from Mrs. Lewis's Journals, 1892–1893*, although she wrote only a brief introduction to the text, which consists of Lewis's first-person diary entries made during their first two journeys. All of these entries had previously been published in the columns of the *Presbyterian Churchman*. In *In the Shadow of Sinai: A Story of Travel and Research from 1895 to 1897* (1898), Lewis published the accounts of their journeys in 1895, 1896, and 1897, along with photographs they took of people and places. The first journey was undertaken to check the palimpsest, the second to buy manuscripts, and the last to verify the correctness of the Palestinian Syriac lectionary.

Their journeys were uncomfortable: the weather conditions were often harsh, as they endured rain, hail, wind, and extremes of temperature, and they frequently encountered unreliable, inhospitable, and even threatening people. Although Thomas Cook and others made arrangements for most travelers and certainly for women, the sisters hired their own dragomen, as they had done on the first trip. Lewis comments that Baedeker underestimated the cost and number of camels needed for the journey, and she describes the sleeping and eating arrangements of the sisters: they traveled with one tent for dining and another for sleeping; at night the cook and dragoman slept in the one used for dining. She also notes the unreliability of the Syrians' statements about their religious customs, which are not uniform because they practice different faiths – Muslim, Protestant, and Catholic. Lewis provides a brief lesson in Arabic pronunciation; describes the habits of the Bedouin, including their penchant for temper tantrums; and even gives instructions on using water filters and on riding (by using pillows to fill up the center of one's saddle) and mounting a camel: *"Observe what your dragoman has got for himself, and adopt it."* Lewis also displays her sense of humor about such means of transport: "Messr. Cook and Gaze, ought to design a camel which will not be so hap-hazard in its structure."

In 1899 Lewis received the first of many honorary degrees to be bestowed upon her and Gibson. Their last visit to Sinai was in 1906, when they were sixty-three years old. In 1915 Lewis received the Triennial Gold Medal of the Royal Asiatic Society for her research, in particular for that which appeared in *The Commentaries of Isho'dad of Merv* (1911–1916). The sisters continued to live at Castlebrae, which housed many of the priceless treasures they had collected and was where they entertained Cambridge fellows and students, often for lunch. Among their friends was Mary Kingsley, another famous lady traveler, who reportedly said of them, "they stroll to and fro in the Sinai desert as though it were Sauchiehall Street."

The sisters were to donate the land on which Westminster College was erected, and they also contributed funds to the modernizing of Saint Catherine's. Yet in spite of their philanthropy, their scholarship, and the value of their travel writing to a history of the times, the "Giblews" were forgotten until their extraordinary contributions were recounted by the Price biography in 1985 and Billie A. Melman in 1992.

Biography:

Whigham A. Price, *The Ladies of Castlebrae: A Story of Nineteenth-Century Travel and Research* (Gloucester: Sutton, 1985).

Reference:

Billie A. Melman, *Women's Orients: English Women and the Middle East, 1718–1918* (Ann Arbor: University of Michigan Press, 1992).

Papers:

Lewis and Gibson's papers, notebooks, photographs, and lantern slides are maintained by Westminster College, Cambridge.

Elinor Mordaunt
(Evelyn May Clowes Wiehe)

(1872? – 25 June 1942)

Colleen Hobbs
Rutgers University

BOOKS: *The Garden of Contentment* (London: Heinemann, 1902);

Rosemary: That's for Remembrance (Melbourne: Lothian, 1909; London: Scott, 1909);

On the Wallaby through Victoria, as E. M. Clowes (London: Heinemann, 1911);

A Ship of Solace (London: Heinemann, 1911; New York: Sturgis & Walton, 1911);

The Cost of It (London: Heinemann, 1912; New York: Sturgis & Walton, 1912);

The Kid and the Captain (London: Heinemann, 1912);

Lu of the Ranges (London: Heinemann, 1913; New York: Sturgis & Walton, 1913);

Simpson (London: Methuen, 1913; Boston & New York: Houghton Mifflin, 1913);

Bellamy (London: Methuen, 1914; New York: John Lane, 1914);

The Island (London: Heinemann, 1914);

The Family (London: Methuen, 1915; New York: John Lane, 1915);

The Rose of Youth (London & New York: Cassell, 1915; New York: John Lane, 1915);

Shoe and Stocking Stories (New York: John Lane, 1915; London: A. & C. Black, 1926);

The Park Wall (London & New York: Cassell, 1916);

Before Midnight (London & New York: Cassell, 1917);

The Pendulum (London & New York: Cassell, 1918);

The Processionals (London & New York: Cassell, 1918);

Old Wine in New Bottles (London: Hutchinson, 1919);

While There's Life (New York: Holt, 1919);

The Little Soul: A Novel (London: Hutchinson, 1920; New York: McCann, 1921);

Laura Creichton (London: Hutchinson, 1921; Boston: Small, Maynard, 1922);

Alas, That Spring! (London: Hutchinson, 1922; Boston: Small, Maynard, 1923);

Short Shipments (London: Hutchinson, 1922);

Reputation (London: Hutchinson, 1923; Boston: Small, Maynard, 1924);

People, Houses, & Ships (London: Hutchinson, 1924);

The Real Sally (London: Hutchinson, 1925);

The Venture Book (London: John Lane, 1926; New York & London: Century, 1926);

The Further Venture Book (London: John Lane, 1926; New York & London: Century, 1927);

And Then – ? Tales of Land and Sea (London: Hutchinson, 1927);

The Dark Fire (London: Hutchinson, 1927; New York & London: Century, 1927);

Father and Daughter (London: Hutchinson, 1928); republished as *Too Much Java* (New York: Payson & Clarke, 1928);

These Generations (London: Hutchinson, 1930; New York: Brewer & Warren, 1930);

Full Circle (London: Secker, 1931); republished as *Gin and Bitters,* as A. Riposte (New York: Farrar & Rinehart, 1931);

Cross Winds (London: Secker, 1932; New York: Day, 1932);

Purely for Pleasure (London: Secker, 1932); republished as *Rich Tapestry* (New York: Farrar & Rinehart, 1932);

Mrs. Van Kleek (London: Secker, 1933; New York: Day, 1933);

Traveller's Pack (London: Secker, 1933);

The Tales of Elinor Mordaunt (London: Secker, 1934);

Prelude to Death (London: Secker & Warburg, 1936);

Royals Free (London: M. Joseph, 1937);

Sinabada (London: M. Joseph, 1937; New York: Greystone, 1938);

Pity of the World (London: M. Joseph, 1938; New York: Greystone, 1939);

Death It Is and Other Stories (London: Hutchinson, 1939);

Roses in December (London: M. Joseph, 1939);

Hobby Horse (London & Melbourne: Hutchinson, 1940);

Judge Not (London & Melbourne: Hutchinson, 1940);

Return to Spring (New York: Greystone, 1940);

Blitz Kids (London & New York: Oxford University Press, 1941);

Here Too Is Valour (London: Muller, 1941);

Tropic Heat (London & Melbourne: Hutchinson, 1941);

This Was Our Life (London & Melbourne: Hutchinson, 1942);

To Sea! To Sea! (London: Muller, 1943).

PLAY PRODUCTION: *Mrs. Van Kleek,* London, The Playhouse, 17 March 1938.

SELECTED PERIODICAL PUBLICATIONS – UNCOLLECTED: "The Natives in Kenya," *Times* (London), 30 July 1926, p. 10;

"The White Settlers in Kenya: Mrs. Mordaunt's Vindication," *Times* (London), 6 September 1927, p. 8;

"Kenya Coffee," *Times* (London), 19 April 1930, p. 12.

The travel writing of Elinor Mordaunt, the pen name of Evelyn May Clowes Wiehe, provides a wry, breathless, often lyrical description of a world that delights the reader with its variations. Mordaunt's description of Polynesia reflects her passion for travel and illustrates how she was drawn to a life of adventure: "The dropping of the anchor outside new islands – islands and islands and islands, no two ever alike, – ever-changing languages and ever-changing peoples.... That, for me, is life."

As she sailed the ocean in leaky cargo boats often operated by disreputable crews, Mordaunt had a quiet disdain for the complacent, comfort-loving tourists who "go to the same English seaside resort every summer of their lives, and they are – well, that sort of people." In her voyages to Australia, Central and South America, the Dutch East Indies, Southeast Asia, and Africa, Mordaunt's meager budget strained her frail constitution and reinforced her scorn for luxuries and, often, even for necessities. A dust jacket for *Too Much Java* (1928; published in England as *Father and Daughter*), a novel set in the Dutch East Indies, compares Mordaunt to Joseph Conrad "because they both know and write about the same things ... [and] they both tell the truth about the things they know best." Like Conrad, Mordaunt sometimes casts a critical eye on British colonialism, but her travelogues are shaped by humor, a sharp wit, and her awareness of the novelty of her position as a unmarried woman traveling alone. Her accounts provide glimpses of colonies and colonials during the waning days of the British Empire and insights into opportunities that travel offered a woman writer to define herself by rules quite different from those of home.

Evelyn May Clowes's birth at Cotgrove, Nottinghamshire, was casually unrecorded. She claimed to have been born in 1872; when asked the month and day, she characteristically stated that she could not remember. Her parents, St. John Legh Clowes and Elizabeth Bingham Clowes, produced a large, active family of six boys and two girls. As a member of a family in which Mordaunt says she did not "even [do] my hair for myself," she received only a mediocre education provided by "a succession of inefficient governesses."

Mordaunt's autobiography, *Sinabada* (1937), indicates that even this privileged existence was somewhat tenuous. Her family made it clear that she would need to marry well because they could offer her no financial support. Life in rural Victorian England ended for Mordaunt when her fiancé, her childhood sweetheart, died while leading an exploration on the Zambezi River funded by Cecil Rhodes. Heartbroken and seeking change, she accompanied a cousin to the island of Mauritius, where in 1898 she married Maurice Wiehe, a sugar planter.

The marriage appears to have been an unmitigated disaster: her fortune-seeking husband even offered her as a stake in a card game. Suffering from malaria on Mauritius, she went home to England after two and a half years, but her stay there was short. In 1902 the exhausted, emaciated Mordaunt sought passage on a vessel bound for Australia, because "the idea of dying in bed was abhorrent to me." Her account of her failed marriage is purposefully vague, and her failure to inform her family suggests that her separation from her husband had invited some censure. She offers few details about the birth of her son, Godfry Weston Wiehe, shortly after her arrival in Melbourne, but through her eight years of working in Australia she struggled to provide for herself and her son. A much later marriage (in 1933) to Robert Rawnsley Bowles, a retired barrister from Gloucestershire, was short-lived. Reflecting on her relationships with men and on her financial situations, Mordaunt wrote in her autobiography that it seems "evident that from the very first the fates planned for me to walk alone and to possess nothing that I have not earned."

Mordaunt's novels draw on her travel experiences to make detailed observations about distant places and their inhabitants. One review by the *Times Literary Supplement* (12 March 1938) observes

Frontispiece and title page for the first American edition of Elinor Mordaunt's narrative of her voyages throughout the Dutch East Indies

that her story in *Pity of the World* (1938), a late novel that is set in Kenya, captures a "sense of the values of a country, with its resulting influence on the lives of human beings." This particular knowledge of a country and its character "gives Mrs. Mordaunt's work distinction." Her first book, *The Garden of Contentment* (1902), was written during her illness on Mauritius. As an imaginary series of letters, it records her memories of England. *A Ship of Solace* (1911), a book praised by sailors for its accuracy, records her original trip to Australia, and her experiences there are recorded in several novels, including *Lu of the Ranges* (1913) and *The Rose of Youth* (1915). *Laura Creichton* (1921) documents the hard labor and poverty she endured there. When she returned to England after almost eight years, Mordaunt worked in a Staffordshire silk factory, an experience she recounted in the industrial novel *Bellamy* (1914).

Mordaunt's first travel book recounts her life in Australia from about 1902 to 1910. This work, *On the Wallaby through Victoria* (1911), is the only volume in which she uses her name, E. M. Clowes, rather than her pseudonym – perhaps because the book evokes a much more staid and impersonal tone than do her later efforts. She offers an early disclaimer that "this is not supposed to be a national or political history of Victoria," and she adds that "I can only write about Victoria as I know it." To show her knowledge of Australia, she recites her extensive job experience as an editor, a decorator, a gardener, and a housekeeper there.

In assessing the Australian economy and social structure, she discusses the trade imbalance between England and Australia, and she informs readers about election expenses, income taxes, land values, and railway miles. She describes in detail the problems created by a government policy that divides large sheep stations to increase population density. Her analyses draw on her experiences to defend the national character of a country that England was still condescendingly condemning for its "lack of enterprise and general slackness," and she reminds readers that Melbourne was founded only in 1835. While Mordaunt makes no claim as a historian, *On the Wallaby through Victoria* thoughtfully depicts the growing pains that Victoria experienced at the turn of the century.

In her descriptions of the Australian countryside Mordaunt discusses displaced wildlife, which was becoming ever more scarce, and forests cleared for roadbeds. The book's accounts of Australian natives sound rather condescending. Mordaunt refers to Aborigines as "black fellows," for example, but she shows sensitivity for a complex culture falling victim to European colonization. She notes that settlers do not respect aboriginal religion, because "the general tendency [is] to so name only our own particular belief, lumping all the others together under the name of 'superstition,' . . . [and] if we would comprehend anything of a people's beliefs, we must know everything that there is to be known of themselves, their surroundings, and their lives." In this early book Mordaunt reveals the attention to native customs and an appreciation for diversity that she continued to show throughout her travels.

A chapter on "The Working Women of Melbourne" shows the perspective that makes Mordaunt's analysis particularly valuable. She describes the lives of young girls and married women who support themselves through strenuous labor, and she notes their salaries, clothes, and even reading habits. The details of their Australian toil and deprivation are in many ways those of Mordaunt, as she supported herself in working-class positions far removed from those of her privileged background. Her portrait of charwomen is drawn from a short distance, and her authority is credible when she notes that deserted wives comprise most of the trade. The strength, competence, and courage of these women, she says, are the very qualities that leave them open to exploitation by scurrilous husbands, for if these women "had not willingly and bravely put their shoulders to the wheel at the time of some crisis, scarcity of employment, or illness, their husbands would never have found out how much more capable they were than they themselves."

The toil of these women is accompanied by an independence unfamiliar to women in Great Britain, and Mordaunt observes the suffrage movement in England from a colony where women already enjoy the right to vote. As women in Victoria hear of the stone-throwing tactics used by British suffragettes, the Australian women are "truly sorry that there seems no prospect of the brains of those in authority being reached in any less forcible fashion; but, then, they literally *cannot* comprehend a woman's side of the question being disregarded, simply because she is a woman." The description of the enfranchisement of Australian women reveals the strengths of *On the Wallaby through Victoria*. In relinquishing its claim to be an official and historical view of Victoria, it reserves a right to note specific, seemingly inconsequential details that give the province its color and character. These details, presumably unimportant to an official historian, reveal the Australian social, economic, and political climate from a perspective that is not usually considered – that of a working woman.

Mordaunt returned with her school-age son to England, where she cared for him and produced more than twenty novels between 1911 and 1923. Many readers impatient with melodramatic works such as *Alas, That Spring!* (1922) are enthusiastic about the novels and short fiction that draw on her travel experiences, including trips to the Balkans, Italy, and Morocco. A reviewer in the *Times Literary Supplement* (6 April 1922) finds that her stories sometimes seem rushed, but the review notes the "excellent tales" of *Short Shipments,* a 1922 collection distinguished by "much cleverness, a vigour that more than once verges on brutality, excitement, and a dash of horror."

She began her travel writing in earnest when, in 1923, she approached the London *Daily News* and proposed that she sail the world in cargo and sailing ships, "anything but passenger boats," and file stories about her journeys. Knowing of a French cargo route from Marseilles to Tahiti, she made her agreement with the *Daily News,* which paid her twenty pounds for each story she filed. The results were collected in two volumes, *The Venture Book* (1926) and *The Further Venture Book* (1926). The former chronicles her trip through the South Pacific – Tahiti, Samoa, Tonga, and Fiji – while the latter volume concentrates on the Dutch East Indies. Mordaunt's finances required her to produce a constant stream of stories, and her behavior earned her the name *Dauvolavola,* "the one who is always writing," from the Fiji islanders. Written, by necessity, in boats, canoes, and crowded marketplaces, the stories are informed by Mordaunt's personal experience and complemented by her sketches and photographs.

Perhaps to prepare readers for the tales, Mordaunt prefaces *The Venture Book* with a thoughtful discussion of her need for adventure. Although she is happy in her home, she acknowledges that "another self" in her is nourished only by travel.

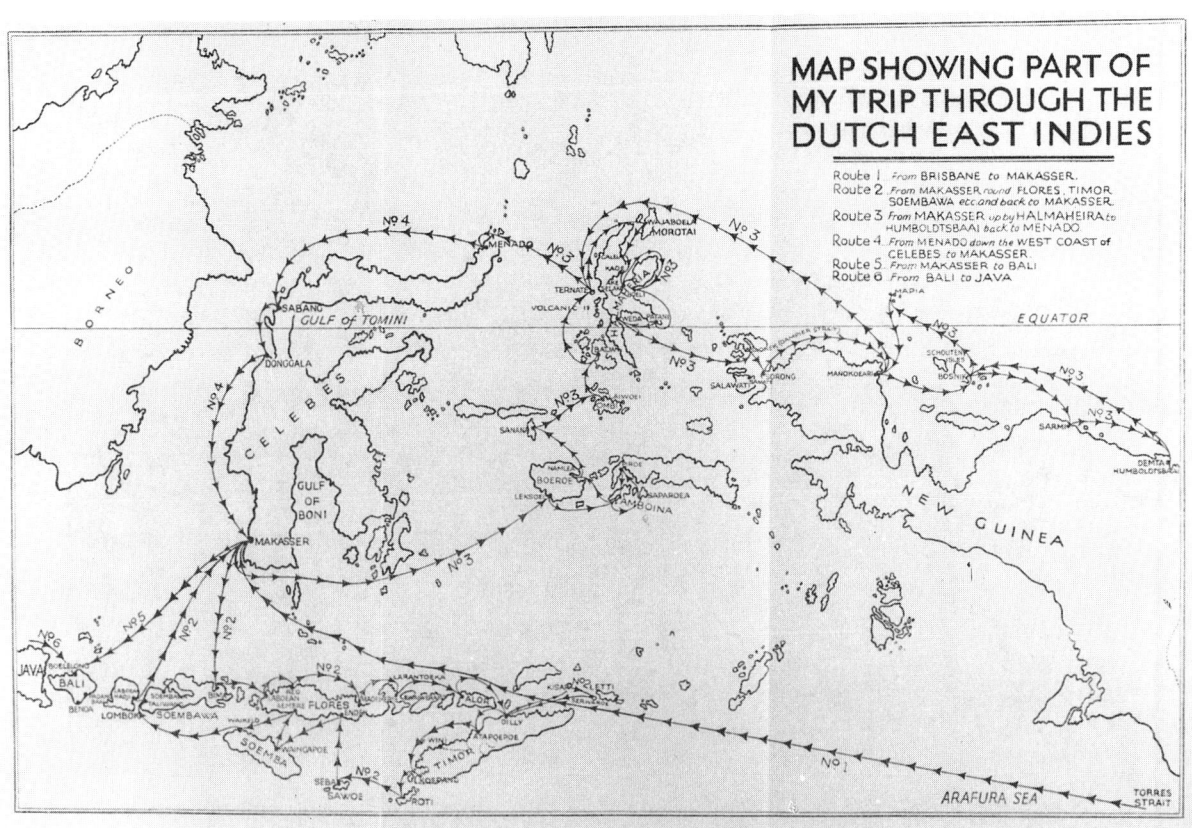

Map of Mordaunt's various journeys, which she financed by writing stories for the London Daily News (from Mordaunt, The Further Venture Book, 1926)

She has no use for "altogether safe and comfortable travelling, in which one is surrounded by everything to which one is accustomed." Her enjoyment arises from danger, from living "by the skin of one's teeth." For a single woman, part of that danger comes from the ability to slip into another identity "quite naturally" as one leaves home. Mordaunt supplies an anecdote in which, while traveling abroad, she is unable to sleep because of a noisy, drunken argument among a group of men. Wearing her dressing gown, she mediates a potentially violent squabble concerning an unfaithful spouse and sends the men back to bed. "That was one of the occasions when, returning to my own room," she recalls, "I found myself wondering if this was, indeed, I."

The ability to shed one's identity entirely, to cast off both the confining mores and the protective taboos of one's culture, is for Mordaunt the most dangerous part of travel. She recognizes that, among the South Pacific islands or any other foreign destination, social, sexual, and moral boundaries can be rewritten if one dares, and in so doing the possibilities are both exhilarating and terrifying.

At one port she almost lost a pack that had no particular importance except that it contained belongings that "might have helped me prove my own identity to myself," she says, and the threat of this loss was "the only time when I ever felt really and truly frightened." Mordaunt quotes a nursery rhyme to illustrate her fear of losing herself in her travels: "If this be I, as I should hope it be, / I've a little dog at home, and he'll know me." The rhyme concludes with the dog barking at a master whom it no longer recognizes.

In its discussion of her interior journey, Mordaunt's account parallels the themes of Conrad, an author who contends that some voyages take one so far that he or she may never return. Mordaunt refuses to disclose specific incidents, but her reflections indicate that she has confronted her own heart of darkness by the end of *The Further Venture Book*. Something indefinable has happened, something that she attempts to explain by reviewing her concept of boundaries. After sailing steadily westward across the Pacific for more than a year, she asks whether "I am now come to the extreme west or the extreme eastern lap of it." The answer she finds al-

A Humboldtsbaai village that Mordaunt visited and photographed for The Further Venture Book

most echoes that of the nihilistic Mr. Kurtz from Conrad's *Heart of Darkness* (1902): east or west, mere geographical locus does not matter. "Not that anything matters, ever will matter again," she concludes. "At the back of my mind I know that there are still other little islands before me . . . but to all this I am profoundly indifferent, for they won't ever be the same. Nothing will ever be the same again."

Mordaunt is left "profoundly indifferent," presumably about her return to the West, after spending more than a year amid the beauty and brutality of the South Pacific. On Bali she witnesses a drama held to celebrate a man's successful lawsuit. The elaborate prelude of graceful dancers and orchestra players lasts more than twelve hours. The loveliness of Fiji and its people touch her so that upon leaving "my heart ached as it had never ached since I saw my own son go away from me." As she details the minute variations among the islanders' dress, food, and customs, Mordaunt is astonished at the "profound depths of stupidity" of westerners who profess that "all these islands are so alike!" Each island has its distinct character and soul, Mordaunt argues, and she offers Ambon as an example. On this Pacific island time seems to have stopped in the seventeenth century, and inhabitants look like subjects in a Rembrandt painting: the natives dress in the black broadcloth and the leather shoes with the upturned, pointed toes of the Dutch settlers.

Mordaunt's recountings of tropical splendors are tempered by accounts of equatorial disease and titled hardship. She suffered recurrent attacks of fever, probably malaria; she describes the volcanic eruption that destroyed Saint-Pierre, Martinique, and killed more than twenty thousand people in 1902. She treats cannibalism, potentially the most gruesome topic, with a surprising amount of sensitivity and humor. Rather than dismissing as merely barbaric the islanders' tales of cannibalism, Mordaunt notes the nuances in cultural taboos. In some islands, for example, eating one's dead relatives is considered good manners; in others, it is unthinkable. In her accounts of general slaughter, however, Mordaunt remarks that manners and morals are "the merest matter of latitude and longitude," and, as a benchmark, compares moral standards of British and Fijian men: "The moral standards of almost any European man would have brought a blush to the cheek of these warriors, with whom incontinence in the young man before marriage was regarded, not as something manly, but as the sign of a feeble, weak-minded person."

Mordaunt's most droll account of cannibalism occurs when she interviews two elderly ladies who had come to Fiji as children. As they tell the story, the women see nothing remarkable in their mother's experience of meeting new neighbors who are shortly murdered and eaten. The women are "surprised" at Mordaunt's suggestion that their mother might have wanted to return to the safety of New Zealand, and they unself-consciously relate another story of a European missionary who was killed after unwittingly having offended a native chief. The missionary was eaten, but only after his feet had been boiled many times. The natives recalled their "shame over having been so uneducated and provincial" as to overlook a crucial fact: the missionary had been wearing boots.

Incongruities such as those that embarrass the Fijian cannibals leave Mordaunt acutely aware of the dangers of making easy generalizations about another culture. On these islands, she says, are "people who are innocent as we can never now be, guilty as we have never yet been; people with their own ideals, their own traditions, so entirely different from ours that we touch a new heaven and a new hell in realizing them." Mordaunt makes no attempt to distill "truths" about South Pacific islanders for westerners; the incongruities are part of the story. She also makes no attempt to explain her motivations and impressions. She leaves readers to draw their own conclusions from the anecdotes she chooses to share. As she passes through the Panama Canal, for instance, she first provides some information about tonnage, geography, and locks. But she quickly abandons any attempt at objectivity and tells her readers to go to a public library for "any more definite, scientific, or practical knowledge of the canal."

Mordaunt does hold some Western biases, and her accounts can gloss over the problems of British colonialism. She notes the differences among French, Dutch, and British attitudes toward their colonies, and in *The Further Venture Book* she grows impatient with criticism of British policy. She writes of Dutch passengers "continually condemning our method of dealing with native labour under the indenture system – and there seems to be a general belief, or a pretence at a belief, that the English make slaves of the Papuans." Mordaunt's response to such criticism ignores the magnitude of British colonial abuse and exploitation; she finds this problem, for example, to lie in the "degenerate, dirty, and diseased" Papuans themselves, who are "no great credit to their conquerors."

In a series of letters to the London *Times* she defended Kenyan settlers against charges of having mistreated African natives and blamed instead the "sentimentalists" who undermined coffee planters by luring away laborers. Even as she railed against misguided policies that transformed "very fine and honest people" into ineffective, "spoilt" workers, Mordaunt was aware of Kenya's cultural complexities. She denounced a critic of white settlers partly for the oversimplifications of his criticism: "he refers to [Kenyans] as though they consisted of one tribe with precisely similar tastes, instead of being many, very different in customs and habits, and sharply antagonistic."

What is constant in Mordaunt's accounts is her adaptability, the ease with which she adjusts to each crisis and modifies the rules as she must. By sheer force of will, she gains passage on a bedraggled schooner, the only vessel bound straight for her intended destination, Samoa. She hounds the captain, who has refused her passage, until he surrenders unconditionally: "I can't refuse you anything. I'm just fair crazy about skinny women!" Though her persistence often succeeds, it can also lead her to lose her temper when a situation demands it, and the presence of a polite British woman turned murderously angry frequently elicits the desired response. The tactic is effective in the Dutch East Indies, where, unbeknownst to Mordaunt, the islanders declare her "King of the Dimdims" – that is, ruler of all the white people in the world. When she attempts to buy a pig for her native guides and is unable to provide the promised animal, the natives' feast is saved only by her temper: "I swear roundly at them all; declaring upon my oath . . . that if I do not get that pig within half an hour I will 'Make Big Government.' " Mordaunt never explains the meaning of this obscure threat, but "The Affair of the Pig," as she calls it, concludes with a bountiful pork dinner.

The details of Mordaunt's travel reveal as much about her as they do of her destinations. For example, she can cast aside Western inhibitions as she watches a frankly sexual dance in the Dutch East Indies. She expresses thankfulness only that "there is no white man here with me," presumably to remind her that the dance, or her presence, is inappropriate. As she describes her newly acquired parrot named Johnny Walker, she reflects that her year of solitary travel has taken its toll on her: "It is really dreadful to be as lonely as I am, for of course I shall grow to love him, and know that will be the inevitable death of him."

Often she writes as if she feels that her embrace of unknown, potentially dangerous adventures is a strategy to avoid the pang of more-familiar venues. As a result, she presents a wistful view of the lush, tropical islands she describes. As her journey closes, she spends a final night aboard ship before her return voyage. This situation, on a ship poised to set sail, seems to her to be a limbo, "a bridge between this world and the next." She thinks about the relative safety that she has felt in the islands and, in contrast to her earlier descriptions, imagines them for a moment as small and easily contained, as "little islands kicked off like baby shoes behind me: all the small familiar things of little islands, with the arms of the sea about them." Mordaunt here rewrites the terrible beauty of the South Pacific to reflect her emotional state. She is poised to strike out for a new world – the familiar world to which she will be returning, but with a changed perspective – and, as she returns, she must kick off her "baby shoes" to make the journey. In such a farewell to the islands Mordaunt blurs the boundaries between personal narrative and travelogue; the result is an emotional geography that is entirely subjective and utterly engaging.

A final volume devoted entirely to travel is *Purely for Pleasure* (1932), a title that Mordaunt intended to be read ironically. Certainly many events she describes – a revolution in Honduras or an attack of dysentery that makes El Salvador "uninteresting" – are not pleasurable. The form of this book is more episodic than that of the two *Venture Book* narratives and covers three separate journeys – to Central America, to equatorial Africa, and to Southeast Asia. This format ensures that the tales have no overarching narrative structure; they lack the emotional tension that carries the reader through the earlier *Venture Book* narratives. Yet the individual portraits that the book paints are perhaps all the more striking because they are contained and distinct. The title of the American edition of the book, *Rich Tapestry*, provides a metaphor for the presentation Mordaunt seeks to make in this work – a tapestry of remembered events.

Purely for Pleasure opens with Mordaunt despairing that it is "impossible for me to write a proper travel book; impossible to get things in their proper order." However, she says, she finds travel books that do so to be boring. As one reviewer for the London *Times* (17 June 1932) observes, this apologia sets her tone for another tale of travel that illustrates "an exquisite curiosity, a perennial eagerness, a lovely sense of humour, [and] a complete inability to 'stay put.'"

As in her earlier texts, Mordaunt's travelogue is a study in extremes. Prerevolutionary Cuba is all grace and charm, "an atmosphere of good cigars and wine." In Honduras, however, the revolution is already under way. Stopping at a family's home for the night, "several men with guns" attempt to force open the door. Mordaunt's calm assessment is that "they are not very much in earnest, for after we have all put our shoulders to it, including the children, they retreat and there is no sound of shooting."

Purely for Pleasure covers a wide expanse of settings, from African American church services in Florida to Kenyan coffee plantations. The most chilling vignette in the author's tapestry is found in "The Seamy Side of Singapore," a chapter that documents traffic in child prostitution. After having seen the tourists' version of the city, Mordaunt hires a guide who can "impersonally show me things which I should not have cared to see in the company of anyone of my own kind." Free from the presence of Western acquaintances, she visits a business that trades in young girls from all over the world – China, Java, the Philippines, or Somalia. These girls are "like flowers, torn up by the roots and sold, casually to anyone: so much delicacy, so much fair young beauty: so much budding womanhood, possible motherhood, tied up in bunches and sold." Without railing against the injustices of the slave system, Mordaunt's intimate description of the tiny girls' faces, their hair, and their laughter clearly reveals the brutality of the market.

Mordaunt concludes her travels by disembarking in Genoa after having toured the coast of the Italian Somaliland, and she closes *Purely for Pleasure* by refusing to analyze the images she has presented. That kind of reflection is found, only briefly, in her foreword, where she observes that travel is "not so much a matter of the body as of the mind." The most complex journey is the inner one, and it cannot be explained simply by following a map and listing ports of call. That focus on her inner journey allows Mordaunt to reflect later that she has "lived so many lives, died so many deaths, been in so many countries, among such divers people . . . that there are times when I feel like a disembodied spirit floating about seeing and not being seen." A white, unmarried, and unaccompanied British woman could hardly be invisible in Java, or Singapore, or Kenya, but Mordaunt's reflection reveals how she imagines herself during her travels and explains why she can have so little regard for her physical discomfort: she is not a body but a spirit, gathering images and memories too rich to be fully understood at the moment.

The journeys documented in *The Venture Book* and *The Further Venture Book* provided Mordaunt with materials that she continued to mine in her fic-

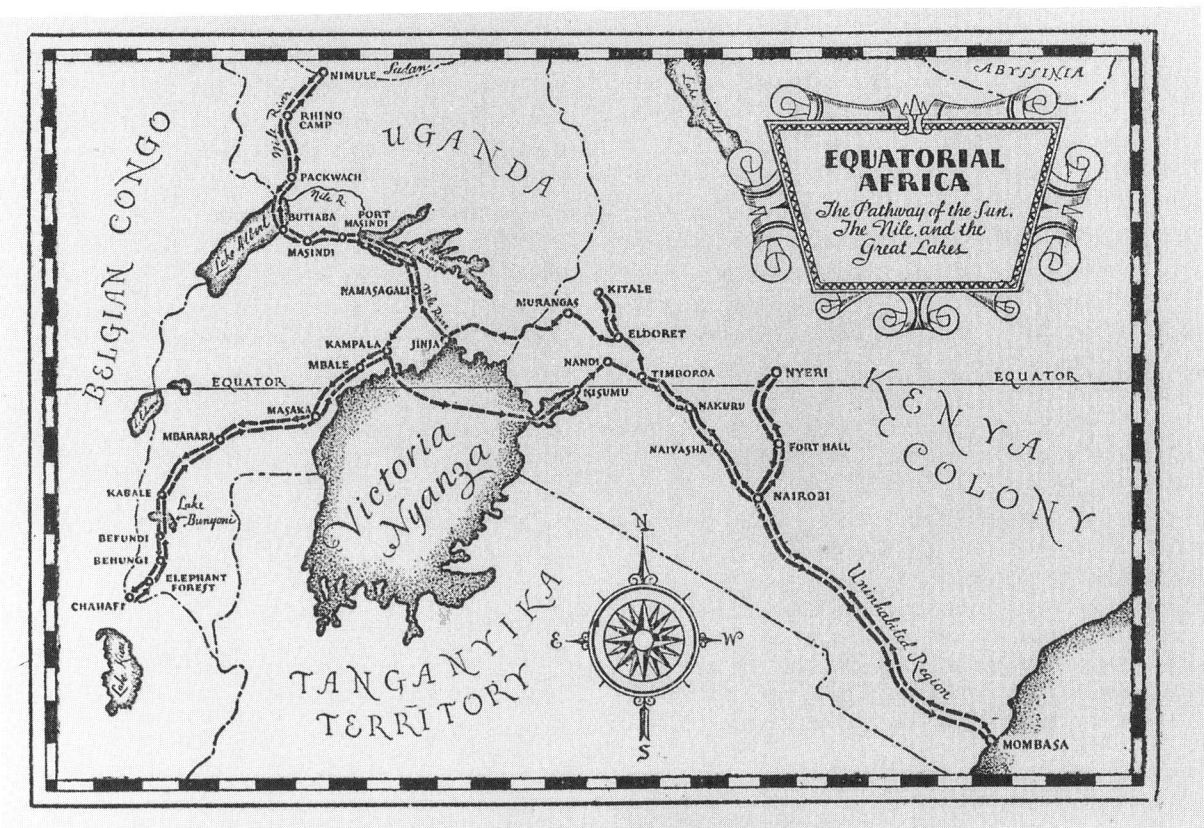

Map showing the African parts of the trips described in Mordaunt's last travel book, Purely for Pleasure *(1932)*

tion. A reviewer in the *Times Literary Supplement* (6 September 1928) wrote that Mordaunt's *Father and Daughter,* which depicts an interracial love affair in the Dutch East Indies, irresistibly reminds one of Conrad. *Mrs. Van Kleek* (1933), one of her best-received novels, is set in a hotel in the tropics that attracts the same misfits, missionaries, and government officials whom Mordaunt met in her journeys. One of her final novels, *This Was Our Life* (1942), includes the story of cannibals boiling a boot-clad missionary. Another review in the *Times Literary Supplement* (1 August 1942) found that such imaginative incidents would be suspect, "had Mrs. Mordaunt not been so well known as an intrepid and imaginative traveller. The reader might be tempted to think she had been drawing a very long bow, but the matter-of-fact manner in which she tells her story carries its own conviction."

Mordaunt's acute observations of human behavior were not limited to her travels abroad. At least once what the *Times Literary Supplement* (7 September 1931) called her "slightly malicious raps . . . in the English literary world" registered offense. W. Somerset Maugham sued Mordaunt for libel because of her 1931 novel *Full Circle,* a story about a popular novelist who unsparingly fictionalizes his acquaintances. The review does not note the outcome of Maugham's lawsuit.

The title of Mordaunt's autobiography, *Sinabada,* translates loosely as "Lady King," the title given her by natives of the Dutch East Indies. Her memoirs provide more insight into the isolation that leads a woman surrounded by friends to feel that "everyone of us is as alone as a chick in its shell." The autobiography also includes details that she might have considered either too revealing or too risqué to include in her earlier volumes. For example, her stop in Djibouti included an erotic dance performance by four naked Somali women. Mordaunt's poverty precluded a potential love affair in Australia, and she casually observes in her memoirs that few people realize "how many women are kept virtuous by shabby underclothes."

In recounting an incident that happened at home, on the Thames River in Wapping, Mordaunt also indicates some difficulties that she must have faced in her travels. When she expresses an interest in sailing on the river wherries, a sailor tells her that such a cruise is impossible. She presses the point until her companion is forced to declare, "You couldn't do it. 'Cause why? You'd 'ave to sleep with the skipper." Mordaunt consistently downplays the difficulties faced by a woman traveler, but her exchange on the Thames shows that minor inconveniences were laced with real sexual intimidation. The incident gives readers a heightened respect for her ability to negotiate various languages, customs, and sexual mores safely for so many years.

Mordaunt's compulsion for travel grew from more than just a need for financial support. As she sailed the oceans in difficult circumstances, Mordaunt gradually found that she no longer "belonged" on land. Sun-blistered and thirsty in the Pacific, where she sleeps on the deck of a schooner loaded with lumber, dynamite, and gasoline, she describes herself as "happy":

> Upon a boat like this, I have somehow or other the sensation of being at home among people who like me, whom I understand and who understand me. I am altogether in my own element, translated by my freedom from what most people call the ordinary life – to me so extraordinary that it misfits me like badly made clothes, stiff and ungainly, in which I never fail to feel an ass.

Among strangers and deprivations Elinor Mordaunt found her family, her comfort, and her home.

Mordaunt is perhaps best known for her treatment of Australian subjects, but her interests and her travels are wide-ranging, and much research remains to be done on her. Her journalistic publications remain scattered in many newspapers and magazines in Australia, England, and the United States: the Australian *Bulletin* and *Lone Hand;* the London *Daily News, Black and White, Pearson's Magazine;* and the American *Metropolitan Magazine.* Her use of pseudonyms makes even more difficult the task of tracing her writing; for example, "John Heron" is identified as the author of her boys' stories published in *Young England.*

Mordaunt's chronicles remain uncollected and out of print, but her writing offers a wealth of information about sea travel and colonial travelers in the first half of twentieth century. At home among strangers as she tackled one adversity after another, Mordaunt is a writer whose stories reward readers with views of the world shaped by her poignant humor. That she saw so much of the world with so few resources attests to her independence, endurance, and ingenuity. She died in the Radcliffe Infirmary, Oxford, in 1942.

Reference:

Dale Spender, *Writing a New World: Two Centuries of Australian Women Writers* (London: Pandora, 1988).

Papers:

The Bodley Head files in the University of Reading Library contain the largest collection of Mordaunt's correspondence, fifty-four letters written from 1923 through 1926. Small collections of letters are owned by the Richmond Central Library, Richmond, Surrey; the University of Newcastle upon Tyne Library; and the state Library of Victoria, Melbourne.

Marianne North
(24 October 1830 – 30 August 1890)

Maura O'Connor
University of Cincinnati

BOOKS: *Recollections of a Happy Life: Being the Autobiography of Marianne North,* 2 volumes, edited by Catherine North Symonds (London: Macmillan, 1892; London & New York: Macmillan, 1893);

Some Further Recollections of a Happy Life, Selected from the Journals of Marianne North, Chiefly between the Years 1859 and 1869, edited by Symonds (London & New York: Macmillan, 1893);

A Vision of Eden: The Life and Work of Marianne North, edited by Graham Bateman (New York: Holt, Rinehart & Winston, 1980; Exeter: Webb & Bower, 1980).

Marianne North was an exceptional Victorian traveler. She had the good fortune of accompanying her father, Frederick North, sometimes a member of Parliament for Hastings, on his professional and leisurely travels abroad. When her mother – whom Marianne North identified only as "the beautiful widow of Robert Shuttleworth of Gawthorpe Hall in Lancashire" and "the eldest daughter of Sir John Marjoribanks, Bart., of Lees, MP for Berwickshire" – was quite ill and near death in 1855, Mrs. North made her daughter promise never to leave her father. This promise ironically proved to be liberating for the young woman, for until his death in 1869 she traveled throughout Europe and the far reaches of the British Empire with her father. After 1869 she set out alone in search of natural wonders. From 1871 until 1882 she continued to travel, as she visited almost every continent and many islands in the Pacific and Atlantic Oceans. As an amateur botanist she has been called an "English botanical explorer" because her adventures were in search of flowers, tropical foliage, trees, and other natural wonders.

She was also a talented painter, as were many middle-class and upper-middle-class British travelers abroad during the nineteenth century. Her legacy, both artistic and scientific, she left to all travelers and lovers of nature. In 1882 the distant botanical wonders of the British Empire, the delights of

Marianne North, photographed by Julia Margaret Cameron in Sri Lanka in 1877

the California mountain ranges, the tropical flowers of South America, and the plants of Japan were made available to the public at the Royal Botanic Gardens at Kew, outside London. Because North agreed to finance the building, 832 of her paintings are displayed in the gallery that she helped design and construct. She supervised the architectural designs of James Fergusson as well as the arrangements of paintings in the gallery, which bears her name. She also painted the frieze and decorations surrounding the doors, and she used pieces of wood collected from her travels, some 246 different kinds, to frame the paintings on the walls. Her gallery attests beautifully to the spirit of adventure, the scientific curiosity, and the artistic talent of a Victorian woman traveler, and it also embodies a legacy of privilege, wealth, and empire.

Marianne North was born on 24 October 1830 in the lush botanical setting of Hastings, and her social background contributed directly to her career as a world traveler and amateur botanist. She descended from well-connected gentry – renowned intellectuals, politicians, and successful businessmen – on both sides of her family. Painting and learning about horticulture on three beautiful estates in Hastings, Norfolk, and Lancashire, she spent her childhood and youth playing with an older half sister, a younger sister, and an older brother. North explains early in her memoirs what prompted her love for botany and travel when she confesses that, as a young adolescent, she found her summer's delight in collecting and painting varieties of British fungi.

Another summer's delight was traveling on the Continent. In fact, when North first set out for the Continent with her family in August 1847, they stayed abroad for three years. She contracted typhoid fever, from which she recovered, and witnessed revolutions in central Europe and France. But she was not interested in political intrigue. She preferred German music and the surrounding scenery to politics. Given the habits and itineraries of many wealthy English families, the extensive and prolonged European traveling that her family enjoyed was customary in the early Victorian period. Her writings and paintings show, however, that the Continent did not hold for her a fascination as great as it held for many of her contemporaries. Perhaps this is partly attributable to her great interest in botanical discovery. The Continent was too cultivated for Marianne North, who sought more-exotic landscapes and new species of plants, flowers, and trees.

Her father's influence profoundly shaped the direction of North's life. After her mother died, North accompanied her father on his travels abroad and took her mother's place as mistress of several large households. Partly because of Frederick's political connections, the North family entertained such well-known nineteenth-century people as Elizabeth Cleghorn Gaskell and Sir Francis Galton and his family. Other influential visitors included Sir Edward Sabine, the president of the Royal Society, and his wife, Lady Sabine, as well as Sir William Hooker, a professor of botany at Glasgow University and the creator and first director of the Royal Botanic Gardens.

When North and her father lived in their town house in London after her mother had died, they frequented the Royal Botanic Gardens, where North began to collect and paint samples of rare plants. She confesses in her memoirs that her interests in these specimens prompted her desire to see tropical lands and to travel in search of new species of plants. At home in England, she writes, she and her father

> rode often to Chiswick Gardens and got specimen flowers to paint; were also often at Kew, and once when there Sir William Hooker gave me a hanging bunch of the *Amherstia nobilis,* one of the grandest flowers in existence. It was the first that had bloomed in England, and made me long more and more to see the tropics.

Although Frederick North enjoyed collecting plants, he was less inspired than his daughter was to travel in search of exotic botanical wonders. When he died North was thus free to establish her travel itinerary to those distant places. Her father's friends and extensive network of political and scientific connections made it possible for her, certainly at first, to travel far off the beaten track, and she acknowledges this as a matter of course in her memoirs. At the beginning of a chapter on her travels through Borneo and Java, for example, she explains how she has made her way there from Singapore: "After a fortnight at Government House, Sir William [Hooker] wrote me letters to the Rajah [Sir James Brooke] and Rani [Lady Margaret Brooke] of Sarawak, and I went on board the little steamer which goes there every week from Singapore."

Economic security, social status, and connections allowed North to venture where few privileged Englishmen had gone in the nineteenth century. Yet privilege does not explain North's desire to travel in search of new plants, flowers, and trees, nor does it explain her willingness to endure physical discomfort to paint rare plant species. Botany interested many European and American women, but North's botanical endeavors were unusual because, without supervision, she journeyed to such remote places to satisfy her scientific curiosity and then rendered her findings with great accuracy. As Susan Morgan notes, North discovered four new species of plants, and her travel writings and paintings reveal continuing Victorian preoccupations about the British Empire.

North's travels and work certainly were connected with imperialism. Of all of her father's friends and acquaintances, perhaps none was so valuable to North as her association with Sir Wil-

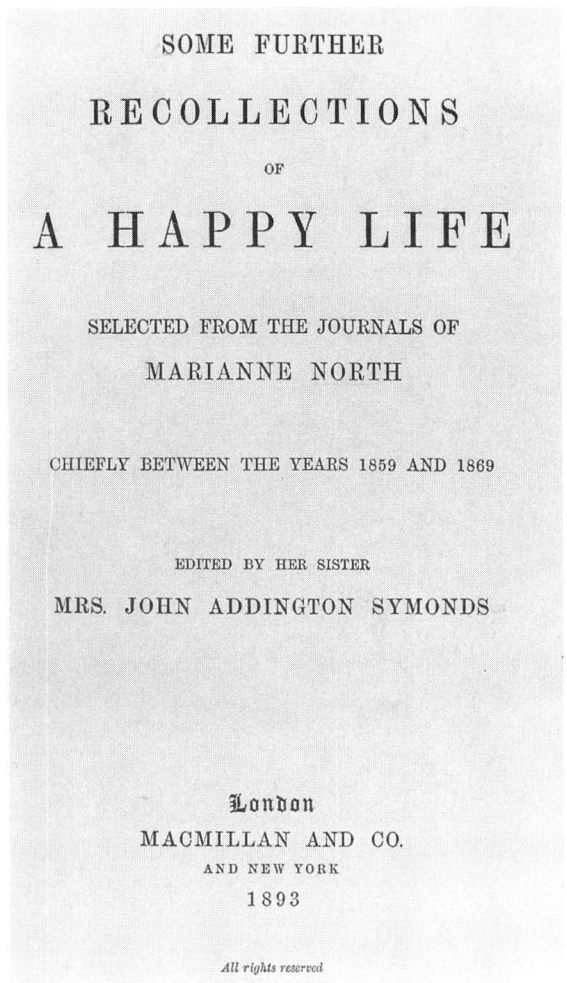

Frontispiece and title page for the second series of North's posthumously published memoirs

liam Hooker and her close friendship with Hooker's son, Sir Joseph, a renowned botanist who succeeded his father as director of the Royal Botanic Gardens. As Lucile H. Brockway has shown, the father and son together set up gardens in many British colonies, and as directors at Kew they had the privilege of nominating and appointing the directors of these colonial botanical gardens from some of the most talented botanists trained at Kew. North's close association with the Hookers gave her access to a scientific community that would otherwise have been closed to her as a woman. It also brought her a special diplomatic role to perform for these two botanists, for she carried messages and greetings from them, as directors at Kew, to their ambassadors in the colonial gardens.

In her travel memoirs composed between 1886 and 1888 and published posthumously, North delights in sharing with her readers the botanical details and discoveries of her scientific adventures. She carefully describes the Canadian wilderness as well as the more exotic landscapes of Brazil, Java, and Borneo. In her travel book North is cheerful, unself-conscious, and amazingly resilient and resourceful, demonstrating qualities that her late-nineteenth-century audience both appreciated and admired. She also displays her prodigious memory, which was to shape her stories. Her sketches, her paintings, and the itineraries of her travels as well as her diary or engagement book were the sources she could use to craft her travel narratives, but the passing of time between North's travel experiences and the writing of her narrative of those experiences shapes and edits what happened.

Although she is not a consistently good storyteller, North knows how to pique her reader's curiosity, as do many good travel writers. A bold, confident, and quietly self-possessed writer, she generally cultivates the manners of her social class and expresses opinions acceptable to her audience. At

North at Alderley in 1887

the same time, however, as an explorer and a traveler recounting her experiences she is stubbornly independent, pays little attention to the narrow dictates of society, holds rather negative views of marriage, and never doubts her talents or abilities. Her social position, privileges, and good manners helped North to travel around the world with a certain ease. Despite being an unconventional woman, she made most people feel comfortable.

When she traveled to the United States, for example, she went to considerable lengths to visit former family servants who had immigrated to the Midwest after the North estate in Hastings had been sold. On this same trip she also visited members of the Charles Francis Adams family of Quincy, Massachusetts, as well as members of other famous Boston families, and she seems to have been perfectly at ease whether traveling by freight train or in a first-class carriage. Generally unself-conscious among people whom she meets on her journey, she reflects very little on social engagements and acquaintances and is more interested in the places and the special botanical wonders that she discovers. Such a fondness for geography and nature makes her writing fascinating; as a travel writer North effectively draws her readers into her adventures.

North is best at describing in her travel memoirs what she is best at painting: the plants, flowers, and trees that betoken her fascination with botany and arouse her curiosity about new species. Her depictions of the people, particularly of British colonies and nonwestern lands, resemble spectacles and are sometimes artificially rendered, in contrast to her descriptions and representations of the glorious autumn foliage of New England, the tropical plants of Borneo and Java, or the dramatic landscapes of the Indian subcontinent. While she is not ill-mannered or willfully harsh toward her Jamaican servants, for example, she clearly presents herself as superior to them in her writings. She never doubts or questions that "these old people" – one of whom is her male servant, "a coal-black mortal with a gray head" – are to serve her, draw her bath, and protect her while she investigates the beauty of an island – its fruits, plants, gorgeous trees, and flowers. She represents the indigenous people of Jamaica only when they get in the way, or at times in the middle, of her botanical landscapes, as they do in her account of a West Indian vista in December 1871, when she recalls

> golden-flowered allamandas, bignonias, and ipomoeas over everything, heliotropes, lemon-verbenas, and geraniums from the long-neglected garden running wild like weeds: over all a giant cotton-tree quite 200 feet high was within sight, standing up like a ghost in its winter nakedness against the forest of evergreen trees, only coloured by the quantities of orchids, wild pines, and other parasites which had lodged themselves in its soft bark and branches. Little negro huts nestled among the 'bush' everywhere, and zigzag paths led in all directions round the house.

North's account, like much Victorian travel writing, devotes a disproportionate amount of space to these natural wonders that she views as hers to enjoy, collect, and cultivate. The only society that she describes in any detail is British society spread out across the world. The Jamaicans, Indians, Indonesians, Brazilians, and even at times the Canadians and United States citizens whom she encounters along the way inhabit her travel account only to ease the difficult conditions of some journeys – to cook, clean, and make sure that she is in the best position to undertake her scientific investigations. Her arrogance is usually couched in good manners, but that arrogance and a sense that she is entitled to such privileges are obvious to readers of her travel books. These stand as evidence of an imperial identity that many Victorian travelers and travel writers such as she helped to forge, an identity predicated on and maintained by the belief that their being white and English made them racially and culturally superior.

Visitors walking through the Marianne North Gallery at Kew Gardens must be struck by how precisely she organized her life's work and traveling records. One large room displays records of her journeys through botanical representations and picturesque scenes of daily life in Australia, New Zealand, Java, Japan, Borneo, Singapore, India, Ceylon, America, Jamaica, Brazil, and Chile. The innermost part of the gallery continues to record her journeys through the Cape of Good Hope, South Africa, and the Seychelles islands. Her meticulous arrangement and aesthetic judgment are certainly apparent. Light from the windows beautifully illuminates the wood samples that she collected and her vivid oil paintings of various species of flowers and plants. Her paintings of the flowers are particularly expert; her paintings of the indigenous peoples of Japan, India, Brazil, and Java – much like her representations of those peoples in her travel memoirs – are simplistic and, not surprisingly, picturesque. The gallery reflects her fondness for plants and exotic fauna and her lack of attention to the world's peoples and their diverse cultures.

North's accomplishments reveal an arrogance bred of privilege and prerogative. Her paintings and travel writings represented and delineated the British Empire to metropolitan readers because she saw that empire as her own – to consume, tame, and finally arrange as she saw fit. As a wealthy, white Englishwoman, she never doubted that the world was her garden. As Morgan observes,

> North could be an amateur botanist because the Royal Botanic Gardens had impressive garden outposts in many tropical countries whose functions were in large part to cultivate plants in order to service and further British imperial policies. North could be an independent woman traveler because upper-class English people lived in elaborate bungalows all around the world governing local laborers ... and they were delighted to put her up as part of their function of disseminating British culture and supporting the expansion of British knowledge.

The Marianne North Gallery at the Royal Botanic Garden and *Recollections of a Happy Life* (1892) are two vivid and concrete reminders of important ways in which this well-connected and independent woman traveler, who escaped the tedious demands of genteel domesticity and society, helped to shape, articulate, and represent an imperial culture that defines nineteenth-century England.

References:

Dea Birkett, *Spinsters Abroad: Victorian Lady Travellers* (London: Blackwell, 1988);

Lucile H. Brockway, *Science and Colonial Expansion: The Role of the British Royal Gardens* (New York: Academic Press, 1979);

Shirley Foster, *Across New Worlds: Nineteenth-Century Women Travellers and Their Writings* (Hempstead: Harvestor Wheatsheaf, 1950);

Barbara McCrimmon, "Marianne North," *Manuscripts*, 41, no. 4 (1989): 283–293;

Billie Melman, *Women's Orients: English Women and the Middle East, 1718–1918* (Ann Arbor: University of Michigan Press, 1992);

Dorothy Middleton, *Victorian Lady Travellers* (New York & London: Routledge & Kegan Paul, 1965);

Sara Mills, *Discourses of Difference: An Analysis of Women's Travel Writing and Colonialism* (London & New York: Routledge, 1991);

Brenda Moon, "Marianne North's *Recollections of a Happy Life*: How They Came to Be Written and Published," *Journal for the Society for the Bibliography of Natural History*, 8, no. 4 (1978): 497–505;

Susan Morgan, *Place Matters: Gendered Geography in Victorian Women's Travel Books about Southeast Asia* (New Brunswick: Rutgers University Press, 1996);

Morgan, Introduction to North's *Recollections of a Happy Life: Being the Autobiography of Marianne North*, 2 volumes (Charlottesville: University of Virginia Press, 1993): I, xi–xl;

Laura Ponsonby, ed., *Marianne North at Kew Gardens* (Exeter: Webb & Bower, 1990).

Papers:

The North family papers are located at Rougham, Norfolk; Marianne North's letters can be found at the Royal Botanic Gardens, Kew.

Frederick Courteney Selous

(31 December 1851 – 4 January 1917)

David C. Judkins
University of Houston

BOOKS: *A Hunter's Wanderings in Africa: Being a Narrative of Nine Years Spent amongst the Game of the Far Interior of South Africa, Containing Accounts of Explorations beyond the Zambesi, on the River Chobe, and in the Matabele and Mashuna Countries, with Full Notes upon the Natural History of All the Large Mammalia* (London: Bentley, 1881; London & New York: Macmillan, 1907);

Travel and Adventure in South-East Africa: Being the Narrative of the Last Eleven Years Spent by the Author on the Zambesi and Its Tributaries; with an Account of the Colonisation of Mashunaland and the Progress of the Gold Industry in That Country (London: Ward, 1893; New York: Arno, 1967);

Sunshine and Storm in Rhodesia: Being a Narrative of Events in Matabeleland Both before and during the Recent Native Insurrection up to the Date of the Disbandment of the Bulawayo Field Force (London: Ward, 1896; New York: Negro Universities Press, 1969);

The War in South Africa: Letters Contributed to the "Times" (London: National Press Agency, 1899);

Reconstruction in South Africa (London: South African Conciliation Committee, 1899);

Sport and Travel, East and West (London & New York: Longmans, Green, 1900);

Recent Hunting Trips in British North America (London: Witherby, 1907); republished as *Hunting Trips in North America* (New York: Scribners, 1907);

African Nature Notes and Reminiscences (London: Macmillan, 1908);

The Gun at Home and Abroad: The Big Game of Africa and Europe, by Selous, J. G. Millais, and Abel Chapman (London: Macmillan, 1914).

OTHER: *The Living Animals of the World: A Popular Natural History with One Thousand Illustrations*, edited by Selous, Charles J. Cornish, Sir Harry Johnston, and others (New York: Dodd, Mead, 1902);

Frederick Courteney Selous at age fifty-nine

John H. Patterson, *The Man-Eaters of the Tsavo, and Other East African Adventures,* with a foreword by Selous (London: Macmillan, 1907);

Hesketh Vernon Prichard, *Hunting Camps in Wood and Wilderness,* with a foreword by Selous (London: Heinemann, 1910);

Sir Arnold Wienholt Hodson, *Trekking the Great Thirst: Travel and Sport in the Kalahari Desert,* with a foreword by Selous (London & Leipzig: Unwin, 1912).

SELECTED PERIODICAL PUBLICATIONS – UNCOLLECTED: "Recent Exportations in Mashuna-land," *Proceedings of the Royal Geo-*

graphical Society, new series 3 (June 1881): 352-358;

"Further Explorations in the Mashuna Country," *Proceedings of the Royal Geographical Society,* new series 5 (May 1883): 268-271;

"Further Explorations in Matabese-land," *Proceedings of the Royal Geographical Society,* new series 10 (May 1888): 293-296;

"Mashunaland and the Mashunas," *Fortnightly Review,* 15 (1 May 1889): 661-666.

Frederick Courteney Selous was a big-game hunter, naturalist, explorer, collector, and writer whose life followed a pattern of adventure, courage, and bravery that many men held as an ideal. The adventures Selous recounts in his books were models for those of H. Rider Haggard's fictional character Alan Quatermain. Establishing a pattern for his life in which Africa was always to be as important as England, Selous first went to Africa at the age of nineteen and returned only briefly to England for visits. On one of his trips to England when he was forty-two years old, Selous married Gladys Maddy, the vivacious twenty-year-old daughter of a Gloustershire clergyman, and the two became parents of two sons.

The Selouses briefly tried life as settlers in Rhodesia, but Selous was destined never to settle anywhere for long. In Surrey they were to own one home, to which they attached a small museum to house Selous's collection of stuffed animals, butterflies, and birds' nests and eggs. Yet he continued to travel, to write accounts of his experiences, and to deliver popular lectures during short respites in England. He frequently addressed the Royal Geographical Society, which in 1893 awarded him a gold medal for his twenty years of exploration in southern central Africa.

Frederick Courteney Selous, one of five children of Frederick Lokes Selous and Ann Sherborn Selous, was born at the Selous home in Regents Park. Frederick Lokes was a member and chairman of the London Stock Exchange, and Frederick Courteney enjoyed a middle-class childhood. He attended Bruce Castle School in Tottenham and later went to Rugby, where he began to show his interests in natural history and collecting. Undisciplined and highly independent, he was a difficult child: he was particularly fond of bird-watching and collecting specimen eggs from birds' nests, although he seldom had the permission of his masters to do so.

At the age of sixteen, one year before he was to complete his studies, Selous left Rugby in 1868 and went to Switzerland, where he was to study French, music, and medicine. He had already announced his career plans, which were to shoot elephants in Africa. His father, however, hoped that Selous would find a conventional profession that would absorb his considerable energy. But after being in Switzerland only a few months, Selous announced that he was not interested in medicine, and for the next three years he drifted about Europe and returned frequently to London, apparently for family consultations. Following minor brushes with authorities, he was threatened with possible imprisonment for having struck an official who had attempted to arrest him for poaching. Selous was sent to South Africa by his father, who no doubt hoped that his son would avoid further trouble. Selous, nineteen years old and still restless, was haunted by his reading of William Charles Baldwin's *African Hunting from Natal to the Zambesi, Including Lake Ngami, the Kalahari Desert, &c. from 1852 to 1860* (1863).

Selous's first book, *A Hunter's Wanderings in Africa: Being a Narrative of Nine Years Spent amongst the Game of the Far Interior of South Africa, Containing Accounts of Explorations beyond the Zambesi, on the River Chobe, and in the Matabele and Mashuna Countries, with Full Notes upon the Natural History of All the Large Mammalia* (1881), provides an autobiographical account of his travels and hunting adventures and offers notes to guide other hunters. Selous arrived in Africa at Port Elizabeth in 1871 with £400, a considerable sum, but he knew no one. He immediately set out northward for the interior to carry out his plans to become an elephant hunter and support himself through ivory selling and perhaps miscellaneous trading.

By this time nearly all the elephants had been killed in the southern, more accessible areas, and, making frequent stops, Selous spent more than a year of slow travel to reach elephant country. He had purchased two lightweight, smoothbore, muzzle-loading, single-shot weapons that fired four-ounce bullets. Although these guns were accurate and successful in bringing down elephants, each weapon had a strong recoil. On one occasion when Selous's piece failed to discharge, his gun bearer, not realizing what had happened, reloaded it with a second charge of powder and bullet. When Selous refired the gun, he recalled that its recoil "lifted me clean from the ground, and turning round in the air, [I] fell with my face in the sand, whilst the gun was carried yards away over my shoulder." Selous remembered being "stunned," and he "soon found that I could not lift my right arm. Besides this, I was covered with blood, which spurted from a deep wound under the right cheek-bone, caused by the stock of the gun as it flew upwards from the vio-

 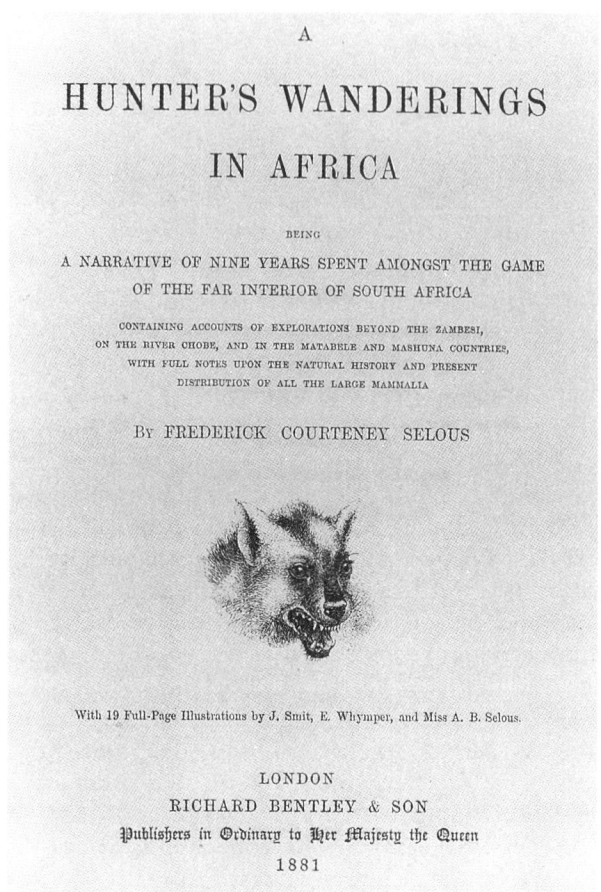

Frontispiece and title page for Selous's first account of his experiences as a hunter and trader in Africa

lence of the recoil." Yet Selous continued to pursue the elephant, which he killed when it charged him again.

Selous and his fellow hunters observed safety standards casually, at best, in their pursuit of elephants. Powder was measured only by the handful before being rammed into the muzzle loader. Native gun bearers were sent around the elephants, where they might be in the line of fire, to drive elephants toward the hunters. When accidents inevitably happened from such arrangements, the European hunters mourned little. "Poor Mendose!" wrote Selous after one hunt, during which another gun bearer's bullet had killed the servant; "he was an obedient, willing servant, and by far the best shot of all our native hunters."

Although readers today recoil in horror at the wholesale slaughter of elephants and other big game, Selous considered himself a conservationist. He did not kill animals simply because they were within his gun sights. He killed only for ivory or for the meat that he and his men needed, and his book describes much game that he did not kill and some elephants with insignificant tusks that he left undisturbed. But Selous notes that a successful hunter might have a large following of hungry people urging him to kill animals so that they might be fed.

From central southern Africa, Selous followed the Zambezi River north and west to reach Victoria Falls in June 1874. In recounting his trip Selous complains that his writing talent does not adequately convey the majesty of the falls before him. "I wish I could give you some idea of their wonderful grandeur and beauty! But the task is far beyond me." He nonetheless begins by trying to convey the size of the river before he describes the falls:

> The river tumbles into a narrow rend in the earth which runs right across its course.... One stands, it must be understood, on the very edge of the chasm [with the river], seemingly as still and quiet as a lake, flowing tranquilly on heedless of its coming danger, till with a crash it leaps in one splendid mass of fleecy, snow-white foam into an abyss four hundred feet in depth.

He summarizes in an uncharacteristic flourish, "Such are the Victoria Falls – one of, if not *the* most transcendently beautiful natural phenomena on this side of Paradise."

Later this same year, although lion hunting had not been his interest, Selous killed his first lion, a female, by chance when his party encountered a small pride in the bush. He never considered himself a great lion hunter, though others called him the greatest to come out of Africa. He seldom went after lions, except to defend his party or to kill a particular specimen. When Selous reached Tati on the Zambezi, he received the first mail he had read in three years and decided to return to England for a year.

Upon returning to Africa in 1876, Selous resumed his life as an elephant hunter. Game was becoming scarce, however, and he joined other hunters to find more remote areas where valuable game had not yet been killed. These areas were seldom visited by armed Europeans because of natural barriers. For instance, some good hunting remained in areas infested by the tsetse fly. Hunters usually traveled by horseback and carried their supplies in wagons drawn by oxen, but neither horses nor oxen could be safely taken into "fly country," as domestic animals would not survive. Thus, porters not only had to carry all supplies but also had to carry out ivory and other trophies.

Another area where Selous hunted was north of the Kalahari Desert. His trek across the desert in 1879 was successful, for he lost no members of his expedition nor any of his animals, but it was a highly risky journey. After four days of travel with no water, Selous wrote that it was a "sudden relief to sight a pool of muddy water that lifted a weight of fear and anxiety from our hearts. Only an hour before it had seemed that I was doomed to lose all my live stock – nearly everything I possessed in the world from thirst; and now the danger was past."

The year 1879 was also that of the Zulu War, and although Selous was not involved, he sympathized with the Boers in Natal and the Transvaal. He recognized that the Zulus were a brave and proud people, and he admired their courage. He also knew that the Zulus and the Boers could not share a geographical region unless one was subordinate to the other. In a letter to his mother he wrote, "You think it [the Zulu War] was unjustifiable, but it was not so, for so long as the military power of the Zulus remained unbroken there could be no peace in South Africa and the white inhabitants of Natal and the Transvaal would have had an assegai constantly dangling over their heads." He saw little

Selous during his early years as a big-game hunter in Africa

glory in the British victory and felt sad for the Zulus, but the events seemed inevitable to him.

When England annexed the Transvaal, Selous remained ambivalent. He recognized that problems were likely when the Boers were to run their own affairs, but he saw little justification for the English position, which was little more than a grab for power and control. By the time armed resistance to British rule was mounted, Selous was back in England preparing his book for publication. *A Hunter's Wanderings in Africa* achieved immediate success and lifted Selous into the ranks of the foremost hunters and naturalists. The book established his reputation as a big-game hunter who recounted interesting tales, but the royalties it brought were small. Unlike writers of typical hunting or fishing tales, Selous does not boast or gloat about his successful kills or narrow escapes but rather tells his stories in a straightforward manner occasionally punctuated with understatement. Yet contemporary readers so criticized him for the amount of game he killed that he had to answer accusations of having indiscriminately slaughtered African game.

In the 1870s no controls existed on the killing of game in Africa. Hunters were not held to quotas, nor were hunting seasons established. No game parks existed as animal refuges. Most hunters, including Selous, believed that game was so abundant that it would last forever. When game was scarce hunters attributed its disappearance to their having driven off, rather than killed off, the animals.

Apart from criticism of the number of animals Selous had shot, his book was greatly admired. It provided readers with a vicarious experience of which they could only dream; as its hero, Selous appeared with all the attendant virtues: courage, tenacity, bravery, and a measure of good luck. Yet his narrative included an unvarnished honesty, sometimes in his understatement but always in a plain-faced truthfulness that was the hallmark of his writing. He became the standard-bearer of British heroism, not as a great general or admiral but as the educated gentleman who took the moral principles he had learned at public school and applied them to a wilderness far removed from the playing fields and classrooms of Rugby. Today's audiences may read *A Hunter's Wanderings in Africa* in ways much different from those of nineteenth-century readers, for it is not a story of cross-cultural understanding; it is an account of cultural domination. Selous never suggests that he learns much from the natives whom he hires to assist him in the hunts. Instead, he attempts to teach them the virtues of hard work, loyalty, and sacrifice, and he complains, sometimes with bitterness but more often with resignation, when the natives want to rest or give up the hunt for the day.

During all the time that Selous spends alone in the bush, readers never find him entertaining any reflections on his life or on the occupation he has chosen. Indeed, he gives the impression that he seldom plans more than a few months ahead and thinks little about the future. Selous does bother to learn enough of the natives' languages to be able to converse with the local chiefs and the porters of his parties. Although he only notes the natural beauty of the Matabele girls and does not mention his relations with native women in his book, Selous kept at least one African woman and fathered two children by her. He was the product of Victorian culture, and, though he may have rebelled against certain restrictive conventions of respectability, he did not emphasize or even acknowledge his breaks in print. He used the flora, the fauna, the people, and indeed the entire land for his own pleasure, but he was highly respected among his peers. He was certainly not a drunken despoiler, a debaucher setting himself up as ruler of some miserable, malnourished tribe in central Africa. Indeed, at his death years later Selous was eulogized as "a moral antiseptic in a country where men are not saints. Anything mean or sordid literally shriveled up in his presence."

Selous returned to South Africa in 1882, but in his letters he almost constantly complains of financial distress. The story of his next years of hunting and exploring is told in his second book, *Travel and Adventure in South-East Africa: Being the Narrative of the Last Eleven Years Spent by the Author on the Zambesi and Its Tributaries; with an Account of the Colonisation of Mashunaland and the Progress of the Gold Industry in That Country* (1893). He resumed hunting more in order to supply trophies and specimens to dealers and museums than to amass ivory. Although his commissions for specimens were constant, the expenses he incurred proved greater than the income he earned from them; he therefore had to borrow money to mount his expeditions, and his income barely paid the interest on his loans. Although his first book was selling well and was to be reprinted several times, Selous earned little money from it because of his financial arrangements with the publisher, Richard Bentley. Selous was more than thirty years old, but he had inherited none of his father's financial acumen. Throughout his life he continued to complain of losing money or of not making enough.

In 1887 he acted as a guide for J. A. Jameson, A. C. Fountaine, and Frank Cooper on a yearlong hunting trip into Mashonaland on the upper Manyami River. During this trip Selous and his party discovered the caves of Sinoia, which he later described in the *Proceedings of the Royal Geographical Society*. But on this expedition he also had serious difficulties with Matabele natives, many of whom were hired as servants and bearers for the hunters. In *Travel and Adventure in South-East Africa* Selous glosses over the incident in which the hunters stood by as natives killed more than 150 other natives for supposedly having violated the Selous party's agreement not to prospect for minerals. It must have been a horrific incident to witness, but Selous provides no information about it in his book or in articles he was writing for the Royal Geographical Society. His biographer J. G. Millais also writes nothing about it except to note that "no complete record seems to have been kept."

After this expedition Selous pushed north of the Zambezi in 1888. Although the natives of this area had bad reputations among the white hunters, the lure of abundant game and a desire to get away from Mashonaland made Selous determined to undertake a risky enterprise. He assembled a core party in Bamangwato and headed north toward his destination, seven hundred miles away. By the time his party crossed the Zambezi, it had grown larger with additional bearers carrying the goods to be bartered. But Selous's native bearers began deserting, and the size of the party was reduced to just twenty-five men.

One night natives from a nearby village attacked and overran the party. Twelve men were killed as six others were wounded, and Selous narrowly escaped with only his rifle and four cartridges. In this hostile land he traveled only at night, and occasionally he would stop at a native village to get directions or other help in order to reach the Zambezi. During one of these meetings his rifle was snatched from him and another attempt was made on his life before he could slip away into the long grass. Finally he reached the Zambezi, where he found natives who were more friendly and survivors of his own party who had independently escaped. Selous wrote that this experience was the nearest he had come to death during his wanderings in Africa, but it did not deter him from making future expeditions.

In the late 1880s this area north of the Zambezi and the other regions where he had traveled with Jameson were attracting new interest from empire builders in western Europe, particularly in England. The mass of land north and south of the Zambezi was lightly populated and generally unmapped, and the interests of empire builders had two aims – first, that of expanding the empire, and second (and more pragmatic), that of acquiring precious metals. Having finished his harrowing hunting trip, Selous emerged from this area and returned to England in 1889.

His last few years in Africa had been financially unsuccessful. Apart from acting as a guide to Jameson and his party, Selous's efforts at specimen collecting, elephant hunting, and trading had been unprofitable, as they had been throughout his nearly twenty years in Africa. He returned to England with a significant debt and only two prospects for paying it off. He could submit to his family, give up Africa, and settle down to manage the family finances – or he could join a group of other adventurers seeking to secure mining rights in the area he had explored. Selous probably never seriously considered the first alternative.

Because Selous was one of the few Englishmen who knew this area called Mashonaland, he found that his presence was often in demand in London, where interest in this area of Africa was high. Selous fanned this interest by writing an article on his latest African adventures for the *Fortnightly Review*. In part a promotion piece, the article summarizes the advantages of the land:

> This is in fact a country where European children would grow up strong and healthy, and our English fruits retain their flavor. The soil is very rich and fertile, and owing to the facilities for irrigation enormous quantities of wheat could be grown (it may be in ten years or it may be in a longer period, but sooner or later), to supply the large centers of population that will spring up in the gold-producing districts to the north and east of the plateau.

The article is also an adventure story. First Selous tells about a man-eating lion that was eventually killed by two courageous natives; then he tells his own hunting story of a large eland that, to the great delight of the natives following him, he eventually killed. Finally, with apparently disarming honesty, Selous asks why he has written this article. The answer is not as candid as his rhetorical question would suggest: Selous was seeking investors. No doubt he believed in what he was promoting, but the article was written by a man with a vested financial interest in promoting a real-estate development.

Selous returned to Africa in May 1889 to lead a prospecting expedition. When he reached Cape Town he found that speculation on new mining ventures was high, and with the help of friends he quickly raised £10,000 and formed the Selous Exploration Syndicate. Soon he and the party he gathered were steaming up the east coast of Africa to the Portuguese port of Quelimane. From there they proceeded up the Zambezi to Mashonaland, where they obtained a treaty for mining rights in the region. When he returned to Cape Town in December 1889, Selous was immediately summoned to Kimberley to meet with Cecil Rhodes, whom he had met in London before having returned to Africa earlier that year. At meetings between the two during the next few months, Rhodes convinced Selous that the mining treaty was worthless and that Selous should come to work for him. Selous seems not to have protected his shareholders, for Rhodes's "merger of interests" contained no compensation for shareholders of the Selous Exploration Syndicate; only Selous received a substantial retainer.

During these negotiations Selous persuaded Rhodes to open a road north for settlers to move into Mashonaland. Selous's argument was that the Portuguese would soon establish a valid claim to the region if the English would not do so, and the only real way to stake a legitimate claim was to populate the region. He further maintained that the

Frontispiece and title page for Selous's book on his experiences as a guide, hunter, explorer, trader, and developer of colonial enterprise

high plateau of Mashonaland, although presently without human habitation, would prove to be wonderful farmland. To cut such a three-hundred-mile road north and east from the camp of the British South Africa Company on the Macloutsi River to Fort Salisbury, in July 1890 Selous led an advance party of several hundred men. Apart from his literary contributions, this road was Selous's most enduring contribution to the development of Africa. He wisely led the settlers on an eastern route that skirted the hostile Matabeleland and brought them safely to the high, fertile Mashonaland plateau.

Making roads, negotiating with local tribes, and assisting in the settlement of the country, Selous continued working for the British South Africa Company until late 1892, when he returned to England. He owned much land in the recently opened territory and hoped to realize a gain from it when the railroad was pushed through and goods could be marketed more easily and inexpensively. During his last year in Africa he had worked on his second book, *Travel and Adventure in South-East Africa,* which was published upon his return to England. He also planned a lecture tour across the United States, a tour that was to culminate in a Rocky Mountain hunting expedition.

This was a time of extraordinary fulfillment for Selous. His fame was widespread as leader of the pioneer group that settled what was to become Rhodesia and is now Zimbabwe. His new book was at the publishers, who firmly believed that it would be a great success. The Royal Geographical Society conferred upon him its highest honor, the Founder's Gold Medal. Yet Selous was nearly forty-one years old when he returned to England before Christmas 1892, and he may never have established a close relationship with a white woman. Early in 1893 he became engaged to Gladys Maddy, the teenage daughter of a clergyman and a houseguest at his mother's home. An extraordinarily attractive and vivacious young lady, she was apparently as attracted to Selous as he was to her. Maddy's parents were apparently not so pleased, however, and plans for the marriage were set far in the future. In all, Selous had returned to a hero's welcome – complete with a fair young damsel.

As he prepared to leave for his tour of the United States, word arrived of a Matabele uprising in the recently opened territory, and without hesitation Selous canceled his tour, put his courtship on hold, and left immediately for South Africa. There he was soon to face the first of two Matabele uprisings. The first, in 1893, was relatively short, but in it Selous was wounded by a bullet to his right side while he galloped to the aid of a wagon train under attack. In December 1893 this revolt was quelled, and Selous returned to England.

On 4 April 1894 he and Gladys married and immediately went abroad to Switzerland, Italy, and Hungary for a honeymoon. Upon returning to England, they bought an estate in Surrey at Worplesdon, southwest of London and just north of Guildford. *Travel and Adventure in South-East Africa* had been published and was selling well. For this book Selous received much more income than he had from his first book, and although books on southern Africa were being published at a furious pace, Selous was recognized as the authority. He was a handsome, even dashing, figure on the lecture platform and could mesmerize audiences with his hunting tales, which he delivered with so much wholesome honesty that one reporter wrote, "You cannot be in his company a minute before you feel that Diogenes would have stopped his search and said: 'Here, at last, is an honest man.'"

Despite this popularity and financial success, less than a year after their marriage he and Gladys returned to Africa with the intent of managing a farm in the country that he had helped to open. The reason for this move is not clear. Selous complained that his income had fallen with the drop in value of his investments. No doubt he also wanted to show Africa to Gladys, and it would have been more surprising had he settled permanently in Surrey.

In 1895 Selous and his wife settled at Essexvale, in Southern Rhodesia. Although the land was peaceful and prosperous at first, it was actually a poor time to be in Rhodesia. Rinderpest, a fatal animal disease, was spreading across the land from the north and killing nearly all the domestic cattle as well as much of the game. A plague of locusts was nearly as destructive to the crops. Finally, and most important, the second Matabele uprising occurred with a fury that was not anticipated after the earlier, disorganized rebellion. The Matabele natives, having acquired Winchester rifles during the peaceful times, were much better armed and more unified than they had been.

Selous gives an account of the second uprising in his third book, *Sunshine and Storm in Rhodesia: Being a Narrative of Events in Matabeleland Both before and during the Recent Native Insurrection up to the Date of the Disbandment of the Bulawayo Field Force* (1896), published shortly after the end of the uprising and when he was back in England. The book is extraordinary in its balanced presentation. Certainly Selous, like almost every Englishman of the day, believed in racial superiority, but unlike armchair analysts, he does not ascribe stereotypical traits to the indigenous peoples. In fact, Selous blames this uprising less on the Matabele than on the mismanagement and poor judgment of the British South Africa Company. He attempts to correct the popular belief that the Matabele are somehow inherently brutal and so savage that they will always be violent and intractable. He points out that they are being treated like a conquered nation: the able-bodied men are required to work for the settlers or in the mines at a fixed rate of pay – ten shillings per month, with food provided. These men had been unaccustomed to regular work and did not take well to it. In addition, the confiscation of their cattle had made the natives unhappy, and when the cattle were confiscated not once but repeatedly, the Matabele were kept in a nearly constant state of agitation. The police protection that the British South Africa Company had provided for the settlers had also been withdrawn when it clearly should have been retained.

Despite the validity of these complaints, Selous assisted the settlers and the army in suppressing the rebellion. The Matabele killed isolated farmers and their families, and Selous quickly moved Gladys from Essexvale into Bulawayo, the main British stronghold. During the next months Selous led patrols from Bulawayo to recover cattle, rescue farmers, and dislodge the Matabele as they moved against the settlers' stronghold. In none of these actions was Selous wounded, though by his own account he had several close calls; Gladys later commented, "he will always be to the fore in everything – but I am proud of having such a husband."

In May 1896 two relief columns of British soldiers were converging upon Bulawayo and killing many Matabele. Although the rebellion was effectively suppressed by the end of June, Selous decided to give up trying to become a settler, and, after another hunting excursion, he returned to England with his wife late in 1896. *Sunshine and Storm in Rhodesia* was released before the end of the year, and a careful reading of the book reveals that Selous was making a final break with the British South Africa Company and with Cecil Rhodes. In addition to his criticisms of their policies, Selous retreats from his earlier optimism about Rhodesia's future. He writes that it will be years, perhaps decades, until profitable farming can be undertaken. He cautions anyone planning to settle in the newly opened area to be prepared for severe hardship and few financial rewards. In short, Selous's brief career in land development and promotion was over.

Having returned to England, Selous resumed his interest in egg and nest collecting, a hobby that he had begun while he was a schoolboy. He traveled as far as Turkey and as near as Reading to locate unusual birds' eggs and nests, and he hunted them with his customary tenacity. In April 1898 his wife gave birth to their first son, Freddie, and late in 1899 their second son, Harold, was born. Throughout this time Selous continued to travel. In 1898 he left for the United States to hunt in the Rocky Mountains. He hoped to meet Theodore Roosevelt, with whom he had been corresponding frequently since Roosevelt had read his first two books in the mid 1890s, but Selous was unable to meet him. Roosevelt respected Selous both for his exploits as a big-game hunter and for his ability to write about those exploits in an open, unadorned style.

The outbreak of the Boer War in 1899 was the most difficult period of Selous's life. His country had initiated a war against a people among whom Selous had lived and whom he much admired. He believed that this war brought out the worst in British character – an arrogant, close-minded superiority that would harvest decades of bitter hatred, anger, and despair. Selous therefore wrote a series of letters to the London *Times* and *Morning Post,* and in these letters he demonstrates a courage that was far greater than what he had shown when he stared down a lion or pushed on after a wounded elephant following the recoil of a double-loaded gun that had nearly killed him.

His letters oppose the British and defend the Boers as a conservative but peaceful people who in no way deserve the treatment that they are receiving from the British. He notes that England is less interested in protecting the English subjects who have settled among the Boers than in seizing the newly discovered gold mines in the Witwatersrand, and he passionately argues that the differences between the British and the Boers can be resolved without warfare. He correctly foresees that a war will last far longer than the six months predicted by the British authorities and that it will create a deepening hatred that will not pass for several generations. Ultimately, he believed, regardless of the outcome of the current war, South Africa would finally be independent from Great Britain. Selous respected the Boer character and knew that it would not be easily broken.

In a 24 October 1899 letter to the *Times* Selous adopts his typical straightforward, personal argument. Speaking of his experience with the Boers, he writes: "I met with no insults nor the least incivility anywhere, nor ever heard any boasting about Boer success over our troops." In denying this, Selous was denying a familiar complaint directed against the Boers by the British. "In common with all who really know the Boers, who have lived amongst them, and not taken their character at second-hand, I have always been struck by their moderation in speaking of their victories over our soldiers." Not content simply to attest to the admirable character of the Boers, he contrasts it with that of the British: "As for the Boers having a contempt for Englishmen as individuals, that is nonsense. They hate the British Government, and knowing their history, I for one think they have ample reason for doing so. But the individual Englishman that they know, they take for his real value."

Selous, like most men who preach moderation and sanity at the beginning of a war when patriotic fever is high, was strongly criticized both by the press and by friends. It was a difficult period for the former hunter and explorer, and he wrote to Roosevelt, "It is a very bitter grief to me not to be able to side with my own country in this war, and I feel it so much that were I a younger man and unmarried I would leave this country and settle in America." In fact, Selous left England, but not permanently.

In spring 1900 he was in Hungary collecting eggs and nests. Later that year while his home in Surrey was being remodeled and enlarged and the museum to house his trophies was also being expanded, Selous went to Canada to hunt moose. In 1901 he returned to North America to hunt in Newfoundland, and after a brief visit home he left for Asia to look for bird eggs and to hunt. Not until November 1902 did he return to Africa – but not South Africa. Instead, he turned his attention to East Africa, as he arrived in Mombasa in late November for a hunting safari.

Selous's first three books were his most successful; his last three were not as financially rewarding. *Sport and Travel, East and West* is difficult to lo-

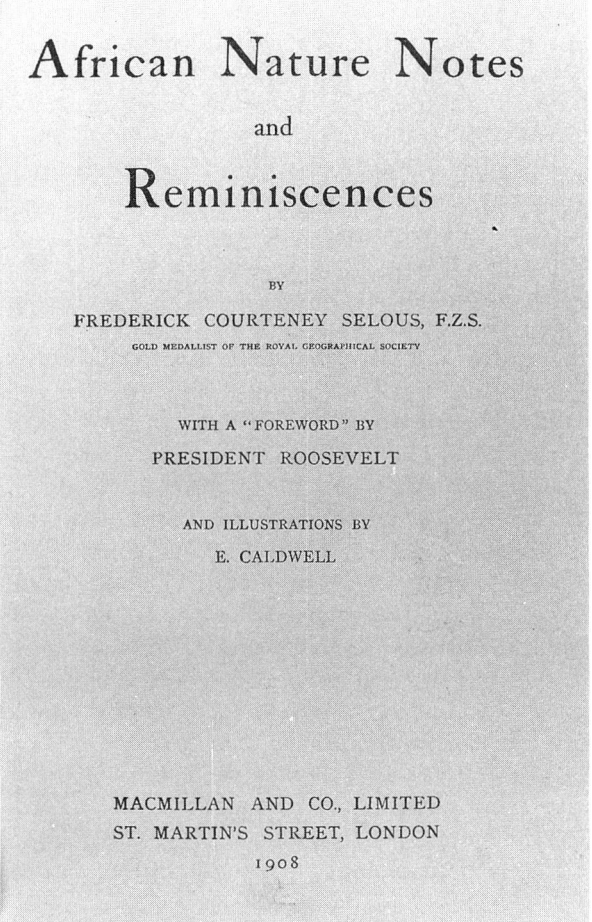

Frontispiece and title page for the last, and least popular, of Selous's African books – a work aimed at imparting information rather than relating adventure stories

cate today and was not much easier to find when it was first published in 1900.

Selous called *Recent Hunting Trips in British North America* (1907) "the record of the few unimportant hunting trips which have been the outcome of my restlessness during the last few years." The book is much more than that. Never satisfied with a typical hunting trip, Selous always sought areas where few others had walked. For instance, while hunting caribou in Newfoundland, he notes, "Between the rivers there are stretches of country which may be said to be absolutely unknown – pathless wastes of marsh and forest, studded with countless little lakes and ponds never yet looked upon by the eye of civilized men." It is this type of country to which Selous is attracted and in which he hunts most successfully. North America was not the same as Africa in the minds of readers, however, and the book did not sell particularly well.

One year later Selous's *African Nature Notes and Reminiscences* (1908), with an introduction by Theodore Roosevelt, was published. Although Selous had adopted a more scientific style and wanted the book to be read as the notes of a naturalist rather than the adventures of a hunter, the public was apparently not prepared for this change. This book, which Roosevelt so admired, was also a poor seller; it was never reprinted and soon was out of print.

Selous's passion for hunting, collecting, and travel nevertheless continued. Perhaps the most notable expedition of his later years was in sailing to Mombasa with Roosevelt in 1909. Roosevelt had just finished his term as president of the United States, and he had sought Selous's advice and counsel on hunting in Africa. Selous had devoted much time to planning such a trip, and, although he and Roosevelt apparently had never actually hunted together, at nearly the last moment it became possible for him to join Roosevelt on an outbound steamer from Naples to Mombasa. Roosevelt had invited him, but Selous privately complained that Roosevelt was accompanied by such a crowd of followers that hunting would lose its appeal. Selous may also have shied away from the trigger-happy former president, who apparently blasted away at nearly anything that moved.

In 1910 Selous was the English representative at the Second International Congress of Field Sports held in Vienna, where Selous worked to pass resolutions to protect migrating birds. The next year he returned to the lower Sudan of Africa, where he looked for a specimen of the great elands for the Natural History Museum. The trip was not particularly successful, and Millais reports that Selous was ill at the end of the trip. Unable to ride his horse, "he had to tramp the whole way to the Nile on foot." Following his return to England and an operation for hemorrhoids, from which he quickly recovered, Selous was back in British East Africa later that year to hunt with his old friend, W. N. MacMillan. On this trip Selous was nearly sixty years old and was beginning to slow down; some encounters with lion and buffalo proved to be particularly dangerous. Selous subsequently spent more time at home, and except for a trip to Iceland to collect eggs, he seems to have confined his restlessness to relatively tame shooting trips around England or visits to offshore islands to collect eggs.

Yet in August 1914 when World War I broke out, the sixty-three-year-old Selous offered his services to his country. England was reluctant to induct Selous as a combat soldier, despite the fact that a major life insurance company had given him a medical report that pronounced him fit and ready for active duty. For a time he had to satisfy his patriotic impulses by serving as a special constable under the Defense of the Realm Act. But by March 1915 he had been appointed as a lieutenant in the Royal Fusiliers, and on 4 May he landed with his men in Mombasa. Little more than a month later Selous and his men were deep in Africa to capture a wireless station from the Germans. As reprinted by Millais, notes left by Selous provide an account of the assault written in his typically understated style:

> Presently Sergeant-Major Bottomly of C Company came across the swamp, and lay down alongside of me, or at least separated from me by just a yard. My black boy Ramazani lying between us, but a little lower down, so that his head was on a line with my hips. I just said a word to Bottomly, and then turned my head away from him again to look in the direction from which the bullets were coming. Almost immediately my black boy Ramazani touched me, and said: "Master, soldier hit, dead." I had never heard a sound, but turning my head I saw poor Bottomly lying on his back, stone dead, with a bullet though his head.

Following this loss, Selous turns to the larger conflict:

> Our ships had now crept right into the Bay of Bukoba, and as they fired on the town, or the enemy's gun positions, their shells came screaming and whistling over us. The machine-guns were going too with their wicked rattle, and bullets from snipers' rifles came with an unpleasant sound, sometimes apparently within a few inches of our bodies, which were just then pressed as close to the ground as possible. I thought, as I lay there only a yard away from the blood-stained corpse of poor Sergeant-Major Bottomly, listening to the peculiar noise of each kind of projectile as it found its invisible course through the air above and around me, that I could recall various half-hours of my life passed amidst much pleasanter surroundings.

Fearing that his reader might find him over-impressed with this relatively insignificant engagement in Africa, Selous hastily adds,

> Yet what a small and miserable thing this was, after all, in the way of battle compared with the titanic combats which have been taking place in Europe ever since the greatest war in history commenced last August. I can well understand how the nerves of any man, however strong, may be shaken to pieces, by the awful clamour of the giant shells and the concentrated fire of many machine-guns, and countless numbers of rifles, and the terrible havoc wrought by these fearful weapons of destruction.

After this engagement neither Selous nor his men saw much action against the enemy. His service consisted of many marches and patrols, and weather conditions were bad. The days were hot, often rainy, and the men seemed constantly wading in a sea of mud. Even though Selous was a captain, he marched with his men and shared their army rations and lack of shelter at night. These miserable conditions took their toll, as the troops were depleted far more by illness than by occasional sniper fire. Again suffering from hemorrhoids, Selous was sent home for an operation. He was in England for only a few weeks and promptly returned to Africa, when he and four hundred new recruits with the Twenty-fifth Royal Fusiliers landed at Mombasa in August 1916. They proceeded inland and then down the coast toward Dar es Salaam. On this journey Selous learned that he had been awarded the

Distinguished Service Order for gallantry and bravery.

In December Selous and his troops, as part of a larger contingent attempting to encircle the German troops and capture the remaining forces in East Africa, reached Kissaki in Tanganyika. On 4 January the Royal Fusiliers were in the forefront of the encirclement and were attempting to close a road that would prevent the Germans from escaping. The country was rough, with heavy bush making it difficult to see the enemy clearly. According to contemporary accounts, the Royal Fusiliers were outnumbered five to one. Gunfire was heavy, and when Selous advanced in order to see the enemy positions better, he was shot in the head. Greatly mourned by his men and apparently by all who knew him, he was buried a short time later at the foot of a tamarind tree.

Selous's body was never returned to England and remains buried in the land he loved. Although he wrote six books and many articles on Africa, he was best known as the living embodiment of the English colonial gentleman who exuded his English resolution and confidence worldwide. Even at a time of great carnage in Europe, the world seemed to pause at the passing of such a notable figure, as people voiced both sorrow for his death and admiration for what he had achieved. His wife expressed apparently widespread sentiments: "It has been a fine ending to a wonderful life."

A huge game preserve in Tanzania encloses the area where he spent the last weeks of his life. Fittingly named the Selous Game Reserve, a full description of the area is provided in Peter Matthessen's *Sand Rivers* (1981).

Biographies:

J. G. Millais, *Life of Frederick Courtenay Selous, D.S.O.* (London: Longmans, Green, 1919);

Stephen Taylor, *The Mighty Nimrod: A Life of Frederick Courteney Selous* (London: Collins, 1989).

References:

R. Blair, "Selous – A Reassessment," *Rhodesiana,* 17 (1967): 31-43;

"Mr. Frederick Courteney Selous," *Review of Reviews,* 8 (March 1893): 256-269;

Bertha Selous Philips and Elizabeth Walsh, "Selous and His Road," *African World Annual,* 47 (1951): 117-122.

Papers:

Selous's journals, letters to his mother, and other papers are housed by the Zimbabwe National Archives, Harare.

Robert Louis Stevenson
(13 November 1850 - 3 December 1894)

Gordon Hirsch
University of Minnesota

See also the Stevenson entries in *DLB 18: Victorian Novelists After 1885; DLB 57: Victorian Prose Writers After 1867; DLB 141: British Children's Writers, 1880-1914; DLB 156: British Short-Fiction Writers, 1880-1914: The Romantic Tradition;* and *DS 13: The House of Scribner, 1846-1904.*

BOOKS: *The Pentland Rising* (Edinburgh: Privately printed, 1866);

An Appeal to the Clergy (Edinburgh & London: Blackwood, 1875);

An Inland Voyage (London: Kegan Paul, 1878; Boston: Roberts, 1883);

Edinburgh: Picturesque Notes, with Etchings (London: Seeley, Jackson & Halliday, 1879; New York: Macmillan, 1889);

Travels with a Donkey in the Cevennes (London: Kegan Paul, 1879; Boston: Roberts, 1879);

Virginibus Puerisque and Other Papers (London: Kegan Paul, 1881; New York: Collier, 1881);

Familiar Studies of Men and Books (London: Chatto & Windus, 1882; New York: Dodd, Mead, 1887);

New Arabian Nights (2 volumes, London: Chatto & Windus, 1882; 1 volume, New York: Holt, 1882);

The Story of a Lie (London: Hayley & Jackson, 1882); republished as *The Story of a Lie and Other Tales* (Boston: Turner, 1904);

The Silverado Squatters (London: Chatto & Windus, 1883; New York: Munro, 1884);

Treasure Island (London: Cassell, 1883; Boston: Roberts, 1884);

A Child's Garden of Verses (London: Longmans, Green, 1885; New York: Scribners, 1885);

More New Arabian Nights: The Dynamiter, by Stevenson and Fanny Van de Grift Stevenson (London: Longmans, Green, 1885; New York: Holt, 1885);

Macaire (Edinburgh: Privately printed, 1885);

Prince Otto: A Romance (London: Chatto & Windus, 1885; Boston: Roberts, 1886);

Robert Louis Stevenson in 1879

The Strange Case of Dr. Jekyll and Mr. Hyde (London: Longmans, Green, 1886; New York: Scribners, 1886);

Kidnapped (London: Cassell, 1886; New York: Scribners, 1886);

Some College Memories (Edinburgh: University Union Committee, 1886; New York: Mansfield & Wessels, 1899);

The Merry Men and Other Tales and Fables (London: Chatto & Windus, 1887; New York: Scribners 1887);

Underwoods (London: Chatto & Windus, 1887; New York: Scribners, 1887);

Memories and Portraits (London: Chatto & Windus, 1887; New York: Scribners, 1887);

Memoir of Fleeming Jenkin (London & New York: Longmans, Green, 1887);

The Misadventures of John Nicholson: A Christmas Story (New York: Lovell, 1887);

The Black Arrow: A Tale of the Two Roses (London: Cassell, 1888; New York: Scribners, 1888);

The Master of Ballantrae: A Winter's Tale (London: Cassell, 1889; New York: Scribners, 1889);

The Wrong Box, by Stevenson and Lloyd Osbourne (London: Longmans, Green, 1889; New York: Scribners, 1889);

Ballads (London: Chatto & Windus, 1890; New York: Scribners, 1890);

Father Damien: An Open Letter to the Reverend Dr. Hyde of Honolulu (London: Chatto & Windus, 1890; Portland, Maine: Mosher, 1897);

Across the Plains: With Other Memories and Essays (London: Chatto & Windus, 1892; New York: Scribners, 1892);

A Footnote to History: Eight Years of Trouble in Samoa (London: Cassell, 1892; New York: Scribners, 1892);

Three Plays: Deacon Brodie, Beau Austin, Admiral Guinea, by Stevenson and W. E. Henley (London: Nutt, 1892; New York: Scribners, 1892);

The Wrecker, by Stevenson and Osbourne (London: Cassell, 1892; New York, 1892);

Island Nights' Entertainments: Consisting of The Beach of Falesá, The Bottle Imp, The Isle of Voices (London: Cassell, 1893; New York: Scribners, 1893);

Catriona: A Sequel to Kidnapped (London: Cassell, 1893; New York: Scribners, 1893);

The Ebb-Tide: A Trio and a Quartette, by Stevenson and Osbourne (Chicago: Stone & Kimball, 1894; London: Heinemann, 1894);

The Body-Snatcher (New York: Merriam, 1895);

The Amateur Emigrant from the Clyde to Sandy Hook (Chicago: Stone & Kimball, 1895; New York: Scribners, 1899);

The Strange Case of Dr. Jekyll and Mr. Hyde, with Other Fables (London: Longmans, Green, 1896);

Weir of Hermiston: An Unfinished Romance (London: Chatto & Windus, 1896; New York: Scribners, 1896);

A Mountain Town in France: A Fragment (New York & London: John Lane, 1896);

Songs of Travel and Other Verses (London: Chatto & Windus, 1896);

In the South Seas (London: Chatto & Windus, 1896; New York: Scribners, 1896);

St. Ives: Being the Adventures of a French Prisoner in England (New York: Scribners, 1897; London: Heinemann, 1898);

The Morality of the Profession of Letters (Gouverneur, N.Y.: Brothers of the Book, 1899);

A Stevenson Medley, edited by S. Colvin (London: Chatto & Windus, 1899);

Essays and Criticism (Boston: Turner, 1903);

Prayers Written at Vailima, With an Introduction by Mrs. Stevenson (New York: Scribners, 1904; London: Chatto & Windus, 1905);

Essays of Travel (London: Chatto & Windus, 1905);

Essays in the Art of Writing (London: Chatto & Windus, 1905);

Essays, edited by W. L. Phelps (New York: Scribners, 1906);

Lay Morals and Other Papers (London: Chatto & Windus, 1911; New York: Scribners, 1911);

Records of a Family of Engineers (London: Chatto & Windus, 1912);

The Waif Woman (London: Chatto & Windus, 1916);

On the Choice of a Profession (London: Chatto & Windus, 1916);

Poems Hitherto Unpublished, 2 volumes, edited by G. S. Hellman (Boston: Bibliophile Society, 1916);

New Poems and Variant Readings (London: Chatto & Windus, 1918);

Robert Louis Stevenson: Hitherto Unpublished Prose Writings, edited by H. H. Harper (Boston: Bibliophile Society, 1921);

When the Devil Was Well, edited by William P. Trent (Boston: Bibliophile Society, 1921);

Confessions of a Unionist: An Unpublished Talk on Things Current, Written in 1888, edited by F. V. Livingstone (Cambridge, Mass.: Privately printed, 1921);

The Best Thing in Edinburgh: An Address to the Speculative Society of Edinburgh in March 1873, edited by K. D. Osbourne (San Francisco: Howell, 1923);

Selected Essays, edited by H. G. Rawlinson (London: Oxford University Press, 1923);

The Castaways of Soledad: A Manuscript by Stevenson Hitherto Unpublished, edited by Hellman (Buffalo: Privately printed, 1928);

Monmouth: A Tragedy, edited by C. Vale (New York: Rudge, 1928);

The Charity Bazaar: An Allegorical Dialogue (Westport, Conn.: Georgian Press, 1929);

The Essays of Robert Louis Stevenson, edited by M. Elwin (London: Macdonald, 1950);

Salute to R L S, edited by F. Holland (Edinburgh: Cousland, 1950);

Tales and Essays, edited by G. B. Stern (London: Falcon, 1950);

Silverado Journal, edited by John E. Jordan (San Francisco: Book Club of California, 1954);

From Scotland to Silverado, edited by James D. Hart (Cambridge, Mass.: Harvard University Press, 1966);

The Amateur Emigrant with Some First Impressions of America, 2 volumes, edited by Swearingen (Ashland, Oreg.: Osborne, 1976–1977);

The Cevennes Journal: Notes on a Journey Through the French Highlands, edited by Gordon Golding (Edinburgh: Mainstream, 1978);

A Newly Discovered Long Story "An Old Song" and a Previously Unpublished Short Story "Edifying Letters of the Rutherford Family," edited by Swearingen (Hamden, Conn.: Archon, 1982; Paisley, Scotland: Wilfion, 1982);

Robert Louis Stevenson and "The Beach of Falesá": A Study in Victorian Publishing with the Original Text, edited by Barry Menikoff (Stanford, Cal.: Stanford University Press, 1984).

Collections: *The Works of R. L. Stevenson,* Edinburgh Edition, 28 volumes, edited by Sidney Colvin (London: Chatto & Windus, 1894–1898);

The Works of Robert Louis Stevenson, Vailima Edition, 26 volumes, edited by Lloyd Osbourne and Fanny Van de Grift Stevenson (London: Heinemann, 1922–1923; New York: Scribners, 1922–1923);

The Works of Robert Louis Stevenson, Tusitala Edition, 35 volumes (London: Heinemann, 1924);

The Works of Robert Louis Stevenson, South Seas Edition, 32 volumes (New York: Scribners, 1925).

A combination of artistic ambitions, a thirst for adventure and discovery, courtship of and eventual marriage to an American, and above all poor health turned Robert Louis Stevenson into a lifelong wanderer. His early travel writing demonstrates his keen eye for visual detail, his interest in particular people and situations encountered along the way, and his relative indifference to conventional tourist attractions. Stevenson is also admired for his ability to create and convey an image of himself as a somewhat bohemian, humorous, and introspective traveler. In his later travel writing, dealing with European immigration to America and describing the impact of European and American trade and settlement on the South Sea islands, Stevenson records significant historic events and social changes and develops an empathy with fellow British expatriates as well as with oppressed native peoples. Readers from Stevenson's time to the present have admired his precise, felicitous style, as well as his engaging personal voice. He may be the nineteenth-century travel writer whose footsteps have been most avidly followed by subsequent writers in this genre.

Robert Louis Stevenson was born on 13 November 1850 to a respectable Edinburgh family – "a family of engineers," as he called them in his autobiographical *Records of a Family of Engineers,* an unfinished work first published in the Edinburgh Edition (1896). His father, Thomas Stevenson, was a civil engineer, whose own father, Robert Stevenson, was a particularly well-known builder of lighthouses and harbors along the Scottish coast. The writer's mother, Isabella Balfour Stevenson, was a cleric's daughter. Stevenson's childhood home was, in fact, animated by Calvinist religious feeling. Not only was Thomas Stevenson a stern practitioner, but young Louis's nurse, Alison Cunningham, "Cummy," also filled his mind with religious literature and stories tinged with her Calvinism.

From infancy Stevenson was a sickly child, suffering from recurrent respiratory illnesses and coughs. These illnesses, which made him feel fragile and isolated as a youth, persisted throughout his life, taking the form of fevers, coughing, bronchial infections, and eventually "Bluidy Jack," a hemorrhage of the lungs that results in blood issuing from the mouth. Stevenson's delicate health dictated his being sent south to warmer climates at an early age, and his contributions to the art of travel writing derive significantly from his poor health and his interminable wanderings in quest of a salutary climate.

Stevenson attended Edinburgh University from 1867 to 1872, beginning with the study of engineering and drifting with little enthusiasm to law. As his studies progressed, he became increasingly bohemian and artistic. He dressed in a velvet jacket, had his first sexual experiences in the brothels of Edinburgh's Old Town, announced his agnosticism to his horrified parents, and in general declared his independence from the middle-class values of his family. He also began to write verse and literary essays, several of which represent his first efforts at travel writing.

In his early twenties, Stevenson traveled in Scotland, England, and France, undertaking frequent walking tours wherever he went. His cousin and official biographer, Graham Balfour, cites a "pencil list of towns in which he had slept, compiled about 1886," giving the totals as forty-six in England (nineteen more than once), fifty in Scotland (twenty-three more than once), seventy-four in France (thirty-one more than once), and forty in the

Stevenson with his father, Thomas, in 1866

rest of Europe (sixteen more than once). Balfour notes as well that "between 1871 and 1876 no less than nine of [Stevenson's essays] deal with travel or the external appearances of places known to him."

These early travel essays comprise an important portion of Stevenson's literary apprenticeship, although during the same period he also published short stories and essays of literary criticism, and he experimented with longer literary forms. Some of these early walking-tour essays – for example, "A Retrospect," written at Dunoon, Argyll, in 1870; "Cockermouth and Keswick," written between 1871 and 1873; and "A Winter's Walk in Carrick and Galloway," written in 1876 – were clearly provisional efforts, remaining as unpublished fragments during Stevenson's lifetime, though many of them were collected and published posthumously in *Essays of Travel* (1905). In many respects they are meditative, personal, or moral essays – less concerned with a particular place than with the moods, thoughts, perceptions, chance encounters, memory, personal history, and identity of the author. Thus, they derive from the tradition of the Romantic essay; the essayist whom Stevenson most admired and imitated was William Hazlitt.

Even though some of these early essays may be considered pedestrian in more ways than one, nearly all include a few striking thoughts or observations. The essay on "Roads," for example, first published in the December 1873 issue of *The Portfolio*, describes the "prolongations of expectancy" that winding roads can foster concerning something glimpsed from afar. The same essay amusingly captures the bewilderment of a rural denizen encountering the city for the first time in that person's observation "that there seemed to be a great deal of meeting thereabouts" as the paths of passersby cross in a crowded, bustling city street.

Several of these papers illustrate the influence of Romanticism in their praise of childhood, nature, imaginative solitude, and chance encounters on the road. "An Autumn Effect" (The Portfolio, April 1875), based on a walking tour of the Chiltern Hills, Buckinghamshire, celebrates agricultural labor, the beauty of the forest canopy in autumn, a visit to a peacock farm, and, above all, the freedom of the pedestrian wanderer "with a sufficiency of money and a knapsack" to "change his mind at every fingerpost, and, where ways meet, follow vague preferences freely." As its title suggests, "On the Enjoyment of Unpleasant Places" (*The Portfolio*, November 1874) commends residence in such "sublime" places as "the bleak and gusty North."

Many of the individuals Stevenson describes in these early travel essays become the butt of his jokes or objects for his condescension – such as the hatmaker and "Evangelist" of contentment, Smethurst, in "Cockermouth and Keswick," or the Dogberry-like country constable encountered in "An Autumn Effect," or the young Ayrshire men in "A Winter's Walk in Carrick and Galloway" who "seemed only eager to get drunk, and to do so speedily." The focus of these essays is consistently on Stevenson's narrative persona – his perceptions, moods, memories, and the lessons he draws from these experiences. In the essay "Walking Tours" (*Cornhill Magazine*, January–February 1876), Stevenson insists that "to be properly enjoyed, a walking tour should be gone upon alone," that one must "be open to all impressions and let your thoughts take colour from what you see." Several of these early essays based on walking tours conclude in a contemplative scene following arrival at an inn; a pipe in mouth, a bit of grog in hand, the young Stevenson indulges in some reading or pleasant conversation.

Four of these early essays are especially important in terms of Stevenson's biography. "Ordered South" (*Macmillan's Magazine,* May 1874) establishes the profound connection in Stevenson's life between illness and travel: he had been sent to the olive groves and cypresses of the Mediterranean region on doctor's orders in the fall of 1873, and this essay explores some of the alienation that "the invalid," as he calls himself, experiences in the lush and beautiful surroundings, from which he feels estranged. There are striking passages in which the young Stevenson, still in his early twenties, identifies the "premature old age" of the invalid, for whose "feeble body" some excursions "prove too long or too arduous," who looks upon "the stir of man's activity . . . with a patriarchal impersonality of interest."

"Forest Notes" (*Cornhill Magazine,* May 1876) describes the scene of Stevenson's stay in the artist colonies around Fontainebleau in the spring of 1875, where the Barbizon school of French painting had been located. In this essay Stevenson provides glimpses of the region's farmers and forest dwellers and shows some awareness of class distinctions and the historical development of the area, but he is particularly interested in describing the artists and models who have settled in the inn run by an M. Siron. A later, more developed and revealing treatment of these artist colonies, to which Stevenson returned in succeeding summers, is "Fontainebleau: Village Communities of Painters" (*The Magazine of Art,* May and June 1884). In this later essay Stevenson discusses more frankly the lure of Fontainebleau for aspiring painters seeking escape from bourgeois society in an artists' colony. The French "love of style" is encountered in Fontainebleau, a locale that combines the virtues of "charm, loveliness, [and] proximity to Paris," connecting culture with natural beauty. It is a perfect spot for a youthful apprenticeship in the arts. In such a setting one may learn the craft of the artist and at the same time experience the "hereditary spell of forests on the mind of man." Stevenson also alludes to the growing number of English and American female art students, one of whom was Fanny Van de Grift Osbourne, the American who was to become his wife.

One other early essay, "El Dorado" (*London,* 11 May 1878), should be considered in the context of Stevenson's travel writing, because its title implies that the endpoint of all journeys is more mythic and personal than objective and real. Although it describes an imaginary journey, this well-known essay captures some of Stevenson's essential feelings about travel and reveals his indebtedness to Romanticism. In it he recounts how "when we have discovered a continent, or crossed a chain of mountains, it is only to find another ocean or another plain upon the farther side." The questing spirit of humankind is insatiable, Stevenson suggests, in its search for the city of gold; the spires of El Dorado are always "but a little way farther." The essay concludes with an impassioned and frequently quoted tribute to the virtues of travel, even – or perhaps especially – travel without a particular goal in mind: "Little do ye know your own blessedness; for to travel hopefully is a better thing than to arrive, and the true success is to labour."

In September 1876 Stevenson and a friend, Walter Simpson, set off on a canoe trip on the rivers and canals of Belgium and northern France. Intending from the first to write about the trip, Stevenson kept a journal, which, rewritten a year later, became Stevenson's first real book, *An Inland Voyage* (1878), published by C. Kegan Paul. Stevenson's previous monograph publications were short pamphlets, the first of which was published at his father's expense when Stevenson was only sixteen. Like Stevenson's subsequent early travel books, *An Inland Voyage* received generally favorable reviews but scant attention from the reading public, selling fewer than five hundred copies in its first year.

Except for what he describes as the "incessant, pitiless, beating rain" and an encounter with trees that had fallen across the flood-swollen Oise and knocked him out of his canoe, Stevenson records a relatively uneventful journey. One of the most remarkable incidents of the voyage – Stevenson's brief detention by the police in France as a suspected German spy – was not fully detailed until ten years after the book was first published. In the travel book Stevenson devotes his space instead to his reception at (and, because of his odd appearance and limited funds, expulsion from) the inns at which he called, the meals served, and the local inhabitants glimpsed along the banks.

Even though the canoe trip itself, as he recounts it, was fairly ordinary, Stevenson shows a remarkable ability to capture memorably the people he encountered. He describes the canal-boat families, aboard ship with their dogs, birds, and flowers; a peddler and his wife, who compete extravagantly in singing the praises of their child; and an old woman who distributes her devotions equally to every shrine in the church at Creil – seeking diversification "like a prudent capitalist." Occasionally one is reminded that Stevenson was a young, unmarried man at the time, as when he records waving at the young girls along the bank as if "to play with possibility and knock in a peg for fancy to hang upon." A group of young women whom Stevenson calls "the graces of Origny" blew kisses flirtatiously at the passing canoeists, inviting them to "come back again!"

In the tradition of the Romantic essay, many incidents provide occasions for Stevenson to moralize or philosophize. Some of these digressions are relatively effective, as when he describes the way the receipt of a bag of letters from home breaks the spell of the voyage or when he discusses the state of Nirvana he achieves as the canoe drifts leisurely seaward. Other meditations seem more commonplace, such as his declaration that "I suppose none of us recognize the great part that is played in life by eating and drinking," or his musings on the power of a great cathedral to stir the sentiments, which move toward the conclusion that "I have never yet heard [a sermon] that was so expressive as a cathedral."

Stevenson concentrates not only on his own impressions and thoughts, but also on the responses of others to him, a young, spindly legged, long-haired bohemian wanderer, wearing "a smoking-cap of Indian work, the gold lace pitifully frayed and tarnished." Mistaken by those he encounters for a spy, peddler, servant, or purveyor of pornographic photographs, Stevenson paints a wry portrait of his own apparent lack of defined social rank and respectability abroad. He occasionally dines with judges and the upper reaches of provincial society but also breaks bread with peddlers and a Communist. On the one hand, Stevenson writes from the position of a young but securely middle-class Scotsman, while on the other he delights in the freedom and rootlessness of the barge families and peddlers and in his own detachment from the scenes he visits and records. He partakes of a willed and temporary separation from ordinary life: "'Out of my country and myself I go.' I wish to take a dive among new conditions for a while, as into another element. I have nothing to do with my friends or my affections for the time." In recounting the trip Stevenson jauntily refers to himself and his fellow-voyager, Simpson, by the names of their boats, the *Arethusa* and the *Cigarette,* respectively. The men are neither more nor less, in other words, than their two canoes. Stevenson is interested in the passing scenes but is also aware that he is only momentarily connected with them.

Throughout the book Stevenson records his impressions and thoughts rather than the scenes or places a tourist might seek out. *An Inland Voyage* is emphatically not a Baedeker or a Murray guidebook. It focuses instead on the particular experiences and thoughts of the voyager. Perhaps one might say that the "inland" of the title is in some important respects the interior of Stevenson, and that the book is most appealing where it is dominated by its sense of freedom, independence, and contingency.

Given this context the reader is perhaps less shocked by Stevenson's sudden decision to end his trip and return to his own world: "I was weary of dipping the paddle; I was weary of living on the skirts of life; I wished to be in the thick of it once more; I wished to get to work; I wished to meet people who understood my own speech, and could meet with me on equal terms, as a man, and no longer as a curiosity." He senses that "the traveller ... is not a traveller everywhere ... and his journey is no more than a siesta by the way on the real march of life." Both the canoe trip and the book end abruptly.

Between 1874 and 1879 Stevenson traveled in Europe, shuttling mostly between Edinburgh, London, and France. He had first met Fanny Osbourne in July 1876 at Grez, where both had gone to study art. Fanny would pose a problem for Stevenson's parents for several reasons: she was an American, ten years older than Stevenson; she was already married to a knockabout American miner and adventurer, Sam Osbourne; and she had her two children, Belle and Lloyd, in tow.

During this period Stevenson also began to see his career as a writer taking shape. Though he earned little money from his efforts, he continued to write and publish essays and stories in literary magazines, including some fanciful tales later collected as *New Arabian Nights* (1882). His second and third commercial books, however, were still volumes of travel writing, or at least books about places.

While residing in France in 1878, Stevenson began writing his second book, *Edinburgh: Picturesque Notes, with Etchings* (1879). Although it is debatable whether or not this work qualifies strictly as travel writing – since it does not attempt to describe a particular journey or visit and since Edinburgh was in fact Stevenson's hometown – it nevertheless marks some interesting developments in Stevenson's writing related to his attitudes toward place, time, and change. While at this time of life Stevenson was still returning regularly to Edinburgh, the book has something of a retrospective texture – dwelling on reminiscences of childhood and anticipating his own travels and eventual exile from Scotland for

Decorative title page for Stevenson's 1878 account of his trip down the Oisé River in France

the sake of health. The book was written partly, Stevenson declares nostalgically at the close, for the sake of Edinburgh's many emigrants, in hopes of calling up a picture for them of a familiar place left behind. It is clear that Stevenson was already beginning to count himself among those emigrants and exiles.

One interesting aspect of *Edinburgh: Picturesque Notes* is the way that Stevenson blends a sense of the town's history and myths with telling details reflecting the city's contemporary urban scene and social structure. The book is less personal and more objective and analytic than *An Inland Voyage*. Stevenson is aware of Edinburgh's expansion into the countryside, transforming what used to be pleasant rural villages into urban sprawl, and he records the failure of the builders of the New Town to preserve views of the hills and countryside nearby. He observes the social inequities of the growing metropolis – the ugly back garrets of High Street, on which thoroughfare decrepit buildings had been known to collapse and fall to the ground. The "broken shutters, wry gables, old palsied houses on the brink of ruin" in the city's Irish quarter are also noted. At several points Stevenson positions himself in the more fashionable parts of town, glancing over a parapet to scan the sunless and disreputable streets on the other side.

Because he associates Edinburgh with the stern Calvinism of his youth, Stevenson takes pains to point out the consequences of this creed – for example, the harsh Puritanism of the two maiden sisters who argue and draw a chalk line down the floor to divide the single room in which they both live, never engaging in conversation thereafter. Nor are rival creeds more magnanimous, to Stevenson's way of thinking, as is apparent in his description of the simultaneous, cacophonous peal of the bells of various denominations on a Sunday morning in Edinburgh, which he sees as expressing an "extremity of zeal" and as a deplorable "outcry of incongruous orthodoxies."

Nevertheless, Edinburgh is clearly the site of many happy, youthful memories for Stevenson. His radical ambivalence toward his birthplace comes through most clearly in his descriptions of Edinburgh's winter and new year. While saluting the snowballs, skating, sliding, whiskey, and shortbread associated with this season, Stevenson also records "the gloom and depression" – the short days, grim skies, and wearying winds – of the Edinburgh winter. Edinburgh was always home for Stevenson, but it was also the site where he experienced his worst health and some of his grimmest moods.

For many of Stevenson's admirers, his third book, *Travels with a Donkey in the Cevennes* (1879), comprises his most picturesque and perfect travel writing. It is an elaboration of a journal he kept while roaming this mountainous region in the south of France for twelve days during September and October 1878. The journal was first published one hundred years later under the title *The Cevennes Journal: Notes on a Journey Through the French Highlands* (1978). Stevenson made this tour on foot, with knapsack and pack, though he purchased a donkey, which he named Modestine, to carry his pack.

Several motifs weave their way through Stevenson's description of this walking tour. One concerns the arduousness of the journey. Stevenson traveled through mountainous terrain, out of the usual paths of tourists, and he frequently struggled to find his way from one village to the next and to find lodgings. After being lost and traveling in circles around the hamlet of Fouzilhic, Stevenson at last reaches his destination of Luc and Cheylard; finding himself hard put to explain why one should even attempt to travel to these out-of-the-way villages, Stevenson advances this well-known rationale for travel: "I travel not to go anywhere, but to go. I travel for travel's sake. The great affair is to move; to feel the needs and hitches of our life more nearly; to come down off this feather-bed of civilisation, and find the globe granite underfoot and strewn with cutting flints." Adventure, the struggle to overcome difficulties and hardship, constitutes the goal of travel. Travel frees one from the artificial comforts of everyday life. In Stevenson's mind travel is not primarily a visit to a famous historic or cultural site. This view is certainly a reaction to the already developing European phenomenon of mass tourism, but more specifically it recalls the Romantic ideal of the walking tour, with its key ingredients of freedom and the accidental.

A second topic that is implicitly present throughout much of the book is Stevenson's consciousness of his separation from Fanny Osbourne, who had returned to America in August 1878 at the insistence of her husband, Sam. Occasionally Stevenson's "strange lack" of Fanny rises to the surface of the text, particularly in his feeling of solitude under the stars: "I wished a companion to lie near me in the starlight, silent and not moving, but ever within touch.... To live out of doors with the woman a man loves is of all lives the most complete and free." Toward the end of the journey he speculates that "perhaps someone was thinking of me in another country."

A third motif, complementary to the second, is Stevenson's relationship with his "mouse-coloured" donkey, Modestine, who, in a displacement of the absent Fanny, is presented as a willful, recalcitrant "lady friend" whom Stevenson must learn to master before she masters him. At first Stevenson is hardly able to get Modestine to move at all, but after an innkeeper supplies him with a goad, a stick with a small pin at the end, "thenceforward Modestine was my slave." Nevertheless, Modestine continues to balk, lose her way, suffer from an unbalanced load, or falter on the difficult terrain. Much of the enjoyment many readers have derived from *Travels with a Donkey in the Cevennes* stems from the author's alternating feelings of exasperation and affection toward his particular "lady friend," although a modern (and especially a female) reader may well not find these tests of will – and the resultant blows and blood – quite so endearing as earlier generations of male readers seem to have done. Nor are such readers likely to enjoy Stevenson's observation that Modestine's "faults were those of her race and sex [though] her virtues were her own." Still,

Frontispiece for Stevenson's Travels with a Donkey in the Cevennes *(1879), illustrated by Walter Crane*

although Stevenson begins his relationship with Modestine by relying on his cane and goad, by the end his predominant feeling is of "my bereavement" at parting, expressed in terms that mingle humor with genuine affection. In any event the relationship between Stevenson and Modestine occupies the center stage of *Travels with a Donkey;* the district in which he traveled occupies only second place, even in the volume's title.

The fourth motif that organizes the book concerns religion. Stevenson was traveling in the region of the Camisards, Protestants who were in armed revolt against the government of Catholic France in the early years of the eighteenth century, following the revocation of the Edict of Nantes, which had granted religious toleration to French Protestants. Stevenson identifies these Protestant martyrs with the Covenanters in Scotland, who rebelled against the English Crown's efforts to impose the Anglican liturgy on the Scots Presbyterians during the reigns of the last Stuart kings.

In the Cevennes, Stevenson not only visits some of the sites of the Camisard rebellion, but also dwells on his sense of identification with these Protestant rebels, as he had identified with the Covenanters during the days of his youth. He engages in religious controversy with some of the proselytizing Catholics he meets, even though he expresses admiration for the monastic life of the Trappists. Throughout these discussions Stevenson seems to advance the view that all faiths pursue similar ends by different means. Because most humans derive their national and religious identities from an accident of birth, it is not desirable for them to renounce their beliefs and traditions. Human virtue resides on both sides of a struggle such as the French Catholics and Protestants underwent, and Stevenson values most the "mutual toleration" he finds now among most Protestants and Catholics in France.

Stevenson's ability to understand these earlier religious controversies at the same time that he – a foreigner in another century – feels objective about them produces a sense of distance and equanimity in the book. Stevenson is affectionately drawn toward the hinterlands of France, but he is also a somewhat bemused and cosmopolitan observer. Many of Stevenson's critics, including some of the book's earliest reviewers, have written of its charm, and it is likely that something of Stevenson's coupling of sympathetic engagement with cosmopolitan detachment is captured by this word. As with *An Inland Voyage,* the reader takes away more an idea of the author's sensibility and the particular incidents that occurred to him than descriptions of memorable places visited. A tour de force of this sort is Stevenson's close observation of his own sensations in the chapter, "A Night Among the Pines," in which he records a night of fitful slumbers outdoors until day gradually arrives.

In August 1879, one year after Fanny had sailed back to America, Stevenson decided to join her in California, despite the pleas of his parents and friends such as William Ernest Henley and Sir Sidney Colvin. He traveled to New York on the steamship *Devonia,* which was thronged with immigrants to the United States – mostly English, Scots, and Irish, but including Germans, Scandinavians, and Russians. After spending a day in New York City, Stevenson crossed the continent by rail. Although ill throughout the journey, he again recorded his impressions while traveling, and when recuperating in California he worked on *The Amateur Emigrant from the Clyde to Sandy Hook* (1895), which was to have been a two-part record of the stages of his journey from Scotland to New York, and thence to California. When Colvin, who had become Stevenson's principal literary adviser, received the manuscript, he disparaged it, perhaps because of the grime, realism, and focus on the working-class steerage passengers who were portrayed in the book, all of which set it apart from Stevenson's earlier travel writing. Stevenson's father, perhaps on Colvin's advice, repurchased the manuscript before it could be published. A condensed version of the second part of Stevenson's journey, "Across the Plains," was published in *Longman's Magazine* in 1883, and Stevenson was revising the first half, describing the ship's voyage, at the time of his death. Although *The Amateur Emigrant* was included in the Edinburgh Edition of Stevenson's works (1895), some of the grittier passages were not restored from the manuscript until James D. Hart's 1966 edition, *From Scotland to Silverado.*

One of the ways in which *The Amateur Emigrant* differs from its predecessors is in its intense focus on social forces and class, which have moved from the periphery to the center of Stevenson's vision. He carefully documents the distinctions made onboard ship between the cabin passengers and those in steerage. Stevenson traveled "second cabin," paying eight guineas rather than six to enjoy this "modified oasis in the very heart of the steerages," but his particular interests in – and, at some critical moments, identification with – the steerage passengers are clear. One of the most bitter scenes in the book describes a visit to the steerage by three tittering cabin passengers and the "insults these people managed to convey by their presence."

Throughout the voyage Stevenson praises many of the working-class men and women of the steerage, sometimes idealizing their proximity to "necessity and nature." He notes the solidarity that develops among the emigrants with the introduction of music, dance, and song aboard ship. Stevenson offers a long disquisition on how gentlemanly behavior is accessible to members of any class and how he found much of it among his fellow passengers in steerage. He has some sense of the emigrants as the innocent victims of social forces beyond their control: "Labouring mankind had in the last years, and throughout Great Britain, sustained a prolonged and crushing series of defeats. I had heard vaguely of these reverses." But many times Stevenson holds the individuals he encounters personally responsible for their failures and misfortunes, and he is skeptical that, given their weaknesses of "drink, idleness, and incompetency," their experience in America will be radically different. Stevenson shares with the high Victorian writers of an earlier generation a belief that the common folk must first reform themselves, make themselves over, before they can expect a change of country or a change of political system to produce success and fulfillment.

The passengers to whom Stevenson gives the most space are those who exemplify, positively or negatively, his social and political beliefs. He devotes a chapter to contrasting two stowaways, one who sets willingly to work to earn his passage after he is discovered on board and another who idly skulks, begs, and gets by. The engineer McKay is also accorded several pages. Although McKay blames his failure in Scotland on "corrupt masters and a corrupt State policy," Stevenson portrays him as a tippler and a crude materialist, interested in machinery but lacking in humanity. If he ever becomes rich, Stevenson worries, he will become "an oppressor of men," making money only to squander it – a particularly dangerous sort of capitalist. On the other hand, Stevenson admires the North Sea fisherman whose politics are radical but who has amassed a fair amount of money. If there is to be a working-class revolution, Stevenson hopes it will be led by a generous and successful laborer like this fisherman – a person who resembles the title character of George Eliot's *Felix Holt* (1866). Although aware of some of the causes that prompt the laboring class to emigrate and sensing the political unrest in this class, Stevenson emphasizes individual qualities and decisions as he contrasts one passenger with another.

Stevenson's own social position on this voyage was equivocal. He was the offspring of middle-class parents, but he was going to California to pursue a married woman without his parents' emotional or financial support. He was a writer, but his writing at this point had yet to earn him a living. Stevenson reveals that he was offered compensation by the ship's officers if he would copy out the list of passengers for them – so the sympathetic crew views the profession of this young, obviously impoverished "writer." Once again, Stevenson frequently played the part of a bohemian, yet he strongly desired to be recognized as a gentleman and was chagrined that, despite "every advantage of speech and manner," he was taken by all to be "the average man of the steerage." Stevenson acknowledges his equivocal status in the title of his work: he is an *amateur* emigrant – like his fellow steerage passengers, yet also set apart from them both socially and as an author, observer, and critic. Nevertheless, Stevenson was part of a moment of great historic change and records it with considerable clarity, capturing the dignity of many of the participants in this great migration.

After ten days' sail, the *Devonia* arrived in New York, where it rained for the entire day Stevenson spent there. Thin to begin with, he had lost fourteen pounds in the crossing and arrived drenched and with a terrible itch. The second part of *The Amateur Emigrant*, "Across the Plains," describes Stevenson's rail journey across the country, in the course of which he became increasingly ill, suffering from fever and diarrhea. Stevenson writes that although he came prepared to find America a promised land of youthful pluck and energy, in fact he discovered a cruelly indifferent railway company that kept him and the other immigrants cold and wet on a drafty platform waiting for the cars for the trip west to be unlocked. He also encountered bullying conductors, cheating company agents, and American immigrants to the West who had both "less talent and less good manners" than the European emigrants with whom he had crossed the sea. Although he delights in the collegiality and bonding that takes place among these railway emigrants as they create their informal syndicates to purchase the necessities for washing or for making breakfast, Stevenson finds less true communal feeling than he had found on board the *Devonia*. He is also fully aware of and depressed by the irony that all the places that had been destination points for the ship's passengers are now the points of origin for those aboard the emigrant train headed west. Once more he recognizes the lowly status of the emigrant, as the cheaper emigrant train yields priority to all others on the tracks, or as it pulls out of stations without giving any warning of its departure to its passengers. With a stark realism Stevenson describes how the Union Pacific cars "in which we had been cooped for more than ninety hours had begun to stink abominably."

Similarly, Stevenson finds the landscape of midcontinent America depressing. He is overwhelmed by "this huge sameness" and the "gauntness" of Nebraska's plains. Of "the desert of Wyoming" he writes that the "train was the one piece of life in all the deadly land . . . this paralysis of man and nature."

One of the most interesting developments in *The Amateur Emigrant* is Stevenson's ability to empathize with what he calls the "despised races" – the Chinese, Native Americans, and African Americans – he encounters. Though his Caucasian fellow passengers declare that the Chinese are dirty, Stevenson observes their superior cleanliness, and what the Americans see as stupidity is instead an imperfect knowledge of English. Stevenson is yet more

Page from Stevenson's journal of his travels around the Kona coast of Hawaii, 27 April 1889 (Edinburgh City Libraries) outraged by his fellow passengers' treatment of Native Americans: the passengers "danced and jested round them with a truly Cockney baseness. I was ashamed for the thing we call civilization." Introduced to his first African American, Stevenson realizes "that any shows of pity and generous condescension are entirely misplaced"; rather, African Americans must be regarded as equals. Stevenson repeatedly complains about American incivility and lack of consideration for others, but he finds the racism of his fellow travelers particularly incomprehensible: "They seemed never to have looked at [the Chinese], listened to them, or thought of them, but hated them *a priori*."

Although Stevenson's friends and family found *The Amateur Emigrant* so grim as to warrant suppression, it has gained respect in recent years. For one thing it is an important record by an astute observer of the late-nineteenth-century emigrant experience: the Atlantic was plied, after all, by many ships, not just the *Devonia,* and the transcontinental railways hauled many emigrants and Americans west. Second, perhaps because Stevenson's personal journey was itself so difficult, this book lacks the archness and "charm" that sometimes characterize Stevenson's early travel writing. Third, Stevenson attends more closely than before to the issues of social class and more clearly identifies himself with the deracinated emigrants, though at the same time he is aware of his "amateur" status. But the narrative persona he creates has a kind of engaging fluidity: his identity, class identifications, nationality, marital status, and career are all still unsettled. As a result he grapples with his ambivalent feelings about his neighbors in the steerage and their politics, as well as his ambivalence toward the Americans he encounters once he has landed. On the one hand, he is an author, a "Shakespeare" as he dubs himself, who can make a game of withholding his real name from his fellow emigrants. On the other hand, his sympathy extends to the "despised races" and unemployed workers he finds in the United States. Having come with his own high hopes and dreams, Stevenson can sympathize with the aspirations of his fellow emigrants yet also voice his skepticism over whether anything will really change in "'the good country' we had been going to so long."

Upon arriving in California, Stevenson found his relationship with Fanny, for whom he had undertaken his long journey, still unclear. His parents disapproved of his trip and refused to send any funds, so Stevenson was impoverished. His health continued to suffer after the transcontinental rail journey, and blood hemorrhages filled Stevenson's mouth.

At first Stevenson settled alone in Monterey, from August through mid December 1879; then he moved to San Francisco and eventually to Fanny's house in Oakland. As Fanny nursed Stevenson, they drew closer together once again. Fanny finally obtained a divorce from her husband in December 1879, and Fanny and Stevenson were married on 19 May 1880.

Stevenson continued to write journalistic essays about California, partly in a desperate effort to support himself and his new family. The most important of these pieces describe Monterey and San Francisco. They were published in periodicals in 1880 and 1883, respectively, and then at Stevenson's suggestion were grouped together under the title "The Old and New Pacific Capitals" in the Edinburgh Edition of his works. These essays include more scenic description than was usual for Stevenson and attempt to capture the remarkable physical presence – ocean, woods, chilly fogs, mountains, and hills – of these coastal towns.

Stevenson's efforts to analyze American society continue in these essays. In writing about Monterey, he focuses on the steady encroachment of "Americans" on Native American and Mexican American lands and freedom, and he records the newcomers' exploitation of and indifference toward these native peoples. Describing San Francisco, Stevenson is struck by the juxtaposition of different cultures – Chinese, Italian, Mexican, French, and German. "The town is essentially not Anglo-Saxon; still more essentially not American," Stevenson declares after listening to its languages, looking into its shops, and sampling its cuisines. With considerable perspicacity, Stevenson notes that the "hotch-potch of races" in California is much less likely to be assimilated to the predominant Anglo-Saxon culture than in the eastern United States.

Shortly after Stevenson's marriage to Fanny, the couple took a honeymoon journey fifty miles north of the Bay Area to the higher altitudes and brighter skies of the Napa Valley, staying first in the town of Calistoga and eventually finding an abandoned silver-mining camp on the slopes of Mount Saint Helena where they might "squat," or settle without any legal right to do so. As had been his practice for his previous travel books, Stevenson kept a journal of his stay at the Silverado mine, which he later rewrote as *The Silverado Squatters* (1883).

Stevenson's descriptions of the lodgings in which they stayed are memorable. The house,

Frontispiece for Stevenson's account of his stay at an abandoned silver-mining camp on Mount Saint Helena

which had originally been used by the miners, consisted of three ramshackle rooms, one built atop the other. When the Stevensons arrived, however, it was filled with rubbish from the mine, generally lacked windows and doors, and had chinks between the roof and wall boards so that the interior was partially exposed to sun, stars, wind, and rain. Nevertheless, the cottage was "planted close against the hill," in a canyon, and thus was somewhat protected from the elements by an overhanging rock. There was no stove or hearth, though the squatters improvised by using a blacksmith's forge outside. Poison oak surrounded the cottage and crept in through the chinks in the floor, while the rattles of snakes were heard outside. The edifice opened onto a platform commanding a wide view of the valley below, however, and the view became all the more spectacular when sea fogs inundated the Napa Valley overnight, rising gradually in the morning light.

Unlike Stevenson's record of the rail journey across the plains, *The Silverado Squatters* is a generally buoyant and hopeful book, as befits a honeymoon trip on which the author's health also seemed to improve markedly. He jokingly alludes to himself and his bride as the king and queen of their squat; Fanny's son, Lloyd Osbourne, as the crown prince; and their setter/spaniel cross, Chuchu, as the grand duke. Stevenson delights in recording Chuchu's terror at "every whiz" of the snakes' rattles and the dog's general dubiety about the journey to this mountain, given its marked preference for the sofa cushion and ladies' laps.

As with Stevenson's earlier travel books, conventional tourist attractions are given short shrift. Napa's famed vineyards, for example, are cursorily described. Instead, the author devotes considerable space to developing the characters of the people encountered in the valley. His chapter on "The Petrified Forest," for example, records his being "mightily unmoved" by the vaunted tourist attraction in which trees have been turned to stone: "Sight-seeing is the art of disappointment," he confesses. But he quickly adds that "fortunately, Heaven rewards us with many agreeable prospects and adventures by the way" – in this case by the prospect of a Mr. "Petrified Charley" Evans, discoverer and proprietor of the petrified forest. Born in Sweden, Evans was a former sailor who frequently journeyed out of Glasgow on Scottish ships before settling in America. The Swedish Evans, to Stevenson's amusement, made much of Stevenson as a fellow Scot, greeting him warmly as a compatriot of his adopted land.

Stevenson also devotes considerable space to a rather ambivalent portrait of a Jewish shopkeeper in the valley, called Mr. Kelmar in the book, who helped Stevenson find his squat at the mine. On the one hand, Stevenson is clearly grateful for Kelmar's assistance and enjoys the freedom and celebratory feeling that Kelmar has in his peregrinations through the valley. On the other hand, he is disturbed by Kelmar's practice of selling increasingly large amounts of merchandise to his customers on credit, so that eventually they are caught "in various degrees of servitude" to him.

Perhaps the most memorable portraits of all are those of the Hansons, the family of "poor whites," as Stevenson calls them, who reside closest to the Silverado mine. Rufe Hanson, a hunter by profession, nevertheless is motivated chiefly by his sense of "woodland poetry" and simple playfulness – "the pictorial side of his daily business." Rufe's brother-in-law, Irvine Lovelands, is depicted as the less attractive counterpart of Rufe's rural identity. Irvine is a lout – strong but cowardly, handsome but vain, and thoroughly unpleasant. Stevenson at first employs Irvine to cut firewood but eventually discharges him, since Irvine does little work, preferring instead to sit around and talk, chew pine gum, and spit on the cottage floor.

Despite the small number of irritating locals on the slopes of Mount Saint Helena and despite the relative primitivism of the squat, Stevenson's book on his stay at Silverado in many respects describes an idyll. He was married, happy, and relatively healthy during his stay in the Napa Valley, which he left at the end of July 1880. By then Stevenson's parents were reconciled to his marriage, and they supplied sufficient funds for their son and his wife to travel back to Scotland. By the fall the Stevensons were on the move again, this time to Davos, Switzerland, a spot believed to offer some relief to those who might be tubercular; diagnosis was difficult in Stevenson's time, and biographers today are still uncertain about whether Stevenson actually suffered from tuberculosis or merely had other bronchial or lung problems. Certainly the dry, cold climate of Davos, where Stevenson spent the next two winters, did not improve his health. The essays he wrote about his stays there – three of which were collected in *Essays of Travel* – make clear his sense that an invalid's weakness and an Alpine winter combined to make him feel imprisoned: "the mountains are about you like a trap." Even in an essay titled "The Stimulation of the Alps," Stevenson explains how the bright, cold, energizing Alpine winters join with the sanitarium resident's natural lethargy to produce a "sterile joyousness of spirits . . . a sort of intermittent youth, with periods of lassitude." These were neither happy nor healthful days for Stevenson, and they resulted in his most enervated and gloomy travel writing.

After their first winter (1880–1881) in Davos, the Stevensons returned to Scotland and settled successively in small cottages near Pitlochry and Braemar, where Stevenson began to write some of his finest short stories – "The Merry Men," "Thrawn Janet," and "The Body Snatchers." Still more important, Stevenson and his stepson, Lloyd Osbourne, playing with Lloyd's paints, one day produced the map of an island, stirring Stevenson's imagination to conjure up the story of *Treasure Island* (1883), originally serialized weekly in the magazine *Young Folks* from October 1881 to January 1882. The story was well enough received to warrant publication as a book by Cassell and Company in 1883, in an edition of two thousand.

In his early thirties, after more than ten years of trying, Stevenson was at last beginning to achieve modest success and recognition for his writing. It is noteworthy, too, that the origins of Stevenson's first great novel derived from the map of an imaginary island based on his travels. Many readers have thought that the "treasure island" strongly evokes certain locales in California, even though this part of the novel is set in the West Indies. Stevenson's initial success as a novelist, in other words, is intimately linked to his earlier career as a travel writer and visitor to, at least by nineteenth-century European standards, some fairly exotic places. Many of his novels crisscross the seas, just as he had done.

Stevenson spent the next several years living at various locations in Britain and France, seeking temperate climates for his health's sake. His happiest and most productive periods seem to have come at Hyères (1883–1884), in the south of France, and at the house called Skerryvore, which Thomas Stevenson bought as a wedding present for Fanny, at Bournemouth. During the three-year period he lived at Skerryvore (1884–1887), Stevenson produced such fiction as *The Strange Case of Dr. Jekyll and Mr. Hyde* and *Kidnapped* (both published 1886); two books of verse; including his most notable, *A Child's Garden of Verses* (1885); numerous short stories; and plays. The Stevensons remained relatively settled at Hyères and Skerryvore, however, so this was not a period during which Stevenson wrote travel literature.

Stevenson made his final visit to Edinburgh to see his dying father and, after the latter's death in May 1887, to attend to family affairs. Stevenson still sought a more salubrious climate; Fanny wished to return to America; and at last the Stevensons were free to leave Europe altogether, accompanied, as they were most of the time thereafter, by Stevenson's mother. In August 1887 they sailed, but this time Stevenson arrived in the States as a renowned author, thanks to the book that was his first great popular success, *The Strange Case of Dr. Jekyll and Mr. Hyde,* which had sold forty thousand copies in Britain alone during the six months after it first appeared. The first of many stage (and, later, screen) adaptations, starring Richard Mansfield, had already opened in America when the Stevensons arrived.

They settled for a while at Saranac Lake, in the Adirondack Mountains of New York, where, as in Davos, there was a renowned tuberculosis clinic. This setting inspired Stevenson to begin writing *The Master of Ballantrae* (1889), as well as to collaborate with Lloyd Osbourne on *The Wrong Box* (1889). While still at Saranac, Louis began to dream again of warmer climes and of a yacht trip in the Pacific. He was encouraged in these plans by the American magazine publisher S. S. McClure, who negotiated with the *New York Sun* to publish as many as fifty serial letters from Stevenson describing his voyage, and this agreement marked Stevenson's return to travel literature.

Stevenson eventually did make three voyages in the South Seas – on the *Casco* (June 1888 to January 1889), the *Equator* (June to December 1889), and the *Janet Nicoll* (April to July 1890). He kept daily notes of his experiences with the idea of assembling a comprehensive volume on the South Seas – its history, peoples, and cultures – complete with photographs taken by Stevenson, Lloyd Osbourne, and others of their party. Many of the photographs were lost in a fire aboard ship in April 1890, but those remaining were published in Alanna Knight's *Robert Louis Stevenson in the South Seas: An Intimate Photographic Record* (1986). Stevenson never completed his ambitiously comprehensive study of the South Seas, although he did produce thirty-seven newspaper installments, about half of which were collected by Sidney Colvin, following Stevenson's own plan, under the title *In the South Seas* (1896), for the posthumous Edinburgh Edition of Stevenson's works. One or two twentieth-century editors have added some of Stevenson's essays on Hawaii to their editions of *In the South Seas*.

The editors of the *New York Sun* were disappointed by the series. In a letter to Charles Baxter on 4 February 1891, Sidney Colvin wrote that these records of Stevenson's travels were not so much "letters of incident and experience . . . but only the advance sheets of a book and rather a dull book at that." Fanny, too, complained that her husband's essays on the South Seas lacked the personal focus of Stevenson's early travel books such as *Travels with a Donkey;* and Stevenson's friends in Britain complained that he was straying from the material he knew best in trying to write about so remote a corner of the globe.

Stevenson unquestionably recognized that in writing about the South Seas he was producing a text different from his early travel writing. Describing his project, he bragged that "nobody has had such stuff: such wild stories, such beautiful scenes, such singular intimacies, such manners and traditions, so incredible a mixture of the beautiful and horrible, the savage and civilized." Such nineteenth-century writers as Herman Melville and Charles Warren Stoddard had already introduced English-speaking readers to the attractions of the South Pacific in a rather idealized form, but by the time of Stevenson's voyages the impact of European and American commerce and the religious and colonial rivalries being played out in the Pacific made the region seem less paradisaical. Stevenson is less prone to idealize, readier to report and analyze critically, than most of his predecessors.

Stevenson's experiences are recounted in these essays, but they are frequently exemplary of larger concerns, such as the physical geography of the South Pacific, the customs and beliefs of the island peoples, their past and present political situation, and the effects of European and American trade and colonization on the region. Stevenson's emphasis on photography in the work as originally projected is revealing: he saw himself in the role of journalist, scientist, and anthropologist, coming to study and understand the South Seas and its inhabitants in order to report back to Europe and North America both on the historic cultures found there and on the changes that were transforming them.

Writing about Rudyard Kipling's *Kim* (1901) in *Culture and Imperialism,* Edward Said develops the links between late-nineteenth-century anthropology and colonialism, between coming to know a different culture and exercising authority and control over that culture. There can be little doubt that Stevenson, too, is implicated to some extent in the colonial enterprise. In the late nineteenth century, control over the South Pacific was actively contested by Britain, France, Germany, and the United States. Thus, when Stevenson protested German hegemony over Samoa – where he settled and eventually died – and when he openly sided with a rebellious native chief, Mataafa, he carried the banner for native self-determination, but he was at the same time indirectly advancing the interests of Great Britain and the United States, as opposed to Germany, in that part of the world. In his writing, too, Stevenson acknowledges that he can never be anything but an outsider, an observer of native cultures. He records that the natives refer to him as "the Rich Man" or "Owner," nor is he averse to passing himself off as "an intimate personal friend of Queen Victoria's" (or even as her son), who will report to the queen on island conditions immediately upon his return. At various points in his South Seas collection, too, Stevenson adopts the perspective of the superior European, finding in some of the "barbarous islanders" an "image of our race upon its lowest terms, as the partner of beasts, beastly itself." Nevertheless, in his writing Stevenson characteristically maintains an openness to experience and observation, which forms a counterweight to any tendency to condescend toward the Pacific Islanders. He seems content to allow complex and even some-

Page from the original manuscript for In the South Seas *(1896; American Art Association/Anderson Galleries, sale 4249, 8–9 April 1936)* what contradictory attitudes toward them to coexist in his account.

The book-length collection of Stevenson's essays, *In the South Seas,* has four parts in the edition Colvin prepared, each featuring a different island or group of islands and each developing its own theme and emphasis. The first section of the volume, dealing with the Marquesas, is the most fascinating, because of Stevenson's attempt to understand the culture of the cannibals who once resided there. Stevenson begins this part emphasizing his journey outside of traditional Western culture, leaving "the shadow of the Roman empire" and "that comfortable zone of kindred languages" that constitutes European culture. Although Stevenson insists that he has no defense of cannibalism to offer, he rejects the notion that cannibals are particularly cruel peoples and treats the phenomenon as one that can be understood in terms of the relations among different clans and groups on these islands. He explains it as a product of the natives' desire for revenge and retribution from a group rather than an individual and perhaps also as the manifestation of a hunger for animal foods on islands where such food is relatively scarce. Cannibalism is a cultural artifact, Stevenson insists, and a vegetarian might regard a meat-eating European, or a European might regard a person from a culture where dogs are eaten, with no less horror than a European views a cannibal. Cannibal practices, he notes, inform many features of Marquesan art – from tattooing to their songs and dances. Cannibalism, then, is a defining feature of this culture, and one must so understand it.

One of the pleasures of the first part of *In the South Seas* is Stevenson's sympathetic attempt to understand the customs of the Marquesans. He dilates upon the Marquesans' practice of exchanging gifts with visitors, explaining that this is not usually done greedily or competitively, but rather out of a sense that such an exchange is a social duty. He considers the use of taboos on the islands and explains how they contribute to a sense of social organization and order. He notes how an individual's body size correlates with that person's social status, how the Marquesans generally love and even fall under the sway of their children, and how the spirits of the dead are thought to haunt their communities, an idea Stevenson connects with their cannibal past.

Stevenson also offers his strongest criticism of the pernicious effects of European and American influence in this first section of *In the South Seas*. He records the importation of illnesses such as smallpox and syphilis into the islands and their resultant "depopulation." Those natives who survived frequently suffered from despondency and depression, experiencing the empty and meaningless life of the colonial subject, as the activities and features of their earlier culture decayed or were actively proscribed. Stevenson notes how unsurprising it is that, when both white adventurers and missionaries attack and ridicule native beliefs and taboos, the culturally deracinated natives should grow skeptical of European religions and value systems. He also strongly condemns European responsibility for the opium trade – "As a subject of the British crown, I am an unwilling shareholder in the largest opium business under heaven" – which had turned many islanders into thieves and criminals to support their habits. To counteract this new order of native criminals, Stevenson reports, French gendarmes in the South Seas were equipped with thumbscrews in order to wring confessions from the natives they arrested. Looking broadly at this issue of Western influence, Stevenson concludes that Pacific Islanders are best able to survive and thrive where the fewest changes have been introduced by European and American missionaries, traders, and settlers.

Following this sympathetic social analysis, the first section of *In the South Seas* concludes on a more personal and humorous note when Stevenson recounts how one particularly unpleasant former cannibal chieftain, whose "favourite morsel was the human hand, of which he speaks to-day with an ill-favoured lustfulness," bade farewell to Fanny Stevenson as she was leaving the Marquesas while "holding her hand" most attentively and "viewing her [departure] with tearful eyes."

The second section of *In the South Seas* treats another group of islands, the Paumotus, to the south and slightly to the west of the Marquesas. Whereas the Marquesas are high, volcanic islands, the Paumotus are atolls, or low coral islands, containing lagoons. As with the Marquesas, Stevenson ranges over many topics in his discussion of the Paumotus: their physical geography, flora and fauna, the general character of the natives, and more-specific accounts of his personal experiences and adventures on the islands, such as his problems in determining to whom he owed rent for the house in which he stayed. This section develops into a discussion of the various religions and sects found on the Paumotus and thence to consideration of funeral customs and beliefs about the dead found on this archipelago. These islanders, too, had a cannibal past, and they imagined that the spirits of the dead prowl for food and must be mollified or deflected from their malign purposes through such de-

Stevenson (seated, center) with his mother; his stepson, Lloyd Osbourne; his wife, Fanny; his stepdaughter, Isobel Strong; and servants on the porch of Vailima, his estate in Samoa

vices as graveside vigils. The book's second section concludes with an extended chapter on "Graveyard Stories," retelling various anecdotes about ghosts and spirits and displaying Stevenson's taste for macabre folklore.

The third and fourth sections of *In the South Seas* describe two islands in the Gilberts, which are farther west and north of the other two island groups – due north of New Zealand, in fact, but near the equator. Whereas the focus of the first two sections might fairly be called anthropological, with their emphasis on customs and legends, the focus of the latter sections is on the contemporary society and politics of these two islands, Butaritari and Apemama. In these two parts Stevenson considers more particularly than previously the impact of white traders on these island societies and describes in detail the two independent "kinglets" who continued to rule there.

The third section of the book describes Tebureimoa, the king of Butaritari, who is represented as a corpulent, drug-addicted, and generally unattractive monarch who participates in three weeks of alcoholic dissipation at the time of Stevenson's visit.

Stevenson also describes the rivalry among the island's European traders, who contribute to the debauchery as they compete recklessly in the sale of gin in order to assure their share of the trade in copra (dried coconuts).

Characteristically, Stevenson attempts evenhandedness in his account of white influence and local institutions. After recounting instances of the drunken violence white traders have visited on native women, Stevenson nonetheless proclaims at one point that the white trader "often makes a kind and loyal husband" for the native. While on the one hand describing native "king, magistrates, police, and army joining in one common scene of drunkenness," Stevenson also makes a point of describing his warm feelings toward Maka, a virtuous and upright Hawaiian missionary on Butaritari.

The volume's fourth section focuses on another island king, Tembinok' of Apemama, who also has a taste for alcohol, though in his case it is "a beverage known (and labeled) as Hennessy's brandy" but which "is neither Hennessy, nor even brandy; is about the colour of sherry, but is not sherry; tastes of kirsch, and yet neither is it kirsch."

Last photograph of Stevenson, in Samoa in 1894 with a native

Like Stevenson's other native kings, Tembinok' is subjected to a certain amount of criticism: he controls the island's wealth, whether measured in copra or tobacco; as a trader, he relentlessly seeks to add novelties to his possession; he rules a seraglio of wives and women; he is not averse to governing his kingdom by intimidation and, if necessary, force; and he is even prepared to rig card games in his favor by having two hands dealt to himself.

For all this gentle criticism, though, Stevenson also clearly admires Tembinok', not least because he has managed to fend off white encroachment upon his island kingdom. Even the Stevenson party had to take special pains to establish a relationship with Tembinok' in order to win permission to stay awhile on Apemana; after being won over, however, the king allowed Stevenson to build a small town, named "Equator Town" after Stevenson's ship, and protected it from native intrusion by placing it under taboo. Further, Tembinok' displayed an astonishing openness and candor about many things, including an awareness of his power and that "he can give and take, and slay, and allay the scruples of the conscientious" in his land. "He rejoices in the crooked and violent paths of kingship like a strong man to run a race, or like an artist in his art. . . . I never saw a man more patently in the right trade," Stevenson declares, painfully aware that such independent native rule is rapidly vanishing from the South Pacific.

In the South Seas concludes with the departure of the *Equator* from Tembinok's island: "That night the palm-tops of Apemama had dipped behind the sea, and the schooner sailed solitary under the stars." This sentence essentially closes Stevenson's career as a travel writer. His monumental project on the South Seas was never completed, although one additional portion of that projected work took the form of *A Footnote to History* (1892), a polemical work describing then-recent European (especially German) meddling among the native rulers and factions in Samoa, where Stevenson began to build his last home, Vailima, in 1890.

Between 1888 and 1894, the last half-dozen years of his life, while roaming the South Seas and later when settled in Samoa, Stevenson continued to turn out fiction prodigiously. Books written in whole or in part after Stevenson had come to the South Seas include works with a largely Scottish setting, such as *The Master of Ballantrae, Catriona* (1893), and the unfinished *Weir of Hermiston* (1896) and *St. Ives: Being the Adventures of a French Prisoner in England* (1897).

Yet Stevenson also began to write works of fiction set in the South Seas, reflecting both the folklore and the actual lives of inhabitants of the region. Probably the most impressive of these works is the novella *The Beach of Falesá* (published in *Island Nights' Entertainments,* 1893), which exposes the crudeness and rapacity of some of the white traders toward the South Seas natives while exploring the development of empathy and integrity in one particular trader. This novella, in conjunction with some of Stevenson's other late fiction and *In the South Seas,* reveals his growing maturity as a writer, his increasing sensitivity to cultural difference, and his fierce opposition to colonial exploitation.

After he had built his home in Samoa, Stevenson still cherished the hope that he might return to Scotland. Yet in his last years, he found even the climates of Sydney and Hawaii cold and unhealthful, so a return to Edinburgh proved impossible. Stevenson became a world traveler in search of a warm climate to assuage his respiratory complaints. In the end he died young, on 3 December 1894 at the age of forty-four; the cause of death was a cerebral hemorrhage. His coffin was borne up the steep slopes of Mount Vaea, near his beloved Vailima, by sixty Samoans.

For most of the century following Stevenson's death, he was regarded predominantly as a sentimental traveler who had journeyed to some remote corners of Europe and the globe. The "charm," "style," and personal focus of his earliest travel writing, especially *Travels with a Donkey,* were most praised, and subsequent travel writers and admirers retraced his footsteps. In the last quarter-century even Stevenson's early travel writings have seemed to be more complicated and self-aware than originally thought, and he has garnered new praise for the social realism found predominantly in his later books of travel, particularly *The Amateur Emigrant* and *In the South Seas.* Stevenson grew more attuned to the cultural, social, political, and historical complexities of people and places, and he also became more aware of the complex and contradictory elements in his own personality in response to the things he encountered. Stevenson, in other words, became a more sensitive, reflective, and empathic observer of the places he visited as he achieved maturity as a writer.

Letters:
Sidney Colvin, ed., *The Letters of Robert Louis Stevenson to his Family and Friends,* 2 volumes (London: Methuen, 1899; New York: Scribners, 1899);

DeLancey Ferguson and Marshall Waingrow, eds., *R. L. S.: Stevenson's Letters to Charles Baxter* (New Haven: Yale University Press, 1956; London: Oxford University Press, 1956);

Bradford A. Booth and Ernest Mehew, eds., *The Letters of Robert Louis Stevenson,* 8 volumes (New Haven & London: Yale University Press, 1994–1995).

Bibliography:
W. F. Prideaux, *A Bibliography of the Works of Robert Louis Stevenson,* revised edition, edited and supplemented by Mrs. Luther S. Livingston (London: Hollings, 1918).

Biographies:
Graham Balfour, *The Life of Robert Louis Stevenson,* 2 volumes (New York: Scribners, 1901);

Janet Adam Smith, *Robert Louis Stevenson* (London: Duckworth, 1947);

David Daiches, *Robert Louis Stevenson* (Norfolk, Conn.: New Directions, 1947);

J. C. Furnas, *Voyage to Windward: The Life of Robert Louis Stevenson* (New York: Sloane, 1951);

James Pope Hennessy, *Robert Louis Stevenson* (London: Cape, 1974);

Jenni Calder, *RLS: A Life Study* (London: Hamilton, 1980);

Ian Bell, *Dreams of Exile: Robert Louis Stevenson, A Biography* (New York: Holt, 1992);

Frank McLynn, *Robert Louis Stevenson: A Biography* (London: Hutchinson, 1993; New York: Random House, 1994).

References:
Jenni Calder, ed., *Stevenson and Victorian Scotland* (Edinburgh: University of Edinburgh Press, 1981);

G. K. Chesterton, *Robert Louis Stevenson* (London: Hodder & Stoughton, 1927);

Edwin M. Eigner, *Robert Louis Stevenson and Romantic Tradition* (Princeton: Princeton University Press, 1966);

Richard Holmes, *Footsteps: Adventures of a Romantic Biographer* (New York: Viking, 1985);

Anne Roller Issler, *Our Mountain Hermitage: Silverado and Robert Louis Stevenson* (Stanford: Stanford University Press, 1950);

Robert Kiely, *Robert Louis Stevenson and the Fiction of Adventure* (Cambridge, Mass.: Harvard University Press, 1965);

Alanna Knight, ed., *Robert Louis Stevenson in the South Seas: An Intimate Photographic Record* (Edinburgh: Mainstream, 1986);

Paul Maixner, ed., *Robert Louis Stevenson: The Critical Heritage* (London: Routledge & Kegan Paul, 1981);

Barry Menikoff, " 'These Problematic Shores': Robert Louis Stevenson in the South Seas," in *The Ends of the Earth, 1876-1918,* edited by Simon Gattrell (Atlantic Highlands, N.J.: Ashfield, 1990), IV: 141-156;

Roy Nickerson: *Robert Louis Stevenson in California: A Remarkable Courtship* (San Francisco: Chronicle, 1982);

Andrew Noble, Introduction to Stevenson's *From the Clyde to California: Robert Louis Stevenson's Emigrant Journey* (Aberdeen: Aberdeen University Press, 1985), pp. 3-33;

Noble, ed., *Robert Louis Stevenson* (London: Vision, 1983);

Jonathan Raban, Introduction to Stevenson's *The Amateur Emigrant* (London: Hogarth Press, 1984), pp. vii-xii;

Nicholas Rankin, *Dead Man's Chest: Travels After Robert Louis Stevenson* (London: Faber & Faber, 1987);

Janet Adam Smith, ed., *Henry James and Robert Louis Stevenson: A Record of Friendship and Criticism* (London: Hart-Davis, 1948);

Roger G. Swearingen, "'Essays on the Enjoyment of the World': The Place of *Travels with a Donkey* in Stevenson's Work and Literary Career," *Cahiers Victoriens et Edouardiens,* 8 (April 1979): 25-38;

Swearingen, *The Prose Writings of Robert Louis Stevenson: A Guide* (Hamden, Conn.: Archon, 1980).

Papers:
Collections of Robert Louis Stevenson's papers are at the Beinecke Rare Book and Manuscript Library at Yale University; the Pierpont Morgan Library in New York; the Henry E. Huntington Library in San Marino, California; the Widener Library at Harvard University; the Edinburgh Public Library; the Silverado Museum in Saint Helena, California; and the Stevenson House in Monterey, California.

Ella C. Sykes

(? – 23 March 1939)

Scott R. Christianson
Radford University

BOOKS: *Through Persia on a Side-Saddle* (London: Innes, 1898);
The Story-Book of the Shah; or, Legends of Old Persia (London: Macqueen, 1901);
Persia and Its People (London: Methuen, 1910; New York: Macmillan, 1910);
A Home-Help in Canada (London: Smith, Elder, 1912);
Through Deserts and Oases of Central Asia, by Sykes and Sir Percy Molesworth Sykes (London: Macmillan, 1920).

SELECTED PERIODICAL PUBLICATIONS –
UNCOLLECTED: "Domestic Life in Persia," *Scientific American Supplement,* 55 (25 April 1903): 22830–22832;
"A Talk about Persia and Its Women," *National Geographic Magazine,* 21 (October 1910): 847–866;
"The Simple Life on a Poultry Ranch in British Columbia," *Cornhill Magazine,* 111 (February 1915): 214–222;
"At a Y.M.C.A. Hut Somewhere in France," *Cornhill Magazine,* 115 (February 1917): 204–214.

Ella C. Sykes's three books based on her travels through Persia, Canada, and Central Asia are no longer in print. Her name does not appear in the *Dictionary of National Biography* (though her brother's does), and she is not mentioned in critical studies of travel writing or women's writing. Nevertheless, Sykes's writings describe quite remarkable travels and demonstrate that the business of empire was conducted by middle-class families, the men employed by the government and stationed around the world and the women – their wives and sisters – often accompanying them. It was carried out village by village and district by district as well as in the halls of Parliament, the counting houses of commerce, and the fields of battle. Sykes affords a glimpse of late-Victorian and Edwardian colonization through both the political perspective of official representatives of the British Empire and the domestic perspective of the women who accompanied the men on their official travels.

Courtesy of Widener Library, Harvard University

The man on the business of empire to whom Ella Constance Sykes was attached as sister, companion, and domestic organizer was Sir Percy Molesworth Sykes, who was a brigadier general at the time of his retirement from the British army and who was known for organizing the South Persia Rifles, credited with discouraging German and Turkish invasion of Persia during World War I. His reasons for traveling underwrote her journeys in Persia and Central Asia, and his agenda determined hers. Only the date of Ella Sykes's death is recorded in the few encyclopedic sources that even mention her; it is even impossible to ascertain whether she was born earlier or later than his 1869 birthdate. She evidently never married, giving her the freedom to travel twice with her brother, once on her own to Canada, and once to "do her bit" at a YMCA camp in France during World War I. She died on 23 March 1939, on the eve of the next world war, preceding her brother in death by six years.

Beyond these scant biographical details, information about the life of Ella Sykes comes from her writings. On her first journey through Persia, from June 1894 to March 1897, she accompanied her brother, then a captain, as he conducted negotiations for territory, establishment of a new Persian consulate, and rights of way for telegraph lines linking the Indian portions of the British Empire with the Mediterranean – the basic groundwork of colonization. The Sykeses covered a vast area of Persia on horseback – one anonymous reviewer for the *Athenaeum* (7 May 1898) estimated that they traveled more than two thousand miles – covering deserts, mountains, and other rough terrain. Ella Sykes described the journey in her first book, *Through Persia on a Side-Saddle* (1898).

The frontispiece for *Through Persia on a Side-Saddle* depicts a young woman in a long floral dress and seated in a chair. An anonymous reviewer wrote in the *Critic* (January 1899): "The Author, judging from her frontispiece, is a hearty and healthy English girl who enjoys life at the full and can be happy in either Europe or Asia." The reviewer goes on to affirm that *Through Persia on a Side-Saddle* "is not one of the ephemeral volumes dashed off by a society woman from London who 'does' a country for an outing and then comes back to be lionized for one season." On the contrary, the book is "a capital contribution by a woman of culture and reading, who has also a lively interest in everything human. She shows appreciation of the scenery, sympathy with the people, and delight in nature."

Through Persia on a Side-Saddle is the best of Sykes's three travel books, perhaps because of her youthful enthusiasm and sense of newfound freedom. It begins: "The 'gorgeous East' has always possessed a strong fascination for me, and after reading 'Eothen,' that most delightful book of travels, the indescribable attraction of the Orient became, if possible, stronger than before." Sykes's enthusiasm for Persia never wanes throughout the book, in spite of the severe hardships she endured on the trail:

> I can never forget my feelings of joy and exultation when I realised that I was at last in Persia, on the threshold of a new life, which I ardently trusted might have its quantum of adventure. I had been civilised all my days, and now I had a sense of freedom and expansion which quickened the blood and made the pulse beat high.

She maintains throughout her narrative a typically British "keenness" for all that she experiences; she acknowledges that "although many a time I encountered hard facts, quite sufficient to destroy the romantic illusions of most folk, yet they struck against mine powerlessly." She comments frequently on the excellent health and spirits she enjoyed while in Persia:

> I took to Persia and things Persian at once, and never felt better in my life than at Tehran. The climate seemed to exhilarate me in the most delightful way, and to one accustomed to English winters it was a treat that never palled, to wake up morning after morning to a world bathed in brilliant sunshine, with perhaps a covering of crisp white snow on the ground.

Sykes's enjoyments are, however, selective and colonialist. She appreciates the natural scenery wherever she goes, yet every town is a dirty and undramatic disappointment: "Also the national predilection for mud as a building material is not conducive to a fine or enduring architecture, however beautiful may be the tiles with which it is covered." She expresses a love for the Persian people that was no doubt genuine, but she frequently has recourse to the stereotypes that "Orientals" are dirty, dishonest, and childlike, their development having been arrested sometime during the Middle Ages. In dealing with some French-speaking Persians, she says, "I should have managed much better if I could have realised that they were but children mentally, and must be talked to as such." She finds them "rather difficult to get on with," although "these French-speaking Persians . . . were a great improvement on their fellow-countrymen of the old school."

Sykes encountered hardships as extreme as any faced by the most intrepid explorers: arctic temperatures with only canvas as a shelter, precipices to be climbed or skirted at much peril, hostile armed tribesmen. Proud of her endurance of such conditions, she also describes in loving detail her domestic arrangements at each place of residence along the way. That the business of empire was engendered in the home is revealed in the chapter "Housekeeping at Kerman." Sykes describes the division of labor in her household: "We settled into a routine as time went on; my brother went off to the stables and then on to his office after breakfast, and I began my housekeeping – a novel experience into which I entered with all the enthusiasm of a beginner." Sykes's sphere, however, is no less colonialist than her brother's. Much of the chapter is devoted to the "servant question": "We found our Persian servants, from highest to lowest, afflicted with an incurable laziness, and although we had over a dozen men to minister to our various wants, yet three or four good English servants would have done all the work they did and a great deal more besides." She goes on to reflect, ruefully: "It is disagreeable to feel that there is seldom much personal attachment between master and servant, such as is the rule in India; and certainly the Persian domestic's idea of service, which is to purloin as much as he possibly can, is hardly calculated to produce such a feeling." Even more surprising than the lack of attachment between master and servant, Sykes remarks, was "the interest that the Persian gentry took in our servants." She imagines that the "gentry" suspect that her and her brother's "particularly simple life" was "merely a blind to cover deep designs or extraordinary doings on our part," and so they pumped the servants for information about "our inexplicable European characters."

Sykes does express a seemingly genuine sympathy for what what she perceives as the unfortunate status of Persian women. She describes one Persian lady who "was always eager to hear about my life in England, but I fear that my accounts only made her lot the gloomier and she said frequently that she could never understand why I came to Persia, when I could live in such a well-regulated country as the one I had described to her." Yet Sykes unavoidably juxtaposes overt examples of Persian women's subservience with her own, subtler differential status. She is excused from the manly political discussions and negotiations; she details her cumbersome grooming and attire, uncompromised for the sake of comfort during travel; and she rides throughout Persia on a sidesaddle, even though riding astride would have been not only more comfortable but also safer at such treacherous moments as fording a torrential stream or crossing an icy precipice.

Occasionally and briefly Sykes addresses directly the darker side of imperialism. On their arrival at Bunder Nasseri in the latter part of their journey the Sykeses learn that "Mr. Tanfield (Messrs. Lynch's agent at Shuster)" had been attacked by an enraged servant and "had been brought down that day from that fanatical city in a terribly mutilated condition" yet had managed to give evidence "to the authorities of Shuster against the wretch, who was arrested by the Governor and thrown into prison." Sykes concludes: "His heroism and pluck were something wonderful, and made us prouder than ever of the name of Englishman!" Within six pages Sykes is describing the ingenious methods "Messrs. Lynch" employed for exploiting the region's agriculture.

Through Persia on a Side-Saddle ends as it began, with a celebration of Persia and the Sykeses' travels through it. Reviewers seized on the author's enthusiasm for her subject and commented favorably on the book. The reviewer for the *Dial* (1 December 1898) proclaims: "The woman traveller in foreign parts is either what has been comprehensively called since Shakespeare's time a 'good fellow,' or a very great bore. Miss Sykes is a good fellow" who "went 'Through Persia on a Side Saddle,' as her title proclaims, and with her eyes wide open to a series of things which would have escaped the slower-witted man." The reviewer for the *Bookman* (September 1898) detects "a vein of sarcasm, which, combined with a glowing enthusiasm, gives a racy tone to her language, and keeps the reader in good humour." The reviewer for *The Critic,* however, thinks that Sykes's "sense of fun is strong, but she is never sarcastic." That reviewer also observes that "the Land of the Lion and the Sun is one in which the maximum of comfort is reached by the men, for the women have a dull time of it, though doubtless they get more fun out of life than is apparent to English people." In a long review with lengthy quotations from the book, the *Atheneum* (7 May 1898) offers the sole criticism of the book's extensive description and detail "in matters which are comparatively trivial." *Through Persia on a Side-Saddle* was followed by a children's book, *The Story-Book of the Shah* (1901); two long articles, "Domestic Life in Persia" (1903) and "A Talk about Persia and Its Women" (1910); and a third book, *Persia and Its People* (1910).

In 1911, on behalf of the Colonial Intelligence League for Educated Women, Sykes traveled to Canada disguised as a woman seeking a position as a "home-help" – a domestic servant for middle-class housewives – to discover the range of employment opportunities for educated, single, English women in the Dominion. The record of her mission, *A Home-Help in Canada* (1912), features a frontispiece of Sykes, who is older and somewhat stouter than she appeared in *Through Persia on a Side-Saddle*. A primary interest of *A Home-Help in Canada* resides in the reader's attempts to picture the dignified English lady of the frontispiece in the lowly station of a home-help. Unlike servants in the structured En-

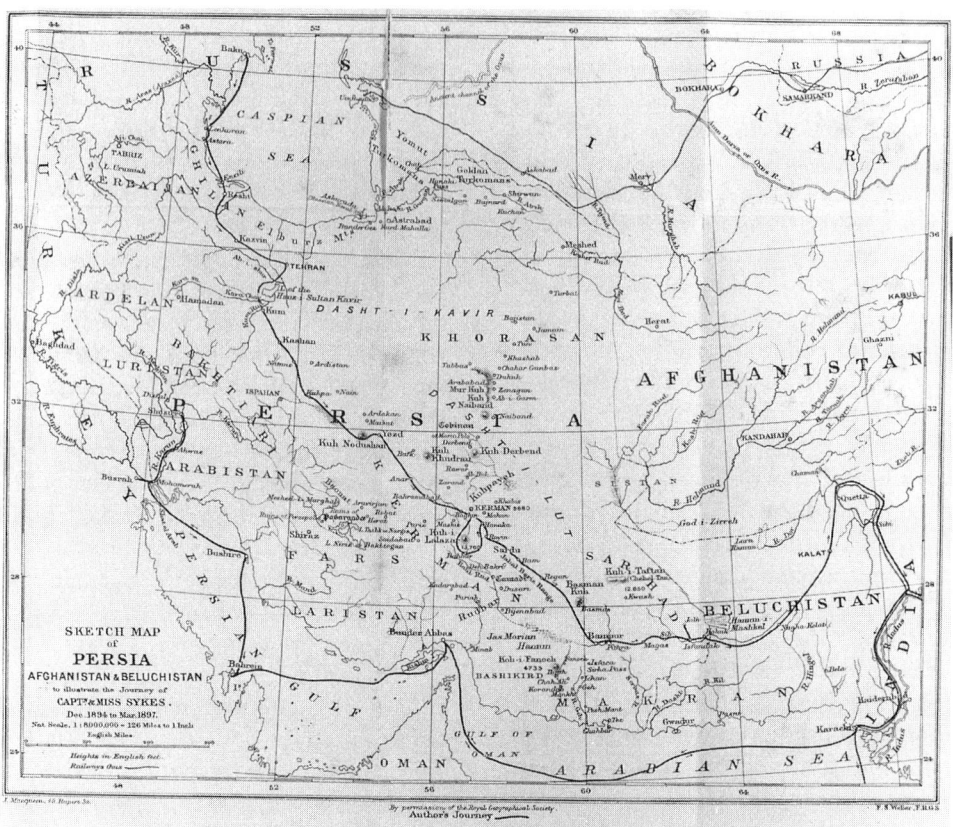

Foldout frontispiece for Sykes's revised edition of Through Persia on a Side-Saddle *(1901)*

glish "upstairs-downstairs" system, the home-help had an ambiguous status: she was a member of the family, yet without quite being so – and was less a professional, in terms of duties performed and remuneration, than her English counterpart.

As she travels across Canada, Sykes offers the kind of "keen" descriptions of scenery and places she delivered in *Through Persia on a Side-Saddle*. She also displays the same condescending sympathy for Canadian women less fortunate than herself that she showed toward Persian women. Her scheme to learn the Canadian employment situation for single Englishwomen by disguising herself as a prospective home-help is not only ludicrously inefficient but embarrassingly condescending. It is also colonialist: a daughter of the empire goes to the Dominion to discover how it may be exploited to solve England's problem of rampant unemployment of single women. Sykes's straightforward narrative lays bare the workings of imperialism – its hierarchy, its exploitation of its own working-class people and its colonies, even its exploitation of upper-class women, such as herself, drafted to "do their bit" for colonization.

In what she describes as an account of her last travels, Sykes recorded her 1915 journey through Chinese and Russian Turkistan in *Through Deserts and Oases of Central Asia* (1920), co-authored with her brother, who at that time held the rank of major. In the preface Percy Sykes notes that his sister wrote the first two-thirds of the text, describing their journey, while he wrote the last third, on historical, geographical, and political matters. Although *Through Deserts and Oases of Central Asia,* like *Through Persia on a Side-Saddle,* lays bare the effects of colonization, the book is perhaps most notable for its almost complete avoidance of immediate geopolitical concerns. The Sykeses made their journey during the first year of World War I, and their book was published two years after its conclusion; yet Major Sykes's treatment of "geographical, historical, and political" material is completely silent about the Russian Revolution, which occurred during the war and affected the region through which the Sykeses traveled.

Through Deserts and Oases of Central Asia is narrated by an older but still enthusiastic Sykes. One of the few women at the time to make it to the "Roof of the World" – the heights of the Russian Pamirs – Sykes again braved the elements, customs, and bad cuisine of a foreign land and rendered an account of her travels that is similar in almost every feature to *Through Persia on a Side-Saddle*. Again she displays her eye for scenery, her exultation in the freedom afforded by travel, her intrepidity in the face of danger, and her keen interest in the minutiae of native domestic life. She also voices similar complaints about the people she finds: "One of the great drawbacks of the Turki is that they never wash." One Kashgari custom, however, "rejoiced" her: "There is a law that, if the husband divorces his wife, the latter may take all the movables in the house"; therefore, the husband "often finds it cheaper to take a second wife."

In spite of the book's neglect of contemporary political issues, the review in the *Bookman* (Spring 1920) says that *Through Deserts and Oases of Central Asia* "should be in the hands of all interested in British policy in the Middle East." Sir George Macartney in the *Asiatic Review* (January 1921) remarks that "It is pathetic to think . . . that the authors of 'Through Deserts and Oases' were about the last Britishers to catch a glimpse of Russian Central Asia before that country became engulfed in the gloom of Bolshevism." Macartney also rebukes Sykes's "rejoicing" in the legal rights of married Kashgari women: "The wish may be the father to the thought; but the thought has not yet assumed the more active form of practice in Kashgar, and if the law exists there at all, it is, alas! more honoured in the breach than the observance." Sykes acknowledged and supported the strictures of patriarchy, even as she celebrated the freedom she herself enjoyed as a traveler. Throughout her writings she feelingly proclaimed the doctrine of "live and let live," all the while applying the "Orientalist" and colonialist stereotypes her readers expected. In short, Sykes displays the personal and feminine face of colonialism.

In *Temperamental Journeys: Essays on the Modern Literature of Travel* (1992) the editor, Michael Kowalewski, criticizes what he sees as the "commodification" of travel; he claims that "stores like the Banana Republic" try to suggest "that wearing uncombed cotton and wool products imported from independent artisans 'itself qualifies as an adventurous feat' analogous to global adventures." Sykes, of course, actually undertook such adventures, and throughout her texts, in language fit for a Banana Republic catalogue, she describes the multifarious gear that accompanies her on her travels – toted, at times, by dozens of indigenous servants. Following the final map in *Through Persia on a Side-Saddle* is a full-page advertisement for "The Sykes Patent Tent and Sleeping Valise," invented by Percy Sykes and described as "invaluable for officers on active service, and for all travellers going abroad or on sporting expeditions." This gear-loving and consumer-oriented aspect of Sykes's texts indicates the link for the European colonialist mind between travel and commerce; between experiencing a foreign place and buying pieces of it to take home (Sykes enumerates such purchases as Persian carpets and Turkistan jade); and between "knowing" and "owning." Sykes's texts are, finally, testimony to the personal and domestic nature of colonization.

Reference:

Michael Kowalewski, ed., *Temperamental Journeys: Essays on the Modern Literature of Travel* (Athens: University of Georgia Press, 1992).

Joseph Thomson
(14 February 1858 – 2 August 1895)

James Barszcz
William Paterson College

BOOKS: *To the Central African Lakes and Back: The Narrative of the Royal Geographical Society's East African Expedition, 1878-80,* 2 volumes (London: Sampson Low, Marston, Searle & Rivington, 1881; Boston: Houghton, Mifflin, 1881);

Through Masăi Land: A Journey of Exploration among the Snowclad Volcanic Mountains and Strange Tribes of Eastern Equatorial Africa (London: Sampson Low, Marston, Searle & Rivington, 1885); republished as *Through Masăi Land: A Journey of Exploration among the Snow Clad Volcanic Mountains and Strange Tribes of Eastern Equatorial Africa* (Boston: Houghton, Mifflin, 1885); revised as *Through Masai Land: A Journey of Exploration among the Snowclad Volcanic Mountains and Strange Tribes of Eastern Equatorial Africa* (London: Sampson Low, Marston, Searle & Rivington, 1887);

From the Equator to the Pole: Adventures of Recent Discovery, by Eminent Travellers, by Thomson, W. W. Graham, and A. H. Markham (London: Isbister, 1887) – includes "In the Heart of Africa," by Thomson;

Ulu: An African Romance, by Thomson and Elizabeth Harris-Smith (London: Sampson Low, Marston, Searle & Rivington, 1888);

Travels in the Atlas and Southern Morocco: A Narrative of Exploration (London: Philip, 1889; New York: Longmans, Green, 1889);

Mungo Park and the Niger (London: Philip, 1890; New York: Dodd, Mead, 1890).

SELECTED PERIODICAL PUBLICATIONS – UNCOLLECTED: "Adventures on the Rovuma: Letters in the Course of an Exploration," *Good Words,* 23 (1882): 240-247;

"East Central Africa and Its Commercial Outlook," *Scottish Geographical Magazine,* 2 (January 1886): 65-79;

"Mohammedanism in Central Africa," *Contemporary Review,* 50 (December 1886): 876-883;

"Downing Street *Versus* Chartered Companies in Africa," *Fortnightly Review,* new series 46 (August 1889): 173-185;

"The Results of European Intercourse with the African," *Contemporary Review,* 57 (March 1890): 339-352;

"To Lake Bangweolo and the Unexplored Region of British Central Africa," *Geographical Journal,* 1 (January 1893): 97-121.

As the author of some of the most popular books of African exploration in the late nineteenth century, Joseph Thomson earned fame not only for his discoveries but also for the humane manner in which he conducted his expeditions. Except for Henry M. Stanley, Thomson traveled more widely in Africa than any other European explorer, and he boasted that he never fired his gun in anger.

The youngest of five sons of William and Agnes Brown Thomson, Joseph Thomson was born on 14 February 1858 in Penpont, in Dumfries, Scotland. William, a stonemason said to have a constitution of iron, moved his family in 1868 to Gatelawbridge, where he ran his own quarry. According to James Thomson, his brother and his first biographer, Joseph early exhibited the traits of character that marked his adult careers as author and explorer: he was intellectually curious and morally upright, deploring the "profanity and indecent talk" of his fellows, and he was a leader among his peers, a lover of romantic landscapes and of geology. He was also interested in the literature of exploration, as he avidly read works by and about David Livingstone and Mungo Park.

Enrolling at Edinburgh University in 1875, Thomson studied geology and chemistry with Sir Archibald Geikie, the eminent Scottish geologist, and he attended T. H. Huxley's lectures in natural history. These lectures, according to a letter James Thomson quotes in his biography of his brother, Thomson found "glorious and sublime." He impressed those around him as an especially active, bold, and intelligent young man. In a letter of recommendation for Thomson, Geikie commented both on the "daring" Thomson showed on his field trips and on his "versatility of adaptation and knowledge of men and manners."

Geikie's letter was written to Keith Johnston, an explorer then planning an expedition to East Central Africa for the Royal Geographical Society. The expedition needed a geologist and botanist, and Thomson, having just graduated, sought and obtained the position. Before this first African sojourn was finished, Thomson needed all his daring, adaptability, and insight. Less than six weeks after leading the company from Zanzibar, Johnston died of dysentery and left Thomson, at the age of twenty, in charge of the expedition. According to the playwright J. M. Barrie, one of Thomson's friends, Johnston's death "made a man in an hour of a stripling." Though the Royal Geographical Society would not have blamed him had he turned back, Thomson pressed on and achieved the original objectives of the trip.

After being received as a hero at home in Scotland, Thomson delivered a paper on the results of the journey at a meeting of the Royal Geographical Society in London. His skill as a public speaker attracted offers for his story from publishers, and in the summer of 1881 *To the Central African Lakes and Back: The Narrative of the Royal Geographical Society's East African Expedition, 1878-80* appeared. The two-volume book was favorably received in both England and the United States. "Considering his youth and want of literary experience," wrote the London *Times* reviewer,

> Mr. Thomson tells the story of his work in Africa with wonderful ease and brightness. He makes no attempt to spin out a long story, but tells in a simple yet vigorous and attractive style the varied results of his observation. And he is a really good and exact observer, with a fund of genuine humour which will certainly amuse the reader.

Similar praise came from the *Dial,* an American journal, which called Thomson's book "in some respects the most valuable of [recent] contributions to modern African research." The reviewer praised Thomson's humor and cited him as one of the best "judges of the Negro character."

In fact, much of Thomson's humor describes Africans and their ways of life with ironically inflated diction, as when he comments on the grooming practices of the Warua people: "Generally, their spare time is occupied in entomological researches, after the manner of monkeys in the Zoological Gardens." Given such passages, a modern reader who wonders at Thomson's reputation for humane attitudes toward Africans must nonetheless remember the intensity of the Europeans' sense of superiority at the time. Such common European values are expressed in Stanley's *How I Found Livingstone* (1872), in which a notorious illustration shows Stanley resolutely pointing a pistol at a porter, who stands up to his neck in water and balances a box on his head; the caption reads, "Look out, you drop that box, I'll shoot you."

For all the sarcasm in his books, Thomson was famous for avoiding violence. Upon his return from this first expedition, he attended a reception at which he delivered a toast that sums up the values he maintained on his expeditions:

> My fondest boast is, not that I have travelled over hundreds of miles hitherto untrodden by the foot of a white man, but that I have been able to do so as a Christian and a Scotchman, carrying everywhere goodwill and

 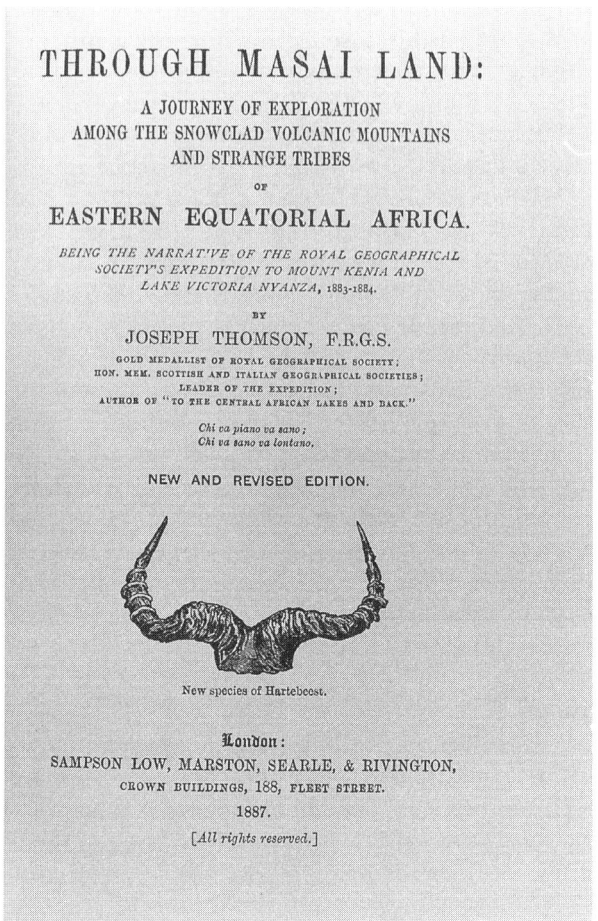

Frontispiece and title page for the revised edition of Thomson's narrative of his second African trip

friendship, finding that a gentle word was more potent than gunpowder, and that it was not necessary, even in Central Africa, to sacrifice the lives of men in order to throw light upon its dark corners.

In 1882 Thomson returned to Africa, where the Royal Geographical Society sent him to open a route from the eastern coast to Lake Victoria. This journey would take him through the territory of the Masai, a people known for the bravery and ferocity of their warriors. Thomson found the Masai to be less fearsome than reported, and he ascribed their reputation to Arab slave traders who exaggerated the dangers of the area in order to scare away Europeans who might want to abolish their business. On his return trip Thomson was tossed by a water buffalo that he had wounded in a hunt, and he was eventually afflicted with dysentery. The illness required him to be carried by porters for part of the journey, a rare violation of his principle that the way to stay healthy in Africa was through unremitting physical exertion.

Thomson's narrative of this expedition, *Through Masäi Land: A Journey of Exploration among the Snowclad Volcanic Mountains and Strange Tribes of Eastern Equatorial Africa,* was published in 1885. Notwithstanding disclaimers in the preface, where he apologizes for the inelegance of the book's style, the narrative seems shaped for the already-established audience of armchair travelers in Europe. Nigel Pavitt notes regretfully that Thomson slights his own important discoveries in natural history in order to provide the reader with "exaggerated and over-dramatized tales of elephant-shooting, since he knew that yarns of big-game hunting would increase the sales of his books." Thomson also describes at length the layout of the villages he visits and the customs he observes, and he often adds dated literary tags, such as "Let us now hurry forward, for the day is big with fate!"

As he does in *To the Central African Lakes and Back,* he also ironically praises the cultures he observes in *Through Masäi Land,* when he describes "the toilet of a swell M-teita damsel," for example. Still, the book registers Thomson's admiration for the Masai, not only for their physical attractiveness, which previous explorers had noted, but also for their intellectual and moral development that he observed nowhere else in Africa. The Masai, he notes, "indulged in none of the obtrusive, vulgar inquisitiveness or aggressive impertinence which make the traveller's life a burden to him among other native tribes."

Through Masăi Land received favorable reviews, as had Thomson's first book. "Mr. Thomson has no need to apologize for his want of practice with the pen," the London *Times* reviewer notes: "... his former narrative proved that he can tell his story quite as well as he leads his expeditions." The reviewer praises the "clear, swinging, vigorous style, rising into eloquence ... and abounding throughout with a sense of humour, or rather rollicking fun, which does not even spare the author's own peculiarities." The reviewer also praises Thomson's forbearance with "the African nature" and his recognition that "the African is only a wayward child whom it is more easy to manage by humouring than by thwarting." Thomson, it is said, acknowledges after all that "the native has some kind of right to his own country" and, as Thomson describes him, is justified in being suspicious of white travelers.

Questions about the African soul – how and whether it can be developed through contact with white culture – are dramatized in *Ulu: An African Romance* (1888), the little-noticed novel Thomson conceived in 1887 and wrote with the assistance of a school friend, Elizabeth Harris-Smith. Thomson wanted to write such a book, in part to correct what he saw as the distorted picture of Africa conveyed in the fiction of H. Rider Haggard, especially in a work such as *She: A History of Adventure* (1886). Published in 1888, Thomson's novel concerns a Scotsman, Tom Gilmour, who has been hurt in love and is disgusted by the "social frauds" and priggish culture of Victorian Britain. Gilmour travels to Africa and, in a Pygmalionesque attempt, decides to take a native girl, the Ulu of the title, as his wife in order to cultivate her. But he soon meets a missionary's daughter, Kate Kennedy, who reminds him of what the book ultimately presents as the superior beauty, moral and physical, of Caucasian womanhood.

Long passages in *Ulu* feature debates on the nature of the soul, the nature of good and evil, and the value of religion. These debates are displaced by action near the end of the book, when the two female characters, having been abducted by Masai warriors, are rescued by Gilmour. Ulu conveniently dies at the end of the adventure, allowing Gilmour, her erstwhile fiancé, to express his love for Kennedy honorably and to receive hers in return. Robert I. Rotberg praises the setting and the action sequences in the book, but he finds the characterization to be thin and the dialogue stilted. Noting similarities between Gilmour and Thomson himself, Rotberg concludes that the book "remains important more for its autobiographical elements than for the story."

After completing but before publishing *Ulu*, Thomson undertook a new journey, this time without institutional support, to the northwestern regions of Africa. The areas he visited were long known to Europeans, though little detailed knowledge about the geography of the interior had been gathered. Thomson added to what was known of the Atlas Mountains, but his book on the journey, *Travels in the Atlas and Southern Morocco: A Narrative of Exploration* (1889), was noticed as much for its depiction of Moroccan culture as for its scientific observations. "Probably no one has ever revealed so plainly the actual facts with regard to the condition of the Jews in the towns of Morocco," the reviewer for the London *Times* remarked. An American reviewer praised Thomson's exposure of the avarice shown by foreign ministers and consuls, including those from the United States, and once again Thomson earned admiration for the vigor and clarity of his writing.

The following year Thomson was asked to write a biography of Mungo Park, the eighteenth-century Scottish explorer, for a series of books on travel, The World's Great Explorers and Explorations. Written in about three months and adding some further biographical and historical material, the biography retells Park's own rendition of his travels on the Niger River. In general, Thomson admires Park for his industry and for his disinterested motives in exploring the region. Thomson also comments on the deplorable social conditions in West Africa, conditions brought on largely by European trade involving slaves, gin, and gunpowder. He concludes by forecasting a glowing future for the region if the Royal Niger Company, a private British trading company that was then becoming established there, proves successful. Although Rotberg considers the book a cursory performance, it is still cited in scholarship on Park.

Illustration from the revised edition of Thomson's Through Masai Land *(1887)*

In addition to his books, Thomson published many articles and essays in both popular and professional journals. Frequently he excerpted portions of books for publication. He also commented, most notably in "Downing Street *Versus* Chartered Companies in Africa" (1889), on what he saw as the failure of British policies in Africa. Thomson deplored the decision of the government to curb its colonial enterprises in Africa and thereby leave other European powers, which Thomson found less well suited than England, to rule over the indigenous peoples of Africa. In the absence of British governmental efforts, Thomson advocated the chartering of private British companies, such as the Royal Niger Company and the British South Africa Company, to maintain British influence.

Thomson's final expedition, which was to obtain trading rights in the interior of southern Africa, was commissioned by the British South Africa Company in 1890. For this endeavor Thomson was to proceed north from South Africa to the western shore of Lake Nyasa. Debilitated by disease, including the cystitis he had contracted in Morocco, he returned home from this expedition in 1892 and recovered sufficiently to present to the Royal Geographical Society a paper based on the trip, "To Lake Bangweolo and the Unexplored Region of British Central Africa," which subsequently appeared in the *Geographical Journal*. He never fully regained his health, however, despite various trips to climates milder than that of England. He died in London on 2 August 1895.

Born in the middle of the nineteenth century, Thomson pursued his career well after the completion of the major African expeditions by Livingstone, Richard Francis Burton, and Stanley. Few major discoveries remained to be made. Nonetheless, Thomson's work helped Europeans add many details to the map of Africa, and he is credited with some important contributions to zoology and botany, most notably the identification of Thomson's gazelle. He has twentieth-century detractors: one reviewer of Rotberg's biography calls Thomson "a middling, provincial, uninspired ... young man." Yet Thomson's contemporaries credited him with reviving the "age of African romance," and some regarded him as the equal of Livingstone in the geniality of his relations with Africans. Today his works are widely considered to be classics for their narrative energy, for their views of the geography and people of Africa, and for the image they convey of a disinterested, courageous, and generous European explorer.

Biographies:

James Thomson, *Joseph Thomson: African Explorer,* 2 volumes (London: Sampson Low, Marston, Searle & Rivington, 1896);

Robert I. Rotberg, *Joseph Thomson and the Exploration of Africa* (London: Chatto & Windus, 1971; New York: Oxford University Press, 1971).

References:

J. M. Barrie, "Joseph Thomson," in his *An Edinburgh Eleven: Pencil Portraits from College Life* (New York: Lovell, Coryell, 1892), pp. 107-114;

Anne Hugon, *The Exploration of Africa: From Cairo to the Cape,* translated by Alexandra Cambell (London: Thames & Hudson, 1993; New York: Abrams, 1993);

John A. Hunter, "Joseph Thomson, African Explorer," in *Tales of the African Frontier,* edited by Hunter and Daniel P. Mannix (New York: Harper, 1954), pp. 44-61;

Frank McLynn, *Hearts of Darkness: The European Exploration of Africa* (New York: Carroll & Graf, 1993);

Nigel Pavitt, *Kenya: The First Explorers* (London: Aurum, 1989; New York: St. Martin's Press, 1989);

Robert I. Rotberg, "Joseph Thomson: Energy, Humanism, and Imperialism," in his *Africa and Its Explorers* (Cambridge, Mass.: Harvard University Press, 1970), pp. 297-320.

Ethel Brilliana Tweedie
(circa 1860 – 15 April 1940)

Josephine A. McQuail
Tennessee Technological University

BOOKS: *A Girl's Ride in Iceland,* as Ethel B. Harley (London: Griffith, Farran, Okeden & Welsh, 1889);

The Oberammergau Passion Play 1890 (London: Kegan Paul, French, Trübner, 1890);

A Winter Jaunt to Norway (London: Bliss, Sands & Foster, 1894);

Danish Versus English Butter Making (London: Cox, 1895);

Wilton, Q. C.; or, Life in a Highland Shooting Box (London: Cox, 1895);

Through Finland in Carts (London: Black, 1897; New York: Macmillan, 1898);

George Harley, F.R.S.: The Life of a London Physician (London: Scientific Press, 1899);

Mexico as I Saw It (London: Hurst & Blackett, 1901; New York: Macmillan, 1901);

Sunny Sicily: Its Rustics and Its Ruins (London & New York: Macmillan, 1904);

Behind the Footlights (London: Hutchinson, 1904; New York: Dodd, Mead, 1904);

Porfirio Diaz: Seven Times President of Mexico (London: Hurst & Blackett, 1906); republished as *The Maker of Modern Mexico: Porfirio Diaz* (London & New York: John Lane, 1906);

Hyde Park: Its History and Romance (London: Nash, 1908; New York: Pott, 1908); republished as *Hyde Park: Its History & Romance* (London: Besant, 1930);

Thirteen Years of a Busy Woman's Life (London & New York: John Lane, 1912);

Busy Days (London: Routledge, 1913);

America as I Saw It; or, America Revisited (London: Hutchinson, 1913; New York: Macmillan, 1913);

Women the World Over: A Sketch Both Light and Gay, Perchance Both Dull and Stupid (London: Hutchinson, 1914; New York: Doran, 1915);

My Table-Cloths: A Few Reminiscences (London: Hutchinson, 1916; New York: Doran, 1917);

Mexico: From Diaz to the Kaiser (London: Hutchinson, 1917; New York: Doran, 1918);

Ethel Brilliana Tweedie in 1922

Women and Soldiers (London & New York: John Lane, 1918);

A Woman on Four Battle-Fronts (Leeds & London: Beck & Inchbold, 1919);

Mainly East (In Prose – Perhaps Prosey) (London: Hutchinson, 1922; New York: Dutton, 1922);

An Adventurous Journey (Russia – Siberia – China) (London: Hutchinson, 1926; revised edition, London: Butterworth, 1929);

Russia As I Saw It (Hastings: Parsons, 1927);

Me and Mine: A Medley of Thoughts and Memories (London: Hutchinson, 1932);

Tight Corners of My Adventurous Life (London: Hutchinson, 1933);
My Legacy Cruise (The Peak Year of My Life) (London: Hutchinson, 1936).

OTHER: *The First College Open to Women, Queen's College, London: Memories and Records of Work Done, 1848-1898,* edited by Tweedie (Edinburgh: Turnbull & Spears, 1898?);
Omnium Gatherum Booklet, edited by Tweedie (London: Simpkin, Marshall, 1909);
William Hickling Prescott, *The Conquest of Mexico,* 2 volumes, introduction by Tweedie (London: Richards / Oxford: Oxford University Press, 1915).

SELECTED PERIODICAL PUBLICATIONS – UNCOLLECTED: "A Chat with Dr. Nansen," *Temple Bar,* 97 (February 1893): 187-198;
"Isthmus of Tehuantepec Railway," *Fortnightly Review,* new series 70 (August 1901): 271-280;
"William Quiller Orchardson," *Fortnightly Review,* new series 87 (June 1910): 1111-1122;
"Women and Work," *Fortnightly Review,* new series 89 (June 1911): 1099-1111;
"Diaz: The Maker of Modern Mexico," *Fortnightly Review,* new series 90 (July 1911): 65-74;
"Eugenics," *Fortnightly Review,* new series 91 (May 1912): 854-865;
"American Women Dissected by an English Writer," *New York Times,* 23 February 1913, VI: 6;
"Mrs. Alec Tweedie Dissects British Women and Ours," *New York Times,* 9 March 1913, VII: 5;
"Europe Can Give America Lessons in Manners," *New York Times,* 16 March 1913, VI: 9;
"Mrs. Alec Tweedie Mourns Our Disappearing Home Life," *New York Times,* 23 March 1913, VI: 5;
"An Englishwoman's Impressions of Our Sleeping Cars," *New York Times,* 30 March 1913, V: 6.

Ethel Brilliana Tweedie, or Mrs. Alec Tweedie, as she signed her work, was a pioneering and prolific female journalist, travel writer, biographer, and historian. She wrote or edited more than twenty books, many of which relate her experiences in traveling – often to exotic places under conditions few women, or men, of her time were prepared to face. She visited the United States, Finland, Iceland, Norway, Mexico, Sicily, the Middle East, India, Egypt, Italy, Greece, China, and a troubled Russia. She was a tireless advocate of women's rights, particularly of a woman's right to vote. Yet some of her other views betray colonialist, sometimes elitist attitudes that, although they might offend modern readers, no doubt indicate well the ways that many of the educated British upper middle classes thought. Tweedie's books are carefully researched and illustrated, sometimes with her own artwork or photographs, and are full of vivid detail.

Her watercolors, sketches, and photographs illustrate the quality of her work as a visual artist in addition to that as a writer. In 1921, after traveling fifty thousand miles on her journey to the Far East, three hundred of her watercolors and sketches were exhibited. She often signed her sketches "E. B. Tweedie," but after her marriage she always used her husband's name to identify her authorship and was upset when a German translation of *The Maker of Modern Mexico: Porfirio Diaz* (1906) identified her not as "Mrs. Alec Tweedie" but as "Alec Tweedie" because the publisher thought this would sell more copies.

Tweedie was born Ethel Brilliana Harley, the daughter of a prominent London physician, George Harley; and his wife, the former Emma Jessie Muspratt of Seaforth Hall, near Liverpool. Harley – who was interested in the chemistry of respiration, the workings of poisons used on arrowheads, and cases of suspended animation induced by drowning, hanging, and other traumas – was the subject of Tweedie's *George Harley, F.R.S.: The Life of a London Physician* (1899). He was an energetic founder of the British Institute of Preventive Medicine, and he also was keenly interested in the English language. In his own book, *The Simplification of English Spelling* (1877), Harley proposed radical revision of the English language and its orthography. In addition to his medical writing, he also contributed "What is a Geyser?" as an appendix to his daughter's first book, *A Girl's Ride in Iceland* (1889).

While Tweedie's books do not mention any details of her education, her trip to Norway with her brother and sister, Vaughn and Olga, was occasioned by Vaughn's illness following his medical studies. (Tweedie was also to go on several journeys with Olga – one to Italy and Sicily, and another to Finland.) As a girl, Tweedie's educational opportunities were limited, and she was probably tutored at home: the first college for women in London had opened in 1848, but not even an upper-class family such as that of Leslie Stephen sent its bright young daughters to college. While Stephen's daughter, Virginia Woolf, made a formidable effort to educate herself in modern and classical languages and literature, Tweedie and her books suffered because she lacked a sophisticated and intimate knowledge of various literary genres. Her love of books and learning expressed in *Thirteen Years of a Busy Woman's Life* (1912) bespeaks a woman whose desire for knowledge has been somehow thwarted: "After one has passed the critical age of twenty . . . the love of books, the real honest pleasure of reading, the insatiable craving for knowledge, takes fast hold of us, and we begin to realise, as we study even one single subject, what a vast field lies open before us." These are the remarks of someone who lacks, and who misses, the general background of knowledge provided by a good formal education.

Pursuing Ethel without permission on her journey to Iceland was Alec Tweedie, the grandson of Dr. Alexander Tweedie and an importunate wooer who was to become her husband. Following their marriage they had two sons, Harley Alec and Leslie Kinloch Tweedie. Tweedie relates how her husband's early death in 1896, six months after losing his fortune, made it necessary for her to turn to writing in order to support herself and her sons. She dedicated *Through Finland in Carts* (1897), her first book after Alexander's death, to her late husband; as well as to her dearest friend, Sir John Eric Ericksen; and her father – all of whom, as she acknowledges in her dedication, had died within a five-month period. Tragedy was to follow Tweedie throughout her life: she lost her youngest son, a lieutenant in World War I, in 1916, and her other son was killed in a flying accident in 1926. Traveling, sketching, and writing were often to be means for Tweedie to use in seeking solace.

Although Tweedie disliked the cheap newspapers and periodicals that began to appear at the end of the nineteenth century, she found opportunities in the new readers educated by board schools that had extended basic literacy and created a demand for entertaining reading material. Tweedie's writing for that public included not only many travel books but also a novel, *Wilton, Q. C; or, Life in a Highland Shooting Box* (1895), which she had published before her husband's death, and works such as *Hyde Park: Its History and Romance* (1908), *My Table-Cloths: A Few Reminiscences* (1916), and *Me and Mine: A Medley of Thoughts and Memories* (1932). Her characteristically chatty, breezy style made all her works easy and amusing reading, but it often invited attack by the critics, who perceived her tone and her domestic topics as "feminine."

Tweedie is comfortable adopting a tone appropriate to domestic genres such as those of the diary, the journal, or the letter. As the title of *My Table-Cloths: A Few Reminiscences* emphasizes, she uses a familiar domestic image of tablecloths to characterize her prose: it is a compilation of her experiences and memories. Dorothy Wordsworth is perhaps the best example of a woman who wrote in a similar fashion – in fragments – because she never could summon that wholeness of vision and confidence that would allow her to write polished masterpieces as her brother, William, wrote. Although Tweedie never seeks to create a disembodied narrative presence in her writing, her learning and culture are apparent in her books, which reflect typical British attitudes toward colonialism and class at the time, as well as feminist views of women and women's rights.

Before her husband died, Tweedie had not earned a living by writing, but she had written several books and articles. Her first article, "Pasteur and His Institute," was published in *Murray's Magazine* after her mother and her husband, both of whom knew Louis Pasteur, had arranged an interview. Tweedie later wrote other articles for the *Pall Mall Gazette* and the *Queen,* and in *Thirteen Years of a Busy Woman's Life* she relates how she came to write for the latter. While visiting Scotland with her husband, who was an avid golfer, Tweedie went to the offices of the *Queen* and proposed that she write articles on sports. At first she met resistance, but when the art editor saw her driving gloves (for holding the reins of horses while driving a carriage), he said that any woman who could drive a pair of horses along big city streets in that season could certainly write an article on sports. Her persistence paid off, and at the time of her death she was called a "pioneer British woman journalist."

Did Tweedie make enough money in her career to support an upper-middle-class life? She certainly appears well dressed in all of the photographs included in her books, whether she is photographed in the wilds of Africa or on the ranches of Mexico. She was one of those tourists who must have high tea to drink and hot water for a bath, no matter where she was. However, she also had to struggle to maintain a decorous existence. In *Thirteen Years of a Busy Woman's Life* she bemoans the insensitivity of friends who, after her husband's death, invited her to visit them or to accompany them to the theater or other entertainments and yet seemed unaware that she could afford neither train fare nor cab fare.

In *America as I Saw It; or, America Revisited* (1913) she comments on the expense of traveling to America: the upper-middle-class Englishwoman has only $500 to spend on vacation, and just the expense of traveling to America takes much of her meager vacation funds. In *Mainly East (In Prose –*

 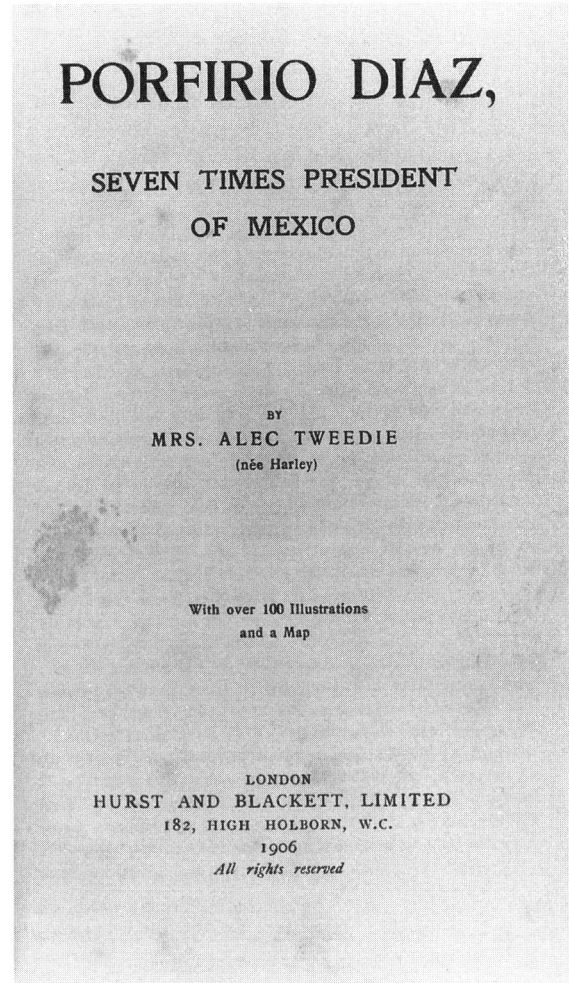

Frontispiece and title page for Tweedie's second book on her travels in Mexico. Díaz, for whom Tweedie had expressed admiration in her first book, invited her to return for the celebration of his seventh election as president.

Perhaps Prosey) (1922) Tweedie is certainly unwilling to spend money carelessly, as she is forced to do when she gets sick and cannot move into a house she has rented, but her steady stream of book and journalistic publications leads one to believe that she made a comfortable living as a writer. She was always able to secure invitations to see important people, to go to significant social events, and to meet dignitaries wherever she traveled.

Her first book, *A Girl's Ride in Iceland,* was prompted by her father's suggestion that she keep a diary of the trip, because posting letters would be impractical. Dr. John Rae, a friend of her father, tried to dissuade her from making the trip, which he said was too arduous an undertaking for ladies. She was not convinced, however, and she departed on the *Camoens,* a steamer that ran between Granton and Reykjavík. Circumnavigating the island by ship, she also made some forays into the interior by horseback.

Such an undertaking raised eyebrows when her book appeared; for example, one reviewer criticized its author's recommendation that women ride astride. Tweedie had seen some hot springs in the interior (near Thingvalla Water, where the Icelandic government had been seated) only because she was willing to attempt the difficult journey by horseback. Her decision to ride astride was undertaken for practical reasons: "Riding man-fashion is less tiring than on a sidesaddle, and I soon found it far more agreeable, especially when traversing rough ground." The frontispiece of Tweedie's book depicts women astride horses, and such features of the book sparked a controversy ("Should Women Ride Astride?") in *The Field* and the *Daily Graphic* in 1890.

Tweedie was later to comment more extensively on riding astride in *Mexico as I Saw It* (1901):

> As I am a warm advocate of riding astride for women, perhaps it may be well to describe why and how I came to adopt that mode. My first long expedition was in Iceland, where on one occasion a girl and I accomplished a distance of 163 miles in three days and a few hours. This was a land where there were no bridges, rivers had to be swum by the ponies, there were no roads, and rough paths and dangerous mountain passes formed the track.... Such rides could never be accomplished by side-saddle, whereas mounted astride the woman is no longer handicapped, and provided she have equal strength with her male companion, can go wherever he goes.

Reviewers called Tweedie immodest and indelicate for her "unfeminine" behavior, but she was not discouraged by such critiques and continued to express her feminist views in her writings.

In Iceland Tweedie visited Akureyri, Saudhárkrókur, Reykir, and Reykjavík, and she regretted not being able to visit the volcano at Hekla. In recounting the history of Iceland, she describes the national character and calls attention to the

> sad, dejected looks of the men and women who surrounded us. There was neither life nor interest depicted on their faces, nothing but stolid indifference. This apathy is no doubt caused by the hard lives these people live, the intense cold they have to endure and the absence of variety in their every-day existence.

Her description of their everyday lives gives an impression of their poverty. In Saudhárkrókur she recounts an auction of a pathetic assembly of items that seem of no value to the foreign tourists but which the Icelanders think are worthy of display and sale.

To complete her account of the northern lands begun in *A Girl's Ride in Iceland,* Tweedie wrote *A Winter Jaunt to Norway* (1894). This work recounts her travels (sometimes on snowshoes, as skis were then called) and encounters with famous people such as Fridtjof Nansen, the Norwegian explorer and humanitarian who won the Nobel Peace Prize in 1922; Henrik Ibsen, the Norwegian dramatist; Bjørnstjerne Bjørnson, the winner of the Nobel Prize for literature in 1903; and Georg Brandes, the Danish literary critic and scholar.

During the unusually severe winter of 1892–1893 Tweedie and her sister attempted to begin their trip to Norway. Tweedie writes that a derisive shipping clerk from whom they were trying to buy tickets "laughed at the idea of two ladies going at all, as none but commercial travellers ever venture there between October and May." Having purchased their tickets, they were turned back on the first leg of their journey, even before having left Victoria Station, because ice had closed the Kiel to Korsør route, they were told. The next day friends abroad assured them by telegram that the route was open, and they began their journey.

Their travel was arduous, however, and in some places the ice had to be dynamited to clear a route. Tweedie describes the special prow of the boat, which was able to "skate" onto the ice and break it up. In Helsingör, Tweedie and her companions later had to travel in a sledge boat drawn over the ice by men. The poor guides were alternately ducked in the freezing water and dragged over the ice by the sledge boat, and the passengers repeatedly had to push it when it became stuck. On the last leg of the journey the sisters traveled by rail through Sweden to arrive in Christiania (Oslo), Norway, where Dr. Vaughn Harley met them.

Tweedie was constantly amazed by methods of travel in the frozen Norwegian lands. She comments on the "powers of endurance and [the] wondrous cat-like agility and sure-footedness" of the horses, and she finds that Norwegian drivers are as determined as the horses are to drive through and over anything that stands in their way. Tweedie also ventured out on skis, and both in 1891 and 1893 she observed the annual ski competitions. Particularly impressive to her were the ski jumps.

Tweedie's friends arranged for her to see sights and towns outside Christiania, and she stayed at a hotel in a town called Sandviken, where local people were amazed, as those in Iceland had been, to see female visitors, or "ladies." At Hönefos she saw a waterfall that was frozen and appeared as a fairyland. She was amazed to encounter some lovers whose lovemaking occurred in an ice cave. Her meditation on love in *A Winter Jaunt to Norway* is followed by lines from Alfred Tennyson's *In Memoriam* (1850): " 'Tis better to have loved and lost / Than never to have loved at all." With the death of her husband in 1896, Tweedie would soon come to know firsthand the poignancy of these lines.

With her brother and some male friends Tweedie also visited a silver mine at Kongsberg, southwest of Christiania, and a *saeter,* a goatherd cabin, where she and her sister again met with scoffing and discouragement: a farmer whom they encountered said that "he still considered it quite impossible for the ladies – and English ladies, too – unaccustomed, of course, to ski, to make the ascent, and even advised the men not to do so." He was

A BIT OF OLD JERUSALEM
From a Water Colour by Mrs. Alec Tweedie.

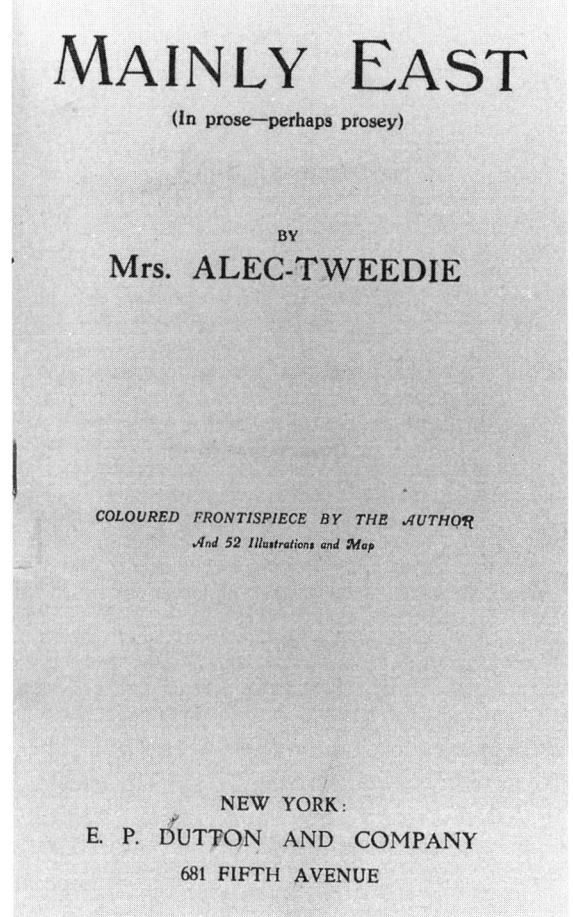

Frontispiece and title page for the American edition of the journey Tweedie took to relieve her grief over the death of her youngest son in World War I

persuaded that these English ladies knew how to ski when he was shown newspaper articles that mentioned them and their prowess. Surrounded by large snowdrifts, they spent three days playing in the snow and talking in the remote cabin, and they were greeted with amazement when they returned to Sandum Farm.

A Winter Jaunt to Norway ends with Tweedie's accounts of three famous Norwegian men and one Dane: Bjørnson, Nansen, Ibsen, and Brandes, respectively. The book's curious arrangement thus undercuts Tweedie's prodigious achievements, for her travels are nothing in comparison to Nansen's Arctic explorations, and her writings pale in comparison to the achievements of these literary figures. Her comments on Ibsen are therefore telling:

> He is a great admirer of the opposite sex, and more affable with them than with his own. Indeed, he has keen appreciation of a pretty face, and says so. Ibsen has done much as a writer for women in his earlier works, advocating their individual development. He thought they should be independent beings, with education and ideas of their own, and not merely the wives of their husbands, the mothers of their children. He advocated that women [*sic*] should be a human thinking being above all else. Latterly he has shifted these views. His women characters have become hysterical and incomprehensible. Clever they are, in their attraction for men, and they often appear capable of achieving great ends, but usually collapse in emancipated shrieks.

In recounting her introduction to Bjørnson — poet, novelist, dramatist, and politician — Tweedie compares "the consular struggle in Norway" to Britain's "struggle with Ireland." She also notes his advocacy for women's rights. Dr. Brandes of Denmark, a literary critic and noted intellectual whom she met on her journey, accompanied her on a visit to a Danish butter factory, and this experience provided the subject of Tweedie's next book, *Danish Versus English Butter Making* (1895). Denmark was to compete with England in producing butter and other dairy products such as eggs. Tweedie remarks

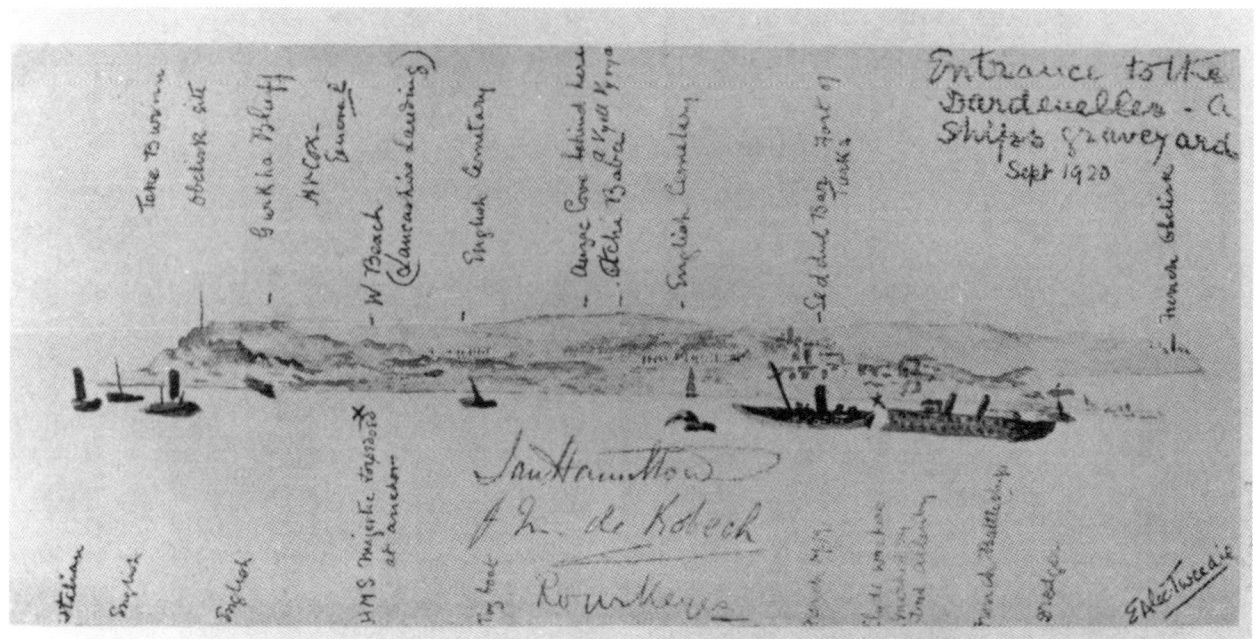

Sketch of the Dardanelles that Tweedie drew during her postwar tour of Mediterranean and Eastern regions (Imperial War Museum, London)

on Danish efficiency; she accuses British farmers, in contrast, of being lethargic.

Tweedie's next travel book, *Through Finland in Carts,* recounts her visits not only to cities such as Helsinki but also to obscure rural locations such as Lake Ladoga. During this Scandinavian trip she journeyed in a tar boat and slept in a peasant's cottage overrun with bugs. She and her sister had to be ready to adapt to the most impromptu abodes – for instance, that of the monastery at Valerno, which was not really prepared to house two ladies. Tweedie remarks on the gravity of the Finns, but she and her sister become laughingstocks to the Finns, who see the two burdened with furs and rugs for the Finnish summer – when Finnish summers are warmer than those of England.

Although she says that Finnish is a difficult language, Tweedie attempts to use some Finnish words, and she condemns British people who never speak any language besides their own. She relates the history of Finland, which for centuries had belonged to Sweden and at the time of her visit is controlled by the Russian czar. She discovers that, in an exercise of censorship, the Russians have tampered with her newspapers from England.

Tweedie found the Finns to be notably democratic. She observes them sharing drinks from a flask in a way that shows no class differences, and such features of Finnish democracy bother her. Visiting a Russian monastery at Lake Ladoga, she meets a young upper-class convert from Moscow. Saying that he has descended both socially and intellectually to the level of those beneath him, she bemoans the loss of his upper-class finish. In another instance, she relates a complaint by the wife of one of Finland's eight governors: having mixed with a soldier's daughter in school, the woman's son began to call her "thou." Tweedie remarks that the upper classes are making a mistake in mixing with the lower, whom they will later have to command.

She claims that sexual equality has always characterized Finland, and in *Women the World Over: A Sketch Both Light and Gay, Perchance Both Dull and Stupid* (1914) she cites Finland as a country where women have rights equal to those of men. Even in *Through Finland in Carts,* published seventeen years earlier, she finds the position of the Finnish woman to be remarkable. Because the Finnish population included thirty-six thousand more women than men, Finnish women had ceased to look upon marriage as their only avenue for fulfillment. The first Finnish women to graduate from the university had done so in 1870. Tweedie carefully emphasizes the morality of the country, because some of her readers believed that female emancipation led to immorality.

Finnish divorce laws at the time were far more liberal than those in England. Tweedie reports that, if a Finnish wife leaves her husband after a year, he may advertise three times for her return in a newspaper. If his wife fails to respond, he may then obtain a divorce, and no scandal is attached to the affair. This was quite unlike the situation in England, where Tweedie finds the process to be a disgrace.

Tweedie characterizes the Finnish temperament as musical and poetical, and she notes that Finns love songs, riddles, puzzles, and stories. She once engages in some storytelling by plunging in medias res into a narrative that convinces her reader that the castle where she is staying is haunted – before she reveals that the story is a dream. Another experimental piece reveals Tweedie's playfulness. In discussing one jaunt, Tweedie engages in some free association: "In Finland ... the correct way to eat wild strawberries was with a pin! The pin reminds us of pricks, and pricks somehow remind us of soap, and soap reminds us of a little incident which may here be mentioned." She then relates a story of how a dwindling supply of precious English soap greatly distresses her traveling companion, her sister. Tweedie taunts her sister by announcing every time she is about to wash her hands with the treasure. One day her sister gleefully returns with a bar of Finnish soap that turns out to be floor soap.

Tweedie's next destination was Mexico, a much different place, which she visited in 1900–1901 on her first trip alone. She traveled more than twenty-five thousand miles by ship, train, and horseback – a prodigious journey, especially for a woman. She visited some major cities as well as ranches in the interior and the great Aztec ruins. Mexico provided the experiences for three books that, if one judges from the number of their reprinted editions, became her most popular works: *Mexico as I Saw It* was published and reprinted in five editions from 1901 through 1911; *Porfirio Diaz: Seven Times President of Mexico* (1906) was published and reprinted in three editions, including a German translation; and *Mexico: From Diaz to the Kaiser* (1917) appeared in two editions. Her first two books were so successful that she was asked to write the introduction for William Hickling Prescott's two-volume *The Conquest of Mexico* (1915).

Her first trip to Mexico established her admiration for its president-dictator, Porfirio Díaz. In 1904 Díaz invited her back to observe his seventh election, after which she began to write her second book on her Mexican travels, *Porfirio Diaz,* which celebrated that leader and his accomplishments. In these two books her attitudes toward Mexico are influenced by her status as British subject and her partiality for Díaz. In a review of the last of her Mexican books, the *Literary Digest International Book Review* (25 May 1918) remarked that her veiled sarcasm toward the United States would irritate most American readers. For instance, in *Mexico: From Diaz to the Kaiser* Tweedie asserts: "There are many people who consider that the United States have no right to interfere in Mexican affairs; that Mexico no more belongs to them than does Canada ... that the Pan-American Union was built not merely for the expansion of trade, but for the final absorption of all the Americas." Her suspicions about the motives of U.S. leaders are evident throughout the book, which asks:

> Has the United States the right to dictate to the whole of Latin America?
> Does the land from the Rio Grande to Cape Horn rightfully come under the United States control?
> Does the United States wish to annex everything, as they annexed Panama from Colombia, and if so, have they the right?

Tweedie shows political acumen in her skepticism about the intentions of the United States.

In *Mexico as I Saw It* Tweedie demonstrates an understanding of the oppressive ranch system in Mexico, which allowed ranchers to oblige workers to contract enormous debts in order to purchase necessities and thereby kept workers virtually enslaved. Tweedie calls this the "peon system." She decries this system as well as the lack of emancipation for the Mexican woman and the cruelty of cockfights and bullfights – which, she asserts, make spectators (especially children) vicious and cruel. She has no faith that the Mexican Indian can be educated and questions Díaz's conviction that "this education will awaken the country, and prove an inestimable boon." Tweedie makes it clear in all of her books that she has little respect for proletarian movements. In *Mexico: From Diaz to the Kaiser,* she condemns such socialist trends as those no doubt behind the Russian Revolution:

> If individuals would only give up talking so much of their rights, their own ridiculous little rights, and think less of themselves and more of other people and their own nation's right, what a blessing it would be.... They only want their rights for themselves, those strikers and revolutionaries; they care nothing of the rights of civilization. Their ethics are small, their stupidity supreme.

Despite Tweedie's skepticism of some of Díaz's aspirations about what education, for example, would do for his proletariat, she was impressed with an Aztec, who declaimed to her: "You have come from a land of great civilisation to visit our

Turkish railway pass used by Tweedie during part of her travels recounted in Mainly East *(1922)*

wild country; but Señora, you must remember that five thousand years ago, when England was unknown, our ancestors raised these ruins." On her next trip, a feeling of awe at what past civilizations have accomplished overwhelmed Tweedie.

Sunny Sicily: Its Rustics and Its Ruins (1904) was the product of Tweedie's next journey abroad. In Sicily and Italy she was struck with awe at the remains of the Roman Empire. The relics of Selinus in the Palermo museums and the remains of amphitheaters and temples impressed her. She remarks that in the face of these ancient wonders, modern civilization, which seems so advanced, seems of little consequence. The sights in Sicily lead her to realize how impermanent even an apparently omnipotent power can be, and meditations on England's colonial power and the strength of the British Empire often preoccupy her.

If Mexican peasants had impressed Tweedie as being childlike and little removed from an animal existence, she concludes in her book on Italy that southern peoples are dishonest. A meal that has been packed for her lunch at her hotel disappears before she can eat it on an excursion, and she comments:

We had our tea, laughed over the incident, and thanked Heaven that Teutonic and Scandinavian peoples are more honest than their Southern Brethren. What a contrast is the honesty of Norway or Finland, the highlands of Scotland, or the mountain districts of Bavaria, with the dishonesty of Spain, Italy, Sicily or Morocco. Does sunshine make thieves, one wonders?

Such sentiments convey typical upper-middle-class British prejudice of this era.

In other respects, however, Tweedie shows sensitivity to Sicilian life. She describes Sicilian poverty and explains it through her social analysis. She bemoans the destruction of forests that have left the Sicilian landscape barren. She shows special sensitivity to the plight of Sicilian women: "Their marriages are arranged for them, for the women have not really emerged from the thrall of unquestioning submission to men. They are wives and mothers, housekeepers and ornaments, but not yet chums and friends."

In describing a meeting with a Sicilian countess, Tweedie reports that her sister asked if the countess drove much. "No, very little," replied the countess. "I potter about in the garden and in the house, but seldom go outside our own grounds." Tweedie concludes, "This seems to be the general life of the ordinary Sicilian lady, to whom the independence, athleticism, and energy of an Englishwoman are surprising." She finds the Sicilian level of energy to be low and repeats a remark from an Sicilian-American woman whose parents had moved back to Sicily from America: "These Sicilians don't live, they crawl, ma'am."

Sunny Sicily got generally good reviews, though not all readers appreciated Tweedie's style. *American Review of Reviews* (March 1905) characterized her writing as an "entertaining style wedded to a real knowledge of how to tell a story." *The New York Times* (4 May 1905) commented on the informality and personal style of the book and found it both informative and entertaining. The reviewer for *The Nation* (9 February 1905) was harsh, admitting that the book was readable and full of practical suggestions, but was offended by Tweedie's flippant tone and called her style slipshod.

Tweedie had first seen the United States on her way to Mexico, so on her second trip to America – on her way to attend the seventh election of Díaz – she gathered material for *America as I Saw It; or, America Revisited*. Like her books on Mexico, her tone and assertions about the United States in this work offended some critics and readers in the United States. For Tweedie, this response only proved her contention that "HYPERSENSITIVENESS IS THE AMERICAN SIN."

Before her visit to United States, Tweedie purposefully avoided reading any books on the country – not even, she notes, anything by Charles Dickens, who had toured the U.S. in 1842 and in 1867–1868. Perhaps Tweedie implicitly aims to compare herself to such a great writer as Dickens, but her books are of more topical interest than his works. In her three trips to the United States in 1900–1901, 1904, and 1912 Tweedie visited various American cities such as New York, Chicago, Boston, Saint Louis, Niagara Falls, El Paso, and Washington, D.C., where she was a guest in the White House and at several embassies. She was also entertained in private homes, and her narrative lists those people whom she wishes to thank for their hospitality. Extracts from her book were published in a series of articles in *The New York Times* in 1913.

She is much impressed with American energy on this visit:

> Americans can work hard and play vigorously but the hour of folded hands and quiet thought is an unknown luxury in their luxurious land.

> A strenuous life lived too strenuously is like an overwound watch – it snaps.

In 1926 Aldous Huxley would make similar observations on the American character, but he would offer more positive conclusions than Tweedie. American vitality to Huxley seemed a psychological product of prosperity. Tweedie, however, sees the Americans' reckless harvest of timber and asserts that "it is about time for the government to intervene, and see that the rainfall and climate of the country is properly protected through its timber."

Tweedie has a keen eye for detail in assessing national character; much of what she says about the United States and its people remains true. For instance, she wonders if any part of the United States will be free of billboards in the future. She complains about the criminality touted in United States newspaper headlines. She reports that Americans are overworked, that sweatshops are everywhere, and that clerks do not get holidays. The almighty dollar reigns, and for this reason the arts are not appreciated and the professions not as respected as in other lands.

Noting that Americans do not sufficiently appreciate or remunerate domestic labor by wives or mothers, Tweedie shows her usual sensitivity to women's issues. She believes that one solution to the oppression of women lies in extending the vote to them, for this alone will give them a say in national affairs. She was mistaken, however, in her conviction that American women would get the franchise before English women, and her feminism seems incongruous with her condemnation of African Americans, Roman Catholics, and Jews, whom she saw as threats or "outside elements."

Women the World Over: A Sketch Both Light and Gay, Perchance Both Dull and Stupid is not, strictly speaking, a travel book, but it demonstrates her playfulness as well as her feminist sympathies and offers insights about women of various nationalities that Tweedie encountered in her travels. "Dedicated to Women the World Over, Wicked and Wise, Learned and Lanky, Dainty and Dauntless, Playful and Pretty, but, Women Still," the book was published in the first year of World War I, when men had gone off to war and women had begun to assume men's jobs. Tweedie reveals an attitude perhaps typical of feminists of those times: while advocating the rights of women, she placates those who might object that women would become "unfeminine" by adopting the roles of men. She frequently emphasizes that woman is man's helpmate and that the woman should dress to please the man.

In the first American edition of the book (1915) she deplores the fact that women are paid less than men even when women do an equal amount of work. She notes that nearly as many women as men work in the great trades in England, and those women are not paid wages equal to those of men who perform the same work. She asserts that every rational person favors woman suffrage, and she anticipates that woman suffrage will be universal before 1925. In 1913 Norway adopted full general suffrage for women, and she praises Norway and Finland for treating women fairly.

However, Tweedie expresses conventional attitudes along with her feminist positions. At some places the book advises men not to marry beneath them and women to wear neat black coats and skirts during the day and simple low-back gowns in the evening to please men. Yet she believes women are capable of entering every profession and calls on them to be everything from balloonists to aviary keepers, lecturers in all branches of science and learning, quality-control inspectors, and doctors. She asserts that women are superior, that women do not make war and do not want war. "Most of the good in the world emanates from women," she claims.

Her writing seems deliberately to cultivate a "feminine" style that appears, in the parenthetical subtitle of *Mainly East (In Prose – Perhaps Prosey)*, to emphasize experimentation. The subtitle of *Women the World Over* also evidenced such a flippant tone – an experimental disposition – and this quality of her style irked reviewers.

Mainly East resulted from a journey, primarily a sketching trip that took two and a half years and covered fifty thousand miles, begun shortly after the armistice to World War I. Tweedie visited places such as Egypt, Palestine, Syria, India, Turkey, southern Sudan, and the islands of Greece. She did not intend to write a book based on her journey, but she had been devastated by the loss of her youngest son in 1916 during the war, and when her spirits remained so, her mother had insisted, "My dear, if you do not do something, you will crack.... You painted quite nicely when you were a little girl. Go and paint." So Tweedie then set off to forget her pain by traveling and painting. She wrote the book by reconstructing her journey from memory, rather than from notes, and this method of composition perhaps explains the conversational style of the narrative – although it may also reflect the commitment to stylistic experimentation suggested by her subtitle. In any case, she addresses her reader as a friend:

Will you take a little jaunt with me, unknown friend – you see, I take it for granted you are a friend. So let us go a jaunt together.

Please understand right away, I'm no "gentleman with a duster" or a "mirror," nor am I a "gentleman" at all; nor, being a woman, has every man I have ever met proposed to me, or twice as many as I have ever met been violently in love with me.... All these items you can find in previous volumes by other people. I'm just an ordinary sort of person who ran away from war surroundings, and myself, in January, 1919, when the word "Armistice" was quite newly on everyone's tongue.

Will you jaunt along some 50,000 miles with me, mainly East, and into war zones?

Tweedie sometimes travels in luxury and at other times receives only the barest of accomodations: "Sometimes we shall live in hovels, sometimes in palaces, sometimes travel in great ocean steamers, and sometimes on river barges; sometimes on a camel, an elephant, a horse or even the humble moke." Her dejection over the loss of her son Leslie weighs on her. Early in her journey one Sunday Tweedie feels strange and dispatches a companion so that she can be alone, for

A curious premonition of something horrible was upon me.

Always being extremely psychic, and never wishing to give way to that dangerous power, I struggled to put it behind me; but it was no good; I could do nothing. My brain was on fire and yet I felt numb.

Later she discovers that her mother has died that same day. Perhaps because of this emphasis on "feminine intuition" in addition to her conversational style, the *Boston Evening Transcript* (24 March 1923) noted her "feminine comments."

Reviews of the book were mixed. The *Literary Review* of the *New York Evening Post* (5 May 1923) pronounced her book of "passing value" and accused her of being "concerned mainly with superficialities," but the *Literary Digest International Book Review* (June 1923) praised her "shrewd comments on the vexed political conditions in the countries she visited."

Among Tweedie's notably prescient observations is her assertion that "Surely we should never have taken over any place unless we intend to keep it," and she adds these provoking precepts on Palestine:

1. We should clear out.
2. The people of the country – the Arabs – should be left to govern themselves. If they fail, it will be their own fault, and not ours.

Invitation drawn by Harry Furniss for an exhibition of Tweedie's sketches (from Mainly East*)*

3. That we should give our aid only when specifically asked for.

4. That the Jews certainly should have a national home if they want one, but they must not plant themselves on somebody else. And they must govern their new home themselves and pay for it themselves.

5. That the "Holy Land" is so holy to many religions that all the holy places should be entirely outside the Government's jurisdiction. They should be treated separately, either by the moral suasion of the League of Nations, or the creation of an international and interreligious council.

6. An alien, although allied, protector of the Church is not possible.

She asks why England should police half the world and points out the enormous cost, both in money and lives, for the role that England plays as "universal policeman." As the cost of this imperialism is reflected in the heavy taxes paid by British citizens, whom Tweedie claims are the most heavily taxed of the day, she points out the appalling cost of war in human sacrifice:

Like so many others, I lost a dear son, and although I feel sad, that sadness is mingled with reverence on Armistice Day and increased loyalty to our wonderful Empire. Having travelled far, from personal experience I can vouch for the marvellous things Britishers have done all over the world. A race to be proud of.

Tweedie must believe in the Empire in order to cope with her own loss.

Despite her reservations about the ill consequences of imperialism, she praises England as the "vast and wonderful Empire." She claims that "Palestine need have no fear in British hands, for Gentiles, Moslems and Jews are safe. The British empire stands for Justice." As she had also wondered in *Mexico: From Diaz to the Kaiser,* she wonders whether "anything [can] be done through the cinema . . . to teach the well-meaning but insensitive," and she regrets that Britain is losing an opportunity to educate the ignorant masses of the East.

Instead of being shown the greatness of the British Empire and of its industrial and colonial benefits, those Eastern masses are subjected to insipid plots that demean their notions of the Englishman. She believes that India should be ruled by force. Alluding to contemporary events there, Tweedie asserts that people who are not present on foreign ground should not be allowed to alter a plan or a mandate that has been formulated abroad. She faults Prime Minister David Lloyd George and For-

A Golden Vase at the Winter Palace, Peking.
[By the Author]
[Frontispiece.

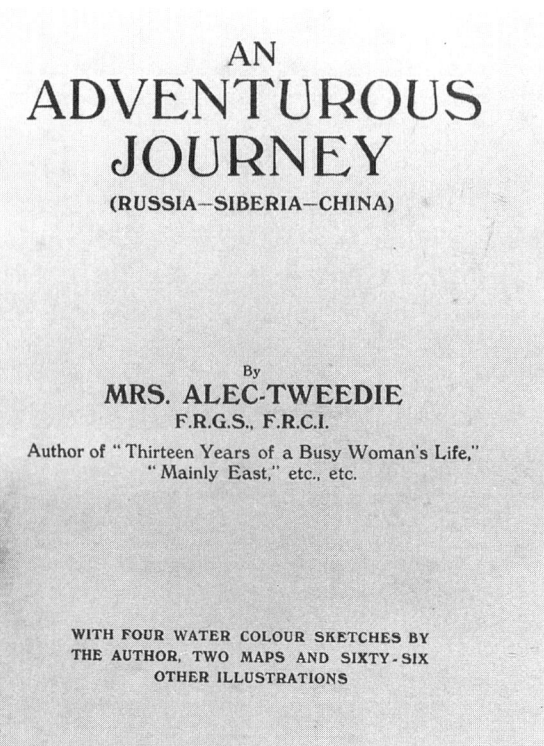

Frontispiece and title page for Tweedie's description of her travels through the Far East in 1924 and 1925

eign Secretary Arthur James Balfour for having made promises that undermine policies established on colonial lands. She specifically condemns the British interference following the April 1919 rioting at Amritsar, India, which resulted after an English woman had been molested and the suspected perpetrators had been punished. The Amritsar massacre showed the extent of Indian unrest:

> Three hundred and seventy of the three thousand or four thousand present were killed.... They were a collection of rebels and of the worst kind, agitators against law and order, and had General [R. E. H.] Dyer's action been upheld by the British Parliament at home, the death penalty of those three hundred seventy men would probably have stopped all further trouble in India, especially if it had been coupled with the arrest of [Mohandas] Gandhi.... In fact, the death roll has become a sad one because that lesson, given to emphasize the necessity of law and order, was blamed instead of praised.

Tweedie's opinion of the Indian natives is not good, and she does not believe that education can relieve the social strains under which native peoples live in a colonial milieu. Of institutions such as the Gordon College, she writes,

> I am villain enough to say that I entirely disapprove.... One finds them scattered all over the Middle East and in India, and the amount of mischief they do, to my mind, is incalculable. It is no good taking a little boy from the wilds, where he is brought up in the simplicity of his native home, and dumping him into a college where he is educated free in a number of things that his brain is quite incapable of grasping, except like a parrot, and then sending him back to his country life, for which he has become utterly unfitted, and the socialistic germ engendered in the small boy's brain grows into rebellion, discontent and far-reaching evil.

Tweedie condemns their ignorance and irresponsibility, qualities in the Indians that the Russians lack:

About twenty-five per cent [of the people] in Russia are literate, only about five per cent in India. They are children, native children, children of the soil, those millions of people in India, with a tiny and sometimes noisy froth at the top who want to reform everyone and everything except themselves apparently, but they are a bagatelle among so many.

She also reiterates the prejudices against southern peoples that she had stated in *Sunny Sicily:* Lacking some "incentive, Southern peoples are idle in the extreme. They seem to have no natural desire for improvement of any sort, and are quite content to laze through the day and do nothing in particular." She notices the same indolence in Constantinople, "where nobody seemed to be busy; everybody seemed to be waiting for something; the true Oriental blood seemed to run in their veins, for verily in the East the saying 'Time waits for no man' is untrue. Time waits on every man and everything. No one bustles."

Tweedie's observations about the status of women are more appealing. In Egypt, when she has met one wife (originally from British Nigeria) of the sultan of Myerno, she comments, "No country ever progresses till its women cease to be slaves. Man has been misnamed the stronger sex because he is not afraid of a mouse – a delightful form of fear he adores in the so-called weaker sex." She observes that in the countries she visits, only a few mosques allow women to enter: one in Hebron, one in Santa Sofia, and another in Damascus. She laments the inability of the woman to participate fully in the life of her culture.

Perhaps Tweedie finds the juxtaposition of squalor and splendor to be the most striking feature of these different cultures she encounters on her journey. On a visit to Bethlehem she observes that

> man has done his best for centuries to destroy history, and religion and nature. It was especially painful, after an interesting drive to Bethlehem, to see the crowding of trumpery lamps and embroideries and mediocre ware in the spot where Christ was born. Alas, also the Holy Sepulchre in Jerusalem itself has been made tawdry and vulgar.

Tweedie's descriptions in *Mainly East* are enhanced by her sketches and paintings of people, buildings, and scenes that she sometimes made at great risk. For instance, the Philae Temple was partly submerged under the waters released by the Aswân Dam, and after having sketched the temple, Tweedie was persuaded by her sister to venture into its interior in their small boat. They squeezed out of the temple only with the greatest difficulty, and they later learned that they could easily have been trapped and drowned inside the temple, for the dam waters had been released while they had been on their boat trip.

Tweedie recounts another exotic sojourn in *An Adventurous Journey: (Russia – Siberia – China)* (1926). She first visited China in January 1924 and then returned in June 1925. As with her travels in *Mainly East,* her intention was to paint, although this time she also intended to write – about China. In fact, she did not intend to write about Russia when she passed through it, but she reported that the country so moved and amazed her she had to write about it. Her account, published within a decade of the Russian Revolution, provides the kind of glimpse that George Orwell was to give about how propaganda has infused the Russian culture and leads children to betray their parents. Tweedie decries the Russian encouragement of free love, which she says has created many young, single mothers – little more than children themselves – who must raise children. She says that she sees not a single happy face in Russia. Tweedie's views seem extreme, but Sybille Bedford writes that in 1927 Huxley concurred in his comments made on his reading of René Fülöp-Miller's *The Mind and Face of Bolshevism: An Examination of Cultural Life in Soviet Russia* (1927):

> It's really humiliating that human beings can be so stupid as the Russians seem to be. They really are the devil. Europeans must join together to resist all the enemies of our civilization – Russians, Americans, orientals – each in their own way a hideous menace.

Tweedie describes a sinister experience on her journey. In traveling by train to China she encounters the daughter of Grigory Yefimovich Rasputin and "Mr. S.," the Russian ambassador to China, both of whom she suspects are spreading anti-British propaganda. Her visit to Russia concludes with a horrible train wreck that she experiences in Siberia, but she arrives safely in China only to find anti-British sentiment at a height, and she almost cancels her visit to Hong Kong.

Despite this dangerous atmosphere Tweedie does go to Hong Kong, but, no doubt because of these attacks on the British, she becomes defensive about the British Empire. China was at this time in the throes of nationalism and was rocked by strikes and demonstrations, the latter of which sometimes resulted in violence against the British. A revised version of *An Adventurous Journey* was published in 1929 with details about continued anti-British ten-

Tweedie sketching at the Lhama Temple, Beijing (from An Adventurous Journey, *1926)*

sions in China that in 1926 had resulted in Britain returning both Wei-hai and the indemnity that the Chinese had been forced to pay following the Boxer uprising in 1901.

In view of these events Tweedie's defensive disposition is understandable, and her rhetoric is imperial: "Who says the British Empire and the Briton are not the greatest product in the world today?" she asks, and she adds that "the more I wander the more I feel we are the most truly democratic people on earth. We are the freest, and we have a crown for our jewel." Perhaps the prime example of her defensiveness about her own country is in her declaration that

> We have the pride and privilege of being part of the far flung British Empire – the Greatest Power on God's earth. That means wider lands and more natural resources than all Americas bunched together. It means more opportunity to prosper for those who work. It means a square deal for all, and the good fortune of living our lives under well-ordered conditions.

Tweedie meets the deposed emperor and empress of China, for whom she has great sympathy, and she takes a trip to the Great Wall. Her personal attendant, My Amah, has had her feet bound, and such observations perhaps lead Tweedie to offer some salient opinions on the position of women in China. She remarks that "China has a far way to go, and she must forgive my saying so, but it is her treatment of her women that is holding her back, and the thraldom of her home life."

Tweedie's descriptions of Russia and China in *An Adventurous Journey* provide a vivid sense of the life and culture of these countries, although this view is necessarily that of an outsider. Tweedie can deftly sketch national character – for instance, in one hilarious episode when she tries to get permission to get access to a building in order to make a sketch, she relates that the Chinese apparently cannot say no. She is never told that she cannot get in, but she also can never enter – until by chance she goes there on a day when it happens to be thrown open to the public: no officials have apprised her of this.

Tweedie dedicated *Tight Corners of My Adventurous Life* (1933), which presents primarily a compilation of incidents from her other books, to her two sons, who had both been tragically killed. Her last book, *My Legacy Cruise (The Peak Year of My Life)* (1936), describes her voyage around the world. She obviously remained energetic in her last years; in fact, her travels had become more ambitious as the years had passed. In *Mexico as I Saw It* Tweedie had quoted with amusement the summary assessment of one reviewer of her early works: "As a girl she 'rode through Iceland,' a little later she 'snow-shoed through Norway' [and] 'therefore in more advanced years we may look for her travels in a bath-chair [wheelchair]!'"

Her views may have been colored by her prejudices, but she was not content to limit her comments to what she saw on her journeys: she attempted to establish relationships with the people of the lands she visited. As a member of the International Council of Women on London in 1899, she was also a well-known philanthropist and civic figure: in 1912 the Italian government officially thanked her for having helped those affected by a massive earthquake in Sicily, and in World War I she contributed significantly by organizing women to help make munitions.

She was also a noted writer who had a feel for national character. She took pains in researching historical and political backgrounds of the places she visited, and she viewed her writing as a product more of hard work than of talent. In fact, she reports that she had written twice, from cover to cover, her 1908 work, *Hyde Park: Its History and Romance*. Writing each of her books caused her to feel anxiety and sleeplessness, as she reports in *Thirteen Years of a Busy Woman's Life,* and her works are all infused with her remarkable personality. They remain informative as period pieces.

References:

Meena Alexander, *Women in Romanticism: Mary Wollstonecraft, Dorothy Wordsworth and Mary Shelley* (Totowa, N.J.: Barnes & Noble, 1989);

Sybille Bedford, *Aldous Huxley: A Biography* (New York: Carroll & Graf, 1974).

Travel Writing, 1876–1909

This checklist of selected travel writing is arranged chronologically, because the names of many travel writers are unfamiliar. It provides a chronicle of events in the history of travel, as many titles identify the destinations and sometimes even the motives for the journeys taken by the authors.

C. G. *Fortnight's Tour amongst Arabs on Mount Lebanon; Including a Visit to Damascus, Ba'albek, the Cedars, Natural Bridge, Etc.* London: Nisbet, 1876.

Madden, Thomas More. *The Principal Health-Resorts of Europe and Africa for the Treatment of Chronic Diseases.* London: Churchill, 1876.

Mazuchelli, Nina Elizabeth. *The Indian Alps and How We Crossed Them: Being a Narrative of Two Years' Residence in the Eastern Himalaya and Two Months' Tour into the Interior.* London: Longmans, Green, 1876.

Moresby, John. *New Guinea and Polynesia.* London: John Murray, 1876.

Barker, Lady Mary Anne. *A Years' Housekeeping in South Africa.* London: Macmillan, 1877.

Bryce, James. *Transcaucasia and Ararat: Being Notes of a Vacation Tour in the Autumn of 1876.* London: Macmlllan, 1877.

Croal, Thomas A. *A Book About Travelling.* London & Edinburgh: Nimmo, 1877.

Fleming, George. *A Nile Novel.* London: Macmillan, 1877.

Lambert, Cowley. *A Trip to Cashmere and Ladak.* London: H. S. King, 1877.

Blunt, Mrs. John Elijah [Fanny Janet]. *The People of Turkey; Twenty Years' Residence among Bulgarians, Greeks, Albanians, Turks and Armenians, by a Consel's Daughter and Wife.* London: John Murray, 1878.

Burton, Richard Francis. *The Gold-Mines of Midian and the Ruined Midianite Cities: A Fortnight's Tour in Northwestern Arabia.* London: Kegan Paul, 1878.

Campion, J. S.. *On the Frontier: Reminiscences of Wild Sports, Personal Adventures, and Strange Scenes.* London: Chapman & Hall, 1878.

Disbrowe, Charlotte Anne Albinia. *Original Letters from Russia 1825–1828.* London: Ladies Printing Press, 1878.

Roche, Harriet A. *On Trek in the Transvaal: Or, Over Berg and Veldt in South Africa.* London: Sampson Low, Marston, Searle & Rivington, 1878.

Stanley, Henry Morton. *Through the Dark Continent,* 2 volumes. London: Sampson Low, Marston, Searle & Rivington, 1878.

Trollope, Anthony. *South Africa,* 2 volumes. London: Chapman & Hall, 1878.

Baker, Samuel White. *Cyprus as I Saw It in 1879.* London: Macmillan, 1879.

Bent, James Theodore. *A Freak of Freedom; or the Republic of San Marino*. London: Longmans, 1879.

Burton, Isabel. *Arabia, Egypt, India: A Narrative of Travel*. London: Mullan, 1879.

Campion, J. S. *On Foot in Spain: A Walk from the Bay of Biscay to the Mediterranean*. London: Chapman & Hall, 1879.

Loftie, W. J. *A Ride in Egypt from Sioot to Luxor in 1879*. London: Macmillan, 1879.

Meredith, Louisa Anne. *Our Island Home: A Tasmanian Sketch Book*. Hobart Town: Walch, 1879.

Murray-Aynsley, Harriet Georginana Maria. *Our Visit to Hindostan, Kashmir, and Ladakh*. London: W. H. Allen, 1879.

Pender, Rose. *No Telegraph: Or, a Trip to Our Unconnected Colonies, 1878*. London: Gilbert & Rivington, 1879.

Stevenson, Mary Esme Gwendoline (Grogan). *Our Home in Cyprus*. London: Chapman & Hall, 1879.

Symonds, J. A. *Sketches and Studies in Italy*. London: Smith, Elder, 1879.

Westminster, Elizabeth Grosvenor, Marchioness of. *Diary of a Tour in Sweden, Norway, and Russia, in 1827, with Letters*. London: Hurst & Blackett, 1879.

Whately, Mary Louisa. *Letters from Egypt to Plain Folks at Home*. London: Seeley, Jackson & Halliday, 1879.

Butler, William Francis. *Far Out: Rovings Retold*. London: Isbister, 1880.

Forde, Gertrude. *A Lady's Tour in Corsica*, 2 volumes. London: Bentley, 1880.

Gray, Mrs. John Henry. *Fourteen Months in Canton*. London: Macmillan, 1880.

Hare, Augustus J. C. *Days Near Paris*. London: George Allen, 1880.

Meredith, Louisa Anne. *Tasmanian Friends and Foes, Feathered, Furred and Finned*. London: Marcus Ward, 1880.

Oliphant, Laurence. *The Land of Gilead, with Excursions in the Lebanon*. Edinburgh & London: Blackwood, 1880.

Whymper, Edward. *The Ascent of the Matterhorn*. London: John Murray, 1880.

Ashbee, Henry Spencer. *A Ride to Peking*. London: Lindsey, 1881.

Blackwood, Lady Alicia. *A Narrative of Personal Experiences and Impressions During a Residence on the Bosphorus throughout the Crimean War*. London: Hatchard, 1881.

Elliot, Frances Dickinson (Minto). *The Diary of an Idle Woman in Sicily*, 2 volumes. London: Bentley, 1881.

Ellis, Tristram James. *On a Raft, and through the Desert*. London: Field & Tuer, 1881.

Hardy, Lady Mary McDowell Duffus. *Through Cities and Prairie Lands*. London: Chapman & Hall, 1881.

Hughes, Mrs. Thomas Francis. *Among the Sons of Han: Notes of a Six Years' Residence in Various Parts of China and Formosa*. London: Tinsley, 1881.

Keane, John F. *My Journey to Medinah: Describing a Pilgrimage to Medinah, Performed by the Author Disguised as a Mohammedan.* London: Tinsley, 1881.

Mazuchelli, Nina Elizabeth. *'Magyarland': Being the Narrative of Our Travels through the Highlands and Lowlands of Hungary,* 2 volumes. London: Sampson Low, Marston, Searle & Rivington, 1881.

Stevenson, Mary Esme Gwendoline (Grogan). *Our Ride through Asia Minor.* London: Chapman & Hall, 1881.

Heckford, Sarah. *A Lady Trader in the Transvaal.* London: Sampson Low, Marston, Searle & Rivington, 1882.

Maxwell, Edward Herbert. *Griffin Ahoy! A Yacht Cruise to the Leavant and Wanderings in Egypt, Syria, the Holy Land, Greece, and Italy in 1881.* London: Hurst & Blackett, 1882.

McDougall, Mrs. Harriette. *Sketches of Our Life at Sarawak.* London: Society for Promoting Christian Knowledge, 1882.

Montgomery, A. N. *Hints about Egypt.* London: Houghton, 1882.

Murray-Aynsley, Harriet Georginana Maria. *An Account of a Three Months' Tour from Simla through Bussahir, Kunowar and Spiti, to Lahoul.* Calcutta: Thacker Spink, 1882.

Oliphant, Laurence. *The Land of Khemi: Up and Down the Middle Nile.* Edinburgh & London: Blackwood, 1882.

Stone, Olivia M. *Norway in June . . . Accompanied by a Sketch May, a Table of Expenses, and a List of Articles Indispensible to the Traveller in Norway.* London: Ward, 1882.

Taylor, Ellen M. *Madeira: Its Scenery, and How to See It. With Letters of a Year's Residence, and Lists of the Trees, Flowers, Ferns, and Seaweeds.* London: Stanford, 1882.

Baxter, Lucy E. (as Scott Leader). *The Renaissance of Art in Italy.* London: Sampson Low, Marston, Searle & Rivington, 1883.

Bridges, F. D. *Journal of a Lady's Travel round the World.* London: John Murray, 1883.

Edgcumbe, Lady Ernestine, and Lady Mary Wood. *Four Months' Cruise in a Sailing Yacht.* London: Hurst & Blackett, 1883.

Hare, Augustus J. C. *Cities of Northern Italy,* 2 volumes. London: Macmillan, 1883.

Hare. *Cities of Southern Italy and Sicily.* London: Smith, Elder, 1883.

Hunt, Shelley Leigh, and Alexander S. Kenny. *Tropical Trials: A Handbook for Women in the Tropics.* London: W. H. Allen, 1883.

Murray-Aynsley, Harriet Georginana Maria. *Our Tour in Southern India.* London: White, 1883.

Stevenson, Mary Esme Gwendoline (Grogan). *On Summer Seas, Including the Mediterranean, the Aegean . . . and a Voyage Down the Danube.* London: Chapman & Hall, 1883.

Symonds, J. A. *Italian Byways.* New York: Holt, 1883.

Whately, Mary Louisa. *Scenes from Life in Cairo.* London: Seeley, Jackson & Halliday, 1883.

Buckland, Anne W. *The World Beyond the Esterelles.* London: Remington, 1884.

Calderwood, Margaret Steuart. *Letters and Journals of Mrs. Calderwood of Polton from England, Holland, and the Low Countries in 1756,* edited by Alexander Fergusson. Edinburgh: David Douglas, 1884.

Elliot, Frances Dickinson (Minto). *The Diary of an Idle Woman in Spain.* London: White, 1884.

Faithfull, Emily. *Three Visits to America.* Edinburgh: David Douglas, 1884.

Hall, Mrs. Cecil. *A Lady's Life on a Farm in Manitoba.* London: W. H. Allen, 1884.

Hare, Augustus J. C. *Cities of Central Italy,* 2 volumes. London: George Allen, 1884.

Hare. *Sketches in Holland & Scandinavia.* Philadelphia: McKay, 1884.

Leonowens, Anna. *Life and Travels in India.* Philadelphia: Porter & Coates, 1884.

Lester, Mary (as Maria Soltera). *A Lady's Ride across Spanish Honduras.* Edinburgh & London: Blackwood, 1884.

McCormick, Robert. *Voyages of Discovery in the Arctic and Antarctic Seas,* 2 volumes. London: Sampson Low, Marston, Searle & Rivington, 1884.

Poole, Annie Sampson. *Mexicans at Home in the Interior.* London: Chapman & Hall, 1884.

Pringle, M. A. *Towards the Mountains of the Moon: A Journey in East Africa.* Edinburgh & London: Blackwood, 1884.

Speedy, Cornelia Mary. *My Wanderings in the Soudan.* London: Bentley, 1884.

Gordon, Charles George. *The Journals of Major-General C. G. Gordon at Khartoum,* 2 volumes, edited by A. Egmont Hake. London: Kegan Paul, Trench, 1885.

Hare, Augustus J. C. *Studies in Russia.* London: Smith, Elder, 1885.

Innes, Emily. *The Chersonese with the Gilding Off,* 2 volumes. London: Bentley, 1885.

Hore, Annie Boyle. *To Lake Tanganyika in a Bath Chair.* London: Sampson Low, Marston, Searle & Rivington, 1886.

Marryat, Florence (later Church, Mrs. Ross, and Lean, Mrs. Francis). *Tom Tiddler's Ground.* London: Swan Sonnenschein, Lowrey, 1886.

Walker, Mary Adelaide. *Eastern Life and Scenery, with Excursions in Asia Minor, Mytilene, Crete, and Roumania.* London: Chapman & Hall, 1886.

Ashbee, Henry, and Alexander Graham. *Travels in Tunisia.* London: Dulau, 1887.

Butler, William Francis. *The Campaign of the Cataracts: Being a Personal Narrative of the Great Nile Expedition of 1884-5.* London: Sampson Low, Marston, Searle & Rivington, 1887.

Forbes, Anna. *Insulinde: Experiences of a Naturalist's Wife in the Eastern Archipelago.* Edinburgh: Blackwood, 1887.

Hort, Dora (Mrs. Alfred). *Via Nicaragua: A Sketch of Travel.* London: Remington, 1887.

Layard, Mrs. Granville (Gertrude). *Through the West Indies.* London: Sampson Low, Marston, Searle & Rivington, 1887.

Layard, Sir Austen Henry. *Early Adventures in Persia, Susiana, and Babylonia.* London: John Murray, 1887.

Oliphant, Laurence. *Haifa: Or, Life in Modern Palestine,* edited by Charles A. Dana. London: Blackwood, 1887.

Stone, Olivia M. *Tenerife and Its Six Satellites: Or, the Canary Islands Past and Present.* London: Ward, 1887.

Crompton-Roberts, Violet and Mildred, and others. *A Jubilee Jaunt to Norway.* London: Griffith, Farran, Okeden & Welsh, 1888.

Gordon, Charles George. *Letters of General C. G. Gordon to His Sister.* London & New York: Macmillan, 1888.

Taylor, Isaac. *Leaves from an Egyptian Note-Book.* London: Kegan Paul, 1888.

Dufferin, Lady Hariot (Blackwood). *Our Viceregal Life in India: Selections from My Journal, 1884–1888,* 2 volumes. London: John Murray, 1889.

Sandwith, F. M. *Egypt as a Winter Resort.* London: Kegan Paul, Trench, 1889.

Wingfield, Lewis Strange. *Wanderings of a Globe-Trotter in the Far East,* 2 volumes. London: Bentley, 1889.

Hare, Augustus J. C. *North-Eastern France.* London: George Allen, 1890.

Kemble, Fanny. *Further Records.* London: Bentley, 1890.

Martin, Annie. *Home Life on an Ostrich Farm.* London: George Philip, 1890.

Stanley, Henry Morton. *In Darkest Africa,* 2 volumes. London: Sampson Low, Marston, Searle & Rivington, 1890.

Dowie, Ménie Muriel. *A Girl in the Karpathians,* third edition. London: George Philip, 1891.

Dufferin, Lady Harriot (Blackwood). *My Canadian Journal 1872–8 . . . Extracts from My Letters Home Written while Lord Dufferin Was Governor-General.* London: John Murray, Spottiswoode, 1891.

Hort, Dora (Mrs. Alfred). *Tahiti: The Garden of the Pacific.* London: Unwin, 1891.

Jerome, Jerome K. *Diary of a Pilgrimage: And Six Essays.* Bristol: Arrowsmith, 1891.

Miller, Ellen E. *Alone through Syria.* London: Kegan Paul, 1891.

Moir, Jane F. *A Lady's Letters from Central Africa: A Journey from Mandala, Shir Highlands to Ujiji, Lake Tanganyika, and Back.* Glasgow: Maclehose, 1891.

Monteiro, Rose. *Delagoa Bay: Its Natives and Natural History.* London: George Philip, 1891.

Whymper, Edward. *Travels Amongst the Great Andes of the Equator.* London: John Murray, 1891.

Elliot, Frances Dickinson (Minto). *Diary of an Idle Woman in Constantinople.* London: John Murray, 1892.

Inglis, Lady Julia Selina. *The Siege of Lucknow: A Diary.* London: Osgood, 1892.

Sheldon, Mary French. *Sultan to Sultan: Adventures among the Masai and Other Tribes of East Africa.* London: Saxon, 1892.

Smith, William, F.S.A.S. *A Yorkshireman's Trip to the United States and Canada.* London: Longmans, Green, 1892.

Whitwell, Mrs. E. R. *Spain: As We Found It in 1891.* London: Eden, Remington, 1892.

Barkly, Fanny A. *Among Boers and Basutos: The Story of Our Life on the Frontier.* London: Remington, 1893.

Blennerhassett, Rose, and Lucy Sleeman. *Adventures in Mashonaland by Two Hospital Nurses.* London: Macmillan, 1893.

Chennell, Ellen. *Recollections of an Egyptian Princess by Her English Governess: Being a Record of Five Years Residence at the Court of Ismail Pasha Khedive,* 2 volumes. Edinburgh & London: Blackwood, 1893.

Colvile, Lady Zlie. *Round the Black Man's Garden.* Edinburgh: Blackwood, 1893.

Curzon, George. "Ladies and the Royal Geographic Society." *Times* (London), 31 May 1893, p. 11.

Landor, Arnold Henry Savage. *Alone with the Hairy Ainu. Or, 3800 Miles on a Pack Saddle in Yezo and a Cruise to the Kurile Islands.* London: John Murray, 1893.

Somerville, E., and V. M. Ross. *Through Connemara in a Governess Cart.* London: Allen, 1893.

Tyacke, Mrs. Richard Humphrey. *How I Shot My Bears: Or, Two Years' Tent Life in Kullu and Lahoul.* London: Sampson Low, Marston, 1893.

Bower, Hamilton. *Diary of a Journey across Tibet.* London: Rivington, Percival, 1894.

Conway, Sir William Martin. *Climbing and Exploration in the Karakoram Himalayas,* 2 volumes. London: Unwin, 1894.

Janeway, Catherine. *Ten Weeks in Egypt and Palestine.* London: Kegan Paul, 1894.

Paton, Margaret Whitecross. *Letters and Sketches from the New Hebrides,* edited by Rev. James Paton. London: Hodder & Stoughton, 1894.

Balfour, Alice Blanche. *Twelve Hundred Miles in a Waggon.* London: Arnold, 1895.

Gardner, Nora Beatrice. *Rifle and Spear with the Rajpoots: Being the Narrative of a Winter's Travel and Sport in Northern India.* London: Chatto & Windus, 1895.

Landor, Arnold Henry Savage. *Corea; or Cho-sen, the Land of Morning Calm,* 2 volumes. London: Heinemann, 1895.

Stanley, Henry Morton. *My Early Travels and Adventures in America and Asia,* 2 volumes. London: Sampson Low, Marston, 1895.

Wilson, Anne Campbell. *After Five Years in India: Or, Life and Work in a Punjab District.* London: Blackie, 1895.

Blaikie, W. G. "Lady Travellers," *Blackwood's Magazine,* 160 (July 1896): 49-66.

Brodrick, Mary, ed. *A Handbook for Travellers in Lower and Upper Egypt.* London: John Murray, 1896.

Browning, Ellen. *A Girl's Wanderings in Hungary.* London: Longmans, Green, 1896.

Fitzgerald, Edward Arthur. *Climbs in the New Zealand Alps: Being an Account of Travel and Discovery*. London: Unwin, 1896.

Whymper, Edward. *Chamonix and the Range of Mont Blanc, a Guide*. London: John Murray, 1896.

Bryce, James. *Impression of South Africa*. London: Macmillan, 1897.

Hare, Augustus J. C. *The Rivieras*. London: George Allen, 1897.

Howard of Glossop, Baroness Winefred Mary. *Journal of a Tour in the United States, Canada and Mexico*. London: Sampson Low, Marston, 1897.

Janeway, Catherine. *Glimpses at Greece, To-Day and Before Yesterday*. London: Kegan Paul, Trench & Trubner, 1897.

Muller, Georgina A. *Letters from Constantinople*. London & New York: Longmans, Green, 1897.

Vivian, Herbert. *Servia, the Poor Man's Paradise*. London: Longmans, Green, 1897.

Landor, Arnold Henry Savage. *In the Forbidden Land*. London: Heinemann, 1898.

Meynell, Alice. *The Spirit of Place and Other Essays*. London: John Lane, 1898.

Stanley, Henry Morton. *Through South Africa*, 2 volumes. London: Sampson Low, Marston, 1898.

Steevens, G. W. *Egypt in 1898*. London: Blackwood, 1898.

Steevens. *With Kitchener to Khartoum*. London: Greenhill, 1898.

Bird, Mary Rebecca. *Persian Women and Their Creed*. London: Church Missionary Society, 1899.

David, Mrs. Edgeworth. *Funafuti, or Three Months on a Coral Island: An Unscientific Account of a Scientific Expedition*. London: John Murray, 1899.

Ellis, Beth. *An English Girl's Impressions of Burmah*. London: Simpkin, Marshall, Hamilton & Kent, 1899.

Neufeld, Charles. *A Prisoner of the Khaleefa: Twelve Years' Captivity at Omdurman*. London: Chapman & Hall, 1899.

Vivian, Herbert. *Tunisia and the Modern Barbary Pirates*. London: Pearson, 1899.

Cook, Frederick A. *Through the First Antarctic Night, 1898-1899*. London: Heinemann, 1900.

Donaldson, Florence Annesley. *Lepcha Land: Or, Six Weeks in the Sikhim Himalayas*. London: Sampson Low, Marston, 1900.

Haldane, J. W. C. *3800 Miles Across Canada*. London: Simpkin, Marshall, Hamilton, Kent, 1900.

Lauder, John. *Journals*, edited by Donald Crawford. Edinburgh: Edinburgh University Press, 1900.

Savory, Isabel. *A Sportswoman in India: Personal Adventures and Experiences of Travel in Known and Unknown India*. London: Hutchinson, 1900.

Thomas, Margaret. *Two Years in Palestine and Syria*. London: Nimmo, 1900.

Whymper, Edward. *The Valley of Zermatt and the Matterhorn.* London: John Murray, 1900.

Barnard, Lady Anne. *Lady Anne Barnard at the Cape of Good Hope 1797–1802,* edited by W. H. Wilkins. London: Smith, Elder, 1901.

Barnard. *South Africa a Century Ago: Letters Written from the Cape of Good Hope 1797–1801,* edited by W. H. Wilkins. London: Smith, Elder, 1901.

Conway, Sir William Martin. *The Bolivian Andes.* London & New York: Harper, 1901.

Davies, Hannah. *Among Hills and Valleys in Western China: Incidents of Missionary Work.* London: Partridge, 1901.

Fenton, Bessie Knox. *The Journal of Mrs. Fenton: A Narrative of Her Life in India, the Isle of France – Mauritius – and Tasmania during the Years 1826–1830.* London: Arnold, 1901.

Gissing, George. *By the Ionian Sea: Notes of a Ramble in Southern Italy.* London: Chapman & Hall, 1901.

Hodgson, Lady Mary Alice. *The Siege of Kumassi.* London: Pearson, 1901.

Landor, Arnold Henry Savage. *China and the Allies,* 2 volumes. London: Heinemann, 1901.

Vivian, Herbert. *Abyssinia: Through the Lion-Land to the Court of the Lion of Judah.* London: Pearson, 1901.

Carey, William. *Travel and Adventure in Tibet: Including the Diary of Miss Annie R. Taylor's Remarkable Journey from Tau-Chau to Ta-Chien-Lu through the Heart of the Forbidden Land.* London: Hodder & Stoughton, 1902.

Collie, J. Norman. *Climbing on the Himalaya and Other Mountain Ranges.* Edinburgh: Douglas, 1902.

Doughty, Marion. *Afoot through the Kashmir Valleys.* London: Sands, 1902.

Durand, Ella R. *An Autumn Tour in Western Persia.* Westminster: Constable, 1902.

Grove, Agnes Geraldine. *Seventy-One Days' Camping in Morocco.* London: Longmans, Green, 1902.

Thomas, Margaret. *Denmark, Past and Present.* London: Anthony Treherne, 1902.

Browning, Oscar. *Impressions of Indian Travel.* London: Hodder & Stoughton, 1903.

Butler, Lady Elizabeth Southerden Thompson. *Letters from the Holy Land.* London: A. & C. Black, 1903.

Landor, Arnold Henry Savage. *Across Crooked Lands,* 2 volumes. New York: Scribners, 1903.

Savory, Isabel. *In the Tail of the Peacock.* London: Hutchinson, 1903.

Shand, Alexander Innes. *Old-Time Travel: Personal Reminiscences of the Continent Forty Years Ago Compared with Experiences of the Present Day.* London: John Murray, 1903.

Stevenson, Margaret Isabella. *From Saranac to the Marquesas and Beyond: Being Letters Written by Mrs. M. I. Stevenson During 1887–88 to Her Sister Jane Whyte Balfour,* edited by Marie Clothilde Balfour. London: Methuen, 1903.

Symons, Arthur. *Cities.* London: Dent, 1903.

Barker, Lady Mary Anne. *Colonial Memories.* London: Smith, Elder, 1904.

Cecil, Mary Rothes Margaret. *Birds Notes from the Nile.* Westminster: Constable, 1904.

Hutton, Edward. *The Cities of Umbria.* London: Dutton, 1904.

Landor, Arnold Henry Savage. *The Gems of the East.* New York & London: Harper, 1904.

Townley, Lady Susan. *My Chinese Note Book.* London: Methuen, 1904.

Wilson, Anne Campbell. *Hints for the First Years of Residence in India.* Oxford: Clarendon Press, 1904.

Armitage, Albert B. *Two Years in the Antarctic: Being a Narrative of the British National Antarctic Expedition.* London: Arnold, 1905.

Goodrich-Freer, Ada M. *In a Syrian Saddle.* London: Methuen, 1905.

Landor, Arnold Henry Savage. *Tibet & Nepal.* London: Black, 1905.

Shaw, Flora Louisa (Lady Lugard). *A Tropical Dependency: An Outline of the Ancient History of the Western Soudan with an Account of the Modern Settlement of Northern Nigeria.* London: Nesbit, 1905.

Waddell, Austine. *Lhasa and Its Mysteries.* London: John Murray, 1905.

Duncan, Jane. *A Summer Ride through Western Tibet.* London: Smith, Elder, 1906.

Hedley, John. *On Tramp among the Mongols.* Shanghai: North-China Daily News & Herald, 1906.

Hutton, Edward. *The Cities of Spain.* London: Methuen, 1906.

Stevenson, Margaret Isabella. *Letters from Samoa 1891–1895,* edited by Marie Clothilde Balfour. London: Metheun, 1906.

Synge, J. M. *The Aran Islands.* Dublin: Maunsel, 1906.

Hall, Mary. *A Woman's Trek from the Cape to Cairo.* London: Methuen, 1907.

Hornby, Emily. *Sinai and Petra: The Journals of Emily Hornby, in 1899 and 1901.* London: Nisbet, 1907.

Hutton, Edward. *Florence and the Cities of Northern Tuscany, with Genoa.* London: Methuen, 1907.

Landor, Arnold Henry Savage. *Across Wildest Africa,* 2 volumes. London: Hurst & Blackett, 1907.

Churchill, Winston. *My African Journey.* London: Hodder & Stoughton, 1908.

Cook, Augusta. *By Way of the East: Or, Gathered Light from Our Travels in Palestine, Egypt, Smyrna, Ephesus, Etc.* London: Banks, 1908.

Cook, Frederick A. *To the Top of the Continent.* New York: Doubleday, Page, 1908.

Hutton, Edward. *Country Walks about Florence.* London: Methuen, 1908.

Kelman, John. *From Damascus to Palmyra.* London: A. & C. Black, 1908.

Larymore, Constance. *A Resident's Wife in Nigeria.* London: Routledge, 1908.

Wilkins, Louisa Jebb. *By Desert Ways to Baghdad*. London: Unwin, 1908.

Benn, Edith Fraser. *An Overland Trek from India by Side-Saddle, Camel, and Rail: The Record of a Journey from Baluchistan to Europe*. London: Longmans, Green, 1909.

Fraser, David. *The Short Cut to India: The Record of a Journey Along the Route of the Baghdad Railway*. Edinburgh: Blackwood, 1909.

Hume-Griffith, M. E. *Behind the Veil in Persia and Turkish Arabia, an Account of an English Woman's Eight Years' Residence amongst the Women of the East*. London: Seeley, 1909.

Jenkins, Lady (Catherine Minna). *Sport and Travel in Both Tibets*. London: Blades, East, & Blades, 1909.

Jones, Henry Festing. *Diversions in Sicily*. London: Rivers, 1909.

Maturin, Mrs. Fred [Edith Money]. *Petticoat Pilgrims on Trek*. London: Nash, 1909.

Books for Further Reading

Adams, Percy G. *Travel Literature and the Evolution of the Novel.* Lexington: University of Kentucky Press, 1983.

Adams, ed. *Travel Literature through the Ages: An Anthology.* New York & London: Garland, 1988.

Adams, William H. Davenport. *Celebrated Women Travellers of the Nineteenth Century.* London: Swan Sonnenschein, 1883.

Aitken, Maria. *A Girdle round the Earth.* London: Constable, 1987.

Allen, Alexandra. *Travelling Ladies.* London: Jupiter, 1980.

Baker, Anne, ed. *Morning Star: Florence Baker's Diary of the Expedition to Put Down the Slave Trade on the Nile, 1870-73.* London: William Kimber, 1972.

Barr, Pat. *The Mensahibs: The Women of Victorian India.* London: Secker & Warburg, 1976.

Bathe, Basil W. *Seven Centuries of Sea Travel.* London: Barrie & Jenkins, 1972.

Behdad, Ali. *Belated Travelers: Orientalism in the Age of Colonial Dissolution.* Durham, N.C.: Duke University Press, 1994.

Birkett, Dea. *Spinsters Abroad: Victorian Lady Explorers.* Oxford & New York: Blackwell, 1989.

Bishop, Peter. *The Myth of Shangri-La: Tibet, Travel Writing and the Western Creation of Sacred Landscape.* Berkeley: University of California Press, 1989.

Brent, Peter. *Far Arabia: Explorers of the Myth.* London: Weidenfeld, 1977.

Brinnin, John Malcolm. *The Sway of the Grand Saloon.* London: Macmillan, 1972.

Burkhardt, A. J., and S. A. Medlik. *Tourism Past, Present, and Future.* London: Heinemann, 1974.

Buzard, James. *The Beaten Track: European Tourism, Literature, and the Ways to "Culture" 1800-1918.* Oxford: Clarendon Press/Oxford University Press, 1993.

Callaway, Helen. *Gender, Culture and Empire: European Women in Colonial Nigeria.* Urbana: University of Illinois Press, 1987.

Clark, Ronald W. *Explorers of the World.* London: Aldus Books, 1964.

Clark. *Men, Myths and Mountains.* London: Weidenfeld, 1975.

Clark. *The Victorian Mountaineers.* London: Batsford, 1953.

Books for Further Reading

Cole, Garold. *Travels in America from the Voyages of Discovery to the Present.* Norman: University of Oklahoma Press, 1984.

Crossley-Holland, Kevin, ed. *The Oxford Book of Travel Verse.* New York: Oxford University Press, 1987.

Davidson, Lilias Campbell. *Hints to Lady Travellers at Home and Abroad.* London: Iliffe, 1889.

de Beer, Gavin Rylands. *Travellers in Switzerland.* London & New York: Oxford University Press, 1949.

Dodd, Philip, ed. *The Art of Travel: Essays on Travel Writing.* London: Frank Cass, 1982.

Dowie, Ménie Muriel, ed. *Women Adventurers.* London: Unwin, 1893.

Eisner, Robert. *Travelers to an Antique Land: The History and Literature of Travel to Greece.* Ann Arbor: University of Michigan Press, 1991.

Faber, Richard. *The Vision and the Need: Late Victorian Imperialist Aims.* London: Faber & Faber, 1966.

Feifer, Maxine. *Tourism in History: From Imperial Rome to the Present.* New York: Stein & Day, 1986.

Foster, Shirley. *Across New Worlds: Nineteenth-Century Women Travellers and Their Writings.* New York: Wheatsheaf-Harvester, 1990.

Fowler, Marian. *Below the Peacock Fan: First Ladies of the Raj.* New York: Viking, 1987.

Fraser, Keath, ed. *Bad Trips.* New York: Vintage Departures, 1991.

Frawley, Maria H. *A Wider Range: Travel Writing by Women in Victorian England.* Rutherford, N.J.: Fairleigh Dickinson, 1994.

Freeth, Zahra, and Victor Winstone. *Explorers of Arabia: From the Renaissance to the Victorian Era.* London: Allen & Unwin, 1978.

Fussell, Paul, ed. *The Norton Book of Travel.* New York: Norton, 1987.

Graham-Brown, Sarah. *Images of Women: The Portrayal of Women in Photography of the Middle East, 1860-1950.* New York: Columbia University Press, 1988.

Greenhill, B., and A. Gifford. *Women under Sail.* London: David & Charles, 1970.

Gregory, Alexis. *The Golden Age of Travel, 1880-1939.* New York: Rizzoli, 1990.

Headrick, Daniel R. *The Tools of Empire: Technology and European Imperialism in the Nineteenth Century.* New York: Oxford University Press, 1981.

Hibbert, Christopher. *The Grand Tour.* London: Spring Books, 1974.

Hobsbawn, Eric J. *The Age of Empire 1875-1914.* London: Weidenfeld & Nicolson, 1987.

Holmes, Winifred. *Seven Adventurous Women.* London: Bell, 1953.

Hopkirk Peter. *Foreign Devils on the Silk Road: The Search for the Lost Cities and Treasures of Chinese Central Asia.* London: John Murray, 1980.

Hopkirk. *Trespassers on the Roof of the World: The Race for Lhasa*. London: John Murray, 1982.

Howarth, Patrick. *When the Riviera Was Ours*. London & Henley: Routledge & Kegan Paul, 1977.

Hutchins, Francis. *The Illusion of Permanence: British Imperialism in India*. Princeton: Princeton University Press, 1967.

Hyam, Ronald. *Empire and Sexuality: The British Experience*. Manchester: Manchester University Press, 1990.

Jameson, Fredric. *The Political Unconscious: Narrative as a Socially Symbolic Act*. Ithaca, N.Y.: Cornell University Press, 1981.

Kabbani, Rana. *Europe's Myths of Orient: Devise and Rule*. London: Macmillan, 1986.

Keay, John. *The Gilgit Game: The Explorers of the Western Himalayas 1865-95*. London: John Murray, 1979.

Keay, ed. *The Royal Geographical Society History of World Exploration*. London: Hamlyn, 1991.

Keay, Julia. *With Passport and Parasol: The Adventures of Seven Victorian Ladies*. London: BBC Books, 1989.

Kiernan, V. G. *The Lords of Human Kind: Black Man, Yellow Man, and White Man in an Age of Empire*. New York: Columbia University Press, 1986.

Lane, Edward William. *Manners and Customs of the Modern Egyptians*. London: Dent, 1936.

Lawrence, Karen R. *Penelope Voyages: Women and Travel in British Literary Tradition*. Ithaca, N.Y.: Cornell University Press, 1994.

Leed, Eric J. *The Mind of the Traveler: From Gilgamesh to Global Tourism*. New York: Basic Books, 1991.

Lochsberg, Winifred. *History of Travel*. Leipzig: Edition Leipzig, 1979.

MacCannell, Earle Dean. *The Tourist, a New Theory of the Leisure Class*. London: Macmillan, 1976.

MacGregor, John. *Tibet: A Chronicle of Exploration*. London: Routledge & Kegan Paul, 1970.

MacKenzie, John. *Imperialism and Popular Culture*. Manchester: Manchester University Press, 1986.

Macmillan, Margaret. *Women of the Raj*. London: Thames & Hudson, 1988.

Mahood, Molly. *The Colonial Encounter: A Reading of Six Novels*. London: Rex Collings, 1977.

Marsden-Smedley, Philip, and Jeffrey Klinke, eds. *Views from Abroad: The "Spectator" Book of Travel Writing*. London: Grafton, 1988.

Massingham, Hugh and Pauline, eds. *The Englishman Abroad*. London: Phoenix House, 1962.

Melman, Billie. *Women's Orients: English Women and the Middle East, 1718-1918*. Ann Arbor: University of Michigan Press, 1992.

Michael, Maurice Albert, ed. *Traveller's Quest: Original Contributions towards a Philosophy of Travel*. Freeport, N.Y.: Books for Libraries, 1950.

Middleton, Dorothy. *Victorian Lady Travellers*. New York: Dutton, 1965.

Mills, Sara. *Discourses of Difference: An Analysis of Women's Travel Writing and Colonialism.* New York & London: Routledge, 1991.

Mitchell, Timothy. *Colonising Egypt.* Cambridge: Cambridge University Press, 1988.

Montgomery, A. N. *Hints about Egypt.* London: Houghton, 1882.

Moorehead, Alan. *The White Nile.* London: Hamish Hamilton, 1960.

Morgan, Susan. *Place Matters: Gendered Geography in Victorian Women's Travel Books about Southeast Asia.* New Brunswick, N.J.: Rutgers University Press, 1996.

Morris, Mary, ed. *Maiden Voyages: Writing of Women Travellers.* New York: Vintage, 1993.

Mulvey, Christopher. *Anglo-American Landscapes: A Study of Nineteenth-Century Anglo-American Travel Literature.* London: Cambridge University Press, 1983.

Mulvey. *Transatlantic Manners: Social Patterns in Nineteenth-Century Anglo-American Travel Literature.* London: Cambridge University Press, 1990.

Nevins, Allan. *America through British Eyes.* New York: Oxford University Press, 1948.

Newby, Eric, ed. *A Book of Travellers' Tales.* New York: Penguin, 1985.

Oliver, Caroline. *Western Women in Colonial Africa.* Westport, Conn.: Greenwood Press, 1982.

Oswell, W. Edward. *William Cotton Oswell: Hunter and Explorer,* 2 volumes. New York: Doubleday, Page, 1900.

Ousby, Ian. *The Englishman's England: Taste, Travel, and the Rise of Tourism.* New York: Cambridge University Press, 1991.

Owen, Charles. *The Grand Days of Travel.* Exeter: Webb & Bower, 1979.

Page, Martin. *The Lost Pleasures of the Great Trains.* London: Weidenfeld, 1975.

Pakenham, Valerie. *The Noonday Sun: Edwardians in the Tropics.* London: Methuen, 1985.

Pemble, John. *The Mediterranean Passion: Victorians and Edwardians in the South.* London: Clarendon Press / New York: Oxford University Press, 1987.

Pimlott, J. A. R. *The Englishman's Holiday: A Social History.* London: Faber & Faber, 1947.

Pomeroy, Earl. *In Search of the Golden West: The Tourist in Western America.* New York: Knopf, 1957.

Porter, Dennis. *Haunted Journeys: Desire and Transgression in European Travel Writing.* Princeton: Princeton University Press, 1991.

Pratt, Mary Louise. *Imperial Eyes: Studies in Travel Writing and Transculturation.* London & New York: Routledge, 1992.

Raskin, Jonah. *The Mythology of Imperialism.* New York: Random House, 1971.

Rennie, Neil. *Far-Fetched Facts: The Literature of Travel and the Idea of the South Seas.* Oxford: Oxford University Press, 1996.

Rice, Warner G., ed. *Literature as a Mode of Travel.* New York: New York Public Library, 1963.

Ridley, Hugh. *Images of Imperial Rule.* London: Croom Helm, 1983.

Robinson, Jane, ed. *Unsuitable for Ladies: An Anthology of Women Travellers.* Oxford: Oxford University Press, 1994.

Robinson. *Wayward Women: A Guide to Women Travellers.* Oxford: Oxford University Press, 1990.

Robinson, Ronald, and John Gallagher. *Africa and the Victorians: The Official Mind of Imperialism.* London & New York: St. Martin's Press, 1961.

Russell, Mary. *The Blessings of a Good Thick Skirt: Women Travellers and Their World.* London: Collins, 1986.

Said, Edward. *Culture and Imperialism.* New York: Knopf, 1993.

Said. *Orientalism.* New York: Pantheon, 1978.

Savage-Landor, A. Henry. *Everywhere: The Memoirs of an Explorer.* New York: Stokes, 1924.

Schivelbusch, Wolfgang. *The Railway Journey: The Industrialization of Time and Space in the Nineteenth Century.* Berkeley: University of California Press, 1986.

Severin, Timothy. *The Oriental Adventure: Explorers of the East.* London: Angus & Robertson, 1976.

Sigaux, Gilbert. *History of Tourism,* translated by Joan White. London: Leisure Arts, 1966.

Spurr, David. *The Rhetoric of Empire: Colonial Discourse in Journalism, Travel Writing, and Imperial Administration.* Durham, N.C.: Duke University Press, 1993.

Stefoff, Rebecca. *Women of the World: Women Travelers and Explorers.* New York: Oxford University Press, 1991.

Stevenson, Catherine Barnes. *Victorian Women Travel Writers in Africa.* Boston: Twayne, 1982.

Strobel, Margaret. *European Women and the Second British Empire.* Bloomington: Indiana University Press, 1991.

Swinglehurst, Edmund. *Cook's Tours: The Story of Popular Travel.* Poole, Dorset: Blandford, 1982.

Swinglehurst. *The Romantic Journey: The Story of Thomas Cook and Victorian Travel.* New York & London: Harper, 1974.

Sykes, Christopher. *Four Studies in Loyalty.* London: Collins, 1946.

Tidrick, Kathryn. *Empire and the English Character.* London: Tauris, 1990.

Tidrick. *Heart-Beguiling Araby.* Cambridge: Cambridge University Press, 1981.

Tiltman, Marjorie Hessell. *Women in Modern Adventure.* London: Harrap, 1935.

Tinling, Marion. *Women into the Unknown. A Sourcebook on Women Explorers and Travelers.* New York: Greenwood Press, 1989.

Trease, Robert G. *The Grand Tour.* London: Heinemann, 1967.

Tregaskis, Hugh. *Beyond the Grand Tour: The Levant Lunatics.* London: Ascent Books, 1979.

Trollope, Joanna. *Britannia's Daughters.* London: Hutchinson, 1983.

Van Thal, Herbert. *Victoria's Subjects Travelled.* London: Barker, 1951.

Von Martels, Zwerder. *Travel Fact and Travel Fiction: Studies in Fiction, Literary Tradition, Scholarly Discovery and Observation in Travel Writing.* New York: Brill, 1951.

West, Herbert Faulkner. *The Mind on the Wing: A Book for Readers and Collectors.* New York: Coward-McCann, 1947.

Williams, Cicely. *Women on the Rope: The Feminine Share in Mountain Adventure.* London: Allen & Unwin, 1973.

Woodcock, George. *Into Tibet: The Early British Explorers.* London: Faber & Faber, 1971.

Contributors

J. Allen Barksdale	*Bowling Green State University*
James Barszcz	*William Paterson College*
Kelly Belanger	*Youngstown State University*
Barbara Brothers	*Youngstown State University*
Scott R. Christianson	*Radford University*
Joan Corwin	*Evanston, Illinois*
Julie English Early	*University of Alabama in Huntsville*
Ira Grushow	*Franklin and Marshall College*
John W. M. Hallock	*Temple University*
Gordon Hirsch	*University of Minnesota*
Colleen Hobbs	*Rutgers University*
Syrine C. Hout	*University of Maine at Fort Kent*
Timothy S. Jones	*Augustana College*
David C. Judkins	*University of Houston*
Josephine A. McQuail	*Tennessee Technological University*
Susan Morgan	*Miami University*
Maura O'Connor	*University of Cincinnati*
Keith C. Odom	*Texas Christian University*
Patricia O'Neill	*Hamilton College*
Cynthia Ellen Patton	*Mesa State College*
Brian D. Reed	*Kent State – East Liverpool*
James Rigney	*Roehampton Institute, London*
Catherine Barnes Stevenson	*University of Hartford*
Stephen E. Tabachnick	*University of Oklahoma*
Susan Schoenbauer Thurin	*University of Wisconsin – Stout*
Nancy V. Workman	*Lewis University*

Cumulative Index

Dictionary of Literary Biography, Volumes 1-174
Dictionary of Literary Biography Yearbook, 1980-1995
Dictionary of Literary Biography Documentary Series, Volumes 1-14

Cumulative Index

DLB before number: *Dictionary of Literary Biography*, Volumes 1-174
Y before number: *Dictionary of Literary Biography Yearbook*, 1980-1995
DS before number: *Dictionary of Literary Biography Documentary Series*, Volumes 1-14

A

Abbey Press DLB-49

The Abbey Theatre and Irish Drama, 1900-1945 ....DLB-10

Abbot, Willis J. 1863-1934 ....DLB-29

Abbott, Jacob 1803-1879 ....DLB-1

Abbott, Lee K. 1947- DLB-130

Abbott, Lyman 1835-1922 ....DLB-79

Abbott, Robert S. 1868-1940 ....DLB-29, 91

Abelard, Peter circa 1079-1142 .....DLB-115

Abelard-Schuman DLB-46

Abell, Arunah S. 1806-1888 ....DLB-43

Abercrombie, Lascelles 1881-1938 ...DLB-19

Aberdeen University Press Limited DLB-106

Abish, Walter 1931- DLB-130

Ablesimov, Aleksandr Onisimovich 1742-1783 ....DLB-150

Abraham à Sancta Clara 1644-1709 ....DLB-168

Abrahams, Peter 1919- DLB-117

Abrams, M. H. 1912- DLB-67

Abrogans circa 790-800 ....DLB-148

Abschatz, Hans Aßmann von 1646-1699 ....DLB-168

Abse, Dannie 1923- DLB-27

Academy Chicago Publishers DLB-46

Accrocca, Elio Filippo 1923- DLB-128

Ace Books DLB-46

Achebe, Chinua 1930- DLB-117

Achtenberg, Herbert 1938- DLB-124

Ackerman, Diane 1948- DLB-120

Ackroyd, Peter 1949- DLB-155

Acorn, Milton 1923-1986 ....DLB-53

Acosta, Oscar Zeta 1935?- DLB-82

Actors Theatre of Louisville DLB-7

Adair, James 1709?-1783? DLB-30

Adam, Graeme Mercer 1839-1912 ... DLB-99

Adame, Leonard 1947- DLB-82

Adamic, Louis 1898-1951 ....DLB-9

Adams, Alice 1926- Y-86

Adams, Brooks 1848-1927 ....DLB-47

Adams, Charles Francis, Jr. 1835-1915 ....DLB-47

Adams, Douglas 1952- Y-83

Adams, Franklin P. 1881-1960 ......DLB-29

Adams, Henry 1838-1918 ....DLB-12, 47

Adams, Herbert Baxter 1850-1901 ... DLB-47

Adams, J. S. and C. [publishing house] DLB-49

Adams, James Truslow 1878-1949 ... DLB-17

Adams, John 1735-1826 ....DLB-31

Adams, John Quincy 1767-1848 .....DLB-37

Adams, Léonie 1899-1988 ....DLB-48

Adams, Levi 1802-1832 ....DLB-99

Adams, Samuel 1722-1803 ......DLB-31, 43

Adams, Thomas 1582 or 1583-1652 ....DLB-151

Adams, William Taylor 1822-1897 .. DLB-42

Adamson, Sir John 1867-1950 ......DLB-98

Adcock, Arthur St. John 1864-1930 ....DLB-135

Adcock, Betty 1938- DLB-105

Adcock, Betty, Certain Gifts DLB-105

Adcock, Fleur 1934- DLB-40

Addison, Joseph 1672-1719 ....DLB-101

Ade, George 1866-1944 ....DLB-11, 25

Adeler, Max (see Clark, Charles Heber)

Adonias Filho 1915-1990 ....DLB-145

Advance Publishing Company DLB-49

AE 1867-1935 ....DLB-19

Ælfric circa 955-circa 1010 ....DLB-146

Aesthetic Poetry (1873), by Walter Pater DLB-35

After Dinner Opera Company Y-92

Afro-American Literary Critics: An Introduction DLB-33

Agassiz, Jean Louis Rodolphe 1807-1873 ....DLB-1

Agee, James 1909-1955 ....DLB-2, 26, 152

The Agee Legacy: A Conference at the University of Tennessee at Knoxville Y-89

Aguilera Malta, Demetrio 1909-1981 ....DLB-145

Ai 1947- DLB-120

Aichinger, Ilse 1921- DLB-85

Aidoo, Ama Ata 1942- DLB-117

Aiken, Conrad 1889-1973 ....DLB-9, 45, 102

Aiken, Joan 1924- DLB-161

Aikin, Lucy 1781-1864 ....DLB-144, 163

Ainsworth, William Harrison 1805-1882 ....DLB-21

Aitken, George A. 1860-1917 ......DLB-149

Aitken, Robert [publishing house] ...DLB-49

Akenside, Mark 1721-1770 ....DLB-109

Akins, Zoë 1886-1958 ....DLB-26

Alabaster, William 1568-1640 ......DLB-132

Alain-Fournier 1886-1914 ....DLB-65

Alarcón, Francisco X. 1954- DLB-122

Alba, Nanina 1915-1968 ....DLB-41

Albee, Edward 1928- DLB-7

Albert the Great circa 1200-1280 ...DLB-115

Alberti, Rafael 1902- DLB-108

Albertinus, Aegidius circa 1560-1620 ....DLB-164

Alcott, Amos Bronson 1799-1888 .....DLB-1

Alcott, Louisa May 1832-1888 ....DLB-1, 42, 79; DS-14

Alcott, William Andrus 1798-1859 ....DLB-1

Alcuin circa 732-804 ....DLB-148

Alden, Henry Mills 1836-1919 ......DLB-79

Alden, Isabella 1841-1930 ....DLB-42

Alden, John B. [publishing house] DLB-49

Alden, Beardsley and Company DLB-49

Aldington, Richard
1892-1962 ....DLB-20, 36, 100, 149

Aldis, Dorothy 1896-1966 .... DLB-22

Aldiss, Brian W. 1925- DLB-14

Aldrich, Thomas Bailey
1836-1907 ....DLB-42, 71, 74, 79

Alegría, Ciro 1909-1967 .... DLB-113

Alegría, Claribel 1924- DLB-145

Aleixandre, Vicente 1898-1984 .... DLB-108

Aleramo, Sibilla 1876-1960 .... DLB-114

Alexander, Charles 1868-1923 ..... DLB-91

Alexander, Charles Wesley
[publishing house] DLB-49

Alexander, James 1691-1756 ...... DLB-24

Alexander, Lloyd 1924- DLB-52

Alexander, Sir William, Earl of Stirling
1577?-1640 .... DLB-121

Alexis, Willibald 1798-1871 .... DLB-133

Alfred, King 849-899 .... DLB-146

Alger, Horatio, Jr. 1832-1899 ...... DLB-42

Algonquin Books of Chapel Hill DLB-46

Algren, Nelson
1909-1981 .... DLB-9; Y-81, 82

Allan, Andrew 1907-1974 .... DLB-88

Allan, Ted 1916- DLB-68

Allbeury, Ted 1917- DLB-87

Alldritt, Keith 1935- DLB-14

Allen, Ethan 1738-1789 .... DLB-31

Allen, Frederick Lewis 1890-1954 .. DLB-137

Allen, Gay Wilson
1903-1995 .... DLB-103; Y-95

Allen, George 1808-1876 .... DLB-59

Allen, George [publishing house] ... DLB-106

Allen, George, and Unwin
Limited DLB-112

Allen, Grant 1848-1899 .... DLB-70, 92

Allen, Henry W. 1912- Y-85

Allen, Hervey 1889-1949 ......DLB-9, 45

Allen, James 1739-1808 .... DLB-31

Allen, James Lane 1849-1925 ...... DLB-71

Allen, Jay Presson 1922- DLB-26

Allen, John, and Company DLB-49

Allen, Samuel W. 1917- DLB-41

Allen, Woody 1935- DLB-44

Allende, Isabel 1942- DLB-145

Alline, Henry 1748-1784 .... DLB-99

Allingham, Margery 1904-1966 ...... DLB-77

Allingham, William 1824-1889 ..... DLB-35

Allison, W. L. [publishing house] DLB-49

The *Alliterative Morte Arthure* and
the *Stanzaic Morte Arthur*
circa 1350-1400 .... DLB-146

Allott, Kenneth 1912-1973 .... DLB-20

Allston, Washington 1779-1843 ...... DLB-1

Almon, John [publishing house] DLB-154

Alonzo, Dámaso 1898-1990 .... DLB-108

Alsop, George 1636-post 1673 ...... DLB-24

Alsop, Richard 1761-1815 .... DLB-37

Altemus, Henry, and Company DLB-49

Altenberg, Peter 1885-1919 .... DLB-81

Altolaguirre, Manuel 1905-1959 .... DLB-108

Aluko, T. M. 1918- DLB-117

Alurista 1947- DLB-82

Alvarez, A. 1929- DLB-14, 40

Amadi, Elechi 1934- DLB-117

Amado, Jorge 1912- DLB-113

Ambler, Eric 1909- DLB-77

*America: or, a Poem on the Settlement of the
British Colonies* (1780?), by Timothy
Dwight DLB-37

American Conservatory Theatre DLB-7

American Fiction and the 1930s DLB-9

American Humor: A Historical Survey
East and Northeast
South and Southwest
Midwest
West DLB-11

The American Library in Paris Y-93

American News Company DLB-49

The American Poets' Corner: The First
Three Years (1983-1986) Y-86

American Proletarian Culture:
The 1930s DS-11

American Publishing Company DLB-49

American Stationers' Company DLB-49

American Sunday-School Union DLB-49

American Temperance Union DLB-49

American Tract Society DLB-49

The American Writers Congress
(9-12 October 1981) Y-81

The American Writers Congress: A Report
on Continuing Business Y-81

Ames, Fisher 1758-1808 .... DLB-37

Ames, Mary Clemmer 1831-1884 .... DLB-23

Amini, Johari M. 1935- DLB-41

Amis, Kingsley 1922-
................DLB-15, 27, 100, 139

Amis, Martin 1949- DLB-14

Ammons, A. R. 1926- DLB-5, 165

Amory, Thomas 1691?-1788 ....DLB-39

Anaya, Rudolfo A. 1937- DLB-82

Ancrene Riwle circa 1200-1225 ....DLB-146

Andersch, Alfred 1914-1980 ....DLB-69

Anderson, Margaret 1886-1973 ... DLB-4, 91

Anderson, Maxwell 1888-1959 ....DLB-7

Anderson, Patrick 1915-1979 ....DLB-68

Anderson, Paul Y. 1893-1938 ....DLB-29

Anderson, Poul 1926- DLB-8

Anderson, Robert 1750-1830 ....DLB-142

Anderson, Robert 1917- DLB-7

Anderson, Sherwood
1876-1941 ....DLB-4, 9, 86; DS-1

Andreae, Johann Valentin
1586-1654 ....DLB-164

Andreas-Salomé, Lou 1861-1937 ....DLB-66

Andres, Stefan 1906-1970 ....DLB-69

Andreu, Blanca 1959- DLB-134

Andrewes, Lancelot
1555-1626 .... DLB-151, 172

Andrews, Charles M. 1863-1943 .....DLB-17

Andrews, Miles Peter ?-1814 ....DLB-89

Andrian, Leopold von 1875-1951 ....DLB-81

Andrić, Ivo 1892-1975 ....DLB-147

Andrieux, Louis (see Aragon, Louis)

Andrus, Silas, and Son DLB-49

Angell, James Burrill 1829-1916 ......DLB-64

Angell, Roger 1920- DLB-171

Angelou, Maya 1928- DLB-38

Anger, Jane flourished 1589 ....DLB-136

Angers, Félicité (see Conan, Laure)

Anglo-Norman Literature in the Development
of Middle English Literature DLB-146

The Anglo-Saxon Chronicle
circa 890-1154 ....DLB-146

The "Angry Young Men" DLB-15

Angus and Robertson (UK)
Limited DLB-112

Anhalt, Edward 1914- DLB-26

Anners, Henry F. [publishing house] ...DLB-49

Annolied between 1077 and 1081 ....DLB-148

Anselm of Canterbury 1033-1109 ...DLB-115

Anstey, F. 1856-1934 ....DLB-141

Anthony, Michael 1932- DLB-125	Der arme Hartmann ?-after 1150 .... DLB-148	Aubert de Gaspé, Phillipe-Ignace-François 1814-1841 ....DLB-99
Anthony, Piers 1934- DLB-8	Armed Services Editions DLB-46	Aubert de Gaspé, Phillipe-Joseph 1786-1871 ....DLB-99
Anthony Burgess's *99 Novels:* An Opinion Poll Y-84	Armstrong, Richard 1903- DLB-160	Aubin, Napoléon 1812-1890 ....DLB-99
Antin, David 1932- DLB-169	Arndt, Ernst Moritz 1769-1860 ..... DLB-90	Aubin, Penelope 1685-circa 1731 ....DLB-39
Antin, Mary 1881-1949 ....Y-84	Arnim, Achim von 1781-1831 ...... DLB-90	Aubrey-Fletcher, Henry Lancelot (see Wade, Henry)
Anton Ulrich, Duke of Brunswick-Lüneburg 1633-1714 ....DLB-168	Arnim, Bettina von 1785-1859 ...... DLB-90	Auchincloss, Louis 1917- DLB-2; Y-80
Antschel, Paul (see Celan, Paul)	Arno Press DLB-46	Auden, W. H. 1907-1973 ....DLB-10, 20
Anyidoho, Kofi 1947- DLB-157	Arnold, Edwin 1832-1904 .... DLB-35	Audio Art in America: A Personal Memoir Y-85
Anzaldúa, Gloria 1942- DLB-122	Arnold, Matthew 1822-1888 ..... DLB-32, 57	Auerbach, Berthold 1812-1882 .....DLB-133
Anzengruber, Ludwig 1839-1889 ...DLB-129	Arnold, Thomas 1795-1842 .... DLB-55	Auernheimer, Raoul 1876-1948 ......DLB-81
Apodaca, Rudy S. 1939- DLB-82	Arnold, Edward [publishing house] DLB-112	Augustine 354-430 ....DLB-115
Apple, Max 1941- DLB-130	Arnow, Harriette Simpson 1908-1986 .... DLB-6	Austen, Jane 1775-1817 ....DLB-116
Appleton, D., and Company DLB-49	Arp, Bill (see Smith, Charles Henry)	Austin, Alfred 1835-1913 ....DLB-35
Appleton-Century-Crofts DLB-46	Arreola, Juan José 1918- DLB-113	Austin, Mary 1868-1934 ....DLB-9, 78
Applewhite, James 1935- DLB-105	Arrowsmith, J. W. [publishing house] DLB-106	Austin, William 1778-1841 ....DLB-74
Apple-wood Books DLB-46	Arthur, Timothy Shay 1809-1885 ....DLB-3, 42, 79; DS-13	Author-Printers, 1476–1599 ....DLB-167
Aquin, Hubert 1929-1977 ....DLB-53	The Arthurian Tradition and Its European Context DLB-138	The Author's Apology for His Book (1684), by John Bunyan DLB-39
Aquinas, Thomas 1224 or 1225-1274 ....DLB-115	Artmann, H. C. 1921- DLB-85	An Author's Response, by Ronald Sukenick Y-82
Aragon, Louis 1897-1982 ....DLB-72	Arvin, Newton 1900-1963 .... DLB-103	Authors and Newspapers Association DLB-46
Arbor House Publishing Company DLB-46	As I See It, by Carolyn Cassady DLB-16	Authors' Publishing Company DLB-49
Arbuthnot, John 1667-1735 ....DLB-101	Asch, Nathan 1902-1964 .... DLB-4, 28	Avalon Books DLB-46
Arcadia House DLB-46	Ash, John 1948- DLB-40	Avancini, Nicolaus 1611-1686 ......DLB-164
Arce, Julio G. (see Ulica, Jorge)	Ashbery, John 1927- DLB-5, 165; Y-81	Avendaño, Fausto 1941- DLB-82
Archer, William 1856-1924 ....DLB-10	Ashendene Press DLB-112	Averroës 1126-1198 ....DLB-115
The Archpoet circa 1130?-? DLB-148	Asher, Sandy 1942- Y-83	Avery, Gillian 1926- DLB-161
Archpriest Avvakum (Petrovich) 1620?-1682 ....DLB-150	Ashton, Winifred (see Dane, Clemence)	Avicenna 980-1037 ....DLB-115
Arden, John 1930- DLB-13	Asimov, Isaac 1920-1992 .... DLB-8; Y-92	Avison, Margaret 1918- DLB-53
Arden of Faversham DLB-62	Askew, Anne circa 1521-1546 ...... DLB-136	Avon Books ....DLB 46
Ardis Publishers Y-89	Asselin, Olivar 1874-1937 .... DLB-92	Awdry, Wilbert Vere 1911- DLB-160
Ardizzone, Edward 1900-1979 ......DLB-160	Asturias, Miguel Angel 1899-1974 .... DLB-113	Awoonor, Kofi 1935- DLB-117
Arellano, Juan Estevan 1947- DLB-122	Atheneum Publishers DLB-46	Ayckbourn, Alan 1939- DLB-13
The Arena Publishing Company DLB-49	Atherton, Gertrude 1857-1948 .... DLB-9, 78	Aymé, Marcel 1902-1967 ....DLB-72
Arena Stage DLB-7	Athlone Press DLB-112	Aytoun, Sir Robert 1570-1638 ......DLB-121
Arenas, Reinaldo 1943-1990 ....DLB-145	Atkins, Josiah circa 1755-1781 ...... DLB-31	Aytoun, William Edmondstoune 1813-1865 ....DLB-32, 159
Arensberg, Ann 1937- Y-82	Atkins, Russell 1926- DLB-41	
Arguedas, José María 1911-1969 ....DLB-113	The Atlantic Monthly Press DLB-46	
Argueta, Manilio 1936- DLB-145	Attaway, William 1911-1986 .... DLB-76	**B**
Arias, Ron 1941- DLB-82	Atwood, Margaret 1939- DLB-53	
Arland, Marcel 1899-1986 ....DLB-72	Aubert, Alvin 1930- DLB-41	B. V. (see Thomson, James)
Arlen, Michael 1895-1956 .. DLB-36, 77, 162		Babbitt, Irving 1865-1933 ....DLB-63
Armah, Ayi Kwei 1939- DLB-117		

Cumulative Index

Babbitt, Natalie 1932- DLB-52

Babcock, John [publishing house] ... DLB-49

Baca, Jimmy Santiago 1952- DLB-122

Bache, Benjamin Franklin
  1769-1798 .... DLB-43

Bachmann, Ingeborg 1926-1973 ..... DLB-85

Bacon, Delia 1811-1859 .... DLB-1

Bacon, Francis 1561-1626 .... DLB-151

Bacon, Roger circa
  1214/1220-1292 .... DLB-115

Bacon, Sir Nicholas
  circa 1510-1579 .... DLB-132

Bacon, Thomas circa 1700-1768 .... DLB-31

Badger, Richard G.,
  and Company DLB-49

Bage, Robert 1728-1801 .... DLB-39

Bagehot, Walter 1826-1877 .... DLB-55

Bagley, Desmond 1923-1983 .... DLB-87

Bagnold, Enid 1889-1981 ...... DLB-13, 160

Bagryana, Elisaveta 1893-1991 ..... DLB-147

Bahr, Hermann 1863-1934 ..... DLB-81, 118

Bailey, Alfred Goldsworthy
  1905- DLB-68

Bailey, Francis [publishing house] ... DLB-49

Bailey, H. C. 1878-1961 .... DLB-77

Bailey, Jacob 1731-1808 .... DLB-99

Bailey, Paul 1937- DLB-14

Bailey, Philip James 1816-1902 ..... DLB-32

Baillargeon, Pierre 1916-1967 .... DLB-88

Baillie, Hugh 1890-1966 .... DLB-29

Baillie, Joanna 1762-1851 .... DLB-93

Bailyn, Bernard 1922- DLB-17

Bainbridge, Beryl 1933- DLB-14

Baird, Irene 1901-1981 .... DLB-68

Baker, Augustine 1575-1641 .... DLB-151

Baker, Carlos 1909-1987 .... DLB-103

Baker, David 1954- DLB-120

Baker, Herschel C. 1914-1990 ..... DLB-111

Baker, Houston A., Jr. 1943- DLB-67

Baker, Samuel White 1821-1893 ... DLB-166

Baker, Walter H., Company
  ("Baker's Plays") DLB-49

The Baker and Taylor Company DLB-49

Balaban, John 1943- DLB-120

Bald, Wambly 1902- DLB-4

Balde, Jacob 1604-1668 .... DLB-164

Balderston, John 1889-1954 .... DLB-26

Baldwin, James
  1924-1987 .... DLB-2, 7, 33; Y-87

Baldwin, Joseph Glover
  1815-1864 .... DLB-3, 11

Baldwin, Richard and Anne
  [publishing house] DLB-170

Baldwin, William
  circa 1515-1563 .... DLB-132

Bale, John 1495-1563 .... DLB-132

Balestrini, Nanni 1935- DLB-128

Ballantine Books DLB-46

Ballantyne, R. M. 1825-1894 .... DLB-163

Ballard, J. G. 1930- DLB-14

Ballerini, Luigi 1940- DLB-128

Ballou, Maturin Murray
  1820-1895 .... DLB-79

Ballou, Robert O.
  [publishing house] DLB-46

Balzac, Honoré de 1799-1855 ...... DLB-119

Bambara, Toni Cade 1939- DLB-38

Bancroft, A. L., and
  Company DLB-49

Bancroft, George
  1800-1891 .... DLB-1, 30, 59

Bancroft, Hubert Howe
  1832-1918 .... DLB-47, 140

Bangs, John Kendrick
  1862-1922 .... DLB-11, 79

Banim, John 1798-1842 ... DLB-116, 158, 159

Banim, Michael 1796-1874 .... DLB-158, 159

Banks, John circa 1653-1706 .... DLB-80

Banks, Russell 1940- DLB-130

Bannerman, Helen 1862-1946 ..... DLB-141

Bantam Books DLB-46

Banville, John 1945- DLB-14

Baraka, Amiri
  1934- DLB-5, 7, 16, 38; DS-8

Barbauld, Anna Laetitia
  1743-1825 ...... DLB-107, 109, 142, 158

Barbeau, Marius 1883-1969 .... DLB-92

Barber, John Warner 1798-1885 ..... DLB-30

Bàrberi Squarotti, Giorgio
  1929- DLB-128

Barbey d'Aurevilly, Jules-Amédée
  1808-1889 .... DLB-119

Barbour, John circa 1316-1395 ..... DLB-146

Barbour, Ralph Henry
  1870-1944 .... DLB-22

Barbusse, Henri 1873-1935 .... DLB-65

Barclay, Alexander
  circa 1475-1552 .... DLB-132

Barclay, E. E., and Company DLB-49

Bardeen, C. W.
  [publishing house] DLB-49

Barham, Richard Harris
  1788-1845 .... DLB-159

Baring, Maurice 1874-1945 .... DLB-34

Baring-Gould, Sabine 1834-1924 .... DLB-156

Barker, A. L. 1918- DLB-14, 139

Barker, George 1913-1991 .... DLB-20

Barker, Harley Granville
  1877-1946 .... DLB-10

Barker, Howard 1946- DLB-13

Barker, James Nelson 1784-1858 .....DLB-37

Barker, Jane 1652-1727 .... DLB-39, 131

Barker, Lady Mary Anne
  1831-1911 .... DLB-166

Barker, William
  circa 1520-after 1576 .... DLB-132

Barker, Arthur, Limited DLB-112

Barkov, Ivan Semenovich
  1732-1768 .... DLB-150

Barks, Coleman 1937- DLB-5

Barlach, Ernst 1870-1938 ...... DLB-56, 118

Barlow, Joel 1754-1812 .... DLB-37

Barnard, John 1681-1770 .... DLB-24

Barne, Kitty (Mary Catherine Barne)
  1883-1957 .... DLB-160

Barnes, Barnabe 1571-1609 .... DLB-132

Barnes, Djuna 1892-1982 ...... DLB-4, 9, 45

Barnes, Julian 1946- Y-93

Barnes, Margaret Ayer 1886-1967 .....DLB-9

Barnes, Peter 1931- DLB-13

Barnes, William 1801-1886 .... DLB-32

Barnes, A. S., and Company DLB-49

Barnes and Noble Books DLB-46

Barnet, Miguel 1940- DLB-145

Barney, Natalie 1876-1972 .... DLB-4

Barnfield, Richard 1574-1627 .... DLB-172

Baron, Richard W.,
  Publishing Company DLB-46

Barr, Robert 1850-1912 .... DLB-70, 92

Barral, Carlos 1928-1989 .... DLB-134

Barrax, Gerald William
  1933- DLB-41, 120

Barrès, Maurice 1862-1923 .... DLB-123

Barrett, Eaton Stannard 1786-1820 .... DLB-116

Barrie, J. M. 1860-1937 .... DLB-10, 141, 156

Barrie and Jenkins DLB-112

Barrio, Raymond 1921- DLB-82

Barrios, Gregg 1945- DLB-122

Barry, Philip 1896-1949 .... DLB-7

Barry, Robertine (see Françoise)

Barse and Hopkins DLB-46

Barstow, Stan 1928- DLB-14, 139

Barth, John 1930- DLB-2

Barthelme, Donald 1931-1989 .... DLB-2; Y-80, 89

Barthelme, Frederick 1943- Y-85

Bartholomew, Frank 1898-1985 .... DLB-127

Bartlett, John 1820-1905 .... DLB-1

Bartol, Cyrus Augustus 1813-1900 .... DLB-1

Barton, Bernard 1784-1849 .... DLB-96

Barton, Thomas Pennant 1803-1869 .... DLB-140

Bartram, John 1699-1777 .... DLB-31

Bartram, William 1739-1823 .... DLB-37

Basic Books DLB-46

Basille, Theodore (see Becon, Thomas)

Bass, T. J. 1932- Y-81

Bassani, Giorgio 1916- DLB-128

Basse, William circa 1583-1653 ..... DLB-121

Bassett, John Spencer 1867-1928 ..... DLB-17

Bassler, Thomas Joseph (see Bass, T. J.)

Bate, Walter Jackson 1918- DLB-67, 103

Bateman, Christopher [publishing house] DLB-170

Bateman, Stephen circa 1510-1584 .... DLB-136

Bates, H. E. 1905-1974 .... DLB-162

Bates, Katharine Lee 1859-1929 ..... DLB-71

Batsford, B. T. [publishing house] DLB-106

Battiscombe, Georgina 1905- DLB-155

The Battle of Maldon circa 1000 ...... DLB-146

Bauer, Bruno 1809-1882 .... DLB-133

Bauer, Wolfgang 1941- DLB-124

Baum, L. Frank 1856-1919 .... DLB-22

Baum, Vicki 1888-1960 .... DLB-85

Baumbach, Jonathan 1933- Y-80

Bausch, Richard 1945- DLB-130

Bawden, Nina 1925- DLB-14, 161

Bax, Clifford 1886-1962 .... DLB-10, 100

Baxter, Charles 1947- DLB-130

Bayer, Eleanor (see Perry, Eleanor)

Bayer, Konrad 1932-1964 .... DLB-85

Baynes, Pauline 1922- DLB-160

Bazin, Hervé 1911- DLB-83

Beach, Sylvia 1887-1962 .... DLB-4

Beacon Press DLB-49

Beadle and Adams DLB-49

Beagle, Peter S. 1939- Y-80

Beal, M. F. 1937- Y-81

Beale, Howard K. 1899-1959 .... DLB-17

Beard, Charles A. 1874-1948 .... DLB-17

A Beat Chronology: The First Twenty-five Years, 1944-1969 .... DLB-16

Beattie, Ann 1947- Y-82

Beattie, James 1735-1803 .... DLB-109

Beauchemin, Nérée 1850-1931 ...... DLB-92

Beauchemin, Yves 1941- DLB-60

Beaugrand, Honoré 1848-1906 ..... DLB-99

Beaulieu, Victor-Lévy 1945- DLB-53

Beaumont, Francis circa 1584-1616 and Fletcher, John 1579-1625 ... DLB-58

Beaumont, Sir John 1583?-1627 .... DLB-121

Beaumont, Joseph 1616–1699 ...... DLB-126

Beauvoir, Simone de 1908-1986 .... DLB-72; Y-86

Becher, Ulrich 1910- DLB-69

Becker, Carl 1873-1945 .... DLB-17

Becker, Jurek 1937- DLB-75

Becker, Jurgen 1932- DLB-75

Beckett, Samuel 1906-1989 .... DLB-13, 15; Y-90

Beckford, William 1760-1844 .... DLB-39

Beckham, Barry 1944- DLB-33

Becon, Thomas circa 1512-1567 .... DLB-136

Beddoes, Thomas 1760-1808 .... DLB-158

Beddoes, Thomas Lovell 1803-1849 .... DLB-96

Bede circa 673-735 .... DLB-146

Beecher, Catharine Esther 1800-1878 .... DLB-1

Beecher, Henry Ward 1813-1887 .... DLB-3, 43

Beer, George L. 1872-1920 .... DLB-47

Beer, Johann 1655-1700 .... DLB-168

Beer, Patricia 1919- DLB-40

Beerbohm, Max 1872-1956 ..... DLB-34, 100

Beer-Hofmann, Richard 1866-1945 .... DLB-81

Beers, Henry A. 1847-1926 .... DLB-71

Beeton, S. O. [publishing house] DLB-106

Bégon, Elisabeth 1696-1755 .... DLB-99

Behan, Brendan 1923-1964 .... DLB-13

Behn, Aphra 1640?-1689 ... DLB-39, 80, 131

Behn, Harry 1898-1973 .... DLB-61

Behrman, S. N. 1893-1973 .... DLB-7, 44

Belaney, Archibald Stansfeld (see Grey Owl)

Belasco, David 1853-1931 .... DLB-7

Belford, Clarke and Company DLB-49

Belitt, Ben 1911- DLB-5

Belknap, Jeremy 1744-1798 ...... DLB-30, 37

Bell, Clive 1881-1964 .... DS-10

Bell, Gertrude Margaret Lowthian 1868-1926 .... DLB-174

Bell, James Madison 1826-1902 ..... DLB-50

Bell, Marvin 1937- DLB-5

Bell, Millicent 1919- DLB-111

Bell, Quentin 1910- DLB-155

Bell, Vanessa 1879-1961 .... DS-10

Bell, George, and Sons DLB-106

Bell, Robert [publishing house] DLB-49

Bellamy, Edward 1850-1898 .... DLB-12

Bellamy, John [publishing house] ... DLB-170

Bellamy, Joseph 1719-1790 .... DLB-31

Bellezza, Dario 1944- DLB-128

La Belle Assemblée 1806-1837 .... DLB-110

Belloc, Hilaire 1870-1953 .... DLB-19, 100, 141, 174

Bellow, Saul 1915- DLB-2, 28; Y-82; DS-3

Belmont Productions DLB-46

Bemelmans, Ludwig 1898-1962 ..... DLB-22

Bemis, Samuel Flagg 1891-1973 ..... DLB-17

Bemrose, William [publishing house] DLB-106

Benchley, Robert 1889-1945 .... DLB-11

Benedetti, Mario 1920- DLB-113

Benedictus, David 1938- DLB-14

Benedikt, Michael 1935- DLB-5

Benét, Stephen Vincent 1898-1943 .... DLB-4, 48, 102

Benét, William Rose 1886-1950 ..... DLB-45

Benford, Gregory 1941- Y-82

Benjamin, Park 1809-1864 .... DLB-3, 59, 73

Benlowes, Edward 1602-1676 ...... DLB-126

Benn, Gottfried 1886-1956 .... DLB-56

Benn Brothers Limited DLB-106

Bennett, Arnold
1867-1931 .... DLB-10, 34, 98, 135

Bennett, Charles 1899- DLB-44

Bennett, Gwendolyn 1902- DLB-51

Bennett, Hal 1930- DLB-33

Bennett, James Gordon 1795-1872 ... DLB-43

Bennett, James Gordon, Jr.
1841-1918 .... DLB-23

Bennett, John 1865-1956 .... DLB-42

Bennett, Louise 1919- DLB-117

Benoit, Jacques 1941- DLB-60

Benson, A. C. 1862-1925 .... DLB-98

Benson, E. F. 1867-1940 ..... DLB-135, 153

Benson, Jackson J. 1930- DLB-111

Benson, Robert Hugh 1871-1914 ... DLB-153

Benson, Stella 1892-1933 ...... DLB-36, 162

Bent, James Theodore 1852-1897 .. DLB-174

Bent, Mabel Virginia Anna ?-? DLB-174

Bentham, Jeremy 1748-1832 ... DLB-107, 158

Bentley, E. C. 1875-1956 .... DLB-70

Bentley, Richard
[publishing house] DLB-106

Benton, Robert 1932- and Newman,
David 1937- DLB-44

Benziger Brothers DLB-49

Beowulf circa 900-1000
or 790-825 .... DLB-146

Beresford, Anne 1929- DLB-40

Beresford, John Davys
1873-1947 .... DLB-162

Beresford-Howe, Constance
1922- DLB-88

Berford, R. G., Company DLB-49

Berg, Stephen 1934- DLB-5

Bergengruen, Werner 1892-1964 .... DLB-56

Berger, John 1926- DLB-14

Berger, Meyer 1898-1959 .... DLB-29

Berger, Thomas 1924- DLB-2; Y-80

Berkeley, Anthony 1893-1971 ..... DLB-77

Berkeley, George 1685-1753 .... DLB-31, 101

The Berkley Publishing
Corporation DLB-46

Berlin, Lucia 1936- DLB-130

Bernal, Vicente J. 1888-1915 .... DLB-82

Bernanos, Georges 1888-1948 .... DLB-72

Bernard, Harry 1898-1979 .... DLB-92

Bernard, John 1756-1828 .... DLB-37

Bernard of Chartres
circa 1060-1124? DLB-115

Bernhard, Thomas
1931-1989 .... DLB-85, 124

Bernstein, Charles 1950- DLB-169

Berriault, Gina 1926- DLB-130

Berrigan, Daniel 1921- DLB-5

Berrigan, Ted 1934-1983 .... DLB-5, 169

Berry, Wendell 1934- DLB-5, 6

Berryman, John 1914-1972 .... DLB-48

Bersianik, Louky 1930- DLB-60

Berthelet, Thomas
[publishing house] DLB-170

Bertolucci, Attilio 1911- DLB-128

Berton, Pierre 1920- DLB-68

Besant, Sir Walter 1836-1901 ...... DLB-135

Bessette, Gerard 1920- DLB-53

Bessie, Alvah 1904-1985 .... DLB-26

Bester, Alfred 1913-1987 .... DLB-8

The Bestseller Lists: An Assessment Y-84

Betham-Edwards, Matilda Barbara (see
Edwards, Matilda Barbara Betham-)

Betjeman, John 1906-1984 ..... DLB-20; Y-84

Betocchi, Carlo 1899-1986 .... DLB-128

Bettarini, Mariella 1942- DLB-128

Betts, Doris 1932- Y-82

Beveridge, Albert J. 1862-1927 ...... DLB-17

Beverley, Robert
circa 1673-1722 .... DLB-24, 30

Beyle, Marie-Henri (see Stendhal)

Bianco, Margery Williams
1881-1944 .... DLB-160

Bibaud, Adèle 1854-1941 .... DLB-92

Bibaud, Michel 1782-1857 .... DLB-99

Bibliographical and Textual Scholarship
Since World War II Y-89

The Bicentennial of James Fenimore
Cooper: An International
Celebration Y-89

Bichsel, Peter 1935- DLB-75

Bickerstaff, Isaac John
1733-circa 1808 .... DLB-89

Biddle, Drexel [publishing house] DLB-49

Bidermann, Jacob
1577 or 1578-1639 .... DLB-164

Bidwell, Walter Hilliard
1798-1881 .... DLB-79

Bienek, Horst 1930- DLB-75

Bierbaum, Otto Julius 1865-1910 .... DLB-66

Bierce, Ambrose
1842-1914? DLB-11, 12, 23, 71, 74

Bigelow, William F. 1879-1966 ...... DLB-91

Biggle, Lloyd, Jr. 1923- DLB-8

Biglow, Hosea (see Lowell, James Russell)

Bigongiari, Piero 1914- DLB-128

Billinger, Richard 1890-1965 .... DLB-124

Billings, John Shaw 1898-1975 ..... DLB-137

Billings, Josh (see Shaw, Henry Wheeler)

Binding, Rudolf G. 1867-1938 .... DLB-66

Bingham, Caleb 1757-1817 .... DLB-42

Bingham, George Barry
1906-1988 .... DLB-127

Bingley, William
[publishing house] DLB-154

Binyon, Laurence 1869-1943 .... DLB-19

Biographia Brittanica DLB-142

Biographical Documents I Y-84

Biographical Documents II Y-85

Bioren, John [publishing house] DLB-49

Bioy Casares, Adolfo 1914- DLB-113

Bird, Isabella Lucy 1831-1904 ...... DLB-166

Bird, William 1888-1963 .... DLB-4

Birken, Sigmund von 1626-1681 .... DLB-164

Birney, Earle 1904- DLB-88

Birrell, Augustine 1850-1933 .... DLB-98

Bisher, Furman 1918- DLB-171

Bishop, Elizabeth 1911-1979 ..... DLB-5, 169

Bishop, John Peale 1892-1944 .. DLB-4, 9, 45

Bismarck, Otto von 1815-1898 ..... DLB-129

Bisset, Robert 1759-1805 .... DLB-142

Bissett, Bill 1939- DLB-53

Bitzius, Albert (see Gotthelf, Jeremias)

Black, David (D. M.) 1941- DLB-40

Black, Winifred 1863-1936 .... DLB-25

Black, Walter J.
[publishing house] DLB-46

The Black Aesthetic: Background DS-8

The Black Arts Movement, by
    Larry Neal DLB-38

Black Theaters and Theater Organizations in
    America, 1961-1982:
    A Research List DLB-38

Black Theatre: A Forum
    [excerpts] DLB-38

Blackamore, Arthur 1679-? DLB-24, 39

Blackburn, Alexander L. 1929- Y-85

Blackburn, Paul 1926-1971 ....DLB-16; Y-81

Blackburn, Thomas 1916-1977 ......DLB-27

Blackmore, R. D. 1825-1900 ....DLB-18

Blackmore, Sir Richard
    1654-1729 ....DLB-131

Blackmur, R. P. 1904-1965 ....DLB-63

Blackwell, Basil, Publisher DLB-106

Blackwood, Algernon Henry
    1869-1951 ....DLB-153, 156

Blackwood, Caroline 1931- DLB-14

Blackwood, William, and
    Sons, Ltd. DLB-154

Blackwood's Edinburgh Magazine
    1817-1980 ....DLB-110

Blair, Eric Arthur (see Orwell, George)

Blair, Francis Preston 1791-1876 .....DLB-43

Blair, James circa 1655-1743 ....DLB-24

Blair, John Durburrow 1759-1823 ....DLB-37

Blais, Marie-Claire 1939- DLB-53

Blaise, Clark 1940- DLB-53

Blake, Nicholas 1904-1972 ....DLB-77
    (see Day Lewis, C.)

Blake, William
    1757-1827 .... DLB-93, 154, 163

The Blakiston Company DLB-49

Blanchot, Maurice 1907- DLB-72

Blanckenburg, Christian Friedrich von
    1744-1796 ....DLB-94

Blaser, Robin 1925- DLB-165

Bledsoe, Albert Taylor
    1809-1877 ....DLB-3, 79

Blelock and Company DLB-49

Blennerhassett, Margaret Agnew
    1773-1842 ....DLB-99

Bles, Geoffrey
    [publishing house] DLB-112

Blessington, Marguerite, Countess of
    1789-1849 ....DLB-166

The Blickling Homilies
    circa 971 ....DLB-146

Blish, James 1921-1975 ....DLB-8

Bliss, E., and E. White
    [publishing house] DLB-49

Bliven, Bruce 1889-1977 ....DLB-137

Bloch, Robert 1917-1994 ....DLB-44

Block, Rudolph (see Lessing, Bruno)

Blondal, Patricia 1926-1959 ....DLB-88

Bloom, Harold 1930- DLB-67

Bloomer, Amelia 1818-1894 ......DLB-79

Bloomfield, Robert 1766-1823 ......DLB-93

Bloomsbury Group DS-10

Blotner, Joseph 1923- DLB-111

Bloy, Léon 1846-1917 ....DLB-123

Blume, Judy 1938- DLB-52

Blunck, Hans Friedrich 1888-1961 ... DLB-66

Blunden, Edmund
    1896-1974 ....DLB-20, 100, 155

Blunt, Lady Anne Isabella Noel
    1837-1917 ....DLB-174

Blunt, Wilfrid Scawen
    1840-1922 .... DLB-19, 174

Bly, Nellie (see Cochrane, Elizabeth)

Bly, Robert 1926- DLB-5

Blyton, Enid 1897-1968 ....DLB-160

Boaden, James 1762-1839 ....DLB-89

Boas, Frederick S. 1862-1957 ......DLB-149

The Bobbs-Merrill Archive at the
    Lilly Library, Indiana University ...Y-90

The Bobbs-Merrill Company DLB-46

Bobrov, Semen Sergeevich
    1763?-1810 ....DLB-150

Bobrowski, Johannes 1917-1965 .....DLB-75

Bodenheim, Maxwell 1892-1954 .. DLB-9, 45

Bodenstedt, Friedrich von
    1819-1892 ....DLB-129

Bodini, Vittorio 1914-1970 ....DLB-128

Bodkin, M. McDonnell
    1850-1933 .... DLB-70

Bodley Head DLB-112

Bodmer, Johann Jakob 1698-1783 ... DLB-97

Bodmershof, Imma von 1895-1982 .. DLB-85

Bodsworth, Fred 1918- DLB-68

Boehm, Sydney 1908- DLB-44

Boer, Charles 1939- DLB-5

Boethius circa 480-circa 524 ....DLB-115

Boethius of Dacia circa 1240-? DLB-115

Bogan, Louise 1897-1970 ...... DLB-45, 169

Bogarde, Dirk 1921- DLB-14

Bogdanovich, Ippolit Fedorovich
    circa 1743-1803 ....DLB-150

Bogue, David [publishing house] ...DLB-106

Böhme, Jakob 1575-1624 ....DLB-164

Bohn, H. G. [publishing house] DLB-106

Bohse, August 1661-1742 ....DLB-168

Boie, Heinrich Christian
    1744-1806 ....DLB-94

Bok, Edward W. 1863-1930 ....DLB-91

Boland, Eavan 1944- DLB-40

Bolingbroke, Henry St. John, Viscount
    1678-1751 ....DLB-101

Böll, Heinrich 1917-1985 ...... Y-85, DLB-69

Bolling, Robert 1738-1775 ....DLB-31

Bolotov, Andrei Timofeevich
    1738-1833 ....DLB-150

Bolt, Carol 1941- DLB-60

Bolt, Robert 1924- DLB-13

Bolton, Herbert E. 1870-1953 ....DLB-17

Bonaventura DLB-90

Bonaventure circa 1217-1274 ......DLB-115

Bond, Edward 1934- DLB-13

Bond, Michael 1926- DLB-161

Boni, Albert and Charles
    [publishing house] DLB-46

Boni and Liveright DLB-46

Robert Bonner's Sons DLB-49

Bontemps, Arna 1902-1973 ......DLB-48, 51

The Book League of America DLB-46

Book Reviewing in America: I Y-87

Book Reviewing in America: II Y-88

Book Reviewing in America: III Y-89

Book Reviewing in America: IV Y-90

Book Reviewing in America: V Y-91

Book Reviewing in America: VI Y-92

Book Reviewing in America: VII Y-93

Book Reviewing in America: VIII Y-94

Book Reviewing in America and the
    Literary Scene Y-95

Book Supply Company DLB-49

The Book Trade History Group Y-93

The Booker Prize
    Address by Anthony Thwaite,
    Chairman of the Booker Prize Judges
    Comments from Former Booker
    Prize Winners Y-86

Boorde, Andrew circa 1490-1549 ...DLB-136

Boorstin, Daniel J. 1914- DLB-17

Cumulative Index

Booth, Mary L. 1831-1889 .... DLB-79

Booth, Philip 1925- Y-82

Booth, Wayne C. 1921- DLB-67

Borchardt, Rudolf 1877-1945 .... DLB-66

Borchert, Wolfgang
    1921-1947 .... DLB-69, 124

Borel, Pétrus 1809-1859 .... DLB-119

Borges, Jorge Luis
    1899-1986 .... DLB-113; Y-86

Börne, Ludwig 1786-1837 .... DLB-90

Borrow, George
    1803-1881 .... DLB-21, 55, 166

Bosch, Juan 1909- DLB-145

Bosco, Henri 1888-1976 .... DLB-72

Bosco, Monique 1927- DLB-53

Boston, Lucy M. 1892-1990 .... DLB-161

Boswell, James 1740-1795 ..... DLB-104, 142

Botev, Khristo 1847-1876 .... DLB-147

Botta, Anne C. Lynch 1815-1891 ......DLB-3

Bottomley, Gordon 1874-1948 ...... DLB-10

Bottoms, David 1949- DLB-120; Y-83

Bottrall, Ronald 1906- DLB-20

Boucher, Anthony 1911-1968 ....DLB-8

Boucher, Jonathan 1738-1804 .... DLB-31

Boucher de Boucherville, George
    1814-1894 .... DLB-99

Boudreau, Daniel (see Coste, Donat)

Bourassa, Napoléon 1827-1916 ..... DLB-99

Bourget, Paul 1852-1935 .... DLB-123

Bourinot, John George 1837-1902 ... DLB-99

Bourjaily, Vance 1922- DLB-2, 143

Bourne, Edward Gaylord
    1860-1908 .... DLB-47

Bourne, Randolph 1886-1918 .... DLB-63

Bousoño, Carlos 1923- DLB-108

Bousquet, Joë 1897-1950 .... DLB-72

Bova, Ben 1932- Y-81

Bovard, Oliver K. 1872-1945 .... DLB-25

Bove, Emmanuel 1898-1945 .... DLB-72

Bowen, Elizabeth 1899-1973 .... DLB-15, 162

Bowen, Francis 1811-1890 .... DLB-1, 59

Bowen, John 1924- DLB-13

Bowen, Marjorie 1886-1952 .... DLB-153

Bowen-Merrill Company DLB-49

Bowering, George 1935- DLB-53

Bowers, Claude G. 1878-1958 ...... DLB-17

Bowers, Edgar 1924- DLB-5

Bowers, Fredson Thayer
    1905-1991 .... DLB-140; Y-91

Bowles, Paul 1910- DLB-5, 6

Bowles, Samuel III 1826-1878 .... DLB-43

Bowles, William Lisles 1762-1850 ... DLB-93

Bowman, Louise Morey
    1882-1944 .... DLB-68

Boyd, James 1888-1944 .... DLB-9

Boyd, John 1919- DLB-8

Boyd, Thomas 1898-1935 .... DLB-9

Boyesen, Hjalmar Hjorth
    1848-1895 .... DLB-12, 71; DS-13

Boyle, Kay
    1902-1992 .... DLB-4, 9, 48, 86; Y-93

Boyle, Roger, Earl of Orrery
    1621-1679 .... DLB-80

Boyle, T. Coraghessan 1948- Y-86

Brackenbury, Alison 1953- DLB-40

Brackenridge, Hugh Henry
    1748-1816 ....DLB-11, 37

Brackett, Charles 1892-1969 .... DLB-26

Brackett, Leigh 1915-1978 ....DLB-8, 26

Bradburn, John
    [publishing house] DLB-49

Bradbury, Malcolm 1932- DLB-14

Bradbury, Ray 1920- DLB-2, 8

Bradbury and Evans DLB-106

Braddon, Mary Elizabeth
    1835-1915 ....DLB-18, 70, 156

Bradford, Andrew 1686-1742 .... DLB-43, 73

Bradford, Gamaliel 1863-1932 ..... DLB-17

Bradford, John 1749-1830 .... DLB-43

Bradford, Roark 1896-1948 .... DLB-86

Bradford, William 1590-1657 ....DLB-24, 30

Bradford, William III
    1719-1791 .... DLB-43, 73

Bradlaugh, Charles 1833-1891 ...... DLB-57

Bradley, David 1950- DLB-33

Bradley, Marion Zimmer 1930- DLB-8

Bradley, William Aspenwall
    1878-1939 .... DLB-4

Bradley, Ira, and Company DLB-49

Bradley, J. W., and Company DLB-49

Bradstreet, Anne
    1612 or 1613-1672 .... DLB-24

Bradwardine, Thomas circa
    1295-1349 .... DLB-115

Brady, Frank 1924-1986 .... DLB-111

Brady, Frederic A.
    [publishing house] DLB-49

Bragg, Melvyn 1939- DLB-14

Brainard, Charles H.
    [publishing house] DLB-49

Braine, John 1922-1986 .... DLB-15; Y-86

Braithwait, Richard 1588-1673 .....DLB-151

Braithwaite, William Stanley
    1878-1962 .... DLB-50, 54

Braker, Ulrich 1735-1798 ....DLB-94

Bramah, Ernest 1868-1942 ....DLB-70

Branagan, Thomas 1774-1843 ....DLB-37

Branch, William Blackwell
    1927- DLB-76

Branden Press DLB-46

Brassey, Lady Annie (Allnutt)
    1839-1887 ....DLB-166

Brathwaite, Edward Kamau
    1930- DLB-125

Brault, Jacques 1933- DLB-53

Braun, Volker 1939- DLB-75

Brautigan, Richard
    1935-1984 .... DLB-2, 5; Y-80, 84

Braxton, Joanne M. 1950- DLB-41

Bray, Anne Eliza 1790-1883 ....DLB-116

Bray, Thomas 1656-1730 ....DLB-24

Braziller, George
    [publishing house] DLB-46

The Bread Loaf Writers'
    Conference 1983 .... Y-84

The Break-Up of the Novel (1922),
    by John Middleton Murry DLB-36

Breasted, James Henry 1865-1935 ....DLB-47

Brecht, Bertolt 1898-1956 ...... DLB-56, 124

Bredel, Willi 1901-1964 ....DLB-56

Breitinger, Johann Jakob
    1701-1776 ....DLB-97

Bremser, Bonnie 1939- DLB-16

Bremser, Ray 1934- DLB-16

Brentano, Bernard von
    1901-1964 ....DLB-56

Brentano, Clemens 1778-1842 ....DLB-90

Brentano's DLB-49

Brenton, Howard 1942- DLB-13

Breton, André 1896-1966 ....DLB-65

Breton, Nicholas
    circa 1555-circa 1626 ....DLB-136

The Breton Lays
    1300-early fifteenth century DLB-146

Brewer, Warren and Putnam DLB-46

Brewster, Elizabeth 1922- DLB-60

Bridgers, Sue Ellen 1942- DLB-52

Bridges, Robert 1844-1930 ....DLB-19, 98

Bridie, James 1888-1951 ....DLB-10

Briggs, Charles Frederick
1804-1877 ....DLB-3

Brighouse, Harold 1882-1958 ....DLB-10

Bright, Mary Chavelita Dunne
(see Egerton, George)

Brimmer, B. J., Company DLB-46

Brines, Francisco 1932- DLB-134

Brinley, George, Jr. 1817-1875 .....DLB-140

Brinnin, John Malcolm 1916- DLB-48

Brisbane, Albert 1809-1890 ....DLB-3

Brisbane, Arthur 1864-1936 ....DLB-25

British Academy DLB-112

The British Library and the Regular
Readers' Group Y-91

The British Critic 1793-1843 ....DLB-110

*The British Review and London
Critical Journal* 1811-1825 ....DLB-110

Brito, Aristeo 1942- DLB-122

Broadway Publishing Company DLB-46

Broch, Hermann 1886-1951 .....DLB-85, 124

Brochu, André 1942- DLB-53

Brock, Edwin 1927- DLB-40

Brockes, Barthold Heinrich
1680-1747 ....DLB-168

Brod, Max 1884-1968 ....DLB-81

Brodber, Erna 1940- DLB-157

Brodhead, John R. 1814-1873 ....DLB-30

Brodkey, Harold 1930- DLB-130

Broeg, Bob 1918- DLB-171

Brome, Richard circa 1590-1652 .....DLB-58

Brome, Vincent 1910- DLB-155

Bromfield, Louis 1896-1956 .... DLB-4, 9, 86

Broner, E. M. 1930- DLB-28

Bronk, William 1918- DLB-165

Bronnen, Arnolt 1895-1959 ....DLB-124

Brontë, Anne 1820-1849 ....DLB-21

Brontë, Charlotte 1816-1855 ....DLB-21, 159

Brontë, Emily 1818-1848 ....DLB-21, 32

Brooke, Frances 1724-1789 .....DLB-39, 99

Brooke, Henry 1703?-1783 ....DLB-39

Brooke, L. Leslie 1862-1940 ....DLB-141

Brooke, Margaret, Ranee of Sarawak
1849-1936 ....DLB-174

Brooke, Rupert 1887-1915 ....DLB-19

Brooker, Bertram 1888-1955 ....DLB-88

Brooke-Rose, Christine 1926- DLB-14

Brookner, Anita 1928- Y-87

Brooks, Charles Timothy
1813-1883 ....DLB-1

Brooks, Cleanth 1906-1994 ....DLB-63; Y-94

Brooks, Gwendolyn
1917- DLB-5, 76, 165

Brooks, Jeremy 1926- DLB-14

Brooks, Mel 1926- DLB-26

Brooks, Noah 1830-1903 .....DLB-42; DS-13

Brooks, Richard 1912-1992 ....DLB-44

Brooks, Van Wyck
1886-1963 ....DLB-45, 63, 103

Brophy, Brigid 1929- DLB-14

Brossard, Chandler 1922-1993 ......DLB-16

Brossard, Nicole 1943- DLB-53

Broster, Dorothy Kathleen
1877-1950 ....DLB-160

Brother Antoninus (see Everson, William)

Brougham and Vaux, Henry Peter
Brougham, Baron
1778-1868 ....DLB-110, 158

Brougham, John 1810-1880 ....DLB-11

Broughton, James 1913- DLB-5

Broughton, Rhoda 1840-1920 ....DLB-18

Broun, Heywood 1888-1939 ....DLB-29, 171

Brown, Alice 1856-1948 ....DLB-78

Brown, Bob 1886-1959 ....DLB-4, 45

Brown, Cecil 1943- DLB-33

Brown, Charles Brockden
1771-1810 ....DLB-37, 59, 73

Brown, Christy 1932-1981 ....DLB-14

Brown, Dee 1908- Y-80

Brown, Frank London 1927-1962 ...DLB-76

Brown, Fredric 1906-1972 ....DLB-8

Brown, George Mackay
1921- DLB-14, 27, 139

Brown, Harry 1917-1986 ....DLB-26

Brown, Marcia 1918- DLB-61

Brown, Margaret Wise
1910-1952 ....DLB-22

Brown, Morna Doris (see Ferrars, Elizabeth)

Brown, Oliver Madox
1855-1874 ....DLB-21

Brown, Sterling
1901-1989 ....DLB-48, 51, 63

Brown, T. E. 1830-1897 ....DLB-35

Brown, William Hill 1765-1793 .....DLB-37

Brown, William Wells
1814-1884 ....DLB-3, 50

Browne, Charles Farrar
1834-1867 ....DLB-11

Browne, Francis Fisher
1843-1913 ....DLB-79

Browne, Michael Dennis
1940- DLB-40

Browne, Sir Thomas 1605-1682 ....DLB-151

Browne, William, of Tavistock
1590-1645 ....DLB-121

Browne, Wynyard 1911-1964 ....DLB-13

Browne and Nolan DLB-106

Brownell, W. C. 1851-1928 ....DLB-71

Browning, Elizabeth Barrett
1806-1861 ....DLB-32

Browning, Robert
1812-1889 ....DLB-32, 163

Brownjohn, Allan 1931- DLB-40

Brownson, Orestes Augustus
1803-1876 ....DLB-1, 59, 73

Bruccoli, Matthew J. 1931- DLB-103

Bruce, Charles 1906-1971 ....DLB-68

Bruce, Leo 1903-1979 ....DLB-77

Bruce, Philip Alexander
1856-1933 ....DLB-47

Bruce Humphries
[publishing house] DLB-46

Bruce-Novoa, Juan 1944- DLB-82

Bruckman, Clyde 1894-1955 ....DLB-26

Bruckner, Ferdinand 1891-1958 ....DLB-118

Brundage, John Herbert (see Herbert, John)

Brutus, Dennis 1924- DLB-117

Bryant, Arthur 1899-1985 ....DLB-149

Bryant, William Cullen
1794-1878 ....DLB-3, 43, 59

Bryce Echenique, Alfredo
1939- DLB-145

Bryce, James 1838-1922 ....DLB-166

Brydges, Sir Samuel Egerton
1762-1837 ....DLB-107

Bryskett, Lodowick 1546?-1612 ....DLB-167

Buchan, John 1875-1940 ... DLB-34, 70, 156

Buchanan, George 1506-1582 ......DLB-132

Buchanan, Robert 1841-1901 ....DLB-18, 35

Buchman, Sidney 1902-1975 .... DLB-26

Buchner, Augustus 1591-1661 ..... DLB-164

Büchner, Georg 1813-1837 .... DLB-133

Bucholtz, Andreas Heinrich
1607-1671 .... DLB-168

Buck, Pearl S. 1892-1973 .... DLB-9, 102

Bucke, Charles 1781-1846 .... DLB-110

Bucke, Richard Maurice
1837-1902 .... DLB-99

Buckingham, Joseph Tinker 1779-1861 and
Buckingham, Edwin
1810-1833 .... DLB-73

Buckler, Ernest 1908-1984 .... DLB-68

Buckley, William F., Jr.
1925- DLB-137; Y-80

Buckminster, Joseph Stevens
1784-1812 .... DLB-37

Buckner, Robert 1906- DLB-26

Budd, Thomas ?-1698 .... DLB-24

Budrys, A. J. 1931- DLB-8

Buechner, Frederick 1926- Y-80

Buell, John 1927- DLB-53

Buffum, Job [publishing house] DLB-49

Bugnet, Georges 1879-1981 .... DLB-92

Buies, Arthur 1840-1901 .... DLB-99

Building the New British Library
at St Pancras Y-94

Bukowski, Charles
1920-1994 .... DLB-5, 130, 169

Bulger, Bozeman 1877-1932 .... DLB-171

Bullein, William
between 1520 and 1530-1576 ... DLB-167

Bullins, Ed 1935- DLB-7, 38

Bulwer-Lytton, Edward (also Edward Bulwer)
1803-1873 .... DLB-21

Bumpus, Jerry 1937- Y-81

Bunce and Brother DLB-49

Bunner, H. C. 1855-1896 .... DLB-78, 79

Bunting, Basil 1900-1985 .... DLB-20

Bunyan, John 1628-1688 .... DLB-39

Burch, Robert 1925- DLB-52

Burciaga, José Antonio 1940- DLB-82

Bürger, Gottfried August
1747-1794 .... DLB-94

Burgess, Anthony 1917-1993 ...... DLB-14

Burgess, Gelett 1866-1951 .... DLB-11

Burgess, John W. 1844-1931 .... DLB-47

Burgess, Thornton W.
1874-1965 .... DLB-22

Burgess, Stringer and Company DLB-49

Burick, Si 1909-1986 .... DLB-171

Burk, John Daly circa 1772-1808 .... DLB-37

Burke, Edmund 1729?-1797 .... DLB-104

Burke, Kenneth 1897-1993 ......DLB-45, 63

Burlingame, Edward Livermore
1848-1922 .... DLB-79

Burnet, Gilbert 1643-1715 .... DLB-101

Burnett, Frances Hodgson
1849-1924 ...... DLB-42, 141; DS-13, 14

Burnett, W. R. 1899-1982 .... DLB-9

Burnett, Whit 1899-1973 and
Martha Foley 1897-1977 .... DLB-137

Burney, Fanny 1752-1840 .... DLB-39

Burns, Alan 1929- DLB-14

Burns, John Horne 1916-1953 .... Y-85

Burns, Robert 1759-1796 .... DLB-109

Burns and Oates DLB-106

Burnshaw, Stanley 1906- DLB-48

Burr, C. Chauncey 1815?-1883 ...... DLB-79

Burroughs, Edgar Rice 1875-1950 .... DLB-8

Burroughs, John 1837-1921 .... DLB-64

Burroughs, Margaret T. G.
1917- DLB-41

Burroughs, William S., Jr.
1947-1981 .... DLB-16

Burroughs, William Seward
1914- DLB-2, 8, 16, 152; Y-81

Burroway, Janet 1936- DLB-6

Burt, Maxwell S. 1882-1954 .... DLB-86

Burt, A. L., and Company DLB-49

Burton, Hester 1913- DLB-161

Burton, Isabel Arundell
1831-1896 .... DLB-166

Burton, Miles (see Rhode, John)

Burton, Richard Francis
1821-1890 ....DLB-55, 166

Burton, Robert 1577-1640 .... DLB-151

Burton, Virginia Lee 1909-1968 ..... DLB-22

Burton, William Evans
1804-1860 .... DLB-73

Burwell, Adam Hood 1790-1849 .... DLB-99

Bury, Lady Charlotte
1775-1861 .... DLB-116

Busch, Frederick 1941- DLB-6

Busch, Niven 1903-1991 .... DLB-44

Bushnell, Horace 1802-1876 ....DS-13

Bussieres, Arthur de 1877-1913 ..... DLB-92

Butler, Juan 1942-1981 ....DLB-53

Butler, Octavia E. 1947- DLB-33

Butler, Robert Owen 1945- DLB-173

Butler, Samuel 1613-1680 ..... DLB-101, 126

Butler, Samuel 1835-1902 ... DLB-18, 57, 174

Butler, William Francis
1838-1910 ....DLB-166

Butler, E. H., and Company DLB-49

Butor, Michel 1926- DLB-83

Butter, Nathaniel
[publishing house] DLB-170

Butterworth, Hezekiah 1839-1905 ....DLB-42

Buttitta, Ignazio 1899- DLB-114

Byars, Betsy 1928- DLB-52

Byatt, A. S. 1936- DLB-14

Byles, Mather 1707-1788 ....DLB-24

Bynneman, Henry
[publishing house] DLB-170

Bynner, Witter 1881-1968 ....DLB-54

Byrd, William circa 1543-1623 ......DLB-172

Byrd, William II 1674-1744 .... DLB-24, 140

Byrne, John Keyes (see Leonard, Hugh)

Byron, George Gordon, Lord
1788-1824 .... DLB-96, 110

C

Caballero Bonald, José Manuel
1926- DLB-108

Cabañero, Eladio 1930- DLB-134

Cabell, James Branch
1879-1958 .... DLB-9, 78

Cabeza de Baca, Manuel
1853-1915 ....DLB-122

Cabeza de Baca Gilbert, Fabiola
1898- DLB-122

Cable, George Washington
1844-1925 .... DLB-12, 74; DS-13

Cabrera, Lydia 1900-1991 ....DLB-145

Cabrera Infante, Guillermo
1929- DLB-113

Cadell [publishing house] DLB-154

Cady, Edwin H. 1917- DLB-103

Caedmon flourished 658-680 ....DLB-146

Caedmon School circa 660-899 ....DLB-146

Cahan, Abraham
1860-1951 .... DLB-9, 25, 28

Cain, George 1943- DLB-33

Caldecott, Randolph 1846-1886 .....DLB-163

Calder, John (Publishers), Limited DLB-112

Caldwell, Ben 1937- DLB-38

Caldwell, Erskine 1903-1987 ...... DLB-9, 86

Caldwell, H. M., Company DLB-49

Calhoun, John C. 1782-1850 .... DLB-3

Calisher, Hortense 1911- DLB-2

A Call to Letters and an Invitation to the Electric Chair, by Siegfried Mandel DLB-75

Callaghan, Morley 1903-1990 ...... DLB-68

Callaloo Y-87

Calmer, Edgar 1907- DLB-4

Calverley, C. S. 1831-1884 .... DLB-35

Calvert, George Henry 1803-1889 .... DLB-1, 64

Cambridge Press DLB-49

Cambridge Songs (Carmina Cantabrigensia) circa 1050 .... DLB-148

Cambridge University Press DLB-170

Camden, William 1551-1623 .... DLB-172

Camden House: An Interview with James Hardin Y-92

Cameron, Eleanor 1912- DLB-52

Cameron, George Frederick 1854-1885 .... DLB-99

Cameron, Lucy Lyttelton 1781-1858 .... DLB-163

Cameron, William Bleasdell 1862-1951 .... DLB-99

Camm, John 1718-1778 .... DLB-31

Campana, Dino 1885-1932 .... DLB-114

Campbell, Gabrielle Margaret Vere (see Shearing, Joseph, and Bowen, Marjorie)

Campbell, James Dykes 1838-1895 .... DLB-144

Campbell, James Edwin 1867-1896 .... DLB-50

Campbell, John 1653-1728 .... DLB-43

Campbell, John W., Jr. 1910-1971 .... DLB-8

Campbell, Roy 1901-1957 .... DLB-20

Campbell, Thomas 1777-1844 .... DLB-93, 144

Campbell, William Wilfred 1858-1918 .... DLB-92

Campion, Edmund 1539-1581 ...... DLB-167

Campion, Thomas 1567-1620 .... DLB-58, 172

Camus, Albert 1913-1960 .... DLB-72

Canby, Henry Seidel 1878-1961 ..... DLB-91

Candelaria, Cordelia 1943- DLB-82

Candelaria, Nash 1928- DLB-82

Candour in English Fiction (1890), by Thomas Hardy DLB-18

Canetti, Elias 1905-1994 .... DLB-85, 124

Canham, Erwin Dain 1904-1982 .... DLB-127

Canitz, Friedrich Rudolph Ludwig von 1654-1699 .... DLB-168

Cankar, Ivan 1876-1918 .... DLB-147

Cannan, Gilbert 1884-1955 .... DLB-10

Cannell, Kathleen 1891-1974 .... DLB-4

Cannell, Skipwith 1887-1957 ...... DLB-45

Canning, George 1770-1827 ...... DLB-158

Cannon, Jimmy 1910-1973 .... DLB-171

Cantwell, Robert 1908-1978 .... DLB-9

Cape, Jonathan, and Harrison Smith [publishing house] DLB-46

Cape, Jonathan, Limited DLB-112

Capen, Joseph 1658-1725 .... DLB-24

Capes, Bernard 1854-1918 .... DLB-156

Capote, Truman 1924-1984 .... DLB-2; Y-80, 84

Caproni, Giorgio 1912-1990 .... DLB-128

Cardarelli, Vincenzo 1887-1959 .... DLB-114

Cárdenas, Reyes 1948- DLB-122

Cardinal, Marie 1929- DLB-83

Carew, Jan 1920- DLB-157

Carew, Thomas 1594 or 1595-1640 .... DLB-126

Carey, Henry circa 1687-1689-1743 .... DLB-84

Carey, Mathew 1760-1839 ...... DLB-37, 73

Carey and Hart DLB-49

Carey, M., and Company DLB-49

Carlell, Lodowick 1602-1675 .... DLB-58

Carleton, William 1794-1869 ...... DLB-159

Carleton, G. W. [publishing house] DLB-49

Carlile, Richard 1790-1843 .... DLB-110, 158

Carlyle, Jane Welsh 1801-1866 ..... DLB-55

Carlyle, Thomas 1795-1881 .... DLB-55, 144

Carman, Bliss 1861-1929 .... DLB-92

Carmina Burana circa 1230 .... DLB-138

Carnero, Guillermo 1947- DLB-108

Carossa, Hans 1878-1956 .... DLB-66

Carpenter, Humphrey 1946- DLB-155

Carpenter, Stephen Cullen ?-1820? DLB-73

Carpentier, Alejo 1904-1980 .... DLB-113

Carrier, Roch 1937- DLB-53

Carrillo, Adolfo 1855-1926 .... DLB-122

Carroll, Gladys Hasty 1904- DLB-9

Carroll, John 1735-1815 .... DLB-37

Carroll, John 1809-1884 .... DLB-99

Carroll, Lewis 1832-1898 ...... DLB-18, 163

Carroll, Paul 1927- DLB-16

Carroll, Paul Vincent 1900-1968 ..... DLB-10

Carroll and Graf Publishers DLB-46

Carruth, Hayden 1921- DLB-5, 165

Carryl, Charles E. 1841-1920 .... DLB-42

Carswell, Catherine 1879-1946 ...... DLB-36

Carter, Angela 1940-1992 .... DLB-14

Carter, Elizabeth 1717-1806 .... DLB-109

Carter, Henry (see Leslie, Frank)

Carter, Hodding, Jr. 1907-1972 ..... DLB-127

Carter, Landon 1710-1778 .... DLB-31

Carter, Lin 1930- Y-81

Carter, Martin 1927- DLB-117

Carter and Hendee DLB-49

Carter, Robert, and Brothers DLB-49

Cartwright, John 1740-1824 .... DLB-158

Cartwright, William circa 1611-1643 .... DLB-126

Caruthers, William Alexander 1802-1846 .... DLB-3

Carver, Jonathan 1710-1780 .... DLB-31

Carver, Raymond 1938-1988 .... DLB-130; Y-84, 88

Cary, Joyce 1888-1957 .... DLB-15, 100

Cary, Patrick 1623?-1657 .... DLB-131

Casey, Juanita 1925- DLB-14

Casey, Michael 1947- DLB-5

Cassady, Carolyn 1923- DLB-16

Cassady, Neal 1926-1968 .... DLB-16

Cassell and Company DLB-106

Cassell Publishing Company DLB-49

Cassill, R. V. 1919- DLB-6

Cassity, Turner 1929- DLB-105

The Castle of Perserverance circa 1400-1425 .... DLB-146

Castellano, Olivia 1944- DLB-122

Castellanos, Rosario 1925-1974 .... DLB-113

Castillo, Ana 1953- DLB-122

Castlemon, Harry (see Fosdick, Charles Austin)

Caswall, Edward 1814-1878 .... DLB-32

Catacalos, Rosemary 1944- DLB-122

Cather, Willa
  1873-1947 .... DLB-9, 54, 78; DS-1

Catherine II (Ekaterina Alekseevna), "The
  Great," Empress of Russia
  1729-1796 .... DLB-150

Catherwood, Mary Hartwell
  1847-1902 .... DLB-78

Catledge, Turner 1901-1983 .... DLB-127

Cattafi, Bartolo 1922-1979 .... DLB-128

Catton, Bruce 1899-1978 .... DLB-17

Causley, Charles 1917- DLB-27

Caute, David 1936- DLB-14

Cavendish, Duchess of Newcastle,
  Margaret Lucas 1623-1673 .... DLB-131

Cawein, Madison 1865-1914 .... DLB-54

The Caxton Printers, Limited DLB-46

Caxton, William
  [publishing house] DLB-170

Cayrol, Jean 1911- DLB-83

Cecil, Lord David 1902-1986 ...... DLB-155

Celan, Paul 1920-1970 .... DLB-69

Celaya, Gabriel 1911-1991 .... DLB-108

Céline, Louis-Ferdinand
  1894-1961 .... DLB-72

The Celtic Background to Medieval English
  Literature DLB-146

Center for Bibliographical Studies and
  Research at the University of
  California, Riverside Y-91

The Center for the Book in the Library
  of Congress Y-93

Center for the Book Research Y-84

Centlivre, Susanna 1669?-1723 ..... DLB-84

The Century Company DLB-49

Cernuda, Luis 1902-1963 .... DLB-134

Cervantes, Lorna Dee 1954- DLB-82

Chacel, Rosa 1898- DLB-134

Chacón, Eusebio 1869-1948 .... DLB-82

Chacón, Felipe Maximiliano
  1873-? DLB-82

Chadwyck-Healey's Full-Text Literary Databases: Editing Commercial Databases of
  Primary Literary Texts Y-95

Challans, Eileen Mary (see Renault, Mary)

Chalmers, George 1742-1825 .... DLB-30

Chaloner, Sir Thomas
  1520-1565 .... DLB-167

Chamberlain, Samuel S.
  1851-1916 .... DLB-25

Chamberland, Paul 1939- DLB-60

Chamberlin, William Henry
  1897-1969 .... DLB-29

Chambers, Charles Haddon
  1860-1921 .... DLB-10

Chambers, W. and R.
  [publishing house] DLB-106

Chamisso, Albert von
  1781-1838 .... DLB-90

Champfleury 1821-1889 .... DLB-119

Chandler, Harry 1864-1944 .... DLB-29

Chandler, Norman 1899-1973 ..... DLB-127

Chandler, Otis 1927- DLB-127

Chandler, Raymond 1888-1959 .... DS-6

Channing, Edward 1856-1931 .... DLB-17

Channing, Edward Tyrrell
  1790-1856 .... DLB-1, 59

Channing, William Ellery
  1780-1842 .... DLB-1, 59

Channing, William Ellery, II
  1817-1901 .... DLB-1

Channing, William Henry
  1810-1884 .... DLB-1, 59

Chaplin, Charlie 1889-1977 .... DLB-44

Chapman, George
  1559 or 1560 - 1634 .... DLB-62, 121

Chapman, John DLB-106

Chapman, William 1850-1917 .... DLB-99

Chapman and Hall DLB-106

Chappell, Fred 1936- DLB-6, 105

Chappell, Fred, A Detail
  in a Poem DLB-105

Charbonneau, Jean 1875-1960 ...... DLB-92

Charbonneau, Robert 1911-1967 .... DLB-68

Charles, Gerda 1914- DLB-14

Charles, William
  [publishing house] DLB-49

The Charles Wood Affair:
  A Playwright Revived Y-83

Charlotte Forten: Pages from
  her Diary DLB-50

Charteris, Leslie 1907-1993 .... DLB-77

Charyn, Jerome 1937- Y-83

Chase, Borden 1900-1971 .... DLB-26

Chase, Edna Woolman
  1877-1957 .... DLB-91

Chase-Riboud, Barbara 1936- DLB-33

Chateaubriand, François-René de
  1768-1848 .... DLB-119

Chatterton, Thomas 1752-1770 ..... DLB-109

Chatto and Windus DLB-106

Chaucer, Geoffrey 1340?-1400 ..... DLB-146

Chauncy, Charles 1705-1787 .... DLB-24

Chauveau, Pierre-Joseph-Olivier
  1820-1890 .... DLB-99

Chávez, Denise 1948- DLB-122

Chávez, Fray Angélico 1910- DLB-82

Chayefsky, Paddy
  1923-1981 .... DLB-7, 44; Y-81

Cheever, Ezekiel 1615-1708 .... DLB-24

Cheever, George Barrell
  1807-1890 .... DLB-59

Cheever, John
  1912-1982 .... DLB-2, 102; Y-80, 82

Cheever, Susan 1943- Y-82

Cheke, Sir John 1514-1557 .... DLB-132

Chelsea House DLB-46

Cheney, Ednah Dow (Littlehale)
  1824-1904 .... DLB-1

Cheney, Harriet Vaughn
  1796-1889 .... DLB-99

Cherry, Kelly 1940 .... Y-83

Cherryh, C. J. 1942- Y-80

Chesnutt, Charles Waddell
  1858-1932 .... DLB-12, 50, 78

Chester, Alfred 1928-1971 .... DLB-130

Chester, George Randolph
  1869-1924 .... DLB-78

The Chester Plays circa 1505-1532;
  revisions until 1575 .... DLB-146

Chesterfield, Philip Dormer Stanhope,
  Fourth Earl of 1694-1773 .... DLB-104

Chesterton, G. K.
  1874-1936 .... DLB-10, 19, 34, 70, 98, 149

Chettle, Henry
  circa 1560-circa 1607 .... DLB-136

Chew, Ada Nield 1870-1945 ...... DLB-135

Cheyney, Edward P. 1861-1947 ..... DLB-47

Chicano History DLB-82

Chicano Language DLB-82

Child, Francis James
  1825-1896 .... DLB-1, 64

Child, Lydia Maria
  1802-1880 .... DLB-1, 74

Child, Philip 1898-1978 ....DLB-68

Childers, Erskine 1870-1922 ....DLB-70

Children's Book Awards
 and Prizes DLB-61

Children's Illustrators,
 1800-1880 ....DLB-163

Childress, Alice 1920-1994 ....DLB-7, 38

Childs, George W. 1829-1894 ....DLB-23

Chilton Book Company DLB-46

Chinweizu 1943- DLB-157

Chitham, Edward 1932- DLB-155

Chittenden, Hiram Martin
 1858-1917 ....DLB-47

Chivers, Thomas Holley
 1809-1858 ....DLB-3

Chopin, Kate 1850-1904 ....DLB-12, 78

Chopin, Rene 1885-1953 ....DLB-92

Choquette, Adrienne 1915-1973 .....DLB-68

Choquette, Robert 1905- DLB-68

The Christian Publishing
 Company DLB-49

Christie, Agatha 1890-1976 .....DLB-13, 77

Christus und die Samariterin
 circa 950 ....DLB-148

Chulkov, Mikhail Dmitrievich
 1743?-1792 ....DLB-150

Church, Benjamin 1734-1778 ....DLB-31

Church, Francis Pharcellus
 1839-1906 ....DLB-79

Church, William Conant
 1836-1917 ....DLB-79

Churchill, Caryl 1938- DLB-13

Churchill, Charles 1731-1764 ......DLB-109

Churchill, Sir Winston
 1874-1965 ....DLB-100

Churchyard, Thomas
 1520?-1604 ....DLB-132

Churton, E., and Company DLB-106

Chute, Marchette 1909-1994 ......DLB-103

Ciardi, John 1916-1986 ....DLB 5; Y 86

Cibber, Colley 1671-1757 ....DLB-84

Cima, Annalisa 1941- DLB-128

Cirese, Eugenio 1884-1955 ....DLB-114

Cisneros, Sandra 1954- DLB-122, 152

City Lights Books DLB-46

Cixous, Hélène 1937- DLB-83

Clampitt, Amy 1920-1994 ....DLB-105

Clapper, Raymond 1892-1944 ....DLB-29

Clare, John 1793-1864 ....DLB-55, 96

Clarendon, Edward Hyde, Earl of
 1609-1674 ....DLB-101

Clark, Alfred Alexander Gordon
 (see Hare, Cyril)

Clark, Ann Nolan 1896- DLB-52

Clark, Catherine Anthony
 1892-1977 ....DLB-68

Clark, Charles Heber
 1841-1915 ....DLB-11

Clark, Davis Wasgatt 1812-1871 ....DLB-79

Clark, Eleanor 1913- DLB-6

Clark, J. P. 1935- DLB-117

Clark, Lewis Gaylord
 1808-1873 ....DLB-3, 64, 73

Clark, Walter Van Tilburg
 1909-1971 ....DLB-9

Clark, C. M., Publishing
 Company DLB-46

Clarke, Austin 1896-1974 ....DLB-10, 20

Clarke, Austin C. 1934- DLB-53, 125

Clarke, Gillian 1937- DLB-40

Clarke, James Freeman
 1810-1888 ....DLB-1, 59

Clarke, Pauline 1921- DLB-161

Clarke, Rebecca Sophia
 1833-1906 ....DLB-42

Clarke, Robert, and Company DLB-49

Clarkson, Thomas 1760-1846 ......DLB-158

Claudius, Matthias 1740-1815 ......DLB-97

Clausen, Andy 1943- DLB-16

Claxton, Remsen and
 Haffelfinger DLB-49

Clay, Cassius Marcellus
 1810-1903 ....DLB-43

Cleary, Beverly 1916- DLB-52

Cleaver, Vera 1919- and
 Cleaver, Bill 1920-1981 ....DLB-52

Cleland, John 1710-1789 ....DLB-39

Clemens, Samuel Langhorne
 1835-1910 ....DLB-11, 12, 23, 64, 74

Clement, Hal 1922- DLB-8

Clemo, Jack 1916- DLB-27

Cleveland, John 1613-1658 ....DLB-126

Cliff, Michelle 1946- DLB-157

Clifford, Lady Anne 1590-1676 ....DLB-151

Clifford, James L. 1901-1978 ......DLB-103

Clifford, Lucy 1853?-1929 ....DLB-135, 141

Clifton, Lucille 1936- DLB-5, 41

Clode, Edward J.
 [publishing house] DLB-46

Clough, Arthur Hugh 1819-1861 ....DLB-32

Cloutier, Cécile 1930- DLB-60

Clutton-Brock, Arthur
 1868-1924 ....DLB-98

Coates, Robert M.
 1897-1973 ....DLB-4, 9, 102

Coatsworth, Elizabeth 1893- DLB-22

Cobb, Charles E., Jr. 1943- DLB-41

Cobb, Frank I. 1869-1923 ....DLB-25

Cobb, Irvin S.
 1876-1944 ....DLB-11, 25, 86

Cobbett, William 1763-1835 ....DLB-43, 107

Cobbledick, Gordon 1898-1969 ....DLB-171

Cochran, Thomas C. 1902- DLB-17

Cochrane, Elizabeth 1867-1922 ......DLB-25

Cockerill, John A. 1845-1896 ....DLB-23

Cocteau, Jean 1889-1963 ....DLB-65

Coderre, Emile (see Jean Narrache)

Coffee, Lenore J. 1900?-1984 ....DLB-44

Coffin, Robert P. Tristram
 1892-1955 ....DLB-45

Cogswell, Fred 1917- DLB-60

Cogswell, Mason Fitch
 1761-1830 ....DLB-37

Cohen, Arthur A. 1928-1986 ....DLB-28

Cohen, Leonard 1934- DLB-53

Cohen, Matt 1942- DLB-53

Colden, Cadwallader
 1688-1776 ....DLB-24, 30

Cole, Barry 1936- DLB-14

Cole, George Watson
 1850-1939 ....DLB-140

Colegate, Isabel 1931- DLB-14

Coleman, Emily Holmes
 1899-1974 ....DLB-4

Coleman, Wanda 1946- DLB-130

Coleridge, Hartley 1796-1849 ....DLB-96

Coleridge, Mary 1861-1907 ......DLB-19, 98

Coleridge, Samuel Taylor
 1772-1834 ....DLB-93, 107

Colet, John 1467-1519 ....DLB-132

Colette 1873-1954 ....DLB-65

Colette, Sidonie Gabrielle (see Colette)

Colinas, Antonio 1946- DLB-134

Collier, John 1901-1980 ....DLB-77

Collier, Mary 1690-1762 ....DLB-95

349

Collier, Robert J. 1876-1918 .... DLB-91

Collier, P. F. [publishing house] DLB-49

Collin and Small DLB-49

Collingwood, W. G. 1854-1932 .... DLB-149

Collins, An floruit circa 1653 ...... DLB-131

Collins, Merle 1950- DLB-157

Collins, Mortimer 1827-1876 .... DLB-21, 35

Collins, Wilkie 1824-1889 .. DLB-18, 70, 159

Collins, William 1721-1759 ...... DLB-109

Collins, William, Sons and Company DLB-154

Collins, Isaac [publishing house] DLB-49

Collyer, Mary 1716?-1763? DLB-39

Colman, Benjamin 1673-1747 ...... DLB-24

Colman, George, the Elder 1732-1794 .... DLB-89

Colman, George, the Younger 1762-1836 .... DLB-89

Colman, S. [publishing house] DLB-49

Colombo, John Robert 1936- DLB-53

Colquhoun, Patrick 1745-1820 ..... DLB-158

Colter, Cyrus 1910- DLB-33

Colum, Padraic 1881-1972 .... DLB-19

Colvin, Sir Sidney 1845-1927 ...... DLB-149

Colwin, Laurie 1944-1992 .... Y-80

Comden, Betty 1919- and Green, Adolph 1918- DLB-44

Comi, Girolamo 1890-1968 .... DLB-114

The Comic Tradition Continued [in the British Novel] DLB-15

Commager, Henry Steele 1902- DLB-17

The Commercialization of the Image of Revolt, by Kenneth Rexroth DLB-16

Community and Commentators: Black Theatre and Its Critics DLB-38

Compton-Burnett, Ivy 1884?-1969 .... DLB-36

Conan, Laure 1845-1924 .... DLB-99

Conde, Carmen 1901- DLB-108

Conference on Modern Biography Y-85

Congreve, William 1670-1729 .... DLB-39, 84

Conkey, W. B., Company DLB-49

Connell, Evan S., Jr. 1924- ... DLB-2; Y-81

Connelly, Marc 1890-1980 ..... DLB-7; Y-80

Connolly, Cyril 1903-1974 .... DLB-98

Connolly, James B. 1868-1957 .... DLB-78

Connor, Ralph 1860-1937 .... DLB-92

Connor, Tony 1930- DLB-40

Conquest, Robert 1917- DLB-27

Conrad, Joseph 1857-1924 .... DLB-10, 34, 98, 156

Conrad, John, and Company DLB-49

Conroy, Jack 1899-1990 .... Y-81

Conroy, Pat 1945- DLB-6

The Consolidation of Opinion: Critical Responses to the Modernists DLB-36

Constable, Henry 1562-1613 .... DLB-136

Constable and Company Limited DLB-112

Constable, Archibald, and Company DLB-154

Constant, Benjamin 1767-1830 ..... DLB-119

Constant de Rebecque, Henri-Benjamin de (see Constant, Benjamin)

Constantine, David 1944- DLB-40

Constantin-Weyer, Maurice 1881-1964 .... DLB-92

Contempo Caravan: Kites in a Windstorm Y-85

A Contemporary Flourescence of Chicano Literature Y-84

The Continental Publishing Company DLB-49

A Conversation with Chaim Potok Y-84

Conversations with Editors Y-95

Conversations with Publishers I: An Interview with Patrick O'Connor Y-84

Conversations with Publishers II: An Interview with Charles Scribner III Y-94

Conversations with Publishers III: An Interview with Donald Lamm Y-95

Conversations with Rare Book Dealers I: An Interview with Glenn Horowitz Y-90

Conversations with Rare Book Dealers II: An Interview with Ralph Sipper Y-94

The Conversion of an Unpolitical Man, by W. H. Bruford DLB-66

Conway, Moncure Daniel 1832-1907 .... DLB-1

Cook, Ebenezer circa 1667-circa 1732 .... DLB-24

Cook, Edward Tyas 1857-1919 .... DLB-149

Cook, Michael 1933- DLB-53

Cook, David C., Publishing Company DLB-49

Cooke, George Willis 1848-1923 .... DLB-71

Cooke, Increase, and Company DLB-49

Cooke, John Esten 1830-1886 .... DLB-3

Cooke, Philip Pendleton 1816-1850 .... DLB-3, 59

Cooke, Rose Terry 1827-1892 .... DLB-12, 74

Coolbrith, Ina 1841-1928 .... DLB-54

Cooley, Peter 1940- DLB-105

Cooley, Peter, Into the Mirror DLB-105

Coolidge, Susan (see Woolsey, Sarah Chauncy)

Coolidge, George [publishing house] DLB-49

Cooper, Giles 1918-1966 .... DLB-13

Cooper, James Fenimore 1789-1851 ... DLB-3

Cooper, Kent 1880-1965 .... DLB-29

Cooper, Susan 1935- DLB-161

Cooper, William [publishing house] DLB-170

Coote, J. [publishing house] DLB-154

Coover, Robert 1932- DLB-2; Y-81

Copeland and Day DLB-49

Copland, Robert 1470?-1548 .... DLB-136

Coppard, A. E. 1878-1957 .... DLB-162

Coppel, Alfred 1921- Y-83

Coppola, Francis Ford 1939- DLB-44

Corazzini, Sergio 1886-1907 .... DLB-114

Corbett, Richard 1582-1635 .... DLB-121

Corcoran, Barbara 1911- DLB-52

Corelli, Marie 1855-1924 ...... DLB-34, 156

Corle, Edwin 1906-1956 .... Y-85

Corman, Cid 1924- DLB-5

Cormier, Robert 1925- DLB-52

Corn, Alfred 1943- DLB-120; Y-80

Cornish, Sam 1935- DLB-41

Cornish, William circa 1465-circa 1524 .... DLB-132

Cornwall, Barry (see Procter, Bryan Waller)

Cornwallis, Sir William, the Younger circa 1579-1614 .... DLB-151

Cornwell, David John Moore (see le Carré, John)

Corpi, Lucha 1945- DLB-82

Corrington, John William 1932- DLB-6

Corrothers, James D. 1869-1917 .... DLB-50

Corso, Gregory 1930- DLB-5, 16

Cortázar, Julio 1914-1984 .... DLB-113

Cortez, Jayne 1936- DLB-41

Corvinus, Gottlieb Siegmund 1677-1746 ....DLB-168

Corvo, Baron (see Rolfe, Frederick William)

Cory, Annie Sophie (see Cross, Victoria)

Cory, William Johnson 1823-1892 ....DLB-35

Coryate, Thomas 1577?-1617 ....DLB-151, 172

Cosin, John 1595-1672 ....DLB-151

Cosmopolitan Book Corporation DLB-46

Costain, Thomas B. 1885-1965 ....DLB-9

Coste, Donat 1912-1957 ....DLB-88

Costello, Louisa Stuart 1799-1870 ..DLB-166

Cota-Cárdenas, Margarita 1941- DLB-122

Cotter, Joseph Seamon, Sr. 1861-1949 ....DLB-50

Cotter, Joseph Seamon, Jr. 1895-1919 ....DLB-50

Cottle, Joseph [publishing house] ...DLB-154

Cotton, Charles 1630-1687 ....DLB-131

Cotton, John 1584-1652 ....DLB-24

Coulter, John 1888-1980 ....DLB-68

Cournos, John 1881-1966 ....DLB-54

Cousins, Margaret 1905- DLB-137

Cousins, Norman 1915-1990 .....DLB-137

Coventry, Francis 1725-1754 ....DLB-39

Coverdale, Miles 1487 or 1488-1569 ....DLB-167

Coverly, N. [publishing house] DLB-49

Covici-Friede DLB-46

Coward, Noel 1899-1973 ....DLB-10

Coward, McCann and Geoghegan DLB-46

Cowles, Gardner 1861-1946 ....DLB-29

Cowles, Gardner ("Mike"), Jr. 1903-1985 ....DLB-127, 137

Cowley, Abraham 1618-1667 ....DLB-131, 151

Cowley, Hannah 1743-1809 ....DLB-89

Cowley, Malcolm 1898-1989 .... DLB-4, 48; Y-81, 89

Cowper, William 1731-1800 ....DLB-104, 109

Cox, A. B. (see Berkeley, Anthony)

Cox, James McMahon 1903-1974 ....DLB-127

Cox, James Middleton 1870-1957 ....DLB-127

Cox, Palmer 1840-1924 ....DLB-42

Coxe, Louis 1918-1993 .... DLB-5

Coxe, Tench 1755-1824 ....DLB-37

Cozzens, James Gould 1903-1978 ....DLB-9; Y-84; DS-2

Crabbe, George 1754-1832 ....DLB-93

Crackanthorpe, Hubert 1870-1896 ....DLB-135

Craddock, Charles Egbert (see Murfree, Mary N.)

Cradock, Thomas 1718-1770 ....DLB-31

Craig, Daniel H. 1811-1895 ....DLB-43

Craik, Dinah Maria 1826-1887 ....DLB-35, 136

Cranch, Christopher Pearse 1813-1892 ....DLB-1, 42

Crane, Hart 1899-1932 ....DLB-4, 48

Crane, R. S. 1886-1967 ....DLB-63

Crane, Stephen 1871-1900 ...DLB-12, 54, 78

Crane, Walter 1845-1915 ....DLB-163

Cranmer, Thomas 1489-1556 ......DLB-132

Crapsey, Adelaide 1878-1914 ....DLB-54

Crashaw, Richard 1612 or 1613-1649 ....DLB-126

Craven, Avery 1885-1980 ....DLB-17

Crawford, Charles 1752-circa 1815 ....DLB-31

Crawford, F. Marion 1854-1909 ....DLB-71

Crawford, Isabel Valancy 1850-1887 ....DLB-92

Crawley, Alan 1887-1975 ....DLB-68

Crayon, Geoffrey (see Irving, Washington)

Creamer, Robert W. 1922- DLB-171

Creasey, John 1908-1973 ....DLB-77

Creative Age Press DLB-46

Creech, William [publishing house] DLB-154

Creede, Thomas [publishing house] DLB-170

Creel, George 1876-1953 ....DLB-25

Creeley, Robert 1926- DLB-5, 16, 169

Creelman, James 1859-1915 ....DLB-23

Cregan, David 1931- DLB-13

Creighton, Donald Grant 1902-1979 ....DLB-88

Cremazie, Octave 1827-1879 ....DLB-99

Crémer, Victoriano 1909?- DLB-108

Crescas, Hasdai circa 1340-1412? DLB-115

Crespo, Angel 1926- DLB-134

Cresset Press DLB-112

Cresswell, Helen 1934- DLB-161

Crèvecoeur, Michel Guillaume Jean de 1735-1813 ....DLB-37

Crews, Harry 1935- DLB-6, 143

Crichton, Michael 1942- Y-81

A Crisis of Culture: The Changing Role of Religion in the New Republic DLB-37

Crispin, Edmund 1921-1978 ....DLB-87

Cristofer, Michael 1946- DLB-7

"The Critic as Artist" (1891), by Oscar Wilde DLB-57

"Criticism In Relation To Novels" (1863), by G. H. Lewes DLB-21

Crnjanski, Miloš 1893-1977 ....DLB-147

Crockett, David (Davy) 1786-1836 ....DLB-3, 11

Croft-Cooke, Rupert (see Bruce, Leo)

Crofts, Freeman Wills 1879-1957 ....DLB-77

Croker, John Wilson 1780-1857 ....DLB-110

Croly, George 1780-1860 ....DLB-159

Croly, Herbert 1869-1930 ....DLB-91

Croly, Jane Cunningham 1829-1901 ....DLB-23

Crompton, Richmal 1890-1969 .....DLB-160

Crosby, Caresse 1892-1970 ....DLB-48

Crosby, Caresse 1892-1970 and Crosby, Harry 1898-1929 ....DLB-4

Crosby, Harry 1898-1929 ....DLB-48

Cross, Gillian 1945- DLB-161

Cross, Victoria 1868-1952 ....DLB-135

Crossley-Holland, Kevin 1941- DLB-40, 161

Crothers, Rachel 1878-1958 ....DLB-7

Crowell, Thomas Y., Company DLB-49

Crowley, John 1942- Y-82

Crowley, Mart 1935- DLB-7

Crown Publishers DLB-46

Crowne, John 1641-1712 ....DLB-80

Crowninshield, Edward Augustus 1817-1859 ....DLB-140

Crowninshield, Frank 1872-1947 ....DLB-91

Croy, Homer 1883-1965 ....DLB-4

Crumley, James 1939- Y-84

Cruz, Victor Hernández 1949- DLB-41

Cumulative Index

Csokor, Franz Theodor
1885-1969 .... DLB-81

Cuala Press DLB-112

Cullen, Countee 1903-1946 ... DLB-4, 48, 51

Culler, Jonathan D. 1944- DLB-67

The Cult of Biography
Excerpts from the Second Folio Debate:
"Biographies are generally a disease of
English Literature" – Germaine Greer,
Victoria Glendinning, Auberon Waugh,
and Richard Holmes Y-86

Cumberland, Richard 1732-1811 .... DLB-89

Cummings, Constance Gordon
1837-1924 .... DLB-174

Cummings, E. E. 1894-1962 ..... DLB-4, 48

Cummings, Ray 1887-1957 .... DLB-8

Cummings and Hilliard DLB-49

Cummins, Maria Susanna
1827-1866 .... DLB-42

Cundall, Joseph
[publishing house] DLB-106

Cuney, Waring 1906-1976 .... DLB-51

Cuney-Hare, Maude 1874-1936 ..... DLB-52

Cunningham, Allan
1784-1842 .... DLB-116, 144

Cunningham, J. V. 1911- DLB-5

Cunningham, Peter F.
[publishing house] DLB-49

Cunquiero, Alvaro 1911-1981 ..... DLB-134

Cuomo, George 1929- Y-80

Cupples and Leon DLB-46

Cupples, Upham and Company DLB-49

Cuppy, Will 1884-1949 .... DLB-11

Curll, Edmund
[publishing house] DLB-154

Currie, James 1756-1805 .... DLB-142

Currie, Mary Montgomerie Lamb Singleton,
Lady Currie (see Fane, Violet)

Cursor Mundi circa 1300 .... DLB-146

Curti, Merle E. 1897- DLB-17

Curtis, Anthony 1926- DLB-155

Curtis, Cyrus H. K. 1850-1933 ..... DLB-91

Curtis, George William
1824-1892 .... DLB-1, 43

Curzon, Robert 1810-1873 .... DLB-166

Curzon, Sarah Anne 1833-1898 ..... DLB-99

Cynewulf circa 770-840 .... DLB-146

Czcpko, Daniel 1605-1660 .... DLB-164

D

D. M. Thomas: The Plagiarism
Controversy Y-82

Dabit, Eugène 1898-1936 .... DLB-65

Daborne, Robert circa 1580-1628 .... DLB-58

Dacey, Philip 1939- DLB-105

Dacey, Philip, Eyes Across Centuries:
Contemporary Poetry and "That
Vision Thing" DLB-105

Dach, Simon 1605-1659 .... DLB-164

Daggett, Rollin M. 1831-1901 .... DLB-79

D'Aguiar, Fred 1960- DLB-157

Dahl, Roald 1916-1990 .... DLB-139

Dahlberg, Edward 1900-1977 ..... DLB-48

Dahn, Felix 1834-1912 .... DLB-129

Dale, Peter 1938- DLB-40

Daley, Arthur 1904-1974 .... DLB-171

Dall, Caroline Wells (Healey)
1822-1912 .... DLB-1

Dallas, E. S. 1828-1879 .... DLB-55

The Dallas Theater Center DLB-7

D'Alton, Louis 1900-1951 .... DLB-10

Daly, T. A. 1871-1948 .... DLB-11

Damon, S. Foster 1893-1971 .... DLB-45

Damrell, William S.
[publishing house] DLB-49

Dana, Charles A. 1819-1897 ..... DLB-3, 23

Dana, Richard Henry, Jr
1815-1882 .... DLB-1

Dandridge, Ray Garfield DLB-51

Dane, Clemence 1887-1965 .... DLB-10

Danforth, John 1660-1730 .... DLB-24

Danforth, Samuel, I 1626-1674 ..... DLB-24

Danforth, Samuel, II 1666-1727 ..... DLB-24

Dangerous Years: London Theater,
1939-1945 .... DLB-10

Daniel, John M. 1825-1865 .... DLB-43

Daniel, Samuel
1562 or 1563-1619 .... DLB-62

Daniel Press DLB-106

Daniells, Roy 1902-1979 .... DLB-68

Daniels, Jim 1956- DLB-120

Daniels, Jonathan 1902-1981 ..... DLB-127

Daniels, Josephus 1862-1948 ..... DLB-29

Dannay, Frederic 1905-1982 and
Manfred B. Lee 1905-1971 ..... DLB-137

Danner, Margaret Esse 1915- DLB-41

Danter, John [publishing house] DLB-170

Dantin, Louis 1865-1945 .... DLB-92

Danzig, Allison 1898-1987 .... DLB-171

D'Arcy, Ella circa 1857-1937 .... DLB-135

Darley, George 1795-1846 .... DLB-96

Darwin, Charles 1809-1882 .... DLB-57, 166

Darwin, Erasmus 1731-1802 .... DLB-93

Daryush, Elizabeth 1887-1977 .... DLB-20

Dashkova, Ekaterina Romanovna
(née Vorontsova) 1743-1810 .... DLB-150

Dashwood, Edmée Elizabeth Monica
de la Pasture (see Delafield, E. M.)

Daudet, Alphonse 1840-1897 .... DLB-123

d'Aulaire, Edgar Parin 1898- and
d'Aulaire, Ingri 1904- DLB-22

Davenant, Sir William
1606-1668 .... DLB-58, 126

Davenport, Guy 1927- DLB-130

Davenport, Robert ?-? DLB-58

Daves, Delmer 1904-1977 .... DLB-26

Davey, Frank 1940- DLB-53

Davidson, Avram 1923-1993 .... DLB-8

Davidson, Donald 1893-1968 .... DLB-45

Davidson, John 1857-1909 .... DLB-19

Davidson, Lionel 1922- DLB-14

Davie, Donald 1922- DLB-27

Davie, Elspeth 1919- DLB-139

Davies, Sir John 1569-1626 .... DLB-172

Davies, John, of Hereford
1565?-1618 .... DLB-121

Davies, Rhys 1901-1978 .... DLB-139

Davies, Robertson 1913- DLB-68

Davies, Samuel 1723-1761 .... DLB-31

Davies, Thomas 1712?-1785 ... DLB-142, 154

Davies, W. H. 1871-1940 ...... DLB-19, 174

Davies, Peter, Limited DLB-112

Daviot, Gordon 1896?-1952 .... DLB-10
(see also Tey, Josephine)

Davis, Charles A. 1795-1867 .... DLB-11

Davis, Clyde Brion 1894-1962 .... DLB-9

Davis, Dick 1945- DLB-40

Davis, Frank Marshall 1905-? DLB-51

Davis, H. L. 1894-1960 .... DLB-9

Davis, John 1774-1854 .... DLB-37

Davis, Lydia 1947- DLB-130

Davis, Margaret Thomson 1926- ... DLB-14

Davis, Ossie 1917- DLB-7, 38

Davis, Paxton 1925-1994 ....Y-94

Davis, Rebecca Harding
1831-1910 ....DLB-74

Davis, Richard Harding
1864-1916 ....DLB-12, 23, 78, 79; DS-13

Davis, Samuel Cole 1764-1809 ......DLB-37

Davison, Peter 1928- DLB-5

Davys, Mary 1674-1732 ....DLB-39

DAW Books DLB-46

Dawson, Ernest 1882-1947 ....DLB-140

Dawson, Fielding 1930- DLB-130

Dawson, William 1704-1752 ....DLB-31

Day, Angel flourished 1586 ....DLB-167

Day, Benjamin Henry 1810-1889 ....DLB-43

Day, Clarence 1874-1935 ....DLB-11

Day, Dorothy 1897-1980 ....DLB-29

Day, Frank Parker 1881-1950 ....DLB-92

Day, John circa 1574-circa 1640 .....DLB-62

Day, John [publishing house] DLB-170

Day Lewis, C. 1904-1972 ....DLB-15, 20
(see also Blake, Nicholas)

Day, Thomas 1748-1789 ....DLB-39

Day, The John, Company DLB-46

Day, Mahlon [publishing house] DLB-49

Deacon, William Arthur
1890-1977 ....DLB-68

Deal, Borden 1922-1985 ....DLB-6

de Angeli, Marguerite 1889-1987 ....DLB-22

De Angelis, Milo 1951- DLB-128

De Bow, James Dunwoody Brownson
1820-1867 ....DLB-3, 79

de Bruyn, Günter 1926- DLB-75

de Camp, L. Sprague 1907- DLB-8

The Decay of Lying (1889),
by Oscar Wilde [excerpt] DLB-18

Dedication, *Ferdinand Count Fathom* (1753),
by Tobias Smollett DLB-39

Dedication, *The History of Pompey the Little*
(1751), by Francis Coventry DLB-39

Dedication, *Lasselia* (1723), by Eliza
Haywood [excerpt] DLB-39

Dedication, *The Wanderer* (1814),
by Fanny Burney DLB-39

Dee, John 1527-1609 ....DLB-136

Deeping, George Warwick
1877-1950 ....DLB 153

Defense of *Amelia* (1752), by
Henry Fielding DLB-39

Defoe, Daniel 1660-1731 ....DLB-39, 95, 101

de Fontaine, Felix Gregory
1834-1896 ....DLB-43

De Forest, John William
1826-1906 ....DLB-12

DeFrees, Madeline 1919- DLB-105

DeFrees, Madeline, The Poet's Kaleidoscope:
The Element of Surprise in the Making
of the Poem DLB-105

de Graff, Robert 1895-1981 ....Y-81

de Graft, Joe 1924-1978 ....DLB-117

De Heinrico circa 980? DLB-148

Deighton, Len 1929- DLB-87

DeJong, Meindert 1906-1991 ....DLB-52

Dekker, Thomas
circa 1572-1632 ....DLB-62, 172

Delacorte, Jr., George T.
1894-1991 ....DLB-91

Delafield, E. M. 1890-1943 ....DLB-34

Delahaye, Guy 1888-1969 ....DLB-92

de la Mare, Walter
1873-1956 ....DLB-19, 153, 162

Deland, Margaret 1857-1945 ....DLB-78

Delaney, Shelagh 1939- DLB-13

Delany, Martin Robinson
1812-1885 ....DLB-50

Delany, Samuel R. 1942- DLB-8, 33

de la Roche, Mazo 1879-1961 ....DLB-68

Delbanco, Nicholas 1942- DLB-6

De León, Nephtal 1945- DLB-82

Delgado, Abelardo Barrientos
1931- DLB-82

De Libero, Libero 1906-1981 ......DLB-114

DeLillo, Don 1936- DLB-6, 173

de Lisser H. G. 1878-1944 ....DLB-117

Dell, Floyd 1887-1969 ....DLB-9

Dell Publishing Company DLB-46

delle Grazie, Marie Eugene
1864-1931 ....DLB-81

Deloney, Thomas died 1600 ....DLB-167

del Rey, Lester 1915-1993 ....DLB-8

Del Vecchio, John M. 1947- DS-9

de Man, Paul 1919-1983 ....DLB-67

Demby, William 1922- DLB-33

Deming, Philander 1829-1915 ......DLB-74

Demorest, William Jennings
1822-1895 ....DLB-79

De Morgan, William 1839-1917 ....DLB-153

Denham, Henry
[publishing house] DLB-170

Denham, Sir John
1615-1669 ....DLB-58, 126

Denison, Merrill 1893-1975 ....DLB-92

Denison, T. S., and Company DLB-49

Dennie, Joseph
1768-1812 ....DLB-37, 43, 59, 73

Dennis, John 1658-1734 ....DLB-101

Dennis, Nigel 1912-1989 ....DLB-13, 15

Dent, Tom 1932- DLB-38

Dent, J. M., and Sons DLB-112

Denton, Daniel circa 1626-1703 .....DLB-24

DePaola, Tomie 1934- DLB-61

De Quincey, Thomas
1785-1859 ....DLB-110, 144

Derby, George Horatio
1823-1861 ....DLB-11

Derby, J. C., and Company DLB-49

Derby and Miller DLB-49

Derleth, August 1909-1971 ....DLB-9

The Derrydale Press DLB-46

Derzhavin, Gavriil Romanovich
1743-1816 ....DLB-150

Desaulniers, Gonsalve
1863-1934 ....DLB-92

Desbiens, Jean-Paul 1927- DLB-53

des Forêts, Louis-Rene 1918- DLB-83

DesRochers, Alfred 1901-1978 ......DLB-68

Desrosiers, Léo-Paul 1896-1967 ....DLB-68

Destouches, Louis-Ferdinand
(see Céline, Louis-Ferdinand)

De Tabley, Lord 1835-1895 ....DLB-35

Deutsch, Babette 1895-1982 ....DLB-45

Deutsch, André, Limited DLB-112

Deveaux, Alexis 1948- DLB-38

The Development of the Author's Copyright
in Britain DLB-154

The Development of Lighting in the Staging
of Drama, 1900-1945 ....DLB-10

de Vere, Aubrey 1814-1902 ....DLB-35

Devereux, second Earl of Essex, Robert
1565-1601 ....DLB-136

The Devin-Adair Company DLB-46

De Voto, Bernard 1897-1955 ....DLB-9

De Vries, Peter 1910-1993 ......DLB-6; Y-82

Dewdney, Christopher 1951- DLB-60

Dewdney, Selwyn 1909-1979 ....DLB-68

DeWitt, Robert M., Publisher DLB-49

DeWolfe, Fiske and Company DLB-49

Dexter, Colin 1930- DLB-87

de Young, M. H. 1849-1925 .... DLB-25

Dhlomo, H. I. E. 1903-1956 ...... DLB-157

Dhuoda circa 803-after 843 .... DLB-148

The Dial Press DLB-46

Diamond, I. A. L. 1920-1988 ..... DLB-26

Di Cicco, Pier Giorgio 1949- DLB-60

Dick, Philip K. 1928-1982 .... DLB-8

Dick and Fitzgerald DLB-49

Dickens, Charles
 1812-1870 ..... DLB-21, 55, 70, 159, 166

Dickinson, Peter 1927- DLB-161

Dickey, James
 1923- DLB-5; Y-82, 93; DS-7

Dickey, William 1928-1994 .... DLB-5

Dickinson, Emily 1830-1886 .... DLB-1

Dickinson, John 1732-1808 .... DLB-31

Dickinson, Jonathan 1688-1747 ..... DLB-24

Dickinson, Patric 1914- DLB-27

Dickinson, Peter 1927- DLB-87

Dicks, John [publishing house] DLB-106

Dickson, Gordon R. 1923- DLB-8

*Dictionary of Literary Biography
 Yearbook* Awards Y-92, 93

The Dictionary of National Biography
 .... DLB-144

Didion, Joan 1934- ...DLB-2, 173; Y-81, 86

Di Donato, Pietro 1911- DLB-9

Diego, Gerardo 1896-1987 .... DLB-134

Digges, Thomas circa 1546-1595 ... DLB-136

Dillard, Annie 1945- Y-80

Dillard, R. H. W. 1937- DLB-5

Dillingham, Charles T.,
 Company DLB-49

The Dillingham, G. W.,
 Company DLB-49

Dilly, Edward and Charles
 [publishing house] DLB-154

Dilthey, Wilhelm 1833-1911 .... DLB-129

Dingelstedt, Franz von
 1814-1881 .... DLB-133

Dintenfass, Mark 1941- Y-84

Diogenes, Jr. (see Brougham, John)

DiPrima, Diane 1934- DLB-5, 16

Disch, Thomas M. 1940- DLB-8

Disney, Walt 1901-1966 .... DLB-22

Disraeli, Benjamin 1804-1881 ....DLB-21, 55

D'Israeli, Isaac 1766-1848 .... DLB-107

Ditzen, Rudolf (see Fallada, Hans)

Dix, Dorothea Lynde 1802-1887 ..... DLB-1

Dix, Dorothy (see Gilmer,
 Elizabeth Meriwether)

Dix, Edwards and Company DLB-49

Dixie, Florence Douglas
 1857-1905 .... DLB-174

Dixon, Paige (see Corcoran, Barbara)

Dixon, Richard Watson
 1833-1900 .... DLB-19

Dixon, Stephen 1936- DLB-130

Dmitriev, Ivan Ivanovich
 1760-1837 .... DLB-150

Dobell, Sydney 1824-1874 .... DLB-32

Döblin, Alfred 1878-1957 .... DLB-66

Dobson, Austin
 1840-1921 .... DLB-35, 144

Doctorow, E. L.
 1931- DLB-2, 28, 173; Y-80

Documents on Sixteenth-Century
 Literature DLB-167, 172

Dodd, William E. 1869-1940 .... DLB-17

Dodd, Anne [publishing house] DLB-154

Dodd, Mead and Company DLB-49

Doderer, Heimito von 1896-1968 .... DLB-85

Dodge, Mary Mapes
 1831?-1905 .... DLB-42, 79; DS-13

Dodge, B. W., and Company DLB-46

Dodge Publishing Company DLB-49

Dodgson, Charles Lutwidge
 (see Carroll, Lewis)

Dodsley, Robert 1703-1764 .... DLB-95

Dodsley, R. [publishing house] DLB-154

Dodson, Owen 1914-1983 .... DLB-76

Doesticks, Q. K. Philander, P. B.
 (see Thomson, Mortimer)

Doheny, Carrie Estelle
 1875-1958 .... DLB-140

Domínguez, Sylvia Maida
 1935- DLB-122

Donahoe, Patrick
 [publishing house] DLB-49

Donald, David H. 1920- DLB-17

Donaldson, Scott 1928- DLB-111

Donleavy, J. P. 1926- DLB-6, 173

Donnadieu, Marguerite (see Duras,
 Marguerite)

Donne, John 1572-1631 .... DLB-121, 151

Donnelley, R. R., and Sons
 Company DLB-49

Donnelly, Ignatius 1831-1901 ....DLB-12

Donohue and Henneberry DLB-49

Donoso, José 1924- DLB-113

Doolady, M. [publishing house] DLB-49

Dooley, Ebon (see Ebon)

Doolittle, Hilda 1886-1961 .... DLB-4, 45

Doplicher, Fabio 1938- DLB-128

Dor, Milo 1923- DLB-85

Doran, George H., Company DLB-46

Dorgelès, Roland 1886-1973 ....DLB-65

Dorn, Edward 1929- DLB-5

Dorr, Rheta Childe 1866-1948 ......DLB-25

Dorset and Middlesex, Charles Sackville,
 Lord Buckhurst,
 Earl of 1643-1706 .... DLB-131

Dorst, Tankred 1925- DLB-75, 124

Dos Passos, John
 1896-1970 .... DLB-4, 9; DS-1

Doubleday and Company DLB-49

Dougall, Lily 1858-1923 ....DLB-92

Doughty, Charles M.
 1843-1926 .... DLB-19, 57, 174

Douglas, Gavin 1476-1522 ....DLB-132

Douglas, Keith 1920-1944 ....DLB-27

Douglas, Norman 1868-1952 ....DLB-34

Douglass, Frederick
 1817?-1895 .... DLB-1, 43, 50, 79

Douglass, William circa
 1691-1752 ....DLB-24

Dourado, Autran 1926- DLB-145

Dove, Rita 1952- DLB-120

Dover Publications DLB-46

Doves Press DLB-112

Dowden, Edward 1843-1913 ... DLB-35, 149

Dowell, Coleman 1925-1985 ....DLB-130

Dowland, John 1563-1626 ....DLB-172

Downes, Gwladys 1915- DLB-88

Downing, J., Major (see Davis, Charles A.)

Downing, Major Jack (see Smith, Seba)

Dowriche, Anne
 before 1560-after 1613 ....DLB-172

Dowson, Ernest 1867-1900 ..... DLB-19, 135

Doxey, William [publishing house] DLB-49

Doyle, Sir Arthur Conan 1859-1930 ....DLB-18, 70, 156

Doyle, Kirby 1932- DLB-16

Drabble, Margaret 1939- DLB-14, 155

Drach, Albert 1902- DLB-85

The Dramatic Publishing Company DLB-49

Dramatists Play Service DLB-46

Drant, Thomas early 1540s?-1578 ....DLB-167

Draper, John W. 1811-1882 ....DLB-30

Draper, Lyman C. 1815-1891 ....DLB-30

Drayton, Michael 1563-1631 ....DLB-121

Dreiser, Theodore 1871-1945 ....DLB-9, 12, 102, 137; DS-1

Drewitz, Ingeborg 1923-1986 ....DLB-75

Drieu La Rochelle, Pierre 1893-1945 ....DLB-72

Drinkwater, John 1882-1937 .... DLB-10, 19, 149

Droste-Hülshoff, Annette von 1797-1848 ....DLB-133

The Drue Heinz Literature Prize Excerpt from "Excerpts from a Report of the Commission," in David Bosworth's *The Death of Descartes* An Interview with David Bosworth.....................Y-82

Drummond, William Henry 1854-1907 ....DLB-92

Drummond, William, of Hawthornden 1585-1649 ....DLB-121

Dryden, Charles 1860?-1931 ....DLB-171

Dryden, John 1631-1700 .. DLB-80, 101, 131

Držić, Marin circa 1508-1567 ....DLB-147

Duane, William 1760-1835 ....DLB-43

Dubé, Marcel 1930- DLB-53

Dubé, Rodolphe (see Hertel, François)

Dubic, Norman 1945- DLB-120

Du Bois, W. E. B. 1868-1963 .... DLB-47, 50, 91

Du Bois, William Pène 1916- DLB-61

Dubus, Andre 1936- DLB-130

Ducharme, Réjean 1941- DLB-60

Dučić, Jovan 1871-1943 ....DLB-147

Duck, Stephen 1705?-1756 ....DLB-95

Duckworth, Gerald, and Company Limited DLB-112

Dudek, Louis 1918- DLB-88

Duell, Sloan and Pearce DLB-46

Duff Gordon, Lucie 1821-1869 .....DLB-166

Duffield and Green DLB-46

Duffy, Maureen 1933- DLB-14

Dugan, Alan 1923- DLB-5

Dugard, William [publishing house] DLB-170

Dugas, Marcel 1883-1947 ....DLB-92

Dugdale, William [publishing house] DLB-106

Duhamel, Georges 1884-1966 ....DLB-65

Dujardin, Edouard 1861-1949 .....DLB-123

Dukes, Ashley 1885-1959 ....DLB-10

Du Maurier, George 1834-1896 ....DLB-153

Dumas, Alexandre, *père* 1802-1870 ....DLB-119

Dumas, Henry 1934-1968 ....DLB-41

Dunbar, Paul Laurence 1872-1906 ....DLB-50, 54, 78

Dunbar, William circa 1460-circa 1522 ...... DLB-132, 146

Duncan, Norman 1871-1916 ....DLB-92

Duncan, Quince 1940- DLB-145

Duncan, Robert 1919-1988 .... DLB-5, 16

Duncan, Ronald 1914-1982 ....DLB-13

Duncan, Sara Jeannette 1861-1922 ....DLB-92

Dunigan, Edward, and Brother DLB-49

Dunlap, John 1747-1812 ....DLB-43

Dunlap, William 1766-1839 ....DLB-30, 37, 59

Dunn, Douglas 1942- DLB-40

Dunn, Stephen 1939- DLB-105

Dunn, Stephen, The Good, The Not So Good DLB-105

Dunne, Finley Peter 1867-1936 ....DLB-11, 23

Dunne, John Gregory 1932- Y-80

Dunne, Philip 1908-1992 ....DLB-26

Dunning, Ralph Cheever 1878-1930 ....DLB-4

Dunning, William A. 1857-1922 .... DLB-17

Duns Scotus, John circa 1266-1308 ....DLB-115

Dunsany, Lord (Edward John Moreton Drax Plunkett, Baron Dunsany) 1878-1957 ....DLB-10, 77, 153, 156

Dunton, John [publishing house] ... DLB-170

Dupin, Amantine-Aurore-Lucile (see Sand, George)

Durand, Lucile (see Bersianik, Louky)

Duranty, Walter 1884-1957 ....DLB-29

Duras, Marguerite 1914- DLB-83

Durfey, Thomas 1653-1723 ....DLB-80

Durrell, Lawrence 1912-1990 ....DLB-15, 27; Y-90

Durrell, William [publishing house] DLB-49

Dürrenmatt, Friedrich 1921-1990 ....DLB-69, 124

Dutton, E. P., and Company DLB-49

Duvoisin, Roger 1904-1980 ....DLB-61

Duyckinck, Evert Augustus 1816-1878 ....DLB-3, 64

Duyckinck, George L. 1823-1863 .....DLB-3

Duyckinck and Company DLB-49

Dwight, John Sullivan 1813-1893 .....DLB-1

Dwight, Timothy 1752-1817 ....DLB-37

Dybek, Stuart 1942- DLB-130

Dyer, Charles 1928- DLB-13

Dyer, George 1755-1841 ....DLB-93

Dyer, John 1699-1757 ....DLB-95

Dyer, Sir Edward 1543-1607 ....DLB-136

Dylan, Bob 1941- DLB-16

E

Eager, Edward 1911-1964 ....DLB-22

Eames, Wilberforce 1855-1937 .....DLB-140

Earle, James H., and Company DLB-49

Earle, John 1600 or 1601-1665 .....DLB-151

Early American Book Illustration, by Sinclair Hamilton DLB-49

Eastlake, William 1917- DLB-6

Eastman, Carol ?- DLB-44

Eastman, Max 1883-1969 ....DLB-91

Eaton, Daniel Isaac 1753-1814 .....DLB-158

Eberhart, Richard 1904- DLB-48

Ebner, Jeannie 1918- DLB-85

Ebner-Eschenbach, Marie von 1830-1916 ....DLB-81

Ebon 1942- DLB-41

Ecbasis Captivi circa 1045 ....DLB-148

Ecco Press DLB-46

Eckhart, Meister circa 1260-circa 1328 ....DLB-115

The Eclectic Review 1805-1868 ...... DLB-110

Edel, Leon 1907- DLB-103

Edes, Benjamin 1732-1803 .... DLB-43

Edgar, David 1948- DLB-13

Edgeworth, Maria
 1768-1849 .... DLB-116, 159, 163

The Edinburgh Review 1802-1929 .... DLB-110

Edinburgh University Press DLB-112

The Editor Publishing Company DLB-49

Editorial Statements DLB-137

Edmonds, Randolph 1900- DLB-51

Edmonds, Walter D. 1903- DLB-9

Edschmid, Kasimir 1890-1966 ...... DLB-56

Edwards, Amelia Anne Blandford
 1831-1892 .... DLB-174

Edwards, Jonathan 1703-1758 ...... DLB-24

Edwards, Jonathan, Jr. 1745-1801 ... DLB-37

Edwards, Junius 1929- DLB-33

Edwards, Matilda Barbara Betham-
 1836-1919 .... DLB-174

Edwards, Richard 1524-1566 .... DLB-62

Edwards, James
 [publishing house] DLB-154

Effinger, George Alec 1947- DLB-8

Egerton, George 1859-1945 .... DLB-135

Eggleston, Edward 1837-1902 ...... DLB-12

Eggleston, Wilfred 1901-1986 ...... DLB-92

Ehrenstein, Albert 1886-1950 ...... DLB-81

Ehrhart, W. D. 1948- DS-9

Eich, Günter 1907-1972 .... DLB-69, 124

Eichendorff, Joseph Freiherr von
 1788-1857 .... DLB-90

1873 Publishers' Catalogues DLB-49

Eighteenth-Century Aesthetic
 Theories DLB-31

Eighteenth-Century Philosophical
 Background DLB-31

Eigner, Larry 1927- DLB-5

Eikon Basilike 1649 .... DLB-151

Eilhart von Oberge
 circa 1140-circa 1195 .... DLB-148

Einhard circa 770-840 .... DLB-148

Eisenreich, Herbert 1925-1986 ...... DLB-85

Eisner, Kurt 1867-1919 .... DLB-66

Eklund, Gordon 1945- Y-83

Ekwensi, Cyprian 1921- DLB-117

Eld, George
 [publishing house] DLB-170

Elder, Lonne III 1931- DLB-7, 38, 44

Elder, Paul, and Company DLB-49

Elements of Rhetoric (1828; revised, 1846),
 by Richard Whately [excerpt] DLB-57

Elie, Robert 1915-1973 .... DLB-88

Elin Pelin 1877-1949 .... DLB-147

Eliot, George 1819-1880 ......DLB-21, 35, 55

Eliot, John 1604-1690 .... DLB-24

Eliot, T. S. 1888-1965 ..... DLB-7, 10, 45, 63

Eliot's Court Press DLB-170

Elizabeth I 1533-1603 .... DLB-136

Elizondo, Salvador 1932- DLB-145

Elizondo, Sergio 1930- DLB-82

Elkin, Stanley 1930- DLB-2, 28; Y-80

Elles, Dora Amy (see Wentworth, Patricia)

Ellet, Elizabeth F. 1818?-1877 .... DLB-30

Elliot, Ebenezer 1781-1849 .... DLB-96

Elliot, Frances Minto (Dickinson)
 1820-1898 .... DLB-166

Elliott, George 1923- DLB-68

Elliott, Janice 1931- DLB-14

Elliott, William 1788-1863 .... DLB-3

Elliott, Thomes and Talbot DLB-49

Ellis, Edward S. 1840-1916 .... DLB-42

Ellis, Frederick Staridge
 [publishing house] DLB-106

The George H. Ellis Company DLB-49

Ellison, Harlan 1934- DLB-8

Ellison, Ralph Waldo
 1914-1994 .... DLB-2, 76; Y-94

Ellmann, Richard
 1918-1987 .... DLB-103; Y-87

The Elmer Holmes Bobst Awards in Arts
 and Letters Y-87

Elyot, Thomas 1490?-1546 .... DLB-136

Emanuel, James Andrew 1921- DLB-41

Emecheta, Buchi 1944- DLB-117

The Emergence of Black Women
 Writers DS-8

Emerson, Ralph Waldo
 1803-1882 ....DLB-1, 59, 73

Emerson, William 1769-1811 .... DLB-37

Emin, Fedor Aleksandrovich
 circa 1735-1770 .... DLB-150

Empson, William 1906-1984 .... DLB-20

The End of English Stage Censorship,
 1945-1968 ....DLB-13

Ende, Michael 1929- DLB-75

Engel, Marian 1933-1985 ....DLB-53

Engels, Friedrich 1820-1895 ....DLB-129

Engle, Paul 1908- DLB-48

English Composition and Rhetoric (1866),
 by Alexander Bain [excerpt] DLB-57

The English Language:
 410 to 1500 ....DLB-146

The English Renaissance of Art (1908),
 by Oscar Wilde DLB-35

Enright, D. J. 1920- DLB-27

Enright, Elizabeth 1909-1968 ....DLB-22

L'Envoi (1882), by Oscar Wilde DLB-35

Epps, Bernard 1936- DLB-53

Epstein, Julius 1909- and
 Epstein, Philip 1909-1952 ....DLB-26

Equiano, Olaudah
 circa 1745-1797 .... DLB-37, 50

Eragny Press DLB-112

Erasmus, Desiderius 1467-1536 .....DLB-136

Erba, Luciano 1922- DLB-128

Erdrich, Louise 1954- DLB-152

Erichsen-Brown, Gwethalyn Graham
 (see Graham, Gwethalyn)

Eriugena, John Scottus
 circa 810-877 ....DLB-115

Ernest Hemingway's Toronto Journalism
 Revisited: With Three Previously
 Unrecorded Stories Y-92

Ernst, Paul 1866-1933 .... DLB-66, 118

Erskine, Albert 1911-1993 .... Y-93

Erskine, John 1879-1951 .... DLB-9, 102

Ervine, St. John Greer 1883-1971 ....DLB-10

Eschenburg, Johann Joachim
 1743-1820 ....DLB-97

Escoto, Julio 1944- DLB-145

Eshleman, Clayton 1935- DLB-5

Espriu, Salvador 1913-1985 ....DLB-134

Ess Ess Publishing Company DLB-49

Essay on Chatterton (1842), by
 Robert Browning DLB-32

Essex House Press DLB-112

Estes, Eleanor 1906-1988 ....DLB-22

Estes and Lauriat DLB-49

Etherege, George 1636-circa 1692 ....DLB-80

Ethridge, Mark, Sr. 1896-1981 ......DLB-127

Ets, Marie Hall 1893- DLB-22

Etter, David 1928- DLB-105

Ettner, Johann Christoph
1654-1724 ....DLB-168

Eudora Welty: Eye of the Storyteller Y-87

Eugene O'Neill Memorial Theater
Center DLB-7

Eugene O'Neill's Letters: A Review Y-88

Eupolemius
flourished circa 1095 ....DLB-148

Evans, Caradoc 1878-1945 ....DLB-162

Evans, Donald 1884-1921 ....DLB-54

Evans, George Henry 1805-1856 .....DLB-43

Evans, Hubert 1892-1986 ....DLB-92

Evans, Mari 1923- DLB-41

Evans, Mary Ann (see Eliot, George)

Evans, Nathaniel 1742-1767 ....DLB-31

Evans, Sebastian 1830-1909 ....DLB-35

Evans, M., and Company DLB-46

Everett, Alexander Hill
790-1847 ....DLB-59

Everett, Edward 1794-1865 ....DLB-1, 59

Everson, R. G. 1903- DLB-88

Everson, William 1912-1994 ......DLB-5, 16

Every Man His Own Poet; or, The
Inspired Singer's Recipe Book (1877),
by W. H. Mallock DLB-35

Ewart, Gavin 1916- DLB-40

Ewing, Juliana Horatia
1841-1885 ....DLB-21, 163

The Examiner 1808-1881 ....DLB-110

Exley, Frederick
1929-1992 ....DLB-143; Y-81

Experiment in the Novel (1929),
by John D. Beresford DLB-36

Eyre and Spottiswoode DLB-106

Ezzo ? after 1065 ....DLB-148

F

"F. Scott Fitzgerald: St. Paul's Native Son
and Distinguished American Writer":
University of Minnesota Conference,
29-31 October 1982 ....Y-82

Faber, Frederick William
1814-1863 ....DLB-32

Faber and Faber Limited DLB-112

Faccio, Rena (see Aleramo, Sibilla)

Fagundo, Ana María 1938- DLB-134

Fair, Ronald L. 1932- DLB-33

Fairfax, Beatrice (see Manning, Marie)

Fairlie, Gerard 1899-1983 ....DLB-77

Fallada, Hans 1893-1947 ....DLB-56

Fancher, Betsy 1928- Y-83

Fane, Violet 1843-1905 ....DLB-35

Fanfrolico Press DLB-112

Fanning, Katherine 1927 ....DLB-127

Fanshawe, Sir Richard
1608-1666 ....DLB-126

Fantasy Press Publishers DLB-46

Fante, John 1909-1983 ....DLB-130; Y-83

Al-Farabi circa 870-950 ....DLB-115

Farah, Nuruddin 1945- DLB-125

Farber, Norma 1909-1984 ....DLB-61

Farigoule, Louis (see Romains, Jules)

Farjeon, Eleanor 1881-1965 ....DLB-160

Farley, Walter 1920-1989 ....DLB-22

Farmer, Penelope 1939- DLB-161

Farmer, Philip José 1918- DLB-8

Farquhar, George circa 1677-1707 ...DLB-84

Farquharson, Martha (see Finley, Martha)

Farrar, Frederic William
1831-1903 ....DLB-163

Farrar and Rinehart DLB-46

Farrar, Straus and Giroux DLB-46

Farrell, James T.
1904-1979 ....DLB-4, 9, 86; DS-2

Farrell, J. G. 1935-1979 ....DLB-14

Fast, Howard 1914- DLB-9

Faulkner, William 1897-1962
........DLB-9, 11, 44, 102; DS-2; Y-86

Faulkner, George
[publishing house] DLB-154

Fauset, Jessie Redmon 1882-1961 ...DLB-51

Faust, Irvin 1924- DLB-2, 28; Y-80

Fawcett Books DLB-46

Fearing, Kenneth 1902-1961 ....DLB-9

Federal Writers' Project DLB-46

Federman, Raymond 1928- Y-80

Feiffer, Jules 1929- DLB-7, 44

Feinberg, Charles E. 1899-1988 ......Y-88

Feind, Barthold 1678-1721 ....DLB-168

Feinstein, Elaine 1930- DLB-14, 40

Feldman, Irving 1928- DLB-169

Felipe, Léon 1884-1968 ....DLB-108

Fell, Frederick, Publishers DLB-46

Felltham, Owen 1602?-1668 ...DLB-126, 151

Fels, Ludwig 1946- DLB-75

Felton, Cornelius Conway
1807-1862 ....DLB-1

Fennario, David 1947- DLB-60

Fenno, John 1751-1798 ....DLB-43

Fenno, R. F., and Company DLB-49

Fenton, Geoffrey 1539?-1608 .....DLB-136

Fenton, James 1949- DLB-40

Ferber, Edna 1885-1968 ....DLB-9, 28, 86

Ferdinand, Vallery III (see Salaam, Kalamu ya)

Ferguson, Sir Samuel 1810-1886 .....DLB-32

Ferguson, William Scott
1875-1954 ....DLB-47

Fergusson, Robert 1750-1774 ......DLB-109

Ferland, Albert 1872-1943 ....DLB-92

Ferlinghetti, Lawrence 1919- DLB-5, 16

Fern, Fanny (see Parton, Sara Payson Willis)

Ferrars, Elizabeth 1907- DLB-87

Ferré, Rosario 1942- DLB-145

Ferret, E., and Company DLB-49

Ferrier, Susan 1782-1854 ....DLB-116

Ferrini, Vincent 1913- DLB-48

Ferron, Jacques 1921-1985 ....DLB-60

Ferron, Madeleine 1922- DLB-53

Fetridge and Company DLB-49

Feuchtersleben, Ernst Freiherr von
1806-1849 ....DLB-133

Feuchtwanger, Lion 1884-1958 ......DLB-66

Feuerbach, Ludwig 1804-1872 .....DLB-133

Fichte, Johann Gottlieb
1762-1814 ....DLB-90

Ficke, Arthur Davison 1883-1945 ....DLB-54

Fiction Best-Sellers, 1910-1945 ....DLB-9

Fiction into Film, 1928-1975: A List of Movies
Based on the Works of Authors in
British Novelists, 1930-1959 ....DLB-15

Fiedler, Leslie A. 1917- DLB-28, 67

Field, Edward 1924- DLB-105

Field, Edward, The Poetry File DLB-105

Field, Eugene
1850-1895 ......DLB-23, 42, 140; DS-13

Field, John 1545?-1588 ....DLB-167

Field, Marshall, III 1893-1956 ......DLB-127

Field, Marshall, IV 1916-1965 ......DLB-127

Field, Marshall, V 1941- DLB-127

Field, Nathan 1587-1619 or 1620 ....DLB-58

Field, Rachel 1894-1942 ....DLB-9, 22

A Field Guide to Recent Schools of American
  Poetry Y-86

Fielding, Henry
  1707-1754 .... DLB-39, 84, 101

Fielding, Sarah 1710-1768 .... DLB-39

Fields, James Thomas 1817-1881 ...... DLB-1

Fields, Julia 1938- DLB-41

Fields, W. C. 1880-1946 .... DLB-44

Fields, Osgood and Company DLB-49

Fifty Penguin Years Y-85

Figes, Eva 1932- DLB-14

Figuera, Angela 1902-1984 .... DLB-108

Filmer, Sir Robert 1586-1653 ...... DLB-151

Filson, John circa 1753-1788 .... DLB-37

Finch, Anne, Countess of Winchilsea
  1661-1720 .... DLB-95

Finch, Robert 1900- DLB-88

Findley, Timothy 1930- DLB-53

Finlay, Ian Hamilton 1925- DLB-40

Finley, Martha 1828-1909 .... DLB-42

Finn, Elizabeth Anne (McCaul)
  1825-1921 .... DLB-166

Finney, Jack 1911- DLB-8

Finney, Walter Braden (see Finney, Jack)

Firbank, Ronald 1886-1926 .... DLB-36

Firmin, Giles 1615-1697 .... DLB-24

First Edition Library/Collectors'
  Reprints, Inc. Y-91

First International F. Scott Fitzgerald
  Conference Y-92

First Strauss "Livings" Awarded to Cynthia
  Ozick and Raymond Carver
  An Interview with Cynthia Ozick
  An Interview with Raymond
  Carver Y-83

Fischer, Karoline Auguste Fernandine
  1764-1842 .... DLB-94

Fish, Stanley 1938- DLB-67

Fishacre, Richard 1205-1248 ...... DLB-115

Fisher, Clay (see Allen, Henry W.)

Fisher, Dorothy Canfield
  1879-1958 .... DLB-9, 102

Fisher, Leonard Everett 1924- DLB-61

Fisher, Roy 1930- DLB-40

Fisher, Rudolph 1897-1934 ..... DLB-51, 102

Fisher, Sydney George 1856-1927 ... DLB-47

Fisher, Vardis 1895-1968 .... DLB-9

Fiske, John 1608-1677 .... DLB-24

Fiske, John 1842-1901 .... DLB-47, 64

Fitch, Thomas circa 1700-1774 ...... DLB-31

Fitch, William Clyde 1865-1909 ...... DLB-7

FitzGerald, Edward 1809-1883 ...... DLB-32

Fitzgerald, F. Scott
  1896-1940 ..... DLB-4, 9, 86; Y-81; DS-1

Fitzgerald, Penelope 1916- DLB-14

Fitzgerald, Robert 1910-1985 ....Y-80

Fitzgerald, Thomas 1819-1891 ...... DLB-23

Fitzgerald, Zelda Sayre 1900-1948 ....Y-84

Fitzhugh, Louise 1928-1974 .... DLB-52

Fitzhugh, William
  circa 1651-1701 .... DLB-24

Flanagan, Thomas 1923- Y-80

Flanner, Hildegarde 1899-1987 ..... DLB-48

Flanner, Janet 1892-1978 .... DLB-4

Flaubert, Gustave 1821-1880 ...... DLB-119

Flavin, Martin 1883-1967 .... DLB-9

Fleck, Konrad (flourished circa 1220)
  .... DLB-138

Flecker, James Elroy 1884-1915 ..DLB-10, 19

Fleeson, Doris 1901-1970 .... DLB-29

Fleißer, Marieluise 1901-1974 ...DLB-56, 124

Fleming, Ian 1908-1964 .... DLB-87

Fleming, Paul 1609-1640 .... DLB-164

The Fleshly School of Poetry and Other
  Phenomena of the Day (1872), by Robert
  Buchanan DLB-35

The Fleshly School of Poetry: Mr. D. G.
  Rossetti (1871), by Thomas Maitland
  (Robert Buchanan) DLB-35

Fletcher, Giles, the Elder
  1546-1611 .... DLB-136

Fletcher, Giles, the Younger
  1585 or 1586-1623 .... DLB-121

Fletcher, J. S. 1863-1935 .... DLB-70

Fletcher, John (see Beaumont, Francis)

Fletcher, John Gould 1886-1950 ...DLB-4, 45

Fletcher, Phineas 1582-1650 ...... DLB-121

Flieg, Helmut (see Heym, Stefan)

Flint, F. S. 1885-1960 .... DLB-19

Flint, Timothy 1780-1840 .... DLB-734

Florio, John 1553?-1625 .... DLB-172

Foix, J. V. 1893-1987 .... DLB-134

Foley, Martha (see Burnett, Whit, and
  Martha Foley)

Folger, Henry Clay 1857-1930 ..... DLB-140

Folio Society DLB-112

Follen, Eliza Lee (Cabot) 1787-1860 ... DLB-1

Follett, Ken 1949- Y-81, DLB-87

Follett Publishing Company DLB-46

Folsom, John West
  [publishing house] DLB-49

Fontane, Theodor 1819-1898 ...... DLB-129

Fonvisin, Denis Ivanovich
  1744 or 1745-1792 .... DLB-150

Foote, Horton 1916- DLB-26

Foote, Samuel 1721-1777 .... DLB-89

Foote, Shelby 1916- DLB-2, 17

Forbes, Calvin 1945- DLB-41

Forbes, Ester 1891-1967 .... DLB-22

Forbes and Company DLB-49

Force, Peter 1790-1868 .... DLB-30

Forché, Carolyn 1950- DLB-5

Ford, Charles Henri 1913- DLB-4, 48

Ford, Corey 1902-1969 .... DLB-11

Ford, Ford Madox
  1873-1939 .... DLB-34, 98, 162

Ford, Jesse Hill 1928- DLB-6

Ford, John 1586-? DLB-58

Ford, R. A. D. 1915- DLB-88

Ford, Worthington C. 1858-1941 .... DLB-47

Ford, J. B., and Company DLB-49

Fords, Howard, and Hulbert DLB-49

Foreman, Carl 1914-1984 .... DLB-26

Forester, Frank (see Herbert, Henry William)

Fornés, María Irene 1930- DLB-7

Forrest, Leon 1937- DLB-33

Forster, E. M.
  1879-1970 .... DLB-34, 98, 162; DS-10

Forster, Georg 1754-1794 .... DLB-94

Forster, John 1812-1876 .... DLB-144

Forster, Margaret 1938- DLB-155

Forsyth, Frederick 1938- DLB-87

Forten, Charlotte L. 1837-1914 ..... DLB-50

Fortini, Franco 1917- DLB-128

Fortune, T. Thomas 1856-1928 ...... DLB-23

Fosdick, Charles Austin
  1842-1915 .... DLB-42

Foster, Genevieve 1893-1979 ....DLB-61

Foster, Hannah Webster
  1758-1840 .... DLB-37

Foster, John 1648-1681 .... DLB-24

Foster, Michael 1904-1956 .... DLB-9

Foulis, Robert and Andrew / R. and A.
  [publishing house] DLB-154

Fouqué, Caroline de la Motte
    1774-1831 ....DLB-90

Fouqué, Friedrich de la Motte
    1777-1843 ....DLB-90

Four Essays on the Beat Generation,
    by John Clellon Holmes DLB-16

Four Seas Company DLB-46

Four Winds Press DLB-46

Fournier, Henri Alban (see Alain-Fournier)

Fowler and Wells Company DLB-49

Fowles, John 1926- DLB-14, 139

Fox, John, Jr. 1862 or
    1863-1919 .... DLB-9; DS-13

Fox, Paula 1923- DLB-52

Fox, Richard Kyle 1846-1922 ....DLB-79

Fox, William Price 1926- DLB-2; Y-81

Fox, Richard K.
    [publishing house] DLB-49

Foxe, John 1517-1587 ....DLB-132

Fraenkel, Michael 1896-1957 ....DLB-4

France, Anatole 1844-1924 ....DLB-123

France, Richard 1938- DLB-7

Francis, Convers 1795-1863 ....DLB-1

Francis, Dick 1920- DLB-87

Francis, Jeffrey, Lord 1773-1850 ....DLB-107

Francis, C. S. [publishing house] DLB-49

François 1863-1910 ....DLB-92

François, Louise von 1817-1893 ....DLB-129

Francke, Kuno 1855-1930 ....DLB-71

Frank, Bruno 1887-1945 ....DLB-118

Frank, Leonhard 1882-1961 .....DLB-56, 118

Frank, Melvin (see Panama, Norman)

Frank, Waldo 1889-1967 ....DLB-9, 63

Franken, Rose 1895?-1988 ....Y-84

Franklin, Benjamin
    1706-1790 .... DLB-24, 43, 73

Franklin, James 1697-1735 ....DLB-43

Franklin Library DLB-46

Frantz, Ralph Jules 1902-1979 ....DLB-4

Franzos, Karl Emil 1848-1904 ......DLB-129

Fraser, G. S. 1915-1980 ....DLB-27

Fraser, Kathleen 1935- DLB-169

Frattini, Alberto 1922- DLB-128

Frau Ava ?-1127 ....DLB-148

Frayn, Michael 1933- DLB-13, 14

Frederic, Harold
    1856-1898 .... DLB-12, 23; DS-13

Freeling, Nicolas 1927- DLB-87

Freeman, Douglas Southall
    1886-1953 .... DLB-17

Freeman, Legh Richmond
    1842-1915 .... DLB-23

Freeman, Mary E. Wilkins
    1852-1930 .... DLB-12, 78

Freeman, R. Austin 1862-1943 ...... DLB-70

Freidank circa 1170-circa 1233 ..... DLB-138

Freiligrath, Ferdinand 1810-1876 ... DLB-133

French, Alice 1850-1934 ..... DLB-74; DS-13

French, David 1939- DLB-53

French, James [publishing house] DLB-49

French, Samuel [publishing house] ... DLB-49

Samuel French, Limited DLB-106

Freneau, Philip 1752-1832 .... DLB-37, 43

Freni, Melo 1934- DLB-128

Freshfield, Douglas W.
    1845-1934 .... DLB-174

Freytag, Gustav 1816-1895 .... DLB-129

Fried, Erich 1921-1988 .... DLB-85

Friedman, Bruce Jay 1930- DLB-2, 28

Friedrich von Hausen
    circa 1171-1190 .... DLB-138

Friel, Brian 1929- DLB-13

Friend, Krebs 1895?-1967? DLB-4

Fries, Fritz Rudolf 1935- DLB-75

Fringe and Alternative Theater
    in Great Britain DLB-13

Frisch, Max 1911-1991 .... DLB-69, 124

Frischmuth, Barbara 1941- DLB-85

Fritz, Jean 1915- DLB-52

Fromentin, Eugene 1820-1876 ..... DLB-123

From *The Gay Science*, by
    E. S. Dallas DLB-21

Frost, A. B. 1851-1928 ....DS-13

Frost, Robert 1874-1963 ...... DLB-54; DS-7

Frothingham, Octavius Brooks
    1822-1895 .... DLB-1

Froude, James Anthony
    1818-1894 .... DLB-18, 57, 144

Fry, Christopher 1907- DLB-13

Fry, Roger 1866-1934 ....DS-10

Frye, Northrop 1912-1991 .... DLB-67, 68

Fuchs, Daniel
    1909-1993 .... DLB-9, 26, 28; Y-93

Fuentes, Carlos 1928- DLB-113

Fuertes, Gloria 1918- DLB-108

The Fugitives and the Agrarians:
    The First Exhibition Y-85

Fulbecke, William 1560-1603? DLB-172

Fuller, Charles H., Jr. 1939- DLB-38

Fuller, Henry Blake 1857-1929 ..... DLB-12

Fuller, John 1937- DLB-40

Fuller, Roy 1912-1991 .... DLB-15, 20

Fuller, Samuel 1912- DLB-26

Fuller, Sarah Margaret, Marchesa
    D'Ossoli 1810-1850 ....DLB-1, 59, 73

Fuller, Thomas 1608-1661 .... DLB-151

Fullerton, Hugh 1873-1945 .... DLB-171

Fulton, Len 1934- Y-86

Fulton, Robin 1937- DLB-40

Furbank, P. N. 1920- DLB-155

Furman, Laura 1945- Y-86

Furness, Horace Howard
    1833-1912 .... DLB-64

Furness, William Henry 1802-1896 ... DLB-1

Furthman, Jules 1888-1966 .... DLB-26

The Future of the Novel (1899), by
    Henry James DLB-18

Fyleman, Rose 1877-1957 .... DLB-160

G

The G. Ross Roy Scottish Poetry
    Collection at the University of
    South Carolina Y-89

Gaddis, William 1922- DLB-2

Gág, Wanda 1893-1946 .... DLB-22

Gagnon, Madeleine 1938- DLB-60

Gaine, Hugh 1726-1807 .... DLB-43

Gaine, Hugh [publishing house] DLB-49

Gaines, Ernest J.
    1933- DLB-2, 33, 152; Y-80

Gaiser, Gerd 1908-1976 .... DLB-69

Galarza, Ernesto 1905-1984 .... DLB-122

Galaxy Science Fiction Novels DLB-46

Gale, Zona 1874-1938 .... DLB-9, 78

Gall, Louise von 1815-1855 .... DLB-133

Gallagher, Tess 1943- DLB-120

Gallagher, Wes 1911- DLB-127

Gallagher, William Davis
    1808-1894 .... DLB-73

Gallant, Mavis 1922- DLB-53

Gallico, Paul 1897-1976 ....DLB-9, 171

Galsworthy, John
    1867-1933 .... DLB-10, 34, 98, 162

Galt, John 1779-1839 .... DLB-99, 116

Galton, Sir Francis 1822-1911 ..... DLB-166

Galvin, Brendan 1938- DLB-5

Gambit DLB-46

Gamboa, Reymundo 1948- DLB-122

Gammer Gurton's Needle DLB-62

Gannett, Frank E. 1876-1957 .... DLB-29

Gaos, Vicente 1919-1980 .... DLB-134

García, Lionel G. 1935- DLB-82

García Lorca, Federico
    1898-1936 .... DLB-108

García Márquez, Gabriel
    1928- DLB-113

Gardam, Jane 1928- DLB-14, 161

Garden, Alexander
    circa 1685-1756 .... DLB-31

Gardiner, Margaret Power Farmer (see
    Blessington, Marguerite, Countess of)

Gardner, John 1933-1982 ...... DLB-2; Y-82

Garfield, Leon 1921- DLB-161

Garis, Howard R. 1873-1962 .... DLB-22

Garland, Hamlin
    1860-1940 .... DLB-12, 71, 78

Garneau, Francis-Xavier
    1809-1866 .... DLB-99

Garneau, Hector de Saint-Denys
    1912-1943 .... DLB-88

Garneau, Michel 1939- DLB-53

Garner, Alan 1934- DLB-161

Garner, Hugh 1913-1979 .... DLB-68

Garnett, David 1892-1981 .... DLB-34

Garnett, Eve 1900-1991 .... DLB-160

Garraty, John A. 1920- DLB-17

Garrett, George
    1929- DLB-2, 5, 130, 152; Y-83

Garrick, David 1717-1779 .... DLB-84

Garrison, William Lloyd
    1805-1879 .... DLB-1, 43

Garro, Elena 1920- DLB-145

Garth, Samuel 1661-1719 .... DLB-95

Garve, Andrew 1908- DLB-87

Gary, Romain 1914-1980 .... DLB-83

Gascoigne, George 1539?-1577 .... DLB-136

Gascoyne, David 1916- DLB-20

Gaskell, Elizabeth Cleghorn
    1810-1865 .... DLB-21, 144, 159

Gaspey, Thomas 1788-1871 .... DLB-116

Gass, William Howard 1924- DLB-2

Gates, Doris 1901- DLB-22

Gates, Henry Louis, Jr. 1950- DLB-67

Gates, Lewis E. 1860-1924 .... DLB-71

Gatto, Alfonso 1909-1976 .... DLB-114

Gaunt, Mary 1861-1942 .... DLB-174

Gautier, Théophile 1811-1872 ..... DLB-119

Gauvreau, Claude 1925-1971 .... DLB-88

The *Gawain*-Poet
    flourished circa 1350-1400 ..... DLB-146

Gay, Ebenezer 1696-1787 .... DLB-24

Gay, John 1685-1732 .... DLB-84, 95

The Gay Science (1866), by E. S. Dallas
    [excerpt] DLB-21

Gayarré, Charles E. A. 1805-1895 ... DLB-30

Gaylord, Edward King
    1873-1974 .... DLB-127

Gaylord, Edward Lewis 1919- DLB-127

Gaylord, Charles
    [publishing house] DLB-49

Geddes, Gary 1940- DLB-60

Geddes, Virgil 1897- DLB-4

Gedeon (Georgii Andreevich Krinovsky)
    circa 1730-1763 .... DLB-150

Geibel, Emanuel 1815-1884 .... DLB-129

Geis, Bernard, Associates DLB-46

Geisel, Theodor Seuss
    1904-1991 .... DLB-61; Y-91

Gelb, Arthur 1924- DLB-103

Gelb, Barbara 1926- DLB-103

Gelber, Jack 1932- DLB-7

Gelinas, Gratien 1909- DLB-88

Gellert, Christian Fuerchtegott
    1715-1769 .... DLB-97

Gellhorn, Martha 1908- Y-82

Gems, Pam 1925- DLB-13

A General Idea of the College of Mirania (1753),
    by William Smith [excerpts] DLB-31

Genet, Jean 1910-1986 .... DLB-72; Y-86

Genevoix, Maurice 1890-1980 .... DLB-65

Genovese, Eugene D. 1930- DLB-17

Gent, Peter 1942- Y-82

Geoffrey of Monmouth
    circa 1100-1155 .... DLB-146

George, Henry 1839-1897 .... DLB-23

George, Jean Craighead 1919- DLB-52

Georgslied 896? DLB-148

Gerhardie, William 1895-1977 ...... DLB-36

Gerhardt, Paul 1607-1676 .... DLB-164

Gérin, Winifred 1901-1981 .... DLB-155

Gérin-Lajoie, Antoine 1824-1882 ..... DLB-99

German Drama 800-1280 .... DLB-138

German Drama from Naturalism
    to Fascism: 1889-1933 .... DLB-118

German Literature and Culture from
    Charlemagne to the Early Courtly
    Period DLB-148

German Radio Play, The DLB-124

German Transformation from the Baroque
    to the Enlightenment, The DLB-97

The Germanic Epic and Old English Heroic
    Poetry: *Widseth, Waldere,* and *The
    Fight at Finnsburg* DLB-146

Germanophilism, by Hans Kohn DLB-66

Gernsback, Hugo 1884-1967 .... DLB-8, 137

Gerould, Katharine Fullerton
    1879-1944 .... DLB-78

Gerrish, Samuel [publishing house] ...DLB-49

Gerrold, David 1944- DLB-8

Gersonides 1288-1344 .... DLB-115

Gerstäcker, Friedrich 1816-1872 .... DLB-129

Gerstenberg, Heinrich Wilhelm von
    1737-1823 .... DLB-97

Gervinus, Georg Gottfried
    1805-1871 .... DLB-133

Geßner, Salomon 1730-1788 .... DLB-97

Geston, Mark S. 1946- DLB-8

Al-Ghazali 1058-1111 .... DLB-115

Gibbon, Edward 1737-1794 .... DLB-104

Gibbon, John Murray 1875-1952 ..... DLB-92

Gibbon, Lewis Grassic (see Mitchell,
    James Leslie)

Gibbons, Floyd 1887-1939 .... DLB-25

Gibbons, Reginald 1947- DLB-120

Gibbons, William ?-? DLB-73

Gibson, Charles Dana 1867-1944 ..... DS-13

Gibson, Charles Dana 1867-1944 ..... DS-13

Gibson, Graeme 1934- DLB-53

Gibson, Margaret 1944- DLB-120

Gibson, Margaret Dunlop
    1843-1920 .... DLB-174

Gibson, Wilfrid 1878-1962 .... DLB-19

Gibson, William 1914- DLB-7

Gide, André 1869-1951 .... DLB-65

Giguère, Diane 1937- DLB-53
Giguère, Roland 1929- DLB-60
Gil de Biedma, Jaime 1929-1990 ....DLB-108
Gil-Albert, Juan 1906- DLB-134
Gilbert, Anthony 1899-1973 ....DLB-77
Gilbert, Michael 1912- DLB-87
Gilbert, Sandra M. 1936- DLB-120
Gilbert, Sir Humphrey
  1537-1583 ....DLB-136
Gilchrist, Alexander
  1828-1861 ....DLB-144
Gilchrist, Ellen 1935- DLB-130
Gilder, Jeannette L. 1849-1916 ...DLB-79
Gilder, Richard Watson
  1844-1909 ....DLB-64, 79
Gildersleeve, Basil 1831-1924 ....DLB-71
Giles, Henry 1809-1882 ....DLB-64
Giles of Rome circa 1243-1316 ....DLB-115
Gilfillan, George 1813-1878 ......DLB-144
Gill, Eric 1882-1940 ....DLB-98
Gill, William F., Company DLB-49
Gillespie, A. Lincoln, Jr.
  1895-1950 ....DLB-4
Gilliam, Florence ?-? DLB-4
Gilliatt, Penelope 1932-1993 .....DLB-14
Gillott, Jacky 1939-1980 ....DLB-14
Gilman, Caroline H. 1794-1888 ...DLB-3, 73
Gilman, W. and J.
  [publishing house] DLB-49
Gilmer, Elizabeth Meriwether
  1861-1951 ....DLB-29
Gilmer, Francis Walker
  1790-1826 ....DLB-37
Gilroy, Frank D. 1925- DLB-7
Gimferrer, Pere (Pedro) 1945- DLB-134
Gingrich, Arnold 1903-1976 ....DLB-137
Ginsberg, Allen 1926- DLB-5, 16, 169
Ginzkey, Franz Karl 1871-1963 ....DLB-81
Gioia, Dana 1950- DLB-120
Giono, Jean 1895-1970 ....DLB-72
Giotti, Virgilio 1885-1957 ....DLB-114
Giovanni, Nikki 1943- DLB-5, 41
Gipson, Lawrence Henry
  1880-1971 ....DLB-17
Girard, Rodolphe 1879-1956 ....DLB-92
Giraudoux, Jean 1882-1944 ....DLB-65
Gissing, George 1857-1903 ....DLB-18, 135

Giudici, Giovanni 1924- DLB-128
Giuliani, Alfredo 1924- DLB-128
Gladstone, William Ewart
  1809-1898 ....DLB-57
Glaeser, Ernst 1902-1963 ....DLB-69
Glanville, Brian 1931- DLB-15, 139
Glapthorne, Henry 1610-1643? DLB-58
Glasgow, Ellen 1873-1945 ....DLB-9, 12
Glaspell, Susan 1876-1948 .....DLB-7, 9, 78
Glass, Montague 1877-1934 ....DLB-11
Glassco, John 1909-1981 ....DLB-68
Glauser, Friedrich 1896-1938 .....DLB-56
F. Gleason's Publishing Hall DLB-49
Gleim, Johann Wilhelm Ludwig
  1719-1803 ....DLB-97
Glendinning, Victoria 1937- DLB-155
Glover, Richard 1712-1785 ....DLB-95
Glück, Louise 1943- DLB-5
Glyn, Elinor 1864-1943 ....DLB-153
Gobineau, Joseph-Arthur de
  1816-1882 ....DLB-123
Godbout, Jacques 1933- DLB-53
Goddard, Morrill 1865-1937 ....DLB-25
Goddard, William 1740-1817 ....DLB-43
Godden, Rumer 1907- DLB-161
Godey, Louis A. 1804-1878 ....DLB-73
Godey and McMichael DLB-49
Godfrey, Dave 1938- DLB-60
Godfrey, Thomas 1736-1763 ....DLB-31
Godine, David R., Publisher DLB-46
Godkin, E. L. 1831-1902 ....DLB-79
Godolphin, Sidney 1610-1643 ......DLB-126
Godwin, Gail 1937- DLB-6
Godwin, Mary Jane Clairmont
  1766-1841 ....DLB-163
Godwin, Parke 1816-1904 ....DLB-3, 64
Godwin, William
  1756-1836 ... DLB-39, 104, 142, 158, 163
Godwin, M. J., and Company DLB-154
Goering, Reinhard 1887-1936 ......DLB-118
Goes, Albrecht 1908- DLB-69
Goethe, Johann Wolfgang von
  1749-1832 ....DLB-94
Goetz, Curt 1888-1960 ....DLB-124
Goffe, Thomas circa 1592-1629 ....DLB-58
Goffstein, M. B. 1940- DLB-61

Gogarty, Oliver St. John
  1878-1957 ....DLB-15, 19
Goines, Donald 1937-1974 ....DLB-33
Gold, Herbert 1924- DLB-2; Y-81
Gold, Michael 1893-1967 ....DLB-9, 28
Goldbarth, Albert 1948- DLB-120
Goldberg, Dick 1947- DLB-7
Golden Cockerel Press DLB-112
Golding, Arthur 1536-1606 ....DLB-136
Golding, William 1911-1993 ....DLB-15, 100
Goldman, William 1931- DLB-44
Goldsmith, Oliver
  1730?-1774 ...DLB-39, 89, 104, 109, 142
Goldsmith, Oliver 1794-1861 ......DLB-99
Goldsmith Publishing Company DLB-46
Gollancz, Victor, Limited DLB-112
Gómez-Quiñones, Juan 1942- DLB-122
Gomme, Laurence James
  [publishing house] DLB-46
Goncourt, Edmond de 1822-1896 ...DLB-123
Goncourt, Jules de 1830-1870 .....DLB-123
Gonzales, Rodolfo "Corky"
  1928- DLB-122
González, Angel 1925- DLB-108
Gonzalez, Genaro 1949- DLB-122
Gonzalez, Ray 1952- DLB-122
González de Mireles, Jovita
  1899-1983 ....DLB-122
González-T., César A. 1931- DLB-82
Goodbye, Gutenberg? A Lecture at
  the New York Public Library,
  18 April 1995 ....Y-95
Goodison, Lorna 1947- DLB-157
Goodman, Paul 1911-1972 ....DLB-130
The Goodman Theatre DLB-7
Goodrich, Frances 1891-1984 and
  Hackett, Albert 1900- DLB-26
Goodrich, Samuel Griswold
  1793-1860 ....DLB-1, 42, 73
Goodrich, S. G. [publishing house] ...DLB-49
Goodspeed, C. E., and Company DLB-49
Goodwin, Stephen 1943- Y-82
Googe, Barnabe 1540-1594 ....DLB-132
Gookin, Daniel 1612-1687 ....DLB-24
Gordon, Caroline
  1895-1981 ....DLB-4, 9, 102; Y-81
Gordon, Giles 1940- DLB-14, 139
Gordon, Lyndall 1941- DLB-155

Cumulative Index

Gordon, Mary 1949- DLB-6; Y-81

Gordone, Charles 1925- DLB-7

Gore, Catherine 1800-1861 ....DLB-116

Gorey, Edward 1925- DLB-61

Görres, Joseph 1776-1848 ....DLB-90

Gosse, Edmund 1849-1928 .....DLB-57, 144

Gosson, Stephen 1554-1624 ....DLB-172

Gotlieb, Phyllis 1926- DLB-88

Gottfried von Straßburg
   died before 1230 ....DLB-138

Gotthelf, Jeremias 1797-1854 .....DLB-133

Gottschalk circa 804/808-869 ......DLB-148

Gottsched, Johann Christoph
   1700-1766 ....DLB-97

Götz, Johann Nikolaus
   1721-1781 ....DLB-97

Gould, Wallace 1882-1940 ....DLB-54

Govoni, Corrado 1884-1965 ....DLB-114

Gower, John circa 1330-1408 ......DLB-146

Goyen, William 1915-1983 .....DLB-2; Y-83

Goytisolo, José Augustín 1928- ...DLB-134

Gozzano, Guido 1883-1916 ....DLB-114

Grabbe, Christian Dietrich
   1801-1836 ....DLB-133

Gracq, Julien 1910- DLB-83

Grady, Henry W. 1850-1889 ....DLB-23

Graf, Oskar Maria 1894-1967 ......DLB-56

Graf Rudolf between circa 1170
   and circa 1185 ....DLB-148

Grafton, Richard
   [publishing house] DLB-170

Graham, George Rex 1813-1894 ....DLB-73

Graham, Gwethalyn 1913-1965 .....DLB-88

Graham, Jorie 1951- DLB-120

Graham, Katharine 1917- DLB-127

Graham, Lorenz 1902-1989 ....DLB-76

Graham, Philip 1915-1963 ....DLB-127

Graham, R. B. Cunninghame
   1852-1936 ....DLB-98, 135, 174

Graham, Shirley 1896-1977 ....DLB-76

Graham, W. S. 1918- DLB-20

Graham, William H.
   [publishing house] DLB-49

Graham, Winston 1910- DLB-77

Grahame, Kenneth
   1859-1932 ....DLB-34, 141

Grainger, Martin Allerdale
   1874-1941 ....DLB-92

Gramatky, Hardie 1907-1979 ....DLB-22

Grand, Sarah 1854-1943 ....DLB-135

Grandbois, Alain 1900-1975 ....DLB-92

Grange, John circa 1556-? DLB-136

Granich, Irwin (see Gold, Michael)

Grant, Duncan 1885-1978 ....DS-10

Grant, George 1918-1988 ....DLB-88

Grant, George Monro 1835-1902 ....DLB-99

Grant, Harry J. 1881-1963 ....DLB-29

Grant, James Edward 1905-1966.....DLB-26

Grass, Günter 1927- DLB-75, 124

Grasty, Charles H. 1863-1924 ....DLB-25

Grau, Shirley Ann 1929- DLB-2

Graves, John 1920- Y-83

Graves, Richard 1715-1804 ....DLB-39

Graves, Robert
   1895-1985 ....DLB-20, 100; Y-85

Gray, Asa 1810-1888 ....DLB-1

Gray, David 1838-1861 ....DLB-32

Gray, Simon 1936- DLB-13

Gray, Thomas 1716-1771 ....DLB-109

Grayson, William J. 1788-1863 ....DLB-3, 64

The Great Bibliographers Series Y-93

The Great War and the Theater, 1914-1918
   [Great Britain] DLB-10

Greeley, Horace 1811-1872 ....DLB-3, 43

Green, Adolph (see Comden, Betty)

Green, Duff 1791-1875 ....DLB-43

Green, Gerald 1922- DLB-28

Green, Henry 1905-1973 ....DLB-15

Green, Jonas 1712-1767 ....DLB-31

Green, Joseph 1706-1780 ....DLB-31

Green, Julien 1900- DLB-4, 72

Green, Paul 1894-1981 ....DLB-7, 9; Y-81

Green, T. and S.
   [publishing house] DLB-49

Green, Timothy
   [publishing house] DLB-49

Greenaway, Kate 1846-1901 ....DLB-141

Greenberg: Publisher DLB-46

Green Tiger Press DLB-46

Greene, Asa 1789-1838 ....DLB-11

Greene, Benjamin H.
   [publishing house] DLB-49

Greene, Graham 1904-1991
   ....DLB-13, 15, 77, 100, 162; Y-85, Y-91

Greene, Robert 1558-1592 .....DLB-62, 167

Greenhow, Robert 1800-1854 ....DLB-30

Greenough, Horatio 1805-1852 ....DLB-1

Greenwell, Dora 1821-1882 ....DLB-35

Greenwillow Books DLB-46

Greenwood, Grace (see Lippincott, Sara Jane
   Clarke)

Greenwood, Walter 1903-1974 ......DLB-10

Greer, Ben 1948- DLB-6

Greflinger, Georg 1620?-1677 ......DLB-164

Greg, W. R. 1809-1881 ....DLB-55

Gregg Press DLB-46

Gregory, Isabella Augusta
   Persse, Lady 1852-1932 ....DLB-10

Gregory, Horace 1898-1982 ....DLB-48

Gregory of Rimini
   circa 1300-1358 ....DLB-115

Gregynog Press DLB-112

Greiffenberg, Catharina Regina von
   1633-1694 ....DLB-168

Grenfell, Wilfred Thomason
   1865-1940 ....DLB-92

Greve, Felix Paul (see Grove, Frederick Philip)

Greville, Fulke, First Lord Brooke
   1554-1628 ....DLB-62, 172

Grey, Lady Jane 1537-1554 ....DLB-132

Grey Owl 1888-1938 ....DLB-92

Grey, Zane 1872-1939 ....DLB-9

Grey Walls Press DLB-112

Grier, Eldon 1917- DLB-88

Grieve, C. M. (see MacDiarmid, Hugh)

Griffin, Bartholomew
   flourished 1596 ....DLB-172

Griffin, Gerald 1803-1840 ....DLB-159

Griffith, Elizabeth 1727?-1793 ...DLB-39, 89

Griffiths, Trevor 1935- DLB-13

Griffiths, Ralph
   [publishing house] DLB-154

Griggs, S. C., and Company DLB-49

Griggs, Sutton Elbert 1872-1930 .....DLB-50

Grignon, Claude-Henri 1894-1976 ...DLB-68

Grigson, Geoffrey 1905- DLB-27

Grillparzer, Franz 1791-1872 ......DLB-133

Grimald, Nicholas
   circa 1519-circa 1562 ....DLB-136

Grimké, Angelina Weld
   1880-1958 ....DLB-50, 54

Grimm, Hans 1875-1959 ....DLB-66

Grimm, Jacob 1785-1863 .... DLB-90

Grimm, Wilhelm 1786-1859 .... DLB-90

Grimmelshausen, Johann Jacob Christoffel von 1621 or 1622-1676 .... DLB-168

Grimshaw, Beatrice Ethel 1871-1953 .... DLB-174

Grindal, Edmund 1519 or 1520-1583 .... DLB-132

Griswold, Rufus Wilmot 1815-1857 .... DLB-3, 59

Gross, Milt 1895-1953 .... DLB-11

Grosset and Dunlap DLB-49

Grossman Publishers DLB-46

Grosseteste, Robert circa 1160-1253 .... DLB-115

Grosvenor, Gilbert H. 1875-1966 .... DLB-91

Groth, Klaus 1819-1899 .... DLB-129

Groulx, Lionel 1878-1967 .... DLB-68

Grove, Frederick Philip 1879-1949 ... DLB-92

Grove Press DLB-46

Grubb, Davis 1919-1980 .... DLB-6

Gruelle, Johnny 1880-1938 .... DLB-22

Grymeston, Elizabeth before 1563-before 1604 .... DLB-136

Gryphius, Andreas 1616-1664 ...... DLB-164

Gryphius, Christian 1649-1706 ..... DLB-168

Guare, John 1938- DLB-7

Guerra, Tonino 1920- DLB-128

Guest, Barbara 1920- DLB-5

Guèvremont, Germaine 1893-1968 .... DLB-68

Guidacci, Margherita 1921-1992 .... DLB-128

Guide to the Archives of Publishers, Journals, and Literary Agents in North American Libraries Y-93

Guillén, Jorge 1893-1984 .... DLB-108

Guilloux, Louis 1899-1980 .... DLB-72

Guilpin, Everard circa 1572-after 1608? DLB-136

Guiney, Louise Imogen 1861-1920 ... DLB-54

Guiterman, Arthur 1871-1943 .... DLB-11

Günderrode, Caroline von 1780-1806 .... DLB-90

Gundulić, Ivan 1589-1638 .... DLB-147

Gunn, Bill 1934-1989 .... DLB-38

Gunn, James E. 1923- DLB-8

Gunn, Neil M. 1891-1973 .... DLB-15

Gunn, Thom 1929- DLB-27

Gunnars, Kristjana 1948- DLB-60

Günther, Johann Christian 1695-1723 .... DLB-168

Gurik, Robert 1932- DLB-60

Gustafson, Ralph 1909- DLB-88

Gütersloh, Albert Paris 1887-1973 ... DLB-81

Guthrie, A. B., Jr. 1901- DLB-6

Guthrie, Ramon 1896-1973 .... DLB-4

The Guthrie Theater DLB-7

Gutzkow, Karl 1811-1878 .... DLB-133

Guy, Ray 1939- DLB-60

Guy, Rosa 1925- DLB-33

Guyot, Arnold 1807-1884 .... DS-13

Gwynne, Erskine 1898-1948 .... DLB-4

Gyles, John 1680-1755 .... DLB-99

Gysin, Brion 1916- DLB-16

H

H. D. (see Doolittle, Hilda)

Habington, William 1605-1654 .... DLB-126

Hacker, Marilyn 1942- DLB-120

Hackett, Albert (see Goodrich, Frances)

Hacks, Peter 1928- DLB-124

Hadas, Rachel 1948- DLB-120

Hadden, Briton 1898-1929 .... DLB-91

Hagedorn, Friedrich von 1708-1754 .... DLB-168

Hagelstange, Rudolf 1912-1984 ..... DLB-69

Haggard, H. Rider 1856-1925 .... DLB-70, 156, 174

Haggard, William 1907-1993 .... Y-93

Hahn-Hahn, Ida Gräfin von 1805-1880 .... DLB-133

Haig-Brown, Roderick 1908-1976 ... DLB-88

Haight, Gordon S. 1901-1985 ...... DLB-103

Hailey, Arthur 1920- DLB-88; Y-82

Haines, John 1924- DLB-5

Hake, Edward flourished 1566-1604 .... DLB-136

Hake, Thomas Gordon 1809-1895 ... DLB-32

Hakluyt, Richard 1552?-1616 ...... DLB-136

Halbe, Max 1865-1944 .... DLB-118

Haldane, J. B. S. 1892-1964 .... DLB-160

Haldeman, Joe 1943- DLB-8

Haldeman-Julius Company DLB-46

Hale, E. J., and Son DLB-49

Hale, Edward Everett 1822-1909 .... DLB-1, 42, 74

Hale, Kathleen 1898- DLB-160

Hale, Leo Thomas (see Ebon)

Hale, Lucretia Peabody 1820-1900 .... DLB-42

Hale, Nancy 1908-1988 ....DLB-86; Y-80, 88

Hale, Sarah Josepha (Buell) 1788-1879 .... DLB-1, 42, 73

Hales, John 1584-1656 .... DLB-151

Haley, Alex 1921-1992 .... DLB-38

Haliburton, Thomas Chandler 1796-1865 .... DLB-11, 99

Hall, Anna Maria 1800-1881 .... DLB-159

Hall, Donald 1928- DLB-5

Hall, Edward 1497-1547 .... DLB-132

Hall, James 1793-1868 .... DLB-73, 74

Hall, Joseph 1574-1656 ....DLB-121, 151

Hall, Samuel [publishing house] DLB-49

Hallam, Arthur Henry 1811-1833 .... DLB-32

Halleck, Fitz-Greene 1790-1867 .... DLB-3

Haller, Albrecht von 1708-1777 .... DLB-168

Hallmann, Johann Christian 1640-1704 or 1716? DLB-168

Hallmark Editions DLB-46

Halper, Albert 1904-1984 .... DLB-9

Halperin, John William 1941- DLB-111

Halstead, Murat 1829-1908 .... DLB-23

Hamann, Johann Georg 1730-1788 ... DLB-97

Hamburger, Michael 1924- DLB-27

Hamilton, Alexander 1712-1756 ..... DLB-31

Hamilton, Alexander 1755?-1804 .... DLB-37

Hamilton, Cicely 1872-1952 .... DLB-10

Hamilton, Edmond 1904-1977 .... DLB-8

Hamilton, Elizabeth 1758-1816 .. DLB-116, 158

Hamilton, Gail (see Corcoran, Barbara)

Hamilton, Ian 1938- DLB-40, 155

Hamilton, Patrick 1904-1962 .... DLB-10

Hamilton, Virginia 1936- DLB-33, 52

Hamilton, Hamish, Limited DLB-112

Hammett, Dashiell 1894-1961 .... DS-6

Dashiell Hammett: An Appeal in TAC Y-91

Hammon, Jupiter 1711-died between 1790 and 1806 ....DLB-31, 50

Hammond, John ?-1663 .... DLB-24

Hamner, Earl 1923- DLB-6

Hampton, Christopher 1946- DLB-13

Handel-Mazzetti, Enrica von 1871-1955 ..... DLB-81

Handke, Peter 1942- DLB-85, 124

Handlin, Oscar 1915- DLB-17

Hankin, St. John 1869-1909 ..... DLB-10

Hanley, Clifford 1922- DLB-14

Hannah, Barry 1942- DLB-6

Hannay, James 1827-1873 ..... DLB-21

Hansberry, Lorraine 1930-1965 ..... DLB-7, 38

Hapgood, Norman 1868-1937 ..... DLB-91

Happel, Eberhard Werner 1647-1690 ..... DLB-168

Harcourt Brace Jovanovich DLB-46

Hardenberg, Friedrich von (see Novalis)

Harding, Walter 1917- DLB-111

Hardwick, Elizabeth 1916- DLB-6

Hardy, Thomas 1840-1928 ...DLB-18, 19, 135

Hare, Cyril 1900-1958 ..... DLB-77

Hare, David 1947- DLB-13

Hargrove, Marion 1919- DLB-11

Häring, Georg Wilhelm Heinrich (see Alexis, Willibald)

Harington, Donald 1935- DLB-152

Harington, Sir John 1560-1612 ..... DLB-136

Harjo, Joy 1951- DLB-120

Harlow, Robert 1923- DLB-60

Harman, Thomas flourished 1566-1573 ..... DLB-136

Harness, Charles L. 1915- DLB-8

Harnett, Cynthia 1893-1981 ..... DLB-161

Harper, Fletcher 1806-1877 ..... DLB-79

Harper, Frances Ellen Watkins 1825-1911 ..... DLB-50

Harper, Michael S. 1938- DLB-41

Harper and Brothers DLB-49

Harraden, Beatrice 1864-1943 ..... DLB-153

Harrap, George G., and Company Limited DLB-112

Harriot, Thomas 1560-1621 ..... DLB-136

Harris, Benjamin ?-circa 1720 ... DLB-42, 43

Harris, Christie 1907- DLB-88

Harris, Frank 1856-1931 ..... DLB-156

Harris, George Washington 1814-1869 ..... DLB-3, 11

Harris, Joel Chandler 1848-1908 ..... DLB-11, 23, 42, 78, 91

Harris, Mark 1922- DLB-2; Y-80

Harris, Wilson 1921- DLB-117

Harrison, Charles Yale 1898-1954 ..... DLB-68

Harrison, Frederic 1831-1923 ..... DLB-57

Harrison, Harry 1925- DLB-8

Harrison, Jim 1937- Y-82

Harrison, Mary St. Leger Kingsley (see Malet, Lucas)

Harrison, Paul Carter 1936- DLB-38

Harrison, Susan Frances 1859-1935 ..... DLB-99

Harrison, Tony 1937- DLB-40

Harrison, William 1535-1593 ..... DLB-136

Harrison, James P., Company DLB-49

Harrisse, Henry 1829-1910 ..... DLB-47

Harsdörffer, Georg Philipp 1607-1658 ..... DLB-164

Harsent, David 1942- DLB-40

Hart, Albert Bushnell 1854-1943 ..... DLB-17

Hart, Julia Catherine 1796-1867 ..... DLB-99

The Lorenz Hart Centenary Y-95

Hart, Moss 1904-1961 ..... DLB-7

Hart, Oliver 1723-1795 ..... DLB-31

Harte, Bret 1836-1902 ... DLB-12, 64, 74, 79

Harte, Edward Holmead 1922- ... DLB-127

Harte, Houston Harriman 1927- ... DLB-127

Hartlaub, Felix 1913-1945 ..... DLB-56

Hartlebon, Otto Erich 1864-1905 ..... DLB-118

Hartley, L. P. 1895-1972 ..... DLB-15, 139

Hartley, Marsden 1877-1943 ..... DLB-54

Hartling, Peter 1933- DLB-75

Hartman, Geoffrey H. 1929- DLB-67

Hartmann, Sadakichi 1867-1944 ..... DLB-54

Hartmann von Aue circa 1160-circa 1205 ..... DLB-138

Harvey, Gabriel 1550?-1631 ..... DLB-167

Harvey, Jean-Charles 1891-1967 ..... DLB-88

Harvill Press Limited DLB-112

Harwood, Lee 1939- DLB-40

Harwood, Ronald 1934- DLB-13

Haskins, Charles Homer 1870-1937 ..... DLB-47

Hass, Robert 1941- DLB-105

The Hatch-Billops Collection DLB-76

Hathaway, William 1944- DLB-120

Hauff, Wilhelm 1802-1827 ..... DLB-90

A Haughty and Proud Generation (1922), by Ford Madox Hueffer DLB-36

Haugwitz, August Adolph von 1647-1706 ..... DLB-168

Hauptmann, Carl 1858-1921 ..... DLB-66, 118

Hauptmann, Gerhart 1862-1946 ..... DLB-66, 118

Hauser, Marianne 1910- Y-83

Hawes, Stephen 1475?-before 1529 ..... DLB-132

Hawker, Robert Stephen 1803-1875 ..... DLB-32

Hawkes, John 1925- DLB-2, 7; Y-80

Hawkesworth, John 1720-1773 ..... DLB-142

Hawkins, Sir Anthony Hope (see Hope, Anthony)

Hawkins, Sir John 1719-1789 ..... DLB-104, 142

Hawkins, Walter Everette 1883-? DLB-50

Hawthorne, Nathaniel 1804-1864 ..... DLB-1, 74

Hay, John 1838-1905 ..... DLB-12, 47

Hayden, Robert 1913-1980 ..... DLB-5, 76

Haydon, Benjamin Robert 1786-1846 ..... DLB-110

Hayes, John Michael 1919- DLB-26

Hayley, William 1745-1820 ..... DLB-93, 142

Haym, Rudolf 1821-1901 ..... DLB-129

Hayman, Robert 1575-1629 ..... DLB-99

Hayman, Ronald 1932- DLB-155

Hayne, Paul Hamilton 1830-1886 ..... DLB-3, 64, 79

Hays, Mary 1760-1843 ..... DLB-142, 158

Haywood, Eliza 1693?-1756 ..... DLB-39

Hazard, Willis P. [publishing house] ...DLB-49

Hazlitt, William 1778-1830 .... DLB-110, 158

Hazzard, Shirley 1931- Y-82

Head, Bessie 1937-1986 ..... DLB-117

Headley, Joel T. 1813-1897 .. DLB-30; DS-13

Heaney, Seamus 1939- DLB-40

Heard, Nathan C. 1936- DLB-33

Hearn, Lafcadio 1850-1904 ..... DLB-12, 78

Hearne, John 1926- DLB-117

Hearne, Samuel 1745-1792 ..... DLB-99

Hearst, William Randolph 1863-1951 ..... DLB-25

Hearst, William Randolph, Jr 1908-1993 .... DLB-127

Heath, Catherine 1924- DLB-14

Heath, Roy A. K. 1926- DLB-117

Heath-Stubbs, John 1918- DLB-27

Heavysege, Charles 1816-1876 ...... DLB-99

Hebbel, Friedrich 1813-1863 .... DLB-129

Hebel, Johann Peter 1760-1826 ...... DLB-90

Hébert, Anne 1916- DLB-68

Hébert, Jacques 1923- DLB-53

Hecht, Anthony 1923- DLB-5, 169

Hecht, Ben 1894-1964
.... DLB-7, 9, 25, 26, 28, 86

Hecker, Isaac Thomas 1819-1888 ..... DLB-1

Hedge, Frederic Henry 1805-1890 .... DLB-1, 59

Hefner, Hugh M. 1926- DLB-137

Hegel, Georg Wilhelm Friedrich 1770-1831 .... DLB-90

Heidish, Marcy 1947- Y-82

Heißenbüttel 1921- DLB-75

Hein, Christoph 1944- DLB-124

Heine, Heinrich 1797-1856 .... DLB-90

Heinemann, Larry 1944- DS-9

Heinemann, William, Limited DLB-112

Heinlein, Robert A. 1907-1988 .... DLB-8

Heinrich Julius of Brunswick 1564-1613 .... DLB-164

Heinrich von dem Türlîn flourished circa 1230 .... DLB-138

Heinrich von Melk flourished after 1160 .... DLB-148

Heinrich von Veldeke circa 1145-circa 1190 .... DLB-138

Heinrich, Willi 1920- DLB-75

Heiskell, John 1872-1972 .... DLB-127

Heinse, Wilhelm 1746-1803 .... DLB-94

Heinz, W. C. 1915- DLB-171

Hejinian, Lyn 1941- DLB-165

Heliand circa 850 .... DLB-148

Heller, Joseph 1923- DLB-2, 28; Y-80

Heller, Michael 1937- DLB-165

Hellman, Lillian 1906-1984 .... DLB-7; Y-84

Hellwig, Johann 1609-1674 .... DLB-164

Helprin, Mark 1947- Y-85

Helwig, David 1938- DLB-60

Hemans, Felicia 1793-1835 .... DLB-96

Hemingway, Ernest 1899-1961
.... DLB-4, 9, 102; Y-81, 87; DS-1

Hemingway: Twenty-Five Years Later Y-85

Hémon, Louis 1880-1913 .... DLB-92

Hemphill, Paul 1936- Y-87

Hénault, Gilles 1920- DLB-88

Henchman, Daniel 1689-1761 ...... DLB-24

Henderson, Alice Corbin 1881-1949 .... DLB-54

Henderson, Archibald 1877-1963 .... DLB-103

Henderson, David 1942- DLB-41

Henderson, George Wylie 1904- DLB-51

Henderson, Zenna 1917-1983 .... DLB-8

Henisch, Peter 1943- DLB-85

Henley, Beth 1952- Y-86

Henley, William Ernest 1849-1903 .... DLB-19

Henniker, Florence 1855-1923 ..... DLB-135

Henry, Alexander 1739-1824 .... DLB-99

Henry, Buck 1930- DLB-26

Henry VIII of England 1491-1547 .... DLB-132

Henry, Marguerite 1902- DLB-22

Henry, O. (see Porter, William Sydney)

Henry of Ghent circa 1217-1229 - 1293 .... DLB-115

Henry, Robert Selph 1889-1970 ..... DLB-17

Henry, Will (see Allen, Henry W.)

Henryson, Robert 1420s or 1430s-circa 1505 ..... DLB-146

Henschke, Alfred (see Klabund)

Hensley, Sophie Almon 1866-1946 .. DLB-99

Henty, G. A. 1832?-1902 ...... DLB-18, 141

Hentz, Caroline Lee 1800-1856 ...... DLB-3

Herbert, Agnes circa 1880-1960 .... DLB-174

Herbert, Alan Patrick 1890-1971 .... DLB-10

Herbert, Edward, Lord, of Cherbury 1582-1648 .... DLB-121, 151

Herbert, Frank 1920-1986 .... DLB-8

Herbert, George 1593-1633 .... DLB-126

Herbert, Henry William 1807-1858 .... DLB-3, 73

Herbert, John 1926- DLB-53

Herbert, Mary Sidney, Countess of Pembroke (see Sidney, Mary)

Herbst, Josephine 1892-1969 .... DLB-9

Herburger, Gunter 1932- DLB-75, 124

Hercules, Frank E. M. 1917- DLB-33

Herder, Johann Gottfried 1744-1803 .... DLB-97

Herder, B., Book Company DLB-49

Herford, Charles Harold 1853-1931 .... DLB-149

Hergesheimer, Joseph 1880-1954 .... DLB-9, 102

Heritage Press DLB-46

Hermann the Lame 1013-1054 ..... DLB-148

Hermes, Johann Timotheus 1738-1821 .... DLB-97

Hermlin, Stephan 1915- DLB-69

Hernández, Alfonso C. 1938- DLB-122

Hernández, Inés 1947- DLB-122

Hernández, Miguel 1910-1942 ..... DLB-134

Hernton, Calvin C. 1932- DLB-38

"The Hero as Man of Letters: Johnson, Rousseau, Burns" (1841), by Thomas Carlyle [excerpt] DLB-57

The Hero as Poet. Dante; Shakspeare (1841), by Thomas Carlyle DLB-32

Heron, Robert 1764-1807 .... DLB-142

Herrera, Juan Felipe 1948- DLB-122

Herrick, Robert 1591-1674 .... DLB-126

Herrick, Robert 1868-1938 ....DLB-9, 12, 78

Herrick, William 1915- Y-83

Herrick, E. R., and Company DLB-49

Herrmann, John 1900-1959 .... DLB-4

Hersey, John 1914-1993 .... DLB-6

Hertel, François 1905-1985 .... DLB-68

Hervé-Bazin, Jean Pierre Marie (see Bazin, Hervé)

Hervey, John, Lord 1696-1743 ..... DLB-101

Herwig, Georg 1817-1875 .... DLB-133

Herzog, Emile Salomon Wilhelm (see Maurois, André)

Hesse, Hermann 1877-1962 .... DLB-66

Hewat, Alexander circa 1743-circa 1824 .... DLB-30

Hewitt, John 1907- DLB-27

Hewlett, Maurice 1861-1923 ....DLB-34, 156

Heyen, William 1940- DLB-5

Heyer, Georgette 1902-1974 .... DLB-77

Heym, Stefan 1913- DLB-69

Heyse, Paul 1830-1914 .... DLB-129

Cumulative Index

Heytesbury, William
circa 1310-1372 or 1373 .... DLB-115

Heyward, Dorothy 1890-1961 ....DLB-7

Heyward, DuBose
1885-1940 .... DLB-7, 9, 45

Heywood, John 1497?-1580? DLB-136

Heywood, Thomas
1573 or 1574-1641 .... DLB-62

Hibbs, Ben 1901-1975 .... DLB-137

Hichens, Robert S. 1864-1950 ..... DLB-153

Hickman, William Albert
1877-1957 .... DLB-92

Hidalgo, José Luis 1919-1947 ..... DLB-108

Hiebert, Paul 1892-1987 .... DLB-68

Hierro, José 1922- DLB-108

Higgins, Aidan 1927- DLB-14

Higgins, Colin 1941-1988 .... DLB-26

Higgins, George V. 1939- DLB-2; Y-81

Higginson, Thomas Wentworth
1823-1911 .... DLB-1, 64

Highwater, Jamake 1942?- ... DLB-52; Y-85

Hijuelos, Oscar 1951- DLB-145

Hildegard von Bingen
1098-1179 .... DLB-148

Das Hildesbrandslied circa 820 .... DLB-148

Hildesheimer, Wolfgang
1916-1991 .... DLB-69, 124

Hildreth, Richard
1807-1865 .... DLB-1, 30, 59

Hill, Aaron 1685-1750 .... DLB-84

Hill, Geoffrey 1932- DLB-40

Hill, "Sir" John 1714?-1775 .... DLB-39

Hill, Leslie 1880-1960 .... DLB-51

Hill, Susan 1942- DLB-14, 139

Hill, Walter 1942- DLB-44

Hill and Wang DLB-46

Hill, George M., Company DLB-49

Hill, Lawrence, and Company,
Publishers DLB-46

Hillberry, Conrad 1928- DLB-120

Hilliard, Gray and Company DLB-49

Hills, Lee 1906- DLB-127

Hillyer, Robert 1895-1961 .... DLB-54

Hilton, James 1900-1954 .... DLB-34, 77

Hilton, Walter died 1396 .... DLB-146

Hilton and Company DLB-49

Himes, Chester
1909-1984 .... DLB-2, 76, 143

Hindmarsh, Joseph
[publishing house] DLB-170

Hine, Daryl 1936- DLB-60

Hingley, Ronald 1920- DLB-155

Hinojosa-Smith, Rolando
1929- DLB-82

Hippel, Theodor Gottlieb von
1741-1796 .... DLB-97

Hirsch, E. D., Jr. 1928- DLB-67

Hirsch, Edward 1950- DLB-120

The History of the Adventures of Joseph Andrews
(1742), by Henry Fielding
[excerpt] DLB-39

Hoagland, Edward 1932- DLB-6

Hoagland, Everett H., III 1942- DLB-41

Hoban, Russell 1925- DLB-52

Hobbes, Thomas 1588-1679 .... DLB-151

Hobby, Oveta 1905- DLB-127

Hobby, William 1878-1964 .... DLB-127

Hobsbaum, Philip 1932- DLB-40

Hobson, Laura Z. 1900- DLB-28

Hoby, Thomas 1530-1566 .... DLB-132

Hoccleve, Thomas
circa 1368-circa 1437 .... DLB-146

Hochhuth, Rolf 1931- DLB-124

Hochman, Sandra 1936- DLB-5

Hodder and Stoughton, Limited DLB-106

Hodgins, Jack 1938- DLB-60

Hodgman, Helen 1945- DLB-14

Hodgskin, Thomas 1787-1869 ..... DLB-158

Hodgson, Ralph 1871-1962 .... DLB-19

Hodgson, William Hope
1877-1918 .... DLB-70, 153, 156

Hoffenstein, Samuel 1890-1947 ..... DLB-11

Hoffman, Charles Fenno
1806-1884 .... DLB-3

Hoffman, Daniel 1923- DLB-5

Hoffmann, E. T. A. 1776-1822 ..... DLB-90

Hoffmanswaldau, Christian Hoffman von
1616-1679 .... DLB-168

Hofmann, Michael 1957- DLB-40

Hofmannsthal, Hugo von
1874-1929 .... DLB-81, 118

Hofstadter, Richard 1916-1970 .... DLB-17

Hogan, Desmond 1950- DLB-14

Hogan and Thompson DLB-49

Hogarth Press DLB-112

Hogg, James 1770-1835 ....DLB-93, 116, 159

Hohberg, Wolfgang Helmhard Freiherr von
1612-1688 .... DLB-168

Hohl, Ludwig 1904-1980 .... DLB-56

Holbrook, David 1923- DLB-14, 40

Holcroft, Thomas
1745-1809 .... DLB-39, 89, 158

Holden, Jonathan 1941- DLB-105

Holden, Jonathan, Contemporary
Verse Story-telling DLB-105

Holden, Molly 1927-1981 .... DLB-40

Hölderlin, Friedrich 1770-1843 .....DLB-90

Holiday House DLB-46

Holinshed, Raphael died 1580 .....DLB-167

Holland, J. G. 1819-1881 .... DS-13

Holland, Norman N. 1927- DLB-67

Hollander, John 1929- DLB-5

Holley, Marietta 1836-1926 .... DLB-11

Hollingsworth, Margaret 1940- DLB-60

Hollo, Anselm 1934- DLB-40

Holloway, Emory 1885-1977 ..... DLB-103

Holloway, John 1920- DLB-27

Holloway House Publishing
Company DLB-46

Holme, Constance 1880-1955 ..... DLB-34

Holmes, Abraham S. 1821?-1908 .... DLB-99

Holmes, John Clellon 1926-1988 ..... DLB-16

Holmes, Oliver Wendell
1809-1894 .... DLB-1

Holmes, Richard 1945- DLB-155

Holroyd, Michael 1935- DLB-155

Holst, Hermann E. von
1841-1904 .... DLB-47

Holt, John 1721-1784 .... DLB-43

Holt, Henry, and Company DLB-49

Holt, Rinehart and Winston DLB-46

Holthusen, Hans Egon 1913- DLB-69

Hölty, Ludwig Christoph Heinrich
1748-1776 .... DLB-94

Holz, Arno 1863-1929 .... DLB-118

Home, Henry, Lord Kames (see Kames, Henry
Home, Lord)

Home, John 1722-1808 .... DLB-84

Home, William Douglas 1912- DLB-13

Home Publishing Company DLB-49

Homes, Geoffrey (see Mainwaring, Daniel)

Honan, Park 1928- DLB-111

Hone, William 1780-1842 ..... DLB-110, 158

Hongo, Garrett Kaoru 1951- DLB-120

Honig, Edwin 1919- DLB-5

Hood, Hugh 1928- DLB-53

Hood, Thomas 1799-1845 ....DLB-96

Hook, Theodore 1788-1841 ....DLB-116

Hooker, Jeremy 1941- DLB-40

Hooker, Richard 1554-1600 ....DLB-132

Hooker, Thomas 1586-1647 ....DLB-24

Hooper, Johnson Jones
  1815-1862 ....DLB-3, 11

Hope, Anthony 1863-1933 .....DLB-153, 156

Hopkins, Gerard Manley
  1844-1889 ....DLB-35, 57

Hopkins, John (see Sternhold, Thomas)

Hopkins, Lemuel 1750-1801 ....DLB-37

Hopkins, Pauline Elizabeth
  1859-1930 ....DLB-50

Hopkins, Samuel 1721-1803 ....DLB-31

Hopkins, John H., and Son DLB-46

Hopkinson, Francis 1737-1791 ......DLB-31

Horgan, Paul 1903- DLB-102; Y-85

Horizon Press DLB-46

Horne, Frank 1899-1974 ....DLB-51

Horne, Richard Henry (Hengist)
  1802 or 1803-1884 ....DLB-32

Hornung, E. W. 1866-1921 ....DLB-70

Horovitz, Israel 1939- DLB-7

Horton, George Moses
  1797?-1883? DLB-50

Horváth, Ödön von
  1901-1938 ....DLB-85, 124

Horwood, Harold 1923- DLB-60

Hosford, E. and E.
  [publishing house] DLB-49

Hoskyns, John 1566-1638 ....DLB-121

Hotchkiss and Company DLB-49

Hough, Emerson 1857-1923 ....DLB-9

Houghton Mifflin Company DLB-49

Houghton, Stanley 1881-1913 ......DLB-10

Household, Geoffrey 1900-1988 ....DLB-87

Housman, A. E. 1859-1936 ....DLB-19

Housman, Laurence 1865-1959 .....DLB-10

Houwald, Ernst von 1778-1845 .....DLB-90

Hovey, Richard 1864-1900 ....DLB-54

Howard, Donald R. 1927-1987 ....DLB-111

Howard, Maureen 1930- Y-83

Howard, Richard 1929- DLB-5

Howard, Roy W. 1883-1964 ....DLB-29

Howard, Sidney 1891-1939 ......DLB-7, 26

Howe, E. W. 1853-1937 ....DLB-12, 25

Howe, Henry 1816-1893 ....DLB-30

Howe, Irving 1920-1993 ....DLB-67

Howe, Joseph 1804-1873 ....DLB-99

Howe, Julia Ward 1819-1910 ....DLB-1

Howe, Percival Presland
  1886-1944 ....DLB-149

Howe, Susan 1937- DLB-120

Howell, Clark, Sr. 1863-1936 ......DLB-25

Howell, Evan P. 1839-1905 ....DLB-23

Howell, James 1594?-1666 ....DLB-151

Howell, Warren Richardson
  1912-1984 ....DLB-140

Howell, Soskin and Company DLB-46

Howells, William Dean
  1837-1920 ....DLB-12, 64, 74, 79

Howitt, William 1792-1879 and
  Howitt, Mary 1799-1888 ......DLB-110

Hoyem, Andrew 1935- DLB-5

Hoyers, Anna Ovena 1584-1655 ...DLB-164

Hoyos, Angela de 1940- DLB-82

Hoyt, Palmer 1897-1979 ....DLB-127

Hoyt, Henry [publishing house] DLB-49

Hrabanus Maurus 776?-856 ....DLB-148

Hrotsvit of Gandersheim
  circa 935-circa 1000 ....DLB-148

Hubbard, Elbert 1856-1915 ....DLB-91

Hubbard, Kin 1868-1930 ....DLB-11

Hubbard, William circa 1621-1704 ..DLB-24

Huber, Therese 1764-1829 ....DLB-90

Huch, Friedrich 1873-1913 ....DLB-66

Huch, Ricarda 1864-1947 ....DLB-66

Huck at 100: How Old Is
  Huckleberry Finn? Y-85

Huddle, David 1942- DLB-130

Hudgins, Andrew 1951- DLB-120

Hudson, Henry Norman
  1814-1886 ....DLB-64

Hudson, W. H.
  1841-1922 ....DLB-98, 153, 174

Hudson and Goodwin DLB-49

Huebsch, B. W.
  [publishing house] DLB-46

Hughes, David 1930- DLB-14

Hughes, John 1677-1720 ....DLB-84

Hughes, Langston
  1902-1967 ....DLB-4, 7, 48, 51, 86

Hughes, Richard 1900-1976 ....DLB-15, 161

Hughes, Ted 1930- DLB-40, 161

Hughes, Thomas 1822-1896 ....DLB-18, 163

Hugo, Richard 1923-1982 ....DLB-5

Hugo, Victor 1802-1885 ....DLB-119

Hugo Awards and Nebula Awards DLB-8

Hull, Richard 1896-1973 ....DLB-77

Hulme, T. E. 1883-1917 ....DLB-19

Humboldt, Alexander von
  1769-1859 ....DLB-90

Humboldt, Wilhelm von
  1767-1835 ....DLB-90

Hume, David 1711-1776 ....DLB-104

Hume, Fergus 1859-1932 ....DLB-70

Hummer, T. R. 1950- DLB-120

Humorous Book Illustration DLB-11

Humphrey, William 1924- DLB-6

Humphreys, David 1752-1818 ....DLB-37

Humphreys, Emyr 1919- DLB-15

Huncke, Herbert 1915- DLB-16

Huneker, James Gibbons
  1857-1921 ....DLB-71

Hunold, Christian Friedrich
  1681-1721 ....DLB-168

Hunt, Irene 1907- DLB-52

Hunt, Leigh 1784-1859 ....DLB-96, 110, 144

Hunt, Violet 1862-1942 ....DLB-162

Hunt, William Gibbes 1791-1833 ....DLB-73

Hunter, Evan 1926- Y-82

Hunter, Jim 1939- DLB-14

Hunter, Kristin 1931- DLB-33

Hunter, Mollie 1922- DLB-161

Hunter, N. C. 1908-1971 ....DLB-10

Hunter-Duvar, John 1821-1899 .....DLB-99

Huntington, Henry E.
  1850-1927 ....DLB-140

Hurd and Houghton DLB-49

Hurst, Fannie 1889-1968 ....DLB-86

Hurst and Blackett DLB-106

Hurst and Company DLB-49

Hurston, Zora Neale
  1901?-1960 ....DLB-51, 86

Husson, Jules-François-Félix (see Champfleury)

Huston, John 1906-1987 ....DLB-26

Hutcheson, Francis 1694-1746 .....DLB-31

Cumulative Index

Hutchinson, Thomas 1711-1780 .... DLB-30, 31

Hutchinson and Company (Publishers) Limited DLB-112

Hutton, Richard Holt 1826-1897 .... DLB-57

Huxley, Aldous 1894-1963 .... DLB-36, 100, 162

Huxley, Elspeth Josceline 1907- ... DLB-77

Huxley, T. H. 1825-1895 .... DLB-57

Huyghue, Douglas Smith 1816-1891 .... DLB-99

Huysmans, Joris-Karl 1848-1907 ... DLB-123

Hyman, Trina Schart 1939- DLB-61

I

Iavorsky, Stefan 1658-1722 .... DLB-150

Ibn Bajja circa 1077-1138 .... DLB-115

Ibn Gabirol, Solomon circa 1021-circa 1058 .... DLB-115

The Iconography of Science-Fiction Art DLB-8

Iffland, August Wilhelm 1759-1814 .... DLB-94

Ignatow, David 1914- DLB-5

Ike, Chukwuemeka 1931- DLB-157

Iles, Francis (see Berkeley, Anthony)

The Illustration of Early German Literary Manuscripts, circa 1150-circa 1300 .... DLB-148

Imbs, Bravig 1904-1946 .... DLB-4

Imbuga, Francis D. 1947- DLB-157

Immermann, Karl 1796-1840 ...... DLB-133

Inchbald, Elizabeth 1753-1821 ... DLB-39, 89

Inge, William 1913-1973 .... DLB-7

Ingelow, Jean 1820-1897 ...... DLB-35, 163

Ingersoll, Ralph 1900-1985 .... DLB-127

The Ingersoll Prizes Y-84

Ingoldsby, Thomas (see Barham, Richard Harris)

Ingraham, Joseph Holt 1809-1860 ..... DLB-3

Inman, John 1805-1850 .... DLB-73

Innerhofer, Franz 1944- DLB-85

Innis, Harold Adams 1894-1952 .... DLB-88

Innis, Mary Quayle 1899-1972 ...... DLB-88

International Publishers Company .. DLB-46

An Interview with David Rabe Y-91

An Interview with George Greenfield, Literary Agent Y-91

An Interview with James Ellroy Y-91

An Interview with Peter S. Prescott Y-86

An Interview with Russell Hoban Y-90

An Interview with Tom Jenks Y-86

Introduction to Paul Laurence Dunbar, Lyrics of Lowly Life (1896), by William Dean Howells DLB-50

Introductory Essay: Letters of Percy Bysshe Shelley (1852), by Robert Browning DLB-32

Introductory Letters from the Second Edition of Pamela (1741), by Samuel Richardson DLB-39

Irving, John 1942- DLB-6; Y-82

Irving, Washington 1783-1859 ..... DLB-3, 11, 30, 59, 73, 74

Irwin, Grace 1907- DLB-68

Irwin, Will 1873-1948 .... DLB-25

Isherwood, Christopher 1904-1986 .... DLB-15; Y-86

The Island Trees Case: A Symposium on School Library Censorship
An Interview with Judith Krug
An Interview with Phyllis Schlafly
An Interview with Edward B. Jenkinson
An Interview with Lamarr Mooneyham
An Interview with Harriet Bernstein Y-82

Islas, Arturo 1938-1991 .... DLB-122

Ivers, M. J., and Company DLB-49

Iyayi, Festus 1947- DLB-157

J

Jackmon, Marvin E. (see Marvin X)

Jacks, L. P. 1860-1955 .... DLB-135

Jackson, Angela 1951- DLB-41

Jackson, Helen Hunt 1830-1885 .... DLB-42, 47

Jackson, Holbrook 1874-1948 .... DLB-98

Jackson, Laura Riding 1901-1991 .... DLB-48

Jackson, Shirley 1919-1965 .... DLB-6

Jacob, Piers Anthony Dillingham (see Anthony, Piers)

Jacobi, Friedrich Heinrich 1743-1819 .... DLB-94

Jacobi, Johann Georg 1740-1841 ..... DLB-97

Jacobs, Joseph 1854-1916 .... DLB-141

Jacobs, W. W. 1863-1943 .... DLB-135

Jacobs, George W., and Company ... DLB-49

Jacobson, Dan 1929- DLB-14

Jaggard, William [publishing house] DLB-170

Jahier, Piero 1884-1966 .... DLB-114

Jahnn, Hans Henny 1894-1959 .... DLB-56, 124

Jakes, John 1932- Y-83

James, C. L. R. 1901-1989 .... DLB-125

James, George P. R. 1801-1860 ..... DLB-116

James, Henry 1843-1916 .... DLB-12, 71, 74; DS-13

James, John circa 1633-1729 .... DLB-24

The James Jones Society Y-92

James, M. R. 1862-1936 .... DLB-156

James, P. D. 1920- DLB-87

James Joyce Centenary: Dublin, 1982 ... Y-82

James Joyce Conference Y-85

James VI of Scotland, I of England 1566-1625 .... DLB-151, 172

James, U. P. [publishing house] DLB-49

Jameson, Anna 1794-1860 ...... DLB-99, 166

Jameson, Fredric 1934- DLB-67

Jameson, J. Franklin 1859-1937 ...... DLB-17

Jameson, Storm 1891-1986 .... DLB-36

Janés, Clara 1940- DLB-134

Jaramillo, Cleofas M. 1878-1956 .... DLB-122

Jarman, Mark 1952- DLB-120

Jarrell, Randall 1914-1965 ...... DLB-48, 52

Jarrold and Sons DLB-106

Jasmin, Claude 1930- DLB-60

Jay, John 1745-1829 .... DLB-31

Jefferies, Richard 1848-1887 .... DLB-98, 141

Jeffers, Lance 1919-1985 .... DLB-41

Jeffers, Robinson 1887-1962 .... DLB-45

Jefferson, Thomas 1743-1826 .... DLB-31

Jelinek, Elfriede 1946- DLB-85

Jellicoe, Ann 1927- DLB-13

Jenkins, Elizabeth 1905- DLB-155

Jenkins, Robin 1912- DLB-14

Jenkins, William Fitzgerald (see Leinster, Murray)

Jenkins, Herbert, Limited DLB-112

Jennings, Elizabeth 1926- DLB-27

Jens, Walter 1923- DLB-69

Jensen, Merrill 1905-1980 .... DLB-17

Jephson, Robert 1736-1803 .... DLB-89

Jerome, Jerome K. 1859-1927 .... DLB-10, 34, 135

Jerome, Judson 1927-1991 .... DLB-105

Jerome, Judson, Reflections: After a Tornado DLB-105

Jerrold, Douglas 1803-1857 .... DLB-158, 159

Jesse, F. Tennyson 1888-1958 .... DLB-77

Jewett, Sarah Orne 1849-1909 .... DLB-12, 74

Jewett, John P., and Company DLB-49

The Jewish Publication Society DLB-49

Jewitt, John Rodgers 1783-1821 ..... DLB-99

Jewsbury, Geraldine 1812-1880 ...... DLB-21

Jhabvala, Ruth Prawer 1927- DLB-139

Jiménez, Juan Ramón 1881-1958 .... DLB-134

Joans, Ted 1928- DLB-16, 41

John, Eugenie (see Marlitt, E.)

John of Dumbleton circa 1310-circa 1349 .... DLB-115

John Edward Bruce: Three Documents DLB-50

John O'Hara's Pottsville Journalism Y-88

John Steinbeck Research Center Y-85

John Webster: The Melbourne Manuscript Y-86

Johns, Captain W. E. 1893-1968 .... DLB-160

Johnson, B. S. 1933-1973 .... DLB-14, 40

Johnson, Charles 1679-1748 .... DLB-84

Johnson, Charles R. 1948- DLB-33

Johnson, Charles S. 1893-1956 ... DLB-51, 91

Johnson, Denis 1949- DLB-120

Johnson, Diane 1934- Y-80

Johnson, Edgar 1901- DLB-103

Johnson, Edward 1598-1672 .... DLB-24

Johnson, Fenton 1888-1958 ...... DLB-45, 50

Johnson, Georgia Douglas 1886-1966 .... DLB-51

Johnson, Gerald W. 1890-1980 ...... DLB-29

Johnson, Helene 1907- DLB-51

Johnson, James Weldon 1871-1938 .... DLB-51

Johnson, John H. 1918- DLB-137

Johnson, Linton Kwesi 1952- DLB-157

Johnson, Lionel 1867-1902 .... DLB-19

Johnson, Nunnally 1897-1977 ...... DLB-26

Johnson, Owen 1878-1952 .... Y-87

Johnson, Pamela Hansford 1912- DLB-15

Johnson, Pauline 1861-1913 .... DLB-92

Johnson, Ronald 1935- DLB-169

Johnson, Samuel 1696-1772 .... DLB-24

Johnson, Samuel 1709-1784 .... DLB-39, 95, 104, 142

Johnson, Samuel 1822-1882 .... DLB-1

Johnson, Uwe 1934-1984 .... DLB-75

Johnson, Benjamin [publishing house] DLB-49

Johnson, Benjamin, Jacob, and Robert [publishing house] DLB-49

Johnson, Jacob, and Company DLB-49

Johnson, Joseph [publishing house] ...DLB-154

Johnston, Annie Fellows 1863-1931 .. DLB-42

Johnston, Basil H. 1929- DLB-60

Johnston, Denis 1901-1984 .... DLB-10

Johnston, George 1913- DLB-88

Johnston, Sir Harry 1858-1927 ..... DLB-174

Johnston, Jennifer 1930- DLB-14

Johnston, Mary 1870-1936 .... DLB-9

Johnston, Richard Malcolm 1822-1898 .... DLB-74

Johnstone, Charles 1719?-1800? DLB-39

Johst, Hanns 1890-1978 .... DLB-124

Jolas, Eugene 1894-1952 .... DLB-4, 45

Jones, Alice C. 1853-1933 .... DLB-92

Jones, Charles C., Jr. 1831-1893 ..... DLB-30

Jones, D. G. 1929- DLB-53

Jones, David 1895-1974 .... DLB-20, 100

Jones, Diana Wynne 1934- DLB-161

Jones, Ebenezer 1820-1860 .... DLB-32

Jones, Ernest 1819-1868 .... DLB-32

Jones, Gayl 1949- DLB-33

Jones, Glyn 1905- DLB-15

Jones, Gwyn 1907- DLB-15, 139

Jones, Henry Arthur 1851-1929 ..... DLB-10

Jones, Hugh circa 1692-1760 .... DLB-24

Jones, James 1921-1977 .... DLB-2, 143

Jones, Jenkin Lloyd 1911- DLB-127

Jones, LeRoi (see Baraka, Amiri)

Jones, Lewis 1897-1939 .... DLB-15

Jones, Madison 1925- DLB-152

Jones, Major Joseph (see Thompson, William Tappan)

Jones, Preston 1936-1979 .... DLB-7

Jones, Rodney 1950- DLB-120

Jones, Sir William 1746-1794 ...... DLB-109

Jones, William Alfred 1817-1900 .... DLB-59

Jones's Publishing House DLB-49

Jong, Erica 1942- DLB-2, 5, 28, 152

Jonke, Gert F. 1946- DLB-85

Jonson, Ben 1572?-1637 .... DLB-62, 121

Jordan, June 1936- DLB-38

Joseph, Jenny 1932- DLB-40

Joseph, Michael, Limited DLB-112

Josephson, Matthew 1899-1978 .... DLB-4

Josiah Allen's Wife (see Holley, Marietta)

Josipovici, Gabriel 1940- DLB-14

Josselyn, John ?-1675 .... DLB-24

Joudry, Patricia 1921- DLB-88

Jovine, Giuseppe 1922- DLB-128

Joyaux, Philippe (see Sollers, Philippe)

Joyce, Adrien (see Eastman, Carol)

Joyce, James 1882-1941 .... DLB-10, 19, 36, 162

Judd, Sylvester 1813-1853 .... DLB-1

Judd, Orange, Publishing Company DLB-49

Judith circa 930 .... DLB-146

Julian of Norwich 1342-circa 1420 .... DLB-1146

Julian Symons at Eighty Y-92

June, Jennie (see Croly, Jane Cunningham)

Jung, Franz 1888-1963 .... DLB-118

Jünger, Ernst 1895- DLB-56

Der jüngere Titurel circa 1275 .... DLB-138

Jung-Stilling, Johann Heinrich 1740-1817 .... DLB-94

Justice, Donald 1925- Y-83

The Juvenile Library (see Godwin, M. J., and Company)

K

Kacew, Romain (see Gary, Romain)

Kafka, Franz 1883-1924 .... DLB-81

Kahn, Roger 1927 .... DLB-171

Kaiser, Georg 1878-1945 .... DLB-124

Kaiserchronik circca 1147 .... DLB-148

Kalechofsky, Roberta 1931- DLB-28

Kaler, James Otis 1848-1912 .... DLB-12

Kames, Henry Home, Lord 1696-1782 .... DLB-31, 104

Cumulative Index

Kandel, Lenore 1932- DLB-16

Kanin, Garson 1912- DLB-7

Kant, Hermann 1926- DLB-75

Kant, Immanuel 1724-1804 .... DLB-94

Kantemir, Antiokh Dmitrievich
  1708-1744 .... DLB-150

Kantor, Mackinlay 1904-1977 ... DLB-9, 102

Kaplan, Fred 1937- DLB-111

Kaplan, Johanna 1942- DLB-28

Kaplan, Justin 1925- DLB-111

Kapnist, Vasilii Vasilevich
  1758?-1823 .... DLB-150

Karadžić, Vuk Stefanović
  1787-1864 .... DLB-147

Karamzin, Nikolai Mikhailovich
  1766-1826 .... DLB-150

Karsch, Anna Louisa 1722-1791 .... DLB-97

Kasack, Hermann 1896-1966 .... DLB-69

Kaschnitz, Marie Luise 1901-1974 ... DLB-69

Kaštelan, Jure 1919-1990 .... DLB-147

Kästner, Erich 1899-1974 .... DLB-56

Kattan, Naim 1928- DLB-53

Katz, Steve 1935- Y-83

Kauffman, Janet 1945- Y-86

Kauffmann, Samuel 1898-1971 .... DLB-127

Kaufman, Bob 1925- DLB-16, 41

Kaufman, George S. 1889-1961 ...... DLB-7

Kavanagh, P. J. 1931- DLB-40

Kavanagh, Patrick 1904-1967 .... DLB-15, 20

Kaye-Smith, Sheila 1887-1956 ...... DLB-36

Kazin, Alfred 1915- DLB-67

Keane, John B. 1928- DLB-13

Keary, Annie 1825-1879 .... DLB-163

Keating, H. R. F. 1926- DLB-87

Keats, Ezra Jack 1916-1983 ...... DLB-61

Keats, John 1795-1821 .... DLB-96, 110

Keble, John 1792-1866 .... DLB-32, 55

Keeble, John 1944- Y-83

Keeffe, Barrie 1945- DLB-13

Keeley, James 1867-1934 .... DLB-25

W. B. Keen, Cooke
  and Company DLB-49

Keillor, Garrison 1942- Y-87

Keith, Marian 1874?-1961 .... DLB-92

Keller, Gary D. 1943- DLB-82

Keller, Gottfried 1819-1890 ..... DLB-129

Kelley, Edith Summers 1884-1956 .... DLB-9

Kelley, William Melvin 1937- DLB-33

Kellogg, Ansel Nash 1832-1886 ..... DLB-23

Kellogg, Steven 1941- DLB-61

Kelly, George 1887-1974 .... DLB-7

Kelly, Hugh 1739-1777 .... DLB-89

Kelly, Robert 1935- DLB-5, 130, 165

Kelly, Piet and Company DLB-49

Kelmscott Press DLB-112

Kemble, Fanny 1809-1893 .... DLB-32

Kemelman, Harry 1908- DLB-28

Kempe, Margery
  circa 1373-1438 .... DLB-146

Kempner, Friederike 1836-1904 .... DLB-129

Kempowski, Walter 1929- DLB-75

Kendall, Claude
  [publishing company] DLB-46

Kendell, George 1809-1867 .... DLB-43

Kenedy, P. J., and Sons DLB-49

Kennedy, Adrienne 1931- DLB-38

Kennedy, John Pendleton 1795-1870 ... DLB-3

Kennedy, Leo 1907- DLB-88

Kennedy, Margaret 1896-1967 ..... DLB-36

Kennedy, Patrick 1801-1873 ...... DLB-159

Kennedy, Richard S. 1920- DLB-111

Kennedy, William 1928- DLB-143; Y-85

Kennedy, X. J. 1929- DLB-5

Kennelly, Brendan 1936- DLB-40

Kenner, Hugh 1923- DLB-67

Kennerley, Mitchell
  [publishing house] DLB-46

Kent, Frank R. 1877-1958 .... DLB-29

Kenyon, Jane 1947- DLB-120

Keough, Hugh Edmund 1864-1912 . DLB-171

Keppler and Schwartzmann DLB-49

Kerner, Justinus 1776-1862 .... DLB-90

Kerouac, Jack 1922-1969 ... DLB-2, 16; DS-3

The Jack Kerouac Revival Y-95

Kerouac, Jan 1952- DLB-16

Kerr, Orpheus C. (see Newell, Robert Henry)

Kerr, Charles H., and Company DLB-49

Kesey, Ken 1935- DLB-2, 16

Kessel, Joseph 1898-1979 .... DLB-72

Kessel, Martin 1901- DLB-56

Kesten, Hermann 1900- DLB-56

Keun, Irmgard 1905-1982 .... DLB-69

Key and Biddle DLB-49

Keynes, John Maynard 1883-1946 .... DS-10

Keyserling, Eduard von 1855-1918 ... DLB-66

Khan, Ismith 1925- DLB-125

Khemnitser, Ivan Ivanovich
  1745-1784 .... DLB-150

Kheraskov, Mikhail Matveevich
  1733-1807 .... DLB-150

Khvostov, Dmitrii Ivanovich
  1757-1835 .... DLB-150

Kidd, Adam 1802?-1831 .... DLB-99

Kidd, William
  [publishing house] DLB-106

Kiely, Benedict 1919- DLB-15

Kieran, John 1892-1981 .... DLB-171

Kiggins and Kellogg DLB-49

Kiley, Jed 1889-1962 .... DLB-4

Kilgore, Bernard 1908-1967 ...... DLB-127

Killens, John Oliver 1916- DLB-33

Killigrew, Anne 1660-1685 .... DLB-131

Killigrew, Thomas 1612-1683 ..... DLB-58

Kilmer, Joyce 1886-1918 .... DLB-45

Kilwardby, Robert
  circa 1215-1279 .... DLB-115

Kincaid, Jamaica 1949- DLB-157

King, Clarence 1842-1901 .... DLB-12

King, Florence 1936 .... Y-85

King, Francis 1923- DLB-15, 139

King, Grace 1852-1932 .... DLB-12, 78

King, Henry 1592-1669 .... DLB-126

King, Stephen 1947- DLB-143; Y-80

King, Woodie, Jr. 1937- DLB-38

King, Solomon [publishing house] DLB-49

Kinglake, Alexander William
  1809-1891 .... DLB-55, 166

Kingsley, Charles
  1819-1875 .... DLB-21, 32, 163

Kingsley, Mary Henrietta
  1862-1900 .... DLB-174

Kingsley, Henry 1830-1876 .... DLB-21

Kingsley, Sidney 1906- DLB-7

Kingsmill, Hugh 1889-1949 ...... DLB-149

Kingston, Maxine Hong
  1940- DLB-173; Y-80

Kingston, William Henry Giles
  1814-1880 .... DLB-163

Kinnell, Galway 1927- DLB-5; Y-87

Kinsella, Thomas 1928- DLB-27

Kipling, Rudyard 1865-1936 .... DLB-19, 34, 141, 156

Kipphardt, Heinar 1922-1982 ......DLB-124

Kirby, William 1817-1906 ....DLB-99

Kircher, Athanasius 1602-1680 .....DLB-164

Kirk, John Foster 1824-1904 ....DLB-79

Kirkconnell, Watson 1895-1977 .....DLB-68

Kirkland, Caroline M. 1801-1864 .... DLB-3, 73, 74; DS-13

Kirkland, Joseph 1830-1893 ....DLB-12

Kirkman, Francis [publishing house].............DLB-170

Kirkpatrick, Clayton 1915- DLB-127

Kirkup, James 1918- DLB-27

Kirouac, Conrad (see Marie-Victorin, Frère)

Kirsch, Sarah 1935- DLB-75

Kirst, Hans Hellmut 1914-1989 ......DLB-69

Kitcat, Mabel Greenhow 1859-1922 ....DLB-135

Kitchin, C. H. B. 1895-1967 ....DLB-77

Kizer, Carolyn 1925- DLB-5, 169

Klabund 1890-1928 ....DLB-66

Klaj, Johann 1616-1656 ....DLB-164

Klappert, Peter 1942- DLB-5

Klass, Philip (see Tenn, William)

Klein, A. M. 1909-1972 ....DLB-68

Kleist, Ewald von 1715-1759 ....DLB-97

Kleist, Heinrich von 1777-1811 ......DLB-90

Klinger, Friedrich Maximilian 1752-1831 ....DLB-94

Klopstock, Friedrich Gottlieb 1724-1803 ....DLB-97

Klopstock, Meta 1728-1758 ....DLB-97

Kluge, Alexander 1932- DLB-75

Knapp, Joseph Palmer 1864-1951 ....DLB-91

Knapp, Samuel Lorenzo 1783-1838 ....DLB-59

Knapton, J. J. and P. [publishing house].............DLB-154

Kniazhnin, Iakov Borisovich 1740-1791 ....DLB-150

Knickerbocker, Diedrich (see Irving, Washington)

Knigge, Adolph Franz Friedrich Ludwig, Freiherr von 1752-1796 ....DLB-94

Knight, Damon 1922- DLB-8

Knight, Etheridge 1931-1992 ....DLB-41

Knight, John S. 1894-1981 .... DLB-29

Knight, Sarah Kemble 1666-1727 .... DLB-24

Knight, Charles, and Company DLB-106

Knight-Bruce, G. W. H. 1852-1896 .... DLB-174

Knister, Raymond 1899-1932 .... DLB-68

Knoblock, Edward 1874-1945 ...... DLB-10

Knopf, Alfred A. 1892-1984 .... Y-84

Knopf, Alfred A. [publishing house] DLB-46

Knorr von Rosenroth, Christian 1636-1689 .... DLB-168

Knowles, John 1926- DLB-6

Knox, Frank 1874-1944 .... DLB-29

Knox, John circa 1514-1572 .... DLB-132

Knox, John Armoy 1850-1906 ...... DLB-23

Knox, Ronald Arbuthnott 1888-1957 .... DLB-77

Kober, Arthur 1900-1975 .... DLB-11

Kocbek, Edvard 1904-1981 .... DLB-147

Koch, Howard 1902- DLB-26

Koch, Kenneth 1925- DLB-5

Koenigsberg, Moses 1879-1945 ..... DLB-25

Koeppen, Wolfgang 1906- DLB-69

Koertge, Ronald 1940- DLB-105

Koestler, Arthur 1905-1983 .... Y-83

Kokoschka, Oskar 1886-1980 ...... DLB-124

Kolb, Annette 1870-1967 .... DLB-66

Kolbenheyer, Erwin Guido 1878-1962 .... DLB-66, 124

Kolleritsch, Alfred 1931- DLB-85

Kolodny, Annette 1941- DLB-67

Komarov, Matvei circa 1730-1812 .... DLB-150

Komroff, Manuel 1890-1974 .... DLB-4

Komunyakaa, Yusef 1947- DLB-120

Konigsburg, E. L. 1930- DLB-52

Konrad von Würzburg circa 1230-1287 .... DLB-138

Konstantinov, Aleko 1863-1897 .... DLB-147

Kooser, Ted 1939- DLB-105

Kopit, Arthur 1937- DLB-7

Kops, Bernard 1926?- DLB-13

Kornbluth, C. M. 1923-1958 .... DLB-8

Körner, Theodor 1791-1813 .... DLB-90

Kornfeld, Paul 1889-1942 .... DLB-118

Kosinski, Jerzy 1933-1991 ..... DLB-2; Y-82

Kosovel, Srečko 1904-1926 ....DLB-147

Kostrov, Ermil Ivanovich 1755-1796 ....DLB-150

Kotzebue, August von 1761-1819 ....DLB-94

Kotzwinkle, William 1938- DLB-173

Kovačić, Ante 1854-1889 ....DLB-147

Kraf, Elaine 1946- Y-81

Kranjčević, Silvije Strahimir 1865-1908 ....DLB-147

Krasna, Norman 1909-1984 ....DLB-26

Kraus, Karl 1874-1936 ....DLB-118

Krauss, Ruth 1911-1993 ....DLB-52

Kreisel, Henry 1922- DLB-88

Kreuder, Ernst 1903-1972 ....DLB-69

Kreymborg, Alfred 1883-1966 .....DLB-4, 54

Krieger, Murray 1923- DLB-67

Krim, Seymour 1922-1989 ....DLB-16

Krleža, Miroslav 1893-1981 ....DLB-147

Krock, Arthur 1886-1974 ....DLB-29

Kroetsch, Robert 1927- DLB-53

Krutch, Joseph Wood 1893-1970 ....DLB-63

Krylov, Ivan Andreevich 1769-1844 ....DLB-150

Kubin, Alfred 1877-1959 ....DLB-81

Kubrick, Stanley 1928- DLB-26

Kudrun circa 1230-1240 ....DLB-138

Kuffstein, Hans Ludwig von 1582-1656 ....DLB-164

Kuhlmann, Quirinus 1651-1689 ....DLB-168

Kuhnau, Johann 1660-1722 ....DLB-168

Kumin, Maxine 1925- DLB-5

Kunene, Mazisi 1930- DLB-117

Kunitz, Stanley 1905- DLB-48

Kunjufu, Johari M. (see Amini, Johari M.)

Kunnert, Gunter 1929- DLB-75

Kunze, Reiner 1933- DLB-75

Kupferberg, Tuli 1923- DLB-16

Kürnberger, Ferdinand 1821-1879 ....DLB-129

Kurz, Isolde 1853-1944 ....DLB-66

Kusenberg, Kurt 1904-1983 ....DLB-69

Kuttner, Henry 1915-1958 ....DLB-8

Kyd, Thomas 1558-1594 ....DLB-62

Kyffin, Maurice circa 1560?-1598 ....DLB-136

Kyger, Joanne 1934- DLB-16

Cumulative Index

Kyne, Peter B. 1880-1957 .... DLB-78

L

L. E. L. (see Landon, Letitia Elizabeth)

Laberge, Albert 1871-1960 .... DLB-68

Laberge, Marie 1950- DLB-60

Lacombe, Patrice (see Trullier-Lacombe, Joseph Patrice)

Lacretelle, Jacques de 1888-1985 .... DLB-65

Lacy, Sam 1903- DLB-171

Ladd, Joseph Brown 1764-1786 ..... DLB-37

La Farge, Oliver 1901-1963 .... DLB-9

Lafferty, R. A. 1914- DLB-8

La Guma, Alex 1925-1985 .... DLB-117

Lahaise, Guillaume (see Delahaye, Guy)

Lahontan, Louis-Armand de Lom d'Arce, Baron de 1666-1715? DLB-99

Laing, Kojo 1946- DLB-157

Laird, Carobeth 1895- Y-82

Laird and Lee DLB-49

Lalonde, Michèle 1937- DLB-60

Lamantia, Philip 1927- DLB-16

Lamb, Charles 1775-1834 .... DLB-93, 107, 163

Lamb, Lady Caroline 1785-1828 ... DLB-116

Lamb, Mary 1764-1874 .... DLB-163

Lambert, Betty 1933-1983 .... DLB-60

Lamming, George 1927- DLB-125

L'Amour, Louis 1908?- Y-80

Lampman, Archibald 1861-1899 .... DLB-92

Lamson, Wolffe and Company DLB-49

Lancer Books DLB-46

Landesman, Jay 1919- and Landesman, Fran 1927- DLB-16

Landon, Letitia Elizabeth 1802-1838 . DLB-96

Landor, Walter Savage 1775-1864 .... DLB-93, 107

Landry, Napoléon-P. 1884-1956 .... DLB-92

Lane, Charles 1800-1870 .... DLB-1

Lane, Laurence W. 1890-1967 ..... DLB-91

Lane, M. Travis 1934- DLB-60

Lane, Patrick 1939- DLB-53

Lane, Pinkie Gordon 1923- DLB-41

Lane, John, Company DLB-49

Laney, Al 1896-1988 .... DLB-4, 171

Lang, Andrew 1844-1912 ...... DLB-98, 141

Langevin, André 1927- DLB-60

Langgässer, Elisabeth 1899-1950 .... DLB-69

Langhorne, John 1735-1779 .... DLB-109

Langland, William circa 1330-circa 1400 .... DLB-146

Langton, Anna 1804-1893 .... DLB-99

Lanham, Edwin 1904-1979 .... DLB-4

Lanier, Sidney 1842-1881 ....DLB-64; DS-13

Lanyer, Aemilia 1569-1645 .... DLB-121

Lapointe, Gatien 1931-1983 .... DLB-88

Lapointe, Paul-Marie 1929- DLB-88

Lardner, John 1912-1960 .... DLB-171

Lardner, Ring 1885-1933 .... DLB-11, 25, 86, 171

Lardner, Ring, Jr. 1915- DLB-26

Lardner 100: Ring Lardner Centennial Symposium Y-85

Larkin, Philip 1922-1985 .... DLB-27

La Roche, Sophie von 1730-1807 .... DLB-94

La Rocque, Gilbert 1943-1984 ...... DLB-60

Laroque de Roquebrune, Robert (see Roquebrune, Robert de)

Larrick, Nancy 1910- DLB-61

Larsen, Nella 1893-1964 .... DLB-51

Lasker-Schüler, Else 1869-1945 .... DLB-66, 124

Lasnier, Rina 1915- DLB-88

Lassalle, Ferdinand 1825-1864 ..... DLB-129

Lathrop, Dorothy P. 1891-1980 ..... DLB-22

Lathrop, George Parsons 1851-1898 .... DLB-71

Lathrop, John, Jr. 1772-1820 .... DLB-37

Latimer, Hugh 1492?-1555 .... DLB-136

Latimore, Jewel Christine McLawler (see Amini, Johari M.)

Latymer, William 1498-1583 .... DLB-132

Laube, Heinrich 1806-1884 .... DLB-133

Laughlin, James 1914- DLB-48

Laumer, Keith 1925- DLB-8

Lauremberg, Johann 1590-1658 .... DLB-164

Laurence, Margaret 1926-1987 ..... DLB-53

Laurentius von Schnüffis 1633-1702 .... DLB-168

Laurents, Arthur 1918- DLB-26

Laurie, Annie (see Black, Winifred)

Laut, Agnes Christiana 1871-1936 ... DLB-92

Lavater, Johann Kaspar 1741-1801 ... DLB-97

Lavin, Mary 1912- DLB-15

Lawes, Henry 1596-1662 .... DLB-126

Lawless, Anthony (see MacDonald, Philip)

Lawrence, D. H. 1885-1930 .... DLB-10, 19, 36, 98, 162

Lawrence, David 1888-1973 .... DLB-29

Lawrence, Seymour 1926-1994 .... Y-94

Lawson, John ?-1711 .... DLB-24

Lawson, Robert 1892-1957 .... DLB-22

Lawson, Victor F. 1850-1925 .... DLB-25

Layard, Sir Austen Henry 1817-1894 .... DLB-166

Layton, Irving 1912- DLB-88

LaZamon flourished circa 1200 ..... DLB-146

Lazarević, Laza K. 1851-1890 ...... DLB-147

Lea, Henry Charles 1825-1909 ...... DLB-47

Lea, Sydney 1942- DLB-120

Lea, Tom 1907- DLB-6

Leacock, John 1729-1802 .... DLB-31

Leacock, Stephen 1869-1944 .... DLB-92

Lead, Jane Ward 1623-1704 .... DLB-131

Leadenhall Press DLB-106

Leapor, Mary 1722-1746 .... DLB-109

Lear, Edward 1812-1888 ... DLB-32, 163, 166

Leary, Timothy 1920-1996 .... DLB-16

Leary, W. A., and Company DLB-49

Léautaud, Paul 1872-1956 .... DLB-65

Leavitt, David 1961- DLB-130

Leavitt and Allen DLB-49

Le Blond, Mrs. Aubrey 1861-1934 .... DLB-174

le Carré, John 1931- DLB-87

Lécavelé, Roland (see Dorgeles, Roland)

Lechlitner, Ruth 1901- DLB-48

Leclerc, Félix 1914- DLB-60

Le Clézio, J. M. G. 1940- DLB-83

Lectures on Rhetoric and Belles Lettres (1783), by Hugh Blair [excerpts] DLB-31

Leder, Rudolf (see Hermlin, Stephan)

Lederer, Charles 1910-1976 .... DLB-26

Ledwidge, Francis 1887-1917 .... DLB-20

Lee, Dennis 1939- DLB-53

Lee, Don L. (see Madhubuti, Haki R.)

Lee, George W. 1894-1976 .... DLB-51

Lee, Harper 1926- DLB-6

Lee, Harriet (1757-1851) and
  Lee, Sophia (1750-1824) DLB-39

Lee, Laurie 1914- DLB-27

Lee, Li-Young 1957- DLB-165

Lee, Manfred B. (see Dannay, Frederic, and
  Manfred B. Lee)

Lee, Nathaniel circa 1645 - 1692 .....DLB-80

Lee, Sir Sidney 1859-1926 ....DLB-149

Lee, Sir Sidney, "Principles of Biography," in
  Elizabethan and Other Essays DLB-149

Lee, Vernon
  1856-1935 .... DLB-57, 153, 156, 174

Lee and Shepard DLB-49

Le Fanu, Joseph Sheridan
  1814-1873 .... DLB-21, 70, 159

Leffland, Ella 1931- Y-84

le Fort, Gertrud von 1876-1971 ......DLB-66

Le Gallienne, Richard 1866-1947 .....DLB-4

Legaré, Hugh Swinton
  1797-1843 .... DLB-3, 59, 73

Legaré, James M. 1823-1859 ....DLB-3

The Legends of the Saints and a Medieval
  Christian Worldview DLB-148

Léger, Antoine-J. 1880-1950 ....DLB-88

Le Guin, Ursula K. 1929- DLB-8, 52

Lehman, Ernest 1920- DLB-44

Lehmann, John 1907- DLB-27, 100

Lehmann, Rosamond 1901-1990 .....DLB-15

Lehmann, Wilhelm 1882-1968 ......DLB-56

Lehmann, John, Limited DLB-112

Leiber, Fritz 1910-1992 ....DLB-8

Leibniz, Gottfried Wilhelm
  1646-1716 ....DLB-168

Leicester University Press DLB-112

Leinster, Murray 1896-1975 ....DLB-8

Leisewitz, Johann Anton
  1752-1806 ....DLB-94

Leitch, Maurice 1933- DLB-14

Leithauser, Brad 1943- DLB-120

Leland, Charles G. 1824-1903 ....DLB-11

Leland, John 1503?-1552 ....DLB-136

Lemay, Pamphile 1837-1918 ....DLB-99

Lemelin, Roger 1919- DLB-88

Lemon, Mark 1809-1870 ....DLB-163

Le Moine, James MacPherson
  1825-1912 ....DLB-99

Le Moyne, Jean 1913- DLB-88

L'Engle, Madeleine 1918- DLB-52

Lennart, Isobel 1915-1971 ....DLB-44

Lennox, Charlotte
  1729 or 1730-1804 ....DLB-39

Lenox, James 1800-1880 ....DLB-140

Lenski, Lois 1893-1974 ....DLB-22

Lenz, Hermann 1913- DLB-69

Lenz, J. M. R. 1751-1792 ....DLB-94

Lenz, Siegfried 1926- DLB-75

Leonard, Elmore 1925- DLB-173

Leonard, Hugh 1926- DLB-13

Leonard, William Ellery
  1876-1944 ....DLB-54

Leonowens, Anna 1834-1914 ... DLB-99, 166

LePan, Douglas 1914- DLB-88

Leprohon, Rosanna Eleanor
  1829-1879 ....DLB-99

Le Queux, William 1864-1927 ......DLB-70

Lerner, Max 1902-1992 ....DLB-29

Lernet-Holenia, Alexander
  1897-1976 ....DLB-85

Le Rossignol, James 1866-1969 .....DLB-92

Lescarbot, Marc circa 1570-1642 ....DLB-99

LeSeur, William Dawson
  1840-1917 ....DLB-92

LeSieg, Theo. (see Geisel, Theodor Seuss)

Leslie, Frank 1821-1880 ....DLB-43, 79

Leslie, Frank, Publishing House DLB-49

Lesperance, John 1835?-1891 ....DLB-99

Lessing, Bruno 1870-1940 ....DLB-28

Lessing, Doris 1919- DLB-15, 139; Y-85

Lessing, Gotthold Ephraim
  1729-1781 ....DLB-97

Lettau, Reinhard 1929- DLB-75

Letter from Japan Y-94

Letter to [Samuel] Richardson on *Clarissa*
  (1748), by Henry Fielding DLB-39

Lever, Charles 1806-1872 ....DLB-21

Leverson, Ada 1862-1933 ....DLB-153

Levertov, Denise 1923- DLB-5, 165

Levi, Peter 1931- DLB-40

Levien, Sonya 1888-1960 ....DLB-44

Levin, Meyer 1905-1981 .... DLB-9, 28; Y-81

Levine, Norman 1923- DLB-88

Levine, Philip 1928- DLB-5

Levis, Larry 1946- DLB-120

Levy, Amy 1861-1889 ....DLB-156

Levy, Benn Wolfe
  1900-1973 ....DLB-13; Y-81

Lewald, Fanny 1811-1889 ....DLB-129

Lewes, George Henry
  1817-1878 ....DLB-55, 144

Lewis, Agnes Smith 1843-1926 .....DLB-174

Lewis, Alfred H. 1857-1914 ....DLB-25

Lewis, Alun 1915-1944 ....DLB-20, 162

Lewis, C. Day (see Day Lewis, C.)

Lewis, C. S. 1898-1963 .... DLB-15, 100, 160

Lewis, Charles B. 1842-1924 ....DLB-11

Lewis, Henry Clay 1825-1850 ....DLB-3

Lewis, Janet 1899- Y-87

Lewis, Matthew Gregory
  1775-1818 ....DLB-39, 158

Lewis, R. W. B. 1917- DLB-111

Lewis, Richard circa 1700-1734 .....DLB-24

Lewis, Sinclair
  1885-1951 .... DLB-9, 102; DS-1

Lewis, Wilmarth Sheldon
  1895-1979 ....DLB-140

Lewis, Wyndham 1882-1957 ....DLB-15

Lewisohn, Ludwig
  1882-1955 .... DLB-4, 9, 28, 102

Lezama Lima, José 1910-1976 ......DLB-113

The Library of America DLB-46

The Licensing Act of 1737 ....DLB-84

Lichfield, Leonard I
  [publishing house] DLB-170

Lichtenberg, Georg Christoph
  1742-1799 ....DLB-94

Lieb, Fred 1888-1980 ....DLB-171

Liebling, A. J. 1904-1963 ....DLB-4, 171

Lieutenant Murray (see Ballou, Maturin
  Murray)

Lighthall, William Douw
  1857-1954 ....DLB-92

Lilar, Françoise (see Mallet-Joris, Françoise)

Lillo, George 1691-1739 ....DLB-84

Lilly, J. K., Jr. 1893-1966 ....DLB-140

Lilly, Wait and Company DLB-49

Lily, William circa 1468-1522 ......DLB-132

Limited Editions Club DLB-46

Lincoln and Edmands DLB-49

Lindsay, Jack 1900- Y-84

Lindsay, Sir David
  circa 1485-1555 ....DLB-132

Lindsay, Vachel 1879-1931 ....DLB-54

Linebarger, Paul Myron Anthony (see Smith, Cordwainer)

Link, Arthur S. 1920- DLB-17

Linn, John Blair 1777-1804 .... DLB-37

Lins, Osman 1924-1978 .... DLB-145

Linton, Eliza Lynn 1822-1898 ...... DLB-18

Linton, William James 1812-1897 ... DLB-32

Lintot, Barnaby Bernard [publishing house] DLB-170

Lion Books DLB-46

Lionni, Leo 1910- DLB-61

Lippincott, Sara Jane Clarke 1823-1904 .... DLB-43

Lippincott, J. B., Company DLB-49

Lippmann, Walter 1889-1974 ...... DLB-29

Lipton, Lawrence 1898-1975 .... DLB-16

Liscow, Christian Ludwig 1701-1760 .... DLB-97

Lish, Gordon 1934- DLB-130

Lispector, Clarice 1925-1977 ...... DLB-113

The Literary Chronicle and Weekly Review 1819-1828 .... DLB-110

Literary Documents: William Faulkner and the People-to-People Program Y-86

Literary Documents II: *Library Journal* Statements and Questionnaires from First Novelists Y-87

Literary Effects of World War II [British novel] DLB-15

Literary Prizes [British] DLB-15

Literary Research Archives: The Humanities Research Center, University of Texas Y-82

Literary Research Archives II: Berg Collection of English and American Literature of the New York Public Library Y-83

Literary Research Archives III: The Lilly Library Y-84

Literary Research Archives IV: The John Carter Brown Library Y-85

Literary Research Archives V: Kent State Special Collections Y-86

Literary Research Archives VI: The Modern Literary Manuscripts Collection in the Special Collections of the Washington University Libraries Y-87

Literary Research Archives VII: The University of Virginia Libraries Y-91

Literary Research Archives VIII: The Henry E. Huntington Library Y-92

"Literary Style" (1857), by William Forsyth [excerpt] DLB-57

Literatura Chicanesca: The View From Without DLB-82

Literature at Nurse, or Circulating Morals (1885), by George Moore DLB-18

Littell, Eliakim 1797-1870 .... DLB-79

Littell, Robert S. 1831-1896 .... DLB-79

Little, Brown and Company DLB-49

Littlewood, Joan 1914- DLB-13

Lively, Penelope 1933- DLB-14, 161

Liverpool University Press DLB-112

The Lives of the Poets DLB-142

Livesay, Dorothy 1909- DLB-68

Livesay, Florence Randal 1874-1953 .... DLB-92

Livings, Henry 1929- DLB-13

Livingston, Anne Howe 1763-1841 .... DLB-37

Livingston, Myra Cohn 1926- DLB-61

Livingston, William 1723-1790 ..... DLB-31

Livingstone, David 1813-1873 ...... DLB-166

Liyong, Taban lo (see Taban lo Liyong)

Lizárraga, Sylvia S. 1925- DLB-82

Llewellyn, Richard 1906-1983 .... DLB-15

Lloyd, Edward [publishing house] DLB-106

Lobel, Arnold 1933- DLB-61

Lochridge, Betsy Hopkins (see Fancher, Betsy)

Locke, David Ross 1833-1888 .... DLB-11, 23

Locke, John 1632-1704 .... DLB-31, 101

Locke, Richard Adams 1800-1871 ... DLB-43

Locker-Lampson, Frederick 1821-1895 .... DLB-35

Lockhart, John Gibson 1794-1854 .... DLB-110, 116, 144

Lockridge, Ross, Jr. 1914-1948 .... DLB-143; Y-80

Locrine and *Selimus* DLB-62

Lodge, David 1935- DLB-14

Lodge, George Cabot 1873-1909 .... DLB-54

Lodge, Henry Cabot 1850-1924 ..... DLB-47

Lodge, Thomas 1558-1625 .... DLB-172

Loeb, Harold 1891-1974 .... DLB-4

Loeb, William 1905-1981 .... DLB-127

Lofting, Hugh 1886-1947 .... DLB-160

Logan, James 1674-1751 ...... DLB-24, 140

Logan, John 1923- DLB-5

Logan, William 1950- DLB-120

Logau, Friedrich von 1605-1655 ....DLB-164

Logue, Christopher 1926- DLB-27

Lohenstein, Daniel Casper von 1635-1683 .... DLB-168

Lomonosov, Mikhail Vasil'evich 1711-1765 .... DLB-150

London, Jack 1876-1916 ...... DLB-8, 12, 78

The London Magazine 1820-1829 ..... DLB-110

Long, Haniel 1888-1956 .... DLB-45

Long, Ray 1878-1935 .... DLB-137

Long, H., and Brother DLB-49

Longfellow, Henry Wadsworth 1807-1882 .... DLB-1, 59

Longfellow, Samuel 1819-1892 .... DLB-1

Longford, Elizabeth 1906- DLB-155

Longley, Michael 1939- DLB-40

Longman, T. [publishing house] DLB-154

Longmans, Green and Company DLB-49

Longmore, George 1793?-1867 ......DLB-99

Longstreet, Augustus Baldwin 1790-1870 .... DLB-3, 11, 74

Longworth, D. [publishing house] DLB-49

Lonsdale, Frederick 1881-1954 ...... DLB-10

A Look at the Contemporary Black Theatre Movement DLB-38

Loos, Anita 1893-1981 .... DLB-11, 26; Y-81

Lopate, Phillip 1943- Y-80

López, Diana (see Isabella, Ríos)

Loranger, Jean-Aubert 1896-1942 .... DLB-92

Lorca, Federico García 1898-1936 ...DLB-108

Lord, John Keast 1818-1872 .... DLB-99

The Lord Chamberlain's Office and Stage Censorship in England DLB-10

Lorde, Audre 1934-1992 .... DLB-41

Lorimer, George Horace 1867-1939 .... DLB-91

Loring, A. K. [publishing house] DLB-49

Loring and Mussey DLB-46

Lossing, Benson J. 1813-1891 ....DLB-30

Lothar, Ernst 1890-1974 .... DLB-81

Lothrop, Harriet M. 1844-1924 ...... DLB-42

Lothrop, D., and Company DLB-49

Loti, Pierre 1850-1923 ....DLB-123

Lott, Emeline ?-? DLB-166

The Lounger, no. 20 (1785), by Henry Mackenzie DLB-39

Lounsbury, Thomas R. 1838-1915 ... DLB-71

Louÿs, Pierre 1870-1925 .... DLB-123

Lovelace, Earl 1935- DLB-125

Lovelace, Richard 1618-1657 .... DLB-131

Lovell, Coryell and Company DLB-49

Lovell, John W., Company DLB-49

Lover, Samuel 1797-1868 .... DLB-159

Lovesey, Peter 1936- DLB-87

Lovingood, Sut (see Harris, George Washington)

Low, Samuel 1765-? DLB-37

Lowell, Amy 1874-1925 .... DLB-54, 140

Lowell, James Russell 1819-1891 .... DLB-1, 11, 64, 79

Lowell, Robert 1917-1977 ...... DLB-5, 169

Lowenfels, Walter 1897-1976 .... DLB-4

Lowndes, Marie Belloc 1868-1947 ... DLB-70

Lownes, Humphrey [publishing house] DLB-170

Lowry, Lois 1937- DLB-52

Lowry, Malcolm 1909-1957 .... DLB-15

Lowther, Pat 1935-1975 .... DLB-53

Loy, Mina 1882-1966 .... DLB-4, 54

Lozeau, Albert 1878-1924 .... DLB-92

Lubbock, Percy 1879-1965 .... DLB-149

Lucas, E. V. 1868-1938 ... DLB-98, 149, 153

Lucas, Fielding, Jr. [publishing house] DLB-49

Luce, Henry R. 1898-1967 .... DLB-91

Luce, John W., and Company DLB-46

Lucie-Smith, Edward 1933- DLB-40

Lucini, Gian Pietro 1867-1914 ..... DLB-114

Ludlum, Robert 1927- Y-82

Ludus de Antichristo circa 1160 .... DLB-148

Ludvigson, Susan 1942- DLB-120

Ludwig, Jack 1922- DLB-60

Ludwig, Otto 1813-1865 .... DLB-129

Ludwigslied 881 or 882 .... DLB-148

Luera, Yolanda 1953- DLB-122

Luft, Lya 1938- DLB-145

Luke, Peter 1919- DLB-13

Lupton, F. M., Company DLB-49

Lupus of Ferrières circa 805-circa 862 .... DLB-148

Lurie, Alison 1926- DLB-2

Luzi, Mario 1914- DLB-128

L'vov, Nikolai Aleksandrovich 1751-1803 .... DLB-150

Lyall, Gavin 1932- DLB-87

Lydgate, John circa 1370-1450 ..... DLB-146

Lyly, John circa 1554-1606 ..... DLB-62, 167

Lynch, Patricia 1898-1972 .... DLB-160

Lynch, Richard flourished 1596-1601 .... DLB-172

Lynd, Robert 1879-1949 .... DLB-98

Lyon, Matthew 1749-1822 .... DLB-43

Lytle, Andrew 1902-1995 ...... DLB-6; Y-95

Lytton, Edward (see Bulwer-Lytton, Edward)

Lytton, Edward Robert Bulwer 1831-1891 .... DLB-32

M

Maass, Joachim 1901-1972 .... DLB-69

Mabie, Hamilton Wright 1845-1916 .... DLB-71

Mac A'Ghobhainn, Iain (see Smith, Iain Crichton)

MacArthur, Charles 1895-1956 .... DLB-7, 25, 44

Macaulay, Catherine 1731-1791 .... DLB-104

Macaulay, David 1945- DLB-61

Macaulay, Rose 1881-1958 .... DLB-36

Macaulay, Thomas Babington 1800-1859 .... DLB-32, 55

Macaulay Company DLB-46

MacBeth, George 1932- DLB-40

Macbeth, Madge 1880-1965 .... DLB-92

MacCaig, Norman 1910- DLB-27

MacDiarmid, Hugh 1892-1978 ...... DLB-20

MacDonald, Cynthia 1928- DLB-105

MacDonald, George 1824-1905 .... DLB-18, 163

MacDonald, John D. 1916-1986 .... DLB-8; Y-86

MacDonald, Philip 1899?-1980 ..... DLB-77

Macdonald, Ross (see Millar, Kenneth)

MacDonald, Wilson 1880-1967 ..... DLB-92

Macdonald and Company (Publishers) DLB-112

MacEwen, Gwendolyn 1941- DLB-53

Macfadden, Bernarr 1868-1955 .... DLB-25, 91

MacGregor, John 1825-1892 .... DLB-166

MacGregor, Mary Esther (see Keith, Marian)

Machado, Antonio 1875-1939 ...... DLB-108

Machado, Manuel 1874-1947 ...... DLB-108

Machar, Agnes Maule 1837-1927 .... DLB-92

Machen, Arthur Llewelyn Jones 1863-1947 .... DLB-36, 156

MacInnes, Colin 1914-1976 .... DLB-14

MacInnes, Helen 1907-1985 .... DLB-87

Mack, Maynard 1909- DLB-111

Mackall, Leonard L. 1879-1937 .... DLB-140

MacKaye, Percy 1875-1956 .... DLB-54

Macken, Walter 1915-1967 .... DLB-13

Mackenzie, Alexander 1763-1820 .... DLB-99

Mackenzie, Compton 1883-1972 .... DLB-34, 100

Mackenzie, Henry 1745-1831 .... DLB-39

Mackey, Nathaniel 1947- DLB-169

Mackey, William Wellington 1937- DLB-38

Mackintosh, Elizabeth (see Tey, Josephine)

Mackintosh, Sir James 1765-1832 .... DLB-158

Maclaren, Ian (see Watson, John)

Macklin, Charles 1699-1797 .... DLB-89

MacLean, Katherine Anne 1925- DLB-8

MacLeish, Archibald 1892-1982 .... DLB-4, 7, 45; Y-82

MacLennan, Hugh 1907-1990 .... DLB-68

Macleod, Fiona (see Sharp, William)

MacLeod, Alistair 1936- DLB-60

Macleod, Norman 1906-1985 .... DLB-4

Macmillan and Company DLB-106

The Macmillan Company DLB-49

Macmillan's English Men of Letters, First Series (1878-1892) DLB-144

MacNamara, Brinsley 1890-1963 .... DLB-10

MacNeice, Louis 1907-1963 ...... DLB-10, 20

MacPhail, Andrew 1864-1938 ...... DLB-92

Macpherson, James 1736-1796 ..... DLB-109

Macpherson, Jay 1931- DLB-53

Macpherson, Jeanie 1884-1946 ...... DLB-44

Macrae Smith Company DLB-46

Macrone, John [publishing house] DLB-106

MacShane, Frank 1927- DLB-111

Macy-Masius DLB-46

Madden, David 1933- DLB-6

Maddow, Ben 1909-1992 ....DLB-44

Maddux, Rachel 1912-1983 ....Y-93

Madgett, Naomi Long 1923- DLB-76

Madhubuti, Haki R.
1942- DLB-5, 41; DS-8

Madison, James 1751-1836 ....DLB-37

Maginn, William 1794-1842 ...DLB-110, 159

Mahan, Alfred Thayer 1840-1914 ...DLB-47

Maheux-Forcier, Louise 1929- DLB-60

Mahin, John Lee 1902-1984 ....DLB-44

Mahon, Derek 1941- DLB-40

Maikov, Vasilii Ivanovich
1728-1778 ....DLB-150

Mailer, Norman
1923- ...DLB-2, 16, 28; Y-80, 83; DS-3

Maillet, Adrienne 1885-1963 ....DLB-68

Maimonides, Moses 1138-1204 ....DLB-115

Maillet, Antonine 1929- DLB-60

Maillu, David G. 1939- DLB-157

Main Selections of the Book-of-the-Month
Club, 1926-1945 ....DLB-9

Main Trends in Twentieth-Century Book
Clubs DLB-46

Mainwaring, Daniel 1902-1977 ....DLB-44

Mair, Charles 1838-1927 ....DLB-99

Mais, Roger 1905-1955 ....DLB-125

Major, Andre 1942- DLB-60

Major, Clarence 1936- DLB-33

Major, Kevin 1949- DLB-60

Major Books DLB-46

Makemie, Francis circa 1658-1708 ...DLB-24

The Making of a People, by
J. M. Ritchie DLB-66

Maksimović, Desanka 1898-1993 ...DLB-147

Malamud, Bernard
1914-1986 ....DLB-2, 28, 152; Y-80, 86

Malet, Lucas 1852-1931 ....DLB-153

Malleson, Lucy Beatrice (see Gilbert, Anthony)

Mallet-Joris, Françoise 1930- DLB-83

Mallock, W. H. 1849-1923 ....DLB-18, 57

Malone, Dumas 1892-1986 ....DLB-17

Malone, Edmond 1741-1812 ....DLB-142

Malory, Sir Thomas
circa 1400-1410 - 1471 ....DLB-146

Malraux, André 1901-1976 ....DLB-72

Malthus, Thomas Robert
1766-1834 ....DLB-107, 158

Maltz, Albert 1908-1985 ....DLB-102

Malzberg, Barry N. 1939- DLB-8

Mamet, David 1947- DLB-7

Manaka, Matsemela 1956- DLB-157

Manchester University Press DLB-112

Mandel, Eli 1922- DLB-53

Mandeville, Bernard 1670-1733 ....DLB-101

Mandeville, Sir John
mid fourteenth century DLB-146

Mandiargues, André Pieyre de
1909- DLB-83

Manfred, Frederick 1912-1994 ....DLB-6

Mangan, Sherry 1904-1961 ....DLB-4

Mankiewicz, Herman 1897-1953 .....DLB-26

Mankiewicz, Joseph L. 1909-1993 ....DLB-44

Mankowitz, Wolf 1924- DLB-15

Manley, Delarivière
1672?-1724 ....DLB-39, 80

Mann, Abby 1927- DLB-44

Mann, Heinrich 1871-1950 ....DLB-66, 118

Mann, Horace 1796-1859 ....DLB-1

Mann, Klaus 1906-1949 ....DLB-56

Mann, Thomas 1875-1955 ....DLB-66

Mann, William D'Alton
1839-1920 ....DLB-137

Manning, Marie 1873?-1945 ....DLB-29

Manning and Loring DLB-49

Mannyng, Robert
flourished 1303-1338 ....DLB-146

Mano, D. Keith 1942- DLB-6

Manor Books DLB-46

Mansfield, Katherine 1888-1923 ....DLB-162

Mapanje, Jack 1944- DLB-157

March, William 1893-1954 ....DLB-9, 86

Marchand, Leslie A. 1900- DLB-103

Marchant, Bessie 1862-1941 ....DLB-160

Marchessault, Jovette 1938- DLB-60

Marcus, Frank 1928- DLB-13

Marden, Orison Swett
1850-1924 ....DLB-137

Marechera, Dambudzo
1952-1987 ....DLB-157

Marek, Richard, Books DLB-46

Mares, E. A. 1938- DLB-122

Mariani, Paul 1940- DLB-111

Marie-Victorin, Frère 1885-1944 .....DLB-92

Marin, Biagio 1891-1985 ....DLB-128

Marincović, Ranko 1913- DLB-147

Marinetti, Filippo Tommaso
1876-1944 ....DLB-114

Marion, Frances 1886-1973 ....DLB-44

Marius, Richard C. 1933- Y-85

The Mark Taper Forum DLB-7

Mark Twain on Perpetual Copyright ...Y-92

Markfield, Wallace 1926- DLB-2, 28

Markham, Edwin 1852-1940 ....DLB-54

Markle, Fletcher 1921-1991 ...DLB-68; Y-91

Marlatt, Daphne 1942- DLB-60

Marlitt, E. 1825-1887 ....DLB-129

Marlowe, Christopher 1564-1593 ....DLB-62

Marlyn, John 1912- DLB-88

Marmion, Shakerley 1603-1639 .....DLB-58

Der Marner
before 1230-circa 1287 ....DLB-138

The *Marprelate* Tracts 1588-1589 .....DLB-132

Marquand, John P. 1893-1960 ...DLB-9, 102

Marqués, René 1919-1979 ....DLB-113

Marquis, Don 1878-1937 ....DLB-11, 25

Marriott, Anne 1913- DLB-68

Marryat, Frederick 1792-1848 ..DLB-21, 163

Marsh, George Perkins
1801-1882 ....DLB-1, 64

Marsh, James 1794-1842 ....DLB-1, 59

Marsh, Capen, Lyon and Webb DLB-49

Marsh, Ngaio 1899-1982 ....DLB-77

Marshall, Edison 1894-1967 ....DLB-102

Marshall, Edward 1932- DLB-16

Marshall, Emma 1828-1899 ....DLB-163

Marshall, James 1942-1992 ....DLB-61

Marshall, Joyce 1913- DLB-88

Marshall, Paule 1929- DLB-33, 157

Marshall, Tom 1938- DLB-60

Marsilius of Padua
circa 1275-circa 1342 ....DLB-115

Marson, Una 1905-1965 ....DLB-157

Marston, John 1576-1634 ....DLB-58, 172

Marston, Philip Bourke 1850-1887 ...DLB-35

Martens, Kurt 1870-1945 ....DLB-66

Martien, William S.
[publishing house] DLB-49

Martin, Abe (see Hubbard, Kin)

Martin, Charles 1942- DLB-120

Martin, Claire 1914- DLB-60

Martin, Jay 1935- DLB-111

Martin, Johann (see Laurentius von Schnüffis)

Martin, Violet Florence (see Ross, Martin)

Martin du Gard, Roger 1881-1958 ...DLB-65

Martineau, Harriet
  1802-1876 ....DLB-21, 55, 159, 163, 166

Martínez, Eliud 1935- DLB-122

Martínez, Max 1943- DLB-82

Martyn, Edward 1859-1923 ....DLB-10

Marvell, Andrew 1621-1678 ....DLB-131

Marvin X 1944- DLB-38

Marx, Karl 1818-1883 ....DLB-129

Marzials, Theo 1850-1920 ....DLB-35

Masefield, John
  1878-1967 ....DLB-10, 19, 153, 160

Mason, A. E. W. 1865-1948 ....DLB-70

Mason, Bobbie Ann
  1940- DLB-173; Y-87

Mason, William 1725-1797 ....DLB-142

Mason Brothers DLB-49

Massey, Gerald 1828-1907 ....DLB-32

Massinger, Philip 1583-1640 ......DLB-58

Masson, David 1822-1907 ....DLB-144

Masters, Edgar Lee 1868-1950 .....DLB-54

Mather, Cotton
  1663-1728 ....DLB-24, 30, 140

Mather, Increase 1639-1723 ....DLB-24

Mather, Richard 1596-1669 ....DLB-24

Matheson, Richard 1926- DLB-8, 44

Matheus, John F. 1887- DLB-51

Mathews, Cornelius
  1817?-1889 ....DLB-3, 64

Mathews, Elkin
  [publishing house] DLB-112

Mathias, Roland 1915- DLB-27

Mathis, June 1892-1927 ....DLB-44

Mathis, Sharon Bell 1937- DLB-33

Matoš, Antun Gustav 1873-1914 ....DLB-147

The Matter of England
  1240-1400 ....DLB-146

The Matter of Rome
  early twelfth to late fifteenth
  century DLB-146

Matthews, Brander
  1852-1929 ....DLB-71, 78; DS-13

Matthews, Jack 1925- DLB-6

Matthews, William 1942- DLB-5

Matthiessen, F. O. 1902-1950 .....DLB-63

Matthiessen, Peter 1927- DLB-6, 173

Maugham, W. Somerset
  1874-1965 .....DLB-10, 36, 77, 100, 162

Maupassant, Guy de 1850-1893 ....DLB-123

Mauriac, Claude 1914- DLB-83

Mauriac, François 1885-1970 ....DLB-65

Maurice, Frederick Denison
  1805-1872 ....DLB-55

Maurois, André 1885-1967 ....DLB-65

Maury, James 1718-1769 ....DLB-31

Mavor, Elizabeth 1927- DLB-14

Mavor, Osborne Henry (see Bridie, James)

Maxwell, H. [publishing house] DLB-49

Maxwell, John [publishing house] ..DLB-106

Maxwell, William 1908- Y-80

May, Elaine 1932- DLB-44

May, Karl 1842-1912 ....DLB-129

May, Thomas 1595 or 1596-1650 ...DLB-58

Mayer, Bernadette 1945- DLB-165

Mayer, Mercer 1943- DLB-61

Mayer, O. B. 1818-1891 ....DLB-3

Mayes, Herbert R. 1900-1987 .....DLB-137

Mayes, Wendell 1919-1992 ....DLB-26

Mayfield, Julian 1928-1984 ....DLB-33; Y-84

Mayhew, Henry 1812-1887 ....DLB-18, 55

Mayhew, Jonathan 1720-1766 ....DLB-31

Mayne, Jasper 1604-1672 ....DLB-126

Mayne, Seymour 1944- DLB-60

Mayor, Flora Macdonald
  1872-1932 ....DLB-36

Mayrocker, Friederike 1924- DLB-85

Mazrui, Ali A. 1933- DLB-125

Mažuranić, Ivan 1814-1890 ....DLB-147

Mazursky, Paul 1930- DLB-44

McAlmon, Robert 1896-1956 .....DLB-4, 45

McArthur, Peter 1866-1924 ....DLB-92

McBride, Robert M., and
  Company DLB-46

McCaffrey, Anne 1926- DLB-8

McCarthy, Cormac 1933- DLB-6, 143

McCarthy, Mary 1912-1989 ....DLB-2; Y-81

McCay, Winsor 1871-1934 ....DLB-22

McClane, Albert Jules 1922-1991 ...DLB-171

McClatchy, C. K. 1858-1936 ....DLB-25

McClellan, George Marion
  1860-1934 ....DLB-50

McCloskey, Robert 1914- DLB-22

McClung, Nellie Letitia 1873-1951 ...DLB-92

McClure, Joanna 1930- DLB-16

McClure, Michael 1932- DLB-16

McClure, Phillips and Company DLB-46

McClure, S. S. 1857-1949 ....DLB-91

McClurg, A. C., and Company DLB-49

McCluskey, John A., Jr. 1944- DLB-33

McCollum, Michael A. 1946 ....Y-87

McConnell, William C. 1917- DLB-88

McCord, David 1897- DLB-61

McCorkle, Jill 1958- Y-87

McCorkle, Samuel Eusebius
  1746-1811 ....DLB-37

McCormick, Anne O'Hare
  1880-1954 ....DLB-29

McCormick, Robert R. 1880-1955 ...DLB-29

McCourt, Edward 1907-1972 ....DLB-88

McCoy, Horace 1897-1955 ....DLB-9

McCrae, John 1872-1918 ....DLB-92

McCullagh, Joseph B. 1842-1896 ...DLB-23

McCullers, Carson
  1917-1967 ....DLB-2, 7, 173

McCulloch, Thomas 1776-1843 ......DLB-99

McDonald, Forrest 1927- DLB-17

McDonald, Walter
  1934- DLB-105, DS-9

McDonald, Walter, Getting Started:
  Accepting the Regions You Own—
  or Which Own You DLB-105

McDougall, Colin 1917-1984 ....DLB-68

McDowell, Obolensky DLB-46

McEwan, Ian 1948- DLB-14

McFadden, David 1940- DLB-60

McFall, Frances Elizabeth Clarke
  (see Grand, Sarah)

McFarlane, Leslie 1902-1977 ......DLB-88

McFee, William 1881-1966 ....DLB-153

McGahern, John 1934- DLB-14

McGee, Thomas D'Arcy
  1825-1868 ....DLB-99

McGeehan, W. O. 1879-1933 ....DLB-25, 171

McGill, Ralph 1898-1969 ....DLB-29

McGinley, Phyllis 1905-1978 ...DLB-11, 48

McGirt, James E. 1874-1930 . . . . DLB-50	Meeke, Mary ?-1816? DLB-116	Metcalf, J. [publishing house] DLB-49
McGlashan and Gill DLB-106	Meinke, Peter 1932- DLB-5	Metcalf, John 1938- DLB-60
McGough, Roger 1937- DLB-40	Mejia Vallejo, Manuel 1923- DLB-113	The Methodist Book Concern DLB-49
McGraw-Hill DLB-46	Melançon, Robert 1947- DLB-60	Methuen and Company DLB-112
McGuane, Thomas 1939- DLB-2; Y-80	Mell, Max 1882-1971 . . . . .DLB-81, 124	Mew, Charlotte 1869-1928 . . . . . DLB-19, 135
McGuckian, Medbh 1950- DLB-40	Mellow, James R. 1926- DLB-111	Mewshaw, Michael 1943- Y-80
McGuffey, William Holmes 1800-1873 . . . . DLB-42	Meltzer, David 1937- DLB-16	Meyer, Conrad Ferdinand 1825-1898 . . . . .DLB-129
McIlvanney, William 1936- DLB-14	Meltzer, Milton 1915- DLB-61	Meyer, E. Y. 1946- DLB-75
McIlwraith, Jean Newton 1859-1938 . . . . DLB-92	Melville, Elizabeth, Lady Culross circa 1585-1640 . . . . DLB-172	Meyer, Eugene 1875-1959 . . . . .DLB-29
McIntyre, James 1827-1906 . . . . DLB-99	Melville, Herman 1819-1891 . . . . . DLB-3, 74	Meyer, Michael 1921- DLB-155
McIntyre, O. O. 1884-1938 . . . . DLB-25	Memoirs of Life and Literature (1920), by W. H. Mallock [excerpt] DLB-57	Meyers, Jeffrey 1939- DLB-111
McKay, Claude 1889-1948 . . . . .DLB-4, 45, 51, 117	Menantes (see Hunold, Christian Friedrich)	Meynell, Alice 1847-1922 . . . . DLB-19, 98
The David McKay Company DLB-49	Mencke, Johann Burckhard 1674-1732 . . . . DLB-168	Meynell, Viola 1885-1956 . . . . .DLB-153
McKean, William V. 1820-1903 . . . . . DLB-23	Mencken, H. L. 1880-1956 . . . . DLB-11, 29, 63, 137	Meyrink, Gustav 1868-1932 . . . . .DLB-81
McKinley, Robin 1952- DLB-52	Mencken and Nietzsche: An Unpublished Excerpt from H. L. Mencken's My Life as Author and Editor Y-93	Michaels, Leonard 1933- DLB-130
McLachlan, Alexander 1818-1896 . . . DLB-99		Micheaux, Oscar 1884-1951 . . . . .DLB-50
McLaren, Floris Clark 1904-1978 . . . DLB-68	Mendelssohn, Moses 1729-1786 . . . . . DLB-97	Michel of Northgate, Dan circa 1265-circa 1340 . . . . .DLB-146
McLaverty, Michael 1907- DLB-15	Méndez M., Miguel 1930- DLB-82	Micheline, Jack 1929- DLB-16
McLean, John R. 1848-1916 . . . . DLB-23	Mercer, Cecil William (see Yates, Dornford)	Michener, James A. 1907?- DLB-6
McLean, William L. 1852-1931 . . . . . DLB-25	Mercer, David 1928-1980 . . . . DLB-13	Micklejohn, George circa 1717-1818 . . . . .DLB-31
McLennan, William 1856-1904 . . . . . DLB-92	Mercer, John 1704-1768 . . . . DLB-31	
McLoughlin Brothers DLB-49	Meredith, George 1828-1909 . . . . DLB-18, 35, 57, 159	Middle English Literature: An Introduction DLB-146
McLuhan, Marshall 1911-1980 . . . . . DLB-88		The Middle English Lyric DLB-146
McMaster, John Bach 1852-1932 . . . . DLB-47	Meredith, Louisa Anne 1812-1895 . . . . DLB-166	Middle Hill Press DLB-106
McMurtry, Larry 1936- DLB-2, 143; Y-80, 87	Meredith, Owen (see Lytton, Edward Robert Bulwer)	Middleton, Christopher 1926- DLB-40
		Middleton, Richard 1882-1911 . . . . .DLB-156
McNally, Terrence 1939- DLB-7	Meredith, William 1919- DLB-5	Middleton, Stanley 1919- DLB-14
McNeil, Florence 1937- DLB-60	Mergerle, Johann Ulrich (see Abraham à Sancta Clara)	Middleton, Thomas 1580-1627 . . . . . .DLB-58
McNeile, Herman Cyril 1888-1937 . . . . DLB-77		Miegel, Agnes 1879-1964 . . . . .DLB-56
	Mérimée, Prosper 1803-1870 . . . . DLB-119	Miles, Josephine 1911-1985 . . . . .DLB-48
McPherson, James Alan 1943- DLB-38	Merivale, John Herman 1779-1844 . . . . DLB-96	Milius, John 1944- DLB-44
McPherson, Sandra 1943- Y-86		Mill, James 1773-1836 . . . . DLB-107, 158
McWhirter, George 1939- DLB-60	Meriwether, Louise 1923- DLB-33	Mill, John Stuart 1806-1873 . . . . .DLB-55
McWilliams, Carey 1905-1980 . . . . . DLB-137	Merlin Press DLB-112	Millar, Kenneth 1915-1983 . . . . .DLB-2; Y-83; DS-6
Mead, L. T. 1844-1914 . . . . DLB-141	Merriam, Eve 1916-1992 . . . . DLB-61	
Mead, Matthew 1924- DLB-40	The Merriam Company DLB-49	Millar, Andrew [publishing house] DLB-154
Mead, Taylor ?- DLB-16	Merrill, James 1926-1995 . . . . .DLB-5, 165; Y-85	
Meany, Tom 1903-1964 . . . . DLB-171		Millay, Edna St. Vincent 1892-1950 . . . . .DLB-45
Mechthild von Magdeburg circa 1207-circa 1282 . . . . DLB-138	Merrill and Baker DLB-49	
	The Mershon Company DLB-49	Miller, Arthur 1915- DLB-7
Medill, Joseph 1823-1899 . . . . DLB-43	Merton, Thomas 1915-1968 . . .DLB-48; Y-81	Miller, Caroline 1903-1992 . . . . .DLB-9
Medoff, Mark 1940- DLB-7	Merwin, W. S. 1927- DLB-5, 169	Miller, Eugene Ethelbert 1950- DLB-41
Meek, Alexander Beaufort 1814-1865 . . . . DLB-3	Messner, Julian [publishing house] . . . DLB-46	Miller, Heather Ross 1939- . . . . . . .DLB 120

Miller, Henry 1891-1980 .... DLB-4, 9; Y-80

Miller, J. Hillis 1928- DLB-67

Miller, James [publishing house] DLB-49

Miller, Jason 1939- DLB-7

Miller, May 1899- DLB-41

Miller, Paul 1906-1991 ....DLB-127

Miller, Perry 1905-1963 ....DLB-17, 63

Miller, Sue 1943- DLB-143

Miller, Walter M., Jr. 1923- DLB-8

Miller, Webb 1892-1940 ....DLB-29

Millhauser, Steven 1943- DLB-2

Millican, Arthenia J. Bates
  1920- DLB-38

Mills and Boon DLB-112

Milman, Henry Hart 1796-1868 .....DLB-96

Milne, A. A.
  1882-1956 .... DLB-10, 77, 100, 160

Milner, Ron 1938- DLB-38

Milner, William
  [publishing house] DLB-106

Milnes, Richard Monckton (Lord Houghton)
  1809-1885 ....DLB-32

Milton, John 1608-1674 ....DLB-131, 151

The Minerva Press DLB-154

Minnesang circa 1150-1280 ....DLB-138

Minns, Susan 1839-1938 ....DLB-140

Minor Illustrators, 1880-1914 ......DLB-141

Minor Poets of the Earlier Seventeenth
  Century DLB-121

Minton, Balch and Company DLB-46

Mirbeau, Octave 1848-1917 ....DLB-123

Mirk, John died after 1414? DLB-146

Miron, Gaston 1928- DLB-60

A Mirror for Magistrates DLB-167

Mitchel, Jonathan 1624-1668 ....DLB-24

Mitchell, Adrian 1932- DLB-40

Mitchell, Donald Grant
  1822-1908 .... DLB-1; DS-13

Mitchell, Gladys 1901-1983 ....DLB-77

Mitchell, James Leslie 1901-1935 ....DLB-15

Mitchell, John (see Slater, Patrick)

Mitchell, John Ames 1845-1918 ......DLB-79

Mitchell, Julian 1935- DLB-14

Mitchell, Ken 1940- DLB-60

Mitchell, Langdon 1862-1935 ....DLB-7

Mitchell, Loften 1919- DLB-38

Mitchell, Margaret 1900-1949 ....DLB-9

Mitchell, W. O. 1914- DLB-88

Mitchison, Naomi Margaret (Haldane)
  1897- DLB-160

Mitford, Mary Russell
  1787-1855 .... DLB-110, 116

Mittelholzer, Edgar 1909-1965 ..... DLB-117

Mitterer, Erika 1906- DLB-85

Mitterer, Felix 1948- DLB-124

Mitternacht, Johann Sebastian
  1613-1679 .... DLB-168

Mizener, Arthur 1907-1988 .... DLB-103

Modern Age Books DLB-46

"Modern English Prose" (1876),
  by George Saintsbury DLB-57

The Modern Language Association of America
  Celebrates Its Centennial Y-84

The Modern Library DLB-46

"Modern Novelists – Great and Small" (1855),
  by Margaret Oliphant DLB-21

"Modern Style" (1857), by Cockburn
  Thomson [excerpt] DLB-57

The Modernists (1932), by Joseph Warren
  Beach DLB-36

Modiano, Patrick 1945- DLB-83

Moffat, Yard and Company DLB-46

Moffet, Thomas 1553-1604 .... DLB-136

Mohr, Nicholasa 1938- DLB-145

Moix, Ana María 1947- DLB-134

Molesworth, Louisa 1839-1921 .... DLB-135

Möllhausen, Balduin 1825-1905 .... DLB-129

Momaday, N. Scott 1934- DLB-143

Monkhouse, Allan 1858-1936 .... DLB-10

Monro, Harold 1879-1932 .... DLB-19

Monroe, Harriet 1860-1936 ..... DLB-54, 91

Monsarrat, Nicholas 1910-1979 .... DLB-15

Montagu, Lady Mary Wortley
  1689-1762 .... DLB-95, 101

Montague, John 1929- DLB-40

Montale, Eugenio 1896-1981 ...... DLB-114

Monterroso, Augusto 1921- DLB-145

Montgomerie, Alexander
  circa 1550?-1598 .... DLB-167

Montgomery, James
  1771-1854 .... DLB-93, 158

Montgomery, John 1919- DLB-16

Montgomery, Lucy Maud
  1874-1942 .... DLB-92; DS-14

Montgomery, Marion 1925- DLB-6

Montgomery, Robert Bruce (see Crispin,
  Edmund)

Montherlant, Henry de 1896-1972 ...DLB-72

The Monthly Review 1749-1844 ...... DLB-110

Montigny, Louvigny de 1876-1955 ...DLB-92

Montoya, José 1932- DLB-122

Moodie, John Wedderburn Dunbar
  1797-1869 ....DLB-99

Moodie, Susanna 1803-1885 ....DLB-99

Moody, Joshua circa 1633-1697 .....DLB-24

Moody, William Vaughn
  1869-1910 ....DLB-7, 54

Moorcock, Michael 1939- DLB-14

Moore, Catherine L. 1911- DLB-8

Moore, Clement Clarke 1779-1863 ...DLB-42

Moore, Dora Mavor 1888-1979 .....DLB-92

Moore, George
  1852-1933 .... DLB-10, 18, 57, 135

Moore, Marianne
  1887-1972 ....DLB-45; DS-7

Moore, Mavor 1919- DLB-88

Moore, Richard 1927- DLB-105

Moore, Richard, The No Self, the Little Self,
  and the Poets DLB-105

Moore, T. Sturge 1870-1944 ....DLB-19

Moore, Thomas 1779-1852 .....DLB-96, 144

Moore, Ward 1903-1978 ....DLB-8

Moore, Wilstach, Keys and
  Company DLB-49

The Moorland-Spingarn Research
  Center DLB-76

Moorman, Mary C. 1905-1994 .....DLB-155

Moraga, Cherríe 1952- DLB-82

Morales, Alejandro 1944- DLB-82

Morales, Mario Roberto 1947- DLB-145

Morales, Rafael 1919- DLB-108

Morality Plays: *Mankind* circa 1450-1500 and
  Everyman circa 1500 ....DLB-146

Mordaunt, Elinor 1872-1942 ......DLB-174

More, Hannah
  1745-1833 ...... DLB-107, 109, 116, 158

More, Henry 1614-1687 ....DLB-126

More, Sir Thomas
  1477 or 1478-1535 ....DLB-136

Moreno, Dorinda 1939- DLB-122

Morency, Pierre 1942- DLB-60

Moretti, Marino 1885-1979 ....DLB-114

Morgan, Berry 1919- DLB-6

Morgan, Charles 1894-1958 .... DLB-34, 100

Morgan, Edmund S. 1916- DLB-17

Morgan, Edwin 1920- DLB-27

Morgan, John Pierpont
 1837-1913 .... DLB-140

Morgan, John Pierpont, Jr.
 1867-1943 .... DLB-140

Morgan, Robert 1944- DLB-120

Morgan, Sydney Owenson, Lady
 1776?-1859 .... DLB-116, 158

Morgner, Irmtraud 1933- DLB-75

Morhof, Daniel Georg
 1639-1691 .... DLB-164

Morier, James Justinian
 1782 or 1783?-1849 .... DLB-116

Mörike, Eduard 1804-1875 .... DLB-133

Morin, Paul 1889-1963 .... DLB-92

Morison, Richard 1514?-1556 ..... DLB-136

Morison, Samuel Eliot 1887-1976 ... DLB-17

Moritz, Karl Philipp 1756-1793 ..... DLB-94

Moriz von Craûn
 circa 1220-1230 .... DLB-138

Morley, Christopher 1890-1957 ......DLB-9

Morley, John 1838-1923 .... DLB-57, 144

Morris, George Pope 1802-1864 .... DLB-73

Morris, Lewis 1833-1907 .... DLB-35

Morris, Richard B. 1904-1989 ...... DLB-17

Morris, William
 1834-1896 ....DLB-18, 35, 57, 156

Morris, Willie 1934- Y-80

Morris, Wright 1910- DLB-2; Y-81

Morrison, Arthur 1863-1945 ... DLB-70, 135

Morrison, Charles Clayton
 1874-1966 .... DLB-91

Morrison, Toni
 1931- DLB-6, 33, 143; Y-81

Morrow, William, and Company ... DLB-46

Morse, James Herbert 1841-1923 .... DLB-71

Morse, Jedidiah 1761-1826 .... DLB-37

Morse, John T., Jr. 1840-1937 ...... DLB-47

Mortimer, Favell Lee 1802-1878 ... DLB-163

Mortimer, John 1923- DLB-13

Morton, Carlos 1942- DLB-122

Morton, John P., and Company DLB-49

Morton, Nathaniel 1613-1685 ...... DLB-24

Morton, Sarah Wentworth
 1759-1846 .... DLB-37

Morton, Thomas
 circa 1579-circa 1647 .... DLB-24

Moscherosch, Johann Michael
 1601-1669 .... DLB-164

Moseley, Humphrey
 [publishing house] DLB-170

Möser, Justus 1720-1794 .... DLB-97

Mosley, Nicholas 1923- DLB-14

Moss, Arthur 1889-1969 .... DLB-4

Moss, Howard 1922-1987 .... DLB-5

Moss, Thylias 1954- DLB-120

The Most Powerful Book Review in America
 [*New York Times Book Review*] Y-82

Motion, Andrew 1952- DLB-40

Motley, John Lothrop
 1814-1877 .... DLB-1, 30, 59

Motley, Willard 1909-1965 ..... DLB-76, 143

Motte, Benjamin Jr.
 [publishing house] DLB-154

Motteux, Peter Anthony
 1663-1718 .... DLB-80

Mottram, R. H. 1883-1971 .... DLB-36

Mouré, Erin 1955- DLB-60

Movies from Books, 1920-1974 .... DLB-9

Mowat, Farley 1921- DLB-68

Mowbray, A. R., and Company,
 Limited DLB-106

Mowrer, Edgar Ansel 1892-1977 .... DLB-29

Mowrer, Paul Scott 1887-1971 ..... DLB-29

Moxon, Edward
 [publishing house] DLB-106

Moxon, Joseph
 [publishing house] DLB-170

Mphahlele, Es'kia (Ezekiel)
 1919- DLB-125

Mtshali, Oswald Mbuyiseni
 1940- DLB-125

Mucedorus DLB-62

Mudford, William 1782-1848 ..... DLB-159

Mueller, Lisel 1924- DLB-105

Muhajir, El (see Marvin X)

Muhajir, Nazzam Al Fitnah (see Marvin X)

Mühlbach, Luise 1814-1873 ...... DLB-133

Muir, Edwin 1887-1959 ....DLB-20, 100

Muir, Helen 1937- DLB-14

Mukherjee, Bharati 1940- DLB-60

Mulcaster, Richard
 1531 or 1532-1611 .... DLB-167

Muldoon, Paul 1951- DLB-40

Müller, Friedrich (see Müller, Maler)

Müller, Heiner 1929- DLB-124

Müller, Maler 1749-1825 ....DLB-94

Müller, Wilhelm 1794-1827 ....DLB-90

Mumford, Lewis 1895-1990 ....DLB-63

Munby, Arthur Joseph 1828-1910 ....DLB-35

Munday, Anthony 1560-1633 .. DLB-62, 172

Mundt, Clara (see Mühlbach, Luise)

Mundt, Theodore 1808-1861 ....DLB-133

Munford, Robert circa 1737-1783 ....DLB-31

Mungoshi, Charles 1947- DLB-157

Munonye, John 1929- DLB-117

Munro, Alice 1931- DLB-53

Munro, H. H. 1870-1916 ...... DLB-34, 162

Munro, Neil 1864-1930 ....DLB-156

Munro, George
 [publishing house] DLB-49

Munro, Norman L.
 [publishing house] DLB-49

Munroe, James, and Company DLB-49

Munroe, Kirk 1850-1930 ....DLB-42

Munroe and Francis DLB-49

Munsell, Joel [publishing house] DLB-49

Munsey, Frank A. 1854-1925 .... DLB-25, 91

Munsey, Frank A., and
 Company DLB-49

Murav'ev, Mikhail Nikitich
 1757-1807 ....DLB-150

Murdoch, Iris 1919- DLB-14

Murdoch, Rupert 1931- DLB-127

Murfree, Mary N. 1850-1922 .... DLB-12, 74

Murger, Henry 1822-1861 ....DLB-119

Murger, Louis-Henri (see Murger, Henry)

Muro, Amado 1915-1971 ....DLB-82

Murphy, Arthur 1727-1805 .... DLB-89, 142

Murphy, Beatrice M. 1908- DLB-76

Murphy, Emily 1868-1933 ....DLB-99

Murphy, John H., III 1916- DLB-127

Murphy, John, and Company DLB-49

Murphy, Richard 1927-1993 ....DLB-40

Murray, Albert L. 1916- DLB-38

Murray, Gilbert 1866-1957 ....DLB-10

Murray, Judith Sargent 1751-1820 ...DLB-37

Murray, Pauli 1910-1985 ....DLB-41

Murray, John [publishing house] DLB-154

Murry, John Middleton
    1889-1957 ....DLB-149

Musäus, Johann Karl August
    1735-1787 ....DLB-97

Muschg, Adolf 1934- DLB-75

The Music of *Minnesang* DLB-138

Musil, Robert 1880-1942 ....DLB-81, 124

Muspilli circa 790-circa 850 ....DLB-148

Mussey, Benjamin B., and
    Company DLB-49

Mwangi, Meja 1948- DLB-125

Myers, Gustavus 1872-1942 ....DLB-47

Myers, L. H. 1881-1944 ....DLB-15

Myers, Walter Dean 1937- DLB-33

N

Nabbes, Thomas circa 1605-1641 ....DLB-58

Nabl, Franz 1883-1974 ....DLB-81

Nabokov, Vladimir
    1899-1977 ..... DLB-2; Y-80, Y-91; DS-3

Nabokov Festival at Cornell Y-83

The Vladimir Nabokov Archive
    in the Berg Collection Y-91

Nafis and Cornish DLB-49

Naipaul, Shiva 1945-1985 .....DLB-157; Y-85

Naipaul, V. S. 1932- DLB-125; Y-85

Nancrede, Joseph
    [publishing house] DLB-49

Naranjo, Carmen 1930- DLB-145

Narrache, Jean 1893-1970 ....DLB-92

Nasby, Petroleum Vesuvius (see Locke, David Ross)

Nash, Ogden 1902-1971 ....DLB-11

Nash, Eveleigh
    [publishing house] DLB-112

Nashe, Thomas 1567-1601? DLB-167

Nast, Conde 1873-1942 ....DLB-91

Nastasijević, Momčilo 1894-1938 ...DLB-147

Nathan, George Jean 1882-1958 ....DLB-137

Nathan, Robert 1894-1985 ....DLB-9

The National Jewish Book Awards Y-85

The National Theatre and the Royal
    Shakespeare Company: The
    National Companies DLB-13

Naughton, Bill 1910- DLB-13

Naylor, Gloria 1950- DLB-173

Nazor, Vladimir 1876-1949 ....DLB-147

Ndebele, Njabulo 1948- DLB-157

Neagoe, Peter 1881-1960 ....DLB-4

Neal, John 1793-1876............ DLB-1, 59

Neal, Joseph C. 1807-1847 ....DLB-11

Neal, Larry 1937-1981 ....DLB-38

The Neale Publishing Company DLB-49

Neely, F. Tennyson
    [publishing house] DLB-49

Negri, Ada 1870-1945 ....DLB-114

"The Negro as a Writer," by
    G. M. McClellan DLB-50

"Negro Poets and Their Poetry," by
    Wallace Thurman DLB-50

Neidhart von Reuental
    circa 1185-circa 1240 ....DLB-138

Neihardt, John G. 1881-1973 ..... DLB-9, 54

Neledinsky-Meletsky, Iurii Aleksandrovich
    1752-1828 ....DLB-150

Nelligan, Emile 1879-1941 ....DLB-92

Nelson, Alice Moore Dunbar
    1875-1935 ....DLB-50

Nelson, Thomas, and Sons [U.S.] ... DLB-49

Nelson, Thomas, and Sons [U.K.] .. DLB-106

Nelson, William 1908-1978 ....DLB-103

Nelson, William Rockhill
    1841-1915 ....DLB-23

Nemerov, Howard 1920-1991 ...DLB-5, 6; Y-83

Nesbit, E. 1858-1924 .... DLB-141, 153

Ness, Evaline 1911-1986 ....DLB-61

Nestroy, Johann 1801-1862 ....DLB-133

Neukirch, Benjamin 1655-1729 ....DLB-168

Neugeboren, Jay 1938- DLB-28

Neumann, Alfred 1895-1952 ....DLB-56

Neumark, Georg 1621-1681 ....DLB-164

Neumeister, Erdmann 1671-1756 ... DLB-168

Nevins, Allan 1890-1971 ....DLB-17

Nevinson, Henry Woodd
    1856-1941 ....DLB-135

The New American Library DLB-46

New Approaches to Biography: Challenges
    from Critical Theory, USC Conference
    on Literary Studies, 1990 ....Y-90

New Directions Publishing
    Corporation DLB-46

A New Edition of *Huck Finn* Y-85

New Forces at Work in the American Theatre:
    1915-1925 ....DLB-7

New Literary Periodicals:
    A Report for 1987 ....Y-87

New Literary Periodicals:
    A Report for 1988 ....Y-88

New Literary Periodicals:
    A Report for 1989 ....Y-89

New Literary Periodicals:
    A Report for 1990 ....Y-90

New Literary Periodicals:
    A Report for 1991 ....Y-91

New Literary Periodicals:
    A Report for 1992 ....Y-92

New Literary Periodicals:
    A Report for 1993 ....Y-93

The New Monthly Magazine
    1814-1884 ....DLB-110

The New *Ulysses* Y-84

The New Variorum Shakespeare Y-85

A New Voice: The Center for the Book's First
    Five Years Y-83

The New Wave [Science Fiction] DLB-8

New York City Bookshops in the 1930s and
    1940s: The Recollections of Walter
    Goldwater Y-93

Newbery, John
    [publishing house] DLB-154

Newbolt, Henry 1862-1938 ....DLB-19

Newbound, Bernard Slade (see Slade, Bernard)

Newby, P. H. 1918- DLB-15

Newby, Thomas Cautley
    [publishing house] DLB-106

Newcomb, Charles King 1820-1894 ...DLB-1

Newell, Peter 1862-1924 ....DLB-42

Newell, Robert Henry 1836-1901 ....DLB-11

Newhouse, Samuel I. 1895-1979 ....DLB-127

Newman, Cecil Earl 1903-1976 .....DLB-127

Newman, David (see Benton, Robert)

Newman, Frances 1883-1928 ....Y-80

Newman, John Henry
    1801-1890 ....DLB 18, 32, 55

Newman, Mark [publishing house] ...DLB-49

Newnes, George, Limited DLB-112

Newsome, Effie Lee 1885-1979 .....DLB-76

Newspaper Syndication of American
    Humor DLB-11

Newton, A. Edward 1864-1940 .....DLB-140

Ngugi wa Thiong'o 1938- DLB-125

The *Nibelungenlied* and the *Klage*
    circa 1200 ....DLB-138

Nichol, B. P. 1944- DLB-53

Nicholas of Cusa 1401-1464 ....DLB-115

Nichols, Dudley 1895-1960 ....DLB-26

Cumulative Index

Nichols, Grace 1950- DLB-157

Nichols, John 1940- Y-82

Nichols, Mary Sargeant (Neal) Gove 1810-1884 .... DLB-1

Nichols, Peter 1927- DLB-13

Nichols, Roy F. 1896-1973 .... DLB-17

Nichols, Ruth 1948- DLB-60

Nicholson, Norman 1914- DLB-27

Nicholson, William 1872-1949 ..... DLB-141

Ní Chuilleanáin, Eiléan 1942- DLB-40

Nicol, Eric 1919- DLB-68

Nicolai, Friedrich 1733-1811 .... DLB-97

Nicolay, John G. 1832-1901 and Hay, John 1838-1905 .... DLB-47

Nicolson, Harold 1886-1968 ... DLB-100, 149

Nicolson, Nigel 1917- DLB-155

Niebuhr, Reinhold 1892-1971 ...... DLB-17

Niedecker, Lorine 1903-1970 .... DLB-48

Nieman, Lucius W. 1857-1935 ...... DLB-25

Nietzsche, Friedrich 1844-1900 .... DLB-129

Niggli, Josefina 1910- Y-80

Nightingale, Florence 1820-1910 ... DLB-166

Nikolev, Nikolai Petrovich 1758-1815 .... DLB-150

Niles, Hezekiah 1777-1839 .... DLB-43

Nims, John Frederick 1913- DLB-5

Nin, Anaïs 1903-1977 .... DLB-2, 4, 152

1985: The Year of the Mystery: A Symposium Y-85

Nissenson, Hugh 1933- DLB-28

Niven, Frederick John 1878-1944 ... DLB-92

Niven, Larry 1938- DLB-8

Nizan, Paul 1905-1940 .... DLB-72

Njegoš, Petar II Petrović 1813-1851 .... DLB-147

Nkosi, Lewis 1936- DLB-157

Nobel Peace Prize
The 1986 Nobel Peace Prize
Nobel Lecture 1986: Hope, Despair and Memory
Tributes from Abraham Bernstein, Norman Lamm, and John R. Silber Y-86

The Nobel Prize and Literary Politics ... Y-86

Nobel Prize in Literature

The 1982 Nobel Prize in Literature
Announcement by the Swedish Academy of the Nobel Prize Nobel Lecture 1982: The Solitude of Latin America Excerpt from *One Hundred Years of Solitude* The Magical World of Macondo A Tribute to Gabriel García Márquez Y-82

The 1983 Nobel Prize in Literature
Announcement by the Swedish Academy Nobel Lecture 1983 The Stature of William Golding Y-83

The 1984 Nobel Prize in Literature
Announcement by the Swedish Academy Jaroslav Seifert Through the Eyes of the English-Speaking Reader
Three Poems by Jaroslav Seifert Y-84

The 1985 Nobel Prize in Literature
Announcement by the Swedish Academy Nobel Lecture 1985 .... Y-85

The 1986 Nobel Prize in Literature
Nobel Lecture 1986: This Past Must Address Its Present Y-86

The 1987 Nobel Prize in Literature
Nobel Lecture 1987 .... Y-87

The 1988 Nobel Prize in Literature
Nobel Lecture 1988 .... Y-88

The 1989 Nobel Prize in Literature
Nobel Lecture 1989 .... Y-89

The 1990 Nobel Prize in Literature
Nobel Lecture 1990 .... Y-90

The 1991 Nobel Prize in Literature
Nobel Lecture 1991 .... Y-91

The 1992 Nobel Prize in Literature
Nobel Lecture 1992 .... Y-92

The 1993 Nobel Prize in Literature
Nobel Lecture 1993 .... Y-93

The 1994 Nobel Prize in Literature
Nobel Lecture 1994 .... Y-94

The 1995 Nobel Prize in Literature
Nobel Lecture 1995 .... Y-95

Nodier, Charles 1780-1844 .... DLB-119

Noel, Roden 1834-1894 .... DLB-35

Nolan, William F. 1928- DLB-8

Noland, C. F. M. 1810?-1858 .... DLB-11

Nonesuch Press DLB-112

Noonday Press DLB-46

Noone, John 1936- DLB-14

Nora, Eugenio de 1923- DLB-134

Nordhoff, Charles 1887-1947 ...... DLB-9

Norman, Charles 1904- DLB-111

Norman, Marsha 1947- Y-84

Norris, Charles G. 1881-1945 .... DLB-9

Norris, Frank 1870-1902 .... DLB-12

Norris, Leslie 1921- DLB-27

Norse, Harold 1916- DLB-16

North, Marianne 1830-1890 .... DLB-174

North Point Press DLB-46

Nortje, Arthur 1942-1970 .... DLB-125

Norton, Alice Mary (see Norton, Andre)

Norton, Andre 1912- DLB-8, 52

Norton, Andrews 1786-1853 .... DLB-1

Norton, Caroline 1808-1877 .... DLB-21, 159

Norton, Charles Eliot 1827-1908 .. DLB-1, 64

Norton, John 1606-1663 .... DLB-24

Norton, Mary 1903-1992 .... DLB-160

Norton, Thomas (see Sackville, Thomas)

Norton, W. W., and Company DLB-46

Norwood, Robert 1874-1932 .... DLB-92

Nossack, Hans Erich 1901-1977 ..... DLB-69

Notker Balbulus circa 840-912 ..... DLB-148

Notker III of Saint Gall circa 950-1022 .... DLB-148

Notker von Zweifalten ?-1095 ..... DLB-148

A Note on Technique (1926), by Elizabeth A. Drew [excerpts] DLB-36

Nourse, Alan E. 1928- DLB-8

Novak, Vjenceslav 1859-1905 ...... DLB-147

Novalis 1772-1801 .... DLB-90

Novaro, Mario 1868-1944 .... DLB-114

Novás Calvo, Lino 1903-1983 ...... DLB-145

"The Novel in [Robert Browning's] 'The Ring and the Book'" (1912), by Henry James DLB-32

The Novel of Impressionism, by Jethro Bithell DLB-66

Novel-Reading: *The Works of Charles Dickens, The Works of W. Makepeace Thackeray* (1879), by Anthony Trollope DLB-21

The Novels of Dorothy Richardson (1918), by May Sinclair DLB-36

Novels with a Purpose (1864), by Justin M'Carthy DLB-21

Noventa, Giacomo 1898-1960 ...... DLB-114

Novikov, Nikolai Ivanovich 1744-1818 .... DLB-150

Nowlan, Alden 1933-1983 .... DLB-53

Noyes, Alfred 1880-1958 .... DLB-20

Noyes, Crosby S. 1825-1908 .... DLB-23

Noyes, Nicholas 1647-1717 .... DLB-24

Noyes, Theodore W. 1858-1946 ..... DLB-29

N-Town Plays circa 1468 to early sixteenth century DLB-146

Nugent, Frank 1908-1965 .... DLB-44
Nugent, Richard Bruce 1906- DLB-151
Nusic, Branislav 1864-1938 .... DLB-147
Nutt, David [publishing house] DLB-106
Nwapa, Flora 1931- DLB-125
Nye, Edgar Wilson (Bill) 1850-1896 .... DLB-11, 23
Nye, Naomi Shihab 1952- DLB-120
Nye, Robert 1939- DLB-14

O

Oakes, Urian circa 1631-1681 .... DLB-24
Oates, Joyce Carol 1938- DLB-2, 5, 130; Y-81
Ober, William 1920-1993 .... Y-93
Oberholtzer, Ellis Paxson 1868-1936 .... DLB-47
Obradović, Dositej 1740?-1811 ..... DLB-147
O'Brien, Edna 1932- DLB-14
O'Brien, Fitz-James 1828-1862 .... DLB-74
O'Brien, Kate 1897-1974 .... DLB-15
O'Brien, Tim 1946- DLB-152; Y-80; DS-9
O'Casey, Sean 1880-1964 .... DLB-10
Ochs, Adolph S. 1858-1935 .... DLB-25
Ochs-Oakes, George Washington 1861-1931 .... DLB-137
O'Connor, Flannery 1925-1964 ..... DLB-2, 152; Y-80; DS-12
O'Connor, Frank 1903-1966 ...... DLB-162
Octopus Publishing Group DLB-112
Odell, Jonathan 1737-1818 .... DLB-31, 99
O'Dell, Scott 1903-1989 .... DLB-52
Odets, Clifford 1906-1963 .... DLB-7, 26
Odhams Press Limited DLB-112
O'Donnell, Peter 1920- DLB-87
O'Donovan, Michael (see O'Connor, Frank)
O'Faolain, Julia 1932- DLB-14
O'Faolain, Sean 1900- DLB-15, 162
Off Broadway and Off-Off Broadway . DLB-7
Off-Loop Theatres DLB-7
Offord, Carl Ruthven 1910- DLB-76
O'Flaherty, Liam 1896-1984 .... DLB-36, 162; Y-84
Ogilvie, J. S., and Company DLB-49
Ogot, Grace 1930- DLB-125

O'Grady, Desmond 1935- DLB-40
Ogunyemi, Wale 1939- DLB-157
O'Hagan, Howard 1902-1982 .... DLB-68
O'Hara, Frank 1926-1966 .... DLB-5, 16
O'Hara, John 1905-1970 .... DLB-9, 86; DS-2
Okara, Gabriel 1921- DLB-125
O'Keeffe, John 1747-1833 .... DLB-89
Okes, Nicholas [publishing house] DLB-170
Okigbo, Christopher 1930-1967 .... DLB-125
Okot p'Bitek 1931-1982 .... DLB-125
Okpewho, Isidore 1941- DLB-157
Okri, Ben 1959- DLB-157
Olaudah Equiano and Unfinished Journeys: The Slave-Narrative Tradition and Twentieth-Century Continuities, by Paul Edwards and Pauline T. Wangman DLB-117
Old English Literature: An Introduction DLB-146
Old English Riddles eighth to tenth centuries DLB-146
Old Franklin Publishing House DLB-49
Old German Genesis and *Old German Exodus* circa 1050-circa 1130 .... DLB-148
Old High German Charms and Blessings DLB-148
The *Old High German Isidor* circa 790-800 .... DLB-148
Older, Fremont 1856-1935 .... DLB-25
Oldham, John 1653-1683 .... DLB-131
Olds, Sharon 1942- DLB-120
Olearius, Adam 1599-1671 .... DLB-164
Oliphant, Laurence 1829?-1888 .... DLB-18, 166
Oliphant, Margaret 1828-1897 .... DLB-18
Oliver, Chad 1928- DLB-8
Oliver, Mary 1935- DLB-5
Ollier, Claude 1922- DLB-83
Olsen, Tillie 1913?- DLB-28; Y-80
Olson, Charles 1910-1970 .... DLB-5, 16
Olson, Elder 1909- DLB-48, 63
Omotoso, Kole 1943- DLB-125
"On Art in Fiction "(1838), by Edward Bulwer DLB-21
On Learning to Write Y-88
On Some of the Characteristics of Modern Poetry and On the Lyrical Poems of Alfred Tennyson (1831), by Arthur Henry Hallam DLB-32

"On Style in English Prose" (1898), by Frederic Harrison DLB-57
"On Style in Literature: Its Technical Elements" (1885), by Robert Louis Stevenson DLB-57
"On the Writing of Essays" (1862), by Alexander Smith DLB-57
Ondaatje, Michael 1943- DLB-60
O'Neill, Eugene 1888-1953 .... DLB-7
Onetti, Juan Carlos 1909-1994 ..... DLB-113
Onions, George Oliver 1872-1961 .... DLB-153
Onofri, Arturo 1885-1928 .... DLB-114
Opie, Amelia 1769-1853 .... DLB-116, 159
Opitz, Martin 1597-1639 .... DLB-164
Oppen, George 1908-1984 .... DLB-5, 165
Oppenheim, E. Phillips 1866-1946 ... DLB-70
Oppenheim, James 1882-1932 .... DLB-28
Oppenheimer, Joel 1930- DLB-5
Optic, Oliver (see Adams, William Taylor)
Orczy, Emma, Baroness 1865-1947 .... DLB-70
Origo, Iris 1902-1988 .... DLB-155
Orlovitz, Gil 1918-1973 .... DLB-2, 5
Orlovsky, Peter 1933- DLB-16
Ormond, John 1923- DLB-27
Ornitz, Samuel 1890-1957 .... DLB-28, 44
Ortiz, Simon 1941- DLB-120
Ortnit and *Wolfdietrich* circa 1225-1250 .... DLB-138
Orton, Joe 1933-1967 .... DLB-13
Orwell, George 1903-1950 .... DLB-15, 98
The Orwell Year Y-84
Ory, Carlos Edmundo de 1923- ... DLB-134
Osbey, Brenda Marie 1957- DLB-120
Osbon, B. S. 1827-1912 .... DLB-43
Osborne, John 1929-1994 .... DLB-13
Osgood, Herbert L. 1855-1918 ...... DLB-47
Osgood, James R., and Company DLB-49
Osgood, McIlvaine and Company DLB-112
O'Shaughnessy, Arthur 1844-1881 .... DLB-35
O'Shea, Patrick [publishing house] DLB-49
Osipov, Nikolai Petrovich 1751-1799 .... DLB-150
Osofisan, Femi 1946- DLB-125

Ostenso, Martha 1900-1963 .... DLB-92

Ostriker, Alicia 1937- DLB-120

Osundare, Niyi 1947- DLB-157

Oswald, Eleazer 1755-1795 .... DLB-43

Otero, Blas de 1916-1979 .... DLB-134

Otero, Miguel Antonio
1859-1944 .... DLB-82

Otero Silva, Miguel 1908-1985 ..... DLB-145

Otfried von Weißenburg
circa 800-circa 875? DLB-148

Otis, James (see Kaler, James Otis)

Otis, James, Jr. 1725-1783 .... DLB-31

Otis, Broaders and Company DLB-49

Ottaway, James 1911- DLB-127

Ottendorfer, Oswald 1826-1900 ..... DLB-23

Otto-Peters, Louise 1819-1895 ..... DLB-129

Otway, Thomas 1652-1685 .... DLB-80

Ouellette, Fernand 1930- DLB-60

Ouida 1839-1908 .... DLB-18, 156

Outing Publishing Company DLB-46

Outlaw Days, by Joyce Johnson DLB-16

Overbury, Sir Thomas
circa 1581-1613 .... DLB-151

The Overlook Press DLB-46

Overview of U.S. Book Publishing,
1910-1945 .... DLB-9

Owen, Guy 1925- DLB-5

Owen, John 1564-1622 .... DLB-121

Owen, John [publishing house] DLB-49

Owen, Robert 1771-1858 ..... DLB-107, 158

Owen, Wilfred 1893-1918 .... DLB-20

Owen, Peter, Limited DLB-112

The Owl and the Nightingale
circa 1189-1199 .... DLB-146

Owsley, Frank L. 1890-1956 .... DLB-17

Oxford, Seventeenth Earl of, Edward de Vere
1550-1604 .... DLB-172

Ozerov, Vladislav Aleksandrovich
1769-1816 .... DLB-150

Ozick, Cynthia 1928- DLB-28, 152; Y-82

P

Pace, Richard 1482?-1536 .... DLB-167

Pacey, Desmond 1917-1975 .... DLB-88

Pack, Robert 1929- DLB-5

Packaging Papa: *The Garden of Eden* Y-86

Padell Publishing Company DLB-46

Padgett, Ron 1942- DLB-5

Padilla, Ernesto Chávez 1944- DLB-122

Page, L. C., and Company DLB-49

Page, P. K. 1916- DLB-68

Page, Thomas Nelson
1853-1922 .... DLB-12, 78; DS-13

Page, Walter Hines 1855-1918 ...DLB-71, 91

Paget, Francis Edward
1806-1882 .... DLB-163

Paget, Violet (see Lee, Vernon)

Pagliarani, Elio 1927- DLB-128

Pain, Barry 1864-1928 .... DLB-135

Pain, Philip ?-circa 1666 .... DLB-24

Paine, Robert Treat, Jr. 1773-1811 ... DLB-37

Paine, Thomas
1737-1809 .... DLB-31, 43, 73, 158

Painter, George D. 1914- DLB-155

Painter, William 1540?-1594 .... DLB-136

Palazzeschi, Aldo 1885-1974 .... DLB-114

Paley, Grace 1922- DLB-28

Palfrey, John Gorham
1796-1881 .... DLB-1, 30

Palgrave, Francis Turner
1824-1897 .... DLB-35

Palmer, Joe H. 1904-1952 .... DLB-171

Palmer, Michael 1943- DLB-169

Paltock, Robert 1697-1767 .... DLB-39

Pan Books Limited DLB-112

Panamaa, Norman 1914- and
Frank, Melvin 1913-1988 .... DLB-26

Pancake, Breece D'J 1952-1979 ..... DLB-130

Panero, Leopoldo 1909-1962 .... DLB-108

Pangborn, Edgar 1909-1976 .... DLB-8

"Panic Among the Philistines": A Postscript,
An Interview with Bryan Griffin Y-81

Panneton, Philippe (see Ringuet)

Panshin, Alexei 1940- DLB-8

Pansy (see Alden, Isabella)

Pantheon Books DLB-46

Paperback Library DLB-46

Paperback Science Fiction DLB-8

Paquet, Alfons 1881-1944 .... DLB-66

Paradis, Suzanne 1936- DLB-53

Pareja Diezcanseco, Alfredo
1908-1993 .... DLB-145

Pardoe, Julia 1804-1862 .... DLB-166

Parents' Magazine Press DLB-46

Parisian Theater, Fall 1984: Toward
A New Baroque Y-85

Parizeau, Alice 1930- DLB-60

Parke, John 1754-1789 ....DLB-31

Parker, Dorothy
1893-1967 .... DLB-11, 45, 86

Parker, Gilbert 1860-1932 ....DLB-99

Parker, James 1714-1770 ....DLB-43

Parker, Theodore 1810-1860 ....DLB-1

Parker, William Riley 1906-1968 ....DLB-103

Parker, J. H. [publishing house] DLB-106

Parker, John [publishing house] DLB-106

Parkman, Francis, Jr.
1823-1893 .... DLB-1, 30

Parks, Gordon 1912- DLB-33

Parks, William 1698-1750 ....DLB-43

Parks, William [publishing house] DLB-49

Parley, Peter (see Goodrich, Samuel Griswold)

Parnell, Thomas 1679-1718 ....DLB-95

Parr, Catherine 1513?-1548 ....DLB-136

Parrington, Vernon L.
1871-1929 .... DLB-17, 63

Parronchi, Alessandro 1914- DLB-128

Partridge, S. W., and Company DLB-106

Parton, James 1822-1891 ....DLB-30

Parton, Sara Payson Willis
1811-1872 .... DLB-43, 74

Pasolini, Pier Paolo 1922- DLB-128

Pastan, Linda 1932- DLB-5

Paston, George 1860-1936 ....DLB-149

The *Paston Letters* 1422-1509 ....DLB-146

Pastorius, Francis Daniel
1651-circa 1720 ....DLB-24

Patchen, Kenneth 1911-1972 .... DLB-16, 48

Pater, Walter 1839-1894 .... DLB-57, 156

Paterson, Katherine 1932- DLB-52

Patmore, Coventry 1823-1896 ... DLB-35, 98

Paton, Joseph Noel 1821-1901 ......DLB-35

Paton Walsh, Jill 1937- DLB-161

Patrick, Edwin Hill ("Ted")
1901-1964 ....DLB-137

Patrick, John 1906- DLB-7

Pattee, Fred Lewis 1863-1950 ....DLB-71

Pattern and Paradigm: History as
Design, by Judith Ryan DLB-75

Patterson, Alicia 1906-1963 ....DLB-127

Patterson, Eleanor Medill 1881-1948 ...... DLB-29

Patterson, Eugene 1923- DLB-127

Patterson, Joseph Medill 1879-1946 ...... DLB-29

Pattillo, Henry 1726-1801 ...... DLB-37

Paul, Elliot 1891-1958 ...... DLB-4

Paul, Jean (see Richter, Johann Paul Friedrich)

Paul, Kegan, Trench, Trubner and Company Limited DLB-106

Paul, Peter, Book Company DLB-49

Paul, Stanley, and Company Limited DLB-112

Paulding, James Kirke 1778-1860 ...... DLB-3, 59, 74

Paulin, Tom 1949- DLB-40

Pauper, Peter, Press DLB-46

Pavese, Cesare 1908-1950 ...... DLB-128

Paxton, John 1911-1985 ...... DLB-44

Payn, James 1830-1898 ...... DLB-18

Payne, John 1842-1916 ...... DLB-35

Payne, John Howard 1791-1852 ...... DLB-37

Payson and Clarke DLB-46

Peabody, Elizabeth Palmer 1804-1894 ...... DLB-1

Peabody, Elizabeth Palmer [publishing house] DLB-49

Peabody, Oliver William Bourn 1799-1848 ...... DLB-59

Peace, Roger 1899-1968 ...... DLB-127

Peacham, Henry 1578-1644? DLB-151

Peacham, Henry, the Elder 1547-1634 ...... DLB-172

Peachtree Publishers, Limited DLB-46

Peacock, Molly 1947- DLB-120

Peacock, Thomas Love 1785-1866 ...... DLB-96, 116

Pead, Deuel ?-1727 ...... DLB-24

Peake, Mervyn 1911-1968 ...... DLB-15, 160

Pear Tree Press DLB-112

Pearce, Philippa 1920- DLB-161

Pearson, H. B. [publishing house] DLB-49

Pearson, Hesketh 1887-1964 ...... DLB-149

Peck, George W. 1840-1916 ...... DLB-23, 42

Peck, H. C., and Theo. Bliss [publishing house] DLB-49

Peck, Harry Thurston 1856-1914 ...... DLB-71, 91

Peele, George 1556-1596 ...... DLB-62, 167

Pegler, Westbrook 1894-1969 ...... DLB-171

Pellegrini and Cudahy DLB-46

Pelletier, Aimé (see Vac, Bertrand)

Pemberton, Sir Max 1863-1950 ...... DLB-70

Penguin Books [U.S.] DLB-46

Penguin Books [U.K.] DLB-112

Penn Publishing Company DLB-49

Penn, William 1644-1718 ...... DLB-24

Penna, Sandro 1906-1977 ...... DLB-114

Penner, Jonathan 1940- Y-83

Pennington, Lee 1939- Y-82

Pepys, Samuel 1633-1703 ...... DLB-101

Percy, Thomas 1729-1811 ...... DLB-104

Percy, Walker 1916-1990 ...... DLB-2; Y-80, 90

Percy, William 1575-1648 ...... DLB-172

Perec, Georges 1936-1982 ...... DLB-83

Perelman, S. J. 1904-1979 ...... DLB-11, 44

Perez, Raymundo "Tigre" 1946- DLB-122

Peri Rossi, Cristina 1941- DLB-145

Periodicals of the Beat Generation DLB-16

Perkins, Eugene 1932- DLB-41

Perkoff, Stuart Z. 1930-1974 ...... DLB-16

Perley, Moses Henry 1804-1862 ...... DLB-99

Permabooks DLB-46

Perrin, Alice 1867-1934 ...... DLB-156

Perry, Bliss 1860-1954 ...... DLB-71

Perry, Eleanor 1915-1981 ...... DLB-44

Perry, Sampson 1747-1823 ...... DLB-158

"Personal Style" (1890), by John Addington Symonds DLB-57

Perutz, Leo 1882-1957 ...... DLB-81

Pesetsky, Bette 1932- DLB-130

Pestalozzi, Johann Heinrich 1746-1827 ...... DLB-94

Peter, Laurence J. 1919-1990 ...... DLB-53

Peter of Spain circa 1205-1277 ...... DLB-115

Peterkin, Julia 1880-1961 ...... DLB-9

Peters, Lenrie 1932- DLB-117

Peters, Robert 1924- DLB-105

Peters, Robert, Foreword to Ludwig of Bavaria DLB-105

Petersham, Maud 1889-1971 and Petersham, Miska 1888-1960 ...... DLB-22

Peterson, Charles Jacobs 1819-1887 ...... DLB-79

Peterson, Len 1917- DLB-88

Peterson, Louis 1922- DLB-76

Peterson, T. B., and Brothers DLB-49

Petitclair, Pierre 1813-1860 ...... DLB-99

Petrov, Gavriil 1730-1801 ...... DLB-150

Petrov, Vasilii Petrovich 1736-1799 ...... DLB-150

Petrović, Rastko 1898-1949 ...... DLB-147

Petruslied circa 854? DLB-148

Petry, Ann 1908- DLB-76

Pettie, George circa 1548-1589 ...... DLB-136

Peyton, K. M. 1929- DLB-161

Pfaffe Konrad flourished circa 1172 ...... DLB-148

Pfaffe Lamprecht flourished circa 1150 ...... DLB-148

Pforzheimer, Carl H. 1879-1957 ...... DLB-140

Phaer, Thomas 1510?-1560 ...... DLB-167

Phaidon Press Limited DLB-112

Pharr, Robert Deane 1916-1992 ...... DLB-33

Phelps, Elizabeth Stuart 1844-1911 ...... DLB-74

Philander von der Linde (see Mencke, Johann Burckhard)

Philip, Marlene Nourbese 1947- DLB-157

Philippe, Charles-Louis 1874-1909 ...... DLB-65

Philips, John 1676-1708 ...... DLB-95

Philips, Katherine 1632-1664 ...... DLB-131

Phillips, Caryl 1958- DLB-157

Phillips, David Graham 1867-1911 ...... DLB-9, 12

Phillips, Jayne Anne 1952- Y-80

Phillips, Robert 1938- DLB-105

Phillips, Robert, Finding, Losing, Reclaiming: A Note on My Poems DLB-105

Phillips, Stephen 1864-1915 ...... DLB-10

Phillips, Ulrich B. 1877-1934 ...... DLB-17

Phillips, Willard 1784-1873 ...... DLB-59

Phillips, William 1907- DLB-137

Phillips, Sampson and Company DLB-49

Phillpotts, Eden 1862-1960 ...... DLB-10, 70, 135, 153

Philosophical Library DLB-46

"The Philosophy of Style" (1852), by Herbert Spencer DLB-57

Phinney, Elihu [publishing house] DLB-49

Phoenix, John (see Derby, George Horatio)

PHYLON (Fourth Quarter, 1950),
The Negro in Literature:
The Current Scene DLB-76

Physiologus
circa 1070-circa 1150 .... DLB-148

Piccolo, Lucio 1903-1969 .... DLB-114

Pickard, Tom 1946- DLB-40

Pickering, William
[publishing house] DLB-106

Pickthall, Marjorie 1883-1922 ...... DLB-92

Pictorial Printing Company DLB-49

Piel, Gerard 1915- DLB-137

Piercy, Marge 1936- DLB-120

Pierro, Albino 1916- DLB-128

Pignotti, Lamberto 1926- DLB-128

Pike, Albert 1809-1891 .... DLB-74

Pilon, Jean-Guy 1930- DLB-60

Pinckney, Josephine 1895-1957 .... DLB-6

Pindar, Peter (see Wolcot, John)

Pinero, Arthur Wing 1855-1934 .... DLB-10

Pinget, Robert 1919- DLB-83

Pinnacle Books DLB-46

Piñon, Nélida 1935- DLB-145

Pinsky, Robert 1940- Y-82

Pinter, Harold 1930- DLB-13

Piontek, Heinz 1925- DLB-75

Piozzi, Hester Lynch [Thrale]
1741-1821 .... DLB-104, 142

Piper, H. Beam 1904-1964 .... DLB-8

Piper, Watty DLB-22

Pisar, Samuel 1929- Y-83

Pitkin, Timothy 1766-1847 .... DLB-30

The Pitt Poetry Series: Poetry Publishing
Today Y-85

Pitter, Ruth 1897- DLB-20

Pix, Mary 1666-1709 .... DLB-80

Plaatje, Sol T. 1876-1932 .... DLB-125

The Place of Realism in Fiction (1895), by
George Gissing DLB-18

Plante, David 1940- Y-83

Platen, August von 1796-1835 ...... DLB-90

Plath, Sylvia 1932-1963 .... DLB-5, 6, 152

Platon 1737-1812 .... DLB-150

Platt and Munk Company DLB-46

Playboy Press DLB-46

Playford, John
[publishing house] DLB-170

Plays, Playwrights, and Playgoers ... DLB-84

Playwrights and Professors, by
Tom Stoppard DLB-13

Playwrights on the Theater DLB-80

Der Pleier flourished circa 1250 .... DLB-138

Plenzdorf, Ulrich 1934- DLB-75

Plessen, Elizabeth 1944- DLB-75

Plievier, Theodor 1892-1955 .... DLB-69

Plomer, William 1903-1973 ..... DLB-20, 162

Plumly, Stanley 1939- DLB-5

Plumpp, Sterling D. 1940- DLB-41

Plunkett, James 1920- DLB-14

Plymell, Charles 1935- DLB-16

Pocket Books DLB-46

Poe, Edgar Allan
1809-1849 .... DLB-3, 59, 73, 74

Poe, James 1921-1980 .... DLB-44

The Poet Laureate of the United States
Statements from Former Consultants
in Poetry Y-86

Pohl, Frederik 1919- DLB-8

Poirier, Louis (see Gracq, Julien)

Polanyi, Michael 1891-1976 .... DLB-100

Pole, Reginald 1500-1558 .... DLB-132

Poliakoff, Stephen 1952- DLB-13

Polidori, John William
1795-1821 .... DLB-116

Polite, Carlene Hatcher 1932- DLB-33

Pollard, Edward A. 1832-1872 ...... DLB-30

Pollard, Percival 1869-1911 .... DLB-71

Pollard and Moss DLB-49

Pollock, Sharon 1936- DLB-60

Polonsky, Abraham 1910- DLB-26

Polotsky, Simeon 1629-1680 ...... DLB-150

Ponce, Mary Helen 1938- DLB-122

Ponce-Montoya, Juanita 1949- DLB-122

Ponet, John 1516?-1556 .... DLB-132

Poniatowski, Elena 1933- DLB-113

Ponsonby, William
[publishing house] DLB-170

Pony Stories DLB-160

Poole, Ernest 1880-1950 .... DLB-9

Poole, Sophia 1804-1891 .... DLB-166

Poore, Benjamin Perley
1820-1887 .... DLB-23

Pope, Abbie Hanscom
1858-1894 .... DLB-140

Pope, Alexander 1688-1744 .... DLB-95, 101

Popov, Mikhail Ivanovich
1742-circa 1790 .... DLB-150

Popular Library DLB-46

Porlock, Martin (see MacDonald, Philip)

Porpoise Press DLB-112

Porta, Antonio 1935-1989 .... DLB-128

Porter, Anna Maria
1780-1832 .... DLB-116, 159

Porter, Eleanor H. 1868-1920 .... DLB-9

Porter, Gene Stratton (see Stratton-Porter, Gene)

Porter, Henry ?-? DLB-62

Porter, Jane 1776-1850 .... DLB-116, 159

Porter, Katherine Anne
1890-1980 ....DLB-4, 9, 102; Y-80; DS-12

Porter, Peter 1929- DLB-40

Porter, William Sydney
1862-1910 .... DLB-12, 78, 79

Porter, William T. 1809-1858 .... DLB-3, 43

Porter and Coates DLB-49

Portis, Charles 1933- DLB-6

Postans, Marianne
circa 1810-1865 .... DLB-166

Postl, Carl (see Sealsfield, Carl)

Poston, Ted 1906-1974 .... DLB-51

Postscript to [the Third Edition of] *Clarissa*
(1751), by Samuel Richardson DLB-39

Potok, Chaim 1929- DLB-28, 152; Y-84

Potter, Beatrix 1866-1943 .... DLB-141

Potter, David M. 1910-1971 .... DLB-17

Potter, John E., and Company DLB-49

Pottle, Frederick A.
1897-1987 .... DLB-103; Y-87

Poulin, Jacques 1937- DLB-60

Pound, Ezra 1885-1972 .... DLB-4, 45, 63

Povich, Shirley 1905- DLB-171

Powell, Anthony 1905- DLB-15

Powers, J. F. 1917- DLB-130

Pownall, David 1938- DLB-14

Powys, John Cowper 1872-1963 ..... DLB-15

Powys, Llewelyn 1884-1939 .... DLB-98

Powys, T. F. 1875-1953 .... DLB-36, 162

Poynter, Nelson 1903-1978 .... DLB-127

The Practice of Biography: An Interview
with Stanley Weintraub Y-82

The Practice of Biography II: An Interview
with B. L. Reid Y-83

The Practice of Biography III: An Interview
with Humphrey Carpenter Y-84

The Practice of Biography IV: An Interview
with William Manchester Y-85

The Practice of Biography V: An Interview
with Justin Kaplan Y-86

The Practice of Biography VI: An Interview
with David Herbert Donald Y-87

The Practice of Biography VII: An Interview
with John Caldwell Guilds Y-92

The Practice of Biography VIII: An Interview
with Joan Mellen Y-94

The Practice of Biography IX: An Interview
with Michael Reynolds Y-95

Prados, Emilio 1899-1962 . . . . DLB-134

Praed, Winthrop Mackworth
1802-1839 . . . . DLB-96

Praeger Publishers DLB-46

Praetorius, Johannes 1630-1680 . . . . DLB-168

Pratt, E. J. 1882-1964 . . . . DLB-92

Pratt, Samuel Jackson 1749-1814 . . . . DLB-39

Preface to *Alwyn* (1780), by
Thomas Holcroft DLB-39

Preface to *Colonel Jack* (1722), by
Daniel Defoe DLB-39

Preface to *Evelina* (1778), by
Fanny Burney DLB-39

Preface to *Ferdinand Count Fathom* (1753), by
Tobias Smollett DLB-39

Preface to *Incognita* (1692), by
William Congreve DLB-39

Preface to *Joseph Andrews* (1742), by
Henry Fielding DLB-39

Preface to *Moll Flanders* (1722), by
Daniel Defoe DLB-39

Preface to *Poems* (1853), by
Matthew Arnold DLB-32

Preface to *Robinson Crusoe* (1719), by
Daniel Defoe DLB-39

Preface to *Roderick Random* (1748), by
Tobias Smollett DLB-39

Preface to *Roxana* (1724), by
Daniel Defoe DLB-39

Preface to *St. Leon* (1799), by
William Godwin DLB-39

Preface to Sarah Fielding's *Familiar Letters*
(1747), by Henry Fielding
[excerpt] DLB-39

Preface to Sarah Fielding's *The Adventures of
David Simple* (1744), by
Henry Fielding DLB-39

Preface to *The Cry* (1754), by
Sarah Fielding DLB-39

Preface to *The Delicate Distress* (1769), by
Elizabeth Griffin DLB-39

Preface to *The Disguis'd Prince* (1733), by
Eliza Haywood [excerpt] DLB-39

Preface to *The Farther Adventures of Robinson
Crusoe* (1719), by Daniel Defoe . . DLB-39

Preface to the First Edition of *Pamela* (1740), by
Samuel Richardson DLB-39

Preface to the First Edition of *The Castle of
Otranto* (1764), by
Horace Walpole DLB-39

Preface to *The History of Romances* (1715), by
Pierre Daniel Huet [excerpts] DLB-39

Preface to *The Life of Charlotta du Pont* (1723),
by Penelope Aubin DLB-39

Preface to *The Old English Baron* (1778), by
Clara Reeve DLB-39

Preface to the Second Edition of *The Castle of
Otranto* (1765), by Horace
Walpole DLB-39

Preface to *The Secret History, of Queen Zarah,
and the Zarazians* (1705), by Delariviere
Manley DLB-39

Preface to the Third Edition of *Clarissa* (1751),
by Samuel Richardson
[excerpt] DLB-39

Preface to *The Works of Mrs. Davys* (1725), by
Mary Davys DLB-39

Preface to Volume 1 of *Clarissa* (1747), by
Samuel Richardson DLB-39

Preface to Volume 3 of *Clarissa* (1748), by
Samuel Richardson DLB-39

Préfontaine, Yves 1937- DLB-53

Prelutsky, Jack 1940- DLB-61

Premisses, by Michael Hamburger . . . DLB-66

Prentice, George D. 1802-1870 . . . . . DLB-43

Prentice-Hall DLB-46

Prescott, William Hickling
1796-1859 . . . . DLB-1, 30, 59

The Present State of the English Novel (1892),
by George Saintsbury DLB-18

Prešeren, France 1800-1849 . . . . DLB-147

Preston, Thomas 1537-1598 . . . . DLB-62

Price, Reynolds 1933- DLB-2

Price, Richard 1723-1791 . . . . DLB-158

Price, Richard 1949- Y-81

Priest, Christopher 1943- DLB-14

Priestley, J. B. 1894-1984
. . . . DLB-10, 34, 77, 100, 139; Y-84

Primary Bibliography: A
Retrospective Y-95

Prime, Benjamin Young 1733-1791 . . . DLB-31

Primrose, Diana
floruit circa 1630 . . . . DLB-126

Prince, F. T. 1912- DLB-20

Prince, Thomas 1687-1758 . . . . . DLB-24, 140

The Principles of Success in Literature (1865), by
George Henry Lewes [excerpt] . . . DLB-57

Printz, Wolfgang Casper
1641-1717 . . . . DLB-168

Prior, Matthew 1664-1721 . . . . DLB-95

Pritchard, William H. 1932- DLB-111

Pritchett, V. S. 1900- DLB-15, 139

Procter, Adelaide-Anne 1825-1864 . . . DLB-32

Procter, Bryan Waller
1787-1874 . . . . DLB-96, 144

The Profession of Authorship:
Scribblers for Bread Y-89

The Progress of Romance (1785), by Clara Reeve
[excerpt] DLB-39

Prokopovich, Feofan 1681?-1736 . . . DLB-150

Prokosch, Frederic 1906-1989 . . . . . . DLB-48

The Proletarian Novel DLB-9

Propper, Dan 1937- DLB-16

The Prospect of Peace (1778), by
Joel Barlow DLB-37

Proud, Robert 1728-1813 . . . . DLB-30

Proust, Marcel 1871-1922 . . . . DLB-65

Prynne, J. H. 1936- DLB-40

Przybyszewski, Stanislaw
1868-1927 . . . . DLB-66

Pseudo-Dionysius the Areopagite floruit
circa 500 . . . . DLB-115

The Public Lending Right in America
Statement by Sen. Charles McC.
Mathias, Jr. PLR and the Meaning
of Literary Property Statements on
PLR by American Writers Y-83

The Public Lending Right in the United Kingdom Public Lending Right: The First Year
in the United Kingdom Y-83

The Publication of English
Renaissance Plays DLB-62

Publications and Social Movements
[Transcendentalism] DLB-1

Publishers and Agents: The Columbia
Connection Y-87

A Publisher's Archives: G. P. Putnam . . . Y-92

Publishing Fiction at LSU Press Y-87

Pückler-Muskau, Hermann von
1785-1871 . . . . DLB-133

Pufendorf, Samuel von
1632-1694 . . . . DLB-168

Cumulative Index

Pugh, Edwin William 1874-1930 ... DLB-135

Pugin, A. Welby 1812-1852 .... DLB-55

Puig, Manuel 1932-1990 .... DLB-113

Pulitzer, Joseph 1847-1911 .... DLB-23

Pulitzer, Joseph, Jr. 1885-1955 ...... DLB-29

Pulitzer Prizes for the Novel, 1917-1945 .... DLB-9

Pulliam, Eugene 1889-1975 .... DLB-127

Purchas, Samuel 1577?-1626 ...... DLB-151

Purdy, Al 1918- DLB-88

Purdy, James 1923- DLB-2

Purdy, Ken W. 1913-1972 .... DLB-137

Pusey, Edward Bouverie 1800-1882 .... DLB-55

Putnam, George Palmer 1814-1872 .... DLB-3, 79

Putnam, Samuel 1892-1950 ....DLB-4

G. P. Putnam's Sons [U.S.] DLB-49

G. P. Putnam's Sons [U.K.] DLB-106

Puzo, Mario 1920- DLB-6

Pyle, Ernie 1900-1945 .... DLB-29

Pyle, Howard 1853-1911 ..... DLB-42; DS-13

Pym, Barbara 1913-1980 ...... DLB-14; Y-87

Pynchon, Thomas 1937- DLB-2, 173

Pyramid Books DLB-46

Pyrnelle, Louise-Clarke 1850-1907 .. DLB-42

Q

Quad, M. (see Lewis, Charles B.)

Quarles, Francis 1592-1644 .... DLB-126

The Quarterly Review 1809-1967 .... DLB-110

Quasimodo, Salvatore 1901-1968 .. DLB-114

Queen, Ellery (see Dannay, Frederic, and Manfred B. Lee)

The Queen City Publishing House .. DLB-49

Queneau, Raymond 1903-1976 ..... DLB-72

Quennell, Sir Peter 1905-1993 ..... DLB-155

Quesnel, Joseph 1746-1809 .... DLB-99

The Question of American Copyright in the Nineteenth Century
Headnote
Preface, by George Haven Putnam
The Evolution of Copyright, by Brander Matthews
Summary of Copyright Legislation in the United States, by R. R. Bowker
Analysis of the Provisions of the Copyright Law of 1891, by George Haven Putnam
The Contest for International Copyright, by George Haven Putnam
Cheap Books and Good Books, by Brander Matthews DLB-49

Quiller-Couch, Sir Arthur Thomas 1863-1944 .... DLB-135, 153

Quin, Ann 1936-1973 .... DLB-14

Quincy, Samuel, of Georgia ?-? DLB-31

Quincy, Samuel, of Massachusetts 1734-1789 .... DLB-31

Quinn, Anthony 1915- DLB-122

Quintana, Leroy V. 1944- DLB-82

Quintana, Miguel de 1671-1748
A Forerunner of Chicano Literature DLB-122

Quist, Harlin, Books DLB-46

Quoirez, Françoise (see Sagan, Francçise)

R

Raabe, Wilhelm 1831-1910 .... DLB-129

Rabe, David 1940- DLB-7

Raboni, Giovanni 1932- DLB-128

Rachilde 1860-1953 .... DLB-123

Racin, Kočo 1908-1943 .... DLB-147

Rackham, Arthur 1867-1939 ...... DLB-141

Radcliffe, Ann 1764-1823 .... DLB-39

Raddall, Thomas 1903- DLB-68

Radiguet, Raymond 1903-1923 ..... DLB-65

Radishchev, Aleksandr Nikolaevich 1749-1802 .... DLB-150

Radványi, Netty Reiling (see Seghers, Anna)

Rahv, Philip 1908-1973 .... DLB-137

Raimund, Ferdinand Jakob 1790-1836 .... DLB-90

Raine, Craig 1944- DLB-40

Raine, Kathleen 1908- DLB-20

Rainolde, Richard circa 1530-1606 .... DLB-136

Rakić, Milan 1876-1938 .... DLB-147

Ralegh, Sir Walter 1554?-1618 ..... DLB-172

Ralph, Julian 1853-1903 ....DLB-23

Ralph Waldo Emerson in 1982 .... Y-82

Ramat, Silvio 1939- DLB-128

Rambler, no. 4 (1750), by Samuel Johnson [excerpt] DLB-39

Ramée, Marie Louise de la (see Ouida)

Ramírez, Sergío 1942- DLB-145

Ramke, Bin 1947- DLB-120

Ramler, Karl Wilhelm 1725-1798 ....DLB-97

Ramon Ribeyro, Julio 1929- DLB-145

Ramous, Mario 1924- DLB-128

Rampersad, Arnold 1941- DLB-111

Ramsay, Allan 1684 or 1685-1758 ....DLB-95

Ramsay, David 1749-1815 ....DLB-30

Ranck, Katherine Quintana 1942- DLB-122

Rand, Avery and Company DLB-49

Rand McNally and Company DLB-49

Randall, David Anton 1905-1975 ....DLB-140

Randall, Dudley 1914- DLB-41

Randall, Henry S. 1811-1876 ....DLB-30

Randall, James G. 1881-1953 ....DLB-17

The Randall Jarrell Symposium: A Small Collection of Randall Jarrells Excerpts From Papers Delivered at the Randall Jarrell Symposium Y-86

Randolph, A. Philip 1889-1979 ......DLB-91

Randolph, Anson D. F. [publishing house] DLB-49

Randolph, Thomas 1605-1635 .. DLB-58, 126

Random House DLB-46

Ranlet, Henry [publishing house] DLB-49

Ransom, John Crowe 1888-1974 .... DLB-45, 63

Ransome, Arthur 1884-1967 ....DLB-160

Raphael, Frederic 1931- DLB-14

Raphaelson, Samson 1896-1983 .....DLB-44

Raskin, Ellen 1928-1984 ....DLB-52

Rastell, John 1475?-1536 ...... DLB-136, 170

Rattigan, Terence 1911-1977 ....DLB-13

Rawlings, Marjorie Kinnan 1896-1953 .... DLB-9, 22, 102

Raworth, Tom 1938- DLB-40

Ray, David 1932- DLB-5

Ray, Gordon Norton 1915-1986 .... DLB-103, 140

Ray, Henrietta Cordelia
 1849-1916 ....DLB-50

Raymond, Henry J. 1820-1869 ...DLB-43, 79

Raymond Chandler Centenary Tributes
 from Michael Avallone, James Elroy, Joe Gores,
 and William F. Nolan Y-88

Reach, Angus 1821-1856 ....DLB-70

Read, Herbert 1893-1968 ....DLB-20, 149

Read, Herbert, "The Practice of Biography," in
 The English Sense of Humour and Other Essays DLB-149

Read, Opie 1852-1939 ....DLB-23

Read, Piers Paul 1941- DLB-14

Reade, Charles 1814-1884 ....DLB-21

Reader's Digest Condensed
 Books DLB-46

Reading, Peter 1946- DLB-40

Reaney, James 1926- DLB-68

Rèbora, Clemente 1885-1957 ....DLB-114

Rechy, John 1934- DLB-122; Y-82

The Recovery of Literature: Criticism in the
 1990s: A Symposium Y-91

Redding, J. Saunders
 1906-1988 ....DLB-63, 76

Redfield, J. S. [publishing house] DLB-49

Redgrove, Peter 1932- DLB-40

Redmon, Anne 1943- Y-86

Redmond, Eugene B. 1937- DLB-41

Redpath, James [publishing house] ...DLB-49

Reed, Henry 1808-1854 ....DLB-59

Reed, Henry 1914- DLB-27

Reed, Ishmael
 1938- DLB-2, 5, 33, 169; DS-8

Reed, Sampson 1800-1880 ....DLB-1

Reed, Talbot Baines 1852-1893 .....DLB-141

Reedy, William Marion 1862-1920 ...DLB-91

Reese, Lizette Woodworth
 1856-1935 ....DLB-54

Reese, Thomas 1742-1796 ....DLB-37

Reeve, Clara 1729-1807 ....DLB-39

Reeves, James 1909-1978 ....DLB-161

Reeves, John 1926- DLB-88

Regnery, Henry, Company DLB-46

Rehberg, Hans 1901-1963 ....DLB-124

Rehfisch, Hans José 1891-1960 .....DLB-124

Reid, Alastair 1926- DLB-27

Reid, B. L. 1918-1990 ....DLB-111

Reid, Christopher 1949- DLB-40

Reid, Forrest 1875-1947 ....DLB-153

Reid, Helen Rogers 1882-1970 ......DLB-29

Reid, James ?-? DLB-31

Reid, Mayne 1818-1883 ....DLB-21, 163

Reid, Thomas 1710-1796 ....DLB-31

Reid, V. S. (Vic) 1913-1987 ......DLB-125

Reid, Whitelaw 1837-1912 ....DLB-23

Reilly and Lee Publishing
 Company DLB-46

Reimann, Brigitte 1933-1973 ....DLB-75

Reinmar der Alte
 circa 1165-circa 1205 ....DLB-138

Reinmar von Zweter
 circa 1200-circa 1250 ....DLB-138

Reisch, Walter 1903-1983 ....DLB-44

Remarque, Erich Maria 1898-1970 ..DLB-56

"Re-meeting of Old Friends": The Jack
 Kerouac Conference Y-82

Remington, Frederic 1861-1909 .....DLB-12

Renaud, Jacques 1943- DLB-60

Renault, Mary 1905-1983 ....Y-83

Rendell, Ruth 1930- DLB-87

Representative Men and Women: A Historical
 Perspective on the British Novel,
 1930-1960 ....DLB-15

(Re-)Publishing Orwell Y-86

Rettenbacher, Simon 1634-1706 ....DLB-168

Reuter, Christian 1665-after 1712 ..DLB-168

Reuter, Fritz 1810-1874 ....DLB-129

Reuter, Gabriele 1859-1941 ....DLB-66

Revell, Fleming H., Company DLB-49

Reventlow, Franziska Gräfin zu
 1871-1918 ....DLB-66

Review of Reviews Office DLB-112

Review of [Samuel Richardson's] *Clarissa*
 (1748), by Henry Fielding DLB-39

The Revolt (1937), by Mary Colum
 [excerpts] DLB-36

Rexroth, Kenneth
 1905-1982 ....DLB-16, 48, 165; Y-82

Rey, H. A. 1898-1977 ....DLB-22

Reynal and Hitchcock DLB-46

Reynolds, G. W. M. 1814-1879 .....DLB-21

Reynolds, John Hamilton
 1794-1852 ....DLB-96

Reynolds, Mack 1917- DLB-8

Reynolds, Sir Joshua 1723-1792 ....DLB-104

Reznikoff, Charles 1894-1976 ....DLB-28, 45

"Rhetoric" (1828; revised, 1859), by
 Thomas de Quincey [excerpt] ...DLB-57

Rhett, Robert Barnwell 1800-1876 ...DLB-43

Rhode, John 1884-1964 ....DLB-77

Rhodes, James Ford 1848-1927 .....DLB-47

Rhys, Jean 1890-1979 .....DLB-36, 117, 162

Ricardo, David 1772-1823 .....DLB-107, 158

Ricardou, Jean 1932- DLB-83

Rice, Elmer 1892-1967 ....DLB-4, 7

Rice, Grantland 1880-1954 .....DLB-29, 171

Rich, Adrienne 1929- DLB-5, 67

Richards, David Adams 1950- DLB-53

Richards, George circa 1760-1814 ...DLB-37

Richards, I. A. 1893-1979 ....DLB-27

Richards, Laura E. 1850-1943 ......DLB-42

Richards, William Carey
 1818-1892 ....DLB-73

Richards, Grant
 [publishing house] DLB-112

Richardson, Charles F. 1851-1913 ...DLB-71

Richardson, Dorothy M.
 1873-1957 ....DLB-36

Richardson, Jack 1935- DLB-7

Richardson, John 1796-1852 ....DLB-99

Richardson, Samuel
 1689-1761 ....DLB-39, 154

Richardson, Willis 1889-1977 ......DLB-51

Riche, Barnabe 1542-1617 ....DLB-136

Richler, Mordecai 1931- DLB-53

Richter, Conrad 1890-1968 ....DLB-9

Richter, Hans Werner 1908- DLB-69

Richter, Johann Paul Friedrich
 1763-1825 ....DLB-94

Rickerby, Joseph
 [publishing house] DLB-106

Rickword, Edgell 1898-1982 ....DLB-20

Riddell, Charlotte 1832-1906 ......DLB-156

Riddell, John (see Ford, Corey)

Ridge, Lola 1873-1941 ....DLB-54

Ridge, William Pett 1859-1930 ....DLB-135

Riding, Laura (see Jackson, Laura Riding)

Ridler, Anne 1912- DLB-27

Ridruego, Dionisio 1912-1975 .....DLB-108

Riel, Louis 1844-1885 ....DLB-99

Riemer, Johannes 1648-1714 ......DLB-168

Riffaterre, Michael 1924- DLB-67

Riis, Jacob 1849-1914 .... DLB-23

Riker, John C. [publishing house] ... DLB-49

Riley, John 1938-1978 .... DLB-40

Rilke, Rainer Maria 1875-1926 ..... DLB-81

Rinehart and Company DLB-46

Ringuet 1895-1960 .... DLB-68

Ringwood, Gwen Pharis
1910-1984 .... DLB-88

Rinser, Luise 1911- DLB-69

Ríos, Alberto 1952- DLB-122

Ríos, Isabella 1948- DLB-82

Ripley, Arthur 1895-1961 .... DLB-44

Ripley, George 1802-1880 ..... DLB-1, 64, 73

The Rising Glory of America:
Three Poems DLB-37

The Rising Glory of America: Written in 1771
(1786), by Hugh Henry Brackenridge and
Philip Freneau DLB-37

Riskin, Robert 1897-1955 .... DLB-26

Risse, Heinz 1898- DLB-69

Rist, Johann 1607-1667 .... DLB-164

Ritchie, Anna Mowatt 1819-1870 .....DLB-3

Ritchie, Anne Thackeray
1837-1919 .... DLB-18

Ritchie, Thomas 1778-1854 .... DLB-43

Rites of Passage
[on William Saroyan] Y-83

The Ritz Paris Hemingway Award Y-85

Rivard, Adjutor 1868-1945 .... DLB-92

Rive, Richard 1931-1989 .... DLB-125

Rivera, Marina 1942- DLB-122

Rivera, Tomás 1935-1984 .... DLB-82

Rivers, Conrad Kent 1933-1968 ..... DLB-41

Riverside Press DLB-49

Rivington, James circa 1724-1802 ... DLB-43

Rivington, Charles
[publishing house] DLB-154

Rivkin, Allen 1903-1990 .... DLB-26

Roa Bastos, Augusto 1917- DLB-113

Robbe-Grillet, Alain 1922- DLB-83

Robbins, Tom 1936- Y-80

Roberts, Charles G. D. 1860-1943 ... DLB-92

Roberts, Dorothy 1906-1993 .... DLB-88

Roberts, Elizabeth Madox
1881-1941 .... DLB-9, 54, 102

Roberts, Kenneth 1885-1957 ....DLB-9

Roberts, William 1767-1849 ...... DLB-142

Roberts Brothers DLB-49

Roberts, James [publishing house] .. DLB-154

Robertson, A. M., and Company DLB-49

Robertson, William 1721-1793 ..... DLB-104

Robinson, Casey 1903-1979 .... DLB-44

Robinson, Edwin Arlington
1869-1935 .... DLB-54

Robinson, Henry Crabb
1775-1867 .... DLB-107

Robinson, James Harvey
1863-1936 .... DLB-47

Robinson, Lennox 1886-1958 .... DLB-10

Robinson, Mabel Louise
1874-1962 .... DLB-22

Robinson, Mary 1758-1800 .... DLB-158

Robinson, Richard
circa 1545-1607 .... DLB-167

Robinson, Therese
1797-1870 .... DLB-59, 133

Robison, Mary 1949- DLB-130

Roblès, Emmanuel 1914- DLB-83

Roccatagliata Ceccardi, Ceccardo
1871-1919 .... DLB-114

Rochester, John Wilmot, Earl of
1647-1680 .... DLB-131

Rock, Howard 1911-1976 .... DLB-127

Rodgers, Carolyn M. 1945- DLB-41

Rodgers, W. R. 1909-1969 .... DLB-20

Rodríguez, Claudio 1934- DLB-134

Rodriguez, Richard 1944- DLB-82

Rodríguez Julia, Edgardo
1946- DLB-145

Roethke, Theodore 1908-1963 ..... DLB-5

Rogers, Pattiann 1940- DLB-105

Rogers, Samuel 1763-1855 .... DLB-93

Rogers, Will 1879-1935 .... DLB-11

Rohmer, Sax 1883-1959 .... DLB-70

Roiphe, Anne 1935- Y-80

Rojas, Arnold R. 1896-1988 .... DLB-82

Rolfe, Frederick William
1860-1913 .... DLB-34, 156

Rolland, Romain 1866-1944 ...... DLB-65

Rolle, Richard
circa 1290-1300 - 1340 .... DLB-146

Rolvaag, O. E. 1876-1931 .... DLB-9

Romains, Jules 1885-1972 .... DLB-65

Roman, A., and Company DLB-49

Romano, Octavio 1923- DLB-122

Romero, Leo 1950- DLB-122

Romero, Lin 1947- DLB-122

Romero, Orlando 1945- DLB-82

Rook, Clarence 1863-1915 ....DLB-135

Roosevelt, Theodore 1858-1919 .....DLB-47

Root, Waverley 1903-1982 ....DLB-4

Root, William Pitt 1941- DLB-120

Roquebrune, Robert de 1889-1978 ...DLB-68

Rosa, João Guimarães
1908-1967 ....DLB-113

Rosales, Luis 1910-1992 ....DLB-134

Roscoe, William 1753-1831 ....DLB-163

Rose, Reginald 1920- DLB-26

Rosegger, Peter 1843-1918 ....DLB-129

Rosei, Peter 1946- DLB-85

Rosen, Norma 1925- DLB-28

Rosenbach, A. S. W. 1876-1952 .....DLB-140

Rosenberg, Isaac 1890-1918 ....DLB-20

Rosenfeld, Isaac 1918-1956 ....DLB-28

Rosenthal, M. L. 1917- DLB-5

Ross, Alexander 1591-1654 ....DLB-151

Ross, Harold 1892-1951 ....DLB-137

Ross, Leonard Q. (see Rosten, Leo)

Ross, Martin 1862-1915 ....DLB-135

Ross, Sinclair 1908- DLB-88

Ross, W. W. E. 1894-1966 ....DLB-88

Rosselli, Amelia 1930- DLB-128

Rossen, Robert 1908-1966 ....DLB-26

Rossetti, Christina Georgina
1830-1894 .... DLB-35, 163

Rossetti, Dante Gabriel 1828-1882 ...DLB-35

Rossner, Judith 1935- DLB-6

Rosten, Leo 1908- DLB-11

Rostenberg, Leona 1908- DLB-140

Rostovsky, Dimitrii 1651-1709 .....DLB-150

Bertram Rota and His Bookshop Y-91

Roth, Gerhard 1942- DLB-85, 124

Roth, Henry 1906?- DLB-28

Roth, Joseph 1894-1939 ....DLB-85

Roth, Philip 1933- ... DLB-2, 28, 173; Y-82

Rothenberg, Jerome 1931- DLB-5

Rotimi, Ola 1938- DLB-125

Routhier, Adolphe-Basile
1839-1920 ....DLB-99

Routier, Simone 1901-1987 ....DLB-88

Routledge, George, and Sons DLB-106

Roversi, Roberto 1923- DLB-128

Rowe, Elizabeth Singer
  1674-1737 .... DLB-39, 95

Rowe, Nicholas 1674-1718 .... DLB-84

Rowlands, Samuel
  circa 1570-1630 .... DLB-121

Rowlandson, Mary
  circa 1635-circa 1678 .... DLB-24

Rowley, William circa 1585-1626 .... DLB-58

Rowse, A. L. 1903- DLB-155

Rowson, Susanna Haswell
  circa 1762-1824 .... DLB-37

Roy, Camille 1870-1943 .... DLB-92

Roy, Gabrielle 1909-1983 .... DLB-68

Roy, Jules 1907- DLB-83

The Royal Court Theatre and the English
  Stage Company DLB-13

The Royal Court Theatre and the New
  Drama DLB-10

The Royal Shakespeare Company
  at the Swan Y-88

Royall, Anne 1769-1854 .... DLB-43

The Roycroft Printing Shop DLB-49

Royster, Vermont 1914- DLB-127

Royston, Richard
  [publishing house] DLB-170

Ruark, Gibbons 1941- DLB-120

Ruban, Vasilii Grigorevich
  1742-1795 .... DLB-150

Rubens, Bernice 1928- DLB-14

Rudd and Carleton DLB-49

Rudkin, David 1936- DLB-13

Rudolf von Ems
  circa 1200-circa 1254 .... DLB-138

Ruffin, Josephine St. Pierre
  1842-1924 .... DLB-79

Ruganda, John 1941- DLB-157

Ruggles, Henry Joseph 1813-1906 .... DLB-64

Rukeyser, Muriel 1913-1980 .... DLB-48

Rule, Jane 1931- DLB-60

Rulfo, Juan 1918-1986 .... DLB-113

Rumaker, Michael 1932- DLB-16

Rumens, Carol 1944- DLB-40

Runyon, Damon 1880-1946 . DLB-11, 86, 171

Ruodlieb circa 1050-1075 .... DLB-148

Rush, Benjamin 1746-1813 .... DLB-37

Rusk, Ralph L. 1888-1962 .... DLB-103

Ruskin, John 1819-1900 .... DLB-55, 163

Russ, Joanna 1937- DLB-8

Russell, B. B., and Company DLB-49

Russell, Benjamin 1761-1845 .... DLB-43

Russell, Bertrand 1872-1970 .... DLB-100

Russell, Charles Edward
  1860-1941 .... DLB-25

Russell, George William (see AE)

Russell, R. H., and Son DLB-49

Rutherford, Mark 1831-1913 .... DLB-18

Ryan, Michael 1946- Y-82

Ryan, Oscar 1904- DLB-68

Ryga, George 1932- DLB-60

Rymer, Thomas 1643?-1713 .... DLB-101

Ryskind, Morrie 1895-1985 .... DLB-26

Rzhevsky, Aleksei Andreevich
  1737-1804 .... DLB-150

S

The Saalfield Publishing
  Company DLB-46

Saba, Umberto 1883-1957 .... DLB-114

Sábato, Ernesto 1911- DLB-145

Saberhagen, Fred 1930- DLB-8

Sacer, Gottfried Wilhelm
  1635-1699 .... DLB-168

Sackler, Howard 1929-1982 .... DLB-7

Sackville, Thomas 1536-1608 ..... DLB-132

Sackville, Thomas 1536-1608
  and Norton, Thomas
  1532-1584 .... DLB-62

Sackville-West, V. 1892-1962 ..... DLB-34

Sadlier, D. and J., and Company DLB-49

Sadlier, Mary Anne 1820-1903 ..... DLB-99

Sadoff, Ira 1945- DLB-120

Saenz, Jaime 1921-1986 .... DLB-145

Saffin, John circa 1626-1710 ...... DLB-24

Sagan, Françoise 1935- DLB-83

Sage, Robert 1899-1962 .... DLB-4

Sagel, Jim 1947- DLB-82

Sagendorph, Robb Hansell
  1900-1970 .... DLB-137

Sahagún, Carlos 1938- DLB-108

Sahkomaapii, Piitai (see Highwater, Jamake)

Sahl, Hans 1902- DLB-69

Said, Edward W. 1935- DLB-67

Saiko, George 1892-1962 .... DLB-85

St. Dominic's Press DLB-112

Saint-Exupéry, Antoine de
  1900-1944 .... DLB-72

St. Johns, Adela Rogers 1894-1988 ... DLB-29

St. Martin's Press DLB-46

St. Omer, Garth 1931- DLB-117

Saint Pierre, Michel de 1916-1987 ... DLB-83

Saintsbury, George
  1845-1933 .... DLB-57, 149

Saki (see Munro, H. H.)

Salaam, Kalamu ya 1947- DLB-38

Salas, Floyd 1931- DLB-82

Sálaz-Marquez, Rubén 1935- DLB-122

Salemson, Harold J. 1910-1988 ...... DLB-4

Salinas, Luis Omar 1937- DLB-82

Salinas, Pedro 1891-1951 .... DLB-134

Salinger, J. D. 1919- DLB-2, 102, 173

Salkey, Andrew 1928- DLB-125

Salt, Waldo 1914- DLB-44

Salter, James 1925- DLB-130

Salter, Mary Jo 1954- DLB-120

Salustri, Carlo Alberto (see Trilussa)

Salverson, Laura Goodman
  1890-1970 .... DLB-92

Sampson, Richard Henry (see Hull, Richard)

Samuels, Ernest 1903- DLB-111

Sanborn, Franklin Benjamin
  1831-1917 .... DLB-1

Sánchez, Luis Rafael 1936- DLB-145

Sánchez, Philomeno "Phil"
  1917- DLB-122

Sánchez, Ricardo 1941- DLB-82

Sanchez, Sonia 1934- DLB-41; DS-8

Sand, George 1804-1876 .... DLB-119

Sandburg, Carl 1878-1967 ..... DLB-17, 54

Sanders, Ed 1939- DLB-16

Sandoz, Mari 1896-1966 .... DLB-9

Sandwell, B. K. 1876-1954 .... DLB-92

Sandy, Stephen 1934- DLB-165

Sandys, George 1578-1644 ..... DLB-24, 121

Sangster, Charles 1822-1893 .... DLB-99

Sanguineti, Edoardo 1930- DLB-128

Sansom, William 1912-1976 .... DLB-139

Santayana, George
  1863-1952 .... DLB-54, 71; DS-13

Santiago, Danny 1911-1988 .... DLB-122

Santmyer, Helen Hooven 1895-1986 .... Y-84

Sapir, Edward 1884-1939 .... DLB-92

Sapper (see McNeile, Herman Cyril)

Sarduy, Severo 1937- DLB-113

Sargent, Pamela 1948- DLB-8

Saro-Wiwa, Ken 1941- DLB-157

Saroyan, William
   1908-1981 .... DLB-7, 9, 86; Y-81

Sarraute, Nathalie 1900- DLB-83

Sarrazin, Albertine 1937-1967 ...... DLB-83

Sarton, May 1912- DLB-48; Y-81

Sartre, Jean-Paul 1905-1980 .... DLB-72

Sassoon, Siegfried 1886-1967 .... DLB-20

Saturday Review Press DLB-46

Saunders, James 1925- DLB-13

Saunders, John Monk 1897-1940 .... DLB-26

Saunders, Margaret Marshall
   1861-1947 .... DLB-92

Saunders and Otley DLB-106

Savage, James 1784-1873 .... DLB-30

Savage, Marmion W. 1803?-1872 ... DLB-21

Savage, Richard 1697?-1743 .... DLB-95

Savard, Félix-Antoine 1896-1982 .... DLB-68

Saville, (Leonard) Malcolm
   1901-1982 .... DLB-160

Sawyer, Ruth 1880-1970 .... DLB-22

Sayers, Dorothy L.
   1893-1957 .... DLB-10, 36, 77, 100

Sayles, John Thomas 1950- DLB-44

Sbarbaro, Camillo 1888-1967 ...... DLB-114

Scannell, Vernon 1922- DLB-27

Scarry, Richard 1919-1994 .... DLB-61

Schaeffer, Albrecht 1885-1950 ...... DLB-66

Schaeffer, Susan Fromberg 1941- .. DLB-28

Schaff, Philip 1819-1893 .... DS-13

Schaper, Edzard 1908-1984 .... DLB-69

Scharf, J. Thomas 1843-1898 ...... DLB-47

Scheffel, Joseph Viktor von
   1826-1886 .... DLB-129

Scheffler, Johann 1624-1677 .... DLB-164

Schelling, Friedrich Wilhelm Joseph von
   1775-1854 .... DLB-90

Scherer, Wilhelm 1841-1886 .... DLB-129

Schickele, René 1883-1940 .... DLB-66

Schiff, Dorothy 1903-1989 .... DLB-127

Schiller, Friedrich 1759-1805 .... DLB-94

Schirmer, David 1623-1687 .... DLB-164

Schlaf, Johannes 1862-1941 .... DLB-118

Schlegel, August Wilhelm
   1767-1845 .... DLB-94

Schlegel, Dorothea 1763-1839 .... DLB-90

Schlegel, Friedrich 1772-1829 .... DLB-90

Schleiermacher, Friedrich
   1768-1834 .... DLB-90

Schlesinger, Arthur M., Jr. 1917- ... DLB-17

Schlumberger, Jean 1877-1968 ...... DLB-65

Schmid, Eduard Hermann Wilhelm (see Edschmid, Kasimir)

Schmidt, Arno 1914-1979 .... DLB-69

Schmidt, Johann Kaspar (see Stirner, Max)

Schmidt, Michael 1947- DLB-40

Schmidtbonn, Wilhelm August
   1876-1952 .... DLB-118

Schmitz, James H. 1911- DLB-8

Schnabel, Johann Gottfried
   1692-1760 .... DLB-168

Schnackenberg, Gjertrud 1953- ... DLB-120

Schnitzler, Arthur 1862-1931 .... DLB-81, 118

Schnurre, Wolfdietrich 1920- DLB-69

Schocken Books DLB-46

Schönbeck, Virgilio (see Giotti, Virgilio)

School Stories, 1914-1960 .... DLB-160

Schönherr, Karl 1867-1943 .... DLB-118

Scholartis Press DLB-112

The Schomburg Center for Research
   in Black Culture DLB-76

Schopenhauer, Arthur 1788-1860 .... DLB-90

Schopenhauer, Johanna 1766-1838 ... DLB-90

Schorer, Mark 1908-1977 .... DLB-103

Schottelius, Justus Georg
   1612-1676 .... DLB-164

Schouler, James 1839-1920 .... DLB-47

Schrader, Paul 1946- DLB-44

Schreiner, Olive 1855-1920 ..... DLB-18, 156

Schroeder, Andreas 1946- DLB-53

Schubart, Christian Friedrich Daniel
   1739-1791 .... DLB-97

Schubert, Gotthilf Heinrich
   1780-1860 .... DLB-90

Schücking, Levin 1814-1883 .... DLB-133

Schulberg, Budd
   1914- DLB-6, 26, 28; Y-81

Schulte, F. J., and Company DLB-49

Schulze, Hans (see Praetorius, Johannes)

Schupp, Johann Balthasar
   1610-1661 .... DLB-164

Schurz, Carl 1829-1906 .... DLB-23

Schuyler, George S. 1895-1977 .. DLB-29, 51

Schuyler, James 1923-1991 ...... DLB-5, 169

Schwartz, Delmore 1913-1966 ... DLB-28, 48

Schwartz, Jonathan 1938- Y-82

Schwarz, Sibylle 1621-1638 .... DLB-164

Schwerner, Armand 1927- DLB-165

Schwob, Marcel 1867-1905 .... DLB-123

Science Fantasy DLB-8

Science-Fiction Fandom and
   Conventions DLB-8

Science-Fiction Fanzines: The Time
   Binders DLB-8

Science-Fiction Films DLB-8

Science Fiction Writers of America and the
   Nebula Awards DLB-8

Scot, Reginald circa 1538-1599 ..... DLB-136

Scotellaro, Rocco 1923-1953 .... DLB-128

Scott, Dennis 1939-1991 .... DLB-125

Scott, Dixon 1881-1915 .... DLB-98

Scott, Duncan Campbell
   1862-1947 .... DLB-92

Scott, Evelyn 1893-1963 .... DLB-9, 48

Scott, F. R. 1899-1985 .... DLB-88

Scott, Frederick George
   1861-1944 .... DLB-92

Scott, Geoffrey 1884-1929 .... DLB-149

Scott, Harvey W. 1838-1910 .... DLB-23

Scott, Paul 1920-1978 .... DLB-14

Scott, Sarah 1723-1795 .... DLB-39

Scott, Tom 1918- DLB-27

Scott, Sir Walter
   1771-1832 .... DLB-93, 107, 116, 144, 159

Scott, William Bell 1811-1890 ...... DLB-32

Scott, Walter, Publishing
   Company Limited DLB-112

Scott, William R.
   [publishing house] DLB-46

Scott-Heron, Gil 1949- DLB-41

Scribner, Charles, Jr. 1921-1995 .... Y-95

Charles Scribner's Sons DLB-49; DS-13

Scripps, E. W. 1854-1926 .... DLB-25

Scudder, Horace Elisha
   1838-1902 .... DLB-42, 71

Scudder, Vida Dutton 1861-1954 ... DLB-71

Scupham, Peter 1933- DLB-40

Seabrook, William 1886-1945 ....DLB-4

Seabury, Samuel 1729-1796 ....DLB-31

Seacole, Mary Jane Grant
  1805-1881 ....DLB-166

The Seafarer circa 970 ....DLB-146

Sealsfield, Charles 1793-1864 ......DLB-133

Sears, Edward I. 1819?-1876 ....DLB-79

Sears Publishing Company DLB-46

Seaton, George 1911-1979 ....DLB-44

Seaton, William Winston
  1785-1866 ....DLB-43

Secker, Martin, and Warburg
  Limited DLB-112

Secker, Martin [publishing house] ...DLB-112

Second-Generation Minor Poets of the
  Seventeenth Century DLB-126

Sedgwick, Arthur George
  1844-1915 ....DLB-64

Sedgwick, Catharine Maria
  1789-1867 ....DLB-1, 74

Sedgwick, Ellery 1872-1930 ....DLB-91

Sedley, Sir Charles 1639-1701 ......DLB-131

Seeger, Alan 1888-1916 ....DLB-45

Seers, Eugene (see Dantin, Louis)

Segal, Erich 1937- Y-86

Seghers, Anna 1900-1983 ....DLB-69

Seid, Ruth (see Sinclair, Jo)

Seidel, Frederick Lewis 1936- Y-84

Seidel, Ina 1885-1974 ....DLB-56

Seigenthaler, John 1927- DLB-127

Seizin Press DLB-112

Séjour, Victor 1817-1874 ....DLB-50

Séjour Marcou et Ferrand, Juan Victor (see
  Séjour, Victor)

Selby, Hubert, Jr. 1928- DLB-2

Selden, George 1929-1989 ....DLB-52

Selected English-Language Little Magazines
  and Newspapers [France,
  1920-1939] DLB-4

Selected Humorous Magazines
  (1820-1950) DLB-11

Selected Science-Fiction Magazines and
  Anthologies DLB-8

Self, Edwin F. 1920- DLB-137

Seligman, Edwin R. A. 1861-1939 ....DLB-47

Selous, Frederick Courteney
  1851-1917 ....DLB-174

Seltzer, Chester E. (see Muro, Amado)

Seltzer, Thomas
  [publishing house] DLB-46

Selvon, Sam 1923-1994 ....DLB-125

Senancour, Etienne de 1770-1846 ..DLB-119

Sendak, Maurice 1928- DLB-61

Senécal, Eva 1905- DLB-92

Sengstacke, John 1912- DLB-127

Senior, Olive 1941- DLB-157

Šenoa, August 1838-1881 ....DLB-147

"Sensation Novels" (1863), by
  H. L. Manse DLB-21

Sepamla, Sipho 1932- DLB-157

Seredy, Kate 1899-1975 ....DLB-22

Sereni, Vittorio 1913-1983 ....DLB-128

Seres, William
  [publishing house] DLB-170

Serling, Rod 1924-1975 ....DLB-26

Serote, Mongane Wally 1944- DLB-125

Serraillier, Ian 1912-1994 ....DLB-161

Serrano, Nina 1934- DLB-122

Service, Robert 1874-1958 ....DLB-92

Seth, Vikram 1952- DLB-120

Seton, Ernest Thompson
  1860-1942 ....DLB-92; DS-13

Settle, Mary Lee 1918- DLB-6

Seume, Johann Gottfried
  1763-1810 ....DLB-94

Seuss, Dr. (see Geisel, Theodor Seuss)

The Seventy-fifth Anniversary of the Armistice:
  The Wilfred Owen Centenary and the
  Great War Exhibit at the University of
  Virginia Y-93

Sewall, Joseph 1688-1769 ....DLB-24

Sewall, Richard B. 1908- DLB-111

Sewell, Anna 1820-1878 ....DLB-163

Sewell, Samuel 1652-1730 ....DLB-24

Sex, Class, Politics, and Religion [in the
  British Novel, 1930-1959] DLB-15

Sexton, Anne 1928-1974 ....DLB-5, 169

Seymour-Smith, Martin 1928- DLB-155

Shaara, Michael 1929-1988 ....Y-83

Shadwell, Thomas 1641?-1692 .....DLB-80

Shaffer, Anthony 1926- DLB-13

Shaffer, Peter 1926- DLB-13

Shaftesbury, Anthony Ashley Cooper,
  Third Earl of 1671-1713 ......DLB-101

Shairp, Mordaunt 1887-1939 ....DLB-10

Shakespeare, William
  1564-1616 ....DLB-62, 172

The Shakespeare Globe Trust Y-93

Shakespeare Head Press DLB-112

Shakhovskoi, Aleksandr Aleksandrovich
  1777-1846 ....DLB-150

Shange, Ntozake 1948- DLB-38

Shapiro, Karl 1913- DLB-48

Sharon Publications DLB-46

Sharp, Margery 1905-1991 ....DLB-161

Sharp, William 1855-1905 ....DLB-156

Sharpe, Tom 1928- DLB-14

Shaw, Albert 1857-1947 ....DLB-91

Shaw, Bernard 1856-1950 ....DLB-10, 57

Shaw, Henry Wheeler 1818-1885 ....DLB-11

Shaw, Joseph T. 1874-1952 ....DLB-137

Shaw, Irwin 1913-1984 .....DLB-6, 102; Y-84

Shaw, Robert 1927-1978 ....DLB-13, 14

Shaw, Robert B. 1947- DLB-120

Shawn, William 1907-1992 ....DLB-137

Shay, Frank [publishing house] DLB-46

Shea, John Gilmary 1824-1892 ......DLB-30

Sheaffer, Louis 1912-1993 ....DLB-103

Shearing, Joseph 1886-1952 ....DLB-70

Shebbeare, John 1709-1788 ....DLB-39

Sheckley, Robert 1928- DLB-8

Shedd, William G. T. 1820-1894 ....DLB-64

Sheed, Wilfred 1930- DLB-6

Sheed and Ward [U.S.] DLB-46

Sheed and Ward Limited [U.K.] DLB-112

Sheldon, Alice B. (see Tiptree, James, Jr.)

Sheldon, Edward 1886-1946 ....DLB-7

Sheldon and Company DLB-49

Shelley, Mary Wollstonecraft
  1797-1851 ....DLB-110, 116, 159

Shelley, Percy Bysshe
  1792-1822 ....DLB-96, 110, 158

Shelnutt, Eve 1941- DLB-130

Shenstone, William 1714-1763 ......DLB-95

Shepard, Ernest Howard
  1879-1976 ....DLB-160

Shepard, Sam 1943- DLB-7

Shepard, Thomas I,
  1604 or 1605-1649 ....DLB-24

Shepard, Thomas II, 1635-1677 .....DLB-24

Shepard, Clark and Brown DLB-49

Cumulative Index

Shepherd, Luke
   flourished 1547-1554 .... DLB-136

Sherburne, Edward 1616-1702 ..... DLB-131

Sheridan, Frances 1724-1766 .... DLB-39, 84

Sheridan, Richard Brinsley
   1751-1816 .... DLB-89

Sherman, Francis 1871-1926 .... DLB-92

Sherriff, R. C. 1896-1975 .... DLB-10

Sherry, Norman 1935- DLB-155

Sherwood, Mary Martha
   1775-1851 .... DLB-163

Sherwood, Robert 1896-1955 ..... DLB-7, 26

Shiel, M. P. 1865-1947 .... DLB-153

Shiels, George 1886-1949 .... DLB-10

Shillaber, B.[enjamin] P.[enhallow]
   1814-1890 .... DLB-1, 11

Shine, Ted 1931- DLB-38

Ship, Reuben 1915-1975 .... DLB-88

Shirer, William L. 1904-1993 .... DLB-4

Shirinsky-Shikhmatov, Sergii Aleksandrovich
   1783-1837 .... DLB-150

Shirley, James 1596-1666 .... DLB-58

Shishkov, Aleksandr Semenovich
   1753-1841 .... DLB-150

Shockley, Ann Allen 1927- DLB-33

Short, Peter
   [publishing house] DLB-170

Shorthouse, Joseph Henry
   1834-1903 .... DLB-18

Showalter, Elaine 1941- DLB-67

Shulevitz, Uri 1935- DLB-61

Shulman, Max 1919-1988 .... DLB-11

Shute, Henry A. 1856-1943 .... DLB-9

Shuttle, Penelope 1947- DLB-14, 40

Sibbes, Richard 1577-1635 .... DLB-151

Sidgwick and Jackson Limited DLB-112

Sidney, Margaret (see Lothrop, Harriet M.)

Sidney, Mary 1561-1621 .... DLB-167

Sidney, Sir Philip 1554-1586 .... DLB-167

Sidney's Press DLB-49

Siegfried Loraine Sassoon: A Centenary Essay
   Tributes from Vivien F. Clarke and
   Michael Thorpe Y-86

Sierra, Rubén 1946- DLB-122

Sierra Club Books DLB-49

Siger of Brabant
   circa 1240-circa 1284 .... DLB-115

Sigourney, Lydia Howard (Huntley)
   1791-1865 .... DLB-1, 42, 73

Silkin, Jon 1930- DLB-27

Silko, Leslie Marmon 1948- DLB-143

Silliman, Ron 1946- DLB-169

Silliphant, Stirling 1918- DLB-26

Sillitoe, Alan 1928- DLB-14, 139

Silman, Roberta 1934- DLB-28

Silva, Beverly 1930- DLB-122

Silverberg, Robert 1935- DLB-8

Silverman, Kenneth 1936- DLB-111

Simak, Clifford D. 1904-1988 .... DLB-8

Simcoe, Elizabeth 1762-1850 .... DLB-99

Simcox, George Augustus
   1841-1905 .... DLB-35

Sime, Jessie Georgina 1868-1958 ..... DLB-92

Simenon, Georges
   1903-1989 ....DLB-72; Y-89

Simic, Charles 1938- DLB-105

Simic, Charles,
   Images and "Images" DLB-105

Simmel, Johannes Mario 1924- DLB-69

Simmes, Valentine
   [publishing house] DLB-170

Simmons, Ernest J. 1903-1972 ...... DLB-103

Simmons, Herbert Alfred 1930- DLB-33

Simmons, James 1933- DLB-40

Simms, William Gilmore
   1806-1870 .... DLB-3, 30, 59, 73

Simms and M'Intyre DLB-106

Simon, Claude 1913- DLB-83

Simon, Neil 1927- DLB-7

Simon and Schuster DLB-46

Simons, Katherine Drayton Mayrant
   1890-1969 .... Y-83

Simpkin and Marshall
   [publishing house] DLB-154

Simpson, Helen 1897-1940 .... DLB-77

Simpson, Louis 1923- DLB-5

Simpson, N. F. 1919- DLB-13

Sims, George 1923- DLB-87

Sims, George Robert
   1847-1922 ....DLB-35, 70, 135

Sinán, Rogelio 1904- DLB-145

Sinclair, Andrew 1935- DLB-14

Sinclair, Bertrand William
   1881-1972 .... DLB-92

Sinclair, Catherine
   1800-1864 .... DLB-163

Sinclair, Jo 1913- DLB-28

Sinclair Lewis Centennial
   Conference Y-85

Sinclair, Lister 1921- DLB-88

Sinclair, May 1863-1946 .... DLB-36, 135

Sinclair, Upton 1878-1968 ....DLB-9

Sinclair, Upton [publishing house] DLB-46

Singer, Isaac Bashevis
   1904-1991 .... DLB-6, 28, 52; Y-91

Singmaster, Elsie 1879-1958 ....DLB-9

Sinisgalli, Leonardo 1908-1981 .....DLB-114

Siodmak, Curt 1902- DLB-44

Sissman, L. E. 1928-1976 ....DLB-5

Sisson, C. H. 1914- DLB-27

Sitwell, Edith 1887-1964 ....DLB-20

Sitwell, Osbert 1892-1969 ....DLB-100

Skármeta, Antonio 1940- DLB-145

Skeffington, William
   [publishing house] DLB-106

Skelton, John 1463-1529 ....DLB-136

Skelton, Robin 1925- DLB-27, 53

Skinner, Constance Lindsay
   1877-1939 ....DLB-92

Skinner, John Stuart 1788-1851 ......DLB-73

Skipsey, Joseph 1832-1903 ....DLB-35

Slade, Bernard 1930- DLB-53

Slater, Patrick 1880-1951 ....DLB-68

Slaveykov, Pencho 1866-1912 ......DLB-147

Slavitt, David 1935- DLB-5, 6

Sleigh, Burrows Willcocks Arthur
   1821-1869 ....DLB-99

A Slender Thread of Hope: The Kennedy
   Center Black Theatre Project DLB-38

Slesinger, Tess 1905-1945 ....DLB-102

Slick, Sam (see Haliburton, Thomas Chandler)

Sloane, William, Associates DLB-46

Small, Maynard and Company DLB-49

Small Presses in Great Britain and Ireland,
   1960-1985 ....DLB-40

Small Presses I: Jargon Society Y-84

Small Presses II: The Spirit That Moves Us
   Press Y-85

Small Presses III: Pushcart Press Y-87

Smart, Christopher 1722-1771 ......DLB-109

Smart, David A. 1892-1957 ....DLB-137

Smart, Elizabeth 1913-1986 ....DLB-88

Smellie, William
   [publishing house] DLB-154

Smiles, Samuel 1812-1904 ....DLB-55

Smith, A. J. M. 1902-1980 ....DLB-88

Smith, Adam 1723-1790 ....DLB-104

Smith, Alexander 1829-1867 .....DLB-32, 55

Smith, Betty 1896-1972................Y-82

Smith, Carol Sturm 1938- Y-81

Smith, Charles Henry 1826-1903 ....DLB-11

Smith, Charlotte 1749-1806 .....DLB-39, 109

Smith, Chet 1899-1973 ....DLB-171

Smith, Cordwainer 1913-1966 ....DLB-8

Smith, Dave 1942- DLB-5

Smith, Dodie 1896- DLB-10

Smith, Doris Buchanan 1934- DLB-52

Smith, E. E. 1890-1965 ....DLB-8

Smith, Elihu Hubbard 1771-1798 ....DLB-37

Smith, Elizabeth Oakes (Prince)
  1806-1893 ....DLB-1

Smith, F. Hopkinson 1838-1915 ...... DS-13

Smith, George D. 1870-1920 ....DLB-140

Smith, George O. 1911-1981 ....DLB-8

Smith, Goldwin 1823-1910 ....DLB-99

Smith, H. Allen 1907-1976 ....DLB-11, 29

Smith, Hazel Brannon 1914- DLB-127

Smith, Henry
  circa 1560-circa 1591 ....DLB-136

Smith, Horatio (Horace)
  1779-1849 ....DLB-116

Smith, Horatio (Horace) 1779-1849 and
  James Smith 1775-1839 ....DLB-96

Smith, Iain Crichton
  1928- DLB-40, 139

Smith, J. Allen 1860-1924 ....DLB-47

Smith, John 1580-1631 ....DLB-24, 30

Smith, Josiah 1704-1781 ....DLB-24

Smith, Ken 1938- DLB-40

Smith, Lee 1944- DLB-143; Y-83

Smith, Logan Pearsall 1865-1946 ....DLB-98

Smith, Mark 1935- Y-82

Smith, Michael 1698-circa 1771 ......DLB-31

Smith, Red 1905-1982 ....DLB-29, 171

Smith, Roswell 1829-1892 ....DLB-79

Smith, Samuel Harrison
  1772-1845 ....DLB-43

Smith, Samuel Stanhope
  1751-1819 ....DLB-37

Smith, Sarah (see Stretton, Hesba)

Smith, Seba 1792-1868 ....DLB-1, 11

Smith, Sir Thomas 1513-1577 .....DLB-132

Smith, Stevie 1902-1971 ....DLB-20

Smith, Sydney 1771-1845 ....DLB-107

Smith, Sydney Goodsir 1915-1975 ...DLB-27

Smith, Wendell 1914-1972 ....DLB-171

Smith, William
  flourished 1595-1597 ....DLB-136

Smith, William 1727-1803 ....DLB-31

Smith, William 1728-1793 ....DLB-30

Smith, William Gardner
  1927-1974 ....DLB-76

Smith, William Henry
  1808-1872 ....DLB-159

Smith, William Jay 1918- DLB-5

Smith, Elder and Company DLB-154

Smith, Harrison, and Robert Haas
  [publishing house] DLB-46

Smith, J. Stilman, and Company DLB-49

Smith, W. B., and Company DLB-49

Smith, W. H., and Son DLB-106

Smithers, Leonard
  [publishing house] DLB-112

Smollett, Tobias 1721-1771 .....DLB-39, 104

Snellings, Rolland (see Touré, Askia
  Muhammad)

Snodgrass, W. D. 1926- DLB-5

Snow, C. P. 1905-1980 ....DLB-15, 77

Snyder, Gary 1930- DLB-5, 16, 165

Sobiloff, Hy 1912-1970 ....DLB-48

The Society for Textual Scholarship and
  TEXT Y-87

The Society for the History of Authorship,
  Reading and Publishing Y-92

Soffici, Ardengo 1879-1964 ....DLB-114

Sofola, 'Zulu 1938- DLB-157

Solano, Solita 1888-1975 ....DLB-4

Sollers, Philippe 1936- DLB-83

Solmi, Sergio 1899-1981 ....DLB-114

Solomon, Carl 1928- DLB-16

Solway, David 1941- DLB-53

Solzhenitsyn and America Y-85

Somerville, Edith Œnone
  1858-1949 ....DLB-135

Song, Cathy 1955- DLB-169

Sontag, Susan 1933- DLB-2, 67

Sorge, Reinhard Johannes
  1892-1916 ....DLB-118

Sorrentino, Gilbert
  1929- DLB-5, 173; Y-80

Sotheby, William 1757-1833 ....DLB-93

Soto, Gary 1952- DLB-82

Sources for the Study of Tudor and Stuart
  Drama DLB-62

Souster, Raymond 1921- DLB-88

The *South English Legendary*
  circa thirteenth-fifteenth
  centuries DLB-146

Southerland, Ellease 1943- DLB-33

Southern Illinois University Press Y-95

Southern, Terry 1924- DLB-2

Southern Writers Between the
  Wars DLB-9

Southerne, Thomas 1659-1746 ......DLB-80

Southey, Caroline Anne Bowles
  1786-1854 ....DLB-116

Southey, Robert
  1774-1843 ....DLB-93, 107, 142

Southwell, Robert 1561?-1595 .....DLB-167

Sowande, Bode 1948- DLB-157

Sowle, Tace
  [publishing house] DLB-170

Soyfer, Jura 1912-1939 ....DLB-124

Soyinka, Wole 1934- ...DLB-125; Y-86, 87

Spacks, Barry 1931- DLB-105

Spalding, Frances 1950- DLB-155

Spark, Muriel 1918- DLB-15, 139

Sparke, Michael
  [publishing house] DLB-170

Sparks, Jared 1789-1866 ....DLB-1, 30

Sparshott, Francis 1926- DLB-60

Späth, Gerold 1939- DLB-75

Spatola, Adriano 1941-1988 ....DLB-128

Spaziani, Maria Luisa 1924- DLB-128

The *Spectator* 1828- DLB-110

Spedding, James 1808-1881 ....DLB-144

Spee von Langenfeld, Friedrich
  1591-1635 ....DLB-164

Speght, Rachel 1597-after 1630 .....DLB-126

Speke, John Hanning 1827-1864 ....DLB-166

Spellman, A. B. 1935- DLB-41

Spence, Thomas 1750-1814 ....DLB-158

Spencer, Anne 1882-1975 ....DLB-51, 54

Spencer, Elizabeth 1921- DLB-6

Spencer, Herbert 1820-1903 ....DLB-57

Spencer, Scott 1945- Y-86

Cumulative Index

Spender, J. A. 1862-1942 . . . . DLB-98

Spender, Stephen 1909- DLB-20

Spener, Philipp Jakob 1635-1705 . . . DLB-164

Spenser, Edmund circa 1552-1599 . . DLB-167

Sperr, Martin 1944- DLB-124

Spicer, Jack 1925-1965 . . . . DLB-5, 16

Spielberg, Peter 1929- Y-81

Spielhagen, Friedrich 1829-1911 . . . DLB-129

"*Spielmannsepen*"
(circa 1152-circa 1500) DLB-148

Spier, Peter 1927- DLB-61

Spinrad, Norman 1940- DLB-8

Spires, Elizabeth 1952- DLB-120

Spitteler, Carl 1845-1924 . . . . DLB-129

Spivak, Lawrence E. 1900- DLB-137

Spofford, Harriet Prescott
1835-1921 . . . . DLB-74

Squibob (see Derby, George Horatio)

Stacpoole, H. de Vere
1863-1951 . . . . DLB-153

Staël, Germaine de 1766-1817 . . . . . DLB-119

Staël-Holstein, Anne-Louise Germaine de
(see Staël, Germaine de)

Stafford, Jean 1915-1979 . . . . DLB-2, 173

Stafford, William 1914- DLB-5

Stage Censorship: "The Rejected Statement"
(1911), by Bernard Shaw
[excerpts] DLB-10

Stallings, Laurence 1894-1968 . . . . DLB-7, 44

Stallworthy, Jon 1935- DLB-40

Stampp, Kenneth M. 1912- DLB-17

Stanford, Ann 1916- DLB-5

Stanković, Borisav ("Bora")
1876-1927 . . . . DLB-147

Stanley, Henry M. 1841-1904 . . . . DS-13

Stanley, Thomas 1625-1678 . . . . . . DLB-131

Stannard, Martin 1947- DLB-155

Stansby, William
[publishing house] DLB-170

Stanton, Elizabeth Cady 1815-1902 . . DLB-79

Stanton, Frank L. 1857-1927 . . . . . . DLB-25

Stanton, Maura 1946- DLB-120

Stapledon, Olaf 1886-1950 . . . . DLB-15

Star Spangled Banner Office DLB-49

Starkey, Thomas circa 1499-1538 . . DLB-132

Starkweather, David 1935- DLB-7

Statements on the Art of Poetry DLB-54

Stationers' Company of
London, The DLB-170

Stead, Robert J. C. 1880-1959 . . . . . . DLB-92

Steadman, Mark 1930- DLB-6

The Stealthy School of Criticism (1871), by
Dante Gabriel Rossetti DLB-35

Stearns, Harold E. 1891-1943 . . . . . . DLB-4

Stedman, Edmund Clarence
1833-1908 . . . . DLB-64

Steegmuller, Francis 1906-1994 . . . . DLB-111

Steel, Flora Annie
1847-1929 . . . . DLB-153, 156

Steele, Max 1922- Y-80

Steele, Richard 1672-1729 . . . . . . DLB-84, 101

Steele, Timothy 1948- DLB-120

Steele, Wilbur Daniel 1886-1970 . . . . DLB-86

Steere, Richard circa 1643-1721 . . . . . DLB-24

Stegner, Wallace 1909-1993 . . . . . DLB-9; Y-93

Stehr, Hermann 1864-1940 . . . . DLB-66

Steig, William 1907- DLB-61

Stieler, Caspar 1632-1707 . . . . DLB-164

Stein, Gertrude 1874-1946 . . . . .DLB-4, 54, 86

Stein, Leo 1872-1947 . . . . DLB-4

Stein and Day Publishers DLB-46

Steinbeck, John 1902-1968 . . . DLB-7, 9; DS-2

Steiner, George 1929- DLB-67

Stendhal 1783-1842 . . . . DLB-119

Stephen Crane: A Revaluation Virginia
Tech Conference, 1989 . . . . Y-89

Stephen, Leslie 1832-1904 . . . . . .DLB-57, 144

Stephens, Alexander H. 1812-1883 . . . DLB-47

Stephens, Ann 1810-1886 . . . . DLB-3, 73

Stephens, Charles Asbury
1844?-1931 . . . . DLB-42

Stephens, James
1882?-1950 . . . . .DLB-19, 153, 162

Sterling, George 1869-1926 . . . . DLB-54

Sterling, James 1701-1763 . . . . DLB-24

Sterling, John 1806-1844 . . . . DLB-116

Stern, Gerald 1925- DLB-105

Stern, Madeleine B. 1912- . . .DLB-111, 140

Stern, Gerald, Living in Ruin DLB-105

Stern, Richard 1928- Y-87

Stern, Stewart 1922- DLB-26

Sterne, Laurence 1713-1768 . . . . DLB-39

Sternheim, Carl 1878-1942 . . . . . DLB-56, 118

Sternhold, Thomas ?-1549 and
John Hopkins ?-1570 . . . . .DLB-132

Stevens, Henry 1819-1886 . . . . .DLB-140

Stevens, Wallace 1879-1955 . . . . .DLB-54

Stevenson, Anne 1933- DLB-40

Stevenson, Lionel 1902-1973 . . . . . . .DLB-155

Stevenson, Robert Louis 1850-1894
. . . . . . DLB-18, 57, 141, 156, 174; DS-13

Stewart, Donald Ogden
1894-1980 . . . . DLB-4, 11, 26

Stewart, Dugald 1753-1828 . . . . .DLB-31

Stewart, George, Jr. 1848-1906 . . . . . .DLB-99

Stewart, George R. 1895-1980 . . . . . . .DLB-8

Stewart and Kidd Company DLB-46

Stewart, Randall 1896-1964 . . . . . . .DLB-103

Stickney, Trumbull 1874-1904 . . . . . .DLB-54

Stifter, Adalbert 1805-1868 . . . . .DLB-133

Stiles, Ezra 1727-1795 . . . . .DLB-31

Still, James 1906- DLB-9

Stirner, Max 1806-1856 . . . . .DLB-129

Stith, William 1707-1755 . . . . .DLB-31

Stock, Elliot [publishing house] DLB-106

Stockton, Frank R.
1834-1902 . . . . DLB-42, 74; DS-13

Stoddard, Ashbel
[publishing house] DLB-49

Stoddard, Richard Henry
1825-1903 . . . . DLB-3, 64; DS-13

Stoddard, Solomon 1643-1729 . . . . . .DLB-24

Stoker, Bram 1847-1912 . . . . DLB-36, 70

Stokes, Frederick A., Company DLB-49

Stokes, Thomas L. 1898-1958 . . . . . . .DLB-29

Stokesbury, Leon 1945- DLB-120

Stolberg, Christian Graf zu
1748-1821 . . . . .DLB-94

Stolberg, Friedrich Leopold Graf zu
1750-1819 . . . . .DLB-94

Stone, Herbert S., and Company DLB-49

Stone, Lucy 1818-1893 . . . . .DLB-79

Stone, Melville 1848-1929 . . . . .DLB-25

Stone, Robert 1937- DLB-152

Stone, Ruth 1915- DLB-105

Stone, Samuel 1602-1663 . . . . .DLB-24

Stone and Kimball DLB-49

Stoppard, Tom 1937- DLB-13; Y-85

Storey, Anthony 1928- DLB-14

Storey, David 1933- DLB-13, 14

Storm, Theodor 1817-1888 . . . . DLB-129

Story, Thomas circa 1670-1742 . . . . . . DLB-31

Story, William Wetmore 1819-1895 . . . DLB-1

Storytelling: A Contemporary
Renaissance Y-84

Stoughton, William 1631-1701 . . . . . . DLB-24

Stow, John 1525-1605 . . . . DLB-132

Stowe, Harriet Beecher
1811-1896 . . . . DLB-1, 12, 42, 74

Stowe, Leland 1899- DLB-29

Stoyanov, Dimitŭr Ivanov (see Elin Pelin)

Strachey, Lytton
1880-1932 . . . . DLB-149; DS-10

Strachey, Lytton, Preface to *Eminent
Victorians* DLB-149

Strahan and Company DLB-106

Strahan, William
[publishing house] DLB-154

Strand, Mark 1934- DLB-5

The Strasbourg Oaths 842 . . . . DLB-148

Stratemeyer, Edward 1862-1930 . . . . . DLB-42

Stratton and Barnard DLB-49

Stratton-Porter, Gene 1863-1924 . . . . . . DS-14

Straub, Peter 1943- Y-84

Strauß, Botho 1944- DLB-124

Strauß, David Friedrich
1808-1874 . . . . DLB-133

The Strawberry Hill Press DLB-154

Streatfeild, Noel 1895-1986 . . . . DLB-160

Street, Cecil John Charles (see Rhode, John)

Street, G. S. 1867-1936 . . . . DLB-135

Street and Smith DLB-49

Streeter, Edward 1891-1976 . . . . DLB-11

Streeter, Thomas Winthrop
1883-1965 . . . . DLB-140

Stretton, Hesba 1832-1911 . . . . DLB-163

Stribling, T. S. 1881-1965 . . . . DLB-9

Der Stricker circa 1190-circa 1250 . . DLB-138

Strickland, Samuel 1804-1867 . . . . DLB-99

Stringer and Townsend DLB-49

Stringer, Arthur 1874-1950 . . . . DLB-92

Strittmatter, Erwin 1912- DLB-69

Strode, William 1630-1645 . . . . DLB-126

Strother, David Hunter 1816-1888 . . . . DLB-3

Strouse, Jean 1945- DLB-111

Stuart, Dabney 1937- DLB-105

Stuart, Dabney, Knots into Webs: Some Autobiographical Sources DLB-105

Stuart, Jesse
1906-1984 . . . . DLB-9, 48, 102; Y-84

Stuart, Lyle [publishing house] DLB-46

Stubbs, Harry Clement (see Clement, Hal)

Stubenberg, Johann Wilhelm von
1619-1663 . . . . DLB-164

Studio DLB-112

The Study of Poetry (1880), by
Matthew Arnold DLB-35

Sturgeon, Theodore
1918-1985 . . . . DLB-8; Y-85

Sturges, Preston 1898-1959 . . . . DLB-26

"Style" (1840; revised, 1859), by
Thomas de Quincey [excerpt] . . . DLB-57

"Style" (1888), by Walter Pater DLB-57

Style (1897), by Walter Raleigh
[excerpt] DLB-57

"Style" (1877), by T. H. Wright
[excerpt] DLB-57

"Le Style c'est l'homme" (1892), by
W. H. Mallock DLB-57

Styron, William 1925- . . . DLB-2, 143; Y-80

Suárez, Mario 1925- DLB-82

Such, Peter 1939- DLB-60

Suckling, Sir John 1609-1641? . . DLB-58, 126

Suckow, Ruth 1892-1960 . . . . DLB-9, 102

Sudermann, Hermann 1857-1928 . . . DLB-118

Sue, Eugène 1804-1857 . . . . DLB-119

Sue, Marie-Joseph (see Sue, Eugène)

Suggs, Simon (see Hooper, Johnson Jones)

Sukenick, Ronald 1932- DLB-173; Y-81

Suknaski, Andrew 1942- DLB-53

Sullivan, Alan 1868-1947 . . . . DLB-92

Sullivan, C. Gardner 1886-1965 . . . . . DLB-26

Sullivan, Frank 1892-1976 . . . . DLB-11

Sulte, Benjamin 1841-1923 . . . . DLB-99

Sulzberger, Arthur Hays
1891-1968 . . . . DLB-127

Sulzberger, Arthur Ochs 1926- . . . DLB-127

Sulzer, Johann Georg 1720-1779 . . . . DLB-97

Sumarokov, Aleksandr Petrovich
1717-1777 . . . . DLB-150

Summers, Hollis 1916- DLB-6

Sumner, Henry A.
[publishing house] DLB-49

Surtees, Robert Smith 1803-1864 . . . . DLB-21

A Survey of Poetry Anthologies,
1879-1960 . . . . DLB-54

Surveys of the Year's Biographies

A Transit of Poets and Others: American
Biography in 1982 . . . . Y-82

The Year in Literary Biography . . . Y-83–Y-95

Survey of the Year's Book Publishing

The Year in Book Publishing Y-86

Survey of the Year's Children's Books

The Year in Children's Books Y-92–Y-95

Surveys of the Year's Drama

The Year in Drama
. . . . Y-82–Y-85, Y-87–Y-95

The Year in London Theatre Y-92

Surveys of the Year's Fiction

The Year's Work in Fiction:
A Survey Y-82

The Year in Fiction: A Biased View Y-83

The Year in
Fiction Y-84–Y-86, Y-89, Y-94, Y-95

The Year in the
Novel Y-87, Y-88, Y-90–Y-93

The Year in Short Stories Y-87

The Year in the
Short Story Y-88, Y-90–Y-93

Survey of the Year's Literary Theory

The Year in Literary Theory Y-92–Y-93

Surveys of the Year's Poetry

The Year's Work in American
Poetry Y-82

The Year in Poetry . . . Y-83–Y-92, Y-94, Y-95

Sutherland, Efua Theodora
1924- DLB-117

Sutherland, John 1919-1956 . . . . DLB-68

Sutro, Alfred 1863-1933 . . . . DLB-10

Swados, Harvey 1920-1972 . . . . DLB-2

Swain, Charles 1801-1874 . . . . DLB-32

Swallow Press DLB-46

Swan Sonnenschein Limited DLB-106

Swanberg, W. A. 1907- DLB-103

Swenson, May 1919-1989 . . . . DLB-5

Swerling, Jo 1897- DLB-44

Swift, Jonathan
1667-1745 . . . . DLB-39, 95, 101

Swinburne, A. C. 1837-1909 . . . . . DLB-35, 57

Swineshead, Richard floruit
circa 1350 . . . . DLB-115

Swinnerton, Frank 1884-1982 . . . . DLB-34

Swisshelm, Jane Grey 1815-1884 .... DLB-43

Swope, Herbert Bayard 1882-1958 .. DLB-25

Swords, T. and J., and Company DLB-49

Swords, Thomas 1763-1843 and
    Swords, James ?-1844 .... DLB-73

Sykes, Ella C. ?-1939 .... DLB-174

Sylvester, Josuah
    1562 or 1563 - 1618 .... DLB-121

Symonds, Emily Morse (see Paston, George)

Symonds, John Addington
    1840-1893 .... DLB-57, 144

Symons, A. J. A. 1900-1941 .... DLB-149

Symons, Arthur
    1865-1945 .... DLB-19, 57, 149

Symons, Julian
    1912-1994 .... DLB-87, 155; Y-92

Symons, Scott 1933- DLB-53

A Symposium on *The Columbia History of
the Novel* Y-92

Synge, John Millington
    1871-1909 .... DLB-10, 19

Synge Summer School: J. M. Synge and the
Irish Theater, Rathdrum, County Wiclow,
Ireland Y-93

Syrett, Netta 1865-1943 .... DLB-135

T

Taban lo Liyong 1939?- DLB-125

Taché, Joseph-Charles 1820-1894 ... DLB-99

Tafolla, Carmen 1951- DLB-82

Taggard, Genevieve 1894-1948 ..... DLB-45

Tagger, Theodor (see Bruckner, Ferdinand)

Tait, J. Selwin, and Sons DLB-49

Tait's Edinburgh Magazine
    1832-1861 .... DLB-110

The Takarazaka Revue Company Y-91

Talander (see Bohse, August)

Tallent, Elizabeth 1954- DLB-130

Talvj 1797-1870 .... DLB-59, 133

Tan, Amy 1952- DLB-173

Taradash, Daniel 1913- DLB-44

Tarbell, Ida M. 1857-1944 .... DLB-47

Tardivel, Jules-Paul 1851-1905 ...... DLB-99

Targan, Barry 1932- DLB-130

Tarkington, Booth 1869-1946 ... DLB-9, 102

Tashlin, Frank 1913-1972 .... DLB-44

Tate, Allen 1899-1979 .... DLB-4, 45, 63

Tate, James 1943- DLB-5, 169

Tate, Nahum circa 1652-1715 .... DLB-80

Tatian circa 830 .... DLB-148

Tavčar, Ivan 1851-1923 .... DLB-147

Taylor, Ann 1782-1866 .... DLB-163

Taylor, Bayard 1825-1878 .... DLB-3

Taylor, Bert Leston 1866-1921 .... DLB-25

Taylor, Charles H. 1846-1921 .... DLB-25

Taylor, Edward circa 1642-1729 ..... DLB-24

Taylor, Elizabeth 1912-1975 .... DLB-139

Taylor, Henry 1942- DLB-5

Taylor, Sir Henry 1800-1886 .... DLB-32

Taylor, Jane 1783-1824 .... DLB-163

Taylor, Jeremy circa 1613-1667 .... DLB-151

Taylor, John
    1577 or 1578 - 1653 .... DLB-121

Taylor, Mildred D. ?- DLB-52

Taylor, Peter 1917-1994 .... Y-81, Y-94

Taylor, William, and Company DLB-49

Taylor-Made Shakespeare? Or Is
"Shall I Die?" the Long-Lost Text
of Bottom's Dream? Y-85

Teasdale, Sara 1884-1933 .... DLB-45

The Tea-Table (1725), by Eliza Haywood
[excerpt] DLB-39

Telles, Lygia Fagundes 1924- DLB-113

Temple, Sir William 1628-1699 .... DLB-101

Tenn, William 1919- DLB-8

Tennant, Emma 1937- DLB-14

Tenney, Tabitha Gilman
    1762-1837 .... DLB-37

Tennyson, Alfred 1809-1892 .... DLB-32

Tennyson, Frederick 1807-1898 ..... DLB-32

Terhune, Albert Payson 1872-1942 ... DLB-9

Terhune, Mary Virginia 1830-1922 ....DS-13

Terry, Megan 1932- DLB-7

Terson, Peter 1932- DLB-13

Tesich, Steve 1943- Y-83

Tessa, Delio 1886-1939 .... DLB-114

Testori, Giovanni 1923-1993 .... DLB-128

Tey, Josephine 1896?-1952 .... DLB-77

Thacher, James 1754-1844 .... DLB-37

Thackeray, William Makepeace
    1811-1863 .... DLB-21, 55, 159, 163

Thames and Hudson Limited DLB-112

Thanet, Octave (see French, Alice)

The Theater in Shakespeare's
Time DLB-62

The Theatre Guild DLB-7

Thegan and the Astronomer
    flourished circa 850 .... DLB-148

Thelwall, John 1764-1834 ...... DLB-93, 158

Theodulf circa 760-circa 821 .... DLB-148

Theriault, Yves 1915-1983 .... DLB-88

Thério, Adrien 1925- DLB-53

Theroux, Paul 1941- DLB-2

Thibaudeau, Colleen 1925- DLB-88

Thielen, Benedict 1903-1965 .... DLB-102

Thiong'o Ngugi wa (see Ngugi wa Thiong'o)

Third-Generation Minor Poets of the
Seventeenth Century DLB-131

Thoma, Ludwig 1867-1921 .... DLB-66

Thoma, Richard 1902- DLB-4

Thomas, Audrey 1935- DLB-60

Thomas, D. M. 1935- DLB-40

Thomas, Dylan
    1914-1953 .... DLB-13, 20, 139

Thomas, Edward
    1878-1917 .... DLB-19, 98, 156

Thomas, Gwyn 1913-1981 .... DLB-15

Thomas, Isaiah 1750-1831 ..... DLB-43, 73

Thomas, Isaiah [publishing house] ...DLB-49

Thomas, Johann 1624-1679 .... DLB-168

Thomas, John 1900-1932 .... DLB-4

Thomas, Joyce Carol 1938- DLB-33

Thomas, Lorenzo 1944- DLB-41

Thomas, R. S. 1915- DLB-27

Thomasîn von Zerclære
    circa 1186-circa 1259 .... DLB-138

Thomasius, Christian 1655-1728 ....DLB-168

Thompson, David 1770-1857 .... DLB-99

Thompson, Dorothy 1893-1961 ..... DLB-29

Thompson, Francis 1859-1907 ...... DLB-19

Thompson, George Selden (see Selden, George)

Thompson, John 1938-1976 .... DLB-60

Thompson, John R. 1823-1873 ... DLB-3, 73

Thompson, Lawrance 1906-1973 ....DLB-103

Thompson, Maurice
    1844-1901 .... DLB-71, 74

Thompson, Ruth Plumly
    1891-1976 .... DLB-22

Thompson, Thomas Phillips
    1843-1933 .... DLB-99

Thompson, William 1775-1833 ..... DLB-158

Thompson, William Tappan
1812-1882 ....DLB-3, 11

Thomson, Edward William
1849-1924 ....DLB-92

Thomson, James 1700-1748 ....DLB-95

Thomson, James 1834-1882 ....DLB-35

Thomson, Joseph 1858-1895 ....DLB-174

Thomson, Mortimer 1831-1875 .....DLB-11

Thoreau, Henry David 1817-1862 ....DLB-1

Thorpe, Thomas Bangs
1815-1878 ....DLB-3, 11

Thoughts on Poetry and Its Varieties (1833),
by John Stuart Mill DLB-32

Thrale, Hester Lynch (see Piozzi, Hester
Lynch [Thrale])

Thümmel, Moritz August von
1738-1817 ....DLB-97

Thurber, James
1894-1961 .... DLB-4, 11, 22, 102

Thurman, Wallace 1902-1934 ....DLB-51

Thwaite, Anthony 1930- DLB-40

Thwaites, Reuben Gold
1853-1913 ....DLB-47

Ticknor, George
1791-1871 .... DLB-1, 59, 140

Ticknor and Fields DLB-49

Ticknor and Fields (revived) DLB-46

Tieck, Ludwig 1773-1853 ....DLB-90

Tietjens, Eunice 1884-1944 ....DLB-54

Tilney, Edmund circa 1536-1610 ...DLB-136

Tilt, Charles [publishing house] DLB-106

Tilton, J. E., and Company DLB-49

Time and Western Man (1927), by Wyndham
Lewis [excerpts] DLB-36

Time-Life Books DLB-46

Times Books DLB-46

Timothy, Peter circa 1725-1782 .....DLB-43

Timrod, Henry 1828-1867 ....DLB-3

Tinker, Chauncey Brewster
1876-1963 ....DLB-140

Tinsley Brothers DLB-106

Tiptree, James, Jr. 1915-1987 ....DLB-8

Titus, Edward William 1870-1952 ....DLB-4

Tlali, Miriam 1933- DLB-157

Todd, Barbara Euphan
1890-1976 ....DLB-160

Tofte, Robert
1561 or 1562-1619 or 1620 .....DLB-172

Toklas, Alice B. 1877-1967 ....DLB-4

Tolkien, J. R. R. 1892-1973 .... DLB-15, 160

Toller, Ernst 1893-1939 ....DLB-124

Tollet, Elizabeth 1694-1754 ....DLB-95

Tolson, Melvin B. 1898-1966 .... DLB-48, 76

Tom Jones (1749), by Henry Fielding
[excerpt] DLB-39

Tomalin, Claire 1933- DLB-155

Tomlinson, Charles 1927- DLB-40

Tomlinson, H. M. 1873-1958 ... DLB-36, 100

Tompkins, Abel [publishing house] ..DLB-49

Tompson, Benjamin 1642-1714 .....DLB-24

Tonks, Rosemary 1932- DLB-14

Tonna, Charlotte Elizabeth
1790-1846 ....DLB-163

Tonson, Jacob the Elder
[publishing house] DLB-170

Toole, John Kennedy 1937-1969 ....Y-81

Toomer, Jean 1894-1967 .... DLB-45, 51

Tor Books DLB-46

Torberg, Friedrich 1908-1979 .....DLB-85

Torrence, Ridgely 1874-1950 ....DLB-54

Torres-Metzger, Joseph V.
1933- DLB-122

Toth, Susan Allen 1940- Y-86

Tottell, Richard
[publishing house] DLB-170

Tough-Guy Literature DLB-9

Touré, Askia Muhammad 1938- ... DLB-41

Tourgée, Albion W. 1838-1905 .....DLB-79

Tourneur, Cyril circa 1580-1626 ....DLB-58

Tournier, Michel 1924- DLB-83

Tousey, Frank [publishing house] ... DLB-49

Tower Publications DLB-46

Towne, Benjamin circa 1740-1793 ... DLB-43

Towne, Robert 1936- DLB-44

The Townely Plays
fifteenth and sixteenth
centuries DLB-146

Townshend, Aurelian
by 1583 - circa 1651 ....DLB-121

Tracy, Honor 1913- DLB-15

Traherne, Thomas 1637?-1674 ....DLB-131

Traill, Catharine Parr 1802-1899 ....DLB-99

Train, Arthur 1875-1945 ....DLB-86

The Transatlantic Publishing
Company DLB-49

Transcendentalists, American DS-5

Translators of the Twelfth Century:
Literary Issues Raised and Impact
Created DLB-115

Travel Writing, 1837-1875 ....DLB-166

Travel Writing, 1876-1909 ....DLB-174

Traven, B.
1882? or 1890?-1969? DLB-9, 56

Travers, Ben 1886-1980 ....DLB-10

Travers, P. L. (Pamela Lyndon)
1899- DLB-160

Trediakovsky, Vasilii Kirillovich
1703-1769 ....DLB-150

Treece, Henry 1911-1966 ....DLB-160

Trejo, Ernesto 1950- DLB-122

Trelawny, Edward John
1792-1881 .... DLB-110, 116, 144

Tremain, Rose 1943- DLB-14

Tremblay, Michel 1942- DLB-60

Trends in Twentieth-Century
Mass Market Publishing DLB-46

Trent, William P. 1862-1939 ....DLB-47

Trescot, William Henry
1822-1898 ....DLB-30

Trevelyan, Sir George Otto
1838-1928 ....DLB-144

Trevisa, John
circa 1342-circa 1402 ....DLB-146

Trevor, William 1928- DLB-14, 139

Trierer Floyris circa 1170-1180 ......DLB-138

Trilling, Lionel 1905-1975 .....DLB-28, 63

Trilussa 1871-1950 ....DLB-114

Trimmer, Sarah 1741-1810 ....DLB-158

Triolet, Elsa 1896-1970 ....DLB-72

Tripp, John 1927- DLB-40

Trocchi, Alexander 1925- DLB-15

Trollope, Anthony
1815-1882 .... DLB-21, 57, 159

Trollope, Frances 1779-1863 ....DLB-21, 166

Troop, Elizabeth 1931- DLB-14

Trotter, Catharine 1679-1749 .....DLB-84

Trotti, Lamar 1898-1952 ....DLB-44

Trottier, Pierre 1925- DLB-60

Troupe, Quincy Thomas, Jr.
1943- DLB-41

Trow, John F., and Company DLB-49

Truillier-Lacombe, Joseph-Patrice
1807-1863 ....DLB-99

Trumbo, Dalton 1905-1976 ....DLB-26

Trumbull, Benjamin 1735-1820 .....DLB-30

Cumulative Index

Trumbull, John 1750-1831 .... DLB-31

Tscherning, Andreas 1611-1659 .... DLB-164

T. S. Eliot Centennial Y-88

Tucholsky, Kurt 1890-1935 .... DLB-56

Tucker, Charlotte Maria
  1821-1893 .... DLB-163

Tucker, George 1775-1861 .... DLB-3, 30

Tucker, Nathaniel Beverley
  1784-1851 .... DLB-3

Tucker, St. George 1752-1827 ...... DLB-37

Tuckerman, Henry Theodore
  1813-1871 .... DLB-64

Tunis, John R. 1889-1975 ..... DLB-22, 171

Tunstall, Cuthbert 1474-1559 ..... DLB-132

Tuohy, Frank 1925- DLB-14, 139

Tupper, Martin F. 1810-1889 .... DLB-32

Turbyfill, Mark 1896- DLB-45

Turco, Lewis 1934- Y-84

Turnbull, Andrew 1921-1970 ..... DLB-103

Turnbull, Gael 1928- DLB-40

Turner, Arlin 1909-1980 .... DLB-103

Turner, Charles (Tennyson)
  1808-1879 .... DLB-32

Turner, Frederick 1943- DLB-40

Turner, Frederick Jackson
  1861-1932 .... DLB-17

Turner, Joseph Addison
  1826-1868 .... DLB-79

Turpin, Waters Edward
  1910-1968 .... DLB-51

Turrini, Peter 1944- DLB-124

Tutuola, Amos 1920- DLB-125

Twain, Mark (see Clemens,
  Samuel Langhorne)

Tweedie, Ethel Brilliana
  circa 1860-1940 .... DLB-174

The 'Twenties and Berlin, by
  Alex Natan DLB-66

Tyler, Anne 1941- DLB-6, 143; Y-82

Tyler, Moses Coit 1835-1900 .... DLB-47, 64

Tyler, Royall 1757-1826 .... DLB-37

Tylor, Edward Burnett 1832-1917 ... DLB-57

Tynan, Katharine 1861-1931 ..... DLB-153

Tyndale, William
  circa 1494-1536 .... DLB-132

U

Udall, Nicholas 1504-1556 .... DLB-62

Uhland, Ludwig 1787-1862 .... DLB-90

Uhse, Bodo 1904-1963 .... DLB-69

Ujević, Augustin ("Tin")
  1891-1955 .... DLB-147

Ulenhart, Niclas
  flourished circa 1600 .... DLB-164

Ulibarrí, Sabine R. 1919- DLB-82

Ulica, Jorge 1870-1926 .... DLB-82

Ulizio, B. George 1889-1969 ..... DLB-140

Ulrich von Liechtenstein
  circa 1200-circa 1275 .... DLB-138

Ulrich von Zatzikhoven
  before 1194-after 1214 .... DLB-138

Unamuno, Miguel de 1864-1936 .... DLB-108

Under the Microscope (1872), by
  A. C. Swinburne DLB-35

Unger, Friederike Helene
  1741-1813 .... DLB-94

Ungaretti, Giuseppe 1888-1970 ..... DLB-114

United States Book Company DLB-49

Universal Publishing and Distributing
  Corporation DLB-46

The University of Iowa Writers' Workshop
  Golden Jubilee Y-86

The University of South Carolina
  Press Y-94

University of Wales Press DLB-112

"The Unknown Public" (1858), by
  Wilkie Collins [excerpt] DLB-57

Unruh, Fritz von 1885-1970 .... DLB-56, 118

Unspeakable Practices II: The Festival of
  Vanguard Narrative at Brown
  University Y-93

Unwin, T. Fisher
  [publishing house] DLB-106

Upchurch, Boyd B. (see Boyd, John)

Updike, John
  1932- DLB-2, 5, 143; Y-80, 82; DS-3

Upton, Bertha 1849-1912 .... DLB-141

Upton, Charles 1948- DLB-16

Upton, Florence K. 1873-1922 ..... DLB-141

Upward, Allen 1863-1926 .... DLB-36

Urista, Alberto Baltazar (see Alurista)

Urzidil, Johannes 1896-1976 .... DLB-85

Urquhart, Fred 1912- DLB-139

The Uses of Facsimile Y-90

Usk, Thomas died 1388 .... DLB-146

Uslar Pietri, Arturo 1906- DLB-113

Ustinov, Peter 1921- DLB-13

Uttley, Alison 1884-1976 .... DLB-160

Uz, Johann Peter 1720-1796 .... DLB-97

V

Vac, Bertrand 1914- DLB-88

Vail, Laurence 1891-1968 .... DLB-4

Vailland, Roger 1907-1965 .... DLB-83

Vajda, Ernest 1887-1954 .... DLB-44

Valdés, Gina 1943- DLB-122

Valdez, Luis Miguel 1940- DLB-122

Valduga, Patrizia 1953- DLB-128

Valente, José Angel 1929- DLB-108

Valenzuela, Luisa 1938- DLB-113

Valeri, Diego 1887-1976 .... DLB-128

Valgardson, W. D. 1939- DLB-60

Valle, Víctor Manuel 1950- DLB-122

Valle-Inclán, Ramón del
  1866-1936 .... DLB-134

Vallejo, Armando 1949- DLB-122

Vallès, Jules 1832-1885 .... DLB-123

Vallette, Marguerite Eymery (see Rachilde)

Valverde, José María 1926- DLB-108

Van Allsburg, Chris 1949- DLB-61

Van Anda, Carr 1864-1945 .... DLB-25

Van Doren, Mark 1894-1972 .... DLB-45

van Druten, John 1901-1957 .... DLB-10

Van Duyn, Mona 1921- DLB-5

Van Dyke, Henry
  1852-1933 .... DLB-71; DS-13

Van Dyke, Henry 1928- DLB-33

van Itallie, Jean-Claude 1936- DLB-7

Van Loan, Charles E. 1876-1919 ... DLB-171

Van Rensselaer, Mariana Griswold
  1851-1934 .... DLB-47

Van Rensselaer, Mrs. Schuyler (see Van
  Rensselaer, Mariana Griswold)

Van Vechten, Carl 1880-1964 ..... DLB-4, 9

van Vogt, A. E. 1912- DLB-8

Vanbrugh, Sir John 1664-1726 .... DLB-80

Vance, Jack 1916?- DLB-8

Vane, Sutton 1888-1963 .... DLB-10

Vanguard Press DLB-46

Vann, Robert L. 1879-1940 .... DLB-29

Vargas, Llosa, Mario 1936- DLB-145

Varley, John 1947- Y-81

Varnhagen von Ense, Karl August 1785-1858 .....DLB-90

Varnhagen von Ense, Rahel 1771-1833 .....DLB-90

Vásquez Montalbán, Manuel 1939- DLB-134

Vassa, Gustavus (see Equiano, Olaudah)

Vassalli, Sebastiano 1941- DLB-128

Vaughan, Henry 1621-1695 .....DLB-131

Vaughan, Thomas 1621-1666 .....DLB-131

Vaux, Thomas, Lord 1509-1556 .....DLB-132

Vazov, Ivan 1850-1921 .....DLB-147

Vega, Janine Pommy 1942- DLB-16

Veiller, Anthony 1903-1965 .....DLB-44

Velásquez-Trevino, Gloria 1949- DLB-122

Veloz Maggiolo, Marcio 1936- DLB-145

Venegas, Daniel ?-? DLB-82

Vergil, Polydore circa 1470-1555 .....DLB-132

Veríssimo, Erico 1905-1975 .....DLB-145

Verne, Jules 1828-1905 .....DLB-123

Verplanck, Gulian C. 1786-1870 .....DLB-59

Very, Jones 1813-1880 .....DLB-1

Vian, Boris 1920-1959 .....DLB-72

Vickers, Roy 1888?-1965 .....DLB-77

Victoria 1819-1901 .....DLB-55

Victoria Press DLB-106

Vidal, Gore 1925- DLB-6, 152

Viebig, Clara 1860-1952 .....DLB-66

Viereck, George Sylvester 1884-1962 .....DLB-54

Viereck, Peter 1916- DLB-5

Viets, Roger 1738-1811 .....DLB-99

Viewpoint: Politics and Performance, by David Edgar DLB-13

Vigil-Piñon, Evangelina 1949- DLB-122

Vigneault, Gilles 1928- DLB-60

Vigny, Alfred de 1797-1863 .....DLB-119

Vigolo, Giorgio 1894-1983 .....DLB-114

The Viking Press DLB-46

Villanueva, Alma Luz 1944- DLB-122

Villanueva, Tino 1941- DLB-82

Villard, Henry 1835-1900 .....DLB-23

Villard, Oswald Garrison 1872-1949 .....DLB-25, 91

Villarreal, José Antonio 1924- DLB-82

Villegas de Magnón, Leonor 1876-1955 .....DLB-122

Villemaire, Yolande 1949- DLB-60

Villena, Luis Antonio de 1951- DLB-134

Villiers de l'Isle-Adam, Jean-Marie Mathias Philippe-Auguste, Comte de 1838-1889 .....DLB-123

Villiers, George, Second Duke of Buckingham 1628-1687 .....DLB-80

Vine Press DLB-112

Viorst, Judith ?- DLB-52

Vipont, Elfrida (Elfrida Vipont Foulds, Charles Vipont) 1902-1992 .....DLB-160

Viramontes, Helena María 1954- DLB-122

Vischer, Friedrich Theodor 1807-1887 .....DLB-133

Vivanco, Luis Felipe 1907-1975 .....DLB-108

Viviani, Cesare 1947- DLB-128

Vizetelly and Company DLB-106

Voaden, Herman 1903- DLB-88

Voigt, Ellen Bryant 1943- DLB-120

Vojnović, Ivo 1857-1929 .....DLB-147

Volkoff, Vladimir 1932- DLB-83

Volland, P. F., Company DLB-46

von der Grün, Max 1926- DLB-75

Vonnegut, Kurt 1922- DLB-2, 8, 152; Y-80; DS-3

Voranc, Prežihov 1893-1950 .....DLB-147

Voß, Johann Heinrich 1751-1826 .....DLB-90

Vroman, Mary Elizabeth circa 1924-1967 .....DLB-33

W

Wace, Robert ("Maistre") circa 1100-circa 1175 .....DLB-146

Wackenroder, Wilhelm Heinrich 1773-1798 .....DLB-90

Wackernagel, Wilhelm 1806-1869 .....DLB-133

Waddington, Miriam 1917- DLB-68

Wade, Henry 1887-1969 .....DLB-77

Wagenknecht, Edward 1900- DLB-103

Wagner, Heinrich Leopold 1747-1779 .....DLB-94

Wagner, Henry R. 1862-1957 .....DLB-140

Wagner, Richard 1813-1883 .....DLB-129

Wagoner, David 1926- DLB-5

Wah, Fred 1939- DLB-60

Waiblinger, Wilhelm 1804-1830 .....DLB-90

Wain, John 1925-1994 .....DLB-15, 27, 139, 155

Wainwright, Jeffrey 1944- DLB-40

Waite, Peirce and Company DLB-49

Wakoski, Diane 1937- DLB-5

Walahfrid Strabo circa 808-849 .....DLB-148

Walck, Henry Z. DLB-46

Walcott, Derek 1930- DLB-117; Y-81, 92

Waldegrave, Robert [publishing house] DLB-170

Waldman, Anne 1945- DLB-16

Waldrop, Rosmarie 1935- DLB-169

Walker, Alice 1944- DLB-6, 33, 143

Walker, George F. 1947- DLB-60

Walker, Joseph A. 1935- DLB-38

Walker, Margaret 1915- DLB-76, 152

Walker, Ted 1934- DLB-40

Walker and Company DLB-49

Walker, Evans and Cogswell Company DLB-49

Walker, John Brisben 1847-1931 .....DLB-79

Wallace, Dewitt 1889-1981 and Lila Acheson Wallace 1889-1984 .....DLB-137

Wallace, Edgar 1875-1932 .....DLB-70

Wallace, Lila Acheson (see Wallace, Dewitt, and Lila Acheson Wallace)

Wallant, Edward Lewis 1926-1962 .....DLB-2, 28, 143

Waller, Edmund 1606-1687 .....DLB-126

Walpole, Horace 1717-1797 .....DLB-39, 104

Walpole, Hugh 1884-1941 .....DLB-34

Walrond, Eric 1898-1966 .....DLB-51

Walser, Martin 1927- DLB-75, 124

Walser, Robert 1878-1956 .....DLB-66

Walsh, Ernest 1895-1926 .....DLB-4, 45

Walsh, Robert 1784-1859 .....DLB-59

Waltharius circa 825 .....DLB-148

Walters, Henry 1848-1931 .....DLB-140

Walther von der Vogelweide circa 1170-circa 1230 .....DLB-138

Walton, Izaak 1593-1683 .....DLB-151

Wambaugh, Joseph 1937- DLB-6; Y-83

Waniek, Marilyn Nelson 1946- DLB-120

Warburton, William 1698-1779 .....DLB-104

Ward, Aileen 1919- DLB-111	Watson, John 1850-1907 .... DLB-156	Weiss, John 1818-1879 .... DLB-1
Ward, Artemus (see Browne, Charles Farrar)	Watson, Sheila 1909- DLB-60	Weiss, Peter 1916-1982 .... DLB-69, 124
Ward, Arthur Henry Sarsfield (see Rohmer, Sax)	Watson, Thomas 1545?-1592 ..... DLB-132	Weiss, Theodore 1916- DLB-5
Ward, Douglas Turner 1930- ... DLB-7, 38	Watson, Wilfred 1911- DLB-60	Weisse, Christian Felix 1726-1804 ... DLB-97
Ward, Lynd 1905-1985 .... DLB-22	Watt, W. J., and Company DLB-46	Weitling, Wilhelm 1808-1871 ...... DLB-129
Ward, Lock and Company DLB-106	Watterson, Henry 1840-1921 ...... DLB-25	Welch, Lew 1926-1971? DLB-16
Ward, Mrs. Humphry 1851-1920 ... DLB-18	Watts, Alan 1915-1973 .... DLB-16	Weldon, Fay 1931- DLB-14
Ward, Nathaniel circa 1578-1652 ... DLB-24	Watts, Franklin [publishing house] ... DLB-46	Wellek, René 1903- DLB-63
Ward, Theodore 1902-1983 .... DLB-76	Watts, Isaac 1674-1748 .... DLB-95	Wells, Carolyn 1862-1942 .... DLB-11
Wardle, Ralph 1909-1988 .... DLB-103	Waugh, Auberon 1939- DLB-14	Wells, Charles Jeremiah circa 1800-1879 .... DLB-32
Ware, William 1797-1852 ....DLB-1	Waugh, Evelyn 1903-1966 ...... DLB-15, 162	Wells, Gabriel 1862-1946 .... DLB-140
Warne, Frederick, and Company [U.S.] DLB-49	Way and Williams DLB-49	Wells, H. G. 1866-1946 .... DLB-34, 70, 156
Warne, Frederick, and Company [U.K.] DLB-106	Wayman, Tom 1945- DLB-53	Wells, Robert 1947- DLB-40
Warner, Charles Dudley 1829-1900 .... DLB-64	Weatherly, Tom 1942- DLB-41	Wells-Barnett, Ida B. 1862-1931 ..... DLB-23
Warner, Rex 1905- DLB-15	Weaver, Gordon 1937- DLB-130	Welty, Eudora 1909- DLB-2, 102, 143; Y-87; DS-12
Warner, Susan Bogert 1819-1885 .... DLB-3, 42	Weaver, Robert 1921- DLB-88	Wendell, Barrett 1855-1921 .... DLB-71
Warner, Sylvia Townsend 1893-1978 .... DLB-34, 139	Webb, Frank J. ?-? DLB-50	Wentworth, Patricia 1878-1961 ...... DLB-77
Warner, William 1558-1609 ...... DLB-172	Webb, James Watson 1802-1884 .... DLB-43	Werder, Diederich von dem 1584-1657 .... DLB-164
Warner Books DLB-46	Webb, Mary 1881-1927 .... DLB-34	Werfel, Franz 1890-1945 .... DLB-81, 124
Warr, Bertram 1917-1943 .... DLB-88	Webb, Phyllis 1927- DLB-53	The Werner Company DLB-49
Warren, John Byrne Leicester (see De Tabley, Lord)	Webb, Walter Prescott 1888-1963 ... DLB-17	Werner, Zacharias 1768-1823 ...... DLB-94
Warren, Lella 1899-1982 .... Y-83	Webbe, William ?-1591 .... DLB-132	Wersba, Barbara 1932- DLB-52
Warren, Mercy Otis 1728-1814 ..... DLB-31	Webster, Augusta 1837-1894 .... DLB-35	Wescott, Glenway 1901- DLB-4, 9, 102
Warren, Robert Penn 1905-1989 .....DLB-2, 48, 152; Y-80, 89	Webster, Charles L., and Company DLB-49	Wesker, Arnold 1932- DLB-13
Die Wartburgkrieg circa 1230-circa 1280 .... DLB-138	Webster, John 1579 or 1580-1634? DLB-58	Wesley, Charles 1707-1788 .... DLB-95
Warton, Joseph 1722-1800 .... DLB-104, 109	Webster, Noah 1758-1843 .... DLB-1, 37, 42, 43, 73	Wesley, John 1703-1791 .... DLB-104
Warton, Thomas 1728-1790 ... DLB-104, 109	Weckherlin, Georg Rodolf 1584-1653 .... DLB-164	Wesley, Richard 1945- DLB-38
Washington, George 1732-1799 ..... DLB-31	Wedekind, Frank 1864-1918 ...... DLB-118	Wessels, A., and Company DLB-46
Wassermann, Jakob 1873-1934 ..... DLB-66	Weeks, Edward Augustus, Jr. 1898-1989 .... DLB-137	*Wessobrunner Gebet* circa 787-815 .... DLB-148
Wasson, David Atwood 1823-1887 ....DLB-1	Weems, Mason Locke 1759-1825 ....DLB-30, 37, 42	West, Anthony 1914-1988 .... DLB-15
Waterhouse, Keith 1929- DLB-13, 15	Weerth, Georg 1822-1856 .... DLB-129	West, Dorothy 1907- DLB-76
Waterman, Andrew 1940- DLB-40	Weidenfeld and Nicolson DLB-112	West, Jessamyn 1902-1984 ..... DLB-6; Y-84
Waters, Frank 1902- Y-86	Weidman, Jerome 1913- DLB-28	West, Mae 1892-1980 .... DLB-44
Waters, Michael 1949- DLB-120	Weigl, Bruce 1949- DLB-120	West, Nathanael 1903-1940 .... DLB-4, 9, 28
Watkins, Tobias 1780-1855 .... DLB-73	Weinbaum, Stanley Grauman 1902-1935 .... DLB-8	West, Paul 1930- DLB-14
Watkins, Vernon 1906-1967 .... DLB-20	Weintraub, Stanley 1929- DLB-111	West, Rebecca 1892-1983 ..... DLB-36; Y-83
Watmough, David 1926- DLB-53	Weise, Christian 1642-1708 ...... DLB-168	West and Johnson DLB-49
Watson, James Wreford (see Wreford, James)	Weisenborn, Gunther 1902-1969 ....DLB-69, 124	Western Publishing Company DLB-46
	Weiß, Ernst 1882-1940 .... DLB-81	*The Westminster Review* 1824-1914 .... DLB-110
		Weston, Elizabeth Jane circa 1582-1612 .... DLB-172

Wetherald, Agnes Ethelwyn 1857-1940 .... DLB-99

Wetherell, Elizabeth (see Warner, Susan Bogert)

Wetzel, Friedrich Gottlob 1779-1819 .... DLB-90

Weyman, Stanley J. 1855-1928 .... DLB-141, 156

Wezel, Johann Karl 1747-1819 ...... DLB-94

Whalen, Philip 1923- DLB-16

Whalley, George 1915-1983 .... DLB-88

Wharton, Edith 1862-1937 ...... DLB-4, 9, 12, 78; DS-13

Wharton, William 1920s?- Y-80

Whately, Mary Louisa 1824-1889 .... DLB-166

What's Really Wrong With Bestseller Lists Y-84

Wheatley, Dennis Yates 1897-1977 .... DLB-77

Wheatley, Phillis circa 1754-1784 .... DLB-31, 50

Wheeler, Anna Doyle 1785-1848? DLB-158

Wheeler, Charles Stearns 1816-1843 .... DLB-1

Wheeler, Monroe 1900-1988 .... DLB-4

Wheelock, John Hall 1886-1978 ..... DLB-45

Wheelwright, John circa 1592-1679 .... DLB-24

Wheelwright, J. B. 1897-1940 .... DLB-45

Whetstone, Colonel Pete (see Noland, C. F. M.)

Whetstone, George 1550-1587 ...... DLB-136

Whicher, Stephen E. 1915-1961 .... DLB-111

Whipple, Edwin Percy 1819-1886 .... DLB-1, 64

Whitaker, Alexander 1585-1617 ..... DLB-24

Whitaker, Daniel K. 1801-1881 ...... DLB-73

Whitcher, Frances Miriam 1814-1852 .... DLB-11

White, Andrew 1579-1656 .... DLB-24

White, Andrew Dickson 1832-1918 .... DLB-47

White, E. B. 1899-1985 .... DLB-11, 22

White, Edgar B. 1947- DLB-38

White, Ethel Lina 1887-1944 .... DLB-77

White, Henry Kirke 1785-1806 ..... DLB-96

White, Horace 1834-1916 .... DLB-23

White, Phyllis Dorothy James (see James, P. D.)

White, Richard Grant 1821-1885 .... DLB-64

White, T. H. 1906-1964 .... DLB-160

White, Walter 1893-1955 .... DLB-51

White, William, and Company DLB-49

White, William Allen 1868-1944 .... DLB-9, 25

White, William Anthony Parker (see Boucher, Anthony)

White, William Hale (see Rutherford, Mark)

Whitechurch, Victor L. 1868-1933 .... DLB-70

Whitehead, Alfred North 1861-1947 .... DLB-100

Whitehead, James 1936- Y-81

Whitehead, William 1715-1785 .... DLB-84, 109

Whitfield, James Monroe 1822-1871 .... DLB-50

Whitgift, John circa 1533-1604 ..... DLB-132

Whiting, John 1917-1963 .... DLB-13

Whiting, Samuel 1597-1679 .... DLB-24

Whitlock, Brand 1869-1934 .... DLB-12

Whitman, Albert, and Company DLB-46

Whitman, Albery Allson 1851-1901 .... DLB-50

Whitman, Alden 1913-1990 .... Y-91

Whitman, Sarah Helen (Power) 1803-1878 .... DLB-1

Whitman, Walt 1819-1892 .... DLB-3, 64

Whitman Publishing Company DLB-46

Whitney, Geoffrey 1548 or 1552?-1601 .... DLB-136

Whitney, Isabella flourished 1566-1573 .... DLB-136

Whitney, John Hay 1904-1982 ..... DLB-127

Whittemore, Reed 1919- DLB-5

Whittier, John Greenleaf 1807-1892 .. DLB-1

Whittlesey House DLB-46

Who Runs American Literature? Y-94

Wideman, John Edgar 1941- ... DLB-33, 143

Widener, Harry Elkins 1885-1912 .... DLB-140

Wiebe, Rudy 1934- DLB-60

Wiechert, Ernst 1887-1950 .... DLB-56

Wied, Martina 1882-1957 .... DLB-85

Wiehe, Evelyn May Clowes (see Mordaunt, Elinor)

Wieland, Christoph Martin 1733-1813 .... DLB-97

Wienbarg, Ludolf 1802-1872 ...... DLB-133

Wieners, John 1934- DLB-16

Wier, Ester 1910- DLB-52

Wiesel, Elie 1928- DLB-83; Y-87

Wiggin, Kate Douglas 1856-1923 .... DLB-42

Wigglesworth, Michael 1631-1705 ... DLB-24

Wilberforce, William 1759-1833 .... DLB-158

Wilbrandt, Adolf 1837-1911 .... DLB-129

Wilbur, Richard 1921- DLB-5, 169

Wild, Peter 1940- DLB-5

Wilde, Oscar 1854-1900 .... DLB-10, 19, 34, 57, 141, 156

Wilde, Richard Henry 1789-1847 .... DLB-3, 59

Wilde, W. A., Company DLB-49

Wilder, Billy 1906- DLB-26

Wilder, Laura Ingalls 1867-1957 ..... DLB-22

Wilder, Thornton 1897-1975 ....DLB-4, 7, 9

Wildgans, Anton 1881-1932 .... DLB-118

Wiley, Bell Irvin 1906-1980 .... DLB-17

Wiley, John, and Sons DLB-49

Wilhelm, Kate 1928- DLB-8

Wilkes, George 1817-1885 .... DLB-79

Wilkinson, Anne 1910-1961 .... DLB-88

Wilkinson, Sylvia 1940- Y-86

Wilkinson, William Cleaver 1833-1920 .... DLB-71

Willard, Barbara 1909-1994 .... DLB-161

Willard, L. [publishing house] DLB-49

Willard, Nancy 1936- DLB-5, 52

Willard, Samuel 1640-1707 .... DLB-24

William of Auvergne 1190-1249 ... DLB-115

William of Conches circa 1090-circa 1154 .... DLB-115

William of Ockham circa 1285-1347 .... DLB-115

William of Sherwood 1200/1205 - 1266/1271 .... DLB-115

The William Chavrat American Fiction Collection at the Ohio State University Libraries Y-92

Williams, A., and Company DLB-49

Williams, Ben Ames 1889-1953 ..... DLB-102

Williams, C. K. 1936- DLB-5

Williams, Chancellor 1905- DLB-76

Williams, Charles 1886-1945 .... DLB-100, 153

Williams, Denis 1923- DLB-117

403

Cumulative Index

Williams, Emlyn 1905- DLB-10, 77

Williams, Garth 1912- DLB-22

Williams, George Washington
1849-1891 .... DLB-47

Williams, Heathcote 1941- DLB-13

Williams, Helen Maria
1761-1827 ....DLB-158

Williams, Hugo 1942- DLB-40

Williams, Isaac 1802-1865 ....DLB-32

Williams, Joan 1928- DLB-6

Williams, John A. 1925- DLB-2, 33

Williams, John E. 1922-1994 ....DLB-6

Williams, Jonathan 1929- DLB-5

Williams, Miller 1930- DLB-105

Williams, Raymond 1921- DLB-14

Williams, Roger circa 1603-1683 .... DLB-24

Williams, Samm-Art 1946- DLB-38

Williams, Sherley Anne 1944- DLB-41

Williams, T. Harry 1909-1979 ..... DLB-17

Williams, Tennessee
1911-1983 ....DLB-7; Y-83; DS-4

Williams, Ursula Moray 1911- ... DLB-160

Williams, Valentine 1883-1946 ..... DLB-77

Williams, William Appleman
1921- DLB-17

Williams, William Carlos
1883-1963 ....DLB-4, 16, 54, 86

Williams, Wirt 1921- DLB-6

Williams Brothers DLB-49

Williamson, Jack 1908- DLB-8

Willingham, Calder Baynard, Jr.
1922- DLB-2, 44

Williram of Ebersberg
circa 1020-1085 .... DLB-148

Willis, Nathaniel Parker
1806-1867 ..... DLB-3, 59, 73, 74; DS-13

Willkomm, Ernst 1810-1886 .... DLB-133

Wilmer, Clive 1945- DLB-40

Wilson, A. N. 1950- DLB-14, 155

Wilson, Angus
1913-1991 ....DLB-15, 139, 155

Wilson, Arthur 1595-1652 .... DLB-58

Wilson, Augusta Jane Evans
1835-1909 .... DLB-42

Wilson, Colin 1931- DLB-14

Wilson, Edmund 1895-1972 ..... DLB-63

Wilson, Ethel 1888-1980 .... DLB-68

Wilson, Harriet E. Adams
1828?-1863? DLB-50

Wilson, Harry Leon 1867-1939 ...... DLB-9

Wilson, John 1588-1667 ....DLB-24

Wilson, John 1785-1854 ....DLB-110

Wilson, Lanford 1937- DLB-7

Wilson, Margaret 1882-1973 ....DLB-9

Wilson, Michael 1914-1978 .... DLB-44

Wilson, Mona 1872-1954 .... DLB-149

Wilson, Thomas
1523 or 1524-1581 .... DLB-132

Wilson, Woodrow 1856-1924 .... DLB-47

Wilson, Effingham
[publishing house] DLB-154

Wimsatt, William K., Jr.
1907-1975 .... DLB-63

Winchell, Walter 1897-1972 .... DLB-29

Winchester, J. [publishing house] DLB-49

Winckelmann, Johann Joachim
1717-1768 .... DLB-97

Winckler, Paul 1630-1686 ....DLB-164

Wind, Herbert Warren 1916- DLB-171

Windet, John [publishing house] DLB-170

Windham, Donald 1920- DLB-6

Wingate, Allan [publishing house] .. DLB-112

Winnifrith, Tom 1938- DLB-155

Winsloe, Christa 1888-1944 .... DLB-124

Winsor, Justin 1831-1897 .... DLB-47

John C. Winston Company DLB-49

Winters, Yvor 1900-1968 .... DLB-48

Winthrop, John 1588-1649 ...... DLB-24, 30

Winthrop, John, Jr. 1606-1676 ...... DLB-24

Wirt, William 1772-1834 .... DLB-37

Wise, John 1652-1725 .... DLB-24

Wiseman, Adele 1928- DLB-88

Wishart and Company DLB-112

Wisner, George 1812-1849 .... DLB-43

Wister, Owen 1860-1938 ....DLB-9, 78

Wither, George 1588-1667 .... DLB-121

Witherspoon, John 1723-1794 .... DLB-31

Withrow, William Henry 1839-1908 .. DLB-99

Wittig, Monique 1935- DLB-83

Wodehouse, P. G.
1881-1975 ....DLB-34, 162

Wohmann, Gabriele 1932- DLB-75

Woiwode, Larry 1941- DLB-6

Wolcot, John 1738-1819 ....DLB-109

Wolcott, Roger 1679-1767 ....DLB-24

Wolf, Christa 1929- DLB-75

Wolf, Friedrich 1888-1953 ....DLB-124

Wolfe, Gene 1931- DLB-8

Wolfe, John [publishing house] DLB-170

Wolfe, Reyner (Reginald)
[publishing house] DLB-170

Wolfe, Thomas
1900-1938 ....DLB-9, 102; Y-85; DS-2

Wolfe, Tom 1931- DLB-152

Wolff, Helen 1906-1994 .... Y-94

Wolff, Tobias 1945- DLB-130

Wolfram von Eschenbach
circa 1170-after 1220 ....DLB-138

Wolfram von Eschenbach's *Parzival*:
Prologue and Book 3 ....DLB-138

Wollstonecraft, Mary
1759-1797 .... DLB-39, 104, 158

Wondratschek, Wolf 1943- DLB-75

Wood, Benjamin 1820-1900 ....DLB-23

Wood, Charles 1932- DLB-13

Wood, Mrs. Henry 1814-1887 ......DLB-18

Wood, Joanna E. 1867-1927 ....DLB-92

Wood, Samuel [publishing house] DLB-49

Wood, William ?-? DLB-24

Woodberry, George Edward
1855-1930 .... DLB-71, 103

Woodbridge, Benjamin 1622-1684 ...DLB-24

Woodcock, George 1912- DLB-88

Woodhull, Victoria C. 1838-1927 ....DLB-79

Woodmason, Charles circa 1720-? ...DLB-31

Woodress, Jr., James Leslie
1916- DLB-111

Woodson, Carter G. 1875-1950 .....DLB-17

Woodward, C. Vann 1908- DLB-17

Woodward, Stanley 1895-1965 ....DLB-171

Wooler, Thomas
1785 or 1786-1853 ....DLB-158

Woolf, David (see Maddow, Ben)

Woolf, Leonard 1880-1969 ....DLB-100; DS-10

Woolf, Virginia
1882-1941 ......DLB-36, 100, 162; DS-10

Woolf, Virginia, "The New Biography," *New York Herald Tribune*, 30 October 1927
.........................DLB-149

Woollcott, Alexander 1887-1943 .....DLB-29

Woolman, John 1720-1772 ....DLB-31

Woolner, Thomas 1825-1892 .... DLB-35

Woolsey, Sarah Chauncy 1835-1905 .... DLB-42

Woolson, Constance Fenimore 1840-1894 .... DLB-12, 74

Worcester, Joseph Emerson 1784-1865 .... DLB-1

Worde, Wynkyn de [publishing house] DLB-170

Wordsworth, Christopher 1807-1885 .... DLB-166

Wordsworth, Dorothy 1771-1855 .... DLB-107

Wordsworth, Elizabeth 1840-1932 .... DLB-98

Wordsworth, William 1770-1850 .... DLB-93, 107

The Works of the Rev. John Witherspoon (1800-1801) [excerpts] DLB-31

A World Chronology of Important Science Fiction Works (1818-1979) DLB-8

World Publishing Company DLB-46

World War II Writers Symposium at the University of South Carolina, 12–14 April 1995 .... Y-95

Worthington, R., and Company DLB-49

Wotton, Sir Henry 1568-1639 ...... DLB-121

Wouk, Herman 1915- Y-82

Wreford, James 1915- DLB-88

Wren, Percival Christopher 1885-1941 .... DLB-153

Wrenn, John Henry 1841-1911 ..... DLB-140

Wright, C. D. 1949- DLB-120

Wright, Charles 1935- DLB-165; Y-82

Wright, Charles Stevenson 1932- ... DLB-33

Wright, Frances 1795-1852 .... DLB-73

Wright, Harold Bell 1872-1944 .... DLB-9

Wright, James 1927-1980 .... DLB-5, 169

Wright, Jay 1935- DLB-41

Wright, Louis B. 1899-1984 .... DLB-17

Wright, Richard 1908-1960 .... DLB-76, 102; DS-2

Wright, Richard B. 1937- DLB-53

Wright, Sarah Elizabeth 1928- DLB-33

Writers and Politics: 1871-1918, by Ronald Gray DLB-66

Writers and their Copyright Holders: the WATCH Project Y-94

Writers' Forum Y-85

Writing for the Theatre, by Harold Pinter DLB-13

Wroth, Lady Mary 1587-1653 ..... DLB-121

Wurlitzer, Rudolph 1937- DLB-173

Wyatt, Sir Thomas circa 1503-1542 .... DLB-132

Wycherley, William 1641-1715 ..... DLB-80

Wyclif, John circa 1335-31 December 1384 .. DLB-146

Wylie, Elinor 1885-1928 .... DLB-9, 45

Wylie, Philip 1902-1971 .... DLB-9

Wyllie, John Cook 1908-1968 ..... DLB-140

Y

Yates, Dornford 1885-1960 ..... DLB-77, 153

Yates, J. Michael 1938- DLB-60

Yates, Richard 1926-1992 ... DLB-2; Y-81, 92

Yavorov, Peyo 1878-1914 .... DLB-147

Yearsley, Ann 1753-1806 .... DLB-109

Yeats, William Butler 1865-1939 .... DLB-10, 19, 98, 156

Yep, Laurence 1948- DLB-52

Yerby, Frank 1916-1991 .... DLB-76

Yezierska, Anzia 1885-1970 .... DLB-28

Yolen, Jane 1939- DLB-52

Yonge, Charlotte Mary 1823-1901 .... DLB-18, 163

The York Cycle circa 1376-circa 1569 .... DLB-146

A Yorkshire Tragedy DLB-58

Yoseloff, Thomas [publishing house] DLB-46

Young, Al 1939- DLB-33

Young, Arthur 1741-1820 .... DLB-158

Young, Dick 1917 or 1918 - 1987 .. DLB-171

Young, Edward 1683-1765 .... DLB-95

Young, Stark 1881-1963 .... DLB-9, 102

Young, Waldeman 1880-1938 ..... DLB-26

Young, William [publishing house] .. DLB-49

Yourcenar, Marguerite 1903-1987 .... DLB-72; Y-88

"You've Never Had It So Good," Gusted by "Winds of Change": British Fiction in the 1950s, 1960s, and After DLB-14

Yovkov, Yordan 1880-1937 .... DLB-147

Z

Zachariä, Friedrich Wilhelm 1726-1777 .... DLB-97

Zamora, Bernice 1938- DLB-82

Zand, Herbert 1923-1970 .... DLB-85

Zangwill, Israel 1864-1926 ..... DLB-10, 135

Zanzotto, Andrea 1921- DLB-128

Zapata Olivella, Manuel 1920- DLB-113

Zebra Books DLB-46

Zebrowski, George 1945- DLB-8

Zech, Paul 1881-1946 .... DLB-56

Zepheria DLB-172

Zeidner, Lisa 1955- DLB-120

Zelazny, Roger 1937-1995 .... DLB-8

Zenger, John Peter 1697-1746 ....DLB-24, 43

Zesen, Philipp von 1619-1689 ..... DLB-164

Zieber, G. B., and Company DLB-49

Zieroth, Dale 1946- DLB-60

Zigler und Kliphausen, Heinrich Anshelm von 1663-1697 .... DLB-168

Zimmer, Paul 1934- DLB-5

Zingref, Julius Wilhelm 1591-1635 .... DLB-164

Zindel, Paul 1936- DLB-7, 52

Zinzendorf, Nikolaus Ludwig von 1700-1760 .... DLB-168

Zola, Emile 1840-1902 .... DLB-123

Zolotow, Charlotte 1915- DLB-52

Zschokke, Heinrich 1771-1848 ...... DLB-94

Zubly, John Joachim 1724-1781 ..... DLB-31

Zu-Bolton II, Ahmos 1936- DLB-41

Zuckmayer, Carl 1896-1977 ....DLB-56, 124

Zukofsky, Louis 1904-1978 ......DLB-5, 165

Župančič, Oton 1878-1949 .... DLB-147

zur Mühlen, Hermynia 1883-1951 ... DLB-56

Zweig, Arnold 1887-1968 .... DLB-66

Zweig, Stefan 1881-1942 .... DLB-81, 118